Early Modern Human Evolution in Central Europe

HUMAN EVOLUTION SERIES

SERIES EDITORS
Russell L. Ciochon, *The University of Iowa*
Bernard A. Wood, *George Washington University*

EDITORIAL ADVISORY BOARD
Leslie Aiello, *University College, London*
Alison Brooks, *George Washington University*
Fred Grine, *State University of New York, Stony Brook*
Andrew Hill, *Yale University*
David Pilbeam, *Harvard University*
Yoel Rak, *Tel-Aviv University*
Mary Ellen Ruvolo, *Harvard University*
Henry Schwarcz, *McMaster University*

African Biogeography, Climate Change, and Human Evolution
edited by Timothy G. Bromage and Friedemann Schrenk

Meat-Eating and Human Evolution
edited by Craig B. Stanford and Henry T. Bunn

The Skull of Australopithecus afarensis
William H. Kimbel, Yoel Rak, and Donald C. Johanson

Early Modern Human Evolution in Central Europe:
The People of Dolní Věstonice and Pavlov
edited by Erik Trinkaus and Jiří Svoboda

Early Modern Human Evolution in Central Europe

The People of Dolní Věstonice and Pavlov

EDITED BY

Erik Trinkaus and Jiří Svoboda
The Dolní Věstonice Studies Volume 12

2006

OXFORD
UNIVERSITY PRESS

Oxford University Press, Inc., publishes works that further
Oxford University's objective of excellence
in research, scholarship, and education.

Oxford   New York
Auckland   Cape Town   Dar es Salaam   Hong Kong   Karachi
Kuala Lumpur   Madrid   Melbourne   Mexico City   Nairobi
New Delhi   Shanghai   Taipei   Toronto

With offices in
Argentina   Austria   Brazil   Chile   Czech Republic   France   Greece
Guatemala   Hungary   Italy   Japan   Poland   Portugal   Singapore
South Korea   Switzerland   Thailand   Turkey   Ukraine   Vietnam

Copyright © 2006 by Oxford University Press, Inc.

Published by Oxford University Press, Inc.
198 Madison Avenue, New York, New York 10016

www.oup.com

Oxford is a registered trademark of Oxford University Press

All rights reserved. No part of this publication may be reproduced,
stored in a retrieval system, or transmitted, in any form or by any means,
electronic, mechanical, photocopying, recording, or otherwise,
without the prior permission of Oxford University Press.

Library of Congress Cataloging-in-Publication Data
Early modern human evolution in Central Europe : The people of Dolní
Věstonice and Pavlov / edited by Erik Trinkaus and Jiří Svoboda.
  p. cm. — (Human evolution series) (The Dolní Věstonice studies ; v. 12)
Includes bibliographical references and index.
ISBN-13 978-0-19-516699-6
ISBN 0-19-516699-X
  1. Dolní Věstonice Site (Czech Republic) 2. Paleolithic period—Czech
Republic—Pavlov Hills. 3. Human remains (Archaeology)—Czech Republic—
Pavlov Hills. 4. Hunting and gathering societies—Czech Republic—Pavlov
Hills. 5. Paleobiology—Czech Republic—Pavlov Hills. 6. Excavations
(Archaeology)—Czech Republic—Pavlov Hills. 7. Pavlov Hills (Czech
Republic)—Antiquities. I. Trinkaus, Erik. II. Svoboda, Jiří, 1953–
III. Series. IV. Dolnovestonické studie ; sv. 12.
GN772.22.C95P34 2005
943.71'02—dc22   2004019755

9 8 7 6 5 4 3 2 1

Printed in the United States of America
on acid-free paper

Preface

In southern Moravia, overlooking the Dyje River to the north, is a long limestone massif known as the Pavlovské Hills. On the loess-covered slopes of those hills, leading down to what was a meandering river basin during the Late Pleistocene, Upper Paleolithic foraging social groups repeatedly stopped, camped, made tools, fashioned figurines in ceramics, wove textiles, and occasionally died. This activity occurred particularly during the period between, minimally, 25,000 and 27,000 years ago (uncalibrated dates), and it resulted in some of the richest Paleolithic archeological sites ever known. These sites, which fall within the jurisdictions of the villages of Dolní Věstonice and Pavlov, have been known since the 1920s. They have justifiably received global recognition for their rich archeological contents, published in a plethora of articles, monographs, and edited volumes by Karel Absolon, Bohuslav Klíma, and Jiří Svoboda. These sites have also yielded an important series of Upper Paleolithic human remains, in fact the largest single sample of human fossil remains available from this time period of the Upper Paleolithic.

The human fossil remains from Pavlov and Dolní Věstonice have been announced in articles by their discoverers, preliminarily described by several of the authors of this volume and other scholars, inventoried in detail, and incorporated into a number of comparative analyses of later Pleistocene late archaic and early modern human fossil remains. However, despite the fact that the preserved human remains from these sites have been known for more than half a century, previously discovered remains having been largely destroyed during World War II, a comprehensive description, comparative analysis, and interpretive consideration of this important sample has yet to be published. The closest approximation to this level of discussion has been two small volumes, one written by and the other edited by one of the contributors to this volume, Emanuel Vlček. In light of this, during the late 1990s, Jiří Svoboda invited Erik Trinkaus "to proceed further than that" with the remains. At the same time, Emanuel Vlček decided that he would not be able to complete his own planned descriptive volume of the fossil sample, and we therefore decided to organize an international team to accomplish this task. Although a number of individuals have worked with the material in the meantime and helped with our knowledge of the sample, it is principally the authors of the chapters in this volume that have done the ultimate task of going through every fragment of bone and assessing them in detail in terms of their comparative morphology and paleobiology.

This preserved sample includes six associated adolescent and adult skeletons from burials, an associated dentition of a child, and a number of isolated skeletal elements and teeth. There are 612 identified skeletal elements, 232 teeth, and a number of fragments retained with the individual skeletons. These fossil human remains—a few of which are curated in the Moravské Zemské Muzeum in Brno, but the majority of which are curated by the Archeologický ústav, Akademie Věd České Republiky in Dolní Věstonice—form the basis for our descriptions and considerations of this prehistoric group of humans. Needless to say, it has been through the hospitality of the curators at the Moravské Zemské Muzeum (Drs. Marta Dočkalová, Martin Oliva, and the late Jan Jelínek) and the ongoing support of the Archeologický ústav of the Akademie Věd České Republiky (Brno) that we have been able to undertake the project at hand. To these individuals and institutions we are most grateful.

It also needs to be mentioned that we have benefited from the access that curators of human fossil and Paleolithic archeological materials across Europe and western Asia, including elsewhere in the Czech Republic, have afforded us, since all assessments of past human remains are, by their very nature, comparative. All of this has been possible through the financial support of the Akademie Věd České Republiky in Dolní Věstonice, the support of our home institutions (see addresses), and grants for these authors from the Wenner-Gren Foundation (J. A. S. and E. T.), the L. S. B. Leakey Foundation (R. G. F. and T. W. H.), and the National Science Foundation (R. G. F., T. W. H., and E. T.). A number of colleagues have also generously shared unpublished comparative data with us. The production of this volume has also benefited from the support of Washington University, the painstaking reading of the chapters by Laura Shackelford, technical assistance in digital photography and data coding by Steven Miller and Joshua Zisson, and the patience and work of the editorial and production staff of Oxford University Press.

It is our hope that this volume will help to fill some of the gaps that remain in our broader understanding of the very successful human populations of the European earlier Upper Paleolithic. It may also convince members of our families that we really were engaged in a serious endeavor and not just the enjoyment of food, wine, beer, countryside, and good friends in and around Dolní Věstonice.

Erik Trinkaus and Jiří A. Svoboda
St. Louis and Dolní Věstonice, June 2004

Contents

1 Introduction *1*
   Erik Trinkaus and Jiří A. Svoboda

2 The Archeological Framework *6*
   Jiří A. Svoboda

3 The Archeological Contexts of the Human Remains *9*
   Jiří A. Svoboda

4 The Burials: Ritual and Taphonomy *15*
   Jiří A. Svoboda

5 The Human Remains: A Summary Inventory *27*
   Trenton W. Holliday, Simon W. Hillson, Robert G. Franciscus, and Erik Trinkaus

6 The Ages at Death *31*
   Simon W. Hillson, Robert G. Franciscus, Trenton W. Holliday, and Erik Trinkaus

7 The Assessment of Sex *46*
   Jaroslav Brůžek, Robert G. Franciscus, Vladimír Novotný, and Erik Trinkaus

8 The Cranial Remains *63*
   Robert G. Franciscus and Emanuel Vlček

9 The Auditory Ossicles *153*
   Petr Lisoněk and Erik Trinkaus

Contents

10 The Mandibular Remains   156
    Robert G. Franciscus, Emanuel Vlček, and Erik Trinkaus

11 Dental Morphology, Proportions, and Attrition   179
    Simon W. Hillson

12 Body Proportions   224
    Trenton W. Holliday

13 Body Length and Body Mass   233
    Erik Trinkaus

14 The Vertebral Columns   242
    Trenton W. Holliday

15. The Costal Skeletons   295
    Trenton W. Holliday

16 The Upper Limb Remains   327
    Erik Trinkaus

17 The Pelvic Remains   373
    Erik Trinkaus

18 The Lower Limb Remains   380
    Erik Trinkaus

19 Skeletal and Dental Paleopathology   419
    Erik Trinkaus, Simon W. Hillson, Robert G. Franciscus, and Trenton W. Holliday

20 The Paleobiology of the Pavlovian People   459
    Erik Trinkaus and Jiří A. Svoboda

    References   467

    Index   485

Contributors

Jaroslav Brůžek
: *Laboratoire d'Anthropologie des Populations du Passé, Université de Bordeaux 1, Avenue des Facultés, 33405 Talence, France; and Katedra antropologie, Fakulta humanitních studií, Západočeská univerzita, Tylova 18, 301 25 Plzeň, Czech Republic*

Robert G. Franciscus
: *Department of Anthropology, Macbride Hall 114, University of Iowa, Iowa City IA 52242, USA*

Simon W. Hillson
: *Institute of Archaeology, University College London, 31-34 Gordon Square, London WC1H 0PY, UK*

Trenton W. Holliday
: *Department of Anthropology, Tulane University, New Orleans LA 70118, USA*

Petr Lisoněk
: *Hanusova 3, 779 00 Olomouc, Czech Republic*

Vladimír Novotný
: *Katedra antropologie, Přírodovědecká fakulta, Masarykova univerzita, Vinařská 5, 60 300 Brno, Czech Republic*

Jiří A. Svoboda
: *Centrum pro Paleolit a Paleoetnologii Dolní Věstonice, Archeologický ústav AV ČR Brno, 691 29 Dolní Věstonice 25, Czech Republic*

Erik Trinkaus
: *Department of Anthropology, Campus Box 1114, Washington University, St. Louis MO 63130, USA*

Emanuel Vlček
: *Oddělení Anthropologie, Národní Muzeum v Praze, Václavské námesti 1, 110 00 Praha 1, Czech Republic*

Early Modern Human Evolution in Central Europe

1

Introduction

Erik Trinkaus and Jiří A. Svoboda

It has long been recognized that human populations and their behaviors underwent a variety of dramatic changes during the middle of the last glacial period, the period also known as the Interpleniglacial or Oxygen Isotope Stage (OIS) 3. This has long been framed in the context of the Middle to Upper Paleolithic transition and as part of the phylogenetic emergence of early modern humans. However, it has become apparent that it involved not only these more generally recognized transitions but also significant human biological, technological, social, and cultural changes within the earlier Upper Paleolithic. For this reason, the Upper Paleolithic complexes of OIS 3 are increasingly divided into an Initial Upper Paleolithic (the Châtelperronian, Bohunician, Szeletian, etc.), the Early Upper Paleolithic (principally the Aurignacian), and the Middle Upper Paleolithic (the Gravettian and its variants in time and space) (Svoboda et al., 1996; Roebroeks et al., 2000; Svoboda & Bar-Yosef, 2003; Svoboda & Sedláčková, 2004). From this comes a recognition, long overdue, of the major amounts of human cultural fluorescence that emerged after 30,000 years B.P. and continued through the middle of the following ten millennia.

Whereas several behavioral changes were elaborated during the Early Upper Paleolithic in Europe—symbolism, representational art, and personal decoration being the archeologically most visible—the Middle Upper Paleolithic involved patterns of variability, complexity, and/or specialization far exceeding those of the earlier Upper Paleolithic. In the Lower Austrian/Moravian region, the earlier Gravettian (Pavlovian) record suggests greater sedentism, the formation of large settlements, the long-distance transport of lithic material, foraging specialization on the largest and smallest mammals (mammoths and the small fur animals), and innovations in technology (e.g., polishing stones, ceramics, and textiles). Therefore, the Gravettian system, compared to the earlier Upper Paleolithic cultural entities, seems not only to be more complex but in some aspects also more labor-expensive. The analysis of the preserved human remains should demonstrate whether and how far these changes are reflected in the anatomy of the humans responsible for it.

At the same time, it is becoming increasingly apparent that there were significant changes in many aspects of human biology during the Interpleniglacial. The first set of these changes involved those associated with the spread of early modern humans across Europe and their variable absorption of populations of late archaic humans (Neandertals) in the process. Focusing on aspects of human biology that reflect principally changes (or stasis) in behavior and adaptation, as opposed to ones that serve mostly as phylogenetic markers, indicates a suite of changes involving manipulative behaviors, locomotor patterns, and decreases in overall stress levels (Trinkaus, 2001).

However, well-dated late Neandertal and early modern human remains from the actual period of transition, the Initial and Early Upper Paleolithic, are extremely rare and have become more so as direct dating of the fossils has proceeded. The available well-dated human skeletal remains between 30,000 and 40,000 years ago in Europe are currently limited to the late Neandertals from Arcy-Renne, Cabezo Gordo, Neandertal, Saint-Césaire, Vindija G_1, and Zafarraya (Mercier et al., 1991; Hublin et al., 1995, 1996; Smith et al., 1999; Walker, 2001; Schmitz et al., 2002) and the early modern humans from Kent's Cavern, Oase, La Quina, and Mladeč (Stringer, 1990; Dujardin, 2003; Svoboda et al., 2002b; Trinkaus et al., 2003a,b), plus isolated teeth from several sites. As a result, the bulk of the comparisons across the late archaic to early modern human biological transition are between earlier Middle Paleolithic Neandertals and Middle Upper Paleolithic early modern humans, with the few diagnostic human remains in between providing a mosaic paleobiological pattern (Trinkaus et al., 1999c, 2003a,b; Trinkaus, 2001).

The second set of changes are those that took place between the Middle and the Late Upper Paleolithic human populations of Europe, alterations in human biology that principally reflect patterns of human behavior and the means of coping biologically and technologically with the environments of glacial Europe (Brennan, 1991; Holliday, 1995; Formicola & Giannecchini, 1999; Churchill et al., 2000; Holt et al., 2000; Trinkaus et al., 2001; Mednikova & Trinkaus, 2001).

In these sets of comparisons, it is evident that it is the Middle Upper Paleolithic human populations which are the pivotal group. They provide the "after" in the consideration of human biological changes across the Middle to Upper Paleolithic transition, and they provide the "before" in the assessment of human biological changes during the Upper Paleolithic. It is therefore appropriate that there should be increasingly detailed assessments of the known human paleontological samples from the Middle Upper Paleolithic of Europe. It is in this context that we present here a description and paleobiological comparative analysis of the largest sample of human remains from the earlier Middle Upper Paleolithic, or Pavlovian, of central Europe, the human remains from the sites of Dolní Věstonice I and II and of Pavlov I.

The Pavlovian People

The human remains from the sites of Dolní Věstonice I and II and Pavlov I in southern Moravia include six partial skeletons, Dolní Věstonice 3, 13–16, and Pavlov 1 (note: the Pavlovian sites have Roman numeral designations, whereas the human remains are individuated by Arabic numerals). There is an associated dentition of an infant, Dolní Věstonice 36, and the isolated remains of several dozen other individuals (see Chapter 5). As such, this sample is approached in terms of completeness only by those from Sunghir (Bader, 1998; Alexeeva et al., 2000), Paglicci (Mallegni & Palma di Cesnola, 1994) and Cussac (Aujoulat et al., 2002) and exceeded in completeness and number of individuals only by the largely lost remains from Předmostí (Matiegka, 1934, 1938; Drozdová, 2001). These fossils have been variably described in the past, through the announcements of their discoveries (Klíma, 1963, 1987a,b,c, 1988, 1990; Svoboda, 1988; Svoboda & Vlček, 1991; Trinkaus et al., 1999b, 2000c) and in variably complete descriptions of their morphology (Malý, 1939;

Jelínek, 1953, 1954; Vlček, 1961, 1991, 1997; Trinkaus & Jelínek, 1997; Trinkaus, 1997b; Trinkaus et al., 1999b, 2000c). Among these publications, Vlček (1991) provided the most complete overall assessment of the then available human remains from the Dolní Věstonice sites, which was followed by a detailed assessment of the Pavlov remains (Vlček, 1997). The articles in Vlček (1992) by Vlček and colleagues provided initial assessments of many aspects of the Dolní Věstonice human remains. And most recently, Sládek et al. (2000) have provided a detailed catalogue of the human remains from these sites, including a detailed inventory of all of the remains and an extensive set of morphometric observations on the remains.

In addition to these publications, the Dolní Věstonice and Pavlov human remains have been increasingly incorporated into comparative analyses of Late Pleistocene human remains, some of which provide original data on the specimens. These publications include a number of doctoral dissertations (e.g., Churchill, 1994; Franciscus, 1995, Holliday, 1995; Schumann, 1995; Pearson, 1997; Holt, 1999; Sládek, 2000; Bailey, 2002) and various paleobiological and/or phylogenetic analyses (e.g., Jelínek, 1989; Alt et al., 1997; Churchill & Formicola, 1997; Holliday, 1997a, 2000b; Formicola & Giannecchini, 1999; Trinkaus et al., 1999c, 2005; Trinkaus & Churchill, 1999; Trinkaus & Ruff, 1999a,b; Churchill & Smith, 2000; Holt et al., 2000; Pearson, 2000; Trinkaus, 2000c; Dobson & Trinkaus, 2001; Shackelford & Trinkaus, 2002; Ruff et al., 2002; Sládek et al., 2002).

As the next step in the process of the detailed description and analysis of this Middle Upper Paleolithic human sample, this volume provides an overall presentation and assessment of the Dolní Věstonice and Pavlov human remains. In addition, it discusses the specific contributions of this sample to our understanding of Late Pleistocene human biology and evolutionary patterns.

Paleobiology versus Phylogeny

The focus of this volume is principally on the paleobiology of these past human populations, assessed from their skeletal and dental remains. As such, it emphasizes those aspects of their remains that permit the evaluation of their functional morphology, paleopathology, and general skeletal correlates of behavioral patterns.

At the same time, it is recognized that a number of aspects of their biology have relevance for phylogenetic considerations of OIS 3 human populations, both in terms of the previous Neandertal to early modern human transition and in terms of the subsequent dynamics of human populations during the earlier Upper Paleolithic. These aspects will be considered as it is deemed appropriate, but it is our contention that these issues need to be focused principally on the regional level given current interpretations of European Late Pleistocene human phylogeny.

The principal, and most contentious, aspect of these phylogenetic considerations is that concerning the spread of early modern humans across Europe. After a decade and a half of animated debate, a consensus is emerging that (1) early modern humans emerged in Africa in the initial Late Pleistocene; (2) sometime after 40,000 years B.P. they spread throughout Europe, fully colonizing the then occupied regions only after 30,000 years B.P.; and (3) in the process they variably absorbed the Neandertal populations resident in Europe (Smith et al., 1989; Duarte et al., 1999; Hublin, 2000; Bräuer, 2001; Stringer, 2001; Wolpoff et al., 2001; Trinkaus & Zilhão, 2002; Trinkaus et al., 2003b; White et al., 2003;). The degree to which those Neandertal populations were integrated into early modern human populations remains uncertain, and it is likely that it will take detailed considerations of regional fossil sequences, employing far larger samples of human remains than are currently available, to resolve differences of interpretation about the degree of admixture between these two human groups. Moreover, it is now apparent that living human genetic variation supports such a paleontological interpretation (Relethford, 2001; Templeton, 2002) and that Late Pleistocene ancient DNA will remain incapable of providing further resolution to the issue (Nordborg, 1998; Tschentscher et al., 2000; Wall, 2000; Pusch & Bachmann, 2004; Serre et al., 2004).

The second area of phylogenetic concern involves the relationships among human biological populations during the Upper Paleolithic and the changes, sometimes dramatic, among the various defined cultural and technological complexes of the Upper Paleolithic (e.g., Svoboda et al., 1996; Svoboda & Bar-Yosef, 2003). It is possible that these archeologically documented shifts involved major displacements of human populations. However, given that all of the human populations were early modern humans and must have overlapped extensively in morphological attributes, a far larger and denser human paleontological record, such as is normally found only with the advent of sedentism and cemeteries in the early Holocene, will be required to address these issues paleontologically.

Consequently, after decades of controversy we may be reaching the point at which the human paleontological record, in its current condition or expected state in the near future, is no longer able to provide further resolution to the phylogenetic debates of the twentieth century. For these reasons, a paleobiological approach to the Pavlovian human remains of southern Moravia, but one that bears in mind the general phylogenetic background of these people, appears to be the more productive one.

General Considerations

Consequently, this presentation of the human remains from the Dolní Věstonice and Pavlov sites will detail their archeological contexts, the burial rituals of the associated skeletons, the paleodemographic parameters of the sample, morphological and paleobiological assessments of the various skeletal and dental anatomical regions, and presentations and evaluations of the multiple pathological lesions on the remains. A brief summary of the skeletal inventory is provided, but the detailed inventory presented in Sládek et al. (2000) is not repeated; only a few minor updates are provided in the relevant anatomical sections. Similarly, the extensive set of tables of morphometric data provided in Sládek et al. (2000) are not repeated; only in cases of new data determined as a result of detailed reassessments are primary data provided. For this volume, and we hope for all subsequent treatments of these fossil human remains, the data published in Sládek et al. (2000), plus any updates in this volume, will supercede all previously published measurements on these remains and thereby provide consistency. It is assumed that the reader will have access to the Sládek et al. (2000) volume, available through the Institute of Archeology of the Czech Academy of Sciences, Brno (Archeologický ústav, Akademie věd České republicky; http://www.iabrno.cz/3cf.htm).

It is therefore hoped that this volume will bring together previously published and new considerations of these Middle Upper Paleolithic human remains from southern Moravia and thereby facilitate their integration into our understanding of the biocultural dynamics of early modern human populations.

2

The Archeological Framework

Jiří A. Svoboda

The lowlands of lower Austria, Moravia, and south Poland form a natural corridor, separating the Bohemian Massif in the west from the Carpathian Mountains in the east. As such, they have allowed movements of both animals and humans from the Danube Valley in the southwest to the north European plain in the northeast.

The Middle Upper Paleolithic, or later Interpleniglacial (terminal OIS 3 and early OIS 2), was a period of global climatic instability leading toward the Last Glacial Maximum (Guthrie & Kolfschoten, 2000). In the loess stratigraphic record of central Europe, this period is represented either by a sedimentary hiatus between the last important pedogenesis and the last loess cover, as in most of the Moravian sites (Klíma et al., 1962), or by shallow loess deposition and initial pedogenesis, as in the Austrian sites (Haesaerts et al., 1996). Wherever stratigraphies are present, the loess/paleosol sequences suggest a dynamic climatic evolution in a "staccato" rhythm. In Moravia, in the absence of such structured vertical sequences, chronological studies are based on the spatial analysis of the large sites combined with series of uncalibrated radiocarbon dates (Svoboda, 1994; Svoboda et al., 2002a). The approximate occupation dynamics appear to fall into three periods, the Early Pavlovian (30,000–27,000 years B.P.), which culminates in the Evolved Pavlovian (27,000–25,000 years B.P., including a large majority of all of the Gravettian radiocarbon dates and the dates of the burials), which then transforms into the Willendorf-Kostenkian, or shouldered-point, horizons (25,000–21,000 years B.P.). Most of the Gravettian sites in this region were settled repeatedly during relatively long time spans, be it millenia, centuries, or seasons of the year (Table 2.1).

The pollen and charcoal analyses from the related stratigraphic layers show that the landscape was partly covered by wooded areas (arboreal pollen usually exceeds 50%), dominated by conifers, and accompanied by decidous trees, including even a few more pretendous species such as oak, beach, and yew (Rybníčková & Rybníček, 1991; Svobodová, 1991a,b; Mason et al., 1994; Opravil, 1994). Geological study of the frost features, supported by evidence from molluscs, in contrast, suggests markedly colder environmental conditions than do the plants—a kind of cold subarctic tundra (Kovanda, 1991). On the basis of discussions of this complex and sometimes contradictory evidence, we would reconstruct a variable and changing environment, both in terms of time and altitude on the landscape, that varies among steppe, shrub and steppe, and partially forested landscapes. This reconstruction is also in accord with the diversity of large mammals in the landscape (West, 2001; Musil, 2002).

A strategic role in the region was played by the so-called geomorphological "gates," topographic places where the river valleys or dry valleys become narrow and the slopes steep. In the southwest, a typical example is the Wachau Gate on the middle Danube River, joining upper and lower Austria. Narrow valleys also occur in southern and central Moravia, along the Dyje and Morava Rivers. In the northeast, another case is the Moravian Gate on Bečva and Odra Rivers, connecting Moravia with south Poland. Existence of such a geographic line is supported by the transport of lithic raw material, flint, from the northeast toward the southwest over distances of hunderds of kilometers.

The Gravettian site distribution (Figure 2.1) copies the axial shape of the territory from the southwest to the northeast along the main rivers of lower Austria, Moravia, and south Poland: Willendorf and Aggsbach in Austria;

Table 2.1 The Gravettian chronological framework based on mean uncalibrated dates, mostly from Groningen.

Years B.P.	Stage	Willendorf Site II	Dolní Věstonice Site I	Dolní Věstonice Site II	Pavlov I: a, b	Předmostí I, II	Brno 2
20,000	Willendorf-Kostenkian						
21,000							
22,000				DV 35: 22,840			
23,000		layer 9: 23,180 23,860					23,680
24,000		24,370 24,910				site II:	
25,000	Evolved Pavlovian	layer 8:	middle and upper part:	DV 16:	25,020–b	25,040	
		24,710 25,800	25,820	**25,570 –25,740**	25,530–b	site I:	
26,000		layer 7: layer 6:	25,950	unit LP: 26,390	26,170–a	26,320	
		26,150 26,500 27,600		DV 13–15: **26,640** majority:	26,620–b 26,650–b 26,730–b	26,870	
27,000	Early Pavlovian	27,620	lower part: 27,250	**26,900 –27,100** unit B: 27,660 lower part:			
28,000			(charcoal layers)	28,300 lower part:			
		layer 5: 27,270					
29,000		30,500	lower part: 29,300	29,000			

Certain questionable dates from other laboratories were omitted. Dates associated with human fossil finds are indicated in bold.

Dolní Věstonice, Pavlov, Boršice, Jarošov, Spytihněv, Předmostí, and Petřkovice in the Czech Republic; and Spadzista in Poland (the "Gravettian landscape"; cf. Otte, 1981: fig. 5; Kozlowski, 1986; Svoboda et al., 1996; Valoch, 1996: carte 6; Oliva, 1998; Škrdla & Lukáš, 2000; Svoboda, 2003). The hunters´ settlements lie in lower altitudes, relative to the Aurignacian or Magdalenian sites (200–300 m a.s.l.), on mid-slopes that are not too high but still control the river valleys or at junctions of a main valley with short, steeply sloping side gullies. This pattern of site location seems to be related to the exploitation of large mammals that follow the river valleys—the mammoths. This presupposition is also supported by the dumps of mammoth osteological material located either inside the settlements or in the adjacent side gullies or as individual pieces scattered in the river floodplain (Svoboda et al., 2005). Such places were also meaningful as areas of human aggregation and communication, with a stimulating effect on technological growth (Soffer, 2000) and symbolism (Klíma, 1989; Svoboda, 1997c; Verpoorte, 2001).

Description and internal analysis of the Dolní Věstonice and Pavlov sites have involved a sequence of monographic presentations, with the tendency toward increasingly interdisciplinary collaboration (Absolon, 1938a,b, 1945; Svoboda, 1991, 1994, 1997b; Klíma, 1963, 1995). The structures of the settlements as reconstructed in these publications and the nature of the activities that have been performed there provide several implications for human anatomy (cf. Trinkaus et al., 2001). First, there is strong evidence of sedentism, supported by indications

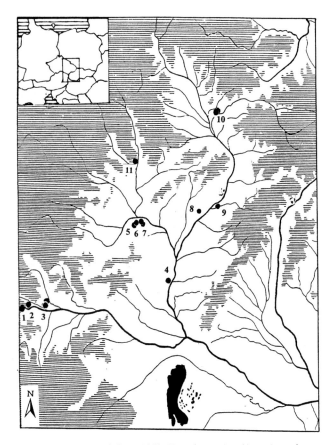

Figure 2.1 Map of the middle Danube region (Austria and Czech Republic), showing the more important Gravettian sites. 1: Aggsbach; 2: Willendorf; 3: Krems-Wachtberg, Langenlois; 4: Stillfried-Grub; 5: Dolní Věstonice II–III; 6: Dolní Věstonice I; 7: Pavlov; 8: Boršice; 9: Jarošov; 10: Předmostí; 11: Brno.

of various seasons of human presence, intensity of the occupation layers and richness of the artifacts, stability of certain dwelling structures, and time-consuming and delicate technologies (microliths, fine ivory carvings, textiles, ceramics, etc.). At the same time, the strong cultural relationships among these sites, long-distance lithic raw material importations, and possible following of animal herds along the rivers suggest a mobility across the landscape. Therefore, we may expect to find both sedentized and mobile individuals in the Gravettian human paleontological sample or individuals suited to both lifestyles. In addition, the labor-expensive activities, such as transport of a considerable weight of lithic raw material to the site or hunting and butchering the largest mammals of that time, may also be reflected in the human anatomy.

3

The Archeological Contexts of the Human Remains

Jiří A. Svoboda

Since 1924 there has been a series of excavations of the complex of Gravettian sites located along the northern slopes of the Pavlovské Hills, above the Dyje River floodplain (Figures 3.1 and 3.2). The chronological course of the excavation process, be it prospective, systematic, or salvage fieldwork, is summarized in Table 3.1.

Based on the sizes of these sites and their relative archeological complexities, there is a visible hierarchy. The large and complex sites, with extended and intensive occupations and numerous art objects, are Dolní Věstonice I and Pavlov I, as well as Předmostí I. Dolní Věstonice II is large in size, but the occupation was less intensive and more structured in time and space than the above sites. In addition, a network of middle-sized and small sites is recorded. Table 3.2 summarizes this hierarchy in the framework of the Dolní Věstonice–Pavlov area (see Chapter 5 for an explanation of the site and human fossil names and numbering).

Human fossils appear as both ritual burials and scattered fragments. Yet, both are an integral component of the settlements, and they are usually concentrated in the central parts of the sites. In the chronological framework (Tables 2.1 and 3.3), the majority of human burials from Dolní Věstonice I and II and Pavlov I, plus Předmostí I, fall into the Evolved Pavlovian stage (27,000–25,000 years B.P.). An earlier Pavlovian horizon, identified at Dolní Věstonice II and dated to around 27,000 years B.P., is related only to the isolated human fragments Dolní Věstonice 33, 36, 39, 47, and 49. This, at least, is the chronological picture given by standard laboratories such as Groningen (GrN). Several younger dates were produced by the Illinois (ISGS) and other laboratories, which do not fit as well into the system and are more problematic.

One later burial, corresponding with the Willendorf-Kostenkian stage of the Gravettian, is Brno 2, 23,700 years B.P. (Pettitt & Trinkaus, 2000). However, this later burial

Table 3.1 Main sites of the Dolní Věstonice–Pavlov area with the years of the excavation campaigns.

Site	Excavator	Type of Fieldwork	Excavation Date
Dolní Věstonice I	K. Absolon	Systematic	1924–1938
	A. Bohmers	Systematic	1939–1942
	K. Žebera a kol.	Prospection	1945–1946
	B. Klíma	Systematic, salvage	1947–1952, 1966, 1971–1979
	J. Svoboda	Control trenching	1990, 1993
Dolní Věstonice II	B. Klíma	Prospection	1959–1960
	B. Klíma-J. Svoboda	Salvage	1985–1991
Dolní Věstonice IIa	J. Svoboda	Salvage	1999
Dolní Věstonice III	P. Škrdla	Prospection	1993–1995
Pavlov I	B. Klíma	Systematic	1952–1965, 1971–1972
Pavlov II	B. Klíma	Systematic	1966–1967

Table 3.2 The hierarchy of the sites in the Dolní Věstonice–Pavlov area.

Large and Complex	Large	Middle	Small	Mammoth Bone Deposits
DV I	DV II	DV IIa	Pavlov III–V	DV I
Pavlov I		DV III		DV II
		Pavlov II		

is unique in two aspects: its location is outside of the typical Pavlovian regions and settlements, and it is unusually rich in grave goods. Another later date obtained directly from the Dolní Věstonice 35 femur from the site of Dolní Věstonice I (22,800 years B.P.) may possibly be contaminated since the majority of the other dates from the Dolní Věstonice I settlement correspond with the Evolved Pavlovian stage (Trinkaus et al., 1999b). However, its original location, and hence archeological associations, within the site complex is not known (it was identified as human during the 1990s in the Moravské Zemské Muzeum).

The only other human remains from the same region and time period are the isolated pieces from Willendorf I and II, which date to the Willendorf-Kostenkian (Teschler-Nicola & Trinkaus, 2001), and two human teeth recently discovered in Grub/Kranawetberg (Antl-Weiser & Fladerer, 2004). The Svitávka 1 partial skeleton, originally considered Gravettian (Vlček, 1967), has been directly dated and shown to be early medieval (Svoboda et al., 2002b). Similarly, the long bones from Krems-Hundssteig,

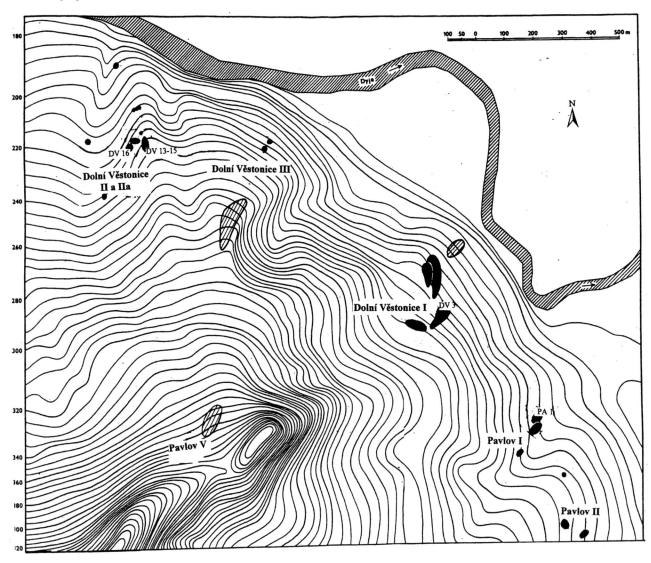

Figure 3.1 Map of the Dolní Věstonice–Pavlov area, showing the locations of Dolní Věstonice I–III and Pavlov I, II, and V and the burials of Dolní Věstonice 3, Dolní Věstonice 13–15, Dolní Věstonice 16, and Pavlov 1.

Figure 3.2 View of the Pavlovské Hills, showing locations of the sites. Photo by M. Novák.

previously assumed to be Gravettian (Jungwirth & Strouhal, 1972), have been shown to be Bronze Age (Trinkaus & Pettitt, 2000). The Brno 1 remains are of questionable antiquity, and the Brno 3 burial, destroyed and unavailable for reanalysis, remains problematic.

Dolní Věstonice I

Spatial Structure and Chronology

This site represents one of the large and complex settlements, with rich evidence of technological development and symbolic activities. The reconstruction presented here combines earlier field observations by Absolon (1938a,b, 1945) with the modern evidence by Klíma (1963, 1981, 1983, 2001) and with still later controlled trenches, aimed mainly at obtaining more radiocarbon dates (Figure 3.3).

The site is separated into the lower, middle, upper, and uppermost parts. Concerning spatial patterning, the upper and uppermost parts appear to be more clearly structured and readable (Klíma, 2001), whereas the middle part has extended but irregular bone, artifact, and charcoal deposits (Klíma, 1981). The problematic lower part includes mostly charcoal areas with poor evidence of occupation. The site is still poorly dated by radiocarbon, even though it was the first one to be discovered and is perhaps the most important one,. The dates from the upper and uppermost parts range to the Evolved Pavlovian (around 25,000–26,000 years B.P.), whereas earlier dates were obtained from the lower part only (27,000–29,000 years B.P., two dates in direct superposition from the 1990 excavation; Svoboda et al., 2002a). Two other dates, one of which comes directly from an isolated human bone, Dolní Věstonice 35 (Trinkaus et al., 1999b), are more recent but not as certain.

Table 3.3 Radiocarbon dates (uncalibrated) directly associated with Gravettian human fossils from Dolní Věstonice (DV) and Pavlov (Pav).

Site	Fossil Number	Lab Number	Material	Context	^{14}C Age (B.P.)
Pavlov I–northwest	Pav 1–3	GrN 20391	Charcoal, exc. 1957	Pavlovian	26,170 ± 450
DV I	DV 35	OxA 8292	Human femur	Gravettian	22,840 ± 200
DV II	DV 13–15	GrN 14831	Associated charcoal	Pavlovian	26,640 ± 110
DV II	DV 13–15	ISGS 1616	Associated charcoal	Pavlovian	24,000 ± 900
DV II	DV 13–15	ISGS 1617	Associated charcoal	Pavlovian	24,970 ± 920
DV II	DV 16	GrN 15276	Associated charcoal	Pavlovian	25,570 ± 280
DV II	DV 16	GrN 15277	Associated hearth	Pavlovian	25,740 ± 210
DV II	DV 16	ISGS 1744	Associated charcoal	Pavlovian	26,390 ± 270
DV II	DV 33	GrN 15324	Associated hearth	Pavlovian	27,070 ± 170
DV II	DV 36, 39, 49	GrN 21122	Associated hearth	Pavlovian	26,970 ± 200
DV II	DV 47	GrN 15279	Associated hearth	Pavlovian	26,920 ± 250
DV II	DV 51–52	GrN 21123	Associated hearth	Pavlovian	26,390 ± 190

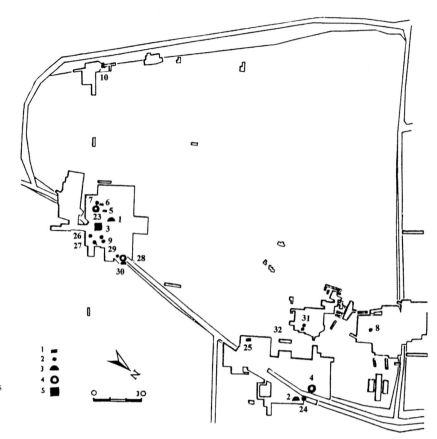

Figure 3.3 Dolní Věstonice I. Plan of the site, with the locations of the human fossils (numbered). 1: smaller skeletal fragments; 2: teeth; 3: calottes; 4: fragments of a burnt calotte; 5: ritual burial. After Klíma (1990).

Archeological Features

Several features deserve special attention. A large mammoth bone deposit ("kjökkenmödding") was located in a terrain furrow along the upper part of the site. A circular structure of mammoth bones, also in the upper part and interpreted as a bone dump by Absolon, was later explained by Klíma as a mammoth bone hut of the eastern European form. The clearest dwelling structure, however, is settlement unit 2 in the uppermost part of the site, with a central hearth, a circular stone alignment, and a few postholes. All other features (such as settlement unit 1, for example) appear as extended charcoal deposits and accumulations of bones, artifacts, and artistic objects, including the ceramic lumps and figurines, and may be of palimpsest character.

Human Fossils

The most important human remains, the burial of the female Dolní Věstonice 3, was located in the upper part of the site at the margin of settlement unit 1, whereas another possible burial, the incomplete remains of the Dolní Věstonice 4 child, lay in the middle part of the site. Neither of these two depositions is directly dated by radiocarbon. Individual skeletal fragments (the calvae of the now lost Dolní Věstonice 1 and 2, plus various cranial fragments) and isolated teeth were scattered over the site (Figure 3.3; Table 5.1; Klíma, 1990; Trinkaus et al., 1999b; Sládek et al., 2000).

Dolní Věstonice II

Spatial Structure and Chronology

Dolní Věstonice II is a well-structured site, both spatially and chronologically (Figure 3.4; Klíma, 1995; Svoboda, 1991, 2001b). Compared to Dolní Věstonice I and Pavlov I, it is the result of short-term but repeated occupations, extending over a considerably larger area (almost 500 m^2) and longer time span (29,000–24,000 years B.P.), with a lower artifact density, a scarcity of decorative objects, and an absence of representational art, but with evidence of certain specialized activities. The first series of radiocarbon dates, all from clearly visible settlement units, are concentrated around 27,000 years B.P., the end of the Early Pavlovian, and interrelations between two activity areas related to this horizon are also revealed by refittings. A later series of dates falls into the time span of 27,000–25,000 years B.P., in the Evolved Pavlovian. Spatial and temporal associations to mammoth bone deposits located in the adjacent gully are evident. In addition, smaller carnivores represent an important part of the faunal material (West, 2001), indicating that systematic fur and hide working was involved, and the use-wear analysis seems to confirm this hypothesis.

Archeological Features

Whereas the individual settlement units cannot be separated in the densely occupied areas at the top of the site, they are clearly visible on the western and northern slopes. In a few cases, the feature appears as a regular shallow depression, with a hearth or hearths in the center (unit 1, northern unit). Others are well-represented hearths, sometimes surrounded by a circle or semicircle of small, kettle-shaped pits (unit 3, eastern unit) but without the visible boundaries to the constructions one would expect around a hearth. Distribution plans of artifacts and decorative objects (Klíma, 1995; Svoboda, 1991: figs. 2.3, 22–23; Svoboda et al., 1993) show concentrations around the settlement units but no specific accumulations related to the burials.

Human Fossils

This site is especially reknowned for the accumulation of both ritual human burials and scattered fragmented human remains in the cultural layer (Klíma, 1987a,b, 1988; Svoboda, 1988; Svoboda & Vlček, 1991; Vlček, 1991; Trinkaus et al., 2000c). If we focus on the dates obtained from charcoal associated with buried human skeletons (the Dolní Věstonice 13–15 triple burial and the Dolní Věstonice 16 adult male burial), the dates from the central hearths of the same settlement units as the human fossils, and the dates obtained directly from human bones (Table 3.3), it appears that a few smaller human fragments such as Dolní Věstonice 33, 36, 39, 47, and 49 fall in the 27,000 years B.P. horizon, whereas the majority of the finds, such as the triple burial, the Dolní Věstonice 16 male burial, and others, follow in the subsequent Evolved Pavlovian stage (27,000–25,000 years B.P.).

Pavlov I

Spatial Structure and Chronology

The site of Pavlov I was excavated systematically by B. Klíma. The excavation recovered a large but spatially more restricted settlement, with archeological accumulations resulting from a variety of activities during a well-defined and restricted time span of 27,000–25,000 years B.P. (the Evolved Pavlovian). The site is separated in the northwestern and southeastern parts (Figure 3.5), showing certain slight differences (as in the flint-radiolarite representation, for example) but with the same radiocarbon dates. The density of features and artifacts is high, and decorative objects and art are associated in the same contexts (Svoboda, 1994, 1997b, 2001c; Verpoorte, 2000).

Archeological Features

The shapes of the features are circular, oval, or irregular, but no regular alignments of larger objects around them were visible. In fact, only feature 5 represents a classic case of a circular, slightly subterranean dwelling with one central hearth. Another circular feature, unusual because

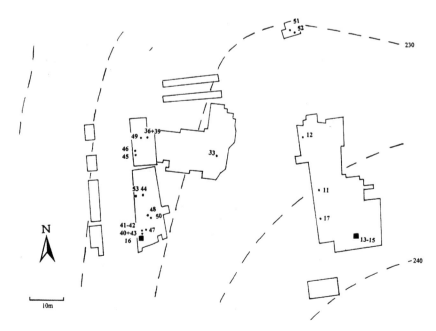

Figure 3.4 Dolní Věstonice II. Plan of the site, with the locations of the human fossils (numbered). Squares: Dolní Věstonice 13–15 and Dolní Věstonice 16 associated skeletons. Dots: individual isolated human remains. The 230 m and 240 m contours above sea level are provided; the hillslope slopes down to the north-northwest. After Trinkaus et al. (2000c).

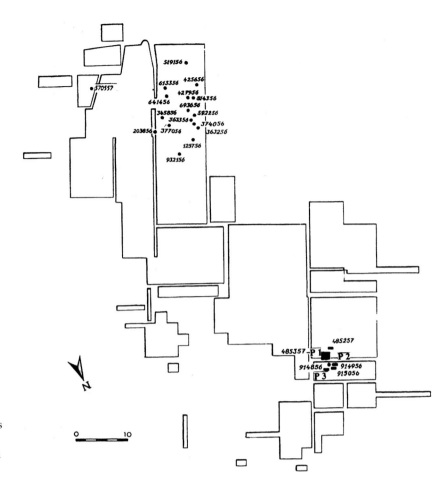

Figure 3.5 Pavlov I. General plan of the site, showing location of the human fossils Pavlov 1–3 and the inventory numbers of the smaller finds. After Klíma (1997), with additions.

of its depth (80 cm) and filled with bones and artifacts (Klíma, 1977), may be interpreted as a storage pit, such as those documented more frequently from the eastern European sites. Coincidence (or spatial overlaps) between the features and the artifact concentrations, as visible on the Surfer graphic presentations, is only very general (cf. the 1953 and 1957 areas), and in certain areas (1954–1956) the artifact densities do not respect the supposed dwellings (Verpoorte, 2000, 2001). Given the intensity of repeated reoccupations and, as a result, the palimpset character of Pavlov I, the reading of these structures at Pavlov I is not always clear.

Human Fossils

The male burial Pavlov 1 was located in the northwestern part of the site, together with portions of a maxilla and two mandibles (Pavlov 2–4). This part of the site has one radiocarbon date, 26,170 years B.P., which, however, is not directly related to the burial. Individual teeth were scattered, especially in the center of the southeastern part, but always in the context of intensive occupation remains and dated with five dates to the 25,000–26,000 years B.P. interval (Vlček, 1997; Sládek et al., 2000).

4

The Burials: Ritual and Taphonomy

Jiří A. Svoboda

Formation and Protection of the Burials

All Gravettian burials of Moravia (in contrast to the cave burials of the Gravettian and Epigravettian in Italy, for example) share the same practice of burying in the open air. However, the details of the deposition show considerable variability as a result of human decisions and mortuary behavior. The human burials within the Dolní Věstonice–Pavlov area have patterns of uniformity, depositing either complete human bodies or small fragments in the central parts of the settled areas. At Předmostí, an accumulation of more than twenty fragmentary bodies was found at one place (Maška, 1895; Absolon & Klíma, 1977; Svoboda, 2001a), and at Brno 2, a male burial was associated with a masculine sculpture and a rich symbolic inventory (Makowsky, 1892; Jelínek et al., 1959; Oliva, 2000). In other cases (Aurignacian, Magdalenian), bodies were dropped through chimneys into vertical karstic cavity systems (Mladeč, Koněprusy; Svoboda, 2000a). These behaviors would cause various effects, such as damage through the deposition of additional bodies into a limited area (Předmostí), exposure of the bodies to current daily activities in settlements (Dolní Věstonice), or additional movements and dispersion of the bodies in underground cave systems (Mladeč, Koněprusy).

All of the burials in the Dolní Věstonice–Pavlov area (Table 4.1) were relatively shallow; some were deposited directly in the cultural layer or slightly below it. The shallow depth is also attested to by extensive root etching on the Dolní Věstonice 13 to 16 burials, indicating that they were relatively close to the surface; the root etching is Pleistocene, given the several meters of loess overlying the remains at the time of their archeological discovery. The lack of pits is recorded from the Moravian settlements in general; the deepest pit in the region, of some 70 cm, was excavated at Pavlov I (Klíma, 1977), and another one, of some 40 cm, was found near the Dolní Věstonice 16 burial at Dolní Věstonice II (Svoboda, 1991). This is in contrast with the Russian sites, where burial pits, storage pits, and other artificial depressions were common components of the settlements (cf. Soffer, 1989). One of the possible explanations may be the larger extension of permafrost in Moravia.

The shallow positions of these burials require protection. The Dolní Věstonice 3 and Pavlov 1 burials were well

Table 4.1 The Dolní Věstonice (DV) and Pavlov ritual burials and their archeological contexts.

Site	Individual	Position	Orientation	Ochre	Items of Decoration
DV I	DV 3	Strongly flexed	NW	Head, upper part of body	10 fox canines
DV II	DV 13	Supine, torded	SSE	Head	20 pierced carnivore teeth, ivory pendants
DV II	DV 14	On belly	S	Head	3 wolf canines, ivory pendants
DV II	DV 15	Supine	S	Head, pelvis	4 pierced fox canines
DV II	DV 16	Flexed	E	Head, pelvis	4 pierced fox canines
Pavlov I	Pavlov 1	Flexed? (disturbed)	SE	—	—

covered by mammoth scapulae. In the latter case, this protection helped to stabilize at least the central part of the body, even when the body was postdepositionally exposed to slope movements. In the cases of Dolní Věstonice 13–15 and 16, we suppose that there were protective constructions built of organic material over the burials—this at least is suggested by the numerous pieces of wooden logs found in the first case, and by the outlines of an archeological feature (a dwelling?) in the second case. The question of the small and scattered fragments of human bones at the same sites remains open; these may be either human remains that were never buried or protected or the remains of disturbed burials.

Ochre

The association of ritual burials with ochre is well documented at these sites, as with many Gravettian burials across Europe (Zilhão & Trinkaus, 2002). Extensive coverage of the remains was observed on Dolní Věstonice 3 (on the upper part of the body, especially on the skull), Dolní Věstonice 13–15 (on the heads), and Dolní Věstonice 15 and 16 (head and pelvis). The usual type of ochre is earthy hematites of dark red color; outcrops for this kind of ochre were located by Klíma (1963) in the variagated marls of the Ždánice Flysh, not far from the site. Another type of hematite ochre is represented by fragments of red-brown to steel-grey color with a heavy polish. Following X-ray analysis, the matter is an iron or a composite of hematite and quartz with a slight admixture of pyrite. The outcrops appear to be at the eastern margins of the Bohemian Massif, most probably along the upper Morava River (Přichystal, 2002).

An ochre sample from the human skull Dolní Věstonice 14 was analyzed palynologically by Svobodová (1991a). She estimated a higher proportion of water algae (*Pediastrum integrum*, *P. boryanum*), suggesting that during processing the ochre could have been diluted with water. Stone plates and pebbles that are still covered by red pigment are frequently found in the related cultural layers. One of these objects, a red-colored plate, was found in square Aa-20 near the head of Dolní Věstonice 16, suggesting a direct relationship of the coloration process to the Dolní Věstonice 16 burial.

Associated Artifacts

One of the major problems in interpreting Paleolithic burials in general is the status of the artifacts found in association (Vanhaeren & d'Errico, 2002). Whereas the relationship is clear in cases of items of body decoration or extraordinary tools and weapons, it is far less clear in cases of common objects such as lithics and faunal remains. In the case of Brno 2, with its rich grave goods, which is an exceptional case in Moravia, it should be recalled that the date of 23,700 years B.P. points to the later Gravettian (Willendorf-Kostenkian). As such, this unique burial is contemporary with the other richly equipped burials of Italy and Russia—Arene Candide: 23,400 years B.P.; Sunghir: 22,930, 23,830, and 24,100 years B.P. (Bader, 1998; Pettitt & Bader, 2000; Pettitt et al., 2003)]—and thus the richness of the ritual may have a chronological significance.

In the Pavlovian burials, we consider as associated artifacts the perforated carnivore canines (Dolní Věstonice 3, 13–16) and perforated ivory pendants (Dolní Věstonice 13). These items were found directly on the bodies, sometimes still attached to the skulls. Other objects found lying around usually do not differ from the content of the cultural layer used as the filling and coverage of the burial. This has been confirmed by the statistical analysis of the Dolní Věstonice 16 sample (Svoboda, 1989). Selecting "nice blades" from these assemblages and placing them into a relationship with the burial events usually lacks justification. Nevertheless, one outstanding artifact, found near the triple burial of Dolní Věstonice II, a siltstone with regular engravings (Figure 4.10), may well represent a special mortuary object.

In terms of taphonomy, all of the burials were subjected to the effects of postdepositional disturbances such as geological processes, carnivore activities, and later human occupation.

Geological Disturbance

Given the geographic and geological setting of the Dolní Věstonice–Pavlov area, two effects are especially important: slope movements and the pressure of loess that was deposited relatively rapidly after the end of the Pavlovian occupation. Both are able to offer us simple taphonomic responses to questions that would otherwise be explained as the effects of complex ritual, violence, or other deliberate human activities.

The sites are located on slopes, so that movements of the blocks of sediments, layers, groups of objects, or individual pieces should be expected. First, the sediments at the Dolní Věstonice I and II sites were affected by step-like landsliding of whole blocks of loess on the clayish and slippery Tertiary subsoil, following vertical and diagonal fissures (Klíma, 1963, 1995). Inside the loess deposit, various types of deformation of the cultural layers by frost were observed, such as the effects of solifluction, cryoturbation, ice wedges, and deformation of the (originally circular or oval-shaped) pits and hearths. The result of these processes is redeposition of a shallow layer of loess over most of the pits, as well as displacement of artifacts and bones, especially small objects such as microliths (Svoboda, 1991, 2001c; Klíma, 1995; Verpoorte, 2000). Nevertheless, the plans of the excavated areas in the Dolní Věstonice–Pavlov sites show that both the features and the artifact accumulations still preserve certain regular forms. Therefore, we should expect the deformation of layers and displacement of objects in the distance of centimeters or tens of centimeters, maximally, and only a slight deformation of the features. This conclusion is reinforced by the human burials from both Dolní Věstonice I (Dolní Věstonice 3) and Dolní Věstonice II (Dolní Věstonice 13 to 16), which show some slight displacement of the larger bones (lower extremities of Dolní Věstonice 14, for example) and of the small bones (e.g., absence of some phalanges).

A more deformed case was the Pavlov 1 burial. The northwestern part of this site was postdepositionally cut by an erosional channel, and the burial appeared on the edge of it (Klíma, 1997; Vlček, 1997). Even if a large portion of the skeleton was preserved, partly because of the protection by a mammoth scapula, the individual bones were displaced.

Other types of geological disturbance may be expected from the effect of the weight of overlying loess deposits on human bodies after decomposition of the soft tissues. Whereas the period of Pavlovian occupation (30,000–25,000 years B.P.) saw only limited or no loess deposition, the following Last Glacial Maximum (around 20,000 years B.P.) caused a rapid and massive sedimentation of several meters of loess. Effects of this pressure on the partly decomposed human bodies may be the compaction of the skeletons, as in the case of Dolní Věstonice 3 (Trinkaus & Jelínek, 1997), and fractures of the skulls, as in the triple burial (Dolní Věstonice 14). This explanation is more likely than the intentional or even ritual human action: secondary burial in the case of Dolní Věstonice 3 or fatal injuries in the case of Dolní Věstonice 13–15.

Carnivore Activity

Alternatively, it is possible that the distribution and postmortem damage patterns of the isolated human remains reflect the activities of carnivores, scavenging human bodies from shallow graves (Trinkaus et al., 2000c). Fox, the most abundant carnivore represented in the faunal record of the sites (Musil, 1994, 1997; West, 2001), has difficulty breaking the bones of medium-sized mammals, including the major bones of humans. Wolves are capable of breaking human bones into spiral fractures, as documented at Dolní Věstonice II by West, and the large number of wolves represented at these sites suggests that this carnivore could be a contributor to the bone breakage and damage patterns observed. Other carnivores are rare in the Gravettian faunal record but are known to alter carcasses. Hyenas can produce (among other damage patterns) bowl-like calottes from human crania as a result of accessing the brain tissue through the cranial base. Hyenas easily crush human bones and, more so than wolves, splinter bones, and they can also excavate, gnaw, destroy, and disperse human bones. In addition, wolverines can scavenge and modify bone, and bears will consume carcasses when available. However, even though the damage patterns observed on the small and scattered human fossil fragments are compatible with carnivore activity, West observed no distinctive carnivore-related damage patterns on these human remains.

Postdepositional Human Activity

Some of the large and complex settlements with repeated occupations and activities clearly have a palimpsest character. In these areas, we should also take into account disturbance by later human activities.

Given all of these phenomena, the good state of preservation of the burials from Dolní Věstonice is surprising. This is especially striking in comparison with the Předmostí burial area, where only parts of bodies or isolated human bones were found in a spatially restricted concentration (Svoboda, 2000b).

Dolní Věstonice 3

Depositional Context

Dolní Věstonice was discovered in 1949 by Klíma (1950). The crushed skull was removed and extensively reconstructed in the Moravské Zemské Muzeum, with missing pieces filled with plaster and painted to resemble the ochre staining on the skull (compare in situ figures in Klíma, 1963, with its published and current condition; Jelínek, 1954; Chapter 8). The postcranial remains were removed en bloc, and they were fully excavated by Trinkaus in 1994 and 1995 (Trinkaus & Jelínek, 1997). The burial was located at the periphery of settlement unit 1, in the following situation: below a shallow but compact horizon of limestone debris mixed with clay and charcoal, there were two large mammoth scapulae that completely covered the upper and lower parts of the strongly flexed body below (Figure 4.1). Following Klíma (1963), another large mammoth bone nearby, the pelvis, was also associated with the burial.

Position

The skeleton was laid on its right side, with the head oriented toward the northwest, in a strongly flexed

Figure 4.1 View of the Dolní Věstonice 3 burial during excavation, showing the covering mammoth scapula (left) and the underlying burial (right). After Klíma (1963).

position, so that the knees were attached to the trunk (Figures 4.1 and 4.2). The postdepositional compaction of the skeleton resulted in a hyperflexed position of the legs and, to a lesser extent, the arms (Trinkaus & Jelínek, 1997).

Ochre and Artifacts

Ochre was dispersed across the upper part of the body and concentrated on the skull, including the lower parts of the skull and the adjacent loess (Figure 4.2). During decomposition the sediment around the humeri in particular became encrusted with ochre, resulting in partial natural molds of the dorsal surfaces of the bones. The large mammoth scapula, used to cover the upper part of the body and the skull, showed a series of parallel striations up to 20 cm long, oriented in two directions, on the lower surface. From the burial area Klíma (1963) recorded the presence of a pointed blade in front of the face, as well as two other blades and a backed microblade near the legs, all made of flint. A few more lithic chips were discovered in the area of the legs during the excavation of the postcrania by Trinkaus (Trinkaus & Jelínek, 1997). The following faunal remains were associated with it: fingers and part of the pelvis of a polar fox and, around the right hand bones in the pelvic area, ten canines of a polar fox and additional incisors.

Interpretation

The strong flexion of the whole body has sometimes been explained as an effect of secondary burial. However, it may more likely be due to pressure of the overlying loess deposits, which were formed shortly after the end of the Gravettian occupation.

Dolní Věstonice 4

Depositional Context

On October 29, 1927, E. Dania recorded in his diary the discovery of a fragmented and incomplete mammoth scapula (Figure 4.3; cf. Absolon, 1929; Klíma, 1990: 10–11). Below one of the mammoth scapular fragments were eleven fragments of a human skull. The whole configuration was inside an extended concentration of charcoal and red-burnt clay, interpreted by Dania as a hearth.

Position

The neck vertebrae were adjacent to the skull, oriented toward the edge of the charcoal area. Further search for human bones in the vicinity was unsuccessful.

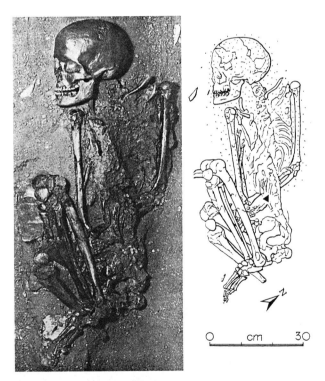

Figure 4.2 Dolní Věstonice I. Photograph and plan of the crouched skeleton Dolní Věstonice 3 after the completed conservation but before excavation of the postcranial remains in the laboratory. The photograph of the skull was taken after reconstruction. The dotted area in the plan shows extension of the ochre; the triangular arrow points to the accumulation of fox teeth in the right hand/pelvis area. After Klíma (1963), with additions.

The Burials: Ritual and Taphonomy

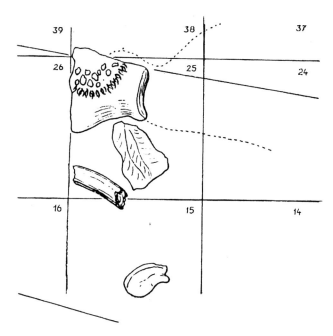

Figure 4.3 Dolní Věstonice I. A child's cranial fragments from Dolní Věstonice 4, with the pattern of opposed pierced fox canines, arranged in couples, as a decoration. After the field diaries by E. Dania.

Ochre and Artifacts

The skull was covered by red ochre. Associated were forty-two fox canines, ordered in a pattern of opposite-oriented pairs (Figure 4.3.); some of the teeth were damaged and some were partly burnt. Absolon (1929: 80) interpreted them as a "necklace," but they may well have been a head decoration, as is usual in other Gravettian burials.

Interpretation

The situation suggested to Dania that the skull had been incompletely burned. In the secondary literature, the find was usually quoted as evidence of early partial cremation. However, this part of the site was characterized by extended charcoal deposits, so that burning the bones—as well as other contents of the cultural layers—may well have occurred subsequently and not as part of a ritual.

Isolated Finds from Dolní Věstonice I

During Absolon's excavation, two human calvaria, Dolní Věstonice 1 and 2, were found in 1925 and 1930 (Absolon, 1938b: 36–38, with anthropological notes by G. M. Morant). Absolon added further cranial vault fragments, evidently of other individuals: Dolní Věstonice 5 and 6, 23 to 25, 28, and 30, and individual teeth numbered Dolní Věstonice 7, 8, and 29. A femoral shaft, Dolní Věstonice 35, was also excavated by Absolon in the late 1930s and rediscovered in the Moravské Zemské Muzeum collections and dated by Trinkaus et al. (1999b).

Klíma's excavation, because of a more systematic floating of the sediments, provided more individual teeth, numbered Dolní Věstonice 9, 10, 26, 27, 31, 32, 37, and 38 (Klíma, 1990; Sládek et al., 2000; Table 5.1). Most of them were found in a concentration in the upper part of the site, around the Dolní Věstonice 3 burial. A smaller and more dispersed cluster was in the middle part of the site, and an isolated tooth lay in the uppermost part (Figure 3.3).

Dolní Věstonice 13 to 15

Depositional Context

On the evening of August 13, 1986, three human skeletons were discovered at the southern margin of the settlement concentration on the top of the Dolní Věstonice II site. They were associated with an unusually large amount of charcoal, particles of burnt wood, and irregular spots of red-burnt loess.

Position

The bodies were lying, more or less, on the ancient surface, oriented with the heads against the slope, to the south (Dolní Věstonice 14 and 15) or south-southeast (Dolní Věstonice 13) (Figures 4.4 to 4.8). Given the slope declination and the horizontal position of the bodies, the legs were located within the cultural layer, whereas the three heads were buried into the underlying loess. There was no burial pit as such, but the upper parts of the bodies were located slightly deeper than the slope surface. The originally shallow location of the bodies is also indicated by the root etching visible on the superior surfaces of a number of the bones.

The scenario of body deposition is visible from the illustrations (Figures 4.4 and 4.5): the central individual, Dolní Věstonice 15, on its back; the right male, Dolní Věstonice 14, on his belly (with his left arm superimposed over the left arm of Dolní Věstonice 15, showing a later deposition of this body); and the left male, Dolní Věstonice 13, on his back, slightly twisted toward Dolní Věstonice 15, with both arms directed toward the latter's pelvis.

The wooden logs, deposited mostly in the same direction as the bodies, are of spruce that is no more than 8–10 cm thick.

Ochre and Artifacts

Ochre was present in a powdered state and as compact lumps in the whole burial area and its direct vicinity, but

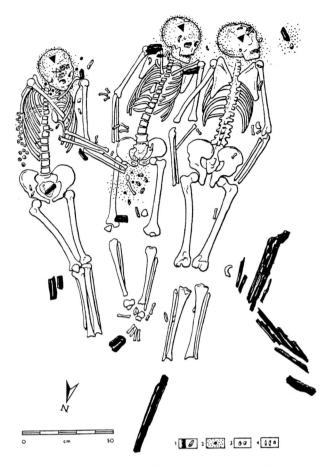

Figure 4.4 Dolní Věstonice II. Plan of the triple burial Dolní Věstonice 13–15. 1: compact charcoal pieces and selected artifacts; 2: ochre and compact pieces of ochre; 3: molluscs (*Melanopsis* and *Arianta arbustorum alpicola*); 4: human teeth, pierced animal teeth, and pierced ivory pendants (exaggerated by the triangular arrows). After Klíma (1995), with additions.

Figure 4.5 Dolní Věstonice II. Photo of the triple burial Dolní Věstonice 13–15.

Figure 4.6 Detail of the skull of Dolní Věstonice 13 during excavation.

The Burials: Ritual and Taphonomy

Figure 4.7 Detail of the skull of Dolní Věstonice 14 during excavation.

compact plastered crusts were attached to the skulls. Another suspicious concentration was in the pelvic area of the central individual. In the frontal part of the Dolní Věstonice 13 cranium, the ochreous coverage formed a thick red crust, with inserted perforated carnivore teeth ordered in three rows of four, five, and eleven teeth, and they were partially selected according to size (Figure 4.9); twenty-seven lumps of ochre were around the neck. Dolní Věstonice 14 had only partial remains of the compact ochreous crust, and three perforated wolf canines were found dispersed nearby. In the case of Dolní Věstonice 15, only four perforated fox canines were found in the vicinity of the skull. In addition, four small, oval-shaped pendants of ivory were dispersed around both of the outside (Dolní Věstonice 13 and 14) skulls.

Figure 4.8 Detail of the skull of Dolní Věstonice 15 during excavation.

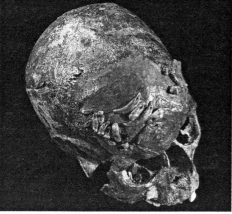

Figure 4.9 Dolní Věstonice 13: compact ochre coverage on the frontal bone, with embedded, pierced decorative objects.

The inventory from the burial area, both from excavation and from floating, includes 4 other perforated fox canines, 6 lumps of burnt clay, nonperforated Tertiary shells, an assemblage of 116 lithic implements and 1 bone awl, and small fragments of bones, some of which were burnt. In the mouth area of Dolní Věstonice 15, there was a piece of animal bone with cutmarks. At the left margin of the burial area, V. Ložek recognized a high—but evidently natural—concentration of molluscs (*Arianta arbustorum alpicola* Fér).

A large hearth was located about 1 m west from the burial area. Five fragments of a longitudinal piece of siltstone, with rhythmically ordered longer and shorter incisions, were found there (Figure 4.10). The incisions were interpreted as callendric by Emmerling and colleagues (1993). Whatever the interpretation, it is an unusual artifact from the broader burial context, and it is therefore worth noting.

Artificial Injuries

Dolní Věstonice 13 has two healed injuries on the skull, first described by Vlček (1991) (see Chapter 19). However,

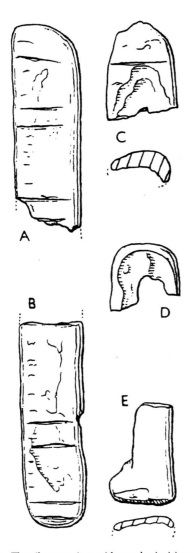

Figure 4.10 The siltstone piece with regular incisions, found west of the burial of Dolní Věstonice 14. After Emmerling et al. (1993).

some presumed fatal injuries, described by the same author, are more problematic. The fractures on the exposed back of the skull of Dolní Věstonice 14 (lying on his abdomen) are probably due to the pressure of overlying loess deposited after decomposition of the brain. The wooden log found near the pelvis of Dolní Věstonice 13, and previously interpreted as a killing weapon, may be just one of the sticks scattered all around the burial.

Interpretation

Deformation of the cultural layer by bioturbation, small-scale slope movements, and cryoturbation, as visible on the geological section, are probably responsible for the disappearance of the smaller bones of the hand and foot skeletons. This type of movement is easily visible in the knee area of Dolní Věstonice 14, for example.

The generally good state of preservation and anatomical connections of the three skeletons, all of which were unburied and appeared almost on the level of the settlement layer, suggest that the bodies were protected against carnivores and other types of disturbance by some kind of a superstructure. This hypothesis is supported by the complex situation, with numerous fragments of wooden sticks scattered around, which may be interpreted as architectural remains of such a hypothetical protective structure.

As a result, the postdepositional changes of the whole area are minimal, and the location of the bodies was authentic and deliberate, so that the positions (e.g., the reverse position of Dolní Věstonice 14) and gestures (the hands of Dolní Věstonice 13 near the pelvis of Dolní Věstonice 15) may have symbolic meaning. It seems that the Dolní Věstonice 15 individual, whatever its gender may have been (see Chapter 7) or, just because of the gender uncertainty (cf. Hollimon, 2001, with ethnographic references) and because of its obvious congenital abnormalities (Chapter 19), played a central role in the mortuary ritual.

Dolní Věstonice 16

Depositional Context

The burial was found inside settlement unit 1, in the western slope of Dolní Věstonice II and in direct spatial and stratigraphic relationship to the central hearth (Figures 4.11 and 4.12). This unit is a shallow depression, about 4.5 m long, and includes further traces of fires and pits filled with stone and bone objects. Unfortunately, the western part of this unique feature was destroyed prior to the rescue excavation at the western slope. The central hearth was relatively small (about 1 m in diameter) but thick (0.35 m), composed of several thin layers of charcoal and burnt loess, and filled with sharp-edged limestone blocks.

Position

The skeleton was buried in the southern part of the depression and oriented in an east-west direction, with the head toward the east and against the slope. He was lying on his right side, with strongly flexed knees; the corpse faced the fire and his knees were only 0.2 m from, and at almost the same level as, the central hearth. Although the head of the skeleton was covered by the loess, the rest of the grave, as well as the rest of the whole depression, lay within a thick cultural layer. This is the same pattern as was recorded for Dolní Věstonice 13–15.

The Burials: Ritual and Taphonomy

Figure 4.11 Dolní Věstonice II. View of settlement unit 1, with the central hearth (center, unexcavated) and the skeleton Dolní Věstonice 16 (right).

Figure 4.12 Dolní Věstonice II, western slope. Plan of settlement unit 1, showing the location of the crouched skeleton Dolní Věstonice 16. A and E: pits; B and C: concentration of objects; D: central hearth of settlement unit 1; 1: red-burnt loess; 2: charcoal accumulation; 3: dispersed chacoal; 4: compact charcoal; 5: red ochre on the body. The triangular arrows indicate locations of the pierced fox canines.

Ochre and Artifacts

The skeleton's head and pelvic area were plastered by ochre. The accompanying inventory found with the skeleton included four perforated carnivore canines, two of them found in the pelvis area and two at the elbow. Further components of the cultural layer from the immediate vicinity of the body were analyzed in detail: a sample from the area of 0.7 x 1.4 m around the skeleton yielded 123 lithic artifacts, 1 Tertiary shell, 1 piece of ochre, and 1 pebble. Statistically, this sample corresponds to the usual content of the cultural layer elsewhere, so a direct association of these artifacts to the burial cannot be demonstrated.

Artificial Injuries

Vlček (1991) described two healed injuries on the skull of Dolní Věstonice 16, but there are multiple lesions on the remains (see Chapter 19).

Interpretation

The male Dolní Věstonice 16 was buried in association with a hearth, the thermal effect of which had been increased and possibly prolonged by the use of limestone blocks. We expect that the whole of settlement unit 1, the eastern outline of which we could follow spatially during the excavation, was sheltered by a construction. If correct, this hypothesis would also solve the problem of protection of the human body and address the fact that it was found in an intact position.

Dolní Věstonice 36

An associated infant's upper and lower dentition, Dolní Věstonice 36, was discovered near the central hearth of unit 4 at the western slope, square F9, but only identified among the faunal remains by West in 1997. The find may represent an infant burial whose skeletal elements were destroyed through chemical disintegration and sediment compaction. Distribution plans of artifacts, ochre, and shells in this area were published by Svoboda and colleagues (1993). However, they demonstrated the usual concentration of artifacts related to settlement unit 4 in general, rather than to the locus of the teeth.

Isolated Finds from Dolní Věstonice II

On June 14, 1986, excavations between settlement concentrations at the upper part of the Dolní Věstonice II site revealed an isolated calotte of an adult individual (Dolní Věstonice 11) and, six days later, a frontal bone fragment of the same individual (Dolní Věstonice 12) in one of the settlement concentrations. Additional, smaller finds and fragments were recorded during the excavation in the whole area (Dolní Věstonice 17, 33, and 34; Klíma, 1990) or during subsequent laboratory processing of the archeozoological material from the western slope of the site by Trinkaus and Fišáková in 1998 (Dolní Věstonice 39 to 53; Sládek et al., 2000; Trinkaus et al., 2000c).

At the western slope, most of the finds were located within a radius of approximately 1 m from the central hearths of the individual settlement units (Dolní Věstonice 33, 39, 40–43, 47, 49, 51, and 52); these are either teeth or postcranial remains. Two of the finds (the Dolní Věstonice 40 and 43 femoral pieces) lay about 1 m from the Dolní Věstonice 16 skeleton within settlement unit 1. Finally, several postcranial elements derived from the peripheral areas (Dolní Věstonice 44–46, 48, 50, and 53; Figure 3.4).

Pavlov I

Depositional Context

The burial was discovered on September 16, 1957, in the northwestern part of the site, north (downslope) from settlement units 12 and 13 (Klíma, 1959). The human body was covered and partly protected by associated mammoth bones: two scapulae, a long bone, and a molar (Figures 4.13 and 4.14). However, this part of the settlement area had been subsequently eroded by shallow furrows. This probably happened during or shortly after the human occupation because the resulting depressions were quickly filled up by redeposited sediments, with parts of the cultural layer, and the whole situation was then covered by the last Pleniglacial loess. During this process, the burial was initially on the edge of one of these furrows and was redeposited downslope.

Ochre and Artifacts

Klíma mentioned no ochre or artifacts related to the burial. The mammoth scapula used as coverage showed numerous striations on the lower face, as in case of the Dolní Věstonice 3 burial. During analysis of the 1957 area inventory (Svoboda, 1997a: 195), we compared the material of the adjacent sector 02, but neither the distribution pattern nor any specific character of the lithic industries can be correlated with the location of the human burial.

Interpretation

Downslope redeposition of the burial took place in the following manner: the group of human bones covered by the mammoth scapula remained more or less in situ or were slightly redeposited en bloc; all unprotected

Figure 4.13 Pavlov I. Excavation of the Pavlov 1 burial, showing the mammoth scapula coverage over the human bones.

bones, especially the skull (even more easily because of its rounded shape), were removed separately.

Isolated Finds from Pavlov I

Adjacent to the Pavlov I burial were a maxilla, Pavlov 2, and two mandibles, Pavlov 3 and 4 (Vlček, 1997; Sládek et al., 2000). All of these finds, together with a few smaller fragments, were clustered in the northwestern part of the site at the margins of the furrowed terrain.

Another concentration is represented by the human teeth dispersed in the southeastern part of the site. This area shows evidence of a dense (almost palimpsest) occupation, with accumulation of features and artifacts. One isolated tooth was found at the southeastern periphery of the site (Figure 3.5).

Human Bones with Artificial Modifications

Absolon (1929, 1938b: figs. 44–46) discovered two human calvae, broken into the shape of apparent "bowls" (Dolní Věstonice 1 and 2). Also, the Dolní Věstonice 11 specimen discovered in 1986 shows an even clearer pattern that could be interpreted as intentional shaping by alternate blows, as described by Vlček (1991). It should also be noted that this last individual, during its lifetime, suffered a serious healed injury to the frontal region, similar to but more severe than those of Dolní Věstonice 13 and 16 (Chapter 19).

However, the rough breakage patterns of the Dolní Věstonice 11 marginal shaping stands in contrast with the otherwise careful bone-working technology of the Gravettian, especially in the case of a precious (and possibly symbolic) raw material such as a human skull. Moreover, the details of the breakage pattern on the Dolní Věstonice 11 calotte (and the general pattern of preservation of the Dolní Věstonice 1 and 2 remains, which were destroyed in 1945) resemble those on many Pleistocene (and recent) human crania subjected to sediment pressure, displacement, and general movement after initial decomposition has taken place. The missing portions—the face, the cranial base, and the nuchal plane—are simply more fragile than the bones of the cranial vault. Moreover, similar breakage patterns are inflicted on human crania by

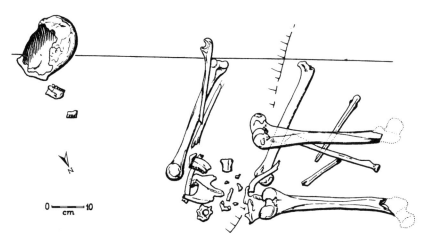

Figure 4.14 Pavlov I. Plan of the burial Pavlov 1. After Klíma (1997), with the slope declination added.

hyenas for brain extraction (Sutcliffe, 1970; Horowitz & Smith, 1988). Therefore, although it is possible that the Dolní Věstonice 11/12 (and Dolní Věstonice 1 and 2) cranial remains were intentionally modified by humans, we should not exclude the possibility of natural breakage and displacement or the opening of the skull for the brain, either by humans or hyenas (see above).

Another type of modified human bone is represented by two perforated human teeth. The first find from Dolní Věstonice (Dolní Věstonice 8) was published by Absolon (1935). The second case (Pavlov 25) was found in 1956 at Pavlov I, square 116, in the context of six perforated animal teeth and three Tertiary shells, suggesting that the human fossils may have been a component of a more complex decorative object.

These cases suggest that certain human remains may have been used postmortem as a part of the Gravettian cultural inventory. The usage is clear in the case of the perforated human teeth. It remains problematic for the other remains.

5

The Human Remains: A Summary Inventory

Trenton W. Holliday, Simon W. Hillson, Robert G. Franciscus, and Erik Trinkaus

The human remains have been inventoried almost completely and in detail by Sládek et al. (2000). However, to provide a quick reference for the material in the context of its discussion in this volume, we present below a summary inventory of the remains. This starts with an overall list of the remains discovered since 1925 at the sites of Dolní Věstonice I (Table 5.1); Pavlov I (Table 5.2); and Dolní Věstonice II, top and upper western slope, western

Table 5.1 Human remains from Dolní Věstonice I.

Specimen	Anatomical Unit(s)	Date of Discovery	Discoverer/ Identifier	Location
DV 1	Calvaria	1925	Absolon	Lost
DV 2	Calvaria	1930	Absolon	Lost
DV 3	Associated skeleton	1949	Klíma	MZM
DV 4	Cranial vault fragments	1927	Absolon	Lost
DV 5	Cranial vault fragment	1925	Absolon	Lost
DV 6	Cranial vault fragments	1925	Absolon	Lost
DV 7	2 teeth	1927	Absolon	Lost
DV 8	Tooth	1934	Absolon	Lost
DV 9	3 teeth	1949	Klíma	MZM
DV 10	Tooth	1951	Klíma	MZM
DV 23	2 cranial vault fragments	1927	Absolon	Lost
DV 24	Cranial vault section	1936	Absolon	Lost
DV 25	Cranial fragment	1936	Absolon	Lost
DV 26	Tooth	1948	Klíma	Lost
DV 27	Tooth	1948	Klíma	MZM
DV 28	2 cranial vault fragments	1924	Absolon	Lost
DV 29	Tooth	1924	Absolon	Lost
DV 30	Cranial fragment	1924	Absolon	Lost
DV 31	3 teeth	1974	Klíma	AV ČR
DV 32	Tooth	1974	Klíma	AV ČR
DV 35*	Femur shaft	1930	Absolon	MZM
DV 37	Tooth	1974	Klíma	AV ČR
DV 38	Tooth	1974	Klíma	AV ČR

Information from Vlček (1971), Klíma (1990), and our personal observations. For the Location, MZM: Moravské Zemské Muzeum (Brno); AV ČR: Archeologický ústav, Akademie Věd České Republiky (Dolní Věstonice).
*Indicates remains identified from the excavated fauna.

Table 5.2 Human remains from Pavlov I.

Specimen	Anatomical Unit(s)
Pavlov 1	Associated skeleton
Pavlov 2	Maxilla with teeth
Pavlov 3	Mandible with teeth
Pavlov 4	Mandible fragment
Pavlov 5	2 teeth
Pavlov 6	2 teeth
Pavlov 7	Tooth
Pavlov 8	Tooth
Pavlov 9	Tooth
Pavlov 10	Tooth
Pavlov 11	Tooth
Pavlov 12	Tooth
Pavlov 13	Tooth
Pavlov 14	Tooth
Pavlov 15	Tooth
Pavlov 16	Tooth
Pavlov 17	Tooth
Pavlov 18	Tooth
Pavlov 19	Tooth
Pavlov 20	Tooth
Pavlov 21	Tooth
Pavlov 22	Tooth
Pavlov 23	Tooth
Pavlov 24	Tooth
Pavlov 25	Tooth
Pavlov 26	Tooth
Pavlov 27	Tooth
Pavlov 28	Tooth

All remains are located at the Archeologický ústav, Akademie Věd České Republiky (Dolní Věstonice) and were discovered by Klíma. Pavlov 1 was discovered in 1954, but the remainder of the remains were found in 1957.

slope, and northern slope (Table 5.3). These are followed by tables listing the remains preserved for the six associated skeletons (Tables 5.4 to 5.9). For details on the identifications, see Sládek et al. (2000) and the appropriate morphological chapters below.

The human remains excavated by Absolon and Klíma from Dolní Věstonice I, which have now been lost, are listed among the remains in Table 5.1, but they are not considered further in the discussion of the human remains from these sites. The same applies to the few fragmentary mammalian remains from Dolní Věstonice II, which were originally considered to be human (Klíma, 1990; Vlček, 1991) but are now recognized to be either nonhuman or sufficiently undiagnostic to cast doubt on whether they are human. And the impressions of human dermatoglyphics present on a number of clay and ceramic objects from these sites (Vlček, 1951, 1991; Králík et al., 2002) are not addressed here.

A Note on Names and Numbers

The human remains of primary consideration here derive from several sites on the slopes of the Pavlovské Hills overlooking the now-dammed and flooded Dyje River Valley. As is common throughout much of central and eastern Europe, the sites are named after the town or village on whose territory they lie, followed by a number if there are multiple sites within the jurisdiction of that village. Hence the sites of concern here fall within the jurisdictions of two adjoining villages, Pavlov to the east and Dolní Věstonice to the west. They therefore have the names of Pavlov and Dolní Věstonice, followed by a Roman numeral to designate the actual site (Pavlov I and II and Dolní Věstonice I, II, IIa, and III). These sites nonetheless occupy adjoining areas of the same hillside and, through the Late Pleistocene, served similar purposes to the human foraging groups that stopped there (see Chapters 2 and 3).

The human remains from the Pavlov sites all derive from Pavlov I, and they are therefore given Arabic numbers to designate the individuals or probable individuals in the order in which they were discovered in the field and/or recognized in the laboratory. A number of the isolated teeth from Pavlov I were not originally given individual numbers, and the ones provided here are based on the recent reassessment of these teeth (Sládek et al., 2000).

The human remains from Dolní Věstonice, however, derive from the two large site complexes, Dolní Věstonice I and II, and those from Dolní Věstonice II come from different portions of that site complex (see Tables 5.1 and 5.3). They are given the designation of Dolní Věstonice plus an Arabic number, but they were also numbered in the sequence in which they were found during excavation and/or recognized in the laboratory, regardless of which Dolní Věstonice site produced them. Even though most of the excavation of Dolní Věstonice I was completed before serious excavation took place at Dolní Věstonice II, a number of the former site's human remains were not recognized or properly catalogued until later. This has resulted in a sequence of numbered specimens that derive from one or the other of these two sites, with little archeological rationale (only an historical one). Tables 5.1 and 5.3 should serve to clarify the origins of these remains, and further details are provided in Sládek et al. (2000).

Table 5.3 Human remains from Dolní Věstonice II, top and upper western slope, western slope, and northern slope.

Specimen	Anatomical Unit(s)	Date of Discovery	Discoverer/Identifier
Dolní Věstonice II, Top and Upper Western Slope			
DV 11	Calotte	1986	Klíma
DV 12	Cranial vault fragment	1986	Klíma
DV 13	Associated skeleton	1986	Klíma
DV 14	Associated skeleton	1986	Klíma
DV 15	Associated skeleton	1986	Klíma
DV 17	2 cranial vault fragments	1986	Klíma
DV 33	Tooth	1987	Klíma
DV 34	Hand phalanx	1987	Klíma
Dolní Věstonice II, Western Slope			
DV 16	Associated skeleton	1987	Svoboda
DV 36*	9 teeth	1997, 1998	Svoboda/West, Trinkaus
DV 39*	Navicular	1998	Svoboda/Trinkaus
DV 40*	Femur shaft	1998	Svoboda/Trinkaus
DV 41*	Humerus shaft	1998	Svoboda/Trinkaus
DV 42*	Fibula shaft	1998	Svoboda/Trinkaus
DV 43*	Femur shaft	1998	Svoboda/Trinkaus
DV 44*	Metatarsal	1998	Svoboda/Trinkaus
DV 45*	Rib fragment	1998	Svoboda/Trinkaus
DV 46*	Cuneiform	1998	Svoboda/Trinkaus
DV 47*	Metatarsal	1998	Svoboda/Trinkaus
DV 48*	Fibula shaft	1998	Svoboda/Trinkaus
DV 49*	Metatarsal	1998	Svoboda/Trinkaus
DV 50*	Radius	1998	Svoboda/Trinkaus
DV 53*	Hand phalanx	1998	Svoboda/Fišáková
Dolní Věstonice II, Northern Slope			
DV 51*	Rib fragment	1998	Svoboda/Trinkaus
DV 52*	Foot phalanx	1998	Svoboda/Trinkaus

All remains are located at the Archeologický ústav, Akademie Věd České Republiky (Dolní Věstonice).
*Indicates that the remains were identified from the excavated fauna. Note that DV 18 to 22 are not included since they have been determined to be nonhuman or are insufficiently preserved to indicate whether they are human (Trinkaus et al., 2000c).

Table 5.4 Dolní Věstonice 3 remains (number of bones/number of pieces).

Cranium (1/1)	Claviculae (2/2)	Os coxae (2/2)
Mandible (1/1)	Scapulae (2/3)	Femora (2/2)
Dentition (31/31)	Humeri (2/2)	Patellae (2/2)
Cervical vertebrae (7/7)	Ulnae (2/2)	Tibiae (2/2)
Thoracic vertebrae (12/11)	Radii (2/2)	Fibulae (2/2)
Lumbar vertebrae (5/5)	Carpals (9/9)	Tarsals (7/7)
Sacrum (1/3)	Metacarpals (10/10)	Metatarsals (4/4)
Ribs (many fragments)	Phalanges (13/13)	Phalanges (6/6)
		Sesamoidea (4/4)

Table 5.5 Dolní Věstonice 13 remains (number of bones/number of pieces).

Cranium (1/1)	Claviculae (2/2)	Os coxae (2/2)
Mandible (1/1)	Scapulae (2/2)	Femora (2/2)
Dentition (26/26)	Humeri (2/2)	Patellae (2/2)
Cervical vertebrae (7/7)	Ulnae (2/2)	Tibiae (2/2)
Thoracic vertebrae (12/12)	Radii (2/2)	Fibulae (2/2)
Lumbar vertebrae (5/5)	Metacarpals (3/3)	Tarsals (2/2)
Sacrum (1/1)	Phalanges (8/8)	
Sternum (2/2)		
Ribs (24/24)		

Table 5.6 Dolní Věstonice 14 remains (number of bones/number of pieces).

Cranium (1/4)	Claviculae (2/2)	Os coxae (2/2)
Auditory ossicles (2/2)	Scapulae (2/3)	Femora (2/2)
Mandible (1/1)	Humeri (2/2)	Patellae (2/2)
Dentition (32/32)	Ulnae (2/3)	Tibiae (2/2)
Cervical vertebrae (7/8)	Radii (2/3)	Fibulae (2/2)
Thoracic vertebrae (12/15)	Carpals (5/5)	Tarsals (5/5)
	Metacarpal (1/1)	Metatarsals (8/8)
Lumbar vertebrae (5/7)	Phalanges (6/6)	Phalanx (1/1)
Sacrum (1/1)		
Ribs (23/29)		

Table 5.7 Dolní Věstonice 15 remains (number of bones/number of pieces).

Cranium (1/1)	Claviculae (2/2)	Os coxae (2/2)
Auditory ossicles (2/2)	Scapulae (2/2)	Femora (2/2)
Mandible (1/1)	Humeri (2/2)	Patellae (2/2)
Dentition (30/30)	Ulnae (2/2)	Tibiae (2/2)
Cervical vertebrae (7/10)	Radii (2/2)	Fibulae (2/2)
Thoracic vertebrae (12/13)	Carpals (4/4)	Tarsals (7/7)
	Metacarpals (4/4)	Metatarsals (7/7)
Lumbar vertebrae (5/6)	Phalanges (8/8)	Phalanges (3/3)
Sacrum (1/1)		Sesamoidea (1/1)
Sternum (1/1)		
Ribs (23/28)		

Table 5.8 Dolní Věstonice 16 remains (number of bones/number of pieces).

Cranium (1/1)	Claviculae (2/2)	Os coxae (2/9)
Mandible (1/1)	Scapulae (2/4)	Femora (2/5)
Dentition (28/28)	Humeri (2/2)	Patella (1/1)
Cervical vertebrae (7/9)	Ulnae (2/2)	Tibiae (2/2)
	Radii (2/2)	Fibulae (2/2)
Thoracic vertebra (1/1)	Carpals (10/10)	Tarsals (13/13)
	Metacarpals (9/9)	Metatarsals (10/10)
Ribs (12?/12)	Phalanges (20/20)	Phalanges (10/10)

Table 5.9 Pavlov 1 remains (number of bones/number of pieces).

Cranium (1/4)	Claviculae (2/2)	Femora (2/2)
Mandible (1/1)	Scapulae (2/2)	Tibiae (1?/2)
Dentition (27/27)	Humeri (2/2)	Fibula (1/1)
Cervical vertebra (1/1)	Ulnae (2/3)	
Thoracic vertebrae (6?/6)	Radii (2/2)	
Lumbar vertebrae (2?/2)	Metacarpal (1/1)	

6

The Ages at Death

Simon W. Hillson, Robert G. Franciscus, Trenton W. Holliday, and Erik Trinkaus

The assessment of the ages at death of the skeletal and dental remains from Dolní Věstonice and Pavlov involves both the consideration of associated skeletons for whom there are multiple lines of evidence and the evaluation of isolated remains for some of whom it is possible to provide a reasonable age at death other than immature or mature.

The ages at death of the six associated skeletons from Dolní Věstonice and Pavlov, Dolní Věstonice 3 and 13–16, plus Pavlov 1, can be based on combinations of skeletal and dental features. Three of these individuals, Dolní Věstonice 3 and 16 and Pavlov 1, were fully mature at death, and therefore the assessments of their ages at death are based on various forms of skeletal and dental degeneration. The other three individuals were either clearly immature (Dolní Věstonice 14) or were sufficiently young adults (Dolní Věstonice 13 and 15) to allow their ages at death to be assessed in part as a result of developmental processes. The one other associated set of remains that can be added to these partial skeletons is the incomplete set of immature teeth of Dolní Věstonice 36.

Among the isolated remains from the Dolní Věstonice sites, there are isolated or associated teeth, whose ages can be assessed according to development and/or attrition, as well as cranial and postcranial pieces. These dental remains include the Dolní Věstonice 9, 10, 27, 31–33, 37, and 38 teeth, as well as the antemortem shed Dolní Věstonice 10 deciduous tooth. The Dolní Věstonice isolated cranial elements whose ages at death can be evaluated are the Dolní Věstonice 17 immature fragments and the Dolní Věstonice 11/12 calotte. In addition, there is a series of isolated postcranial elements, Dolní Věstonice 34, 39–44, 46–50, and 52, some of whose overall degree of maturity can be assessed.

From the Pavlov I site, there are two associated partial dentitions in jaws, Pavlov 2 and 3, and a series of teeth, Pavlov 5 to 28, for most of whom it should be possible to assign approximate ages at death based principally on tooth identification and dental attrition. For each specimen, as available, these are assessed in terms of cranial, dental, axial, and appendicular indicators of age at death. The scales of reference are, by default, those available for recent, largely European or European-derived populations (e.g., Scheuer & Black, 2000, and references therein).

Dolní Věstonice 3

Cranial Indicators

The Dolní Věstonice 3 ectocranial sutures are difficult to assess because of the dark red to brown ochre and painting overlying much of the surface bone. Nonetheless, as has been noted previously (Jelínek, 1954), the closure and obliteration of the ectocranial sutures is generally advanced, indicating age at death beyond young adulthood. After the attainment of adulthood, cranial vault suture closure is the only criterion after dental attrition that is available for skeletal age estimation in the cranium. The methodology has deep historical roots and was used widely until the 1950s, when its utility was questioned and its application waned (Meindl & Lovejoy, 1985; Byers, 2002). Its viability and use has reemerged in more recent years with methodological refinement and the awareness that it should be used only in combination with other "degenerative" indicators of age, rather than as a stand-alone technique. Ectocranial, endocranial, and palatal suture closure are all potentially informative; however, for a variety of practical reasons, ectocranial suture closure is most widely used.

Except for the midlambdoid suture, which is open, all ectocranial suture locations in Dolní Věstonice 3 show evidence for either significant closure or complete obliteration (Table 6.1). Significant suture closure is found in the lambda, obelion, anterior sagittal, and midcoronal locations, and complete obliteration is found in the bregma, pterion, and sphenofrontal locations. Based on Meindl & Lovejoy's (1985) composite vault scoring system, the estimated age for Dolní Věstonice 3 is 45.2 years; however, the total estimated range (24–75 years) and the estimated interdecile range (31–65 years) are quite broad. The estimated age based on the lateral-anterior composite scoring system provides a somewhat older estimate (56.2 years), with similarly wide range estimates for the total range (34–68 years) and interdecile range (49–65 years). The wide range of estimated ages indicated for Dolní Věstonice 3 is a feature of this aging method, and these results should be viewed in aggregate fashion with the other methods employed here.

Dental Indicators

Dolní Věstonice 3 has a fully developed permanent dentition, with roots complete in all teeth and apices closed (Figure 11.1). Even the latest erupting teeth have well-developed occlusal wear. The lower left third molar was congenitally missing, and the corresponding upper left third molar has much less wear than the other teeth, so it cannot be used for age estimation. The two remaining third molars are, however, fully erupted and in wear, implying that Dolní Věstonice 3 was at least older than the mid-20s, by the standards of development in living people.

Brothwell (1963) established a scheme of wear for Romano-British and Medieval material in the Natural History Museum (London). This was calibrated by com-

Table 6.1 Cranial suture closure scoring.

Vault[a]	Pav 1	DV 3	DV 11/12	DV 13	DV 14	DV 15	DV 16
Midlambdoid	2 (L)	0	1 (R)	0 (L)	0 (R)	0 (L)	0 (L)
Lambda	2 (M)	2 (M)	1 (M)	1 (M)	0 (M)	0 (M)	2 (M)
Obelion	2 (M)	2 (M)	3 (M)	0.17[b] (X)	0 (M)	0 (M)	3 (M)
Anterior sagittal	1 (M)	2 (M)	2 (M)	0? (M)	1 (M)	1 (M)	2 (M)
Bregma	0 (M)	3 (M)	2 (M)	0 (M)	0 (M)	1 (M)	2 (M)
Midcoronal	1 (R)	2	1 (R)	0 (R)	0? (L)	0 (R)	3 (L)
Pterion	3 (L)	3	1.67[b] (X)	0 (R)	0 (L)	0 (L)	3 (L)
Composite score	11	14	12	1	1	2	17
Estimated age range	24–60	24–75	24–75	18–45	18–45	18–45	30–71
Estimated interdecile range	28–44	31–65	31–65	19–44	19–44	19–44	35–60
Estimated mean age	39.4	45.2	45.2	30.5*	30.5*	30.5*	48.8
Lateral anterior[a]							
Sphenofrontal	1.67[b] (X)	3	1[b] (X)	0[b] (X)	0 (L)	0 (L)	2.33[b] (X)
Inferior sphenotemporal	1 (L)	2.67[b] (X)	1[b] (X)	0? (L)	0? (L)	0 (R)	1 (L)
Superior sphenotemporal	1.67[b] (X)	2.67[b] (X)	1[b] (X)	0 (L)	0 (L)	0[b] (X)	2.33[b] (X)
Midcoronal	1 (R)	2	1 (R)	0 (R)	0? (L)	0 (R)	3 (L)
Pterion	3 (L)	3	1[b] (X)	0 (R)	0 (L)	0 (L)	3 (L)
Composite score	8	14	5	0	0	0	12
Estimated age range	32–65	34–68	23–68	<50*	<50*	<50*	34–68
Estimated interdecile range	35–57	49–65	28–52	<43*	<43*	<43*	49–65
Estimated age mean	45.5	56.2	41.1	<50*	<50*	<50*	56.2
Sphenooccipital sychndrosis[c]	(X)	(X)	(X)	Closed (M)	(X)	(X)	(X)
Estimated age				>21			

[a]Following Meindl and Lovejoy (1985): 0 = completely open; 1 = minimal closure; 2 = significant closure; 3 = complete obliteration; L = left side; R = right side; M = midline; X = not assessable; ? = inferred from plastic and soil infilling except where junction with other nondamaged suture allows using nondamaged one; * = given the insensitivity of coding at younger age classes, emphasis should be put on the lower end of the age ranges provided and not on the estimated age mean, or considerably below the threshold indicated. All composite scores are rounded to the nearest whole value.
[b]Values are averaged from preserved sections.
[c]Following Steele and Bramblett (1988)

parison with skeletal age indicators, such as pubic symphysis changes. No published tests are available, but Hillson's own experience has suggested that the scheme provides consistent results for a wide range of material. The Brothwell system is based upon the relative wear on the three molars, which is used to define four stages, labeled 17–25 years, 25–35 years, 35–45 years, and more than 45 years. It seems likely that most remains from Upper Paleolithic contexts had a more rapid wear rate than Medieval dentitions, although this is very difficult to test and, for example, Anglo-Saxon jaws often show very heavy wear. An independent study (Miles, 1958, 1962), which used a seriation technique to arrive at age estimates for a British Anglo-Saxon assemblage, nevertheless showed similar results to the Brothwell scheme. The Miles method has been tested independently (Nowell, 1978; Kieser et al., 1983; Lovejoy, 1985; Lovejoy et al., 1985a), and it performs as well as alternative aging methods for adults. In the case of Dolní Věstonice 16 (below), it has been possible to make a direct comparison with a dental study on living people from a population with a high dental wear rate. For all these reasons, the age ranges of the Brothwell scheme should be regarded as approximate, and possibly as maximum ages when applied to remains such as those of Dolní Věstonice 3. The diagrams of the Brothwell scheme can be translated into the Smith scores used to describe occlusal wear in this monograph (Table 6.2). The Dolní Věstonice 3 molars fit best into the 25–35-year category. In other words, the individual had passed into early middle-aged adulthood. This fits the observation that the third molars were fully erupted (with the exception of the congenitally absent lower left tooth) and established in wear.

Axial Indicators

All of the cervical vertebrae (with the exception of C1, an element that lacks a centrum) preserve at least some of the ring epiphyses, and in every case these epiphyses are fused to their respective centra, indicative of skeletal adult status for the specimen. Whereas many of the elements (especially the thoracic and lumbar ones) are extremely fragmentary and/or remain in matrix, there is no evidence of any degenerative change in the vertebral column of Dolní Věstonice 3. Taken in total, these factors best fit an assessment of the individual as a young adult (third to fourth decade), although a somewhat more advanced age is possible.

Appendicular Indicators

Despite the abundance of appendicular remains preserved for Dolní Věstonice 3, the generally poor preservation of all of the epiphyseal regions means that there are very few indicators of the individual's age at death. The sufficiently preserved epiphyseal-metaphyseal remains for any assessment of maturity consist of the right proximal and distal radius, two proximal manual phalangeal bases, two middle manual phalangeal bases, four distal manual phalanges, the left femoral trochanters, a metatarsal 5 head, a proximal pedal phalanx base, a middle pedal phalanx, and three distal pedal phalanges. All of them exhibit epiphyseal regions in which the fusion lines are completely obliterated externally and are invisible radiographically. These data, as well as the general morphology of the remains, support a fully mature status for the individual.

At the same time, there is essentially no evidence of any degenerative process on any of the preserved articular subchondral bone surfaces (Chapters 16–19). The small amount of pitting on the acetabular weight-bearing areas is not indicative of age.

Summary

The age indicators on the Dolní Věstonice 3 partial skeleton, despite fossilization damage and conservation alteration, are in general agreement in indicating the age at death of a middle-aged adult. The degree of dental attrition and general lack of vertebral body and appendicular degeneration support an age at death around the end of the third decade, whereas the degree of ectocranial suture closure suggests an older age, in the fifth or even sixth decade. However, all of these age estimation techniques are approximate, and the best average estimate may therefore be sometime in the fourth decade. This places Dolní Věstonice 3 close to the 35-year boundary in Table 6.4; the specimen is placed in the slightly older age interval in that table, bearing in mind that she may have died earlier in her fourth decade.

Dolní Věstonice 9

If Dolní Věstonice 9/1 and 9/2 are permanent third molars, they both have fully completed mesial roots but distal roots that are still not quite complete at the apex. This

Table 6.2 Brothwell's system for age estimation.

Range of Smith Wear Scores	About 17–25 Years	25–35 Years	35–45 Years	More than 45 Years
First molars	3–4	5–6	7	Any state of wear more than the previous columns
Second molars	2–3	4–5	6–7	
Third molars	1–2	3–4	5–6	

Occlusal attrition in molars is from Brothwell (1963), with the original diagrams of dentine exposure converted into attrition scores (B. H. Smith, 1984).

makes it difficult to suggest an age. Much of the occlusal surface is fractured away, but enough is visible in Dolní Věstonice 9/2 to show that it is possibly at Smith attrition stage 4. In the Brothwell system (above and Table 6.2), this implies the 25–35-year category. Dolní Věstonice 9/3 is a well-worn upper first incisor, equivalent to Dolní Věstonice 3 or Pavlov 1, which are interpreted here as individuals in middle-age adulthood. All this evidence gives a mixed message about age, but the root formation is probably the most reliable, suggesting an age in the early 20s at most.

Dolní Věstonice 10

Dolní Věstonice 10, a deciduous upper first molar (third premolar), is worn to Smith stage 6 and has its root resorbed close to the base of the crown. It was probably exfoliated from the mouth of a living child, and this takes place in living children at 9 to 10 years of age.

Dolní Věstonice 11/12

The Dolní Věstonice 11/12 calotte exhibits a range of ectocranial suture closure. Minimal closure is evident at the midlambdoid, lambda, and midcoronal locations; significant closure is observed at the anterior sagittal and bregma locations; and complete obliteration is evident at obelion (Table 6.1). Based on Meindl and Lovejoy's (1985) composite vault scoring system, the estimated age for Dolní Věstonice 11/12 is the same as that estimated for Dolní Věstonice 3 (i.e., 45.2 years), with the same broad total (24–75 years) and interdecile (31–65 years) estimated ranges. The estimated age based on the lateral-anterior composite scoring system (with far less scoreable locations) provides a younger age estimate (41.1 years) with a wide associated total range estimate (23–68 years) and somewhat smaller interdecile range (28–52 years).

Dolní Věstonice 13

Cranial Indicators

With the exception of lambda, which shows minimal closure, all of the remaining landmarks in Dolní Věstonice 13 exhibit completely open ectocranial sutures (Table 6.1). Because of methodological scoring mechanics and the primary use of this technique for individuals who have attained complete growth, the estimated age based on mean values for vault components (30.5 years) is clearly biased upward, and the actual age for Dolní Věstonice 13 is likely to fall within the lower limits of both the total (18–45 years) and interdecile (19–44) estimated ranges. A closed sphenooccipital synchrondosis evident on Dolní Věstonice 13 indicates a probable age greater than 21 years (Steele & Bramblett, 1988), and it thus helps to limit the lower age range somewhat. The lowest possible composite score (i.e., zero) for lateral anterior components provides only age estimate values of less than 43 and 50 years, and thus it does not serve to further narrow the young age estimate for Dolní Věstonice 13.

Dental Indicators

Dental radiographs and direct observation show that the roots of all of the teeth are completed, with apices closed, including the permanent third molars. There is no occlusal attrition on the crowns of these last teeth, beyond a slight polishing of the enamel (Figures 11.2 and 11.3). This implies either that eruption had not yet fully brought the third molars into the occlusal plane or that eruption had just occurred. The positions of the upper third molars are reconstructed, but the lower third molars appear to be in their original positions. The lower right tooth is not yet in the occlusal plane, whereas the lower left one protrudes slightly. Occlusion was therefore still in the course of completion, and in most living people, this occurs during the late teenage years and early 20s. Assuming that Dolní Věstonice 13 is a male (Chapter 7), the large study by Mincer et al. (1993) suggests that there is a strong probability (>85%) that he would be over 18 years of age by modern standards. In summary, the dental evidence suggests that Dolní Věstonice 13 represents an individual at a developmental stage equivalent to the early 20s in living humans.

Axial Indicators

Throughout the complete vertebral column and rib cage all epiphyses that are preserved have been fused, and there is no evidence of lost but unfused epiphyses. Thus it is apparent that the specimen had attained axial skeletal adult status. There is no indication of degenerative changes anywhere in the vertebral column, sternum, or rib cage; all elements are free from osteoarthritis. The manubrium and mesosternum retain an articulation at the sternal angle and evince no ossification of costal cartilages. Taken together, these observations are consistent with young adult status—that is, an age at death sometime in the third or fourth decade.

Appendicular Indicators

The Dolní Věstonice 13 appendicular skeleton preserves at least one of each of the long bone epiphyses, in most cases both right and left, plus three metacarpals. The only epiphyseal areas not preserving information on degrees of fusion are the distal right ulna and the distal right fibula. The majority of the epiphyseal lines of Dolní Věstonice 13 are completely fused, with the epiphyseal line obliterated externally and internally (as seen radiographically), but there are some for which the lines are still at least partially evident.

In the upper limb, the sternal epiphyses of the proximal clavicles are fully fused, externally and radiographically. The proximal ends are irregularly convex, rather than the furrowed concave surface seen prior to the epiphyseal fusion. The right humeral head and the right and left distal humeri show no trace of the fusion line. The left humeral head is damaged, so that its degree of fusion cannot be ascertained externally. There is a line radiographically, but it is too far distal to be the epiphyseal fusion line and represents a postmortem crack in the bone. The forearm epiphyseal lines are completely fused externally, but the distal radii and left ulna exhibit faint lines of trabecular density in the locations of the epiphyseal fusion lines that traverse the epiphyses. The left metacarpals 2 and 3 heads are completely fused externally and radiographically, but the right metacarpal 2 exhibits a persistent line externally on the ulnar half of the dorsal head and across the palmar head, which is accompanied by a faint line across the mid–head epiphysis.

In the lower limb, all of the femoral epiphyseal lines (head, trochanteric, and condylar) are completely fused and obliterated. The same applies to the fibular epiphyses and the distal tibial epiphyses. The proximal tibial epiphyses, however, exhibit a fusion line across their posterior surfaces, and the line is evident radiographically on the left one (damage to the right one obscures it radiographically).

The largely complete pelvis shows complete fusion of all of the secondary centers of ossification, including the iliac crest and the ischial tuberosity. There is a hint of a fusion line on the right external ischiopubic ramus, which may be a trace of the fusion line there, but it is hard to differentiate it from a muscle line. The pubic symphysis is damaged and partly obscured, but it is apparent that the ventral rampart had not yet begun to form.

Consequently, in general the appendicular remains of Dolní Věstonice 13 indicate fully mature status, but the remnants of epiphyseal fusion lines on the distal forearms and the proximal tibiae suggest that the individual was not markedly beyond 20 years of age. At the same time, the complete fusion of the proximal clavicular epiphyses and of the secondary pelvic centers suggests an age of at least in the middle of the third decade. Yet, the absence of a ventral rampart on the pubic symphysis makes it unlikely that the individual was beyond the third decade. All of these appendicular indicators, therefore, support an age at death for Dolní Věstonice 13 in the middle of his 20s.

Summary

The degree of appendicular epiphyseal fusion and the status of the axial skeleton of Dolní Věstonice 13 suggest an age in the middle or later third decade, although the dental formation and attrition suggest an age in the earlier part of the third decade. The degree of ectocranial suture closure is not of utility here beyond indicating the generally young age of the skeleton at death. Given the developmental status of the third molars in particular and the remnants of epiphyseal lines, it appears more likely that this individual was on the low side of 25 years at death, and hence it is kept in the 21–25-year category in Table 6.4.

Dolní Věstonice 14

Cranial Indicators

With the exception of the anterior sagittal region, which shows minimal closure, all of the remaining areas in Dolní Věstonice 14 exhibit completely open ectocranial sutures (Table 6.1). As mentioned above with respect to Dolní Věstonice 13, the methodological scoring mechanics and the primary use of this technique for individuals who have attained complete growth mean that the resulting estimated age based on mean values for vault components (30.5 years) is clearly biased upward. The actual age for Dolní Věstonice 14 is likely to fall within the lower limits of both the estimated total range (18–45 years) and estimated interdecile range (19–44). Unfortunately, the sphenooccipital synchrondosis cannot be evaluated in Dolní Věstonice 14 to help limit the lower age range, as was possible for Dolní Věstonice 13. As in Dolní Věstonice 13, the lowest possible composite score (i.e., 0) for lateral anterior components provides only age estimate values of less than 43 and 50 years, and thus it does not serve to further narrow the young age estimate for Dolní Věstonice 14.

Dental Indicators

The roots of all of the teeth except the third molars are completed, with apices closed. With the exception of the third molars, the last teeth to complete root formation are the second molars, which in living people close their apices most commonly between 14 and 16 years of age (Gustafson & Koch, 1974). This implies that the Dolní Věstonice 14 individual was at least in the late teenage years at death. The upper third molars (Figure 11.4) have been reconstructed in a way that makes it difficult to judge their eruption state and to observe the state of closure of their root apices. The lower third molars (Figure 11.5), on the other hand, are still contained within their crypts, and it is clear that they had not yet erupted enough to protrude through the crest of the alveolar process. Their roots appear from radiographs to have developed to a full length, although their apices are still not closed. In terms of living young men and women, they are at "Stage G: Root walls parallel, but apices remain open" (Mincer et al., 1993). At this stage of development, there are about equal probabilities that the individual might be over 18 years of age or younger. Taken altogether, the dental evidence implies that the Dolní Věstonice 14 individual was in the late teenage years at death.

Axial Indicators

The vertebral ring epiphyses are badly damaged in Dolní Věstonice 14, but where they are present they appear to have been fully fused to the centra. Likewise, all spinous and transverse process epiphyses appear to be fused to their respective processes. These secondary epiphyses typically fuse to the processes between the ages of 17 and 25 (Bass, 1987). In contrast to the vertebral evidence, however, three ribs (left and right first ribs and left fourth rib) evince heads with the billowing characteristic of unfused metaphyseal surfaces. These secondary epiphyses, which appear on the heads of ribs in adolescence, typically fuse between the ages of 18 and 24 (Bass, 1987). There is no evidence of any degenerative change in either the vertebrae or ribs. Taken together, these data, particularly the lack of fusion of the rib head epiphyses, are consistent with a late adolescent or very young adult (i.e., probably younger than 24) age at death for this specimen.

Appendicular Indicators

The appendicular remains of Dolní Věstonice 14 have damage to a number of the epiphyseal regions, but the preserved portions are sufficient to indicate the mosaic pattern of epiphyseal fusion expected for an adolescent. The bones are nonetheless large (the femora are among the longest known for an earlier Upper Paleolithic human, being exceeded only by those of Barma Grande 2 and Fanciulli 4), suggesting that the individual should have been close to his full adult stature.

The clavicular ends are too damaged to permit assessment of their developmental status. The distal humeral epiphyses are completely fused and the lines obliterated externally and internally, and the right proximal humerus is too incomplete to assess. However, the left humeral head has an anterior border 16 mm long and a medial border 23 mm long, which are still partly open (at least on their margins) and indicate incomplete fusion of the head epiphysis.

The ulnae exhibit fully fused proximal epiphyses, and their distal ends are too damaged to assess. The radii also have completely fused proximal epiphyses and obliteration of the fusion line radiographically and externally. The distal radial metaphyses are not preserved, but the right distal epiphysis is preserved separately and the medial 10 mm of it retains the completely unfused epiphyseal cartilage surface. The second metacarpal head is fully fused, as indicated radiographically, and all of the manual phalangeal bases appear to be fully fused as well.

The juncture of the three bones of the os coxae is completely fused, but there are indicators of immature status elsewhere on the pelvic remains. The iliac crests are incomplete and partly reconstructed, but the left one shows a line of fusion internally along the middle third of the crest, and the dorsal third of the crest is lost in a manner suggesting incomplete fusion of the crest onto the blade. The left ischial tuberosity has a trace of the fusion line for 12.5 mm along the external margin of the main portion of the tuberosity; the right one is broken in a steplike fashion from the tuberosity onto the ischiopubic ramus, which suggests incomplete fusion of the more ventral portion of the secondary ossification center. The auricular surfaces are eroded and the pubic symphyses are partially abraded, limiting observations about them. However, the symphyseal surfaces exhibit clear evidence of transverse billowing and no development of a ventral rampart.

The femoral head epiphyses all appear to be fully fused externally, although damage to the head margins obscures details. However, there is a hint of a fusion line through the femoral heads, especially on the left side. The trochanters are also largely fused, but there is a trace of a fusion line on the anterior half of the lateral side for 9.5 mm on the right side and 19.0 mm on the left side. The distal femoral condyles are damaged, and at least the right one shows no trace of a fusion line radiographically. Yet, there is a trace of a line across the proximoposterior lateral condyle of the right femur; the same area on the left medial condyle exhibits no evidence of a fusion line.

Both proximal tibiae have extensively damaged proximal epiphyses, but they exhibit, along the distal margins of their tibial tuberosities, evidence of the epiphyseal fusion lines. Externally, the left distal tibia appears fully fused, but radiographically there is still evidence of a fusion line across the lateral 20 mm of the distal right tibia. The fibular heads, despite damage to the bones, appear to completely lack their unfused head epiphyses. The distal fibulae are fused, but they still evince fusion lines medially across the proximal malleolar articulations; the lines elsewhere are obliterated.

The only pedal remains that preserve the relevant area are the first metatarsals. The right one has its base fully fused externally, but the eroded left one shows evidence of the fusion line laterally in the exposed trabeculae.

These data therefore show complete fusion and obliteration of the epiphyses that join in the first half of the second decade but variable degrees of fusion of the ones that close during the second half of the adolescent years. The lack of fusion of the distal radius and probably of the proximal fibulae, the partial fusion of the humeral head, and largely complete fusion of the trochanters and proximal tibia—plus traces of lines in the femoral head, distal tibial, and first metatarsal epiphyses—bracket the individual's age maximally between about 16 years and about 20 years. The degree of closure of the pelvic secondary ossification centers supports such a determination.

Summary

Dolní Věstonice 14 is the one individual with an abundance of developmental indicators of the age at death, and they are consistent with indicating an age in the late second decade. This is particularly indicated by the dentition, the vertebrae and ribs, and the appendicular epiphyses. The ectocranial sutures are of little utility other than confirming the young age at death of this individual.

Dolní Věstonice 15

The Dolní Věstonice 15 individual clearly suffered from a form of dysplasia, which especially affected aspects of the pelvic and thigh region (Chapter 19). In addition, there is a variety of abnormalities of the dentition that are or may be directly associated with the congenital difficulties (Chapter 19). It is not known to what extent these abnormalities of development would have affected the timing, both relative and absolute, of various stages of dental and skeletal development. However, the general pattern of dental and skeletal formation appears to be similar to that expected for a developmentally normal individual. Therefore, the assessment of its age at death is undertaken with the assumption that the developmental scheduling of Dolní Věstonice 15 was indeed within normal recent human ranges of variation.

Cranial Indicators

Unlike both Dolní Věstonice 13 and 14, Dolní Věstonice 15 shows more than one location with minimal ectocranial suture closure (the anterior sagittal and bregma locations). All other suture locations, however, are completely open. Even with this difference, the age estimate and associated ranges for those estimates for Dolní Věstonice 15 are the same as for the other two members of the triple burial because of the methodological factors discussed above (Table 6.1). Moreover, for the same reasons, the actual age of Dolní Věstonice 15 is likely to be closer to the youngest end of the estimated ranges. As with Dolní Věstonice 14, Dolní Věstonice 15's sphenooccipital synchrondrosis could not be evaluated to help bracket the generally young age estimate.

Dental Indicators

The roots and root apices for all of the teeth are completed, including the third molars. The upper third molars have been reconstructed in a position that implies they were not fully erupted, but as there is no supporting bone remaining in the alveolar process, it is not necessarily the position they occupied during life. Their little worn state and lack of any clear mesial approximal facet suggest that they had at least not been at all long in the occlusal plane, if they had, in fact, reached it (Figure 11.6). The lower left third molar also shows only slight occlusal wear and is still in the process of eruption, but it is almost in the occlusal plane (Figure 11.7). The right third molar is still partially contained within the jaw because it has erupted in an anomalous position (Chapter 11). It could never have erupted into normal occlusion, and its eruption state cannot therefore be used for age estimation. A living young man or woman would be over 18 years of age at the state of development shown by the Dolní Věstonice 15 third molars (Mincer et al., 1993). The state of eruption is similar to that of Dolní Věstonice 14, but the dental evidence as a whole implies an age in the early to mid-20s.

Axial Indicators

No unfused epiphyses are evident in the Dolní Věstonice 15 ribs, and wherever the cranial and/or caudal surfaces of the specimen's vertebral bodies are preserved, they evince fused ring epiphyses. Similarly, most of the transverse and spinous processes have fused epiphyses, with one notable exception—the spinous process of L4 has a small unfused portion of the epiphyseal line evident on its right side. These epiphyses typically fuse between the ages of 17 and 25 (Bass, 1987). In terms of degenerative changes, T12 evinces a bony lip running from the cranial margin of the body onto the left pedicle that may be osteoarthritic, and the caudal surface of the L2 body exhibits a dorsal pit that may be a Schmorl's node. Osteoarthritis is not unexpected in this pathological individual, however, and thus it is not a good indicator of advanced age. Thus, the lack of fusion of the L4 spinous process is consistent with a young adult age at death for the specimen, probably younger than 25.

Appendicular Indicators

The Dolní Věstonice 15 appendicular remains appear to be largely mature, but there are several indicators of a developmental process that was still underway.

In the upper limb, the long bone epiphyses are mostly fully fused, with the lines of fusion obliterated externally and internally. On the humeri, the only indication of recent fusion is a hint of a line on both humeri across the head and greater tubercle radiographically. The cubital epiphyseal lines are all completely obliterated. The distal radial ones have a trace of a line radiographically across both bones, which is more apparent on the right side. Externally, it is evident on the medial half of the posterior edge, as a hint on the right side and clearly on the left side. The distal ulnae have evidence of a line posteriorly and laterally externally but no evidence of it radiographically. The metacarpal 2 head is completely fused. The clavicles, however, lack their proximal (sternal) epiphyses, have concave metaphyses, and thus show lack of fusion of the sternal end.

On the pelvis, the ischial tuberosity fusion lines are all effaced, as are the left iliac crest ones. However, the right iliac crest has traces of the fusion lines internally just dorsal of the iliac pillar for 7.5 mm, and externally above the fossa for 25 mm. The pubic symphyses exhibit clear transverse billowing and no evidence of a ventral rampart (Figure 6.1).

On the femora, the proximal epiphyseal lines are completely obliterated, as are most of the ones evident around the condyles. There is only a trace of the fusion line by the proximoposterior left medial condyle. The proximal tibial epiphyses have a hint of the fusion line across the posterior surfaces, and the fusion line is clearly visible radiographically. Distally, there is no trace of the fusion line externally, and there is a shadow of it in the lateral halves of the epiphyses. The proximal fibular epiphyses are completely fused, but the distal ones have traces of the fusion line medially, and that distal line continues anteriorly on the right one. The metatarsal heads are all completely fused.

This pattern therefore indicates recent fusion of the various epiphyses that close during the second half of the second decade, with traces of the fusion lines internally and/or externally for a few of them. When these indicators are combined with the proximal clavicular and pubic symphyseal data, they point to an age not far advanced into the third decade.

Summary

As with Dolní Věstonice 13, the age at death of Dolní Věstonice 15 appears to fall well within the first half of the third decade of life. This is supported by the development and attrition of the dentition, the vertebral and rib epiphysis fusion, the appendicular epiphyses, and the pubic symphyseal surface. However, the persistence of more skeletal fusion lines than are present on Dolní Věstonice 13 indicates a slightly younger age in the third decade, trivial delay due to the individual's congenital abnormalities, or normal variation in the progress of fusion of these skeletal elements.

Dolní Věstonice 16

Cranial Indicators

Dolní Věstonice 16 exhibits a wider range of ectocranial suture closure than the other Pavlovian individuals discussed so far. A completely open suture is evident at the midlambdoid location, minimal closure is evident at the inferior sphenotemporal area, significant closure is evident at the lambda and anterior sagittal locations, and complete obliteration is found at the obelion, midcoronal and pterion regions (Table 6.1). Based on Meindl and Lovejoy's (1985) composite vault scoring system, the estimated age for Dolní Věstonice 16 (48.8 years) is slightly older than those estimated for both Dolní Věstonice 3 and 11/12 (45.2 years), although with narrower total (30–71 vs. 24–75 years) and interdecile (35–60 vs. 31–65 years) estimated ranges. The estimated age of Dolní Věstonice 16, based on the lateral-anterior composite scoring system, provides an even older age estimate (56.2 years), with a narrower associated total range estimate, than

Figure 6.1 Left: symphyseal surface of the Dolní Věstonice 15 right pubic symphysis, showing the persistence of the transverse billowing and the absence of a ventral rampart. Right: portion of the Dolní Věstonice 16 left iliac auricular surface, showing the fine porosity.

Dolní Věstonice 3 and 11/12 (34–68 vs. 23–68 years), as well as a smaller interdecile range (49–65 vs. 28–52 years). In general, Dolní Věstonice 16 is estimated to have been the oldest at death of all of the Pavlovian individuals, based on ectocranial suture closure.

Dental Indicators

The radiographs show that the roots of all teeth are completed, with apices closed. This includes both upper third molars and the lower left third molar. It can therefore be concluded that Dolní Věstonice 16 was a fully mature adult, even though the lower right third molar is missing (congenitally absent) and the lower left third molar is not erupted (an occlusal anomaly). All teeth except for the third molars, none of which came into occlusion, are heavily worn (Figures 11.8 and 11.9). In comparison with Brothwell's system (see Dolní Věstonice 3 above and Table 6.2), the lower first and second molars fit into the 35–45-year category (Smith score 7), but the upper first molars are slightly more worn than would be expected (Smith score 8 or more).

These observations imply that Dolní Věstonice 16 was perhaps a little older that Dolní Věstonice 3, perhaps in the "greater than 45 years" category, if wear proceeded at a similar rate to the medieval assemblages used in Brothwell's scheme. This is by no means necessarily the case because the dentitions of hunter-gatherer groups are known to have been subject to rapid wear. The most readily available comparison is with the East Greenland Inuit, who remained isolated into the early decades of the twentieth century; although their diet did contain some imported foods, it was largely composed of the meat and fat derived from fish, birds, and marine mammals. The dentitions of living Inuit were recorded in 1935 by Davies and Pedersen (1955), who recorded wear in the upper first molars on a three-point scale ("1 = attrition of the enamel: 2 = dentine exposed: 3 = exposure of secondary dentine or very rarely the pulp") and tabulated it against age at the time of examination. In his 30–39-year age group, 74% of first molars were at Pedersen's stage 2 and only 5% at stage 3. The 40–49-year group percentages are not given, but from his graph it is apparent that 25% of first molars were at stage 3 and, for the 50 years or greater group, 50% of first molars. The Dolní Věstonice 16 upper first molars expose substantial areas of the darker secondary dentine. They could fit into Pedersen's 30–39-year group of Inuit dentitions but would fit better into the 40–49-year group or older. Pedersen included a photograph of a dental impression taken from a 54-year-old woman who lived in one of the isolated East Greenland villages, and the teeth are clearly less worn than those of Dolní Věstonice 16. In conclusion, it seems likely on the basis of the comparisons available at present that Dolní Věstonice 16 was an adult in late middle age at the time of death.

Axial Indicators

The Dolní Věstonice 16 ribs are too badly damaged to assess whether their secondary epiphyses had been fused at the time of death. However, despite their damaged state, all vertebrae that preserve bodies evince fully fused ring epiphyses, and all of the preserved transverse and spinous processes show fusion of their epiphyses. These suggest a minimum age for the specimen of 17 to 25 years (Bass, 1987). However, there is extensive osteoarthritis in the cervical and upper thoracic regions of the specimen, with osteophytic lipping evident on every element from C4 to T1. Although trauma may be responsible for osteoarthritis in younger individuals, the more parsimonious explanation is that Dolní Věstonice 16 was at least in his fourth decade at the time of death, and he could have been much older.

Appendicular Indicators

The appendicular remains of Dolní Věstonice 16, to the extent that they are preserved (many are eroded and damaged), all indicate a fully mature individual. There is remarkably little degeneration of the articulations and little formation of osteophytes or enthesopathies (Chapter 19). The one marked articular degeneration, in the scaphoid-trapezium articulations bilaterally, appears to be idiopathic and not particularly related to age changes (Chapter 19).

There is one small area of the auricular surface preserved from the left side. The retained surface has a fine porosity and is minimally granular in appearance (Figure 6.1). This suggests a minimum age in the late fourth decade, but it may indicate an age either in the fifth decade (Lovejoy et al., 1985b) or anywhere between the late fourth and the sixth decade (Kobayashi, 1967).

Summary

Dolní Věstonice 16 is clearly the oldest of the known Pavlovian human remains, including those from Předmostí (Matiegka, 1934); this is reflected in his advanced ectocranial suture closure and extensive dental occlusal attrition, and it is supported by his vertebral osteoarthritis and auricular surface metamorphosis. These age indicators generally support an age at death at least well into the fifth decade, but it remains uncertain how much older than 40 to 45 the individual may have been. It is placed in the 45+ year category in Table 6.4, but that assessment may suggest an age rather younger than the actual age at death of this older, male individual.

Dolní Věstonice 17

Dolní Věstonice 17 consists of two small fragments of cranial vault bone, which on the basis of their thinness are considered to be immature. Their size would fit best with a juvenile (6 to 10 years old), but this is a very approximate assessment.

Dolní Věstonice 27

Dolní Věstonice 27, a deciduous upper canine, is worn to Smith stage 5, and it has a root that appears to be broken at about half its length. It is therefore not strongly resorbed and presumably was still in the jaw of the child at the time of death. Without direct evidence of resorption, it is difficult to estimate age, but the wear implies that the tooth had been in the mouth for some time. Eruption takes place between 1 and 2 years of age, and the root is complete around 3 years. Resorption of the root starts around 7 years, and the tooth is exfoliated about 10 to 11 years. All this would indicate an age at death somewhere in the middle of the bracket of 2 to 10 years.

Dolní Věstonice 31

The fully completed root apices of Dolní Věstonice 31, an isolated third molar, imply an age of at least 18 years (Mincer et al., 1993). The presence of occlusal wear shows that the tooth had erupted into the mouth. The wear had, however, caused only a slight polishing of the enamel, which exposed no dentine, and this would place the tooth in the 17–25-year-old category of Brothwell's scheme (Table 6.2). Taken together, the evidence implies that Dolní Věstonice 31 was from a young adult.

Dolní Věstonice 32

This isolated third molar, Dolní Věstonice 32, has fully completed root apices and slight occlusal wear of a very similar type to Dolní Věstonice 31. The age is therefore also likely to be similar, in the early 20s.

Dolní Věstonice 33

The uncertain identification of Dolní Věstonice 33 (Chapter 11) makes it impossible to arrive at an age estimate.

Dolní Věstonice 36

The assemblage of isolated, partly developed teeth of Dolní Věstonice 36 includes deciduous first incisors, first molars (third premolars), and second molars (fourth premolars), as well as a permanent first incisor and first molars (Figure 11.12; Table 6.3).

The three most common tables used for estimating age in immature modern humans are those of Schour and Massler (1941, 1944), Gustafson and Koch (1974), and Ubelaker (1978). The two later tables are based on large radiographic studies of living children. The basis of Schour and Massler is less well known, but it nonetheless seems to yield valid age estimates when independently tested (Liversidge, 1994). The works both of Schour and Massler and of Ubelaker consist of diagrams showing the expected state of dental development at six monthly and yearly intervals. They give an idea of variation in age for the attainment of the different stages but do not show variation in the relative development of different teeth, so, in effect, they portray an "average" developmental sequence. It is therefore to be expected that no one child would necessarily fit neatly into any one of the diagrams, and this is the case with Dolní Věstonice 36.

The development of the deciduous teeth fits best with the diagram for 1 year of age both of Schour and Massler and of Ubelaker, whereas the permanent teeth fit best with the 18-month stage. This degree of disparity is not unusual, but it makes for a less precise age estimate than might be desired. The Gustafson and Koch table works in a different way, with vertical bars representing the development of different teeth and its variation. Ages are estimated by placing a ruler across the table in a best-fit position. For Dolní Věstonice 36, this procedure provides a best-fit age estimate of 1 year. To conclude, the development standards of modern children suggest that Dolní Věstonice 36 died at 1 year of age or perhaps a little older.

Table 6.3 State of dental development in Dolní Věstonice 36 (Moorrees et al., 1963a).

	Deciduous Upper and Lower First Incisors	Deciduous Upper First Molar	Deciduous Upper and Lower Second Molars	Permanent Upper First Incisor	Permanent Upper and Lower First Molars
Development stage	Root half complete—$R_{1/2}$	Root quarter complete—$R_{1/4}$	Root formation initiated—R_i	Crown almost half formed—$C_{1/2}$ or slightly less	Crown three-quarters formed—$C_{3/4}$

Dolní Věstonice 37

This isolated, permanent lower second molar, Dolní Věstonice 37, has fully completed root apices and occlusal wear that involves only a slight polishing of the enamel and no exposure of dentine. Second molar roots are completed in the midteenage years in living children (Gustafson & Koch, 1974), so the roots of Dolní Věstonice 37 and its very slight occlusal wear are compatible with the 17–25-year stage of Brothwell's scheme (Table 6.2). Taken together, the best-fitting age estimate is in the late teenage years, perhaps between 17 and 19, in terms of the recent human dental development.

Dolní Věstonice 38

Like Dolní Věstonice 31 and 32, Dolní Věstonice 38 is an isolated third molar, but it has much more occlusal wear. The root apices are fully complete, showing that it is from a fully mature individual. In addition, the presence of hypercementosis suggests an older age and is compatible with the heavy attrition. The attrition stage fits into the 35–45-year stage of Brothwell's scheme (Table 6.2) or even older. This evidence implies that Dolní Věstonice 38 came from an adult in late middle age.

Dolní Věstonice Isolated Postcranial Remains

The isolated postcranial remains from Dolní Věstonice include a variety of elements (Tables 5.1 and 5.3), all of which appear to be fully mature. However, they are variably indicative of the ages at death of the individuals from which they came.

The majority of the remains, other than appearing fully mature from their general condition, provide little confirmation of this. They include the two rib sections (Dolní Věstonice 45 and 51), the pieces of humerus and radius (Dolní Věstonice 41 and 50), the three pieces of femoral diaphysis (Dolní Věstonice 35, 40, and 43), the two fibular diaphyses (Dolní Věstonice 42 and 48), the two tarsals (Dolní Věstonice 39 and 46) and the three metatarsals (Dolní Věstonice 44, 47, and 49). All of these remains appear to have passed at least midadolescence, based on either general morphology and/or the formation of an articular surface, but further age at death resolution is not possible.

There are three phalanges, two middle hand phalanges (Dolní Věstonice 34 and 53), and one hallucal proximal phalanx (Dolní Věstonice 52), which show complete fusion of their proximal epiphyses. These provide minimum ages of the mid–second decade for the hand phalanges and the second half of the second decade for the hallucal phalanx.

As a result, all of these isolated remains are considered to be mature, but they will not be considered further with respect to their probable ages at death.

Pavlov 1

Cranial Indicators

Pavlov 1, like Dolní Věstonice 16, exhibits a wider range of ectocranial suture closure than the other Pavlovian individuals. A completely open suture is evident at bregma; minimal closure is evident at the anterior sagittal, midcoronal, and inferior sphenotemporal locations; significant closure is evident at the midlambdoid, lambda, and obelion locations; and complete obliteration is found in the pterion region (Table 6.1). Based on Meindl and Lovejoy's (1985) composite vault scoring system, the estimated age for Pavlov 1 (39.4 years) is somewhat younger than those estimated for Dolní Věstonice 3, 11/12, and 16. Pavlov 1's total estimated range (24–60 years) and estimated interdecile range (28–44 years) are narrower than those found in the others, and they do not extend as far into the oldest range of years. The estimated age for Pavlov 1 based on the lateral-anterior composite scoring system provides an older age estimate (45.5 years), which is younger than that estimated for Dolní Věstonice 3 (56.2 years) and Dolní Věstonice 16 (56.2 years) but older than that estimated for Dolní Věstonice 11/12 (41.1 years). The total and interdecile age ranges for Pavlov 1 (32–65 and 35–57 years, respectively) generally span the combined values for Dolní Věstonice 3, 11/12, and 16.

Dental Indicators

In this cranium and mandible, the root apices of all of the teeth are completed, including those of the lower left third molar. This implies an age of at least 18 years at death (Mincer et al., 1993). The permanent first molars show heavy occlusal attrition, which fits best in the 35–45-year category of Brothwell's system (Table 6.2). The second molar wear is a little light for this category, as is the wear on the third molar, although this tooth stands a little proud, perhaps because of an occlusal anomaly that might have affected the pattern and rate of wear. Taken altogether, the dental evidence suggests that Pavlov 1 was a middle-aged adult at death (Figures 11.14 and 11.15).

Axial Indicators

The vertebrae of Pavlov 1 provide virtually no clues to its age at death, other than that the neural arches are fused to the bodies, a development that typically occurs by 7 years of age. The only vertebra that preserves a body is

C2, and the caudal surface of its body is covered in plaster, making evaluation of the presence or absence of ring epiphyses impossible. None of the vertebral elements preserves a transverse or spinous process either, and thus one cannot evaluate whether the secondary epiphyses associated with these structures were present and/or fused. Likewise, the rib fragments do not preserve sites with secondary epiphyses. There is no evidence for degenerative changes in the axial skeleton.

Appendicular Indicators

The appendicular remains of Pavlov 1, which include arm bones, a metacarpal, and the femora, are all indicative of fully mature status. There are no adult age-related changes evident on them that allow further precision.

Summary

The age at death of Pavlov 1, other than fully mature, must be based on the ectocranial suture closure and degree of dental attrition, neither one of which is particularly precise at indicating the age at death of this individual. The suture closure pattern suggests a slightly younger age at death than those of Dolní Věstonice 3 and 11/2, but the degree of dental attrition implies an age perhaps slightly more advanced than that of Dolní Věstonice 3. Taken together, these indicators are most in agreement with an age assessment in the late fourth or early fifth decade.

Pavlov 2

The left and right fragments of a maxilla (Pavlov 2a and 2b) include permanent first and second molars with fully completed root apices (Figure 11.16). In living children, the second molar roots are completed in the midteenage years (Gustafson & Koch, 1974), so Pavlov 2 must at least have been older than that. The state of wear in the first molars fits best with the 25–35-year category of Brothwell's system (Table 6.2), although the wear on the second molars is a little lighter than would be expected for this. Altogether, however, it is likely that Pavlov 2 was a young adult at death.

Pavlov 3

Pavlov 3 comprises a fragment of mandible, including canine, premolars, and first molar (Figure 11.17). All of the the teeth have fully formed root apices. Second premolar roots are completed in living children between 12 and 15 years (Gustafson & Koch, 1974), so the Pavlov 3 fragment represents an older individual. The first molar occlusal wear fits best with the 25–35-year category of Brothwell's scheme (Table 6.2). Taken together, this evidence suggests early middle age for Pavlov 3.

Note that the Pavlov 28 isolated third molar (see below) was originally included with Pavlov 3 (Vlček, 1997), but it has been separated because of our inability to confirm this association. If it does indeed derive from Pavlov 3, its degree of wear would be in agreement with the first molar age indications.

Pavlov 4

The fragmentary piece of mandibular corpus of Pavlov 4 appears mature. Little more can be said of its age at death.

Pavlov 5

Pavlov 5 consists of permanent left and right lower incisors, matched by the fit of their mesial approximal wear facets. Their apices are fully complete, which implies that they represent an individual at least over 9 to 11 years of age. The state of occlusal wear is less than that of incisors in Pavlov 1, Dolní Věstonice 3, and Dolní Věstonice 16, but about the same as Dolní Věstonice 13 and 15 and slightly more than Dolní Věstonice 14. The estimates for the ages of Dolní Věstonice 13 and 15 are about the early 20s, with Dolní Věstonice 14 slightly younger. If we bear in mind the possible variation in the wear of anterior teeth relative to cheek teeth, the incisors of Pavlov 5 should represent a young adult.

Pavlov 6

Pavlov 6 includes deciduous upper first and second molars (third and fourth deciduous premolars). Both have heavy occlusal wear and roots resorbed so that only remnants remain. This implies either that the teeth had been exfoliated from a living child or that they were just about to be exfoliated at the time of death in a child whose remains became fragmentary during deposition. Only in the latter case would they provide any estimate of age at death. The roots of deciduous upper first molars start to resorb just before 5 years of age, and those of second molars at around 6 years (Fanning, 1961; Moorrees et al., 1963b; Haavikko, 1973). If Pavlov 6 died before they were shed, the child must at least have been older than 6 years. Conversely, deciduous second molars are exfoliated around 10 to 11 years of age, and first molars usually a little younger. If the teeth were just about to be exfoliated, the child would need to be about 10 years of age for them still to be in position. In view of the presence on the site of several other isolated deciduous molars with root resorption, it seems more likely that Pavlov 6 represents normally exfoliated teeth.

Pavlov 7–12

Pavlov 7–12, are deciduous molars (premolars) in which the roots are resorbed so that only remnants remain. Pavlov 7–10 are all deciduous lower second molars and, as these teeth are exfoliated around 10 to 11 years of age, they must represent either exfoliated teeth or individuals who died at about that age. The simplest explanation is that they were exfoliated. Pavlov 7 may come from the same individual as Pavlov 8, and both may also belong with Pavlov 6. Pavlov 10 has more wear than the others, but its roots are slightly less resorbed—an anomaly that may be explained by normal variation in exfoliation. Pavlov 11 is a deciduous upper first molar, and Pavlov 12 is a deciduous upper second molar. Upper first molars are usually exfoliated around 9 to 10 years of age in today's children, and second molars at 10 to 11 years, so Pavlov 11 could potentially come from the same child as Pavlov 12, but the approximal facets of the teeth do not fit together well. It seems more likely that they represent separate individuals, which may be those represented by Pavlov 6, 7, 8, 9, or 10.

Pavlov 13 and 14

Pavlov 13 and 14 are deciduous upper canines with roots so strongly resorbed that only remnants remain. They have different morphologies, so despite being left and right teeth, they are unlikely to have belonged to the same individual. Root resorption in upper canines starts at about 7 years of age, and the teeth are exfoliated between 10 and 11. It is most likely that Pavlov 13 and 14 were exfoliated before deposition on the site, but if they had not been, the children must have died at around 10 years of age.

Pavlov 15 and 19

Pavlov 15 is a deciduous upper canine, and Pavlov 19 is a deciduous upper first incisor. Both are exceptionally heavily worn for deciduous teeth, but only about half the root has been resorbed. It is likely that, in both cases, they were retained in the mouth after the other deciduous teeth had been replaced by permanent teeth. For this reason, it is not possible to apply age estimates.

Pavlov 16–18

Pavlov 16 is a deciduous upper second incisor, whereas Pavlov 17 and 18 are deciduous lower second incisors. All have partially resorbed roots, which suggests that they were still retained in the jaw at the time of death, but they have more modest wear than Pavlov 15 and 19. This implies that the individuals were still children, undergoing normal replacement of the deciduous dentition. The teeth can therefore be used to estimate age at death. In deciduous upper second incisors, resorption starts around 5 to 6 years of age and the teeth are exfoliated at 7 to 8 years. In deciduous lower second incisors, resorption starts at about 5 years of age, and the teeth are shed at 6 to 7 years. There is not much difference between these timings, which implies that all the individuals involved were between 5 and 7 years of age at death. All three teeth are from different quadrants of the dentition, so they might represent one individual, although their morphology and wear are different.

Pavlov 20

Pavlov 20 is an isolated third molar. The root apices are completed, implying that the individual from which it came was 18 years of age or older (Mincer et al., 1993). There is very little occlusal wear, but a large mesial approximal wear facet, in the normal position, implies that it was normally erupted. Taken together, this evidence suggests an age not much older than 18 years, perhaps extending into the early 20s. This could be a false impression, however, because there is considerable hypercementosis on the roots, which normally develops progressively with age. It is possible that the lack of wear results from an anomaly of occlusion that somehow did not result in an abnormally placed approximal facet or the absence of the opposite tooth. Originally, this specimen was joined to the reconstructed Pavlov 2, but it was separated when further study showed it could not belong to the same dentition (Chapter 11).

Pavlov 21

The identification of Pavlov 21 is uncertain, so it is not possible to use this tooth in age estimation.

Pavlov 22, 23, and 25

Pavlov 22 is an isolated permanent upper second incisor, and it is quite heavily worn—not so heavily as Pavlov 1 and Dolní Věstonice 3, more than Dolní Věstonice 14 and 15, but around the same as Dolní Věstonice 13. If the rates of anterior tooth wear were similar, this would be compatible with an age in the early 20s. Pavlov 23 is an isolated permanent lower left second incisor, with wear also most similar to that of Dolní Věstonice 13. Pavlov 25 is an isolated permanent lower right second incisor with just slightly more occlusal wear, although this is still less than in Pavlov 1 or Dolní Věstonice 3 and 16.

Pavlov 24, 26, and 27

Pavlov 24, 26, and 27 are too fragmentary to use for age estimation.

Pavlov 28

Pavlov 28 is an isolated lower third molar, formerly associated with the Pavlov 3 remains (Chapter 11). The complete apex of the molar implies an age over 18 (Mincer et al., 1993). It also displays hypercementosis, which usually implies maturity, especially as it is not accompanied by heavy wear.

Summary

The human remains to which ages at death can be reasonably assigned, beyond immature versus mature, are listed in Table 6.4. From these assessments, it appears that the sample contains a number of both immature and mature individuals. There are a couple of young children (Dolní Věstonice 36 and probably the now lost Dolní Věstonice 4), several juveniles (Dolní Věstonice 17 and 27 and Pavlov 16–18), and two late adolescents (Dolní Vostonice 14 and 37). There is a number of young adults, pooling together the individuals likely from the third and early fourth decades, which represent almost half of the ageable individuals and almost three-quarters of the mature individuals. The uncertainties regarding the age categories of Dolní Věstonice 9 and Pavlov 20 are between these two younger adult age categories, and reassignment of them to the other possible category would not change this summary. To these groups are added the older middle-aged Dolní Věstonice 3 and 11/2 and Pavlov 1 remains and the older Dolní Věstonice 16 and 38 specimens.

This overall age span does not change whether one looks at the associated remains, including Dolní Věstonice 36, in this group or only at the isolated cranial and dental remains. There are more older adults among the associated remains from burials and more young adults and juveniles among the isolated teeth and cranial pieces. These contrasts are small but sufficient to provide a significant difference between these two samples (exact chi-square $P = 0.003$; Mehta & Patel, 1999). This pattern of small numbers of associated immature individuals is undoubtedly affected by age-related mortuary practices during the Gravettian as a whole (Zilhão & Trinkaus, 2002). The overall pattern of a relative abundance of prime age adults and a dearth of older mature individuals is one that

Table 6.4 Summary assessments of the ages at death (in years) of the Dolní Věstonice (DV) and Pavlov (Pav) human remains.

0–5	6–10	11–15	16–20	21–25	26–35	36–45	45+
Associated Skeletons							
DV 36			DV 14	DV 13		DV 3	DV 16
DV 4?				DV 15		Pav 1	
Isolated Remains							
	DV 17?		DV 37	DV 9?	Pav 2	DV 11/12	DV 38
	DV 27			DV 31	Pav 3		
	Pav 16			DV 32	Pav 20?		
	Pav 17			Pav 5	Pav 25		
	Pav 18			Pav 22	Pav 28		
				Pav 23			

Note that these are approximate, in terms of the patterns of development, degeneration, and attrition of living humans, and provided here as a general summary. The uncertainties inherent to many of these assessments are discussed in the text. The series of deciduous teeth from Pavlov I that are likely to have been exfoliated during life are not included, nor are the isolated postcranial remains from Dolní Věstonice I and II, which can only be assessed as mature.

has been documented for Late Pleistocene archaic humans (Neandertals; Trinkaus, 1995) and may well hold for earlier Pleistocene archaic *Homo* samples (Streeter et al., 2001), but sufficient data on the ages at death of Gravettian human remains from across Europe are not available to determine whether this pattern was maintained into earlier modern human Upper Paleolithic populations. And, if one assesses only the associated human skeletons from these Pavlovian sites, there does not appear to be a bias toward younger adults.

What becomes apparent from these considerations is that the human skeletal and dental sample from these Pavlovian sites contains remains of individuals from all age categories, from young children to old adults. The apparently synchronous death of three adolescent to young adult males, the triple burial, has struck some as unusual and of possible adverse economic and demographic consequences to their social group (e.g., Wolpoff, 1999), and it certainly may have had that short-term effect. Yet, these ages at death are matched by a number of other individuals from these sites, all represented by isolated dental remains, suggesting that mortality in this age period was not unusual for Gravettian populations. It remains to place these paleodemographic considerations into a broader context of European Middle Upper Paleolithic mortality patterns, to assess the demographic and social implications of this Pavlovian pattern.

7

The Assessment of Sex

Jaroslav Brůžek, Robert G. Franciscus, Vladimír Novotný, and Erik Trinkaus

A number of the interpretations of the Dolní Věstonice and Pavlov human remains is dependent in part on the sex of the individuals. We have therefore attempted to assign sex to as many of the remains as is reasonable, using the skeletal elements available for each individual. The assessment of sex from these skeletal remains involves considerations of several aspects of their skeletal morphology. The first, and most important, involves their pelvic remains, which are assessed from qualitative, degenerative, discrete trait and morphometric perspectives. The second is the form of their cranial and mandibular remains, involving both overall dimensions and detailed traits. The third is an evaluation of overall body size, in the context of the levels of skeletal sexual body size dimorphism present in recent humans and inferred for Upper Paleolithic European populations.

Methodological Issues

The assessment of sex in Pleistocene fossil samples is more difficult than among recent skeletal remains as a result of both their frequently fragmentary and incomplete condition and the absence of reference samples of known sex from the same or closely related populations. It is therefore not possible to apply a reference standard that is specific to an extinct population, despite the necessity of sexual assessment for the evaluation of several aspects of fossil human morphology (see Novotný, 1983).

In this assessment, since we are dealing with a sample whose overall morphological pattern is akin to that of extant humans, it is assumed that the overall patterns, if not necessarily all of the details, of sexual dimorphism of recent humans apply to these Late Pleistocene humans. The contrasts between recent humans and the Late Pleistocene humans in sexual dimorphism are likely, given current knowledge of their skeletal morphology, to lie principally within the realms of body size (also an issue among recent human populations) and general skeletal robusticity, as noted half a century ago (Genovés, 1954). It is therefore necessary to consider the biological bases of the criteria used for sexual assessment.

Sexual dimorphism consists of two elements. On the one hand, there is size dimorphism. On the other hand, shape dimorphism is more appropriate for sex determination among early modern humans (e.g., Krantz, 1982; Novotný, 1983; Hager, 1989), especially as applied to the pelvis, given that by 100,000 years ago early modern humans had evolved an essentially modern pelvic morphology (Vandermeersch, 1981; Rosenberg & Trevathan, 2002).

Assessments of sex based on size dimorphism assume that the level of sexual dimorphism in overall measures of body size, whether of mass or of stature (or skeletal reflections thereof), follow a regular pattern of the degree of dimorphism within the species. Although there is variation in size dimorphism across recent human populations (Hamilton, 1975; Trinkaus, 1980) and there is evidence that it can be affected by levels of stress during development (Stini, 1969; Hamilton, 1975), the variation across populations in percent dimorphism (however calculated) is seldom more than a few percent. Previous assessments of sexual size dimorphism among Late Pleistocene humans (e.g., Frayer, 1980; Trinkaus, 1980) indicate that their overall levels of dimorphism in skeletal indicators of body size were either similar to or slightly greater than those normally recorded across recent human skeletal samples.

However, there is considerable variation across samples in average overall body size, as well as in the dimensions of select elements such as the cranium and mandible, indicating that it is necessary to establish as best as possible a scale of reference for the sample in question. This applies whether individual measurements of body size, such as long bone lengths or multivariate analyses (e.g., discriminant function analysis), are applied to the remains of concern (Trinkaus, 1980; Krogman & Iscan, 1986; Sjøvold, 1988; Brůžek, 1991). For the Late Pleistocene, in order to use such a technique, it is necessary to establish an appropriate reference sample of individuals for whom sex can be reasonably assigned on the basis of other (especially pelvic) shape criteria. Moreover, given the significant overlap in body size between males and females in any one recent human population, and undoubtedly in Pleistocene human groups, reliable results can be obtained for individual specimens principally if they fall within the smaller range of the females or the larger range of the males (Trinkaus, 1980).

In contrast, the degree and pattern of sexual dimorphism of the human pelvis appears to be relatively stable and uniform across populations of recent (and, by inference, past) populations of "anatomically modern" humans (Brůžek, 2002). As a result, a methodology based on the os coxae should be applicable to any "modern" human population, provided that a sufficient number of criteria (or variables) are employed to adequately characterize the overall sexual dimorphism of the pelvis (Novotný, 1981; Brůžek, 1991). Similarly, the patterns of sexual dimorphism in the proportions of the pelvis are sufficiently stable across recent human populations that they lose their specificity and can be applied to specimens from a variety of populations. As a result, discriminant function analyses of the pelvis based on a variety of samples are particularly reliable and provide high levels of correct classification within each of the samples (Brůžek, 1992). Nonetheless, the fragility of the pelvis during fossilization frequently limits the applicability of such metric analyses, since their reliability is related to the number of variables that can be included in the analysis.

It should be noted that—although analyses of cranial and mandibular sexual dimorphism frequently provide uncertain results, especially when recent human standards are applied to Pleistocene human remains (Genovés, 1954; Šefčáková et al., 1999; Sládek et al., 2001)—given variation both in overall cephalic dimensions between populations and in the expression of secondary sexual characteristics of the cranium and mandible, it is possible to provide some indication of sex based on the degree of development of a series of secondary sexual characters of the cranium and mandible. In such cases, as with sexual assesments based on overall size, it is essential to carry out the analyses in the context of the ranges of variation present within the sample of concern.

Previous Sexual Assessments of Dolní Věstonice and Pavlov Samples

In the southern Moravian Gravettian sample, sex was previously assessed for Dolní Věstonice 3 and 13 to 16 principally on the basis of the pelvis, along with considerations of the cranial morphology and, to a lesser extent, body size (e.g., Jelínek, 1954, 1992; Vlček, 1991; Trinkaus & Jelínek, 1997; Novotný & Brůžek, 1999; Novotný, 2003), plus in the case of Dolní Věstonice 15, a diagnosis of its pathological condition (Formicola et al., 2001). For the isolated cranial elements (Dolní Věstonice 11/12, as well as the now lost Dolní Věstonice 1 and 2) and the Pavlov 1 partial skeleton that lacks the pelvis, sex was evaluated principally on the basis of cranial morphology. The various isolated pieces from the Dolní Věstonice sites and Pavlov I have been considered indeterminate as to sex (Novotný & Brůžek, 1999) or not assigned sex (Trinkaus et al., 1999b, 2000c).

Disagreements concerning these attributions have focused almost entirely on the pathological Dolní Věstonice 15 partial skeleton (e.g., Vlček, 1991; Jelínek, 1992; Novotný, 1992, 2003; Novotný & Brůžek, 1999; Formicola et al., 2001; Trinkaus et al., 2001). The sex of this individual has been considered as female, male, and ambiguous. Most of the considerations have concerned the apparent mix of male and female characters in the largely complete pelvis, with some assessments of the development of sexually dimorphic characteristics of the cranium. In addition, Formicola et al. (2001) have proposed a pathological diagnosis of X-linked dominant chondrodysplasia punctata, which has been considered as primarily affecting females and presumed to be lethal in infancy in males. Since Dolní Věstonice 15 survived to early maturity, the conclusion was that the remains must be female. However, the diagnosis of the form of dysplasia evident in Dolní Věstonice 15 is open to discussion (see Chapter 19), and there are documented cases of the survival of males with this condition (Milunsky et al., 2003). Consequently, the sex of Dolní Věstonice 15 is open to reanalysis.

Given that most of the sexual attributions of this sample are based on assessments without clearly stated criteria, this chapter presents detailed descriptions of the methods and results of a reassessment of the sexes of these skeletal remains. The excellent preservation of several of the remains permits the application of a series of morphological and metrical comparisons, several of which (for the pelvis) are based on a large and global sample of recent human remains (Houët et al., 1999; Murail et al., 1999; Brůžek, 2002). This applies particularly to the triple burial individuals (Dolní Věstonice 13, 14, and 15) and to a slightly lesser extent to Dolní Věstonice 3 and 16. All five individuals have pelvic remains, largely complete skulls, and postcranial dimensions. The other remains are then incorporated into the framework provided by these five partial skeletons.

Pelvic Assessments of Sex

The assessment of sex of the five Dolní Věstonice partial skeletons that preserve portions of their pelves, Dolní Věstonice 3 and 13–16, starts with a consideration of the relevant aspects of preservation and morphology of the individual pelves. This is followed by a visual evaluation of discrete characters of relevance to sexual assessment and a morphometric analysis of the pelvic remains.

Individual Pelvic Morphology

Dolní Věstonice 3

The pelvic remains of Dolní Věstonice 3 were badly damaged in situ, such that virtually all of the pubic area and much of the ischial region were lost or crushed beyond restoration. In addition, the left ilium was flattened in situ, with multiple cracks and small displacements of the resultant pieces. As a consequence, the superiorly exposed left ilium (cf., Jelínek, 1954) provided the suggestion that the greater sciatic notch was open posteriorly but was not sufficient to confirm it. The subsequent excavation and analysis of the Dolní Věstonice 3 postcranial remains (Trinkaus & Jelínek, 1997) yielded the undeformed right inferior ilium with most of the greater sciatic notch preserved, the base of the acetabulum, and the inferior portion of the iliac fossa (Figure 7.1). The posterior inferior iliac spine is not retained, and the posterior border of the ischium is only present superiorly.

Figure 7.1 Internal view of the Dolní Věstonice 3 inferior right ilium, with the greater sciatic notch. Scale in centimeters.

Despite the postmortem deformation of the left os coxae, the appearance of the bone is gracile, with a relatively flat internal iliac fossa. The acetabulum is small and shallow. The presence of a composite arc—a single arc joining the superior curve of the auricular surface with the anterior/ventral curve of the greater sciatic notch (Brůžek, 2002)—suggests that it is female. In the preauricular area, there is a small depression, 8 mm by 3 to 4 mm and 2 to 3 mm deep, which resembles a preauricular sulcus. However, low-power magnification reveals that it is the result of fossil preparation, as part of an attempt to separate the crushed sacrum from the ilium.

The shape of the right greater sciatic notch, although incomplete, is commensurate with a diagnosis as female, and it would be seriously at variance with an attribution of male to the skeleton. As noted elsewhere (Trinkaus & Jelínek, 1997), the gluteal lines are well marked on the external ilium.

Dolní Věstonice 13

Dolní Věstonice 13 retains a virtually complete pelvis with only minor damage and restoration to minor margins of the ossa coxae (Figure 17.1). As a result, it is possible to examine it for indications of the sex of Dolní Věstonice 13 from a qualitative morphological approach.

The greater sciatic notches of Dolní Věstonice 13 form distinct full semicircles, with only a hint of longer dorsal sides down to the posterior inferior iliac spines (Figure 7.2). There is no trace of a preauricular sulcus or rugosity, and given the age at death of Dolní Věstonice 13 well into the third decade, it is likely that the individual, if female, would have experienced at least one pregnancy and possibly two. The composite arches are distinctly male, in having separate trajectories for the dorsal auricular and anterior greater sciatic notch curvatures.

The pubic bones are short and, especially the superior pubic rami, stout (Figure 7.3). The ventral pubic ramus thicknesses of 13.0 and 13.8 mm are above the means of a recent human range of variation—Amerindians: males: 11.3 ± 1.5 mm, $N = 42$; females: 10.5 ± 1.9 mm, $N = 40$ (Trinkaus, 1983b)—yet within the ranges of variation of both males and females. The obturator foramina are vertically oriented, with the vertical height markedly greater than the breadth (height/breadth: 1.46 and 1.67). The orientation of the obturator foramina relative to the symphyseal plane and within the plane of the foramen can be indicated by the orientation of the major axis through the digitized surface; this angle has values of 94° and 87°, in which 90° is vertical and an angle >90° indicates that the superior end tilts toward the midline. The Dolní Věstonice 13 obturator foramina are therefore geometrically on either side of vertical.

Figure 7.2 External view of the Dolní Věstonice 13 right greater sciatic notch. The articulated right lateral sacrum is also apparent. Scale in centimeters.

The symphyseal body is mediolaterally narrow, and there are no notches (or profile concavities) when viewed anterolaterally between the symphyseal bodies and the ischiopubic rami.

The ischiopubic rami are also stout (Figure 7.3), and they are especially thick along their anteromedioinferior

Figure 7.3 Anterolateral view of the Dolní Věstonice 13 right ischiopubic region, with the acetabulum. Scale in centimeters.

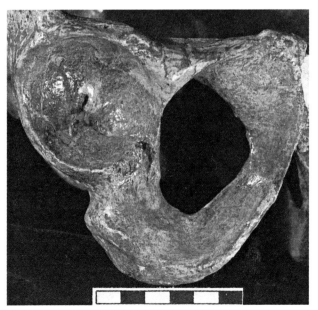

margins. Their inferior margins turn anterolaterally and are displaced anterolaterally relative to each symphyseal body just below the pubic symphysis. However, the subpubic angle (as measured in the plane of the two ischiopubic rami between the right and left sides) of 63° is in the middle of recent male ranges of variation and at least 2.1 standard deviations from the lowest recent human female mean (Table 7.1).

Given the completeness of the Dolní Věstonice 13 pelvis, it is also possible to measure its pelvic aperture diameters. The ratio of its inlet diameters (112.0 mm versus 120.0 mm) provides an index of 93.3, which indicates a relatively round pelvic inlet. This value is relatively high compared to all of the Gravettian specimens providing data except for the female Paglicci 25 (Table 7.1). However, there is an inconsistent pattern of sexual dimorphism in this pattern in modern humans, and it remains within 2 standard deviations of both male and female means. The ischial spines of Dolní Věstonice 13 turn strongly inward, into the pelvic aperture space—an arrangement frequently seen in male pelves.

The Dolní Věstonice 13 pelvis therefore presents a typically "male" morphological pattern in its pelvis, and the characteristics that are generally most indicative of sex (greater sciatic notch shape, short pubis relative to the ischium, the composite arch, and the presence of a *crista phallica*, as well as the obturator foramen shape and orientation, symphyseal body shape and breadth, ischiopubic ramus robusticity, and subpubic angle) all clearly align the pelvis with recent (and Gravettian) male specimens. This is reinforced by the absence of any trace of a preauricular sulcus. The more female proportions, including inlet shape and the anterolateral flaring of the ischiopubic rami, are variable within recent human sexes and do not contradict the diagnosis of the pelvis as male.

Dolní Věstonice 14

The Dolní Věstonice 14 pelvis is also largely complete, even though it sustained more marginal damage than the Dolní Věstonice 13 one (Figure 17.3). The greater sciatic notches form full semicircles, from the posterior inferior iliac spine to the posterior ischium (Figure 7.4). There is no trace of a preauricular sulcus on the preserved left side, but the adolescent age at death of Dolní Věstonice 14 removes any sexual significance of that fact.

The superior pubic rami are short and stout (Figure 7.5), and the ventral thicknesses of the rami (18.6 mm and 17.2 mm) place it well into the male side of the distribution. They are similar to the male Cro-Magnon 1 value (17.5 mm) and above those of the female Caviglione 1 (14.5 mm) and male Paviland 1 (12.4 mm) values. The obturator foramina are vertical and relatively narrow; the ratios of height to breadth are 1.48 and 1.81. The major

Table 7.1 Comparative values for subpubic angle and inlet indices (anteroposterior/transverse diameters) for Dolní Věstonice, Cro-Magnon, Paglicci, Předmostí, and recent human pelves.

	Subpubic Angle		Inlet Index	
	Males	Females	Males	Females
Dolní Věstonice 13	63°		93.3	
Dolní Věstonice 14	62°		72.6	
Dolní Věstonice 15	94°		75.5	
Cro-Magnon 1	—		(77.3)	
Paglicci 25		—		105.5
Předmostí 3	—		(74.2)	
Předmostí 4		—		(81.5)
Předmostí 10		—		(78.2)
Předmostí 14	—		(84.0)	
Recent Humans				
Euro-Americans	63.7° ± 7.8°	88.4° ± 8.5°	79.0 ± 7.9	83.1 ± 10.0
Afro-Americans	65.8° ± 8.7°	85.2° ± 10.4°	81.9 ± 7.9	90.3 ± 11.1
Indian Knoll	73.8° ± 8.2°	98.2° ± 8.4°	82.3 ± 6.7	81.2 ± 5.7
Pecos Pueblo	61.6° ± 8.2°	86.0° ± 10.0°	71.6 ± 6.9	68.5 ± 6.3
Libben	68.8° ± 7.8°	95.2° ± 10.8°	77.4 ± 7.1	72.7 ± 5.7
Haida	65.4° ± 8.2°	93.0° ± 12.3°	78.4 ± 6.6	83.0 ± 9.0

Comparative data from Matiegka (1938), Vallois & Billy (1965), Mallegni et al. (1999), and Tague (1989). The recent human data include two documented cadaver samples (Euro-Americans and Afro-Americans) and four archeological Native American samples. Values in parentheses are indices based on at least one estimated measurement.

axis angles of the foramina are 100° and 93°, indicating a minimal tilt toward the midline. The symphyseal bodies are markedly narrow, and the symphyseal surfaces round evenly onto the anteromedioinferior ischiopubic rami. Yet, the ischiopubic rami are relatively thin and moderately everted. The sexual form of the composite arch is "intermediate." The subpubic angle of 62° is essentially the same as that of Dolní Věstonice 13 and closely aligns Dolní Věstonice 14 with male pelves.

The pelvic inlet appears to be constrained anteroposteriorly. The ratio of its inlet diameters (90.0 mm vs. 124.0 mm) provides an index of 72.6, which confirms the visual impression. This value is moderately low compared to the Gravettian specimens providing data, but it is close to the estimated value for the male Předmostí 3 and not very distant from the values for the male Cro-Magnon 1 and the female Předmostí 10. However, as with Dolní Věstonice 13, the ischial spines (to the extent preserved) turn markedly inward, into the pelvic aperture, a common male characteristic.

The Dolní Věstonice 14 pelvis therefore presents a set of distinctly male features, including the form of the greater sciatic notch, the shape and orientation of the obturator foramen, the shape and breadth of the symphyseal body, the robusticity of the superior pubic ramus, and the small subpubic angle. It is mainly the gracility and eversion of the ischiopubic ramus that indicate a less masculine pelvis. The total pattern of these features, however, argues strongly for a male diagnosis.

Dolní Věstonice 15

The largely complete pelvis of Dolní Věstonice 15 (Figure 17.6) is smaller than any of its contemporaries. Its

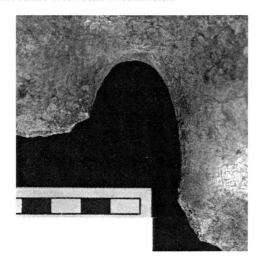

Figure 7.4 External view of the Dolní Věstonice 14 right greater sciatic notch. Scale in centimeters.

The Assessment of Sex

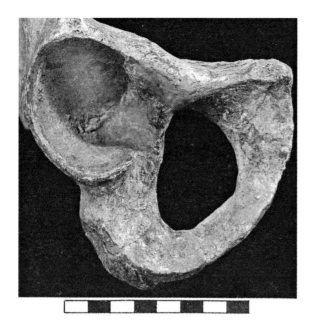

Figure 7.5 Anterolateral view of the Dolní Věstonice 14 right ischiopubic region, with the acetabulum. Scale in centimeters.

Figure 7.6 External view of the Dolní Věstonice 15 right greater sciatic notch. Scale in centimeters.

estimated range of bi-iliac breadth (234–245 mm—given as a range since erosion to the sacroiliac surfaces allows some movement between the ilia and the sacrum when articulated) is below any of the other Gravettian specimens that provide a direct measure of bi-iliac breadth (pooled males and females, including Dolní Věstonice 13 and 14: 272.4 mm ± 12.9 mm, 255.0 – 293.0 mm; $N = 12$). However, its stature estimates of 158.9 cm (male) and 155.2 cm (female) are also low (Chapter 13), such that the index of bi-iliac breadth to stature is 14.7 to 15.4 (male) and 15.1 to 15.8 (female) for Dolní Věstonice 15. The same index is 15.8 ± 0.6 (14.5 – 16.7; $N = 12$) for the Gravettian comparative sample (the low value of 14.5 is for Dolní Věstonice 14). The Dolní Věstonice 15 pelvis is therefore absolutely small, but it is not relatively small compared to its estimated male or female overall body length.

In the context of these small dimensions, the Dolní Věstonice 15 pelvis presents a mosaic of sexually diagnostic features, even though its morphometric proportions clearly align it with males (see below). The greater sciatic notch forms an open semicircle (Figure 7.6), such that the posterior margin descending to the posterior inferior iliac spine is less vertical than those of Dolní Věstonice 13 and 14 (Figures 7.2 and 7.4) but more closed than that of Dolní Věstonice 3. The composite arch arrangement is distinctively male.

The superior pubic ramus is short and stout, and despite the small size of the pelvis overall, it has a ventral ramus thickness (on the left) of 14.8 mm (Figure 7.7). This value is between those of Dolní Věstonice 13 and 14, close to the value for the Caviglione 1 female, and between those of the Cro-Magnon 1 and Paviland 1 males. The symphyseal body is mediolaterally narrow, and there is no concavity in profile between it and the anteroinferior ischiopubic ramus.

The obturator foramina are relatively narrow, as is indicated by height/breadth ratios of 1.45 and 1.79. However, they are more obliquely oriented than those of Dolní Věstonice 13 and 14, providing major axis angles of 117° and 104°. However, the male Paviland 1 pelvis provides an angle of 112°, between the two Dolní Věstonice 15 values, and the late Upper Paleolithic female Obercassel 2 pelvis provides an angle of only 114°.

The ischiopubic rami are moderately everted, similar to the male Dolní Věstonice specimens, and they are

Figure 7.7 Anterolateral view of the Dolní Věstonice 15 right ischiopubic region, with the acetabulum. Scale in centimeters.

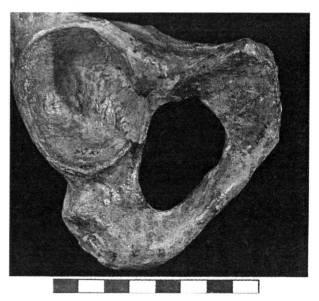

relatively robust for the size of the pelvis. However, the subpubic angle is moderately open, having an angle of 94°, which is in the middle of the recent human female range and quite separate from both modern males and the Dolní Věstonice 13 and 14 values (Table 7.1).

The pelvic inlet is relatively constricted anteroposteriorly (Table 7.1), but not unusually so for either a modern human or Gravettian pelvis. The ischial spines, nonetheless, curve markedly inwardly, much as do those of Dolní Věstonice 13 and 14.

In addition, each posterior ilium exhibits a relatively deep paraglenoidal sulcus, which given damage and restoration, appears to be largely symmetrical (Figure 7.8). On the left side, where it is more complete, the sulcal breadth is about 5 mm superiorly and 3.5 mm inferiorly.

The anteroposterior and transverse inlet diameters (83.0 and 110.0 mm) and outlet diameters (93.5 and 106.0 mm) are well within the ranges of variation of recent human females, many of whom have overall body sizes (as indicated by femoral lengths and femoral head diameters) similar to Dolní Věstonice 15 (Tague, 1989).

Consequently, the detailed morphology of the pelvis of Dolní Věstonice 15 provides mixed messages. Several features are distinctly male, especially those relating to the size and shape of the superior pubic ramus, the symphyseal body, and the ischiopubic ramus. The greater sciatic notch appears to be more male than female, but it is less distinctly so than those of the other members of the triple burial. The subpubic angle is distinctly female, and several of the other features, such as the obturator foramen shape and orientation and the pelvic aperture shape, are ambiguous as to sex.

Dolní Věstonice 16

The Dolní Věstonice 16 partial skeleton preserves nine pieces of its ossa coxae (and none of its sacrum). They indicate a relatively large and/or robust individual with, for example, an iliac crest maximum thickness greater than those of Dolní Věstonice 13 and 14.

In the context of this, it has been possible to place together, in anatomical position, two pieces of the right ischium and ilium, which permits a reconstruction of the greater sciatic notch. It is almost complete, although the ischial spine is broken. The form of the greater sciatic notch (Figure 7.9) is well closed and exhibits a typically male form.

The posterior fragment of the left ilium preserves a portion of the preauricular area, and there is a shallow sulcus between the auricular surface and the posterior greater sciatic notch margin (Figure 7.8). The sulcus is ca. 6.5 mm wide, and it appears to continue to the inferior margin of the bone, but matrix and crushing prevent determination of its full longitudinal extent. The edges of the sulcus, however, are rounded, and the sulcus lacks the characteristic hollowed-out resorption associated with the "bony imprint of pregnancy" (Houghton, 1975). There is no apparent rugosity within or adjacent to the sulcus. It therefore represents the robust impression of the sacroiliac articular capsule, a paraglenoid sulcus (Löhr, 1894; Brůžek, 2002). The remainder of the preauricular surface

Figure 7.8 Internal views of the Dolní Věstonice 15 (left) and 16 (right) preauricular areas, illustrating the paraglenoid sulcus on each of the bones.

Figure 7.9 Internal view of the Dolní Věstonice 16 left greater sciatic notch. Scale in centimeters.

is smooth with a tubercle on its dorsocranial portion. It is similar to those seen on males and older nulliparous females as a result of normal postural and locomotor strains on sacroiliac ligaments.

The morphology of these pelvic remains of Dolní Věstonice 16, especially the form of its greater sciatic notch and the preauricular area, associated with a strong concavity of the iliac blade and an overall robusticity, strongly argues that Dolní Věstonice 16 is male.

Discrete Characters of the Pelvis

Based on these descriptions of the sex-related characteristics of the five Dolní Věstonice pelves, it has been possible to evaluate the observations in the context of five character complexes of the os coxae, the preauricular surface, the greater sciatic notch, the composite arch, the inferior pelvis, and the ischiopubic proportions (Brůžek, 2002). This approach follows the work of Genovés (1954) and Novotný (1981) and provides (among recent humans) a sexual diagnosis reliability of about 95%. Its agreement with other approaches to the sexual assessment of pelvic remains has recently been evaluated by Nepstad-Thornberry et al. (2003). The results are in Table 7.2.

In the context of these evaluations, it should be noted that Dolní Věstonice 13–16 exhibit an unusual configuration of their preauricular surfaces. These configurations consist of grooves, sometimes deep, which turn around the auricular surface. These grooves are never distinctly closed. Two forms of grooves exist in the preauricular region, whose presence and expression are variable and not mutually exclusive. There is a true preauricular groove, which is due to the processes of pregnancy and birth and was first described by Zaajer (1866; see also Houghton, 1974, 1975). It is a closed depression in the form of one or more elongated pits, each with a closed circumference in which the cross-sectional arc is greater than half of the circumference of a circle (Hoshi, 1961). It is produced by endocrinologically mediated skeletal resorption during pregnancy in females, combined with tension on the sacroiliac ligaments during parturition. This is in contrast with a paraglenoid groove, first described by Löhr (1894), which is related to the attachment of the sacroiliac ligaments and capsule (Weisl, 1954). The arc of a paraglenoid groove is open, and it makes up less than half of the circumference of a circle (Hoshi, 1961). This distinction has been formalized methodologically (Novotný, 1981; Brůžek, 2002).

The incomplete nature of the Dolní Věstonice 3 pelvis allows only two characters to be evaluated, and they are female and indeterminate. They therefore support a diagnosis of female but do not by themselves confirm it. The pelves of Dolní Věstonice 13 and 14 have an overwhelming percentage of their characteristics scored as male, with only one preauricular feature being scored as intermediate on one side for Dolní Věstonice 13 and one inferior pelvis feature being scored as intermediate on both sides for Dolní Věstonice 14. The fragmentary state of the Dolní Věstonice 16 pelvis permits only a few assessments, but they are uniformly male.

The scoring of the Dolní Věstonice 15 pelvis, as noted above, provides a certain ambiguity. The depression in

Table 7.2 Pelvic morphology and sexual assessment of the os coxae in the Dolní Věstonice sample, following the method of Brůžek (2002).

	Dolní Věstonice 3	Dolní Věstonice 13		Dolní Věstonice 14		Dolní Věstonice 15		Dolní Věstonice 16
	Left	Right	Left	Right	Left	Right	Left	Left
PS 1	i	m	m	m	m	f	f	m
PS 2	i	m	i	m	m	m	i	m
PS 3	f	m	m	m	m	m	m	m
Sexual form (PS)	o	M	M	M	M	M	o	M
GSN 1	—	m	m	m	m	m	m	m
GSN 2	—	m	m	m	m	m	m	m
GSN 3	—	m	m	m	m	f	f	—
Sexual form (GSN)	—	M	M	M	M	M	M	M
Composite arch's sexual form	F	M	M	o	o	M	M	—
IP 1	—	m	m	m	m	m	m	—
IP 2	—	m	m	i	i	i	i	—
IP 3	—	m	m	m	m	m	m	—
Sexual form (IP)	—	M	M	M	M	M	M	—
Ischiopubic proportion's sexual form	—	M	M	M	M	M	F	—
Sex assessment	Probably female	Male	Male	Male	Male	Male	Male	Probably male

PS: preauricular surface; GSN: greater sciatic notch; IP: inferior pelvis (inferior margin of the os coxae).

the preauricular area, although variable in its scoring, especially on the left side, is relatively open and flat, with a tubercle. It is therefore a sign of ligament attachment, or a paraglenoid sulcus. The greater sciatic notch form, with a scoring of m-m-f, is largely male, occurs in 11.5% of a pooled recent human sample ($N = 1000$), and has a probability of 0.89 of being male. The ischiopubic ramus is largely male and has a clear *crista phallica*. The last two features, the composite arch and the ischiopubic proportions, are male. Therefore, despite some variation in some of the features, the total pattern of the Dolní Věstonice 15 pelvis is predominantly male.

For comparison, the pubic regions of Dolní Věstonice 13 to 15 were assessed by the criteria of Phenice (1969), and the latter two were assessed by the technique of Brůžek et al. (1996) for the sacroiliac region. These two techniques have a reliability of about 80%, since they employ only a portion of the sexual dimorphism of the pelvis.

The pubic criteria of Phenice provide consistent male diagnoses for Dolní Věstonice 13 and 14 (Table 7.3). For Dolní Věstonice 15, the left os coxae is scored consistently as male, whereas the right one provides one male and two intermediate observations. Therefore, although less distinctively male by these criteria than the other two triple burial individuals, Dolní Věstonice 15 remains most likely male in its pubic region.

Evaluation of the sacroiliac regions of Dolní Věstonice 14 and 15 (the Dolní Věstonice 13 surface is obscured by a permanent reassembly of the three pelvic bones) largely supports these conclusions (Table 7.4). Dolní Věstonice 14 has a mostly male and slightly intermediate configuration, whereas Dolní Věstonice 15 has three male scores, two female ones, and one intermediate score.

Pelvic Morphometric Assessment of Sex

A series of measurements of the hip bones of the Dolní Věstonice remains (Table 7.5) was compared to those from a large reference series of recent humans of known sex from four continents (see Table 7.6 for collections). Initially, the individual measurements were compared to the recent human male and female values (Tables 7.6 and 7.7), using probability values derived from adjusted z-scores (following Houët 2001, in which ±1 indicates a difference at the 95% probability level). These were used to assess the positions of the individual measurements relative to the pooled modern human sample.

The estimation of sex was then obtained through the posterior probabilities of the discriminant function analysis derived from the Mahalanobis distances for each specimen relative to the recent human male and female distributions. This approach has been used reliably for archeological samples (Murail et al., 1999) and more recently to confirm the sexual diagnosis of the Magdalenian St. Germain-la-Rivière 4 skeleton (Henry-Gambier et al., 2002).

The application of this technique to the Dolní Věstonice remains (Table 7.8) confirms the female status of Dolní Věstonice 3, with a posterior probability of 0.930 of being female. Its application to the individual hip bones of Dolní Věstonice 13 and 14 provides probabilities of being male of >0.999. A similarly high probability of being male is calculated for the left os coxae of Dolní Věstonice 15, and a trivially lower value of 0.988 is derived from its right side. The less complete pelvic remains of Dolní Věstonice 16 also provide a posterior probability of being male approaching unity. Morphometrically, therefore, there is little doubt that Dolní Věstonice 3 has a female pelvis, and the other four individuals have distinctively male pelvic remains.

The Problem of the Dolní Věstonice 15 Pelvis

The above qualitative and quantitative considerations of the Dolní Věstonice 15 pelvis indicate that it presents a generally overall male morphology but that there is a series of aspects in which it appears female or intermediate between characteristically male and female configurations. The pelvis overall is quite small, even though

Table 7.3 Pelvic morphology and sexual assessment from pubic bone, following the method of Phenice (1969).

	Dolní Věstonice 13		Dolní Věstonice 14		Dolní Věstonice 15	
	Right	Left	Right	Left	Right	Left
Ventral arch	M	M	M	M	M	M
Subpubic concavity	M	M	M	M	o	M
Medial aspect	M	M	M	M	o	M
Sex assessment	Male	Male	Male	Male	Male	Male

Dolní Věstonice 3 and 16 preserve too little of their pubic bones for evaluation.

Table 7.4 Pelvic morphology and sexual assessment of the sacroiliac surface of the iliac bone, following the method of Brůžek et al. (1996).

	Dolní Věstonice 14		Dolní Věstonice 15	
	Right	Left	Right	Left
Segment configuration	o	M	M	o
Surface elevation	M	o	F	F
Retroauricular groove	M	M	M	M
Sex assessment	Male	Male	Male	?

Dolní Věstonice 3 and 16 preserve too little of the sacroiliac surfaces for evaluation, and the Dolní Věstonice 13 pelvis is articulated.

Table 7.5 Pelvic measurements employed for sexing the Dolní Věstonice sample.

	DV 3 L	DV 13 R	DV 13 L	DV 14 R	DV 14 L	DV 15 R	DV 15 L	DV 16 L
Pelvic height (DCOX) M-1[a]	—	226.0	231.0	(226.0)	230.0	197.0	199.0	—
Iliac breadth (SCOX) M-12[a]	(140.0)[g]	153.7	155.0	162.0	160.0	149.5	145.0	—
Horizontal acetabular diam. (HOAC) M-22[a]	(48.0)	53.5	52.3	(55.0)	55.0	(51.0)	(52.0)	—
Acetabulo-symphyseal len. (PUM) M-14[a]	—	68.1	71.0	64.7	66.0	58.2	58.7	—
Cotylo-pubic breadth (SPU)[a]	—	28.0	28.0	30.0	30.0	(26.0)	(29.0)	—
Acetabulo-sciatic breadth (SIS) M-14(1)[a]	(31.0)	37.5	38.5	33.0	33.7	35.5	36.5	(40.8)
Greater sciatic notch breadth (IIMT) M-15(1)[a]	—	33.0	35.0	38.0	39.0	38.0	34.0	(34.5)
Vertical acetabular height (VEAC) M-22	—	55.3	56.0	60.5	59.5	54.8	54.2	(60.0)
Maximum ischial length (ISM)[b]	—	103.1	105.5	105.5	107.0	94.7	95.2	110.0
Postacetabular ischium length (ISMM)[c]	—	112.5	115.3	115.0	116.2	99.5	99.7	—
Spinoauricular length (SA)[d]	—	68.7	67.5	76.8	77.3	70.4	69.2	—
Spinosciatic length (SS)[d]	—	73.2	73.2	74.1	74.7	68.8	70.9	—
Greater sciatic notch breadth (AB)[e]	—	41.0	43.0	38.0	37.0	37.0	38.0	(46.5)
Distance AP of the sciatic notch (AP)[f]	—	20.0	21.0	26.0	28.0	24.0	28.0	(29.0)
Distance BP of the sciatic notch (BP)[f]	—	43.0	46.0	41.0	40.0	37.0	41.0	(41.0)

[a]Measurements are from Sládek et al. (2000). M-## refers to the Martin measurement definitions (Bräuer, 1988). Cotylo-pubic breadth is from Gaillard (1960).
[b]Measurement is from Thieme & Schull (1957).
[c]Measurement is from Schulter-Ellis et al. (1983). Distance is from the most anterior point of the ischial tuberosity to the furthest point on the acetabular border.
[d]Measurements are from Gaillard (1960). Spinoauricular length is the minimum distance between the anterior inferior iliac spine and the auricular point (A). The fixed arm of the caliper is placed on the center of the blunt surface of the anterior inferior iliac spine. Point A is defined as the intersection of the arcuate line with the auricular surface. Spinosciatic length is the minimum distance between the anterior inferior iliac spine and the deepest point in the greater sciatic notch.
[e]Measurements are from Novotný (1975). The distance is from the top of the tubercle of the pyramidal (A) or, when absent, from the posterior inferior iliac spine (A') to the base of the ischial spine (B) measured with a spreading compass "divider."
[f]Measurements are from Brůžek et al. (1994). AP is the distance from point A(A') of the sciatic notch breadth to the deepest point in the greater sciatic notch. BP is the distance from point B of the sciatic notch breadth to the deepest point in the greater sciatic notch.
[g]Measurements in parentheses are estimated on the original specimens because of damage.

Table 7.6 Probability values from adjusted z-scores for each Dolní Věstonice pelvic measurement relative to the distribution of the modern human female global documented sex reference sample, following the procedure of Houët (2001).

	DV 3 L	DV 13 R	DV 13 L	DV 14 R	DV 14 L	DV 15 R	DV 15 L	DV 16 L
DCOX	—	0.011	0.003	0.011	0.004	0.992	0.853	—
SCOX	0.267	0.862	0.767	0.341	0.444	0.826	0.522	—
VEAC	—	0.133	0.082	0.001	0.003	0.182	0.258	0.002
HOAC	0.642	0.137	0.288	0.043	0.043	0.548	0.339	—
PUM	—	0.427	0.809	0.149	0.232	0.007	0.010	—
ISM	—	0.122	0.046	0.046	0.023	0.969	0.956	0.004
SPU	—	0.171	0.171	0.031	0.031	0.559	0.078	—
SIS	.0215	0.540	0.369	0.503	0.638	0.966	0.743	0.120
HMT	—	0.027	0.063	0.186	0.252	0.186	0.042	0.052
SA	—	0.368	0.280	0.750	0.694	0.519	0.409	—
SS	—	0.213	0.205	0.150	0.117	0.757	0.449	—
ISMM	—	0.036	0.009	0.001	0.006	0.790	0.818	—
AB	—	0.292	0.446	0.136	0.102	0.801	0.102	0.658
AP	—	0.012	0.018	0.098	0.170	0.053	0.170	0.219
BP	—	0.038	0.003	0.143	0.249	0.817	0.143	0.143

Note that a value greater than ±1.00 indicates a separation from the reference sample mean with a $P < 0.05$. Modern human $N = 860$. Measurement abbreviations follow Table 7.5. Documented remains are from Collection Olivier, Université de Paris VII; Tamagnini Collection, Universidade de Coimbra; Spitalfields, Natural History Museum (London); Terry Collection, Smithsonian Institution; Hamann-Todd Collection, Cleveland Museum of Natural History; Dart Collection, University of Witwatersrand; and Faculté de Médicine, Chang-Mai, Thailand.

Table 7.7 Probability values from adjusted z-scores for each Dolní Věstonice pelvic measurement relative to the distribution of the modern human male global documented sex reference sample, following the procedure of Houët (2001).

	DV 3 L	DV 13 R	DV 13 L	DV 14 R	DV 14 L	DV 15 R	DV 15 L	DV 16 L
DCOX	—	0.331	0.176	0.331	0.201	0.216	0.278	—
SCOX	0.153	0.876	0.971	0.545	0.673	0.588	0.340	—
VEAC	—	0.764	0.927	0.213	0.343	0.654	0.531	0.273
HOAC	0.036	0.653	0.419	>0.999	>0.999	0.231	0.369	—
PUM	—	0.751	0.798	0.323	0.465	0.023	0.030	—
ISM	—	0.701	0.988	0.988	0.829	0.094	0.110	0.498
SPU	—	0.594	0.594	0.862	0.862	0.216	0.857	—
SIS	0.039	0.646	0.832	0.116	0.162	0.340	0.480	0.721
HMT	—	0.233	0.403	0.765	0.904	0.765	0.310	0.355
SA	—	0.367	0.274	0.690	0.632	0.529	0.411	—
SS	—	0.938	0.953	0.933	0.848	0.387	0.625	—
ISMM	—	0.964	0.643	0.675	0.550	0.058	0.062	—
AB	—	0.842	0.553	0.695	0.556	0.200	0.556	0.695
AP	—	0.075	0.108	0.467	0.708	0.280	0.708	0.842
BP	—	0.177	0.031	0.419	0.592	0.781	0.419	0.419

Note that a value greater than ±1.00 indicates a separation from the reference sample mean with a $P < 0.05$. Modern human $N = 905$. Measurement abbreviations follow Table 7.5. Documented remains are from same collections as in Table 7.6.

comparison of its bi-iliac breadth to its rather small stature (especially for a Gravettian individual—see Chapter 13) does not separate it from the range of variation of those relatively linear Gravettian human populations (see Chapter 12). These small absolute dimensions make it appear rather female in several measurements (Tables 7.6 and 7.7), as do its laterally oriented iliac blades and its open subpubic angle. At the same time, a number of features, some of which are securely based in the sexual development of the pelvis (Moerman, 1982; Budinoff & Tague, 1990), indicate that Dolní Věstonice 15 did not go through the normal developmental processes associated with menarche in females.

Table 7.8 Posterior probabilities for sex estimation for the Dolní Věstonice pelvic remains, using available pelvic measurements (Table 7.5), in the context of a global sample of modern humans of documented sex ($N = 1765$).

	Posterior Probability: Females	Posterior Probability: Males	Sexual Diagnosis
Dolní Věstonice 3 left	0.930	0.070	Female
Dolní Věstonice 13 right	<0.001	>0.999	Male
Dolní Věstonice 13 left	<0.001	>0.999	Male
Dolní Věstonice 14 right	<0.001	>0.999	Male
Dolní Věstonice 14 left	<0.001	>0.999	Male
Dolní Věstonice 15 right	0.012	0.988	Male
Dolní Věstonice 15 left	<0.001	>0.999	Male
Dolní Věstonice 16 left	<0.001	>0.999	Male

At the same time, there are features that are probably related to the dysplasia of the individual. Despite the very open subpubic angle, a mediolateral length of the pubis as short as that of Dolní Věstonice 15 is found in only 1% of the recent human females and 3% of the recent human males. The Dolní Věstonice 15 ischium is also very small, with only 6% of the males having smaller lengths. A variety of forms of dysplasia can affect the proportions of the pelvis, as well as the size and proportions of other aspects of the skeleton. Yet, given the current ambiguities of the paleopathological diagnosis of Dolní Věstonice 15 (Chapter 19), it remains unclear which of these features are likely to have been so affected.

In sum, however, the overwhelming morphological and morphometric indications of the Dolní Věstonice 15 pelvis are that it derives from a very small male, even though such considerations cannot completely exclude the possibility that the skeleton is female. However, the detailed morphological attributes of its pelvis make it unlikely that Dolní Věstonice 15, if female, participated in the normal maturational and reproductive processes and behaviors expected for an individul who had reached the beginning of the third decade of life.

Cranial and Mandibular Indications of Sex

The assessment of sex from cranial and mandibular remains has been carried out on the basis of both the overall dimensions and proportions of the cranium and the relative development of a suite of secondary sexual characteristics. Sexual evaluations based on overall craniomandibular size and proportions, as with skeletal indications of body size (see below), are highly susceptible to the reference sample employed, given between-group variation across recent human samples (e.g., Howells, 1973; Lahr, 1996) and contrasts in overall craniofacial robusticity between Late Pleistocene early modern human samples and late Holocene recent humans (Šefčáková et al., 1999; Wolpoff, 1999; Sládek et al., 2001). Similarly, there is recent human interpopulational variation in the average development of various cranial superstructures (Lahr, 1996), and there have been clear reductions in them, on average, since the earlier Upper Paleolithic (Billy, 1972; Wolpoff, 1999). As a consequence, the use of craniomandibular remains for sexual assessment of the Dolní Věstonice and Pavlov human remains needs to be based on a combination of size, overall proportions, and discrete secondary sexual characteristics, most of them scaled to the extent possible to the levels and patterns of sexual dimorphism evident in the contemporaneous populations of Europe. These assessments are therefore part of the combined analysis of various sexually dimorphic skeletal elements for the five associated skeletons from Dolní Věstonice (Dolní Věstonice 3 and 13–16), but they become primary for the isolated Dolní Věstonice 11/12 cranium and the incomplete Pavlov 1 skeleton.

Overall Craniofacial Size

A multivariate analysis of four dimensions preserved on the six more complete crania provides a first-level assessment of sex based on the size of the crania. The measurements, maximum cranial length (M-1), maximum cranial breadth (M-8), maximum frontal breadth (M-10), and external palate breadth (M-61), were assessed by using a discriminant function analysis of the pooled data ($N = 2{,}524$) from Howells (1973). The analysis provides a Wilks Lambda of 0.644 and a 77.9% correct sexual classification rate for the recent human male and female crania. However, if the criterion for correct classification is raised to 0.80 (as opposed to a binomial one of 0.50), the correctly classified percentage drops to 41.9%, and 53.6% become indeterminant.

When the six Dolní Věstonice and Pavlov crania are entered into the analysis, their posterior probabilities of being male versus female (Table 7.9) place Dolní Věstonice 3 among the females and Dolní Věstonice 13, 14, and 16 and Pavlov 1 among the males. Dolní Věstonice 15, with posterior probabilities of 0.452 and 0.548 for male and female, is ambiguous. Raising the threshold to 0.80 maintains the male attributions of Dolní Věstonice 13, 14, and 16 and Pavlov 1 but makes the attribution of Dolní Věstonice 3 ambiguous.

Table 7.9 Posterior probabilities of being male versus female for Dolní Věstonice and Pavlov crania, relative to the pooled sample of recent human crania of Howells (1973).

	Male	Female
Dolní Věstonice 3	0.290	0.710
Dolní Věstonice 13	0.871	0.129
Dolní Věstonice 14	0.942	0.058
Dolní Věstonice 15	0.452	0.548
Dolní Věstonice 16	0.951	0.049
Pavlov 1	0.976	0.024

Given the ambiguous results for Dolní Věstonice 15 that were based largely on neurocranial comparisons, a further discriminant function analysis was done for this individual on sixteen facial measurements relative to five recent samples from Europe, the Near East, and Africa (total $N = 480$; Table 7.10). A maximum of four facial measurements proved useful in discriminating males and females, depending on the specific reference sample used, with validated percentages of correct classification ranging from 73.2% to 85.7% and Wilks' Lambda ranging from 0.391 to 0.664. Dolní Věstonice 15 comes out as female in four of the five reference samples with posterior probabilities for being female ranging from 0.715 (central Europe) to 0.975 (Mediterranean/Near East). In only one sample, a relatively small and gracile North African sample dominated by dynastic Egyptian crania, does Dolní Věstonice 15 come out as male, with a posterior probability of 0.723.

These results, however, are potentially influenced by a few key factors. First, twenty-one out of the twenty-two recent human reference samples used here consist of individuals whose sex attributions themselves were not documented but rather were estimated, based largely on nonmetric criteria. Although most of these attributions are likely to be correct, it is impossible to know to what extent any misattributed specimens might bias the results for the Pavlovian individuals. Second, it is unclear to what extent contrasts in patterns of a sexual dimorphism and possible contrasts in neurocranial and facial size between Gravettian and recent human samples also influence these results. Given the clear reduction in endocranial capacity related to body mass reduction between this earlier Upper Paleolithic sample and recent humans (Ruff et al., 1997), as well as reductions in facial size and robusticity (Franciscus, 1995; Sládek, 2000; Vlček & Šmahel, 2002), the sexual attributions based on such analyses remain suggestive more than conclusive. Moreover, the overall small size of Dolní Věstonice 15 (see below) undoubtedly contributes to its placement in the middle of the combined recent male and female distributions in neurocranial size and toward the female distributions in terms of facial size.

Table 7.10 Posterior probabilities of being male versus female for Dolní Věstonice 15 based on facial measurements in recent human samples.[a]

Reference Sample	Male N	Female N	Measurements Retained in Stepwise Regression[b]	% Correct Classification[c]	Wilks' Lambda	Probability = Female	Probability = Male
Western Europe	63	49	ZYB	79.6/80.4	0.664	0.719	0.281
Central Europe	56	53	ZYB, IOW, NPH	81.8/73.2	0.604	0.715	0.285
Medit./Near East	37	35	ZYB, NPH, NLB, ALC	84.9/83.8	0.470	0.975	0.025
Northern Africa	38	30	ZYB, IOWS, WNB, ICB	89.9/85.7	0.391	0.277	0.723
Sub-Saharan Bantu	60	59	ZYB, IOWS, WNB	80.0/81.7	0.559	0.747	0.253

[a] Reference sample data are from Franciscus (1995).
[b] Measurements are retained from original set of sixteen facial variables. ZYB: bizygomatic breadth (M-45); IOW: interorbital width (M-50); NPH: nasion-prosthion height (M-48); NLB: nasal breadth (M-54); ALC: alpha chord (Gill et al., 1988); IOWS: interorbital width subtense to nasion (Woo & Morant, 1934); WNB: simotic width (M-57); ICB: intercanine breadth (Glanville, 1969).
[c] First value: initial classification on entire sample; second value: validated classification on one-half of randomized sample.

Craniomandibular Morphology

Scoring for the development of various craniomandibular superstructures is presented in Table 7.11. These include features located on the neurocranial vault, supraorbital and orbital regions, zygomatic bone, mastoid process, and occipital and palatal regions, as well as the mental eminence and gonial regions of the mandible. Scoring of these features followed guidelines based on morphological (i.e., discrete) criteria (Steele & Bramblett, 1988; Buikstra & Ubelaker, 1994; White & Folkens, 2000). It is recognized that such traits, although scored discretely for practical reasons, probably manifest along a metric continuum. Moreover, the potential impact of both regional and temporal shifts in patterns of sexual dimorphism, noted above for overall cranial size, apply here as well.

Dolní Věstonice 3 scores clearly female or probably female across all of the features except two. The first exception is the mandibular gonial angle, which although in the upper range for the Pavlovian sample in terms of acuteness still falls into the male range. This feature, however, is arguably less reliable than most of the other indicators, and it is possibly subject to the pathological remodeling associated with the temporomandibular joint condition for this specimen (Chapter 19). The second feature is a moderately developed supramastoid crest, which scores as indeterminate. Nonetheless, the clear pattern for Dolní Věstonice 3 is female.

Dolní Věstonice 11/12 scores consistently male for all observable features: frontal and parietal eminence absence, a well-developed supraorbital ridge-glabella region, rounded supraorbital margins, and a well-developed nuchal crest.

Dolní Věstonice 13 scores as male in ten of the thirteen traits. Features indicating female are the supraorbital margins, which although not sharp are not as rounded and thick as is usually found in males, and relatively weak development of the nuchal crest. Also, mastoid process development falls in the overlap region for males and females. Nonetheless, the overall pattern indicates male—in fact, the most malelike for all of the three individuals of the triple burial.

Dolní Věstonice 14 scores as male in its absence of frontal and parietal eminences, squared orbital shape, and an acute mandibular gonial angle, the last the most acute of the Pavlovian sample. However, Dolní Věstonice 14 scores female with relatively thin supraorbital margins, a very weakly developed nuchal crest, small occipital condyles, and a palate that is more V-shaped or parabolic. In addition, Dolní Věstonice 14 scores as intermediate in supraorbital ridge-glabellar development, zygomatic robusticity, supramastoid crest development, mastoid process projection, and development of the mental trigone. This result among recent humans with full attainment of adulthood would indicate intermediate or nonsexable status. There is, however, an age issue that needs to be factored in. Unlike the cranial size results presented above for Dolní Věstonice 14, which reflect predominantly gross neurocranial dimensions and are likely to achieve maximum heterochronic values in adolescence or even earlier, the full development of craniomandibular superstructures occurs later in development, especially in males. Dolní Věstonice 14 is the youngest of the three individuals in the triple burial, estimated to be between the ages of 16 and 20 (see Chapter 6), and if its actual age is toward the younger end of this range, then further development of many, if not most, of the features scored as female and indeterminate would be likely. Given these considerations, and the preponderance of evidence in the postcranium and metric cranial comparisons indicating male status for Dolní Věstonice 14, there is nothing substantive in its craniomandibular features to argue against this assessment.

Anthroposcopic sex determination in the cranium is particularly important in Dolní Věstonice 15, given the pathological and other aspects of pelvic anatomy and

Table 7.11 Craniomandibular morphological trait scoring for sexing the Dolní Věstonice and Pavlov sample.

	DV 3	DV 11	DV 13	DV 14	DV 15	DV 16	Pav 1
Frontal eminences[a]	Pres—weak (F?)	Abs (M)	Abs (M)	Abs (M)	Pres—weak (F?)	Abs (M)	Abs (M)
Parietal eminences[a]	Pres—weak (F?)	Abs (M)	Abs (M)	Abs (M)	Pres—weak (F?)	Abs (M)	Abs (M)
Supraorbital ridge-glabella[b]	2 (F)	4 (M)	4 (M)	3 (?)	4 (M)	5 (M)	5+ (M)
Supraorbital margin[b]	1/2 (F)	4 (M)	2 (F)	2 (F)	2 (F)	5 (M)	5+ (M)
Orbital shape[a]	Round (F)	—	Square (M)	Square (M)	Square (M)	Square (M)	Square (M)
Zygomatics[a,c]	Light (F)	—	Heavy (M)	Moderate (?)	Moderate (?)	Heavy (M)	Heavy (M)
Supramastoid crest[a]	Pres—moderate (?)	—	Pres—strong (M)	Pres—weak (?)	Pres—strong (M)	Pres—moderate (?)	Pres—strong (M)
Mastoid process[b]	1/2 (F)	—	3 (?)	3 (?)	4 (M)	4 (M)	5 (M)
Nuchal crest[b]	1/2 (F)	4/5 (M)	2 (F)	*1 (F)	4 (M)	4 (M)	5 (M)
Occipital condyles[a]	Small (F)	—	Large (M)	Small (F)	—	Large (M)	—
Palate[a]	V-shaped (F)	—	U-shaped (M)	V-shaped (F)	U-shaped (M)	U-shaped (M)	V-shaped (F)
Mental eminence[b]	1/2 (F)	—	3+ (M)	3 (?)	3 (?)	4 (M)	5 (M)
Gonial angle[a]	118°–120° (M)	—	117° (M)	114° (M)	118° (M)	115° (M)	116° (M)
Sex assessment	Female	Male	Male	?	Male	Male	Male

[a] Steele & Bramblett (1988: 54); modified from Krogman (1962:115); Abs = absent; Pres = present.
[b] White & Folkens (2000: 364–365); modified from Walker in Buikstra and Ubelaker (1994):1 = female, 2 = probable female, 3 = ambiguous sex, 4 = probable male, 5 = male, + = intermediate between two scores or in excess of highest score, * = see description for clarification or further discussion.
[c] Heavier zygomatics (males) are generally associated with a more laterally arched configuration, whereas lighter zygomatics (females) are associated with more compressed configurations.

craniofacial size detailed above that lead to an ambiguous result, especially in the latter. In seven of twelve traits assessments, Dolní Věstonice 15 comes out as male. These include a well-developed supraorbital ridge-glabella region, square-shaped orbits, strong development of supramastoid crests, well-developed and projecting mastoid processes, a well-developed nuchal crest, a U-shaped palate, and a relatively acute mandibular gonial angle. A fewer number of features indicate intermediate status, including frontal and parietal eminences (although present, only weakly developed) and moderately robust zygomatic and mandibular mental trigone development. Only one feature falls clearly into the female scoring range: the supraorbital margins, which, although thin, are not sharp. The overall pattern of features therefore indicates male rather than female.

Dolní Věstonice 16 scores consistently as male across all features except supramastoid crest development, which, although present, is only moderately developed and therefore inconclusive. In supraorbital ridge-glabella development, supraorbital margin thickness, and mandibular mental eminence development, Dolní Věstonice 16 exceeds all of the other individuals assessed as male except Pavlov 1.

Pavlov 1 also scores as overwhelmingly male, with only a V-shaped or parabolic palate indicating female. Pavlov 1, in fact, is hypermale in several features, showing more robusticity or superstructure development than Dolní Věstonice 16 or any of the other individuals assessed as males. This is evident in supraorbital ridge-glabella development, supraorbital margin thickness, mastoid process and nuchal development, and mandibular mental eminence development.

Femoral Length and Sex

Body size, for which femoral bicondylar length is used here, by itself is a minimally adequate means of assessing the sex of past human samples. For example, in thirteen recent human samples (Trinkaus, 1980, 1981), the percent dimorphism in femoral bicondylar length (female mean/male mean) varies between 90.6 and 94.9 (92.7% ± 1.5%); the means of twelve of them are separated by between 0.94 and 1.91 standard deviations (1.49 ± 0.30). Consequently, individuals whose femoral lengths fall within the overlap between the male and female means are difficult to assess with respect to sex, and only the very high and low outliers provide any substantial confidence of sex attribution based on femoral (or other long bone) lengths.

In addition, as mentioned above, the normal variation in mean stature (and hence long bone lengths) across populations means that the sex-specific reference values must be based on the population of derivation of the individuals in question. Although this can be done for large recent cemetery or forensic samples for which long bone length is employed solely to assign sex to fragmentary specimens, it is more difficult to apply to paleontological samples, which must include individuals spanning large geographical regions and multiple millennia. For example, the Gravettian sample employed here, although still limited in size, spans European Russia to Atlantic Britain to peninsular Italy and includes specimens from at least five millennia.

Despite these caveats and serious limitations in using femoral length to evaluate the sexes of the Pavlov and Dolní Věstonice human remains, these comparisons can provide substantiation of attributions made on the basis of their pelves and crania and, for the individual with femora and a cranium but lacking a pelvis (Pavlov 1) and the individual represented by an isolated femur (Dolní Věstonice 35), such an assesement can provide at least tenuous evidence for such an attribution.

The primary reference sample for the assessment of male and female body size for the Dolní Věstonice and Pavlov human remains is therefore the available sample of pelvically sexed Gravettian human remains from Europe. These include remains from Arene Candide, Baousso da Torre, Barma Grande, Caviglione, Cro-Magnon, Fanciulli (Grotte des Enfants), Paglicci, Paviland, Předmostí, Sunghir, and Veneri (Parabita). Most of these remains are male (twelve males vs. nine females). (Their summary statistics are provided in Table 7.12.) These male and female samples have considerable overlap, given the high value of 470.5 mm for the Veneri 2 female and the low values of 444.5 mm and 452.0 mm for the Předmostí 9 and 14 males. They provide a percentage of dimorphism (female mean/male mean) of 90.0%. This dimorphism value is slightly below the range of recent human femoral length dimorphism values (see above) but not significantly so, given the small sample sizes (plus 2 SE_{mean} for the male mean and minus 2 SE_{mean} for the female mean, give a dimorphism of 84.2%). The addition of the four pelvically sexable Dolní Věstonice specimens, Dolní Věstonice 3, 13, 14, and 16, changes the mean values slightly and decreases the degree of dimorphism trivially (90.2%).

However, the female sample remains small, and the standard error of the mean relatively large (7.8 mm without Dolní Věstonice 3 and 7.2 mm with her); this will directly affect the perceived level of dimorphism, given that it is based on a ratio of the male and female means. An alternative approach is to take the average level of dimorphism for the thirteen recent human samples (92.7%, SE_{mean} = 0.4%), use the male earlier Upper Paleolithic mean value to compute a female mean (449.5 mm; 446.8 mm with Dolní Věstonice 13, 14, and 16), and then use the average recent human coefficient of variation in femo-

Table 7.12 Summary statistics for femoral bicondylar length for Gravettian human remains by sex.

	Mean ± SD	Minimum–Maximum	N
Comparative sample			
Males	484.9 ± 25.3	444.5–524.0	12
Females	436.2 ± 23.3	408.6–470.5	9
With Dolní Věstonice 3, 13, 14, 16			
Males	482.6 ± 25.7	442.8–524.0	15
Females	434.4 ± 22.8	408.6–470.5	10

ral length (4.52%, SE_{mean} = 0.3%) to calculate a standard deviation for this hypothetical earlier Upper Paleolithic female mean (± 20.3 mm; ± 20.2 mm with Dolní Věstonice 13, 14, and 16). The use of these estimated female sample statistics is conservative since the use of a lower level of dimorphism than that provided by the pelvically sexable fossil specimens will tend to make male versus female distinctions more tenuous.

The resultant z-scores (value minus mean/standard deviation) for the Dolní Věstonice and Pavlov values (Table 7.13) provide clear results for three of the specimens and more ambiguous results for the other three. Dolní Věstonice 3, in agreement with its pelvic and craniomandibular morphology, is clearly aligned with the females, being well below the male mean values and even moderately low for a female. Dolní Věstonice 14 and the estimated value for the Dolní Věstonice 35 femur (see Chapter 13) are slightly above the male mean and distant from the female mean. Dolní Věstonice 13, despite its male pelvic and craniomandibular morphology (see above), is small for a male although close to the Předmostí 9 and 14 males. The Dolní Věstonice 16 and Pavlov 1 femoral lengths fall between the male and female means, near the upper limit of the documented earlier Upper Paleolithic female range and somewhat closer to the male mean. Since Pavlov 1 lacks pelvic remains, it is its cranial morphology (see above and Chapter 8), in combination with its femoral length, that serve to provide a sexual diagnosis. The sexual assessment of Dolní Věstonice 16 must be based on its pelvic and craniomandibular remains, both of which strongly support its being male; its femoral length is in agreement with this conclusion.

The Dolní Věstonice 15 skeleton had its proportions, and almost certainly its size, altered by its congenital abnormalities, and it is therefore difficult to employ indicators of size to support a sexual diagnosis of the specimen. However, for completeness of comparisons, its left femur bicondylar length of 383.0 mm is 4.03 and 3.85 standard deviations, respectively, below Gravettian male means without and with Dolní Věstonice 13, 14, and 16. It is 3.28 and 3.16 standard deviations below reconstructed Gravettian female means without and with the Dolní Věstonice specimens. The femoral length of 443 mm estimated from its skeletal trunk height (had its femoral-to-trunk proportions been similar to those of other Gravettian individuals; see Chapter 13) still places it well below the Gravettian male means (1.66 and 1.52 standard deviations, respectively) but slightly above the female means (Table 7.12). Consequently, as is also evident from stature and body mass estimations (Chapter 13), Dolní Věstonice 15 has the overall body size of a Gravettian female, and it would be unusual but not completely exceptional for a Gravettian male. However, it must be emphasized that such size-related considerations are not appropriate in the diagnosis of sex for an individual who clearly suffered from some form of childhood growth abnormality.

Summary

These considerations of pelvic proportions and detailed morphology, craniomandibular morphology, and femo-

Table 7.13 Z-scores [(value − mean)/standard deviation] for each of the Dolní Věstonice and Pavlov specimens with a femoral length versus Gravettian samples.

	Femur Bicondylar Length (mm)	Versus Males	Versus Females	Versus Males with Dolní Věstonice	Versus Females with Dolní Věstonice
Dolní Věstonice 3	(417.5)	−2.66	−1.57	−2.51	−1.45
Dolní Věstonice 13	442.8	−1.66	−0.33	−1.74	−0.20
Dolní Věstonice 14	502.0	0.68	2.59	0.82	2.73
Dolní Věstonice 16	(467.5)	−0.69	0.89	−0.59	1.02
Dolní Věstonice 35	(498.0)	0.52	2.39	0.62	2.53
Pavlov 1	(475.0)	−0.39	1.26	−0.27	1.40

Femoral length is the average of the right and left femora, when available; values in parentheses are estimated (Sládek et al., 2000). The "males with Dolní Věstonice" values do not include the individual in question in the reference sample. The female means and standard deviations are estimated from the male means, following the distributions of recent human samples (see text for procedure).

ral indications of body size support several attributions of sex. Dolní Věstonice 3, based on craniomandibular size and shape, pelvic details, and femoral length, is clearly female. Dolní Věstonice 11/12 is clearly male on the basis of its neurocranial morphology. Dolní Věstonice 14, based on overall and specific aspects of pelvic shape and femoral length, is clearly male, with the craniomandibular features that indicate otherwise probably influenced by its relatively young developmental age (Chapter 6). Dolní Věstonice 35 is assessed as male solely because of its femoral length estimate. Dolní Věstonice 13 is definitely male, based on a number of aspects of pelvic shape and craniomandibular morphology, even though its femoral length places it in the overlap zone for the Gravettian males and females. In overall cranial size and craniomandibular morphology, Dolní Věstonice 16 is unambiguously male, and this attribution is further supported by its greater sciatic notch morphology and its femoral lengths. Pavlov 1 lacks diagnostic pelvic remains; however, in overall cranial size and craniomandibular morphology it is also unambiguously male, and it is further supported in this attribution by its femoral lengths. The ever problematical Dolní Věstonice 15 appears female in overall size (assuming little reduction in stature beyond femoral length shortening because of its abnormalities), and several features of its pelvis (including subpubic angle) and aspects of craniofacial size support its being a female. Nevertheless, other features of its pelvis and craniomandibular anatomy, especially those used in the craniomandibular discrete trait and pelvic morphometric comparisons, indicate that it is best viewed as a gracile male who sustained a variety of abnormalities and general reduced growth during development.

This provides a sample that is overwhelmingly male in composition, with one female (Dolní Věstonice 3), six males (Dolní Věstonice 11/12, 13, 14, 16, and 35 and Pavlov 1) and one abnormal male (Dolní Věstonice 15). Given the suite of criteria that are employed to assess sex for these individuals and the documented ranges of variation in sexually dimorphic features among European Gravettian humans, it is unlikely that this is simply a product of the well-noted tendency to make earlier, more craniofacially robust humans differentially male. It is most likely a reflection of the particular social patterns of burial at the Pavlovian sites of Dolní Věstonice I and II and Pavlov I, combined with any possible differential preservation and discovery because of processes that should have been random with respect to sex.

8

The Cranial Remains

Robert G. Franciscus and Emanuel Vlček

The human remains from Dolní Věstonice and Pavlov include a number of relatively complete crania (Chapter 5; Sládek et al., 2000). Specimens with essentially complete neurocrania and splanchnocrania include Dolní Věstonice 3 and 13–16. Less complete specimens include Pavlov 1, which retains a calvarium with small portions of the basicranium and left zygomatic, as well as portions of the left and right maxillae, and Dolní Věstonice 11/12, which retains a less complete calotte. There are also alveolar fragments associated with maxillary teeth that make up Pavlov 2. The Dolní Věstonice 3 and 13–16 crania, although largely complete, were reconstructed from a number of fragments, in most cases suffered ground pressure deformation, and have been restored to varying degrees with insoluble filler and painted. This condition of the crania frequently makes it difficult to ascertain exact bone joins and details of surface anatomy. The last point is particularly true for Dolní Věstonice 3. The neurocrania are generally better preserved than the splanchnocrania throughout the sample, although in some cases basicranial portions are also relatively more incomplete or damaged. All of the cranial specimens except Dolní Věstonice 11/12 also retain relatively complete mandibles, which are considered in Chapter 10.

Materials and Methods

The considerations of these Pavlovian cranial remains involve both detailed morphological descriptions of the bones and morphometric comparisons with relevant samples of Late Pleistocene human remains. The principal comparative sample is of Gravettian associated human remains from Europe and coeval Levantine sites including Arene Candide, Barma Grande, Brno-Francouzská, Combe Capelle, Cro-Magnon, Fanciulli (Grotte-des-Enfants), Nahal Ein Gev, Ohalo, and Pataud. Even though they are geographically, archeologically, and chronologically close to the Dolní Věstonice and Pavlov sample (Svoboda et al., 1996), the Předmostí human remains are included as part of the greater Gravettian sample since the focus here is on the comparative framework of the remains from the Pavlovské Hills. To these Gravettian remains are added the Aurignacian remains from Mladeč and the archeologically unassociated remains from Oase. The more distant comparisons, for considerations of temporal trends through the Late Pleistocene, are to the Middle Paleolithic samples of late archaic and early modern humans. The former consist of the European and Near Eastern Neandertals and include specimens from Amud, La Chapelle-aux-Saints, La Ferrassie, Forbes' Quarry (Gibraltar), Guattari (Circeo), Neandertal (Feldhofer), La Quina, Shanidar, Spy, and Vindija, plus the initial Upper Paleolithic (Châtelperronian) Saint-Césaire specimen. The latter are the Near Eastern remains from Qafzeh and Skhul. Given the dearth of reasonably complete crania, the earlier, Terminal Middle to Initial Late Pleistocene remains from Krapina, Saccopastore, and Tabun are included when possible, although it is realized that they represent a more ancestral pattern. Data are from primary descriptions of the remains (e.g., Fraipont & Lohest, 1887; Verneau, 1906; Boule 1911–1913; H. Martin, 1923; Matiegka, 1934; McCown & Keith, 1939; Sergi, 1942–1946, 1944, 1974; Vallois & Billy, 1965; Suzuki, 1970; Sergi et al., 1974; Billy, 1975; Heim, 1976; F. H. Smith, 1976; Arensburg, 1977; Paoli et al., 1980; Vandermeersch, 1981; Trinkaus, 1983b; Nadel & Hershkovitz, 1991; Condemi, 1992; Hershkovitz et al.,

1995), supplemented by personal analysis of most of the original specimens.

The majority of the morphometric comparisons employs standard osteometrics, using Martin numbers (Bräuer, 1988), and they are listed for the Dolní Věstonice and Pavlov human remains in Sládek et al. (2000). These are augmented, when necessary and possible, by additional measurements defined in Howells (1973, 1989) and Franciscus (1995). Most of the linear measurements are tabulated in Sládek et al. (2000), and the relevant tables in that volume are referenced as appropriate. In a few cases, reexamination of the specimens has produced additional or slightly different values, and these are provided in the text.

The frequent distortion of complete elements, misalignment of reconstructed portions, and inevitable incompleteness of aspects of anatomy in both the Dolní Věstonice and Pavlov and comparative samples preclude meaningful multivariate comparisons. Accordingly, comparisons are limited to univariate and bivariate comparisons.

In addition to metric comparisons, nonmetric or "epigenetic" traits are coded for the Dolní Věstonice and Pavlov sample, using a subset of the extensive series of such traits detailed in Hauser and DeStefano (1989) that have been employed recently for other fossil samples (e.g., DeStefano & Hauser, 1991; Manzi et al., 2000). Beyond basic anthroposcopic description, such nonmetric traits are potentially pertinent to arguments that some of the Pavlovian sample, in particular the triple burial individuals Dolní Věstonice 13, 14, and 15, share particularly close genetic affinity (e.g., Vlček, 1991; Alt et al., 1997). Also, given the importance of craniofacial robusticity patterning in Later Pleistocene human evolution, Lahr's (1996) scoring grades for several aspects of craniofacial robusticity are used here for comparison. These include cranial vault, supraorbital, infraglabellar, orbitozygomatic and occipital features. Her grades were specifically operationalized for relatively robust east Asian modern and fossil samples; however, her comparisons did include Western Old World samples from Europe and Africa and are thus incorporated here.

Since discussions of Later Pleistocene human evolution have inevitably involved discussion regarding the degree of morphological continuity versus noncontinuity between late Neandertals and early modern humans, particularly in central Europe (e.g., Bräuer & Broeg, 1998, vs. Wolpoff, 1999), the presence/absence of Neandertal lineage traits in the Pavlovian sample is also considered. Since some craniofacial traits have been viewed as either autapomorphies or as features that occur in higher frequencies among Neandertals, their distributions in the Dolní Věstonice and Pavlov sample is noted when relevant.

Dolní Věstonice 3

Preservation

The Dolní Věstonice 3 has the neurocranium and the splanchnocranium preserved. The bones of the cranium have been reconstructed from a number of fragments (Klíma, 1963), and therefore the original positions and connections of some of the pieces are difficult to substantiate. The reconstructed parts, as well as some of the fragments of the original bone, have been painted dark red to brown in color, to match the ochre staining of the cranium. This makes it difficult to assess aspects of the reassembly and restoration of the cranium, and it provides a false sense of the completeness and integrity of portions of the cranium. Moreover, the current state of preservation makes distinctions between antemortem and postmortem alteration, as well as the scoring of minute anatomical details and epigenetic information, particularly difficult.

The neurocranium has preserved the frontal bone, left and right parietal bones, left and right temporal bones, the occipital bone, and some portions of the sphenoid bone. The neurocranial damage is mainly to the parietal bones (on the right side the area of the sphenoidal angle is damaged, and on the left side both the sphenoidal angle and mastoid angle are damaged). Also the nuchal plane of the occipital bone and both temporal squamous portions were partially lost. The splanchnocranium consists of the left and right zygomatic bones, the left and right maxillae, and a small portion of the nasal bones. The left zygomatic bone is damaged, a large portion of it is reconstructed, and it also sustained an antemortem traumatic injury (Chapter 19). The reconstruction influences perceptions of the shape of the bone. Also, both maxillae are reconstructed along the dorsal portions of the bones, the maxillary tuberosities, and a small portion of the alveolar process. The nasal bones present only a small portion near nasion.

It should be noted that the Dolní Věstonice 3 cranium has marked asymmetries of the occipital nuchal plane, the supraorbital area, the zygomatic bones, and the infratemporal fossae, as well as somewhat less marked asymmetries in the infraorbital and nasal regions. The occipital asymmetry might be the result of endocranial (cerebral and/or cerebellar asymmetries), whereas the upper facial asymmetries have been related to the injury and degeneration of the left temporomandibular joint (see below and Chapter 19). However, the degree to which the facial asymmetries are due to pathological changes, as opposed to deformation in situ and during reconstruction, cannot be easily assessed. The connection between the neurocranium and splanchnocranium is well preserved in the contacts of the splanchnocranium with the frontal bone.

On the other hand, the connection with the temporal bone (left and right zygomatic arches) is not preserved and is partly reconstructed.

Anterior View

Frontal Region

The coronal suture is obscured by filler and paint and cannot be evaluated for aging or epigenetic traits (Figure 8.1). The frontal squamous has no frontal keeling present in midline. There are very weakly expressed, nearly absent frontal eminences lateral to metopion. A shallow, antemortem healed lesion is located 30 mm posterior to metopion on the midline that is 16.5 mm in maximum width, 13.0 mm in anteroposterior length, and less than 1.0 mm in depth (Chapter 19). The least frontal breadth of Dolní Věstonice 3 (92 mm), as well as the maximum frontal breadth (119 mm), lie at the smallest end of the range for the Dolní Věstonice and Pavlov sample. Although smaller in overall size for these variables, Dolní Věstonice 3 is nonetheless very similar to the rest of the Pavlovian sample in the overall shape of the frontal bone between these two variables (Figure 8.2; Table 8.1). The relationship between these two variables is very similar in this and the European earlier Upper Paleolithic samples overall, with the Qafzeh-Skhul and Neandertal samples showing relatively greater least frontal breadth values. In this, Dolní Věstonice 14 (see below) shows the relatively greatest least frontal breadth and is more similar to the Qafzeh-Skhul sample, although this could well be due to a postmortem narrowing of its maximum frontal breadth value.

Supraorbital Region

The supraorbital region of Dolní Věstonice 3 shows a distinct separation of medial, midorbital, and lateral components. The glabellar region is pronounced in midline, and this robust midline projection extends to the medial third of both right and left supraorbital areas, each superciliary arch extending superiorly and laterally for ca. 20 mm prior to inflecting inferiorly and then fading out onto the midorbital regions. The total length of this medial component (encompassing glabella) for both sides combined is ca. 60 mm. The lateral-most supraorbital margins, the supraorbital trigone on either side, are both ca. 5 to 6 mm in thickness, and they are distinctly separated from the medial superciliary arch by the obliquely flattened supraorbital sulcus (or a sulcus producing segmentation between the superciliary arch and the supraorbital trigone; Sládek et al., 2002), the latter of which is 4 mm in thickness on either side. The flattening of the supraorbital sulcus results in a "pinched" appearance that has been noted

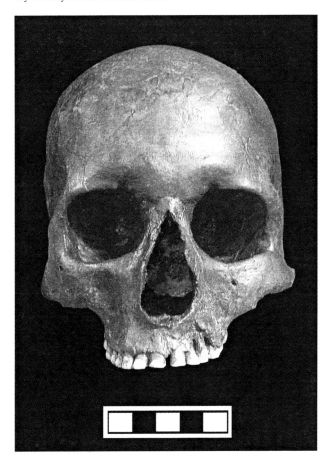

Figure 8.1 Anterior view of the Dolní Věstonice 3 cranium. Note the left-right zygomaticoorbital, infraorbital, and dental asymmetry. Scale in centimeters.

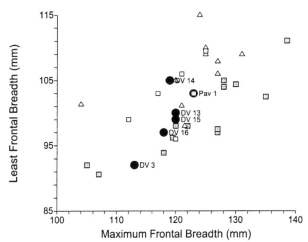

Figure 8.2 Least frontal breadth versus maximum frontal breadth. Open triangles: Neandertals; open squares: Qafzeh-Skhul; gray squares: earlier Upper Paleolithic; solid circles: Dolní Věstonice and Pavlov individuals.

Table 8.1 Summary statistics for neurocranial measurements in the Dolní Věstonice/Pavlov (DV/Pav) and comparative samples [mean ± standard deviation (N)].

Measurement	DV/Pav	EUP	Q-S	Neandertals
Glabello-occipital length (1)	196.3 ± 7.9 (7)	195.2 ± 8.6 (19)	201.2 ± 8.2 (5)	201.7 ±10.1 (12)
Nasio-occipital length (1d)	191.7 ± 7.8 (6)	189.7 ± 11.3 (6)	184.0, 193.0	196.2 ± 10.8 (8)
Basion-nasion length (5)	106.0 ± 3.6 (4)	101.2 ± 7.4 (12)	98.0	113.2 ± 9.4 (5)
Foramen magnum length (7)	38.0 ± 2.9 (5)	35.0, 36.5	37.0, 40.0	34.5, 41.0, 41.9
Maximum cranial breadth (8)	136.7 ± 3.9 (7)	139.8 ± 6.5 (19)	145.1 ± 2.8 (6)	149.8 ± 6.7 (12)
Least frontal breadth (9)	99.3 ± 4.6 (6)	99.5 ± 5.5 (15)	102.4 ± 4.9 (6)	106.2 ± 5.0 (10)
Maximum frontal breadth (10)	118.8 ± 3.3 (6)	123.2 ± 8.9 (17)	119.0 ± 4.8 (5)	122.5 ± 7.1 (12)
Bistephanic breadth (10b)	103.0	120.6 ± 7.5 (7)	111.0	117.6 ± 7.9 (7)
Biasterionic breadth (12)	113.4 ± 4.8 (7)	109.7 ± 5.9 (14)	120.7 ± 7.1 (6)	120.9 ± 5.5 (11)
Mastoid width (13a)	15.0	15.2 ± 3.0 (6)	15.0	11.5 ± 2.9 (6)
Foramen magnum breadth (16)	29.3 ± 1.5 (4)	30.5, 31.0	29.0	27.0, 28.5, 28.9
Basion-bregma height (17)	132.8 ± 3.5 (4)	130.0, 140.0	128.0, 129.0, 130.0	133.0 ± 4.5 (5)
Mastoid height (19a)	26.8 ± 2.6 (5)	30.7 ± 2.9 (6)	28.0	28.6 ± 5.0 (6)
Nasion-bregma chord (29)	116.4 ± 6.5 (7)	114.3 ± 6.1 (17)	114.0 ± 3.2 (5)	109.8 ± 7.4 (11)
Nasion-bregma subtense (29b)	26.5 ± 1.5 (6)	28.8 ± 1.5 (6)	25.0, 28.0	19.7 ± 2.3 (7)
Bregma-lambda chord (30)	119.6 ± 3.4 (7)	120.2 ± 8.6 (18)	118.0 ± 9.2 (4)	109.5 ± 4.9 (11)
Bregma-lambda subtense (30a)	24.3 ± 3.5 (7)	23.3 ± 3.1 (6)	27.0	18.5 ± 4.2 (7)
Lambda-opisthion chord (31)	97.3 ± 5.6 (6)	98.9 ± 4.6 (14)	93.4 ± 5.4 (5)	96.7 ± 7.0 (6)
Lambda-opisthion subtense (31a)	31.2 ± 3.1 (5)	33.0, 35.0, 37.0	26.0, 30.0	33.1 ± 3.7 (4)

Martin (Bräuer, 1988) numbers follow the measurement names. EUP: earlier Upper Paleolithic; Q-S: Qafzeh-Skhul. Estimated Dolní Věstonice 11/12 values are included in the glabello-occipital length and maximum cranial breadth statistics.

frequently in the midorbital region of European Upper Paleolithic specimens (Smith & Ranyard, 1980).

The supraorbital sulcus is distinctly separated both mediolaterally and superoinferiorly from the superciliary arch and visible as a distinct entity for virtually the entire length of the supraorbital margin. The supraorbital trigone is inflated and widened anteroposteriorly, exceeding 5 mm (8.0 mm) on the less reconstructed right side, but retains a smooth lateral surface. The sulcus and trigone therefore correspond most closely to Lahr's (1996) grade 3. This is a conservative score; the left side appears wider and more robust, but it is reconstructed and obscured with paint. The distinct separation of medial, midorbital, and lateral components of the supraorbital region in Dolní Věstonice 3 is present in all of the Dolní Věstonice and Pavlov individuals, and it is fundamentally different from the uniformly rounded supraorbital region lacking these distinct elements found in most Neandertals (Table 8.2; Sládek et al., 2002).

Orbital Region

Interorbital breadth in Dolní Věstonice 3 is smaller than in the male Dolní Věstonice individuals and other earlier

Table 8.2 Occurrence of Neandertal frontal bone and upper splanchnocranial traits in the Dolní Věstonice (DV) and Pavlov (Pav) sample.

Neandertal Configuration	DV 3	DV 11	DV 13	DV 14	DV 15	DV 16	Pav 1
Rounded supraorbital torus without distinct elements	Absent	Absent	Absent	Absent	Absent	Absent	Absent
High, rounded orbits	Absent	—	Absent	Absent	Absent	Absent	Absent
Midfacial prognathism[a]	Absent	—	Absent	Absent	Absent	Absent	—
Infraorbital region horizontally flat or convex[b]	Absent	—	Absent	Absent	Absent	Absent	—
Horizontal nasomaxillary suture	—	—	Absent	Absent	Absent	Absent	—
Square-shaped piriform aperture	Absent	—	Absent	Present	Present	Absent	—
Depressed nasal floor with stage 5 narial margin	Absent	—	Absent	Absent	Absent	Absent	Absent
Posterior rooting of the facial crest[c]	Absent	—	Absent	Absent	Absent	Absent	—
	P^4/M^1		M^1	M^1	M^1/M^2	M^1	

[a]Low subspinale angle; low nasiofrontal angle; large difference between first molar alveolus and zygomaxillary radii.
[b]Obliquely receding in alignment with the anterolateral surface of the zygomatic.
[c]Posterior to M^2 or at M^3.

Upper Paleolithic individuals in absolute terms, and it is also small relative to the width of the upper face at orbital level (Figure 8.3). The Dolní Věstonice interorbital breadths are roughly similar to those found in the other comparative samples. It is interesting to note that the Qafzeh-Skhul and especially Neandertal values for interorbital breadth are not relatively wider than these Pavlovian ones. This probably reflects a cold climate response in the upper internal nose (Franciscus, 2003a).

There is a large degree of asymmetry in the left and right orbital region in Dolní Věstonice 3. The right orbit is square to round and essentially normal in appearance, whereas the left orbit is normal in its medial aspect but vertically compressed in its lateral aspect. The vertical compression is due entirely to the relatively superior placement of the inferolateral border, which is also unusually posteriorly repositioned. The lateral orbital alteration and resulting asymmetry is probably due to a significant traumatic alteration, discussed in Chapter 19. It is also possible that the orbital alteration has influenced to some degree the relationship of interorbital breadth to bifrontal breadth in Dolní Věstonice 3, discussed above. This degree of asymmetry, although visually apparent, is less clearly reflected in standard orbital metrics since the points used in the latter do not correspond to the localities that are most altered. Thus, even with this orbital shape asymmetry in Dolní Věstonice 3, it shares with the other Dolní Věstonice individuals and modern specimens orbits that are lower relative to orbital breadth compared to Neandertals, which have relatively higher values (Figure 8.4; Tables 8.3 and 8.4). The superior orbital margins are moderately sharp in their more medial aspects lateral to each supraorbital notch, and they are more rounded in

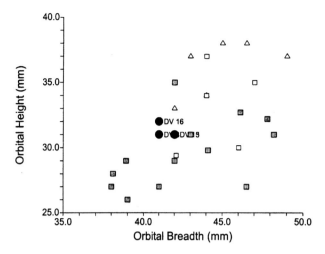

Figure 8.4 Orbital height versus orbital breadth. Left-side values used except where missing. Open triangles: Neandertals; open squares: Qafzeh-Skhul; gray squares: earlier Upper Paleolithic; solid circles: Dolní Věstonice individuals.

each lateral aspect. The right and left supraorbital notches are large, and each is medially positioned (Table 8.3).

Internally, the right orbit contains a large amount of intact bone in its lateral, medial, superior, and inferior aspects, although it is covered with paint and filler to the point where details of anatomy are obscured. The superior and inferior ethmoidal sutures are present, but details cannot be discerned. The left orbit also contains a large amount of internal bone; however, it is even more obscured by consolidant and paint. Some observations, however, are possible that are relevant to the pathological alteration of the left side of the face. The left internal orbit is mediolaterally compressed, and although its superior aspect is identical to the right, the inferior floor of the orbit is more inferiorly depressed than the right. The most posterior portion of the orbit where the frontal bone and the lesser and greater wings of the sphenoid bone form the superior orbital fissure is preserved, and it also appears to be remodeled. If the reconstruction is accurate, it is possible that the asymmetrical distortion of the left orbit is the healed result of an antemortem force trauma that occurred on the left side of the face, damaging the zygomatic, the mandibular condylar neck, and the internal orbit (see below and Chapters 10 and 19).

The inferolateral border of the right orbit is robust in that it is rounded rather than sharp. However, the border is also raised in relation to the floor of the orbits and does not attain the most robust grade; it therefore corresponds most closely to Lahr's (1996) grade 2, the most frequent form in her recent human sample. There is no development of a zygomaxillary tubercle between the orbital and free margin of the zygomatic, which corresponds to a grade 1 degree of robusticity development

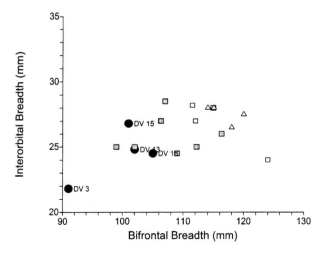

Figure 8.3 Interorbital breadth versus bifrontal breadth. Open triangles: Neandertals; open squares: Qafzeh-Skhul; gray squares: earlier Upper Paleolithic; solid circles: Dolní Věstonice individuals.

Table 8.3 Epigenetic traits of the frontal bone and anterior splanchnocranium for the Dolní Věstonice (DV) and Pavlov (Pav) sample.

Trait	DV 3	DV 11/12	DV 13	DV 14	DV 15	DV 16	Pav 1
Metopism (1,2,3)	Absent	Absent	Absent	Present? *Possible incomplete parietal part*	Absent	Absent	Absent
Frontal grooves (1, 2, 3)	Absent?	—	Absent?	Absent?	Absent?	Present? *Possible right side*	Absent?
Supranasal suture (1)	?	Present *Partially obliterated*	Present *Partially obliterated*	Present *Trace*	Present *Partially obliterated*	Present *Trace*	Present *Partially obliterated*
Supraorbital notch & related structures (1, 2, 3)	Present *Bilateral, large notches on medial aspect*	Present *Right side: large acute medial notch*	Present *Right side: well-expressed, moderate-sized medial nutrient foramen Left side: two small notches*	Present *Left side: very large, blurred medial notch*	Present *Bilateral, large notches on medial aspect*	Present *Bilateral, medium notches on medial aspect*	Present *Bilateral, very large, blurred medial notch*
Ethmoidal foramina (1)	?	—	Present *One medium-sized extrasutural foramen*	—	—	—	—
Infraorbital foramen (1, 2)	?	—	—	—	—	—	—
Zygomaxillary tubercle (1, 3)	Absent	—	Present *Right side: medium expression Left side: strong expression*	Present *Left side: medium expression*	Present *Right side: medium expression Left side: strong expression*	Present *Right side: strong expression Left side: strong expression*	—

Traits as employed by 1: Hauser and DeStefano (1989), 2: DeStefano and Hauser (1991), and 3: Manzi et al. (2000).

Table 8.4 Summary statistics for splanchnocranial measurements in the Dolní Věstonice/Pavlov (DV/Pav) and comparative samples [mean ± standard deviation (N)].

Measurement	DV/Pav	EUP	Q-S	Neandertals
Basion-prosthion Length (40)	104.7 ± 4.5 (4)	102.5 ± 7.7 (13)	111.5	116.2 ± 7.1 (5)
Bifrontal breadth (43a)	102.8 ± 6.8 (6)	106.4 ± 6.9 (10)	111.5, 112.0, 124.0	113.0 ± 5.7 (9)
Bizygomatic breadth (45)	120.7 ± 23.8 (6)	131.6 ± 19.5 (16)	147.8 ± 8.6 (4)	143.3 ± 9.1 (10)
Nasion-prosthion height (48)	68.6 ± 6.7 (6)	66.8 ± 5.4 (20)	75.7 ± 2.0 (5)	84.5 ± 4.7 (11)
Superior ethmoidal breadth (49.1)	20.3	27.0 ± 2.4 (5)	28.0	31.1 ± 3.2 (6)
Inferior ethmoidal breadth (49.2)	31.5	33.5, 34.6, 43.3	—	43.2 ± 10.0 (5)
Interorbital breadth (49a)	24.5 ± 2.1 (4)	25.3 ± 1.9 (16)	24.0, 27.0, 28.2	28.0 ± 1.8 (8)
Nasodacryal subtense (49b)	12.5, 13.0	10.7 ± 2.1 (13)	13.0	12.0 ± 1.1 (7)
Orbital breadth (51)	40.5 ± 3.8 (6)	42.7 ± 3.6 (13)	44.6 ± 1.9 (5)	44.9 ± 2.5 (6)
Orbital height (52)	30 ± 2 (6)	29.6 ± 2.7 (13)	33.1 ± 3.3 (5)	36.4 ± 2.1 (7)
Nasal breadth (54)	26.5 ± 3.8 (6)	26.3 ± 1.8 (24)	30.8 ± 1.8 (5)	32.4 ± 3.2 (14)
Nasal height (55)	52.6 ± 3.8 (6)	50.2 ± 3.3 (20)	53.5 ± 1.6 (5)	61.8 ± 3.8 (10)
Simotic chord (57)	9.2, 12.0, 12.8	9.5 ± 2.5 (17)	7.3	14.7 ± 2.1 (7)
Inferior nasal bone width (57.3)	18.5 ± 3.7 (5)	17.7 ± 1.4 (10)	18.0	25.4 ± 3.7 (9)
Simotic subtense (57a)	3.8	3.9 ± 1.2 (14)	2.3	4.9 ± 1.1 (7)
External palate length (60)	53.7 ± 3.4 (6)	55.5 ± 4.5 (6)	61.5, 65.5	64.5 ± 3.8 (4)
External palate breadth (61)	60.5 ± 4.4 (6)	65.2 ± 4.5 (5)	69.5, 73.0	75 ± 2.8 (4)
Internal palate breadth (63)	36.3 ± 3.2 (6)	37.9 ± 2.7 (5)	46.0	52.0, 53.0
Alpha chord length (ALC)	32.6 ± 5.4 (4)	33.2 ± 3.4 (12)	39.0, 46.5	45.2 ± 3.4 (7)
Basion-subspinale length (BSS)	103.1 ± 3.9 (4)	99.6 ± 9.0 (11)	103.0	111.4 ± 6.0 (5)
Intercanine breadth (ICB)	24.2 ± 3.8 (4)	26.1 ± 2.1 (16)	30.2 ± 1.8 (4)	28.6 ± 2.3 (6)
Interorbital width (IOW)	100.3 ± 6.1 (5)	101.5 ± 6.1 (14)	107.6, 111.2, 114.0	111.9 ± 3.6 (9)
Interorbital width subtense (IOWS)	19.0 ± 1.2 (5)	17.2 ± 3.3 (14)	15.9, 19.0, 24.8	24.4 ± 4.0 (9)
Midorbital breadth (MOB)	58.0 ± 8.0 (4)	57.9 ± 4.7 (12)	67.8, 69.4	71.5 ± 2.7 (8)
Zygoorbitale-inf nasomaxillary suture (ZINMS)	25.9 ± 3.9 (5)	24.4 ± 3.1 (13)	27.5, 31.0	32.7 ± 1.9 (8)
Zygoorbitale-inf nasomaxillary subtense (ZINMSS)	5.2 ± 1.7 (5)	4.5 ± 1.3 (13)	3.9, 4.1	3.1 ± 1.4 (8)

Martin (Brauer, 1988) numbers and letter designations (Franciscus, 1995) follow the measurement names. EUP: earlier Upper Paleolithic; Q-S: Qafzeh-Skhul.

(Lahr, 1996). Just inferior to the inferolateral corner of the orbit on the zygomatic bone, there is a small raised ridge that corresponds to a grade 2 degree of robusticity development. Note that Lahr referred to this feature as the "zygomaxillary (malar) tuberosity." However, the use of this term for this feature is confusing because the "zygomaxillary tuberosity" usually refers to the inferior projection of a tubercle at the inferior-most point of the zygomaxillary suture (Hauser & DeStefano, 1989). Therefore, the term "zygomaxillary ridge" will be used here instead.

The lower orbital margin on the right side is more complete and undamaged, and even though it is covered with paint, the zygoorbitale point can be reliably located at the endpoint of the superior course of the zygomaxillary suture. Zygoorbitale is difficult to locate on the lower orbital border of the pathologically altered left side; however, using the right side distance of this point to midline allows the approximation of the left side point. The resulting breadth across these points relative to facial height (Figure 8.5) is roughly similar across all samples. Dolní Věstonice 3 falls at the bottom of the scatter, and along with the Gravettian Barma Grande 3 adolescent female, shows a relatively narrow midorbital distance, as does, interestingly, the very large Shanidar 5 Neandertal male.

The inferomedial and medial orbital rim is present on the right side. However, the lacrimal canal appears to have been filled in or otherwise obliterated. The inferomedial and especially medial orbital rim on the left side are completely covered or restored, and no bone is visible. There is a trochlear spine with medium expression on the internal aspect of both the left and right internal medial orbital areas just superior to the nasofrontal suture.

Infraorbital Region

The infraorbital region on the right side is damaged postmortem in the precise area of the infraorbital foramen; it has been filled and reconstructed. Nonetheless, enough actual bone exists on the adjacent lateral and medial component to accurately infer the topography of the infraorbital region; the pathologically altered left side cannot be used to infer normal anatomy. Based on the right side, a pronounced infraorbital depression was present. Also, the infraorbital process of the zygomatic bone projects

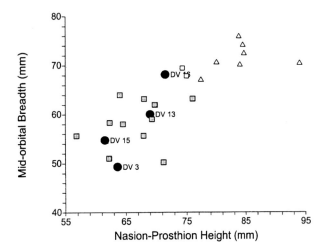

Figure 8.5 Midorbital breadth versus nasion-prosthion height. Open triangles: Neandertals; open squares: Qafzeh-Skhul; gray squares: earlier Upper Paleolithic; solid circles: Dolní Věstonice individuals.

slightly anteriorly over the depressed infraorbital region just lateral to the damaged and filled area, creating a horizontal shelf that further accentuates the infraorbital depression in a horizontal transect. The lateral aspect of the infraorbital region is oriented predominantly in the coronal plane, and the medial aspect adjacent to the piriform aperture is oriented in a parasagittal plane. Therefore, in the transverse plane, the infraorbital region shows the typically modern two-plane configuration that is different from the horizontally flat or convex plane configuration typically found in Neandertals (Table 8.2).

On the pathologically altered left side, the medial aspect of the infraorbital region is inferiorly positioned (consistent with the vertically lowered orbital floor described above) relative to the right side, such that the anterior extent of the insertion for masseter is about 5 mm lower than the right side. Moreover, the lateral left cheek height is considerably diminished relative to the right side, by 25% (17.5 mm vs. 23.2 mm). This diminished aspect is entirely relegated to the masseter origin area and suggests major resorption of the zygomatic bone due to diminished or absent masseter muscle tension for some time.

Nasal Region

The frontal processes of both the right and left maxillae are complete and retain the nasomaxillary sutural edge. The left sutural edge is completely intact, and the right is intact except for its most inferior portion. Only the superior quarters of the nasal bones on either side are preserved. Nasion is superiorly set relative to the superior nasomaxillary points. The frontal processes show a moderate degree of parasagittal eversion and a resulting moderately deep alpha point between the inferior nasomaxillary suture location and zygoorbitale. The alpha point location on the left side was determined both by its present location and by checking it against the right-side distance of this point to midline, given the alteration of the left side of the face. The distance between the estimated alpha points relative to facial height in Dolní Věstonice 3 falls toward the low end of the distribution among the Dolní Věstonice and other earlier Upper Paleolithic modern humans, and they all tend to have relative alpha chord values that are similar to, if smaller than, the Neandertal sample (Figure 8.6). The area between the nasomaxillary suture and inferomedial orbital area is uniformly smooth, lacking the roughened furrow or slight elevation frequently found in both immature and mature archaic *Homo* (Franciscus, unpublished data). The orientation of the nasomaxillary suture is inferiorly inflected rather than horizontally, as is found in many Neandertals (Table 8.2), and this is even more so for Dolní Věstonice 3 compared to the other Dolní Věstonice individuals.

Dolní Věstonice 3 has a narrow nasal breadth (22 mm), trivially larger than the value for Dolní Věstonice 15 (21.5 mm), and both of these values are considerably narrower than the nasal breadths found in the other Dolní Věstonice individuals. Yet, nasal breadths in both Dolní Věstonice 3 and 15 scale with their respective values for nasal height in a fashion that is basically similar across the Dolní Věstonice distribution (Table 8.4; Figure 8.7; Sládek et al., 2000: table 26). If, however, the Dolní Věstonice sample is combined with the other earlier Upper Paleolithic individuals, then both Dolní Věstonice 3 and 15 appear to have relatively narrower nasal breadths

Figure 8.6 Alpha chord length versus nasion-prosthion height. Open triangles: Neandertals; open squares: Qafzeh-Skhul; gray squares: earlier Upper Paleolithic; solid circles: Dolní Věstonice individuals.

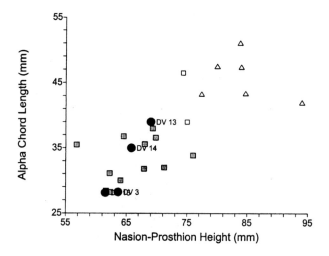

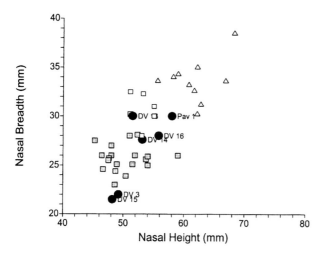

Figure 8.7 Nasal breadth versus nasal height. Open triangles: Neandertals; open squares: Qafzeh-Skhul; gray squares: earlier Upper Paleolithic; solid circles: Dolní Věstonice and Pavlov individuals.

relative to height, and Dolní Věstonice 13 and Combe Capelle 1 have wider than expected nasal breadths. In either case, the Neandertals show higher than expected nasal breadths, given their nasal heights, compared to the combined earlier Upper Paleolithic modern human sample, with the Qafzeh-Skhul sample intermediate between the two.

The lateral margin of the piriform aperture on the right side in Dolní Věstonice 3 is complete and very sharp for its entire course. The left-side lateral margin is asymmetrically altered, with a markedly elevated inferior margin and a thicker border. It is unclear whether this is due to pathological alteration or faulty reconstruction. Given the other pathological alterations, it would seem that the former is more likely, especially since the degree of asymmetry is external and superficial and would not have affected the otherwise functionally constrained nasal capsule (Franciscus, 1995, 2003b). There appears to be a very small inner crest/outer margin separation, such as is occasionally seen on the superior lateral margins of the piriform aperture in recent and fossil *Homo* (Franciscus, 2002). As reconstructed, the internal lateral walls of the nasal fossa are complete from the piriform aperture margin posteriorly to the area of the posterior choanae. However, much of this internal anatomy is reconstructed and/or consolidated. Some of the more subtle features such as the inferior conchal crests are apparent, suggesting that consolidation efforts overlie actual bone, rather than restoration of missing bone. The conchal crests show horizontally positioned anterior components and obliquely positioned posterior components. They are visible in anterior view, although they do not markedly project medially into the nasal cavity.

Based on the undistorted right side (Figure 8.8), the narial margin at the junction of the internal nasal floor and subnasoalveolar clivus is composed of a single sharp crest that is made up of fused lateral, spinal, and turbinal crests (category 1, following De Villiers, 1968; Franciscus, 1995, 2003b). It is this fused single crest or sill that demarcates the internal nasal floor from the subnasoalveolar clivus. However, this sill is not raised to any substantial degree and demarcates the narial margin into a level internal nasal floor (as coded in Franciscus, 2003b); this is particularly evident on the right side (Figure 8.8). In combination with its narial margin pattern, the level internal nasal floor presents a pattern that is clearly different from the last glacial Neandertal pattern. The anterior nasal spine is too damaged to assess, although there is no evidence for the pronounced development of spatulate lateral shelving on either side of the anterior nasal spine in Dolní Věstonice 3 that often accompanies a well-developed anterior nasal spine.

Subnasoalveolar Clivus

The subnasoalveolar clivus in Dolní Věstonice 3 is well preserved on its right side. Its height appears to be relatively short, and it remains uncertain whether this is due to alveolar alterations (see Chapters 11 and 19). The amount of tooth root visibility is not unusual for her estimated age or degree of dental attrition (Chapter 19). The overall shape of the anterior aspect of the clivus in Dolní Věstonice 3 is parabolic, and it exhibits moderate canine

Figure 8.8 Lower border of the piriform aperture in Dolní Věstonice 3.

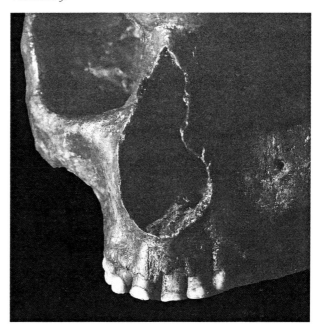

and incisor eminences commensurate with her relatively smaller anterior dentition.

Overall Facial Size and Proportion

Overall facial height and breadth in Dolní Věstonice 3 generally lie at the smallest end of the distribution for both measures among both the Dolní Věstonice and wider earlier Upper Paleolithic samples (Figures 8.3, 8.5, and 8.6; Sládek et al., 2000: tables 23 and 24), although she is comparable to Dolní Věstonice 15 in upper facial height.

Left Lateral View

Frontal Region

In left lateral view (Figure 8.9), the frontal bone of the Dolní Věstonice 3 shows a near vertical frontal inclination with marked projection at metopion relative to the frontal chord. She has the most angulation and the highest subtense values in both absolute and relative terms compared to the rest of the Dolní Věstonice and Pavlov sample and wider earlier Upper Paleolithic sample, with the sole exception of the Aurignacian-associated Mladeč 2 female (Figure 8.10; Sládek et al., 2000: table 21). The high frontal subtense values found in the Pavlovian sample overall are characteristic of all modern human samples and contrast with the lower values found among Neandertals (Table 8.1).

The glabellar region is mildly projecting, with both a very slight supraglabellar inflection and a moderate infraglabellar notch. The degree of infraglabellar notch development in Dolní Věstonice 3 most closely approximates Lahr's (1996) category 2 or 3, and it is similar to the pattern found in all of the Dolní Věstonice and Pavlov individuals, other Upper Paleolithic individuals, and to a large degree, the Neandertal sample. The greatest degree of infraglabellar notch development, as indicated by comparisons of nasio-occipital and glabello-occipital lengths (Figure 8.11), is found in Skhul 4, Skhul 5, and Mladeč 5. The distinct separation of medial, midorbital, and lateral components, described above in anterior view, can be seen and further visualized in lateral view. It includes the distinct separation of the supraorbital trigone from the superciliary arch by the obliquely flattened supraorbital sulcus and the inflated and anteroposteriorly wide supraorbital trigone. The anterior portion of the temporal line on the frontal bone is well marked and rugose. The overall form of the supraorbital region in the Dolní Věstonice and Pavlov sample is remarkably similar between the males and the one female despite differences in overall craniofacial size.

Parietal Region

In left lateral view, the vertex of the cranium is located ca. 27 mm posterior to bregma. There is a moderately acute angle between the midline superior neurocranial

Figure 8.9 Lateral (left) view of the Dolní Věstonice 3 cranium. Scale in centimeters.

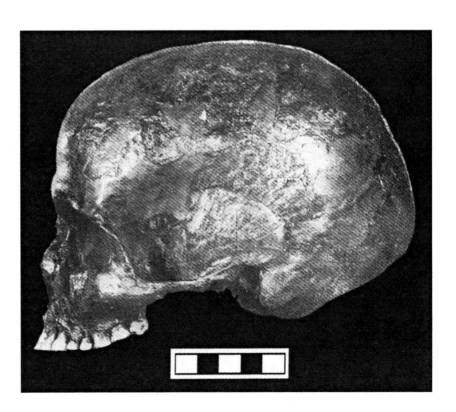

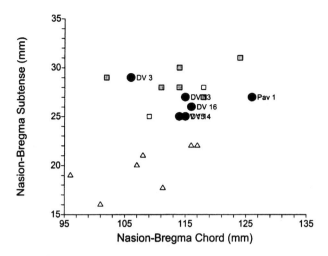

Figure 8.10 Nasion-bregma subtense versus nasion-bregma chord. Open triangles: Neandertals; open squares: Qafzeh-Skhul; gray squares: earlier Upper Paleolithic; solid circles: Dolní Věstonice and Pavlov individuals.

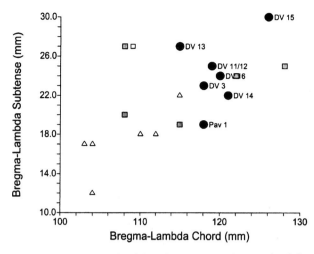

Figure 8.12 Bregma-lambda subtense versus bregma-lambda chord. Open triangles: Neandertals; open squares: Qafzeh-Skhul; gray squares: earlier Upper Paleolithic; solid circles: Dolní Věstonice and Pavlov individuals.

profile and the posterior parietal plane, as is reflected in the bregma-lambda chord subtense. Among Pavlovian individuals, Dolní Věstonice 3 is most similar proportionately to Dolní Věstonice 11, 15, and 16, who have somewhat smaller relative subtenses than does Dolní Věstonice 13 and higher relative values than Dolní Věstonice 14 and Pavlov 1. The early modern human samples show higher values for both the chord and the subtense than the Neandertals but generally similar proportions (Figure 8.12; Table 8.1; Sládek et al., 2000: table 21).

Figure 8.11 Glabello-occipital length versus nasio-occipital length. Infraglabellar notch development is reflected in the comparison of these two distances as greater values for the former relative to those of the latter. Open triangles: Neandertals; open squares: Qafzeh-Skhul; gray squares: earlier Upper Paleolithic; solid circles: Dolní Věstonice and Pavlov individuals.

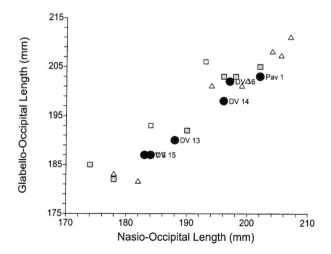

The left parietal has been reconstructed from a number of pieces, and it is filled in and painted over in many areas, obscuring surface detail. The temporal line, to whatever degree it was originally manifested, is difficult to see in the reconstruction; it is even difficult to discern it in its most anterior manifestation at the frontal bone, where it is usually most prominent.

There is a very large area of damaged bone on the left parietal bone just posterior to the coronal suture and superior to where the temporal line would have been. This area is ca. 38 mm in anteroposterior length and ca. 34 mm in mediolateral width, with an uneven surface appearance and differential levels of remodeling. For much of this area, it is difficult to distinguish between true remodeled bone and consolidant. However, in the most anterior portion it more closely approximates actual antemortem remodeling (Chapter 19).

Pterion Region

The junction of the frontal, parietal, temporal, and sphenoid bones in the pterion region has been heavily reconstructed, with a substantial amount of filler on the left side. Although elements of all four bones are present, no useful morphological information can be gleaned from this region.

Temporal Region

The temporal bone is largely intact except for a few pieces that were glued onto the posterior portion. The mastoid process, zygomatic process, and suprameatal crest region are all intact. The fit onto the cranium is excellent;

it appears that it was never separated. The temporal bone is unremarkable, with a moderately developed suprameatal crest that continues from the zygomatic process and extends well beyond porion posterosuperiorly to meet up with the posterior aspect of the temporal line on the parietal. There is a slight depression in the area of asterion. The mastoid process is smaller than that of the male Dolní Věstonice and Pavlov individuals (24.5 mm in oblique length from porion relative to the Frankfurt Horizontal line; 11.0 mm in oblique mediolateral width; 24.3 mm in anteroposterior width from its anterior extent to the mastoid foramen orthogonal to the Frankfurt Horizontal). No anterior mastoid tubercle is present. The mastoid process appears to have a somewhat unnatural inflection into the juxtamastoid eminence without the usual separation (see posterior view), and therefore assessing its extension beyond the eminence with certainty is difficult. However it appears to project inferior to the eminence. Although muscle scarring on the mastoid is present, it is not particularly robust (Table 8.5).

Occipital Region

In left lateral view, Dolní Věstonice 3 shows a degree of occipital bun development. Inferiorly, there is a slight projection of the external occipital protuberance and the superior nuchal line, creating a slight inflection relative to the nuchal plane, but the superior projection near lambda, although present, is not well developed. It corresponds most closely to Lahr's (1996) grade 4 for occipital torus development. The degree of occipital bun development is greater than that found in Dolní Věstonice 13 and 14, less than that found in Dolní Věstonice 15 and Pavlov 1, and comparable to that found in Dolní Věstonice 2, 3, and 11/12. The occipital chord and subtense are both small in Dolní Věstonice 3, with only the somewhat distorted Skhul 4 cranium having smaller values. In relative terms, Dolní Věstonice 3 has a lambda-opisthion subtense that is small compared to the chord. Note that three out of four Neandertals fall in the early modern human distribution, with La Chapelle-aux-Saints 1 being the outlier (Figure 8.13; Table 8.1; Sládek et al., 2000: table 21).

Zygomatic Region

The left zygomatic bone is dramatically altered and unfortunately painted over, with virtually no surface detail visible. There is substantial medial inflection at the junction of the base of the frontal process and zygomatic body that is probably a healed fracture from an antemortem blunt force impact. The body of the zygomatic is also substantially narrowed superoinferiorly and somewhat superiorly positioned; this is most evident in the unnaturally vertically elevated lateral inferior orbital margin on the left side (Chapter 19)

There is weak development of a marginal tubercle that is 10.9 mm in maximum anteroposterior length.

Orbital Region

In left lateral view, the lateral margin of the orbit is posteriorly positioned relative to nasion and other midfacial measures, to a greater degree than for the right. This is almost certainly due to the pathological alteration described above. Based on the nonaltered right side (see below), the degree of midfacial projection would have been rather modest for Dolní Věstonice 3 relative to other early modern (Brno 2) and also some Neandertal (La Quina 5 and La Ferrassie 1) individuals (Figure 8.14, Table 8.2).

Infraorbital Region

Despite the pathological alteration, heavy paint covering and other possible postmortem deformation, the left infraorbital region nonetheless provides some information on the topography of this region.

It is not clear whether the canine fossa depression, evident on the right side, is also present on the left. The left side seems instead to have been a noninflected, flat surface. The horizontal shelf of the zygomatic bone noted on the right side in anterior view (that projects anteriorly over the depressed infraorbital region, creating a horizontal shelf that accentuates the infraorbital depression) does not appear to be present on the left side. The most likely explanation is that the infraorbital flattening on this side is tied to the subsequent posttraumatic remodeling. Despite these pathological changes, the lateral aspect of the infraorbital region is oriented predominantly in the coronal plane, whereas the medial aspect adjacent to the piriform aperture is oriented in a parasagittal plane, resulting in the typically modern two-plane configuration. However, in the sagittal plane the lateral aspect of the infraorbital region is not as obliquely oriented as the right side, so that the inferior margin of the zygomatic process of the maxilla is not as posteriorly positioned relative to the inferior orbital; again this is probably due to the pathological alteration. The anterior root of the zygomatic process of the maxilla is located between the second premolar and first molar (more or less the same as for the right side) rather than posterior to the second molar or even with the third molar, as is often found in Neandertals (Table 8.2; Trinkaus, 1987).

Nasal Region

In the nasal region, there is a moderate degree of parasagittal eversion and a resulting moderately deep alpha

Table 8.5 Occurrence of Neandertal occipitomastoid traits in the Dolní Věstonice (DV) and Pavlov (Pav) sample.

Neandertal Configuration	DV 3	DV 11/12	DV 13	DV 14	DV 15	DV 16	Pav 1
Laterally flattened mastoid process (right)	Present	—	Present *Mild*	Present	Present *Mild*	Absent	—
Laterally flattened mastoid process (left)	Abnormal	—	Absent	Absent	Present *Mild*	Absent	Absent
Anterior mastoid tubercle (right)	Present?	—	Absent *Tubercle is in the middle, not anterior, and is very diffuse*	Present *Diffuse*	Absent	Present *Diffuse, and two others posterior and inferior also present*	—
Anterior mastoid tubercle (left)	Abnormal	—	Absent *Tubercle is in the middle, not anterior, and is very diffuse*	Absent	Absent	Absent	Absent *Tubercle is in the middle, not anterior, and is diffuse*
Mastoid process that does not project significantly below juxtamastoid eminence; large juxtamastoid eminence (right)	Absent	—	Absent	Absent	Absent	Absent	Absent
Mastoid process that does not project significantly below juxtamastoid eminence; large juxtamastoid eminence (left)	Abnormal	—	Absent	Absent	Absent	Absent	Absent
Fully developed suprainiac fossa associated with a bilaterally protruding occipital torus	Absent	Absent	Absent	Absent	Absent	Absent	Absent

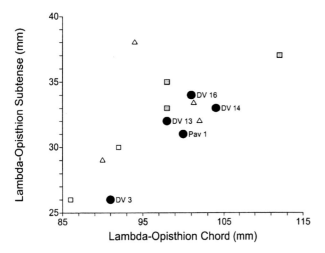

Figure 8.13 Lambda-opisthion subtense versus lambda-opisthion chord. Open triangles: Neandertals; open squares: Qafzeh-Skhul; gray squares: earlier Upper Paleolithic; solid circles: Dolní Věstonice and Pavlov individuals.

point between the inferior nasomaxillary suture location and zygoorbitale. The degree of eversion and alpha point subtense depth in Dolní Věstonice 3 is slightly less than that found in Dolní Věstonice 16, and both of them have much smaller subtense values than Dolní Věstonice 14 and 15 (Figure 8.15). Dolní Věstonice 3's value for the subtense (3.3 mm) is very close to the Neandertal mean and well below the mean for modern humans (Table 8.4), with only Oase 2 and Ohalo 2 having smaller values. It can be seen particularly well in this view that the orientation of the nasomaxillary suture relative to the Frankfurt Horizontal is inferiorly, rather than horizontally,

Figure 8.14 Interorbital width subtense versus interorbital width. Open triangles: Neandertals; open squares: Qafzeh-Skhul; gray squares: earlier Upper Paleolithic; solid circles: Dolní Věstonice individuals.

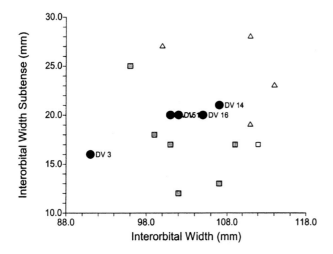

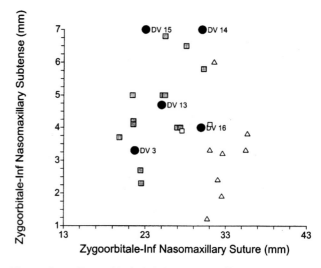

Figure 8.15 Zygoorbitale-inferior nasomaxillary suture subtense versus zygoorbitale-inferior nasomaxillary suture length. Open triangles: Neandertals; open squares: Qafzeh-Skhul; gray squares: earlier Upper Paleolithic; solid circles: Dolní Věstonice individuals.

inflected, as is true for non-Neandertal extinct and extant *Homo* (Table 8.2). The lateral margin of the piriform aperture is slightly more posteriorly positioned (pathologically altered) on the left side than on the unaltered right side, and the lack of anterior nasal spine development, probably due to damage, is also evident.

Subnasoalveolar Clivus and Alveolar Region

There is a moderate inflection at subspinale in lateral view. The clivus also exhibits a mild degree of lower prognathism from this view relative to the other Dolní Věstonice and Pavlov individuals, although well within the range of all of the comparative samples (Figure 8.16). A comparison of the lower facial projection at prosthion and the upper facial projection at nasion (Figure 8.17) also shows Dolní Věstonice 3 to have a similar degree of relative lower facial projection as the other comparative samples.

Overall Size and Proportion in Lateral View

In lateral view, both the height and the length of the neurocranium in Dolní Věstonice 3 are smaller than the Pavlovian males for which complete measurements are available (Figure 8.18; Sládek et al., 2000: tables 18 and 19). However, Dolní Věstonice 3 has the same relationship of height to length as the Pavlovian male sample, with all four individuals falling along a single line. Although the Dolní Věstonice sample and other earlier Upper Paleolithic individuals have cranial heights that are

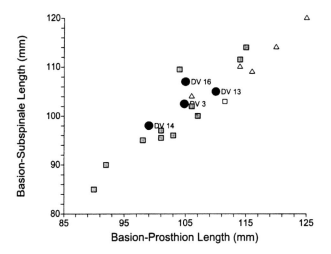

Figure 8.16 Basion-subspinale length versus basion-prosthion length. Open triangles: Neandertals; open squares: Qafzeh-Skhul; gray squares: earlier Upper Paleolithic; solid circles: Dolní Věstonice individuals.

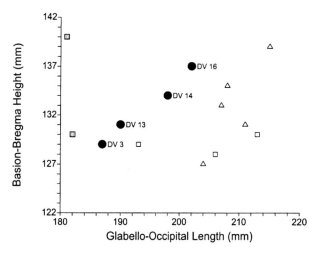

Figure 8.18 Basion-bregma height versus glabello-occipital length. Open triangles: Neandertals; open squares: Qafzeh-Skhul; gray squares: earlier Upper Paleolithic; solid circles: Dolní Věstonice individuals.

comparable to those of Neandertals, their cranial lengths are shorter, resulting in the characteristically modern globular shape (Table 8.1). It should be noted that these comparisons are unfortunately limited to the relatively small number of comparative specimens that retain a complete enough basicranium that standard neurocranial height can be measured (Figure 8.18).

Overall facial height and facial length in Dolní Věstonice 3 is clearly smaller than for Dolní Věstonice 16. However, the former shows roughly comparable values to the males from the triple burial (Figures 8.19 and 8.20; Sládek et al., 2000: tables 22 and 24). Thus, relative to two of the other Dolní Věstonice individuals, Dolní Věstonice 3 has a somewhat smaller neurocranium but a roughly comparable overall splanchnocranial size. Neandertals, not surprisingly, have larger values for facial height and length than the modern human samples, especially when sorted by sex.

Right lateral View

Gross aspects of neurocranial and splanchnocranial morphology that are evident from either side are covered in the preceding section on left lateral view. The following additional features revealed in right lateral view (Figure 8.21) are covered in this section.

Figure 8.17 Basion-nasion length versus basion-prosthion length. Open triangles: Neandertals; open squares: Qafzeh-Skhul; gray squares: earlier Upper Paleolithic; solid circles: Dolní Věstonice individuals.

Figure 8.19 Nasion-prosthion height versus basion-prosthion length. Open triangles: Neandertals; open squares: Qafzeh-Skhul; gray squares: earlier Upper Paleolithic; solid circles: Dolní Věstonice individuals.

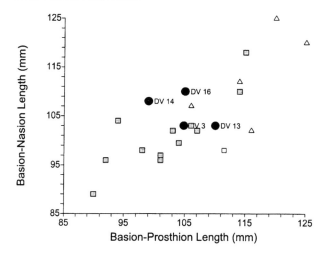
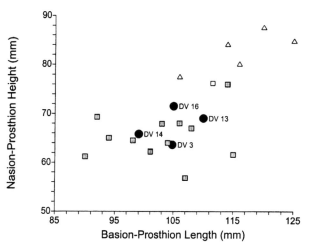

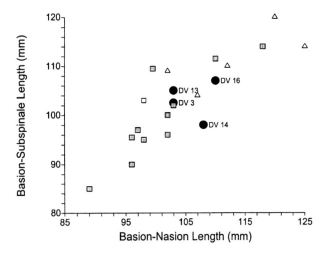

Figure 8.20 Basion-subspinale length versus basion-nasion length. Open triangles: Neandertals; open squares: Qafzeh-Skhul; gray squares: earlier Upper Paleolithic; solid circles: Dolní Věstonice individuals.

Frontal Region

There is a possible healed lesion on the right lateral frontal squamous just anterior of the coronal suture along the temporalis line. It is ca. 15 mm in maximum length and 9 mm maximum width, and it appears to have somewhat delimited edges in its reconstructed state. However, it is located in an area coincident with a clear postmortem break that makes its interpretation difficult (compare figures 6 and 7 in Jelínek, 1954).

Temporal Region

The mastoid process on the right side is laterally flattened and appears to project little below the juxtamastoid eminence (but see posterior view, below).

Parietal Region

There is a surface irregularity on the right parietal along the posterosuperior portion of the temporal line that is ca. 20 mm or less in diameter. As with the frontal surface feature noted above in this view, it is also located in an area coincident with a clear postmortem break that makes its interpretation difficult (compare figures 6 and 7 in Jelínek, 1954). Comparison of the pre- and postreconstruction state of the cranium makes it more likely to be a postmortem artifact than even the previous feature.

Zygomatic Region

As noted in anterior view, the inferolateral border of the right orbit is robust in that it is rounded rather than sharp. However, the border is also raised in relation to the floor of the orbit and does not attain the most robust grade; it therefore corresponds to Lahr's (1996) grade 2. There is

Figure 8.21 Lateral (right) view of the Dolní Věstonice 3 cranium. Scale in centimeters.

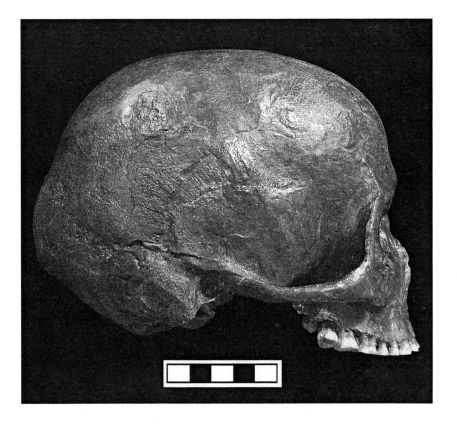

no development of a zygomaxillary ridge between the orbital and free margins of the zygomatic, which corresponds to a grade 1 degree of robusticity development (Lahr, 1996). In terms of epigenetic traits from this view (Table 8.6), there appears to be a single zygomaticofacial foramen of medium size, and there is weak development of the marginal tubercle (10.8 mm maximum anteroposterior length).

Inferior View

Occipital Region

In inferior view (Figure 8.22), the occipital region of Dolní Věstonice 3 has been reconstructed with filler in some areas, especially immediately lateral to the foramen magnum area. The alignment of the foramen magnum and basilar occipital portion deviates to the left. The breadth of the foramen magnum in Dolní Věstonice 3 is identical to those for Dolní Věstonice 14 and 16. Dolní Věstonice 3, like Dolní Věstonice 16, appears to have a more rounded foramen magnum than Dolní Věstonice 14, who has a considerably longer foramen magnum. Dolní Věstonice 13 has the narrowest value for foramen magnum breadth among all of the comparative samples (Figure 8.23; Sládek et al., 2000: tables 18 and 19), and this, with its modest value for length, gives it the superficial appearance of being elongated (Table 8.7).

A weakly developed median occipital crest posterior to the foramen magnum is visible, and its orientation is in correct alignment with the foramen magnum and basilar portion. It is therefore off alignment with the rest of the anterior perspective. The occipital crest spans the weakly expressed inferior nuchal lines and becomes more diffuse posterior to the inferior nuchal line. The anterior portion of the occipital crest is fairly visible, but it is not sharp or prominent, as in Dolní Věstonice 13, for example; the development of the occipital crest most closely approximates category 1 in Lahr (1996). A weakly developed superior nuchal line is evident for most of its course along the occipital, and in midline it joins the *tuberculum linearum* (Hublin, 1978). A weakly expressed supreme nuchal line is present. The right side shows a well-developed sulcus for the posterior digastric muscle and to a much lesser degree the areas for the rectus capitis posterior major, superior oblique, and semispinalis capitis muscles. The left side has major portions filled, and it is difficult to determine the pattern of nuchal muscle scarring. There is a clear asymmetry in the external area coincident with the left internal cerebellar fossa; it projects as a bulge to a noticeably greater degree on the left (pathological) side.

The occipital condyles are well preserved and show clear asymmetry. The right condyle slightly overlies the medial edge of the foramen magnum, and the left condyle is positioned lateral to the medial edge so that the entire edge of the foramen magnum is visible on the left side. The area just lateral to both condyles in the area of the rectus capitis lateralis insertion is damaged and largely reconstructed. The right hypoglossal canal is present on the right side, but the left hypoglossal canal and the posterior condylar fossae are packed with filler and little can be gleaned about the morphology. The basilar portion has a prominent pharyngeal depression (Table 8.8), and it is complete just beyond this feature, where it breaks away to trabecular bone.

Sphenoid Region

The pterygoid processes, posterior nasal choanae, clivus, and the area around hormion are all destroyed. The infratemporal surfaces lateral to the pterygoid processes are preserved to some degree but provide little morphological detail.

Temporal Region

The inferior aspects of the temporal region are well preserved on both sides. The right temporal bone has a well-developed groove for the digastric muscle and an occipital groove medial to the mastoid process; this area on the left side has been filled in. The left temporal bone retains an 8.8 mm length of styloid process, and the right a 9.0 mm length; their original lengths cannot be determined. A stylomastoid foramen can be seen on each side, although they are filled with matrix and consolidant. The areas immediately medial to both styloid processes posterior to the tympanic region are preserved.

The glenoid fossae are well preserved. The right glenoid fossa appears completely normal, with a fossa that is deep relative to the anterior and posterior articular eminences (6.8 mm) with no signs of remodeling. The left glenoid fossa was substantially remodeled antemortem (Chapter 19). The anterior and posterior articular eminences have been completely remodeled away, with only 1.1 mm maximum glenoid fossa depth. The remodeled surface has become a nearly circular, almost flat surface measuring 23.0 mm in maximum width and length. The remodeled glenoid surface is nonuniform, with irregular texture. A transverse line that nearly bisects the two halves of the flattened surface may correspond to the original division between the anterior eminence and the posterior glenoid fossa.

Whereas the normal right infratemporal foramen area shows the usual infratemporal fossa configuration, the pathologically altered left side shows an extreme shorting of the infratemporal fossa length in its superior aspect (left: 33.2 mm; right: 41.6 mm for the maximum

Table 8.6 Epigenetic traits scored in lateral view for the Dolní Věstonice (DV) and Pavlov (Pav) sample.

Trait	DV 3	DV 11/12	DV 13	DV 14	DV 15	DV 16	Pav 1
Zygomaticofacial foramen (1, 2)	Present *Right side: medium-sized foramen Left side: ?*	—	Present *Bilateral, medium-sized foramina*	Absent	Present *Bilateral, medium-sized foramina*	Present *Bilateral, medium- to large-sized foramina*	Present *Left side: two small foramina*
Marginal tubercle (1, 2, 3)	Present *Bilateral weak development*	—	Present *Right side: medium to strong development Left side: weak development*	Present *Bilateral, weak development*	Present *Bilateral, strong development*	Present *Right side: medium development Left side: strong development*	Present *Left side: strong development*
Parietal process of temporal squama (middle meningeal artery emissaries) (1)	?	Present? *Right side*	—	—	Absent *Left side*	Present? *Left side*	—
Suprameatal spine and depression (1, 2, 3)	?	—	Present *Bilateral small spines with slight depressions*	Present *Left side: small spine and depression*	Present *Bilateral, small spines with no depressions*	Present *Bilateral, small spines with moderate depressions*	Present *Bilateral, very small spine*
Squamomastoid suture (1, 3)	?	—	Absent	Absent	Absent?	Absent	Present? *Left side: trace*
Mastoid foramen (1, 2)	Present *Right side: one medium-sized foramen Left side: ?*	—	Present? *Right side: two foramina? Left side: one large foramen*	Present *Right side: one large foramen*	Present *Bilateral, medium-sized foramina*	Present *Right side: two medium-sized foramina Left side: one large foramen and possible second, medium-sized foramen*	Present *Right side: one large foramen*

Traits as employed by 1: Hauser and DeStefano (1989), 2: DeStefano and Hauser (1991), and 3: Manzi et al. (2000).

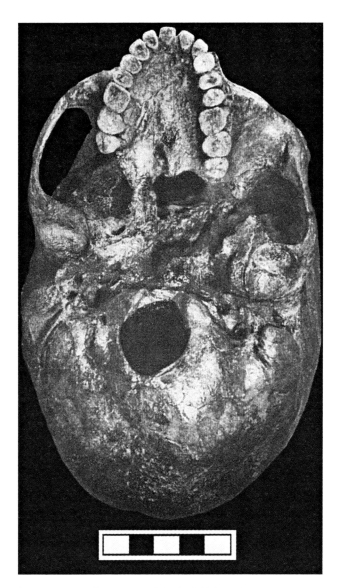

Figure 8.22 Inferior view of the Dolní Věstonice 3 cranium. Note the normal right temporal glenoid fossa and the substantially remodeled and enlarged left glenoid fossa. Note also the marked shortening of the left intratemporal fossa length compared to its right counterpart. Scale in centimeters.

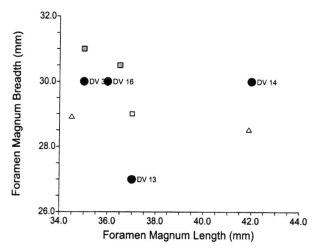

Figure 8.23 Foramen magnum breadth versus foramen magnum length. Open triangles: Neandertals; open squares: Qafzeh-Skhul; gray squares: earlier Upper Paleolithic; solid circles: Dolní Věstonice individuals.

anteroposterior dimension), and the fossa is medially obliquely reoriented compared to the normal right side. This is associated with a dramatic reorientation of the anterior border of the infratemporal fossa (i.e., the confluence of the posterior maxillary zygomatic process, external posterior wall of the maxillary sinus, and anterior temporal surface of the zygomatic bone) on the left side. It is nearly horizontally rather than vertically oriented (Chapter 19).

From inferior view, the zygomatic arches show a high degree of shape asymmetry. The right arch is smoothly curved along its course, showing normal morphology. The left arch shows a medial inflection in the region of the zygomaticotemporal suture. This inflection is part of an overall abnormal shape of the zygomatic bone that is visible in all anatomical normae; it probably reflects a healed fracture sustained by this bone and other bones on this side of the face, resulting from a blunt force impact (Chapter 19). Roughening of the inferior surface for the attachment for masseter is visible on both sides.

Inferior Palatal Region

The inferior palate is complete from the orale anteriorly to the posterior nasal spine in midline, and it shows the degree to which consolidation in situ seems to have saved very fragile bone (see the discussion of orbits above). The right half of the palate is better preserved than the left side, being complete to the palatine suture area, whereas the left is missing about two-thirds of the posterior lateral palate. There appears to be a mild palatine torus running along midline that would have been most expressed in the more posterior palate, but peripheral damage inhibits conclusive scoring. The palate is moderately arched, and the right lateral portions of the palate in the molar region shows some spiculation and relief, indicating that minor palatine bridging was possibly present. From this view, there appears to be asymmetry in the vertical orientation of the lingual alveolar plane on the right and left sides, with the right side being more mediolaterally obliquely angled than the more vertically aligned left side. Whether this is the result of missing bone on the palatal aspect of the left side, postmortem deformation, or a remodeled alteration associated with the other pathological masticatory alterations is unclear.

Table 8.7 Occurrence of Neandertal basicranial traits in the Dolní Věstonice and Pavlov sample.

Neandertal Configuration	DV 3	DV 13	DV 14	DV 15	DV 16	Pav 1
Flat articular eminence, right	Absent	Absent	—	Absent	Absent	Present
Flat articular eminence, left	Pathological	Absent	Present	Absent	Absent	Absent
Elongated foramen magnum	Absent	Absent / Appears elongated because of very narrow breadth	Present Mild	Absent	Absent	—

The anteroposterior axis of the palate, complete all the way to the posterior nasal spine, shows a misalignment relative to that of the foramen magnum and basilar portion of the occipital and a slight misalignment of the dental arcade relative to the plane of the face along the orbital margins that seems to approximate the same degree of asymmetry found in the mandibular dental arcade with respect to the overall mandible. The articulation of the mandibular dentition with that of the maxilla is correct anteroposteriorly, and there is good contact with the anterior dentition when the nonpathological right side mandibular condyle is positioned in the right glenoid fossa. However, beginning with second premolar, the maxillary and mandibular teeth begin to show increasing nonocclusion, which reaches its greatest gap of about 3.7 mm at the second to third molars on both sides; it could indicate a slightly improperly rotated splanchnocranial hafting around the frontal bone contacts. In this position, with the right side mandibular condyle/glenoid fossa articulation in proper alignment, the heavily remodeled left mandibular condyle is positioned anteriorly on the most anterior extent of the left remodeled glenoid fossa (i.e., the area that represented the anterior eminence prior to the pathological alteration). The matching mandibular and maxillary dental arcade similarities noted above indicate that the reconstruction reflects actual remodeled anatomy due to the pathological condition rather than a poor reconstructive effort. The wear on the maxillary dentition mirrors that found on the mandible (Chapter 11).

The Dolní Věstonice 3 palate is similar in length to those of Dolní Věstonice 13, 14, and 16, all of which are slightly longer than in Dolní Věstonice 15 and Pavlov 1. The palatal breadth in Dolní Věstonice 3, in contrast, is the smallest of the Pavlovian individuals (Figure 8.24; Sládek et al., 2000: table 27). The Pavlovian sample as a whole has smaller palates overall than the Qafzeh-Skhul and Neandertal samples (Figure 8.24; Table 8.4).

Posterior View

In posterior view (Figure 8.25), the contour of the lateral vault in Dolní Věstonice 3 is relatively straight-sided, and the parietal tubers and the midline posterior vertex are not prominent. Dolní Věstonice 3 thus does not show the marked pentagonal configuration that is found on some other specimens (e.g., Dolní Věstonice 13). Nonetheless, the greatest breadth of the cranium is found relatively high up at the level of the weakly expressed parietal tubers, and the relatively straight-sided contour in Dolní Věstonice 3 is similar to that found in modern humans and very different from the "en-bombe" Neandertal configuration. Maximum cranial breadth and height are absolutely smaller in Dolní Věstonice 3 than in the Dolní Věstonice males (Sládek et al., 2000: table 19), and the relationship between these two dimensions in Dolní Věstonice 3 places it far from the Neandertal scatter and clearly aligns it with the earlier Upper Paleolithic individuals (Figure 8.26; Table 8.1). There is no evidence for sagittal keeling. This view also shows the extension of both mastoid processes below the level of the more medially positioned juxtamastoid eminences, the usual modern human condition; this is especially evident on the right side, where the morphology medial to the mastoid process is more complete.

The left posterior parietal bone has a large area of filled-in reconstruction, and the right side has a large postmortem fracture located about 22 mm superior to lambda. From this view, the posterior portion of the sagittal suture is visible, and lambda can be determined by using the confluence with the lambdoidal suture. The sagittal suture deviates to the left slightly as it approaches lambda, and thus this point appears somewhat off of the midline to the left side. The right lambdoidal suture is relatively better preserved than the left, although both are visible along their course. The left lambdoidal suture is asymmetrically lower (more inferiorly placed) than the right along its course. There are, in fact, several features that show a more inferior placement on the left side in occipital view than the right; these include the nuchal lines, the asterion point, and the tip of the mastoid processes. There is also a distinct globular protrusion or inferior bulging of the left nuchal planum inferior to the superior nuchal line. Several unresolved issues relating to this last feature are discussed in detail in Chapter 19.

The superior nuchal lines are present on both sides and meet in midline at the *tuberculum linareum*, the point

Table 8.8 Epigenetic traits scored in basal view for the Dolní Věstonice and Pavlov sample.

Trait	DV 3	DV 13	DV 14	DV 15	DV 16	Pav 1
Pharyngeal tubercle (1, 3)	?	Present	Present	—	Present	Present
		Weakly expressed	*Weakly expressed*		*Trace*	*Weakly expressed*
Pharyngeal foveola (1, 3)	Present	Present	Present	—	Present	Present
Craniopharyngeal canal (1)	?	Absent	—	—	—	—
Foramen ovale, spinosum, and Vesalius (1,3)	Present	Present	Present	Present	Present	Present
	Right: foramina ovale and spinosum	*Bilateral, foramina ovale and spinosum*	*Left: foramina ovale and spinosum*	*Left: foramina ovale and spinosum*	*Right: foramina ovale and spinosum*	*Right: foramen spinosum*
Palatine torus (1,2,3)	Present	Present	Absent	Absent	Absent	Absent
	Mild development	*One large greater palatine foramen present*				
Hypoglossal canal bridging (1, 2, 3)	Present	Present	Absent	—	Absent	Absent?
	Bilateral, single unbifurcated canals	*Complete bridging on right side*	*Left: single undivided canal*		*Bilateral, single unbifurcated canals*	
Paracondylar process (1, 3)	?	Present	Present	—	—	Present
		Bilateral, weak expression	*Left: weak expression*			*Left: two weakly developed, unfused processes*

Traits as employed by 1: Hauser & DeStefano (1989), 2: DeStefano & Hauser (1991), and 3: Manzi et al. (2000).

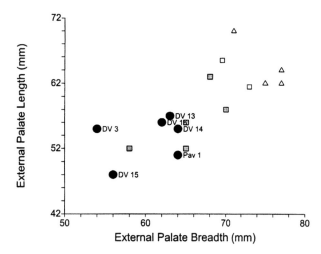

Figure 8.24 External palate length versus external palate breadth. Open triangles: Neandertals; open squares: Qafzeh-Skhul; gray squares: earlier Upper Paleolithic; solid circles: Dolní Věstonice and Pavlov individuals.

frequently below the external occipital protuberance where the superior nuchal lines meet (Hublin, 1978). The supreme nuchal lines are visible, especially on the right side, coursing apart from the superior nuchal lines, which are converging more as they reach the *tuberculum linearum*. An inferior roughened line corresponds to the area that often encompasses an iniac depression; however, this area is not as well defined or pronounced as that found on other specimens (e.g., Pavlov 1 and Dolní Věstonice 14–16). On some of the other Pavlovian specimens (e.g., Dolní Věstonice 14–16), the inferior depression at the external occipital protuberance is matched by an additional distinct depression (usually the larger of the two in surface area) that is more superiorly positioned just below lambda and well above the supreme nuchal line (see occipital discussions for those specimens below). In Dolní Věstonice 3, the superior depression is present but not as rugose or well defined.

It should be emphasized that these depressions above inion, found to varying degrees in these Pavlovian individuals (even in those that are more distinct and pronounced), are not the fully developed, mediolaterally expanded suprainiac fossae associated with a bilaterally protruding occipital torus found in most Neandertals, some Middle Pleistocene European humans, and some of the later Middle Pleistocene African specimens (Hublin, 1978; Santa Luca, 1978; Arsuaga et al., 1997; Rougier, 2003; Haile-Selassie et al., 2004; Trinkaus, 2004). Although Neandertal-like suprainiac fossae are present in some recent humans, they are not present in these specimens.

Figure 8.25 Posterior view of the Dolní Věstonice 3 cranium. Scale in centimeters.

Superior View

In superior view (Figure 8.27) the cranium is dolicocephalic; its greatest width lies not at the parietal tubers in the posterior third of the neurocranium, as in some individuals (Dolní Věstonice 13), but rather more anteriorly near the midpoint. In both maximum breadth and maximum length, it is among the smallest of the comparative early modern human crania, and its cranial breadth is narrow relative to its length, putting it well into the dolichochephalic range (Figure 8.28; Sládek et al., 2000: tables 18 and 19). The cranial capacity for Dolní Věstonice 3 is the smallest of the Dolní Věstonice and Pavlov sample (Table 8.9), although large enough to preclude any postorbital constriction. In oblique superior view, the distinction of the superciliary arch from both the frontal squama and the laterally positioned supraorbital trigone on either side by the obliquely flattened supraorbital sulcus is apparent, although not to the degree that is evident in the more robust male specimens. Relative to the long axis of the cranium from this view, the more posterior placement of the left facial plane and zygomatic region relative to

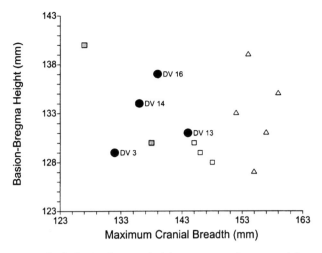

Figure 8.26 Basion-bregma height versus maximum cranial breadth. Open triangles: Neandertals; open squares: Qafzeh-Skhul; gray squares: earlier Upper Paleolithic; solid circles: Dolní Věstonice individuals.

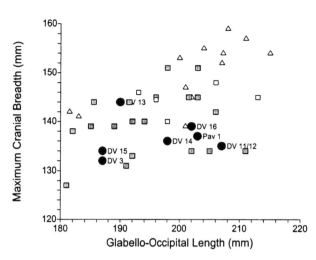

Figure 8.28 Maximum cranial breadth versus glabello-occipital length. Open triangles: Neandertals; open squares: Qafzeh-Skhul; gray squares: earlier Upper Paleolithic; solid circles: Dolní Věstonice and Pavlov individuals.

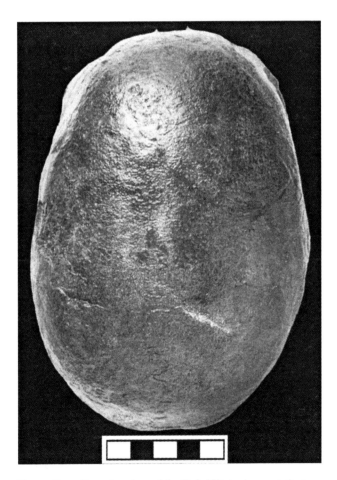

Figure 8.27 Superior view of the Dolní Věstonice 3 cranium. Scale in centimeters.

the right side is evident, as is the medial inflection of the pathologically altered left zygomatic bone.

The sagittal suture is more visible than the coronal suture, and as mentioned in the posterior view, does not follow a linear trajectory but deviates mediolaterally along its course. The large area of damaged bone described in detail in left lateral view on the left parietal just posterior to the coronal suture is evident from this view.

Table 8.9 Cranial capacity estimates for the Dolní Věstonice and Pavlov neurocrania with comparative data for other Late Pleistocene samples.

	Cranial Capacity (cc)
Dolní Věstonice 3	1285
Dolní Věstonice 13	1481
Dolní Věstonice 14	1538
Dolní Věstonice 15	1378
Dolní Věstonice 16	1547
Pavlov 1	1472
Earlier Upper Paleolithic	1514 ± 120 (16)
Qafzeh-Skhul	1501 ± 111 (6)
Neandertals	1498 ± 167 (14)

The Dolní Věstonice and Pavlov values were determined by water displacement of endocranial casts; the comparative data are derived from the original literature and were calculated by different techniques. Mean ± standard deviation (N) are provided for the comparative samples.

Dolní Věstonice 11/12

Preservation

The Dolní Věstonice 11/12 individual retains a calotte with the squamous portions of the frontal bone, the left and right parietal bones, and the occipital bone (catalogued as Dolní Věstonice 11), as well as a separate supraorbital section of the frontal bone that retains a fragment of the right nasal bone (catalogued as Dolní Věstonice 12). A direct connection between the two elements is not present; however, by matching the pattern of a healed compressed fracture that was located in the original conjoin area of the two elements, we can be virtually certain that they belong to the same individual (Figure 8.29; Klíma, 1987b; Vlček, 1991; Sládek et al., 2000).

The bone of the calotte is in good condition. There is minor root etching on the endocranial surface of the frontal squamous on either side of midline, but it is otherwise well preserved. There is also minor root etching on the exocranial surface, but most anatomical details are visible. The bone preservation on Dolní Věstonice 12 is also good, although there is moderate to heavy root etching on the bone surfaces, especially internally.

In the following description, the Dolní Věstonice 11/12 designation will be used when referring to anatomy and measurements that include both elements and to each element in isolation when discussing them separately.

Anterior View

Frontal Region

The frontal bone of Dolní Věstonice 11 consists mainly of the right squamous portion, from near the right stephanion to the supraglabellar area on the midline ca. 68 mm from bregma (Figure 8.30). The coronal suture is evident from the right stephanion to ca. 42 mm left of bregma. There is no evidence of a metopic suture (Table 8.3), no evidence of keeling, and no frontal eminence. Just lateral of the midline, there is a beveling of the external half of the bony edge down to the postmortem break, representing the superior edge of the large anterior frontal trauma. The edge of healed bone is 17 mm in mediolateral diameter and a maximum of 7 mm in its superoinferior extent. It has rounded edges of diploë with vascular canals, and it is clearly distinct from the adjacent postmortem bone breakage (Chapter 19). Minimal suture closure is evident

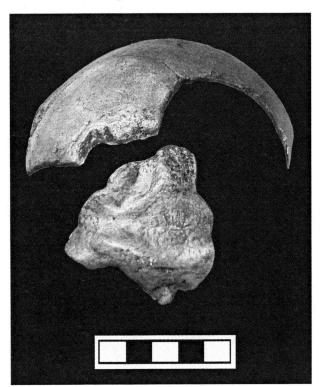

Figure 8.29 Anterior view of composite cast of Dolní Věstonice 11 and 12, showing the relative positions of the two elements from a single individual and the single large traumatic lesion on the frontal region.

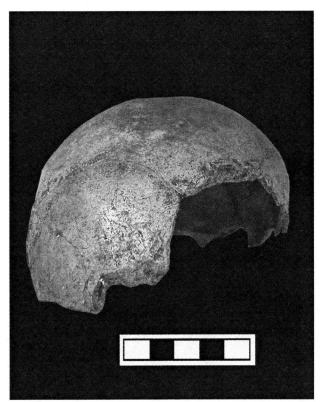

Figure 8.30 Anterior view of Dolní Věstonice 11. Scale in centimeters.

at the midcoronal location, and significant closure is observed at bregma (Chapter 6).

Supraorbital Region

The Dolní Věstonice 12 supraorbital section of the frontal bone is a piece that retains nasion, a fragment of the superior-most right nasal bone, the medial and middle portion of the right upper orbit with its supraorbital elements, and a portion of the mid- to right frontal squamous (Figure 8.31). The maximum preserved height of the specimen is 64 mm, and its maximum preserved breadth is 63 mm.

The supraorbital region of Dolní Věstonice 12 shows a distinct separation of medial and midorbital components; the area of the lateral component is missing. The glabellar region is pronounced in midline. It extends approximately to the medial third of the right supraorbital area (superciliary arch), and it extends superiorly and laterally onto the more diminutive and flattened midorbital region. The total length of this medial component (encompassing glabella) for both sides combined, based on doubling the right side value, is ca. 60 mm. The lateralmost supraorbital margin (the supraorbital trigone) is not preserved, but the flattened supraorbital sulcus of the midorbital region is clearly present. The flattening of the supraorbital sulcus results in the "pinched" appearance noted above (Table 8.2). In addition to these shape considerations, Dolní Věstonice 12 has an overall well-developed supraorbital ridge-glabella region, in size and robusticity consistent with its designation as a male (Chapter 7).

Figure 8.31 Anterior view of Dolní Věstonice 12. Note the large traumatic lesion. Scale in centimeters.

There is a partially obliterated zigzag-shaped supranasal suture on Dolní Věstonice 12 (Hauser & DeStefano, 1989: h, plate VI); it is the most prominent occurrence of this feature among the Dolní Věstonice and Pavlov sample (Table 8.3).

There is a large healed depressed trauma to the frontal squamous, from just lateral of the midline across to the right lateral break of the bone centered ca. 30 mm superior to the supraorbital notch. The edges are rounded and well healed. The depression of the middle of the injured area ectocranially resulted in an endocranial displacement of the internal table by 1 to 2 mm over an area of at least 17 by 17 mm. The superolateral aspects of the injury on this bone match certain details of the anatomically adjacent portion of the right frontal region described above on Dolní Věstonice 11 and are discussed in detail in Chapter 19.

Orbital Region

The right superior orbital margin in Dolní Věstonice 12 is rounded, not sharp (Chapter 7), and it has a large, acute, medial supraorbital notch (Table 8.3; Hauser & DeStefano, 1989: b, plate VIII).

Left Lateral View

Parietal Region

The left parietal bone is missing both inferior angles, as well as the squamosal suture. The inferior temporal line is prominent. There is a relatively acute angle between the midline superior neurocranial profile and the steep posterior parietal plane in left lateral view (Figure 8.32), reflected in the bregma-lambda chord subtense. Dolní Věstonice 11, along with the rest of the Pavlovian sample and other early modern humans, evinces higher absolute values for both the chord and the subtense than Neandertals, but there is little separation of the samples in the relative proportions of these two parietal variables (Figure 8.12; Table 8.1; Sládek et al., 2000: table 21).

Occipital Region

In left lateral view, Dolní Věstonice 11 shows mild occipital bun development. Inferiorly, there is a marked projection of the external occipital protuberance and the superior nuchal line, creating a slight inflexion relative to the nuchal plane, and there is a superior projection near lambda as well.

Overall Size in Lateral View

Only the length of the cranium can provide some information on overall size in lateral view, and even with this

Figure 8.32 Lateral (left) view of Dolní Věstonice 11. Scale in centimeters.

measurement some estimation is required. By approximating the join of Dolní Věstonice 11 and 12, a glabello-occipital length of ca. 207 mm (Vlček, 1997) was obtained. This estimated cranial length is 8.9% longer than the smallest unambiguous male in the Dolní Věstonice sample, Dolní Věstonice 13 (190 mm). Yet, it is very similar to Pavlov 1 (204 mm), who is only 1.5% shorter in length.

Right Lateral View

Parietal Region

In right lateral view (Figure 8.33), the parietal bone is more complete than its left counterpart, almost to the extent of its inferior angles, and it retains the squamosal suture. As on the left side, the right inferior temporal line

Figure 8.33 Lateral (right) view of Dolní Věstonice 11. Note the enlarged enthesopathy at the external occipital protuberance. Scale in centimeters.

is prominent. There is a possible parietal process of the temporal squamous in the posterior moiety of the squamosal suture. This feature is usually related to the presence of middle meningeal artery emissaries; however, the posterior location is not as epigenetically relevant as those found in a more anterior position (Hauser & DeStefano, 1989).

Posterior View

The occipital bone preserves the occipital plane and up to 9 mm of the nuchal plane below the *tuberculum linearum*, primarily on the left side (Figure 8.34). The lambdoidal suture, which exhibits minimal closure at the midlambdoid location and at lambda, is preserved for 72 mm on the right side of lambda and 70 mm on the left (see Table 8.10 for detailed suture scoring). The maximum preserved breadth of the calotte is 135 mm, which most likely approximates the maximum original cranial breadth. This falls at the lower end the distribution of cranial breadths for the Dolní Věstonice and Pavlov sample males, between the values for Dolní Věstonice 15 and 14 (range: 134–144 mm; Sládek et al., 2000: table 19), and above the smaller value for Dolní Věstonice 3 (130 mm). It is modest for an earlier Upper Paleolithic specimen but within 1 standard deviation of that sample's mean (Table 8.1).

The shape of the neurocranium in posterior view, although incomplete in its inferior-most extent, nonetheless shows a straight-sided-to-pentagonal configuration (despite the absence of parietal eminences) that is typically modern. There is a well-developed nuchal crest in Dolní Věstonice 11, consistent with other features that indicate its male designation (Chapter 7). The external occipital protuberance is very prominent, and it may instead be an enlarged enthesopathy (Chapter 19). There is no evidence for a suprainiac fossa of the type that is usually found in Neandertals (see above and summary). The superior nuchal region also has depressions on the right side, just lateral of the midline, whose etiology is unclear (Chapter 19).

Internal (Endocranial) View

In internal (endocranial view (Figure 8.35), there are two broad Pacchionian depressions with rounded edges just posterior to the coronal suture, one on each parietal bone, at the superior end of the anterior meningeal sulcus. The one on the right is ca. 17 mm mediolaterally and ca. 14 mm anteroposteriorly, and the one on the left is ca. 15 mm mediolaterally and ca. 15 mm anteroposteriorly. In addition, there are several smaller distinct depressions. On the right side, there is a pit whose center is ca. 38 mm from the coronal suture and ca. 19 mm from the sagittal suture; it is 4.0 to 4.5 mm in diameter and 1.9 mm deep. There are also two smaller pits just anterior of the larger pit. On the left side, there is a pronounced pit ca. 7 mm in diameter and 2.5 mm deep, with its center ca. 75 mm from the coronal suture and ca. 13 mm from the sagittal suture. The larger Pacchionian depressions, in particular, are consistent with arachnoid granulation enlargement that is associated with midlife to advanced age, and they are not pathological in etiology.

Figure 8.34 Posterior view of Dolní Věstonice 11. Scale in centimeters.

Superior View

Approximating the length of the cranium by conjoining the Dolní Věstonice 11 and 12 elements produces a relatively long cranium (ca. 207 mm), and combining this with estimates of its maximum cranial breadth (ca. 135 mm) results in a very dolicocephalic cranium that lies closest to Pavlov 1 among the Pavlovian individuals and near the limits of the range across all comparative samples (Figure 8.28). In superior view (Figure 8.36), it also appears that the greatest width of the neurocranium in Dolní Věstonice 11 lies more in the middle portion of the neurocranium (as in Dolní Věstonice 3 and 14), rather than in the posterior third (as in Dolní Věstonice 13 and 15 and Pavlov 1). To some extent, this apparent position might be visually influenced by the lack of the most anterior aspect of the frontal region in Dolní Věstonice 11. Also, given the postmortem distortion and other sources of asymmetry in many of these neurocrania, some caution is required in interpreting subtle differences or similarities in these and other aspects of gross neurocranial shape.

Table 8.10 Cranial suture scoring as epigenetic traits for the Dolní Věstonice (DV) and Pavlov (Pav) sample.

Suture and Location	Criteria	DV 11	DV 13	DV 14	DV 15	DV 16	Pav 1
Coronal (bregmatic)	Maximum sutural shape extension	3, 4	—	4	—	—	3
	Basic configurations	2^1	—	2^1	—	—	2^1
	Secondary protrusions	2	—	2	—	—	1
Coronal (complicated)	Maximum sutural shape extension	4, 5	6	—	5	5	4
	Basic configurations	3^1	2^1	—	2^1	3^1	2^1
	Secondary protrusions	2	1, 2	—	1, 2	2	2
Coronal (temporal)	Maximum sutural shape extension	—	—	2	—	4, 5	—
	Basic configurations	—	—	1	—	3^d	—
	Secondary protrusions	—	—	1	—	4	—
Sagittal (bregmatic)	Maximum sutural shape extension	3, 4	—	4	3	4	5
	Basic configurations	3^1	—	2^1	2^1	2^1	2^1
	Secondary protrusions	2, 3	—	2	2	2	2
Sagittal (vertex)	Maximum sutural shape extension	6	—	5, 6	—	5, 6	5
	Basic configurations	2^1	—	2^1	—	2^1	2^1
	Secondary protrusions	2	—	2	—	2	2
Sagittal (obelic)	Maximum sutural shape extension	—	—	6	5	5, 6	—
	Basic configurations	—	—	2^1	2^1	2^1	—
	Secondary protrusions	—	—	2	2	2	—
Sagittal (lambdic)	Maximum sutural shape extension	5	—	—	5	6	5
	Basic configurations	2^1	—	—	2^1	2^1	3^1
	Secondary protrusions	2	—	—	2	2	2
Lambdoid (lambdic)	Maximum sutural shape extension	6	5, 6	4, 5	3	6	6
	Basic configurations	2^1	2^1	2^1	2^1	2^1	2^1
	Secondary protrusions	2	2	2	1, 2	2	2
Lambdoid (intermediate)	Maximum sutural shape extension	6	6	5, 6	5	5, 6	6
	Basic configurations	2^1	2^1	2^1	2^1	3^1	2^1
	Secondary protrusions	2	2	2	2	2, 3	2
Lambdoid (asteric)	Maximum sutural shape extension	6	5, 6	5, 6	—	6	6
	Basic configurations	2^1	2^1	3^1	—	2^1	2^1
	Secondary protrusions	2	2	2	—	2	2

Maximum sutural shape extension: 1: absent, 2: trace (<1 mm), 3: small (1–3 mm), 4: medium (3–6 mm), 5: large (6–10 mm), 6: excessive (>10 mm); basic configurations: 1: simple, 2^d: widely dentate, 2^1: widely looped, 3^d: narrow dentate, 3^1: narrow looped; secondary protrusions: 1: absent, 2: weakly expressed, 3: well expressed, 4: strongly expressed. See Hauser and De Stefano (1989: 90).

There is a modestly depressed area along the sagittal suture just posterior of the coronal suture on the right parietal. It measures 11.3 mm mediolaterally and 14.5 mm anteroposteriorly and exhibits distinct medial and anterior margins with less distinct posterior and lateral edges; it is of uncertain etiology (Chapter 19). There are no unambiguous parietal foramina on either side. There are four or five very small openings, but these are unlikely to be true parietal foramina by the criteria in Hauser & DeStefano (1989).

Dolní Věstonice 13 Cranium

Preservation

In Dolní Věstonice 13, the neurocranium and most of the splanchnocranium are preserved. The cranium is stained overall with red ochre and exhibits root etching that is most prominent on the frontal bone. The neurocranium consists of the frontal bone, the left and right parietal bones, the occipital bone, the left and right temporal bones, and portions of the sphenoid bone. It has been reconstructed from the largely intact frontal bone and large sections of the other neurocranial bones. Some of the joins between the bones and their constituent fragment elements are close fitting, whereas others are separated by up to 11.5 mm of insoluble filler between bone sutures. The reconstructed cranium exhibits a mild degree of deformational asymmetry because of taphonomic ground pressure distortions and difficulty in rejoining some cranial neurocranial elements; these are particularly evident in occipital and inferior views.

In occipital view, there is a mild inferior displacement of the right half of the neurocranium. In inferior view, the left temporal bone and greater wing of the sphenoid bone

The Cranial Remains

Figure 8.35 Endocranial view of Dolní Věstonice 11. Scale in centimeters.

views. In anterior view, the splanchnocranium is slightly skewed to the right side, which is evident in the ca. 10% smaller right orbital height (28 mm) than the left orbital height (31 mm). This skewing is also evident when the undistorted mandible is occluded onto the splanchnocranium; the left mandibular condyle, even when taking into account the temporal lateral misalignment, is inferiorly displaced from the left temporal glenoid fossa relative to the right side.

In lateral view, there remains a question as to the degree to which the splanchnocranium is anteriorly rotated along its contact hinge points, and consequently the facial angulation and the lower facial projection are artificially exaggerated. This can be observed in lateral view by comparing the obliquity of a line connecting the superior and inferior orbital margins relative to the Frankfurt Horizontal, so that the inferior orbital margin in the vicinity of zygoorbitale extends anteriorly relative to the superior orbital margin on both sides. This artificially induced lower facial prognathism can also be seen in basilar view by the rather large anteroposterior distance

are displaced too far laterally relative to the midline by ca. 10 mm and rotated slightly posteriorly. This is particularly evident in the greater distance of the left temporal mastoid process to the occipital condyle and the base of the left sphenoidal pterygoid fossa to that of the basioccipital midline. Moreover, it is confirmed by the lack of proper articulation of the undistorted mandible to the temporal glenoid fossae along the coronal transect. In frontal view, it is also evident that the right and particularly the left temporal squama are not well fitted to the articulating portion of the parietals along the squamosal suture, being displaced laterally by a maximum of ca. 5 mm on the left side.

The splanchnocranium consists of most of the right and left zygomatic bones, right and left maxillae, and a portion of the right palatine bone. The right and left infraorbital areas in the vicinity of the canine fossae are damaged and reconstructed with filler. However, enough of the peripheral morphology is preserved to reliably estimate the general form of the region. The splanchnocranium has been joined to the neurocranium along four contact points, the right and left frontomaxillary sutures and the right and left frontozygomatic sutures. The hafting of the splanchnocranium onto the neurocranium, using these four points, results in a misalignment that is clearly visible in anterior, basilar, lateral, and occipital

Figure 8.36 Superior view of Dolní Věstonice 11. Scale in centimeters.

of the inferior infratemporal fossa (52.0 mm) on the less distorted right side compared with the other large Dolní Věstonice specimens (Dolní Věstonice 16 right: 43.5 mm; left: 36.5 mm; Dolní Věstonice 14 left: 43.7 mm). Articulating the left mandibular condyle with the temporal glenoid fossa produces an incisor overbite of ca. 7 mm, along with posterior tooth nonocclusion, which is rectified by positioning the left condyle anteriorly onto the articular eminence. This further verifies the artificial anterior placement of the maxillary dentition and misaligned obliquity of the facial plane.

In posterior view, the mediolateral misalignment of the splanchnocranium is particularly evident by the incongruity of the dental arcade plane relative to that of the basicranium across the plane of the mastoid processes and occipital condyles. In anterior view, the same can be seen by comparing the plane of the internal nasal floor through the piriform aperture relative to the plane of the occipital condyles.

Anterior View

Frontal Region

The frontal bone is the most complete and undistorted bone in the cranium (Figure 8.37). To some degree, the root etching found throughout on this bone obscures very fine surface detail. Nonetheless, most external features are clearly preserved. Despite some postmortem damage along the coronal suture and infilling with reconstructive filler, it is clear that the coronal suture is largely completely open ectocranially (Chapter 6).

There is a slight hint of frontal keeling in midline that is apparent from an oblique (inferofrontal) view. Although weakly expressed, the keeling is especially evident in the area of metopion. There is no evidence of a metopic suture, and there are no frontal grooves on the lateral squamous portions (Table 8.3), although this is difficult to assess because of the extensive root etching. There is a partially obliterated zigzag-shaped supranasal suture. Approximately 35 mm lateral to metopion on either side are very weakly expressed frontal eminences. There is a surface irregularity located 40 mm superior to the right superomedial orbital region between the right frontal eminence and midline in the vicinity of metopion that appears to be an antemortem healed lesion (Chapter 19).

The least frontal breadth of Dolní Věstonice 13 (100 mm) falls between the highest value for the Dolní Věstonice males, 105 mm (Dolní Věstonice 14) and the lowest value, 97 mm (Dolní Věstonice 16), and the maximum frontal breadth for Dolní Věstonice 13 (120 mm) falls between the high value for the Pavlov 1 (122 mm) and the

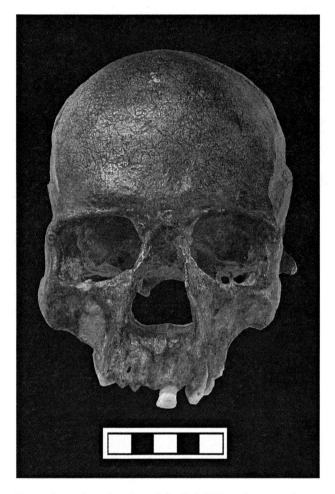

Figure 8.37 Anterior view of the Dolní Věstonice 13 cranium. Note the healed lesion between the right frontal eminence and the midline in the vicinity of metopion. Scale in centimeters.

lower value (118 mm) for Dolní Věstonice 16. The relationship between these two variables is very similar in the Pavlovian and European earlier Upper Paleolithic samples (Figure 8.2), with the Qafzeh-Skhul and Neandertal samples showing relatively greater least frontal breadth values.

Supraorbital Region

The supraorbital region of Dolní Věstonice 13 shows a distinct separation of medial, midorbital, and lateral components. The glabellar region is pronounced in midline, and this robust midline projection extends to the medial third of both right and left supraorbital areas (superciliary arches), each extending superiorly and laterally for ca. 19 mm prior to inflecting inferiorly and then fading out onto the midorbital regions. The total length of this medial component (encompassing glabella) for both sides combined is ca. 63 mm. The lateral-most supraorbital margins, the supraorbital trigone on either side, are both ca. 6 mm in thickness. They are separated from the medial superciliary arch by the obliquely flattened supraorbital sulcus, the latter of which is 3–4 mm in thickness on either side. The flattening of the supraorbital sulcus results in the "pinched" appearance that has been noted frequently in Upper Paleolithic early modern human specimens. The supraorbital sulcus is separated both mediolaterally and superoinferiorly from the superciliary arch and is visible as a distinct entity for virtually the entire length of the supraorbital margin. The supraorbital trigone is inflated and widened anteroposteriorly, but it retains a smooth lateral surface. The supraorbital region corresponds to Lahr's (1996) grade 3.

Orbital Region

Interorbital breadth in Dolní Věstonice 13 is slightly lower than the value for Dolní Věstonice 15, very similar to the value for Dolní Věstonice 16, and expectedly higher than that of Dolní Věstonice 3. When plotted against values for bifrontal breadth (Figure 8.3), their interorbital breadths are roughly similar to those found in the other comparative samples.

As noted above, there is a degree of asymmetry in the orbital region in Dolní Věstonice 13 (right orbital index: 65.1; left: 73.8) not seen in the rest of the Dolní Věstonice sample, and the right orbit is more square than the left one, both resulting from the present splanchnocranial hafting (Sládek et al., 2000). Nonetheless, even with this orbital shape asymmetry in Dolní Věstonice 13, it shares with the other Pavlovian individuals and modern specimens orbits that are lower relative to orbital breadth than those of Neandertals (Figure 8.4; Tables 8.2 and 8.4). The superior orbital margins in Dolní Věstonice 13 are rounded, not sharp (Chapter 7). The right superior orbital margin has one well-expressed, moderately sized nutrient foramen on the medial aspect. The left superior margin has one small notch present medially and one even smaller (5.0 mm) laterally from the first (Table 8.3).

Internally, the superior orbital roof is largely intact in both sides, and there is no indication of thinning or porosity. The superior ethmoidal suture is present on the left side, including the anterior ethmoidal foramen (Table 8.3), but this region is missing on the right side. Neither orbit retains any of the medial wall below the ethmoidal suture. The most posterior portion of the orbit is missing in both sides, although the left orbit retains a large portion of the greater wing of the sphenoid, almost to the level of the inferior orbital fissure. Both orbits preserve some of the lateral internal orbital surfaces contained on the zygomatic bones, and the left orbit retains a small portion of the orbital floor contained on the maxillary bone just lateral to the infraorbital groove and foramen.

The inferolateral borders of both orbits are robust in that they are rounded rather than sharp. However, the borders are also raised in relation to the floor of the orbits, and therefore they do not attain the most robust grade. They correspond most closely to Lahr's (1996) grade 2, her most frequent recent human form. Just inferior to the inferolateral corner of the orbit on the zygomatic bone, there is a small raised ridge, the zygomaxillary ridge (see the comment on terminology above), that corresponds to a grade 2 degree of robusticity development.

The midpoints of the lower orbital margins on both sides are damaged and have been filled. However, the two zygoorbitale points can be reliably located at the endpoints of the superior course of the zygomaxillary sutures. The breadth across these points relative to facial height (Figure 8.5) is similar across all of the samples, with Dolní Věstonice 13 falling roughly in the middle of the earlier Upper Paleolithic scatter. The inferomedial and medial orbital rims are present, even though both of the taphonomically fragile lacrimal bones are missing.

Infraorbital Region

The infraorbital regions on both sides are damaged in the precise area of the infraorbital foramen and have been reconstructed. Nonetheless, enough actual bone exists on the lateral and medial components adjacent to each area to accurately infer the topography of the infraorbital region. Both sides show a clear infraorbital depression that would have been coincident with the depressed infraorbital fossa. Inferior to where the infraorbital foramen would have been, there is a concavity that is similar to the more pronounced concavity found on Dolní Věstonice 15 (Figure 8.38); this feature is also found on the left side of Dolní Věstonice 14, where it is less pronounced than for Dolní Věstonice 15. Its pronounced expression on Dolní Věstonice 15 and lesser expression on Dolní Věstonice 13 and 14 might be evidence for a close genetic relationship among the members of the triple burial.

The infraorbital process of the zygomatic bone projects anteriorly over the depressed infraorbital region just lateral to the damaged and filled area, creating a horizontal shelf that further accentuates the infraorbital depression in a horizontal transect. The lateral aspect of the infraorbital region is oriented predominantly in the coronal plane on both sides, and the medial aspect adjacent to the piriform aperture is oriented in a parasagittal plane on either side. Therefore, in the transverse plane, the infraorbital region shows the typically modern human, two-plane configuration that is different from the horizontally flat or convex plane configuration usually

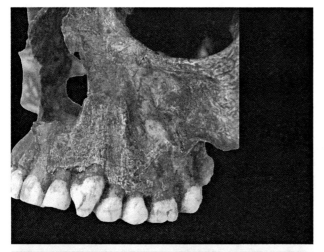
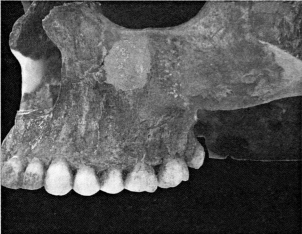
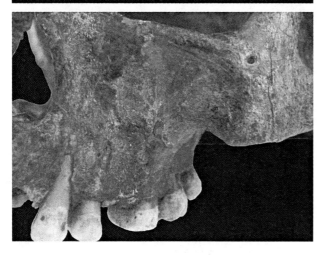

Figure 8.38 Infraorbital regions of Dolní Věstonice 13 (bottom), 14 (middle), and 15 (top). Dolní Věstonice 15's image is reversed to facilitate comparison. Note the deeply depressed infraorbital region, with a sharp vertical sill on the lateral aspect of the medial infraorbital facies in Dolní Věstonice 15 that gives way to a rather recessed posterior area just medial to where the canine fossa is usually found. Note also the more muted but similar morphology found in Dolní Věstonice 13 and 14.

found in Neandertals and some other archaic *Homo* crania (Table 8.2).

Nasal Region

The frontal process of the left maxilla is more complete than the right one and shows the entire course of the left nasomaxillary suture, although none of the nasal bones for either side is preserved. The more complete left-side frontal process exhibits a fairly strong degree of parasagittal eversion and a resulting moderately deep alpha point between the inferior nasomaxillary suture location and zygoorbitale (see lateral view). The distance between the alpha points relative to facial height is roughly comparable among the Dolní Věstonice and other earlier Upper Paleolithic modern humans, and they are also similar to those of the Neandertal sample (Figure 8.6). The area between the nasomaxillary suture and inferomedial orbital area is uniformly smooth, lacking the roughened furrow or slight elevation frequently found in both immature and mature archaic *Homo* (Franciscus, unpublished data). The orientation of the nasomaxillary suture is inferiorly positioned, rather than the horizontally inflected condition found in many Neandertals (Table 8.2).

Dolní Věstonice 13 has the highest nonestimated nasal breadth in the Dolní Věstonice sample (29.2 mm), its value nearly matches the large value for Mladeč 8 (30.0 mm), and it is within measurement rounding of the highest value recorded by Matiegka (1934) for the Předmostí sample (Předmostí 7: 29.0 mm). In contrast to its nasal breadth, the nasal height of Dolní Věstonice 13 (51.5 mm) lies well within the Dolní Věstonice distribution (Table 8.4; Figure 8.7; Sládek et al., 2000: table 26). Combining the Dolní Věstonice sample with the other earlier Upper Paleolithic individuals indicates that Dolní Věstonice 13 does have a wider than expected nasal breadth, although not so much as many Neandertals or even a couple of the Qafzeh-Skhul individuals.

The nasomaxillary suture is continuous with the completely preserved frontonasal suture on the frontal bone, the latter allowing the precise location of nasion. Nasion is superiorly set relative to the superior nasomaxillary points. The lateral margin of the piriform aperture is complete on the left side, and the right side is complete along its inferior two-thirds. The lateral walls of the nasal aperture are sharp for most of their course, but they become slightly thicker and duller in the inferolateral corners of the piriform aperture. There is no evidence of the inner crest/outer margin separation occasionally seen on the superior lateral margins of the piriform aperture in recent and fossil *Homo* (Franciscus, 2002). The internal lateral walls of the nasal fossa are complete from the piriform aperture margin posteriorly to the area bounding the nasolacrimal grooves on both sides, although the grooves themselves are either not present or are obstructed by filler. The ethmoidal crests are difficult to discern. However, the more inferiorly located conchal crests are well preserved, and both of them evince predominantly horizontally positioned anterior components and obliquely positioned posterior components. The conchal crests are visible in anterior view, although they do not markedly project medially into the nasal cavity. The turbinal crest on either side is also visible from anterior view inferior to the conchal crests, and it contributes to a degree of mediolateral thickening in this area that is often associated with turbinal crest development in extant and fossil *Homo*.

The narial margin at the junction of the internal nasal floor and subnasoalveolar clivus is composed of two crests (Figure 8.39). The turbinal crests on either side are fused with the spinal crests, and they are separated from the lateral crest by a maximum of 5 mm in the inferolateral corner of the piriform aperture (category 3, following De Villiers, 1968; Franciscus, 1995, 2003b). It is the fused turbinal and spinal crest, and not the lateral crest, that demarcates the internal nasal floor from the subnasoalveolar clivus. The lateral crest on the right side extends onto the subnasoalveolar clivus but does not extend further than the roots of the lateral incisor (category 1, following De Villiers, 1968; Franciscus, 1995). On the left side, the lateral crest appears to swing back up from the lateral incisor root position to partially fuse with the combined spinal and turbinal crest in the vicinity of the anterior nasal spine. However, the medial portion of the crest is diffuse and barely discernable. The area encompassed by these two lines is shallow and does not reach the depth or clarity associated with a true triangular *fossa intranasalis* found in the category 5 narial margin pattern (De Villiers, 1968) and characteristically found in Neandertals (Franciscus, 1995, 2003b).

The anterior nasal spine is well developed (category 4 or marked; Franciscus, 1995). It is likely that the very tip of the anterior nasal spine is broken and that it would have originally attained the most projecting category designation of 5. This is further supported by the pronounced development of spatulate lateral shelving on either side of the anterior nasal spine, which frequently accompanies the pronounced overall development found in category 5.

The internal nasal floor is preserved from the posterior narial margin and incisive crest to the palatomaxillary suture. The floor is completely level with the narial margin as coded in Franciscus (2003b), contrasting with the last glacial Neandertal pattern. The right maxilla preserves the posteroinferior portion of the maxillary sinus, measuring approximately 19 mm in mediolateral breadth. The fused root complex of the right third molar can be seen extruding into the sinus by several millimeters.

Subnasoalveolar Clivus

The subnasoalveolar clivus in Dolní Věstonice 13 is well preserved. Its height appears to be relatively short, although there is no evidence for substantial alveolar resorption (Chapter 19). There is a small 2.5 by 1.5 mm fenestration superior to the right lateral incisor alveolus that is postmortem damage. The overall shape of the anterior aspect of the clivus in Dolní Věstonice 13 is parabolic; it is not as rounded as in Dolní Věstonice 14 or as flattened as in Dolní Věstonice 15. The anterior aspect of the clivus in Dolní Věstonice 13 also exhibits prominent canine eminences, and to a lesser degree this is also true for the medial incisor eminences.

Overall Facial Size and Proportion

Overall facial height and breadth in Dolní Věstonice 13 generally lie in the middle to upper end of the distributions for both measures among the Dolní Věstonice sample (Figures 8.3, 8.5, & 8.6; Sládek et al., 2000: table 23).

Left Lateral View

Frontal Region

In left lateral view (Figure 8.40) the frontal bone of the Dolní Věstonice 13 shows relatively steep frontal inclination with marked projection at metopion. In this, it is similar to Dolní Věstonice 14–16, with Dolní Věstonice 3 showing a higher frontal subtense relative to the chord, and Pavlov 1 showing the relatively lowest value (Figure 8.10; Sládek et al., 2000: table 21), all of them contrasting with the lower values found among Neandertals (Table 8.1).

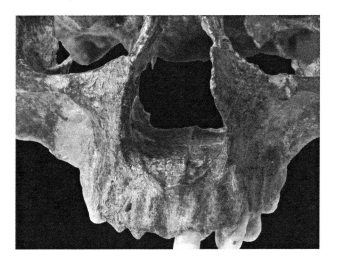

Figure 8.39 Lower border of the piriform aperture in Dolní Věstonice 13.

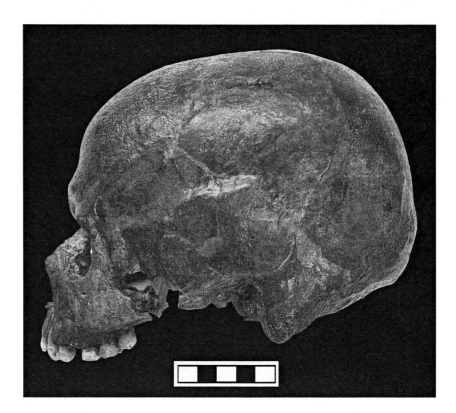

Figure 8.40 Lateral (left) view of the Dolní Věstonice 13 cranium. Note the postmortem break, along the superior temporal line on the parietal posterior to the coronal suture. Scale in centimeters.

The glabellar region is markedly projecting, with both a supraglabellar inflection and an infraglabellar notch. The supraglabellar inflection in Dolní Věstonice 13 is not the pronounced supraglabellar sulcus found in Neandertals and other late archaic humans, but it is similar to that found in other earlier Upper Paleolithic specimens (see anterior view description above). The degree of infraglabellar notch development in Dolní Věstonice 13, with allowance for missing nasal bone articulation with nasion, most closely approximates Lahr's (1996) category 2 or 3 and is similar to the pattern found in all of the Dolní Věstonice and Pavlov individuals, other Upper Paleolithic individuals, and to a large degree the Neandertal sample. The greatest degree of infraglabellar notch development is found in Skhul 4 and 5 and Mladeč 5, which appear as modest outliers (Figure 8.11).

The distinct separation of medial, midorbital, and lateral components, described above in anterior view, can be seen in lateral view, including the distinct separation of the supraorbital trigone from the superciliary arch by the obliquely flattened supraorbital sulcus and the inflated and anteroposteriorly wide supraorbital trigone. The anterior portion of the temporal line on the frontal bone is well marked and rugose.

Parietal Region

In left lateral view, the vertex of the cranium is located approximately 37 mm posterior to bregma. There is a relatively acute angulation between the midline superior neurocranial profile and the steep posterior parietal plane in Dolní Věstonice 13, which is reflected in its bregma-lambda chord subtense. Dolní Věstonice 13, along with the Dolní Věstonice and Pavlov specimens and other modern humans, shows absolutely higher values for the chord and subtense than Neandertals, but the samples are proportionately similar (Figure 8.12; Table 8.1; Sládek et al., 2000: table 21). Dolní Věstonice 13, however, has one of the highest relative subtense values compared to both the Neandertals and other earlier Upper Paleolithic specimens.

The left parietal bone has been reconstructed from eleven major pieces and, as presently reconstructed, produces an unnatural angulation at approximately mid-parietal between the lateral and posterior aspects relative to its opposite side in the area of the parietal tuber. The rugose anterior portion of the temporal line found on the frontal fades and becomes less marked on its continuation along the anterior and posterior parietal. It is not possible to clearly discern the distinction between the superior and inferior temporal lines in their superoposterior parietal portions, the location where they usually become most separated. The temporal line, however, does become more robust in its most inferoposterior segment before crossing to the temporal bone.

There is a large postmortem break just posterior to the left coronal suture and midway between the sagittal and squamosal suture. The break cleaves through the outer table and diploë, and it barely penetrates the inner table as minute fenestra. This break and a similarly damaged area located in almost the same area on the other side of this cranium are conspicuous in both taphonomy and preparation, and they are discussed further in Chapter 19.

Pterion Region

The pterion region has been heavily reconstructed, with a substantial amount of filler on the left side. Although elements of all four bones are present, no useful morphological information can be gleaned from this region.

Temporal Region

The temporal bone has been reconstructed from at least six relatively large pieces. These are confined to the temporal squama; the mastoid process, zygomatic process, and suprameatal crest region are all intact. The fit among the other pieces is accurate, but the fit of the reconstructed temporal onto the parietal has resulted in the somewhat laterally displaced position mentioned above. This poor fit and the slight skew of the attached splanchnocranium result in a misalignment of the anterior and superior portions of the zygomatic arch in the mediolateral plane and, to a lesser degree, in the superoinferior plane. There is a 15 mm gap between the preserved portions of the zygomatic arch that is partly due to improper alignment and partly due to missing bone. The fit of the temporal bone with the greater wing of the sphenoid is arguably more accurate.

The temporal bone is robust with a well-developed suprameatal crest that continues from the zygomatic process and extends well beyond porion posterosuperiorly to meet up with the posterior aspect of the temporal line on the parietal. There is a small, crest-type suprameatal spine with a very small depression (Table 8.5; Hauser & DeStefano, 1989: figure 31 closest to c), and there is no evidence of a squamomastoid suture. The mastoid process is large, measuring 28.8 mm in oblique length from porion relative to the Frankfurt Horizontal line, 16.9 mm in oblique mediolateral width, and 31.7 mm in anteroposterior width from anterior extent to mastoid foramen orthogonal to the Frankfurt Horizontal line. A trace tubercle is present on the mastoid process; however, it is not anteriorly positioned on the mastoid, as in many Neandertals, but rather in the middle of the process anteroposteriorly. The mastoid process extends well below the plane of the juxtamastoid eminence and shows no lateral flattening. In all of these features, the mastoid process of Dolní Věstonice 13 is modern, rather than Neandertal-like (Table 8.5). In addition to its large size, the mastoid process is rugose, especially in its inferoposterior portion, and there is a single medium-sized mastoid foramen that is filled with matrix. As in Dolní Věstonice 14 and 15 and Pavlov 1, it is positioned so that it is more visible in posterior than lateral view.

Occipital Region

In left lateral view, Dolní Věstonice 13 shows virtually no occipital bun development. Inferiorly, there is a slight projection of the external occipital protuberance and the superior nuchal line, creating a slight inflexion relative to the nuchal plane, but the superior projection near lambda is almost imperceptible. The occipital chord and subtense configuration is similar in Dolní Věstonice 13 to the other Pavlovian males, with moderately large values for each, but Dolní Věstonice 3 has considerably smaller values (Figure 8.13; Table 8.1; Sládek et al., 2000: table 21).

Zygomatic Region

The left zygomatic is well preserved. There is an oblique vertical break along the body that leads to a filled-in area on the anterolateral surface near the zygomatic foramen, but the reconstruction has produced an undistorted and accurate configuration of the zygomatic bone. There is some damage along the frontozygomatic suture, but the join is fairly accurate. The join to the maxilla along the zygomaxillary suture involves a fair amount of filler, but the result shows no unusual distortions or asymmetry. The sutural edge for the zygomatic process of the temporal is intact, although as noted above, its mediolateral alignment with the zygomatic process of the temporal is incorrect, and there is a 15.0 mm gap between the two. There is a small elevated ridge just inferior to the inferolateral corner of the orbit, referred to here as the zygomaxillary ridge, that corresponds to a grade 2 degree of robusticity development (Lahr, 1996). There is a single zygomaticofacial foramen of medium size (Table 8.6), and there is weak development of a marginal tubercle that is 13.3 mm in maximum anteroposterior length (Hauser & DeStefano, 1989: figure 36, most similar to b).

Orbital Region

In lateral view of the orbital region, the lateral margin of the orbit in Dolní Věstonice 13 is posteriorly positioned relative to nasion and other midfacial measures, indicating some degree of midfacial projection. In this feature, it is similar to that of the other Dolní Věstonice and Pavlov males, although it is not as pronounced as that seen in Neandertals on average (Figure 8.14; Table 8.2). From this perspective, the degree to which the infraorbital margin projects further anteriorly than the supraorbital margin in the area of the supraorbital sulcus is particularly apparent. This forms an oblique alignment of the orbital borders in the sagittal plane that is in part due to an overrotation of the splanchnocranium anteriorly along its contact hinge points so that the degree of facial angulation and lower facial projection is artificially exaggerated (see the general description above). Nonetheless, the lateral placement of the lateral orbital border would be relatively unaffected by this misalignment.

Infraorbital Region

The left infraorbital region, despite its damage and reconstruction in the infraorbital foramen area, provides reliable information on its topography. There is a clear infraorbital depression on Dolní Věstonice 13 that would have accompanied a depressed infraorbital fossa. As noted in anterior view, the infraorbital process of the zygomatic bone projects anteriorly over the depressed infraorbital region, creating a horizontal shelf that further accentuates the infraorbital depression. In lateral view, it can again be seen that the lateral aspect of the infraorbital region is oriented predominantly in the coronal plane, whereas the medial aspect adjacent to the piriform aperture is oriented in a parasagittal plane, resulting in the typically modern two-plane configuration. Moreover, from this view, it is also evident that in a sagittal plane the lateral aspect of the infraorbital region is obliquely oriented so that the inferior margin of the zygomatic process of the maxilla is posteriorly positioned relative to the inferior orbital. This oblique inflection of the infraorbital region is also a feature of modern humans, in contrast to the vertical or oblique extension of the infraorbital plane that is found in archaic *Homo*. The anterior root of the zygomatic process of the maxilla is located at the anterior edge of the first molar in Dolní Věstonice 13 rather than more posteriorly, as in the Neandertals (Table 8.2).

Nasal Region

In the nasal region, there is a break along the frontal process of the left maxilla that has been refit, which faithfully captures the original form of the frontal process and shows the entire course of the left nasomaxillary suture. As noted earlier, there is a fairly strong degree of parasagittal eversion and a resulting moderately deep alpha point between the inferior nasomaxillary suture location and zygoorbitale. The degree of eversion and alpha point subtense depth in Dolní Věstonice 13 is slightly more pronounced than in Dolní Věstonice 3 and 16, although not so pronounced as in the more extreme values for Dolní Věstonice 14 and 15 (Figure 8.15). It can be seen particularly well in this view that the orientation of the nasomaxillary suture relative to the Frankfurt Horizontal is inferiorly, rather than horizontally, inflected, as is true for non-Neandertal extinct and extant *Homo* (Table 8.2). The lateral margin of the piriform aperture is complete on the left side, and the marked projection of the anterior nasal spine can be seen in lateral view at the level of the narial margin, along with its lateral spatulate development (category 4 or marked; Franciscus, 1995; see also Figure 8.39). As noted above, it is likely that the very tip of the anterior nasal spine is broken and that it may have originally attained the most projecting category designation (5).

Subnasoalveolar Clivus and Alveolar Region

There is a clear inflection at subspinale in lateral view that is somewhat obscured by the pronounced left medial incisor eminence. The clivus also exhibits a mild degree of lower prognathism relative to the other Pavlovian individuals, although well within the range for individuals among all of the comparative samples (Figure 8.16). A comparison of the lower facial projection at prosthion and the upper facial projection at nasion (Figure 8.17) provides some additional confirmation of the overrotation of the splanchnocranium in Dolní Věstonice 13 anteriorly along its contact hinge points, such that its lower facial projection is artificially exaggerated (see above).

Overall Size and Proportion in Lateral View

In lateral view, both the height and the length of the neurocranium in Dolní Věstonice 13 are smaller than those of the other Dolní Věstonice males (Dolní Věstonice 14 and 16) for which complete measurements are available but slightly larger than the Dolní Věstonice 3 female (Figure 8.18; Sládek et al., 2000: tables 18 and 19). The relative shorter lengths of the Dolní Věstonice sample and other earlier Upper Paleolithic individuals result in the characteristically modern human globular shape (Table 8.1).

Overall facial height and facial length in Dolní Věstonice 13 is large among the Dolní Věstonice individuals, being surpassed only by Dolní Věstonice 16 (Figures 8.19 and 8.20; Sládek et al., 2000: tables 22 and 24). Both Dolní Věstonice 13 and 16 lie toward the upper end of the range for early modern humans in general for these measures. Neandertals, not surprisingly, have larger values for facial height and length, especially when sorted by sex.

Right Lateral View

Gross aspects of neurocranial and splanchnocranial morphology that are evident from either side are covered in the preceding section in left lateral view. The following additional features are revealed in right lateral view (Figure 8.41).

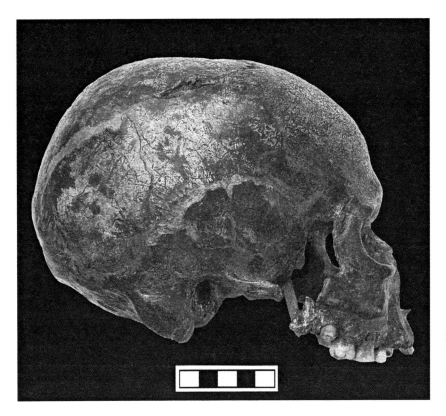

Figure 8.41 Lateral (right) view of the Dolní Věstonice 13 cranium. Note the postmortem break, along the superior temporal line on the parietal posterior to the coronal suture. Scale in centimeters.

Frontal Region

There is a minor post-bregmatic depression evident in right lateral view of the frontal region that is partly exaggerated by a pathological lesion detailed elsewhere (see superior view and Chapter 19). The right parietal has been reconstructed from far fewer pieces than the left side (three versus eleven pieces, respectively), and this results in a more accurate reconstruction of the morphology of the parietal tuber. As with the left side, the rugose anterior portion of the temporal line found on the frontal fades and becomes less marked on its continuation along the anterior and posterior parietal. It is very difficult to discern the distinction between the superior and inferior temporal lines, which usually become most separated in their superoposterior parietal portions. The temporal line, however, does become more robust in its most inferoposterior segment before crossing to the temporal bone.

There is a large postmortem break beginning about 23 mm posterior to the coronal suture and about midway between the sagittal and squamosal suture (Chapter 19).

Pterion Region

As with the left side, the right pterion region has been heavily reconstructed with filler, and it is not possible to decipher the configuration of this region.

Temporal Region

The right temporal bone has been reconstructed from at least four relatively large pieces. These are confined to the temporal squama; the mastoid process, zygomatic process, and suprameatal crest region are all intact. The fit among the other pieces is quite good; relative to the left side, the fit of the right temporal bone onto the parietal bone is close to its original configuration, despite consolidant fill along the squamosal suture. Thus the whole right temporal region is considerably better aligned with its adjacent morphology. There is a 17 mm gap between the preserved portions of the zygomatic arch, which is partly due to slightly anteriorly rotated splanchnocranial pivot at the frontal bone connections, as described above, and partly due to missing bone along the right zygomatic arch. Despite some consolidant, the fit of the temporal with the greater wing of the sphenoid appears to be accurate. Like its counterpart, the right temporal bone is robust, with a well-developed suprameatal crest that continues from the zygomatic process and extends well beyond porion posterosuperiorly to meet up with the posterior aspect of the temporal line on the parietal. As with the left side, there is a small, crest-type suprameatal spine with a very small depression (Table 8.6; Hauser & DeStefano, 1989: figure 31, closest to c), and there no evidence of a squamomastoid suture.

The mastoid process is large (30 mm in oblique length from porion relative to the Frankfurt Horizontal line,

16 mm in oblique mediolateral width, and 30 mm in anteroposterior width from anterior extent to mastoid foramen orthogonal to the Frankfurt Horizontal line). As on the left side, a trace tubercle is present on the right mastoid process that is not anteriorly positioned, as in many Neandertals, but rather in the middle of the process anteroposteriorly. The mastoid process also extends well below the plane of the juxtamastoid eminence. However, unlike its left counterpart, it shows mild flattening (Table 8.5). In addition to its large size, the mastoid process is rugose, especially in its inferoposterior portion.

Occipital Region

In right lateral view, Dolní Věstonice 13 shows virtually no occipital bun development. Compared to the left lateral view, there is a similar degree of projection of the external occipital protuberance and the superior nuchal line, creating a slight inflexion relative to the nuchal plane. The right lateral view shows a slightly greater superior projection near lambda than the left one. However, relative to the marked projection in some other specimens (e.g., Pavlov 1 and Dolní Věstonice 15 and 16), it is almost imperceptible.

Zygomatic Region

The right zygomatic bone is well preserved. There is some damage along the frontozygomatic suture, and the join to the frontal bone is not precisely aligned. Unlike the left side, the join of the right zygomatic to the maxilla along the zygomaxillary suture involves bone contact rather than filler. The sutural edge for the right zygomatic process of the temporal is intact, although its mediolateral alignment with the zygomatic process of the temporal is more correctly aligned than on the left side. There is a 17.0 mm anteroposterior gap between the two. As with the left side, there is a small elevated ridge just inferior to the inferolateral corner of the orbit, the zygomaxillary ridge, that corresponds to a grade 2 degree of robusticity development (Lahr, 1996). A single zygomaticofacial foramen of medium size is present on the body (Table 8.6) and medium to strong development of a marginal tubercle that is 14.3 mm in maximum anteroposterior length (Hauser & DeStefano, 1989: figure 36, most similar to c/d).

Orbital Region

In right lateral view, the position of the lateral margin of the orbit relative to nasion, details of infraorbital and supraorbital margin alignment, and all aspects of facial hafting are essentially identical to that apparent in left lateral view.

Infraorbital Region

The topography of the right-side infraorbital region is virtually identical in all salient aspects as that found on the left side in lateral view.

Nasal Region

In the nasal region, the right-side frontal process of the maxilla is less complete than that on the left side. Nonetheless, the orientation of the nasomaxillary suture, the degree of parasagittal eversion of the frontal process, and the associated alpha point position are very similar to that detailed for the left side. In addition, the lateral margin of the piriform aperture is complete on the right side, and the marked projection of the anterior nasal spine with associated lateral spatulate development found on the left side is apparent from the right side as well.

Subnasoalveolar Clivus and Alveolar Region

The right lateral view of the subnasoalveolar clivus and alveolar region is virtually identical in all salient features on the left side of this region.

Inferior View

Occipital Region

The occipital region (Figure 8.42) has been reconstructed with filler in some areas, including a large area on the right-side medial to the mastoid process and across to the foramen magnum. Large portions of the left side of the foramen magnum have been reconstructed with filler. The alignment of the foramen magnum is clearly off center, such that it is oriented in an oblique direction with the anterior portion deviating toward the right side. Whether this is due only to the restoration or is part of a pattern of asymmetry in the occipital condyles (see below) is not clear. The breadth of the foramen magnum in Dolní Věstonice 13 is narrower than in the other Dolní Věstonice individuals that preserve it, and it has the narrowest value for all comparative samples (Figure 8.23; Sládek et al., 2000: tables 18 and 19). This gives it the superficial appearance of being elongated (Table 8.7), even though this is not the case.

A mildly developed median occipital crest posterior to the foramen magnum is visible, and its orientation is in correct alignment. The occipital crest spans the rather poorly developed inferior nuchal lines. Its anterior portion is fairly sharp and more prominently developed than the portion posterior to the inferior nuchal line; the development of the occipital crest most closely approximates category 2 in Lahr (1996). A fairly well-developed

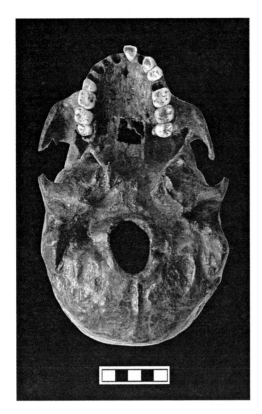

Figure 8.42 Inferior View of the Dolní Věstonice 13 cranium. Scale in centimeters.

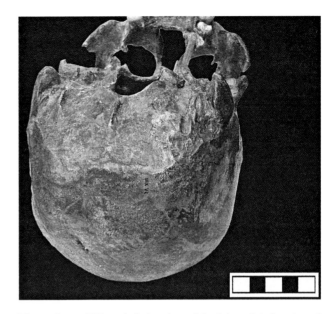

Figure 8.43 Oblique inferior view of the left occipital region of Dolní Věstonice 13. Note the raised area of new bone associated with a minor injury along the superior nuchal line just posterior of the left mastoid process. Scale in centimeters.

superior nuchal line is evident for its entire course along the occipital, and in midline it joins at the *tuberculum linearum* (Hublin, 1978). The lateral-most portions of the superior nuchal lines closest to the posterior mastoid process are particularly prominent. Although damaged somewhat on the right side, these prominent portions appear to be bilaterally symmetrical (Figures 8.43 and 8.44). On the less heavily reconstructed left side, there is a very pronounced ridge that separates the rectus capitis posterior major from the superior oblique muscle attachments. On both left and right sides, the scarring for semispinalis capitis is visible, although it is not pronounced.

The occipital condyles are well preserved and show both size and shape asymmetry (Chapter 19). The area just lateral to the condyle on the right side in the region of the rectus capitis lateralis insertion is damaged and largely missing. This same area lateral to the left side occipital condyle appears to have been reconstructed. The hypoglossal canals and the posterior condylar fossae have been packed with reconstructive filler. There appears to be complete bridging on the right-side hypoglossal canal (Hauser & DeStefano, 1989: Figure 18, corresponding to 5), with one large orifice laterally and one small orifice medially (Table 8.8). The left-side hypoglossal canal is too obscured with matrix to score. On the left side there appears to be weak expression of unfused medial and lateral paracondylar processes, although it is unclear whether the lateral process might be a part of the juxtamastoid eminence. There also appears to be a weakly developed paracondylar process on the right side, although it may also constitute a part of the juxtamastoid eminence (Table 8.8).

Figure 8.44 Oblique inferior view of the right occipital region of Dolní Věstonice 13. Scale in centimeters.

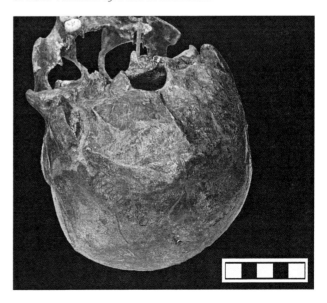

Anterior of the foramen magnum on the inferior basilar surface, a weakly expressed pharyngeal tubercle is present (Hauser & DeStefano, 1989: figure 23, a), and slightly more anteriorly a pharyngeal fovea with a rhombic-shaped margin and shallow tub-shaped bottom is visible (see Table 8.8; Hauser & DeStefano 1989: p. 138, methodology). Lateral to the pharyngeal tubercle, there are well-developed attachment protuberances for longus capitis on both sides. Just anterior to these, the faint line for the fused sphenooccipital synchondrosis can be observed (Chapter 6), especially on the less damaged left aspect.

Sphenoid Region

Just anterior of the fused sphenooccipital synchondrosis, there is a small area of elevated bone that is either the posterior alae of the vomer or a small elevation on the inferior clivus that often occurs just posterior to the vomeral alae in the region near hormion. The sphenoid anterior to this point is missing. Lateral to this point on the right side, the base of the pterygoid process is preserved with the basal portions of the lateral and medial pterygoid laminae (there is a wooden support rod glued from the pterygoid fossa to the left inferior lateral pterygoid lamina); the medial one is better preserved, attaining a height of ca. 12 mm. The scaphoid fossa is present, and lateral to this point the outline for foramen ovale is present, although it is filled in.

The sphenoidal elements on the left side are reconstructed further from midline than on the right by at least 5 mm, as are the articulated left temporal elements. This can be verified by the inability to articulate the left condyle and glenoid fossa when the right mandibular condyle and glenoid fossa are articulated. The base of the pterygoid process is preserved, although less completely than on the right side. The lateral pterygoid lamina is slightly better preserved on the left. The scaphoid fossa is present, and lateral to this point the outlines for both the foramen ovale and the foramen spinosum are clearly present, although filled in. There is no evidence for a foramen of Vesalius on either side (Table 8.8).

Anterior to the pterygoid process on the left side, the greater wing of the sphenoid is preserved superolaterally on its temporal surface to a point near pterion, a region poorly preserved on this cranium. Medially, this part of the left side of the sphenoid also retains a well-developed infratemporal crest, and anteriorly a portion of the orbital surface described in frontal view is also preserved. The right side of the sphenoid, although better aligned with respect to midline than the left, is nonetheless less complete, barely attaining the anterior extent of the infratemporal crest. Small portions of the greater wing of the sphenoid remain.

Temporal Region

The inferior aspects of the temporal region are well preserved on both sides. As noted above, the left temporal bone is reconstructed 5 mm or more too far laterally from midline. Both temporal bones show an occipital groove just medial to the mastoid processes. The left temporal retains an 11 mm length of styloid process, although its original length cannot be determined. A stylomastoid foramen cannot be seen, although it may be filled in. The left-side styloid process is missing, although its base is present. The areas immediately medial to both styloid processes posterior to the tympanic region are damaged and infilled. The posterior glenoid eminences and medial tympanic portions are preserved on both sides.

The glenoid fossae are well preserved, and they show no signs of remodeling. The right glenoid fossa has remnants of dried glue that was at some point used to fix the mandible into articulation on this side. When the condyle was removed a small (7.5 mm) slip of bone detached from the right lateral anterior condyle of the mandible and remains glued to the anterior glenoid eminence. It superficially appears to resemble degenerative remodeling, but it is clearly the result of removing the glued mandible.

The roots of the zygomatic processes extend for 5 to 6 mm beyond the plane of the anterior-most part of the anterior glenoid eminences on either side; neither of these show any roughening of the inferior surface, indicating that the most posterior fibers of the origin for masseter are located further anteriorly. The alignment of the two portions of the zygomatic arch on the right side is fairly good, with about 16 mm of missing bone in the noncontact zone. The left-side zygomatic arch alignment, however, is poor because of the lateral displacement of the temporal region, as noted before; the zygomatic process of the temporal is positioned 11.5 mm lateral to the temporal process of the zygomatic bone. The right-side anterior infratemporal fossa region is accurately positioned in the mediolateral direction relative to the posterior portion, but it is too far anteriorly positioned by at least 4 mm, based on the mandible. In addition to the marked mediolateral misalignment on the left-side infratemporal fossa, the left-side anterior infratemporal fossa is also slightly too far anterior. Given the misalignment of the temporal bone noted earlier, the combined result is that when properly occluded, the left mandibular condyle is ca. 18 mm anterior to the glenoid fossa, resting well anterior to the anterior eminence.

The attachment areas for the masseter muscle are moderately developed, less rugose than for Dolní Věstonice 16, and extend anteriorly to the point where the lateral and anterior aspect of the zygomatic region inflects in the acute angle near zygomaxillare, characteristic of modern humans. The masseter muscle attachment does not extend onto the inferior maxillary surface. Posteriorly, given the incompleteness of the zygomatic arches, it cannot be determined whether the muscle scarring extends beyond the base of the zygomaticotemporal suture on either side or remains anterior to this point, as in Dolní Věstonice 16. The right anterior vertical wall of the temporal fossa is slightly more complete than the left side.

Palatal Region

The palate of Dolní Věstonice 13 is complete anteriorly, with a moderate- to large-sized, nonbridged incisive fossa. The intermaxillary suture area is reconstructed with filler, and the palate lateral to each side of the intermaxillary suture is complete on both sides to the approximate area of the palatomaxillary suture. Except for a small posterolateral portion on the right side, including the greater palatine foramen, the palatine bones are missing. The palate is moderately arched, and there is no evidence for midpalatal (intermaxillary) or alveolar tori (Table 8.8). The lateral portions of the palate in the molar region show some spiculation and relief, indicating that some minor palatine bridging was possibly present. The overall dimensions of the Dolní Věstonice palate are among the largest for the Pavlovian sample (Figure 8.24; Sládek et al., 2000: table 27), although the Pavlovian sample as a whole has smaller palates overall than the Qafzeh-Skhul and Neandertal samples (Figure 8.24; Table 8.4).

Inferior Frontal View

The inferior frontal has relatively complete orbital plates on both sides. Medially, the nasofrontal suture and a portion of the frontal nasal spine are present. The area just lateral to where the superior-most roofs of ethmoidal cells (the ethmoidal notch) would be is also present, although there are no remains of the ethmoid itself. The posterior-most portion of the ethmoidal notch is covered over with filler, but the anterior-most area is open, showing both the internal calvarium and the openings into the frontal sinuses. The left frontal sinus opening shows expansion into midline near the area of glabella but no apparent expansion laterally on the left side beyond the glabellar area. The right frontal sinus opening appears smaller and is filled in with matrix that obscures further visual examination. Radiographic examination (Vlček, 1991: plate 5 and figure 53) confirms these observations; there is a relatively small frontal sinus delimited to the left side only that does not expand beyond the glabellar region. Interestingly, all three members of the triple burial show this left-side delimited frontal sinus pattern (see below). A single supraorbital foramen is visible on the left internal orbital surface associated with the same feature that is evident externally from anterior view. From inferior view, the left superior margin has one small notch medially and one even smaller 5 mm laterally from the first, also visible in anterior view. There is no obvious porosity on the superior internal orbital margins.

Posterior View

In posterior view (Figure 8.45), the posterior portion of the sagittal suture is largely filled in with consolidant, although lambda can be determined by using the confluence with the lambdoidal suture. The left lambdoidal suture is relatively better preserved than the right one, although both are visible along their course. Detailed scoring results for the lambdoidal suture are presented in Table 8.10. The left posterior parietal has been reconstructed from a few large pieces, and the fit in the left tuber region is misaligned, thus slightly exaggerating the prominence of the tuber. There is also a slightly greater posterior projection or prominence in the area of the right superior lambdoidal suture than in the left one that is also a result of reconstruction. The lateral parietal and temporal squama are straight-sided (right side) and superolaterally inclined (left side). Together with the parietal tubers and the posterior vertex, these form the typical pentagonal configuration that is associated with modern humans.

Consistent with the pentagonal shape, the greatest breadth across the parietals is relatively superiorly positioned. Maximum cranial breadth for Dolní Věstonice 13 in posterior view is relatively wide, whereas its cranial height is relatively low compared to the other Dolní

Figure 8.45 Posterior view of the Dolní Věstonice 13 cranium. Scale in centimeters.

Věstonice and Pavlov individuals (Sládek et al., 2000: table 19). This makes Dolní Věstonice 13 somewhat more similar to the Qafzeh-Skhul sample in its neurocranial shape in posterior view and therefore somewhat intermediate between other modern humans and Neandertals in this regard (Figure 8.26). Moderate sagittal keeling is present in Dolní Věstonice 13 (category 2 in Lahr, 1996).

Inferiorly, the posterior view shows the disparity in mediolateral distance between the foramen magnum/occipital condyle area and the mastoid processes. The distance on the left is too large, which is consistent with the misalignment detailed above. The superior nuchal lines are present on both sides and meet in midline at the external occipital protuberance. The supreme nuchal lines are visible, especially on the left side, coursing nearly 15 mm apart from the superior nuchal lines at their widest extent lateral to the midline, and narrowing to ca. 8 mm as they converge at the *tuberculum linearum*.

A 15.5 mm horizontally roughened line in the midline lies just above the most posterior extent of the occipital crest. This roughened line corresponds to the area that often encompasses an iniac depression above inion; however, this area is not as well defined as found on other specimens (e.g., Pavlov 1 and Dolní Věstonice 14–16). On some of the other Dolní Věstonice and Pavlov specimens (e.g., Dolní Věstonice 14–16 and, to a lesser extent, Dolní Věstonice 3 and Pavlov 1), the inferior depression at the external occipital protuberance is matched by an additional distinct depression (usually the larger of the two) that is more superiorly positioned just below lambda and well above the supreme nuchal line. In Dolní Věstonice 13, the superior depression is only vaguely defined and to some extent is affected by reconstruction, especially on the left side (Figure 8.44). As noted above, these depressions above inion are not the suprainiac fossae found in many Neandertals (Table 8.5).

The posterior view also shows the extension of both mastoid processes below the level of the more medially positioned juxtamastoid eminence, the usual modern human condition (Table 8.5).

Superior View

In superior view (Figure 8.46), the less damaged and more accurately reconstructed right side of the neurocranium is evident. This is particularly true in the area of the left parietal tuber, where the join along several breaks is uneven and somewhat distorted. Nonetheless, a reasonably accurate assessment of overall neurocranial form is possible.

The cranium is metrically dolicocephalic, although not extremely so. It lies at the upper limits of the range for maximum cranial breadth relative to length, not only among the Dolní Věstonice and Pavlov individuals (Sládek et al., 2000: tables 18 and 19), but also for the in-

Figure 8.46 Superior view of the Dolní Věstonice 13 cranium. Note the possible healed lesion on the anteromedial aspect of the right parietal bone near bregma. Scale in centimeters.

dividuals across all comparative samples (Figure 8.28). Its greatest width at the parietal tubers is located at the posterior third of the neurocranium. Commensurate with its large cranial capacity estimate of 1481 cc (Table 8.9; Vlček, 1997), there is no postorbital constriction. No unambiguous foramina are present; however, there is minor reconstruction with infilling along the sagittal suture that may have obliterated any foramina that were present. Detailed scoring results for the coronal and sagittal sutures are presented in Table 8.10.

There is a surface irregularity on the right parietal bone located ca. 14 mm lateral to bregma and ca. 12 mm posterior to the coronal suture that appears to be an antemortem healed lesion. This lesion and two enigmatic postmortem breaks, one on each lateral parietal posterior to the coronal suture, are discussed in Chapter 19. In oblique superior view, the occipital region shows a slight posterior asymmetrical projection on the right side along the lambdoidal suture, and then a slight posterior projection on the left side inferior to the lambdoidal suture. It is not clear to what extent these minor asymmetries are naturally occurring or the result of reconstruction. Finally, in oblique superior view, the distinction of the superciliary arch from both the frontal squama and the laterally positioned supraorbital trigone on either side by the obliquely flattened supraorbital sulcus, discussed in various views above, is particularly apparent.

Dolní Věstonice 14

Preservation

Dolní Věstonice 14 includes the neurocranium and much of the splanchnocranium, although there are large areas on both that were originally damaged and missing and have been restored with a consolidant. As with some of the other specimens, root etching is present, and filler and shellac obscure some of the fine surface anatomy.

The neurocranium consists of the frontal bone, the left and right parietal bones, the occipital bone, the left and right temporal bones, and small portions of the sphenoid bone. The neurocranium has been reconstructed from a number of large fragments, and the resulting restoration has several asymmetrical distortions. This is most obvious in anterior and posterior views, which show a marked prominence of the tuber region on the left parietal bone that has been virtually completely reconstituted. The right-side posterior parietal bone has also been reconstructed from several large pieces altered in situ by medially inflected ground pressure, which has resulted in deviation from its original form. In addition, other missing or damaged areas that have been reconstituted with filler include the central area of the rear vault, the infratemporal fossae, the right temporal fossa, and portions of the supraorbital region.

The splanchnocranium has preserved portions of the zygomatic bones, a fragment of right maxillary frontal process, and fragments of the left and right maxillae. The most damaged areas of the splanchnocranium are the infraorbital regions, particularly the right side, and the associated infraorbital margins. The fragment of the right zygomatic bone is not connected with other bones of the splanchnocranium, but it is well connected to the neurocranium. The connection between the splanchnocranium and the neurocranium is not well preserved, and it has been partly estimated. Articulation of the right mandibular condyle into its glenoid fossa produces a moderately good fit along the dentition. The occlusion of the right-side canines through to the second molars and the left-side first molars is good; and although there is mandibular dentition displacement to the right by 1 to 2 mm, facial alignment based on occlusion is probably close to the antemortem alignment.

Anterior View

Frontal Region

The frontal bone (Figure 8.47) is reconstructed from at least twelve major pieces with large areas of filler, and it shows marked postmortem deformation. The right frontolateral squama has been medially compressed, and the supraorbital region is off of coronal alignment, so the left side is more posteriorly positioned. There is root etching

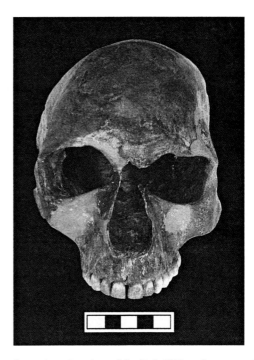

Figure 8.47 Anterior view of the Dolní Věstonice 14 cranium. Scale in centimeters.

on this bone, as well as filler and shellac. There is considerable damage along the coronal suture on the right side, although some details are visible on the left side.

There is no frontal keeling in midline. There is no evidence of a metopic suture, and there are no frontal grooves on the lateral squamous portions (Table 8.3), although root etching and reconstructive efforts preclude confirmation of the latter with certainty. The trace remnant of a supranasal suture is present, (Hauser & DeStefano, 1989; plate VI, h). Inferolaterally to metopion, ca. 28 mm to the left of midline on the left side, there is a very weakly expressed frontal eminence. There also appears to be one in the same general location on the right side, although the greater amount of deformation on this side affects the morphology.

The least frontal breadth of Dolní Věstonice 14 (105 mm) is the largest for the Dolní Věstonice and Pavlov males, indeed, falling in more with the Qafzeh-Skhul and Neandertal samples, but its maximum frontal breadth, on the other hand, falls well within the Pavlovian male distribution (Figure 8.2; Table 8.1). In this, Dolní Věstonice 14 shows the relatively greatest least frontal breadth, but this could well be due to a postmortem-induced reduction of its maximum frontal breadth value.

Supraorbital Region

The supraorbital region of Dolní Věstonice 14 is reconstructed in several places, including the area superiorly and superomedially to the right orbit and the superolateral portion of the right orbit. Fortunately, all portions of the supraorbital anatomy are present on at least one side.

The region shows a distinct separation of medial, midorbital, and lateral components. The glabellar region is pronounced in midline, and this robust midline projection extends to the medial third of both right and left supraorbital areas (superciliary arches), each extending superiorly and laterally for ca. 19 mm prior to inflecting inferiorly and then fading out onto the midorbital regions. The total estimated length of this medial component (encompassing glabella) for both sides combined is ca. 52.4 mm.

The supraorbital trigone on the preserved right side is ca. 5 mm in thickness, smaller than the other Pavlovian males. It is distinctly separated from the more medial superciliary arch by the obliquely flattened supraorbital sulcus, the latter of which is 2.6 mm in thickness on the right side. The supraorbital sulcus is distinctly separated both mediolaterally and superoinferiorly from the superciliary arch and is visible as a distinct entity for virtually the entire length of the supraorbital margin, although the degree of distinctness is less marked on this individual than for any other Pavlovian specimen. The supraorbital trigones are inflated and widened anteroposteriorly, but they retain a smooth lateral surface; they therefore correspond most closely to Lahr's (1996) grade 3. Dolní Věstonice 14, therefore, also has the distinct separation of medial, midorbital, and lateral components of the supraorbital region seen in all of the other Pavlovian individuals.

Orbital Region

The reconstruction of the orbital region in Dolní Věstonice 14 is problematic, especially on the left side. The right orbit has a large reconstructed area in the superomedial portion and along the inferior midorbital region. As reconstructed, the right orbit is narrow superoinferiorly and elongated mediolaterally with a square configuration. The left orbit is reconstructed in the superolateral and lateral aspects, as well as the mid-infraorbital region. In addition, the entire medial rim of the orbit is missing. The superior orbital margins are rounded (Chapter 7). The left orbit has a very large blurred, medially positioned supraorbital notch (larger than any shown in plate VIII in Hauser & DeStefano, 1989; Table 8.3). The right side is too damaged and reconstructed to evaluate this trait. Internally, very little of the superior orbital roof is preserved, and nothing is retained of the medial and inferomedial surfaces. Some lateral internal orbital wall is retained on the left side; however, it is almost completely covered with filler, and surface details are obscured.

The inferolateral borders of both orbits are retained but somewhat reconstructed in parts; they are robust in that they are rounded rather than sharp. However, the borders are also raised in relation to the floor of the orbits and do not attain the most robust grade; they are closest to Lahr's (1996) grade 2, the one most frequently found in her recent human sample. The area just inferior to the inferolateral corner of the orbit on both zygomatic bones is smooth, with no ridge development; this corresponds to a grade 1 degree of robusticity (Lahr, 1996). The midpoints of the lower orbital margins on both sides are damaged, have been filled, and they are not visible. The left zygoorbitale point can be estimated by extrapolating from the endpoints of the superior course of the zygomaxillary suture. The inferomedial and medial orbital rim is present posteriorly, near to the lacrimal groove on the right side; however, the taphonomically fragile lacrimal bone is missing. Nothing of the left medial orbit is preserved.

Infraorbital Region

The infraorbital region on both sides is damaged in the area of the infraorbital foramen and has been reconstructed. The right side in particular is heavily reconstructed; the left retains considerably more surface anatomy and is therefore more useful in describing this region. Both sides originally had a clear infraorbital depression that would have been coincident with a depressed infraorbital fossa. Inferior to where the infraorbital foramen would have been is marked concavity that is very similar to the concavity found on Dolní Věstonice 13 (Figure 8.38), although less extreme than in Dolní Věstonice 15 (see below).

The infraorbital process of the zygomatic bone (based on the right side) projects anteriorly over the depressed infraorbital region, creating a horizontal shelf that further accentuates the infraorbital depression in a horizontal transect. The lateral aspect of the infraorbital region (based on the left side) was oriented predominantly in the coronal plane, and the medial aspect adjacent to the piriform aperture was oriented in a parasagittal plane on each side. Therefore, in the transverse plane, the infraorbital region shows the typically modern two-plane configuration (Table 8.2).

Nasal Region

In the nasal region, the frontal process of the right maxilla is more complete than that on the left side. It preserves the entire course of the left nasomaxillary suture, although neither of the nasal bones for either side is preserved. The more complete right-side frontal process shows a strong degree of parasagittal eversion and a resulting relatively deep alpha point between the inferior nasomaxillary suture location and estimated zygoorbitale (see lateral view). The distance between the estimated alpha points relative to facial height in Dolní Věstonice 14 falls in the middle of the distribution among the Dolní Věstonice and other earlier Upper Paleolithic modern humans, and they all tend to have relative alpha chord values that are similar to those of the Neandertal sample (Figure 8.6). The area between the nasomaxillary suture and inferomedial orbital area is uniformly smooth, lacking the roughened furrow or slight elevation frequently found in both immature and mature archaic *Homo* (Franciscus, unpub. data). The orientation of the nasomaxillary suture is inferiorly oriented, rather than the horizontally inflected condition found in many Neandertals (Table 8.2).

Nasal breadth for Dolní Věstonice 14 (27.6 mm) is similar to that for Dolní Věstonice 16, and both lie between the highest and lowest values for Dolní Věstonice males (Figure 8.7). In contrast to its nasal breadth, the nasal height of Dolní Věstonice 14 (53.1 mm) lies well within the Dolní Věstonice sample distribution (Table 8.4; Figure 8.7; Sládek et al., 2000: table 26 with corrected value noted here). This places it below most Neandertals and Qafzeh-Skhul individuals in relative nasal breadth. The right nasomaxillary suture is discontinuous with the completely preserved frontonasal suture on the frontal bone by 6 mm of filler, the latter allowing the precise location of nasion. Nasion is superiorly set relative to the superior nasomaxillary points.

The lateral margin of the piriform aperture is complete on the left side, whereas the right side is missing 14.5 mm of bone along its margin that extends posterolaterally to the alpha point. The lateral wall of the left nasal aperture is sharp all along its course. There is no evidence of the inner crest/outer margin separation occasionally seen on the superior lateral margins of the piriform aperture in recent and fossil *Homo* (Franciscus, 2002). The internal lateral walls of the nasal fossa are complete from the piriform aperture margin posteriorly to the area bounding the nasolacrimal grooves on both sides (except for the missing area of bone on the right). The ethmoidal crest is visible on the right side, and the more inferiorly located conchal crest is well preserved on the right side. The inferior conchal crest is largely horizontally oriented with a slight obliquity from anterior to posterior portions. The conchal crest is visible in anterior view, although it does not project markedly medially into the nasal cavity.

The narial margin at the junction of the internal nasal floor and subnasoalveolar clivus is composed of two crests (Figure 8.48). The turbinal and spinal crests are fused, and there is partial fusion of the lateral crest to these two that forms a triangular *fossa intranasalis* (maximum separation of crests: 3 mm at the inferolateral corner). This configuration is close to either a category 3 or 5, but in its overall anatomy it is closest to a 5 since a *fossa intranasalis* is technically absent in a stage 3 (De Villiers, 1968; Franciscus, 1995, 2003b). It is the fused turbinal and spinal crest, and not the partially fused lateral crest, that

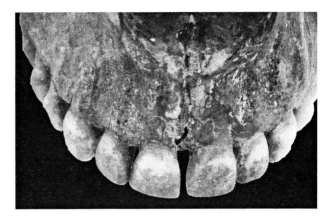

Figure 8.48 Lower border of the piriform aperture in Dolní Věstonice 14.

demarcates the internal nasal floor from the subnasoalveolar clivus. The fossa thus is open anteriorly and is very shallow, rather than opening superiorly. In this respect it differs from category 5 as manifested in Neandertals, in which the lateral margin is as vertically positioned as the fused turbinal and spinal crest and the *fossa intranasalis* is very deep. The anterior nasal spine is very small and nonprojecting, but it appears to be either remodeled or damaged postmortem. The internal nasal floor is preserved, but it is covered with consolidant in the more posterior aspect. Based on what is visible, it appears that the internal nasal floor was completely level as coded in Franciscus (2003b), similar to the other Dolní Věstonice individuals and different from the latest Neandertals.

Subnasoalveolar Clivus

The subnasoalveolar clivus in Dolní Věstonice 14 is well preserved, although there is some postmortem loss of alveolar bone, especially on the anterior aspect of the left first incisor and left canine to second molar. Alveolar bone that is not postmortemly damaged shows virtually no alveolar modification (Chapters 11 and 19). The overall shape of the anterior aspect of the clivus in Dolní Věstonice 14 is parabolic, the most parabolic or rounded in all the Dolní Věstonice sample. The anterior aspect of the clivus in Dolní Věstonice 14 also exhibits moderate canine and incisor eminences.

Overall Facial Size and Proportion

Overall facial height and breadth in Dolní Věstonice 14 generally lie in the middle to upper end of the distribution for both measures among the Dolní Věstonice and Pavlov sample in absolute terms (Figure 8.6; Sládek et al., 2000: tables 23 and 24).

Left Lateral View

Frontal Region

In left lateral view (Figure 8.49), the frontal bone of the Dolní Věstonice 14 shows a mildly steep frontal inclination with moderate projection at metopion relative to the frontal chord (i.e., nasion-bregma subtense). In this, it is similar to Dolní Věstonice 13, 15, and 16, with Dolní Věstonice 3 showing a higher frontal subtense relative to the chord and Pavlov 1 showing the relatively lowest value (Figure 8.10; Sládek et al., 2000: table 21). The high frontal subtense values found in the Dolní Věstonice and Pavlov sample overall are characteristic of all modern human samples, and they contrast with the lower values found among Neandertals (Table 8.1).

The glabellar region is projecting, with both a mild supraglabellar inflection and an even milder infraglabellar notch when taking into account the missing nasal bones. The degree of infraglabellar notch development in Dolní Věstonice 14, with allowance for missing nasal bone articulation with nasion, most closely approximates Lahr's (1996) category 2. Despite the mild supra- and infraglabellar notch development, the actual projection of the point of glabella is similar to the pattern found in all of the Dolní Věstonice and Pavlov individuals, other earlier Upper Paleolithic individuals, and to a large degree, the Neandertal sample. The greatest degree of infraglabellar notch development is found in Skhul 4 and 5 and in Mladeč 5 (Figure 8.11).

The distinct separation of medial, midorbital, and lateral components, described above in anterior view, can be seen in lateral view, including the distinct separation of the supraorbital trigone from the superciliary arch by the obliquely flattened supraorbital sulcus and the inflated and anteroposteriorly wide supraorbital trigone, although these are not as clearly delineated as in other males from this sample. The anterior portion of the temporal line on the frontal bone is weakly expressed.

Parietal Region

In left lateral view, the vertex of the cranium is located approximately 16 mm posterior to bregma. Especially in contrast to Dolní Věstonice 13 and 15, there is a relatively open angulation in Dolní Věstonice 14 between the midline superior neurocranial profile and the steep posterior parietal plane, reflected in its slightly more modest bregma-lambda chord subtense (Table 8.1; Figure 8.12; Sládek et al., 2000: table 21). There is also a distinct and large area of lambdoidal flattening on Dolní Věstonice 14

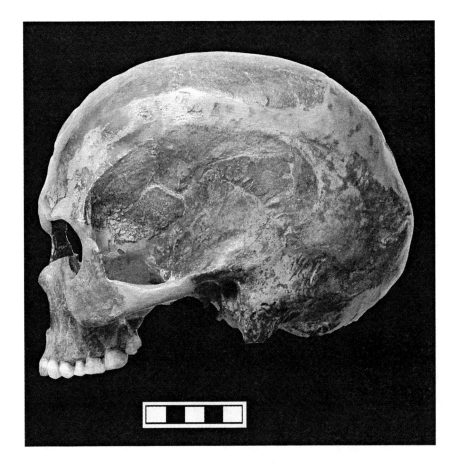

Figure 8.49 Lateral (left) view of the Dolní Věstonice 14 cranium. Scale in centimeters.

that is absent on the other Dolní Věstonice specimens and Pavlov 1 except for Dolní Věstonice 11/12.

The left parietal has been reconstructed from at least eighteen major pieces with a large amount of filler all along the area superior to the temporal line and posteriorly toward lambda. It is difficult to determine whether the left parietal tuber as presently reconstructed is accurate. The weakly developed anterior portion of the temporal line found on the frontal fades and continues along the anterior and posterior parietal as a distinct but weakly marked line (although visible by coloration differences, palpation shows it to have virtually no relief). It is not possible to clearly discern the distinction between the superior and inferior temporal lines that usually become most separated in their superoposterior parietal portions. The temporal line remains weakly developed in its most inferoposterior segment before crossing to the temporal bone.

Pterion Region

The pterion region has been heavily reconstructed on the left side. Although elements of all four bones are present, no useful morphological information can be gleaned from this region.

Temporal Region

The left temporal bone in Dolní Věstonice 14 is largely intact except for the most anterosuperior portion of the squamous portion. The fit of the temporal onto the parietal and occipital bones is accurate. Even though there is a large gap along the zygomatic arch (minimum reconstructed gap: 29 mm), the alignment of the temporal and zygomatic portions of this arch by filler appears more or less sound. The fit of the temporal bone with the remnant of the greater wing of the sphenoid is difficult to assess because of damage to this region.

The temporal bone has a weakly developed suprameatal crest that continues from the zygomatic process and extends well beyond porion posterosuperiorly to meet up with the posterior aspect of the temporal line on the parietal bone. There is a small, crest-type suprameatal spine with a very small depression (Table 8.6; Hauser & DeStefano, 1989: figure 31, closest to c), and there is no evidence of a squamomastoid suture. The mastoid process is large (>30 mm in estimated oblique length from porion relative to the Frankfurt Horizontal line; 14.8 mm in oblique mediolateral width; 30.0 mm in anteroposterior width from its anterior extent to the mastoid foramen orthogonal to the Frankfurt Horizontal). An anterior mastoid tubercle is present, but it is not prominent. The mastoid process extends well below the plane of the juxtamastoid eminence, even though its most inferior tip is missing because of postmortem damage. It shows no lateral flattening. In all of these features, the mastoid process of Dolní Věstonice 14 is modern, rather than Neandertal-like (Table 8.5). In addition to its large size, the mastoid process is rugose, especially in its inferoposterior portion

Occipital Region

In left lateral view, Dolní Věstonice 14 shows virtually no occipital bun development. The occipital chord and subtense configuration in Dolní Věstonice 14 is similar to the other Dolní Věstonice and Pavlov males, with moderately large values for each compared to the considerably smaller values for Dolní Věstonice 3 (Figure 8.13; Table 8.1; Sládek et al., 2000: table 21). It should nonetheless be kept in mind that there is extensive damage and missing bone in this region on Dolní Věstonice 14 (see occipital view).

Zygomatic Region

The left zygomatic bone is not as well preserved as for other Dolní Věstonice and Pavlov specimens. A large area of the frontal process is missing, and the frontozygomatic suture region is completely reconstructed. The mediolateral and anteroposterior alignments of the two roots of the arch appear to be sound. The zygomaticotemporal sutural area is damaged and reconstructed as well. However, the fit of the zygomatic bone onto the maxilla along the zygomaxillary suture is preserved anteriorly and posteriorly, and it is accurate. The area just inferior to the inferolateral corner of the orbit is smooth, with no evidence of a zygomaxillary ridge, that corresponds to a grade 1 degree of robusticity (Lahr, 1996). There is no zygomaticofacial foramen on the body, and there is weak development of a marginal tubercle that is 13.1 mm in maximum anteroposterior length (Table 8.6; Hauser & DeStefano, 1989: figure 36, most similar to b).

Orbital Region

In lateral view, the lateral margin of the orbit is posteriorly positioned relative to nasion, and along with other midfacial measures, this indicates some degree of midfacial projection. In this feature, it is similar to the other Dolní Věstonice males, although not as pronounced as that seen in the Neandertals on average (Figure 8.14; Table 8.2). Given the amount of reconstruction to this area for Dolní Věstonice 14, however, and the asymmetry noted above in anterior view, this observation should be weighed with caution.

Infraorbital Region

The left infraorbital region, despite the minor damage and reconstruction noted above in anterior view, nonetheless provides reliable information on the topography of this region. There is a clear infraorbital depression on Dolní Věstonice 14 that would have accompanied a depressed infraorbital fossa. As noted in anterior view, inferior to where the infraorbital foramen would have been there is marked concavity that is very similar to, albeit less extreme than, the concavity found on Dolní Věstonice 15 (Figure 8.38).

In lateral view, it can again be seen that the lateral aspect of the infraorbital region is oriented predominantly in the coronal plane, whereas the medial aspect adjacent to the piriform aperture is oriented in a parasagittal plane, resulting in the typically modern two-plane configuration. Moreover, from this view, it is also evident that in a sagittal plane the lateral aspect of the infraorbital region is obliquely oriented such that the inferior margin of the zygomatic process of the maxilla is posteriorly positioned relative to the inferior orbit. This oblique inflection of the infraorbital region is also a feature of modern humans, in contrast to the vertical or oblique extension of the infraorbital plane that is found in premodern *Homo*. The anterior root of the zygomatic process of the maxilla is located at the anterior edge of the first molar in Dolní Věstonice 14, similar to other Dolní Věstonice individuals (Table 8.2).

Nasal Region

In the nasal region, there is a break along the frontal process of the left maxilla that makes it appear as though the nasomaxillary suture is horizontally oriented and that the whole orientation of the frontal process is off. This is potentially confusing because the horizontal edge has the superficial appearance of being the nasomaxillary sutural edge. However, the nasolacrimal groove is positioned in anatomically correct positions on both sides, clearly indicating that the horizontal edge is not sutural and is instead a postmortem break above which the frontal process is missing. As noted earlier, there is a fairly strong degree of parasagittal eversion and a resulting relatively deep alpha point between the inferior nasomaxillary suture location and the estimated zygoorbitale point. The degree of eversion and alpha point subtense depth in Dolní Věstonice 14 is more pronounced than in Dolní Věstonice 3 and 16 and similar to that seen in Dolní Věstonice 15, who shows a very everted condition relative to all individuals in all comparative samples (Figure 8.15). It can be seen particularly well in this view that the orientation of the nasomaxillary suture relative to the Frankfurt Horizontal is inferiorly, rather than horizontally, inflected, as is true for non-Neandertal extinct and extant *Homo* (Table 8.2). The lateral margin of the piriform aperture is complete on the left side, and the lack of projection for the damaged anterior nasal spin, as discussed above, can be seen in lateral view at the level of the narial margin.

Subnasoalveolar Clivus and Alveolar Region

There is a mild inflection at subspinale in lateral view in Dolní Věstonice 14 that is not as pronounced as that evident in Dolní Věstonice 13 or 15, although this may be conditioned to some degree by the more complete anterior nasal spines in the latter individuals. The clivus of Dolní Věstonice 14 also exhibits a mild degree of alveolar prognathism from this view, although it is well within the range for individuals among all of the comparative samples (Figure 8.16). A comparison of the lower facial projection at prosthion relative to upper facial projection at nasion (Figure 8.17) indicates the relatively orthognathous face of Dolní Věstonice 14 (despite alveolar prognathism). This is especially true in comparison with Dolní Věstonice 13.

Overall Size and Proportion in Lateral View

The height and the length of the neurocranium in Dolní Věstonice 14 are larger than in Dolní Věstonice 3 and 13, while being slightly smaller than in Dolní Věstonice 16 (Figure 8.18; Sládek et al., 2000: tables 18 and 19), but their proportions remain the same and separate from the Neandertals and a couple of Qafzeh-Skhul specimens.

Overall facial height and facial length in lateral view for Dolní Věstonice 14 are somewhat smaller than for the other Dolní Věstonice individuals, although there are some European earlier Upper Paleolithic individuals who are smaller (Figures 8.19 and 8.20; Sládek et al., 2000: tables 22 and 24). Also, as noted in anterior view, other measures of facial size in Dolní Věstonice 14 lie in the middle to upper end of the distribution among the Dolní Věstonice sample in absolute terms (Figure 8.6; Sládek et al., 2000: tables 23 and 24).

Right Lateral View

Gross aspects of neurocranial and splanchnocranial morphology that are evident from either side are covered in the preceding section. The following additional features are revealed in right lateral view (Figure 8.50).

Parietal Region

The right parietal has been reconstructed from at least thirteen major pieces with a large amount of filler. It is difficult to determine whether the right parietal tuber as presently reconstructed is accurate.

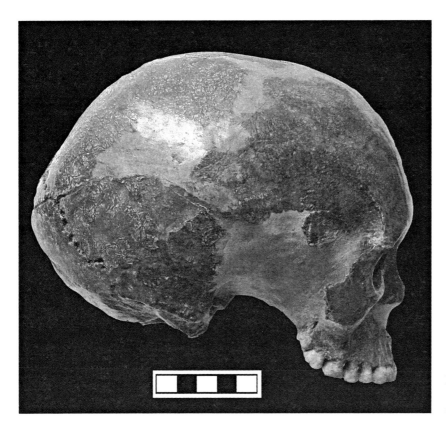

Figure 8.50 Lateral (right) view of the Dolní Věstonice 14 cranium. Scale in centimeters.

Pterion Region

The pterion region has been heavily reconstructed with a substantial amount of filler on the right side, and no useful morphological information can be gleaned from this region.

Temporal Region

The right temporal squamous is heavily damaged, reconstructed, and covered over, and no useful information can be gleaned except a general indication of a medially inflected in situ ground pressure distortion. As on the left side, the squamomastoid suture is absent on this side. There is one large mastoid foramen, filled with matrix, that does not appear to be sutural. It is confined to the temporal bone near asterion at the level of where the missing external acoustic meatus would be, and it is positioned posteriorly such that it is more visible in posterior than in lateral view. In this, it is identical in location to the mastoid foramina found on Dolní Věstonice 13 and 14 and Pavlov 1 (Table 8.6).

Occipital Region

In right lateral view, Dolní Věstonice 14 shows mild occipital bun development, although there are extensive damage and missing bone in this region (see occipital view).

Zygomatic Region

The right zygomatic is not as well preserved as for other Dolní Věstonice and Pavlov specimens. The frontal process and the frontozygomatic suture region are complete. The zygomaticotemporal sutural area, however, is damaged and not reconstructed. The fit of the zygomatic bone onto the maxilla is completely reconstructed, with little useful morphological information. The area just inferior to the inferolateral corner of the orbit is covered with consolidant. It is smooth, with no evidence for a zygomaxillary ridge, which corresponds to a grade 1 degree of robusticity (Lahr, 1996). There is no zygomaticofacial foramen on the body, and there is weak development of a marginal tubercle that is 13.9 mm in maximum anteroposterior length (Table 8.6; Hauser & DeStefano, 1989: figure 36, most similar to b).

Orbital Region

In right lateral view, the lateral margin of the orbit is posteriorly positioned relative to nasion and other midfacial measures, indicating some degree of midfacial projection. However, given the amount of reconstruction to this area and the asymmetry noted above in anterior view, this observation should be weighed with caution.

Infraorbital Region

The infraorbital region on the right side is almost entirely reconstructed.

Nasal Region

All relevant anatomy can be gleaned from the description of the nasal region in anterior view.

Inferior View

Occipital Region

The occipital region has been reconstructed from at least nine major pieces with a large amount of filler superior to the external occipital protuberance. In inferior view (Figure 8.51), large areas of the nuchal plane and peripheral temporal areas show marked ground pressure deformation (a strongly medially inflected right mastoid process, a median crest that deviates markedly to the left, a right nuchal plane that is medially inflected, and an off-center external occipital protuberance). However, there are also features that show a high degree of proper alignment (the foramen magnum and occipital condyles, the basilar portion of the occipital, and the general course of the superior nuchal line except for the most midline component). Thus the postmortem damage and the ensuing reconstruction have combined to produce a rather mixed morphological pattern that makes it difficult to assess actual antemortem symmetry versus asymmetry and any potential pathological alterations. Care is therefore taken here to focus more on presence or absence rather than describing details of anatomy that may or may not be real.

A mildly developed median occipital crest posterior to the foramen magnum is visible, although its orientation is misaligned to the left. The occipital crest does not span the poorly developed (damaged?) inferior nuchal lines. A moderately developed superior nuchal line is evident for much of its course along the occipital; however, the location of midline along the nuchal line is not clear. Specific muscle insertion areas are not clearly differentiated from reconstruction and ground-pressure-induced undulations and fossae.

The occipital condyles are well preserved and show minor asymmetry. The maximum length of the right condyle is 21.0 mm, that of the left 24.5 mm. The width of the right condyle is 12.0 mm (maximum), that of the left 13.5 mm. Compared to other individuals (e.g., Dolní

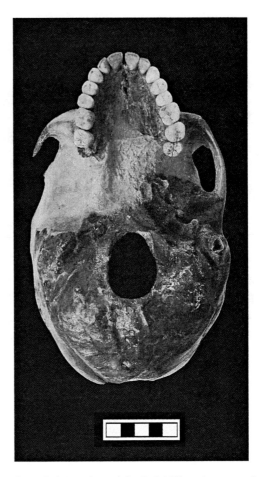

Figure 8.51 Inferior view of the Dolní Věstonice 14 cranium. Scale in centimeters.

Věstonice 13), the level of asymmetry in Dolní Věstonice 14 is not unusual.

The area just lateral to the condyle on the right side in the region of the rectus capitis lateralis insertion is damaged and largely missing. This same area lateral to the left-side occipital condyle appears intact. The hypoglossal canal on the left side is present and open; the one on the right is damaged and filled. The posterior condylar foramina might be present but are not visible under the overlying filler. With respect to hypoglossal canal bridging, there appears to be a single undivided canal on the left side, but the right side is too obscured to evaluate (Table 8.8). On the left side there is weak expression of a fused medial and lateral paracondylar process (Hauser & DeStefano, 1989: figure 20, a). None is evident on the right side.

The breadth of the foramen magnum in Dolní Věstonice 14 at 30 mm is the same as those in Dolní Věstonice 3 and 16, and all three are wider than that in Dolní Věstonice 13. However, in length, the foramen magnum of Dolní Věstonice 14 is much larger than that in the other individuals, making its overall area quite large (Figure

8.23; Sládek et al., 2000: tables 18 and 19). Anterior of the foramen magnum on the inferior basilar surface, there is a weakly expressed pharyngeal tubercle (Hauser & DeStefano, 1989: figure 23, a), and anterior to the tubercle there is also a weakly expressed pharyngeal fovelola (Table 8.8; Hauser & DeStefano, 1989: 138, a, in methodology). Lateral to the pharyngeal tubercle, there are well-developed attachment protuberances for the longus capitis muscle on both sides. Unfortunately, the area anterior to these features that would have contained the sphenooccipital synchondrosis is damaged and/or covered over with filler; given Dolní Věstonice 14's relatively young age (Chapter 6), this synchondrosis might have still been open.

Sphenoid Region

All of the sphenoid is missing or covered with filler, except for the base of the left pterygoid process and a small portion of the basal portion of the greater wing. Foramen ovale is present although filled in, and the outline for foramen spinosum is present (Table 8.8).

Temporal Region

The inferior aspects of the temporal region are well preserved on the left side. However, the right side is missing large portions, including all of the glenoid fossa and lateral portions of the tympanic region. Articulation of the right mandibular condyle into the well-preserved right glenoid fossa produces a moderately good fit along the dentition. The right canine to second molar and left first molar occlude well, but there is a moderate incisor overbite. This may have been due to an actual antemortem overbite rather than postmortem damage and reconstruction, given the minimal wear on the dentition (Chapter 11). There is also some mediolateral mandibular dentition displacement to the right by 1 to 2 mm. Nonetheless, facial alignment based on occlusion is probably close to the antemortem alignment, and the alignments of the palate, basilar occipital, and foramen magnum appear sound. The right mastoid process is medially inflected, although it is difficult to discern whether this is due to ground-pressure distortion or faulty reconstruction. Although mediolateral alignment of the right temporal is off, anteroposterior alignment appears sound.

The left temporal preserves a juxtamastoid eminence medial to the mastoid processes. The left temporal also retains an 8.3 mm length of styloid process, although its original length cannot be determined. A stylomastoid foramen cannot be seen but may be filled in. The right-side styloid process is missing, although its base is present. Additional features in this area that are preserved on the left include the carotid canal, the posterior glenoid eminences, and the tympanic area.

The left glenoid fossa is well preserved. Although healthy in appearance, the fossa does exhibit a horizontally oriented groove at the deepest point. This groove is matched by what appears to be a transversely oriented line across the apex of the left mandibular condyle (there are also a couple of irregularities at either end of the mandibular condyles that could be remodeled bone or dried glue).

The roots of the zygomatic processes are completely missing. The alignment of the two portions of the temporal bone with the left-side zygomatic appears sound, even though most of the zygomatic arch has been reconstructed (see above). Posteriorly, given the incompleteness of the zygomatic arches, masseter muscle scarring cannot be evaluated. Anteriorly, however, it can be seen that masseter did not extend medial of the inferior zygomaxillary suture. None of this anatomy is preserved on the right side. On the left side in inferior view, there is a high degree of angulation between the lateral aspect of the infraorbital and zygomatic root area and the lateral aspect of the inferior zygomatic; this is the typically modern condition that differs from the obtusely angled, or even straight, contour between these two regions that exists in Neandertals.

Inferior Palatal Region

The inferior palate is largely complete anteriorly, although the antemortem form of the incisive fossa has probably been altered with reconstruction. The intermaxillary suture area is reconstructed, and the palate lateral to the intermaxillary suture is complete on both sides to the approximate area of the palatomaxillary suture. Except for a small posterolateral portion on the right side, the palatine bones are missing.

The palate is moderately arched and moderately deep, and there is no evidence for midpalatal (intermaxillary) or alveolar tori (Table 8.8). The overall dimensions of the Dolní Věstonice 14 palate fall toward the large end of the range for the Dolní Věstonice and Pavlov sample (Figure 8.24; Sládek et al., 2000: table 27). The Pavlov and other earlier Upper Paleolithic sample as a whole has smaller palates than the Qafzeh-Skhul and Neandertal samples (Figure 8.24; Table 8.4), a pattern that is evident as well in their mandibular dental arcade breadths (Chapter 10).

Inferior Frontal View

The inferior frontal has relatively incomplete internal orbital plates on both sides, and they are too covered with filler to assess surface detail. Medially, the nasofrontal suture is preserved on the left side more than the right. The ethmoidal notch area is missing. There is a substantial opening into the left medial and left lateral frontal sinus area, showing that the sinuses were expansive laterally beyond the midline glabellar area. Radiographic examination (Vlček, 1991: plate 22 and figure 53) confirms these observations; there is a moderately large frontal sinus delimited to the left glabellar and left lateral side only. The left-side frontal sinus delimitation is also found in the two other members of the triple burial and has further implications regarding their affinity. As noted in anterior view, the left orbit has a very large, blurred, medially positioned supraorbital notch (larger than any shown in plate VIII in Hauser & DeStefano, 1989; Table 8.3).

Internal Calvarium View

Given the damage in the posterior orbital region, it is possible to view the internal calvarial anatomy through the foramen magnum with lighting through the orbital region. However, the internal anatomy is covered with filler in the occipital region and heavily reconstructed between the other bones with a coating of consolidant, obscuring fine surface detail. A general impression of vascular grooving is possible, but detailed observations are precluded.

Posterior View

In posterior view (Figure 8.52), the course of the sagittal suture is clearly seen, and lambda can be determined by using the confluence with the lambdoidal suture. The lambdoidal suture can be seen, although it is damaged and filled in several places. Detailed scoring results for the lambdoidal suture are presented in Table 8.10. The left posterior parietal bone has been heavily reconstructed from several pieces and a large amount of filler. The left tuber region is almost completely reconstituted, and whether the current prominence of the tuber is accurate is unknown. The right posterior parietal bone has also been reconstructed from several large pieces and filler in the lateral-most extent. The parietal tuber is less prominent on this side, but its original form has been influenced by medially inflected ground pressure more anteriorly, making the original form difficult to assess. There is also a greater posterior projection or prominence in the area of the left tuber than in the right, which is almost certainly due to postmortem ground pressure.

Given the present reconstruction, the lateral parietal and temporal squama are straight-sided on the left and more rounded on the right. The left side, with the prominent parietal tuber that is superiorly located and the posterior vertex, form the typical pentagonal configuration that is associated with modern humans, one that is very different from the "en-bombe" Neandertal configuration. The right side is much more rounded, although not typically "en-bombe." Consistent with the pentagonal shape,

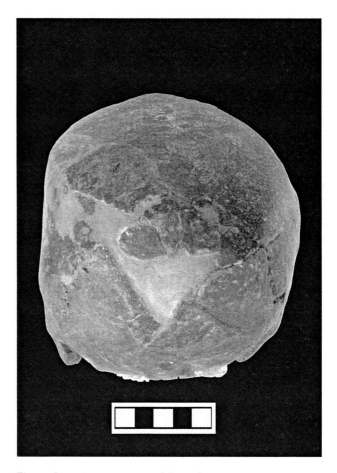

Figure 8.52 Posterior view of the Dolní Věstonice 14 cranium. Scale in centimeters.

the greatest breadth across the parietals is relatively superiorly positioned. Maximum cranial breadth for Dolní Věstonice 14 is smaller than for Dolní Věstonice 16 and especially 13, but its cranial height is relatively higher than in all but Dolní Věstonice 16 (Figure 8.26; Sládek et al., 2000: table 19); however, it must be stressed that the level of neurocranial deformation and reconstitution in this specimen precludes definitive statements about its antemortem shape. There is very modest to no sagittal keeling (category 1; Lahr, 1996).

The superior nuchal lines are present on both sides (Figures 8.53 and 8.54), although they are superoinferiorly asymmetrically positioned (higher on right) because of postmortem damage and meet in midline at the *tuberculum linareum*. The supreme nuchal lines are visible but very modestly developed. A 19.5 mm horizontally roughened line is positioned to the right of the midline that generally corresponds to the external occipital protuberance. Whether it was similar to other Pavlovian specimens is difficult to determine, given the damage. There is a large superiorly located and centrally positioned depression and an inferior central depression that is slightly smaller just

The Cranial Remains

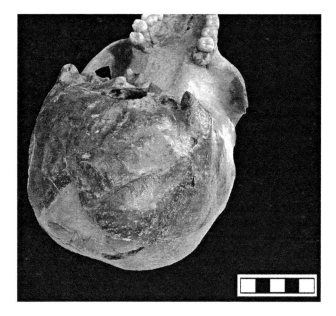

Figure 8.53 Oblique inferior view of the left occipital region of Dolní Věstonice 14. Scale in centimeters.

above the most posterior extent of the occipital crest. The same configuration is found in most of the other specimens from these sites (e.g., Dolní Věstonice 15 and 16 and, to a lesser extent, Dolní Věstonice 3 and Pavlov 1).

The posterior view (Figure 8.52) also shows the extension of the left mastoid processes below the level of the more medially positioned juxtamastoid eminence, the usual modern human condition. Inferiorly, the medially

Figure 8.54 Oblique inferior view of the right occipital region of Dolní Věstonice 14. Scale in centimeters.

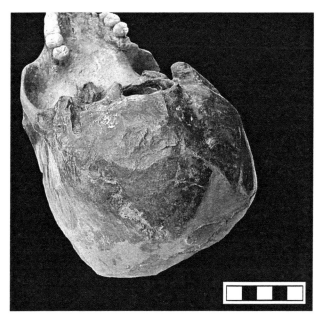

inflected deformation to the right mastoid region can be seen in posterior view.

Superior View

In superior view (Figure 8.55), a medial inflection of the anterior two-thirds of the right side of the neurocranium is evident, associated with a deviation of the sagittal suture to the left. Also evident is a relatively more posteriorly positioned left side of the cranium along the anteroposterior axis (or a right side that is more anteriorly positioned). Whichever is the case, the two sides are clearly off along the anteroposterior axis. Even with this level of postmortem deformation, it is clear that the cranium was dolicocephalic. It lies toward the more dolicocephalic range for maximum cranial breadth relative to length, not only among the Dolní Věstonice and Pavlov individuals (Sládek et al., 2000: tables 18 and 19), but also for the individuals across all of the comparative samples (Figure 8.28). In this, it is very similar to Dolní Věstonice 11/12 and 16 and Pavlov 1. Its greatest width was probably located at the parietal tubers at the posterior third of the neurocranium. Commensurate with its large cranial capacity estimate, 1538 cc (Table 8.9; Vlček,

Figure 8.55 Superior view of the Dolní Věstonice 14 cranium. Scale in centimeters.

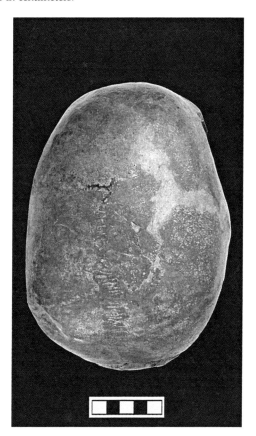

1997), Dolní Věstonice 14 has virtually no postorbital constriction. No parietal foramina appear to be present. Detailed scoring results for the coronal sutures are presented in Table 8.10. In oblique superior view, the distinction of the superciliary arch from both the frontal squama and the laterally positioned supraorbital trigone on either side by the obliquely flattened supraorbital sulcus are apparent, although not as pronounced as for other (and older) male specimens.

Dolní Věstonice 15

Preservation

The neurocranium and splanchnocranium are preserved in Dolní Věstonice 15. The neurocranium is reconstructed from several large fragments that are derived mostly from the calvarium. The neurocranium retains the frontal bone, the left and right parietal bones, the left and right temporal bones, the occipital bone, and the sphenoid bone. The splanchnocranium retains the left and right maxillae and the left and right zygomatic bones. The connection between the neurocranium and splanchnocranium is preserved in four regions. The right frontal and zygomatic bones have a 37 mm connection, and the right side has an 8 mm connection. The right frontal and maxillary bones have a 9 mm connection, and the left side has an 11 mm connection.

There are a number of distortions in the reconstruction. First, the anteroposteior hafting of the splanchnocranium around its hinge contact points is slightly off, probably making the face too orthognathous. Second, left-right anteroposterior asymmetry in the retained basicranial elements is substantial. Third, there is some bone missing along the zygomaticofrontal sutures, especially on the right side, and this slightly shortens the vertical dimensions of the face including the superoinferior maxillary tooth position. The fit of the mandibular condyles into the glenoid fossae, regardless of proper occlusion is good, indicating that at least the relative mediolateral distance between the temporal bones is accurate. In superior view, anteroposterior asymmetry is evident, particularly in the more posterior bulging of the right occipital region. The deviation of the sagittal suture to the left is also evident, which results from the greater degree of medially directed flattening of the right parietal reconstruction.

Anterior View

Frontal Region

The frontal bone (Figure 8.56) is reconstructed from at least five pieces with a small amount of filler between

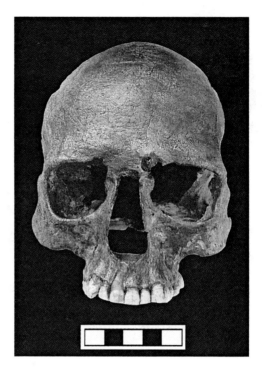

Figure 8.56 Anterior view of the Dolní Věstonice 15 cranium. The opening into the left frontal sinus in the superomedial orbital corner is postmortem damage. Scale in centimeters.

contact points, and it is therefore somewhat more distorted than in Pavlov 1 and Dolní Věstonice 13 and 16, although not as distorted as in Dolní Věstonice 14. The right anterolateral frontal squama is somewhat medially inflected, a postmortem deformation that is more clearly seen from an oblique view. To some degree, as with the other Dolní Věstonice individuals, root etching found throughout on this bone obscures very fine surface detail; nonetheless, most external features are clearly preserved. Much of the coronal suture is damaged and infilled (Chapter 6). There is a possible incomplete (parietal part) metopic suture (Hauser & DeStefano, 1989: plate VI). There are no obvious frontal grooves on the lateral squamous portions (Table 8.3), although this is difficult to assess, given the root etching. There is a partially obliterated zigzag-shaped supranasal suture (Hauser & DeStefano, 1989: plate VI, h). There is no evidence for frontal keeling in midline from either a strict frontal or oblique frontal view. Approximately 25 mm lateral to midline on either side and about 34 mm superior to the superior orbital margins, there are very weakly expressed or trace frontal eminences; they constitute the weakest expression in all of the Pavlovian males.

The least frontal breadth of Dolní Věstonice 15 (99 mm) falls between the highest value for the Pavlovian and Dolní Věstonice males, 105 mm (Dolní Věstonice 14), and the lowest value, 97 mm (Dolní Věstonice 16), whereas the maximum frontal breadth for Dolní Věstonice 15 (120 mm) falls between the highest value, 122 mm (Pavlov 1), and the lowest value, 118 mm (Dolní Věstonice 16). The relationship between these two variables is very similar across the European earlier Upper Paleolithic samples (Figure 8.2), with the Qafzeh-Skhul and Neandertal samples showing relatively greater least frontal breadth values.

Supraorbital Region

The supraorbital region of Dolní Věstonice 15 shows a distinct separation of medial, midorbital, and lateral components. The glabellar region is pronounced in midline, and this robust midline projection extends to the medial third of both right and left supraorbital areas (superciliary arch), each extending superiorly and laterally for a total length across both sides (encompassing glabella) of ca. 57.7 mm. The lateral-most supraorbital margin, the supraorbital trigone on either side, are ca. 5 mm thick, and they are distinctly separated from the medial superciliary arch by the obliquely flattened supraorbital sulcus, the latter of which is 3 mm thick on either side. The supraorbital sulcus is distinctly separated both mediolaterally and superoinferiorly from the superciliary arch, and it is visible as a distinct entity for virtually the entire length of the supraorbital margin. The supraorbital trigone is inflated and widened anteroposteriorly, but it retains a smooth lateral surface; they both correspond most closely to Lahr's (1996) grade 3. The distinct separation of medial, midorbital, and lateral components of the supraorbital region in Dolní Věstonice 15 is thus similar to the other Pavlovian individuals (Table 8.3).

There is a large opening into the left frontal sinus region in the vicinity of the superomedial orbital corner. Internally it is bounded medially by a wall separating this sinus from the other sinus spaces medially. The opening is 11.4 mm wide anteriorly and 18.2 mm anteroposteriorly, and it invades internally into much of the left medial superciliary arch (see further discussion of frontal sinus patterning below). The opening exposing the frontal sinus is most likely a nonpathological feature resulting from postmortem damage (Chapter 19).

Orbital Region

The right orbit is largely intact. The lower border of the orbital margin at zygoorbitale appears to be misaligned, with the zygomatic component positioned slightly anteriorly to the maxillary component. However, it is not unusual to find the zygoorbitale region aligned in this fashion with a resulting small protuberance at zygoorbitale in nondeformed crania. The left orbit is less intact than the right one. In addition to the breakage into the frontal sinus noted above, about 25 mm of the mid–lower orbital margin has been reconstructed with filler. There is a slight degree of asymmetry in orbital shape and positioning in Dolní Věstonice 15. The right orbital breadth (39 mm) is slightly narrower than the left (42 mm), and both sides have the same height (31 mm). Also, the right orbit is slightly more inferiorly positioned than the left. This degree of orbital shape asymmetry is not as pronounced as for Dolní Věstonice 13. Minor orbital shape asymmetry aside, Dolní Věstonice 15 shares with the other Pavlovian individuals and modern specimens orbits that are lower relative to orbital breadth than Neandertals (Figure 8.4, Tables 8.1 and 8.2).

Interorbital breadth in Dolní Věstonice 15 is slightly higher than the similar value for Dolní Věstonice 16, and all are higher than the value for Dolní Věstonice 3. When plotted against values for bifrontal breadth (Figure 8.3), these interorbital breadths are roughly similar to those found in the other comparative samples.

The superior orbital margins in Dolní Věstonice 15 are somewhat sharp, but they do not attain the extreme sharpness found on most gracile females (Chapter 7). Bilateral, large supraorbital notches are present on the very medial aspect of each orbit (Table 8.3; Hauser & DeStefano, 1989: plate VIII, most similar to b). Internally, the superior orbital roof is largely intact on the right side, and there is no indication of thinning or porosity. The left superior orbital surface is less complete, but it also shows no thinning or porosity. The general region of the superior ethmoidal suture area is present on the right side; however, no details are observable. The left-side orbital medial wall is missing. Neither orbit retains any of the medial wall below the ethmoidal suture. The right internal orbit also retains portions of the greater wing of the sphenoid, most of the lateral wall, and the lateral portion of the internal orbital floor. The left orbit retains only a small portion of the anterior lateral wall and internal orbital floor.

The inferolateral borders of both orbits are gracile in that they are sharp rather than rounded; the left side is particularly sharp, but this may be due to reconstruction. The right side is more rounded than the left, but it is considerably less robust and rounded than in Dolní Věstonice 13 and 16. The borders of the inferolateral orbits are also raised in relation to the floor of the orbits; they correspond most closely to Lahr's grade 1, the most common pattern in her recent human sample. Just inferior to the inferolateral corner of the orbits on the zygomatic bones, there is an absence of a raised ridge, the zygomaxillary ridge (see above), which corresponds to a grade 1 degree of robusticity (Lahr, 1996). As mentioned above, the midpoint of the lower orbital margin on the left side is reconstructed and filled in; however, the left zygoorbitale point can be reliably estimated by using the right side. The breadth across these points relative to facial height (Figure 8.5) is similar across all samples, with Dolní Věstonice

15 falling roughly in the middle of the earlier Upper Paleolithic scatter at the lower size range for both measurements. The inferomedial and medial orbital rims are present in Dolní Věstonice 15, but both of the taphonomically fragile lacrimal bones are missing.

Infraorbital Region

The infraorbital region on both sides is damaged in the precise area of the infraorbital foramen and has been filled and reconstructed. Nonetheless, enough bone exists on the lateral and medial components adjacent to each area to accurately infer the topography of the infraorbital region. Both sides show a clear infraorbital depression that would have been coincident with the depressed infraorbital fossa. The infraorbital process of the undamaged right zygomatic bone projects anteriorly over the depressed infraorbital region just lateral to the damaged and filled area, creating a horizontal shelf that further accentuates the infraorbital depression in a horizontal transect.

The lateral aspect of the infraorbital region is oriented entirely in the coronal plane on both sides, and the medial aspect adjacent to the piriform aperture is oriented in a parasagittal plane on either side. Therefore, in the transverse plane, the infraorbital region shows the typically modern two-plane configuration (Table 8.2). In Dolní Věstonice 15, the angulation is particularly acute, with a deeply depressed infraorbital region. Moreover, there is a sharp vertical sill on the lateral aspect of the right medial infraorbital aspect that gives way to a rather recessed posterior area just medial to where the canine fossa is usually found. This markedly depressed area is not pathological; it appears on the left medial infraorbital aspect as well, although the vertical sill is not as sharp on that side. Moreover, the join of the zygomatic bone to the maxilla along the zygomaxillary suture involves a fair amount of filler anteriorly, but posteriorly the zygomaxillary suture and adjacent bone are intact.

These bilaterally expressed depressed areas give the infraorbital region a deeply excavated appearance. This feature is similar to the concavity found on Dolní Věstonice 14 and in more muted expression on Dolní Věstonice 13 (Figure 8.38). The association of this marked concavity with extremely everted frontal processes occurs with some regularity (Franciscus, personal observation). The feature has also been documented recently in other fossil hominids (see Chapter 19), and it may provide evidence for a close genetic relationship among the members of the triple burial.

Nasal Region

The frontal processes of both maxillae are completely preserved, and both preserve the entire course of the nasomaxillary sutures, although neither of the nasal bones on either side is preserved. The frontal process on the left side has been joined to the maxilla along an oblique break just below the plane of the lower orbital margin, and the frontal process and superior piriform aperture margin on the right side have been joined to the maxilla along a horizontal break; however, the breaks are clean, and the joins are accurate. Both frontal processes show a high degree of sagittal eversion that is nearly orthogonal to the coronal plane, and there is a resulting very deep alpha point between the inferior nasomaxillary suture location and zygoorbitale. Dolní Věstonice 14 also shows this highly everted condition with deep alpha points, and possibly so did Dolní Věstonice 3 prior to damage to the region.

The distance between the estimated alpha points relative to facial height in Dolní Věstonice 15 falls in the middle of the distribution among the Dolní Věstonice and other earlier Upper Paleolithic modern humans, and they all tend to have relative alpha chord values that are similar to the Neandertal sample (Figure 8.6). There is a single, moderately developed foramen on the left-side frontal process near the orbital margin; there also appears to be a smaller foramen in the same area on the right side. The area between the nasomaxillary suture and inferomedial orbital area is uniformly smooth, lacking the roughened furrow or slight elevation frequently found in both immature and mature archaic *Homo* (Franciscus, unpublished data).

Dolní Věstonice 15 has the narrowest nasal breadth among the Pavlovian sample (21.5 mm), being smaller even than Dolní Věstonice 3 (22.0 mm) and all other earlier Upper Paleolithic individuals. The nasal height of Dolní Věstonice 15 (49.2 mm) is also the smallest within the Pavlovian sample distribution and among the smallest earlier Upper Paleolithic individuals overall (Table 8.4; Figure 8.7; Sládek et al., 2000: table 26). Yet, when these two measurements are plotted against each other, their proportions fall within the overall Late Pleistocene distribution (Table 8.7), even if Dolní Věstonice 15's nasal breadth is moderately narrow for its height. The orientation of the nasomaxillary suture is inferiorly, rather than horizontally, inflected. The nasomaxillary suture is continuous with the completely preserved frontonasal suture on the frontal bone, the latter allowing the precise location of nasion. Nasion is superiorly set relative to the superior nasomaxillary points.

The lateral margin of the piriform aperture is complete on the left side, with very minor damage along the lateral-most edge. The lateral margin on the right side is missing a small slip of bone of the lateral-most edge (1 to 2 mm) for most of its course. The lateral walls of the nasal aperture are sharp for their entire course. There is only trace evidence of the inner crest/outer margin separation occasionally seen on the superior lateral margins of the piriform aperture in recent and fossil *Homo* (Franciscus,

2002). The internal lateral walls of the nasal fossa are complete from the piriform aperture margin posteriorly to the nasolacrimal grooves on both sides. The ethmoidal crests and the more inferiorly located conchal crests are well preserved. Both inferior conchal crests show predominantly horizontally positioned anterior components and obliquely positioned posterior components. The conchal crests are barely visible in anterior view.

The narial margin at the junction of the internal nasal floor and subnasoalveolar clivus is composed of a single crest that is sharp and well defined (Figure 8.57). Thus all three crests, spinal, turbinal, and lateral, are fused (category 1, following De Villiers, 1968; Franciscus, 1995, 2003b). There is a trace hint of a shallow, separate lateral crest; however, it is most accurately coded as category 1. This single crest demarcates the internal nasal floor from subnasoalveolar clivus, although the sill does not show any real elevation. The anterior nasal spine is moderately developed (category 3; Franciscus, 1995), and there is no evidence for spatulate lateral shelving. The lateral portions of the internal nasal floor are preserved from the posterior narial margin and incisive crest to the general area of the palatomaxillary suture. The floor is completely level with the narial margin.

Subnasoalveolar Clivus

The subnasoalveolar clivus in Dolní Věstonice 15 is well preserved except for the anterior alveolar bone overlying the right medial incisor. The left-side alveolar bone shows less postmortem damage than the right side. Unlike Dolní Věstonice 13, Dolní Věstonice 15 shows very little space between the inferior alveolar margin and the cementoenamel juncture across the dentition (see Chapters 11 and 19). The overall shape of the anterior aspect of the clivus in Dolní Věstonice 15 is not so parabolically rounded as in Dolní Věstonice 14; instead, it is more flattened in its anterior aspect. The anterior aspect of the clivus in Dolní Věstonice 15 also exhibits prominent canine eminences. These shape contrasts may be related to the anomalies of position noted for its dentition (Chapter 11).

Overall Facial Size and Proportion

Overall facial height and breadth in Dolní Věstonice 15 generally lie in the smaller end of the distribution for both measures among the Dolní Věstonice sample in absolute terms (Figures 8.3, 8.5, and 8.6; Sládek et al., 2000: tables 23 and 24).

Left Lateral View

Frontal Region

In left lateral view (Figure 8.58), the frontal bone of the Dolní Věstonice 15 shows relatively steep frontal squama inclination. However, metopion is not as clearly projecting as in some other Pavlovian specimens, and thus there is less angulation between the anterior and posterior metopion planes relative to the frontal chord (i.e., shorter nasion-bregma subtense). In this, it is most similar to Dolní Věstonice 14, with Dolní Věstonice 3 showing a higher frontal subtense relative to the chord and Pavlov 1 showing the relatively lowest value (Figure 8.10; Sládek et al., 2000: table 21).

The glabellar region in Dolní Věstonice 15 is more projecting than in Dolní Věstonice 3, although less than in other Pavlovian specimens. It has both a supraglabellar inflection and an infraglabellar notch. The supraglabellar inflection in Dolní Věstonice 15 is not as pronounced as that found in Neandertals and other premodern *Homo*, but it is similar to that found in other European Upper Paleolithic specimens, albeit at the lower end of the robusticity range (see anterior view description above). The degree of infraglabellar notch development in Dolní Věstonice 15, with allowance for missing nasal bone articulation with nasion, most closely approximates Lahr's (1996) category 2 or 3. It is similar to the pattern found in all of the Pavlovian individuals, other Upper Paleolithic individuals, and to a large degree the Neandertal sample. The greatest degree of infraglabellar notch development is found in Skhul 4 and 5 and Mladeč 5 (Figure 8.11).

The distinct separation of medial, midorbital, and lateral components, described above in anterior view, can be seen and further visualized in lateral view, including the distinct separation of the supraorbital trigone from the superciliary arch, by the obliquely flattened supraorbital sulcus, and the inflated and anteroposteriorly wide supraorbital trigone. The anterior portion of the temporal line on the frontal bone is not as strongly marked as in

Figure 8.57 Lower border of the piriform aperture in Dolní Věstonice 15.

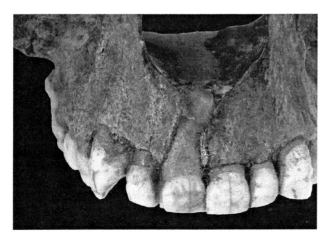

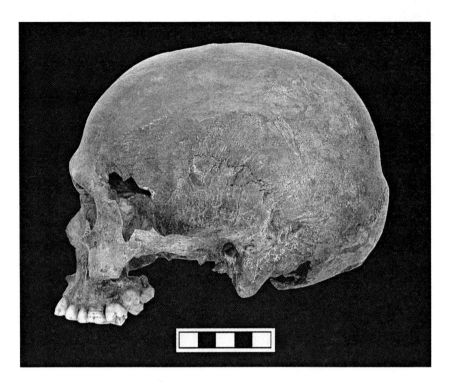

Figure 8.58 Lateral (left) view of the Dolní Věstonice 15 cranium. Scale in centimeters.

other Pavlovian specimens, most closely approximating Dolní Věstonice 14.

Parietal Region

In left lateral view, the vertex of the cranium is located approximately 20 mm posterior to bregma. There is a relatively acute angulation between the midline superior neurocranial profile and the steep posterior parietal plane, reflected in the bregma-lambda chord subtense. Dolní Věstonice 15 has the highest absolute subtense value among all individuals in all comparative samples, and he is also at the high end of the distribution for the bregma-lambda chord (Figure 8.12; Table 8.1; Sládek et al., 2000: table 21). However, proportionately he falls in the middle of the Late Pleistocene distribution, along with Dolní Věstonice 3, 11, and 16.

The left parietal has been reconstructed from at least eight major pieces with varying amounts of filler, and as presently reconstructed it produces an overall reasonable contour along its various curvatures. There are various reconstruction-induced concavities and convexities, although these are minor. The moderate anterior portion of the temporal line found on the frontal fades and becomes even less marked on its continuation along the anterior and posterior parietal. It is not possible to clearly discern the distinction between the superior and inferior temporal lines that usually become most separated in their superoposterior parietal portions. Unlike most of the other Dolní Věstonice and Pavlov specimens, the temporal line does not become more robust in its most inferoposterior segment before crossing to the temporal bone.

Pterion Region

The left pterion region has been heavily reconstructed with a substantial amount of filler, and although elements of all four bones are present, no useful morphological information can be gleaned from it.

Temporal Region

The left temporal bone is virtually intact, and its fit onto the parietal is very good. However, the entire zygomatic region is medially inflected, which accounts for the inability of the zygomatic arch to meet along the zygomaticotemporal suture with a 10 mm mediolateral displacement (i.e., the temporal zygomatic process is too far medially positioned). Also, either the splanchnocranium is hinged too far posteriorly at its contact points or the temporal region is too far anteriorly positioned since there is minimally a 7.5 mm anteroposterior overlap along the area of the zygomaticotemporal suture. The break and misalignment is not actually along the zygomaticotemporal suture but rather in an orthogonal direction to the suture, on the temporal process of the zygomatic bone.

The temporal bone has a moderately well-developed suprameatal crest that continues from the zygomatic process and extends well beyond porion posterosuperiorly to meet up with the poorly developed posterior aspect of the temporal line on the parietal bone. A small, crest-type suprameatal spine with no depression is present (there is no exact analogue shown in Hauser & DeStefano, 1989; Table 8.6). No squamomastoid suture is present, although there is some reconstruction to the mastoid on this side. There is some damage to the mastoid process that has been filled in, but a fair amount of morphology can be determined.

The mastoid process is large in vertical length (27.7 mm in oblique length from porion relative to the Frankfurt Horizontal line), although not as large in other dimensions as the male Pavlovian specimens (12.7 mm in oblique mediolateral width and 26.2 mm in anteroposterior width from its anterior extent to the mastoid foramen orthogonal to the Frankfurt Horizontal). Despite its relatively large vertical height, the process is not particularly rugose. There is no evidence for a tubercle on the mastoid process (anterior or otherwise). The mastoid process extends well below the plane of the juxtamastoid eminence and shows mild lateral flattening. In all of these features, the mastoid process of Dolní Věstonice 13 is modern (Table 8.5). There appears to be a single medium-sized mastoid foramen that is filled in with matrix and does not appear to be sutural; it is basically identical to its right-side counterpart.

Occipital Region

In left lateral view, Dolní Věstonice 15 shows clear occipital bun development. A clear projection begins superiorly just above lambda and continues to just below the external occipital protuberance such that the entire area between projects markedly from the posterior parietal squama and the nuchal plane (Figure 8.59). The projecting region itself is ca. 61 mm in height.

Zygomatic Region

The left zygomatic is relatively well preserved. There is minimal damage along the frontozygomatic suture, but the join is fairly accurate. The join to the maxilla along the zygomaxillary suture appears to involve a fair amount of filler anteriorly, but posteriorly the zygomaxillary suture is intact. There is an unreconstructed break on the zygomatic bone anterior to the zygomaticotemporal suture; the break is in an orthogonal direction to the suture and does not involve any bone posterior to the zygomatic bone (see above).

Just inferior to the inferolateral corner of the orbits on the zygomatic bones, there is an absence of a zygomaxillary ridge, which corresponds to a grade 1 degree of robusticity development (Lahr, 1996). There is a single zygomaticofacial foramen of medium size on the body (Hauser & DeStefano, 1989: figure 35, most similar to a), and strong development of a marginal tubercle that is 16.3 mm in maximum anteroposterior length (figure 36, most similar to d; Table 8.6). The origin for masseter is confined to the inferior border of the zygomatic and does not extend onto the lateral surface as it does in Dolní Věstonice 16.

Orbital Region

In lateral view, the lateral margin of the orbit is posteriorly positioned relative to nasion and other midfacial measures,

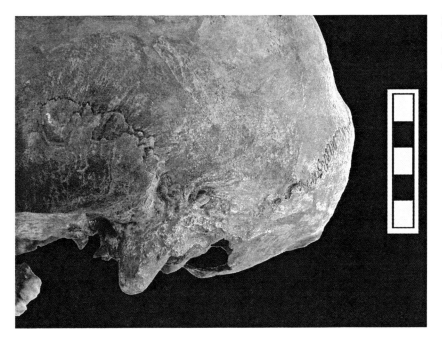

Figure 8.59 Lateral (left) view of the occipital region of Dolní Věstonice 15, showing occipital bunning. Scale in centimeters.

indicating some degree of midfacial projection. In this feature it is similar to the other Pavlovian males, although not as pronounced as that seen in Neandertals on average (Figure 8.14; Table 8.2). The nasolacrimal groove is visible on the medial border of the orbit to its posterior extent.

Infraorbital Region

The left infraorbital region, despite its damage and reconstruction in the anterior infraorbital foramen area, nonetheless provides reliable information on the topography of this region. There was a clear infraorbital depression on Dolní Věstonice 15 that would have accompanied a depressed infraorbital fossa. As noted in anterior view, the infraorbital process of the zygomatic bone has been reconstructed and projects anteriorly over the depressed infraorbital region, creating a horizontal shelf that further accentuates the infraorbital depression; this matches the anatomy on the unreconstructed right side. In lateral view it can again be seen that the lateral aspect of the infraorbital region is oriented entirely in the coronal plane, whereas the medial aspect adjacent to the piriform aperture is oriented in a parasagittal plane, resulting in the typically modern two-plane configuration; the angulation between these two planes in Dolní Věstonice 15 is marked, although normal. Moreover, from this view, it is also evident that in a sagittal plane the lateral aspect of the infraorbital region is obliquely oriented so that the inferior margin of the zygomatic process of the maxilla is posteriorly positioned relative to the inferior orbital. This oblique inflection of the infraorbital region is also a feature of modern humans, in contrast to the vertical or oblique extension of the infraorbital plane that is found in premodern *Homo*.

As can also be seen in anterior view, there is a sharp vertical sill on the lateral aspect of the right medial infraorbital aspect that gives way to a rather recessed posterior area just medial to where the canine fossa is usually found. As noted, this markedly depressed area is not pathological (Figure 8.38). It appears on the left medial infraorbital aspect as well, although the vertical sill is not as sharp on this side. These bilaterally expressed depressed areas give the infraorbital region a deeply excavated appearance (described in greater detail above and in Chapter 19). The anterior root of the zygomatic process of the maxilla is located at the posterior edge of the first molar in Dolní Věstonice 15, as with other Dolní Věstonice crania (Table 8.2).

Nasal Region

In the nasal region, there is a horizontal break along the piriform aperture noted above that can be seen in this view. Nonetheless, the reconstruction faithfully captures the original form of the frontal process. As noted earlier, there is a strong degree of parasagittal eversion of the frontal processes in Dolní Věstonice 15 and a resulting relatively deep alpha point between the inferior nasomaxillary suture location and zygoorbitale. The degree of eversion and alpha point subtense depth in Dolní Věstonice 15, as in Dolní Věstonice 14, is considerably more pronounced than in Dolní Věstonice 3, 13, and 16 (Figure 8.15). It can be seen particularly well in this view that the orientation of the nasomaxillary suture relative to the Frankfurt Horizontal is inferiorly, rather than horizontally, inflected (Table 8.2). The lateral margin of the piriform aperture is complete on the left side except for a small slip of missing bone (ca. 1.6 mm) at the point of the horizontal break. The moderate projection of the anterior nasal spine can be seen in lateral view at the level of the narial margin (category 3; Franciscus, 1995).

Subnasoalveolar Clivus and Alveolar Region

There is a clear inflection at subspinale in lateral view. The clivus also exhibits a mild degree of lower prognathism from this view.

Overall Size and Proportion in Lateral View

The length of the neurocranium in Dolní Věstonice 15 is smaller than in the other Dolní Věstonice and Pavlov males and slightly larger than in Dolní Věstonice 3 (Sládek et al., 2000: table 18). Cranial height is not available for Dolní Věstonice 15 because of the lack of basion. However, the neurocranium is clearly more globular than in the other Pavlovian males. Even with this variation in cranial shape in this sample, they, along with other earlier Upper Paleolithic individuals, although having cranial heights that are comparable to those of Neandertals, exhibit cranial lengths that are shorter, resulting in the characteristically modern globular shape.

Right Lateral View

Gross aspects of neurocranial and splanchnocranial morphology that are evident from either side are covered in the preceding section. The following additional features are revealed in right lateral view (Figure 8.60).

Parietal Region

The right parietal bone has been reconstructed from at least fourteen major pieces with varying amounts of filler, and as reconstructed it produces an overall reasonable contour along its various curvatures. There are various reconstruction-induced concavities and convexities, which to some degree create a greater degree of distortion than in the left side.

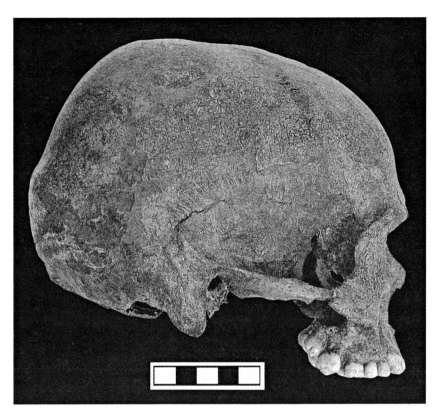

Figure 8.60 Lateral (right) view of the Dolní Věstonice 15 cranium. Note the vertical groove on the temporal bone in the area where the squamomastoid suture is usually found. It has the appearance of a healed antemortem injury rather than a suture; however, its status remains uncertain. Scale in centimeters.

Temporal Region

The right temporal bone is virtually intact, and its fit onto the parietal is good despite the fragmentary makeup of the right parietal. However, there is a poor fit between the zygomatic process of the temporal bone and the zygomatic bone because of the alignment problems described in detail in inferior and left lateral views.

The temporal bone has a moderately well-developed suprameatal crest that continues from the zygomatic process and extends well beyond porion posterosuperiorly to meet up with the poorly developed posterior aspect of the temporal line on the parietal. The right mastoid process is large in vertical length (28.2 mm in oblique length from porion relative to the Frankfurt Horizontal line), although not so large in its other dimensions relative to the male Pavlovian specimens (12.0 mm in oblique mediolateral width and 27.8 mm in anteroposterior width from its anterior extent to mastoid foramen orthogonal to the Frankfurt Horizontal). It should be noted that this last measurement to some degree exaggerates the width of the mastoid because in Dolní Věstonice 15, in contrast to the other Dolní Věstonice males, the actual width of the mastoid is far less than the distance to the mastoid foramen, and the bulk of the mastoid process in Dolní Věstonice 15 is inferiorly narrower compared to its superior distance.

No tubercle is present on the right mastoid process (anterior or otherwise). There is a single medium-sized mastoid foramen; it is filled in with matrix and does not appear to be sutural. It is confined to the temporal bone near asterion at the level of the external acoustic meatus, and it is positioned posteriorly so that it is more visible in posterior than lateral view. It is virtually identical in location to Pavlov 1 and Dolní Věstonice 13 and 14, and it is basically identical to its left-side counterpart. The mastoid process extends well below the plane of the juxtamastoid eminence and shows mild lateral flattening. Unlike the left side, there is no damage to the right mastoid process, and the surface morphology is moderately rugose.

Zygomatic Region

The right zygomatic bone has been reconstructed in its inferoposterior-most aspects with filler, and the most posterior portion of the zygomatic process is missing. There is probably also a minimal amount of bone missing from the frontal process that affects the frontozygomatic suture join and facial hafting by shortening orbital height and allowing play in the hinge angle. The join to the maxilla along the zygomaxillary suture appears to involve a fair amount of filler anteriorly, but in fact, posteriorly, it can be seen that the zygomaxillary suture is completely intact. Details of zygomatic arch misalignment in the mediolateral plane are considered in the inferior view description. In lateral superoinferior view, the alignment is fairly accurate.

Just inferior to the inferolateral corner of the orbit on the right zygomatic bone, there is an absence of a raised ridge, the zygomaxillary ridge, which corresponds to a grade 1 degree of robusticity (Lahr, 1996). There is a single zygomaticofacial foramen of medium size on the body (Table 8.6; Hauser & DeStefano, 1989: figure 35, most similar to a), and there is strong development of the marginal tubercle (16.1 mm maximum anteroposterior length) on the right side (figure 36, most similar to d). It should be noted that the marginal tubercles are really rather large on most Dolní Věstonice individuals, even in otherwise gracile ones such as Dolní Věstonice 15; in fact, Dolní Věstonice 15 has the absolutely largest one.

Infraorbital Region

As can also be seen in anterior view of the infraorbital region, there is a distinct vertical margin above the level of the posterior edge of the second premolar on the lateral aspect of the right medial infraorbital aspect that gives way to a recessed posterior area just medial to where the canine fossa is usually found; it is not as pronounced as on the left side (see discussion above and in Chapter 19). The anterior root of the zygomatic process of the maxilla is located at the posterior first molar/anterior second molar point, and it is therefore somewhat more posteriorly rooted than the left side.

Nasal Region

The lateral margin on the right side of the nasal region is missing a small slip of bone (1 to 2 mm) on the lateralmost edge for most of its course.

Subnasoalveolar Clivus and Alveolar Region

There is a clear inflection at subspinale, but it is hidden in right lateral view by the more lateral alveolar incisor eminences. The clivus exhibits a mild degree of lower prognathism from this view (compare basion-subspinale and basion-prosthion below). The alveolar margin has been damaged postmortem, especially for the molars but less so for the more mesial teeth.

Inferior View

Occipital Region

In inferior view (Figure 8.61) the occipital region is missing large portions of its inferior anatomy, including the

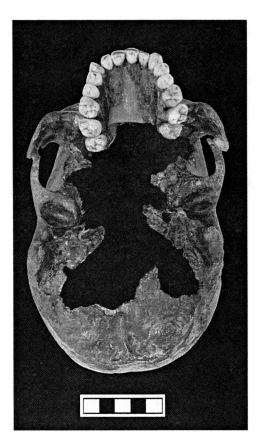

Figure 8.61 Inferior view of the Dolní Věstonice 15 cranium. Scale in centimeters.

foramen magnum, the right- and left-side lateral portions for attachment of rectus capitis posterior major and obliqus superior, the right and left condyles, and the entire basilar portion. As with other basicranial elements in Dolní Věstonice 15, the occipital is asymmetrically aligned because of postmortem compression so that the right asterion point is ca. 14 mm posterior to the left asterion. A median occipital crest posterior to the missing foramen magnum is visible; only the posterior-most portion is well preserved; the rest is missing, along with the foramen magnum, or damaged and filled in. The median occipital crest spans the rather poorly developed inferior nuchal lines. The development of the posterior portion of the occipital crest most closely approximates a category 2 in Lahr (1996). The median crest is not aligned with the long axis of the cranium but instead deviates obliquely through postmortem alteration such that the posterior portion is displaced to the left.

A moderately developed superior nuchal line is evident for its entire course along the occipital, and in midline it joins the *tuberculum linearum*. After taking postmortem deformation into account, the left and right nuchal lines appear to be bilaterally symmetrical. On both left and right sides, the attachment areas for semispinalis capitis are visible; these are broad, although not particularly deeply scalloped. The rectus capitis posterior minor insertions on both sides are also evident and also shallow in relief (Figures 8.62 and 8.63).

The right occipital region superior to the inferior nuchal line bulges further posteriorly than the left side, consistent with the overall direction of basicranial deformation.

Sphenoid Region

A small portion of the inferior-most greater wing of the sphenoid is attached to the left inferior temporal, including the spine of the greater wing. However, none of the adjacent features (e.g., foramina ovale and spinosum) is preserved.

Temporal Region

The inferior aspects of the temporal regions are well preserved on both sides, even though they are highly asymmetrical in alignment. When the mandible is occluded as reconstructed, the right mandibular condyle is 5 to 7 mm too far medial and ca. 10 mm too far inferiorly positioned relative to the glenoid fossa. In occlusion, the left mandibular condyle is 8 to 10 mm too far lateral and ca. 15 mm too far inferiorly positioned relative to the glenoid

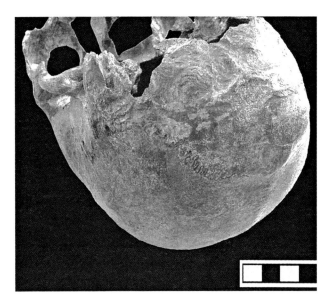

Figure 8.63 Oblique inferior view of the right occipital region of Dolní Věstonice 15. Scale in centimeters.

Figure 8.62 Oblique inferior view of the left occipital region of Dolní Věstonice 15. Scale in centimeters.

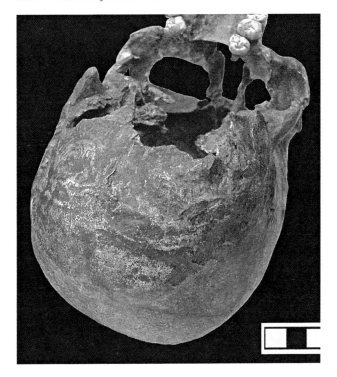

fossa. The mandible is minimally damaged and well reconstructed, so that this lack of occlusal/articular fit is due to at least three factors. First, the anteroposterior hafting of the splanchnocranium around its hinge contact points is slightly off, making the face too orthognathous. Second, the left-right anteroposterior asymmetry in the basicranial elements, including the temporal glenoid fossae is substantial. Third, there is some bone missing along the zygomaticofrontal sutures (especially the right side), and this shortens the vertical dimensions of the face including, slightly, the suproinferior maxillary tooth position. The fit of the mandibular condyles into the glenoid fossae, regardless of proper occlusion, however, is good, indicating that at least the relative mediolateral distance between the temporal bones is accurate.

The left temporal bone shows an occipital groove just medial to the mastoid process, and both temporal bones retain the occipitomastoid sutural edge. Both temporal bones retain the roots and basal elements of the styloid processes. Stylomastoid foramina are present on both sides. The areas immediately medial to both styloid processes are preserved, and the jugular fossae are present. The posterior glenoid eminences and tympanic portions are preserved on both sides. The glenoid fossae are well preserved, and they show no signs of remodeling. The temporal zygomatic processes are completely preserved on both sides. Neither of these show any roughening of the inferior surface posterior to the inferior-most point of the zygomaticotemporal suture, indicating that the most posterior fibers of the origin for masseter are located further anteriorly.

The alignment of the two portions of the zygomatic arch in inferior view is off. There is a 10.0 mm mediolateral displacement (i.e., the temporal zygomatic process is too far medially inflected) on the left side, and on the right there is a similar displacement of 4.6 mm. In addition, either the splanchnocranium is hinged too far posteriorly or the temporal region is too far anteriorly positioned since there is minimally a 7.5 mm anteroposterior overlap along the area of the zygomaticotemporal suture on the left and ca. 3.5 mm of anteroposterior overlap on the right side. As noted above, the break and misalignment are not actually along the zygomaticotemporal suture but instead in an orthogonal direction to the suture and broken on the temporal process of the zygomatic bone.

The attachment areas for masseter along the zygomatic arch are moderately developed and less rugose than for Dolní Věstonice 16. They extend anteriorly to the point where the lateral and anterior aspect of the zygomatic region inflects in an acute angle near zygomaxillare, characteristic of modern humans. Unlike some of the other Pavlovian specimens, the muscle attachment in Dolní Věstonice 15 does extend onto the inferior maxillary surface, ending in a prominent tubercle just medial to the inferior zygomaxillary suture. The right anterior vertical wall of the temporal fossa is more complete than that on the left side.

Inferior Palatal Region

The inferior palate is preserved in its anterior half except along the intermaxillary suture, which is filled in. The posterior half of the maxillary palate and the palatine bones are missing. The palate is fairly shallow, and there is no evidence for midpalatal (intermaxillary) or alveolar tori (Table 8.8). The overall dimensions of the Dolní Věstonice 15 palate are relatively small for the Pavlovian sample (Figure 8.24; Sládek et al., 2000: table 27), and the Pavlovian sample as a whole has smaller palates overall than the Qafzeh-Skhul and Neandertal samples (Figure 8.24; Table 8.4). This is similarly documented for their mandibular dental arcades (Chapter 10).

Inferior Frontal View

In oblique inferior view, bilaterally large supraorbital notches are present on the very medial aspect of each orbit. Internally, the superior orbital roof is largely intact on the right side, and there is no indication of thinning or porosity. The left superior orbital surface is less complete, but it also shows no thinning or porosity. Medially, the nasofrontal suture and a portion of the frontal nasal spine are present. The area just lateral to where the superior-most roofs of ethmoidal cells (the ethmoidal notch) would be is also present on the right side, although there are no remains of the ethmoid bone itself. The anterior-most portion of the ethmoidal notch shows bilateral openings into the midfrontal sinus; this portion is separated from the exposed lateral portion on the left side by a bony partition. Radiographic examination (Vlček, 1991: plate 35 and figure 53) confirms this observation; there is a relatively small frontal sinus delimited to the left side only that does not expand beyond the medial third of the left supraorbital region. All three members of the triple burial show this left-side delimited frontal sinus pattern, with small- to medium-sized, broadly similarly shaped configurations. It is not clear whether a similar pattern is evident in Dolní Věstonice 3 (Jelínek, 1954); however, the left unilateral pattern is clearly different from the bilaterally expanded frontal sinus patterns found in Dolní Věstonice 12 and 16 (Vlček, 1991: plate 56 and figure 53) and the apparent complete lack of any frontal sinus development in Pavlov 1 (Vlček, 1997: figures 33 and 34). As Hauser & DeStefano (1989) point out, there is a well-established pattern, based on extensive family studies, of strong genetic determination of the size and shape of frontal sinuses (Leicher, 1928; Szilvassy, 1982, 1986).

Internal Calvarium View

The large missing area of the sphenoidal and occipital regions allows assessment of some aspects of the internal calvarium. The frontal crest of the frontal bone is present in midline and complete along its full extent. A series of small Pacchionian pits are evident laterally along the midline on both the internal frontal and parietal aspects; none of these are enlarged or grouped together commensurate with Dolní Věstonice 15's relatively young estimated age (Chapter 6). The sulcus for the superior sagittal sinus is well preserved, although not that prominent except in the posterior parietal area. Grooves for the frontal and parietal branches of the middle meningeal vessels are well preserved, especially on the right side. Comparison of the position of the internal occipital protuberance with the external occipital protuberance shows that they are more or less coincident.

Posterior View

In posterior view (Figure 8.64), the sagittal suture can be seen along its entire course. The left lambdoidal suture is better preserved than the right side, and lambda can be determined by using the confluence with the sagittal suture. Detailed scoring results for the lambdoidal suture are presented in Table 8.10. The left and right posterior parietal regions have been reconstructed from a few large pieces, and the fit in the parietal tuber regions is not perfect. Yet, it does not seem to markedly exaggerate or di-

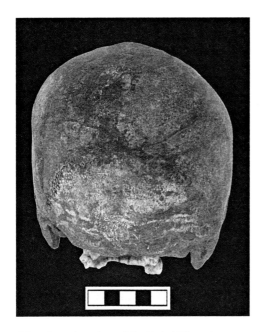

Figure 8.64 Posterior view of the Dolní Věstonice 15 cranium. Scale in centimeters.

minish the original prominence of the tuber. There is a slightly greater posterior projection or prominence in the area of the right superior lambdoidal suture than the left that is also a result of reconstruction and consistent with the anteroposterior asymmetry documented in inferior view. The lateral parietal and temporal squama are straight-sided on the left and slightly more curved on the right side. The left side is also superolaterally inclined and, together with the left parietal tuber and the posterior vertex, form the typical pentagonal configuration that is associated with modern humans. The right side is somewhat more curved along the parietal temporal squama, with the greatest breadth more inferiorly positioned relative to the left. However, it is still qualitatively different from the "en-bombe" Neandertal configuration.

Consistent with the pentagonal shape, the greatest breadth across the parietals is relatively superiorly positioned. Maximum cranial breadth for Dolní Věstonice 15 in posterior view is smaller than in all of the Dolní Věstonice and Pavlov individuals except Dolní Věstonice 3 (Sládek et al., 2000: table 19). Although basion-bregma is not measurable because of a missing foramen magnum area, it can be seen that Dolní Věstonice 15 has a breadth-to-height ratio in the neurocranium that is essentially modern and very different from that found in Neandertals. Moderate sagittal keeling is present in posterior view (category 2; Lahr, 1996), with slight parasagittal depressions that are limited to the anterior sagittal area, not the posterior area. It should be noted that Lahr's sagittal keeling definition was operationalized for extant Asian popula-

tions and does not include parasagittal depressions; it therefore differs from Weidenreich's (1943) definition, which was operationalized for *Homo erectus*.

The superior nuchal lines are present on both sides and meet in midline at the *tuberculum linareum*. The supreme nuchal lines are visible, especially on the left side, coursing nearly 12 mm apart from the superior nuchal lines at their widest extent lateral to the midline and narrowing to ca. 6.5 mm at their maximum narrowest as they converge at the *tuberculum linearum* (Figure 8.62). A 16.5 mm horizontally roughened line in the midline at the *tuberculum linearum* lies just above the most posterior extent of the occipital crest. This roughened line corresponds to the area that encompasses an iniac depression in other Pavlovian specimens.

Also, as on some of the other Pavlovian specimens (e.g., Dolní Věstonice 14 and 16 and, to a lesser extent, Dolní Věstonice 3 and Pavlov 1), there is a depression at the external occipital protuberance that is matched by an additional, distinct depression (the larger of the two) that is more superiorly positioned just below lambda and well above the supreme nuchal line. In Dolní Věstonice 15, the superior depression is well defined and larger than the inferior depression. As with those in the other Pavlovian remains, these depressions in Dolní Věstonice 15 are not the same as the suprainiac fossae found frequently in Neandertals and some other humans since the Dolní Věstonice 15 ones lack the distinctive transverse oval and discretely bordered depressed areas associated with a suprainiac fossa (Hublin, 1978; Trinkaus, 2004). Moreover, in all of the Pavlovian and other earlier Upper Paleolithic individuals that have both depressions (the majority of them), the superior one is always the larger of the two (Franciscus, personal observation). To our knowledge, the broader pattern in other modern humans in this regard is not known.

The posterior view also shows the extension of both mastoid processes below the level of the more medially positioned juxtamastoid eminence, the usual modern human condition (Table 8.5).

Superior View

In superior view (Figure 8.65), the anteroposterior asymmetry discussed above is also evident, particularly in the more posterior bulging of the right occipital region. Also evident is the deviation of the sagittal suture to the left, which results from the greater degree of medially directed flattening of the right parietal reconstruction. Nonetheless, gross assessment of overall neurocranial form indicates that the cranium of Dolní Věstonice 15 is dolicocephalic. Along with Dolní Věstonice 3, it lies in the proportional middle, albeit at the small end, of the range of

Figure 8.65 Superior view of the Dolní Věstonice 15 cranium. Scale in centimeters.

variation of the earlier Upper Paleolithic sample. It is narrower than Dolní Věstonice 13 but rounder than Pavlov 1 and Dolní Věstonice 11/12, 14, and 16. (Figure 8.28; Sládek et al., 2000: tables 18 and 19). The greatest cranial width in Dolní Věstonice 15 is found relatively posteriorly at the parietal tubers in the posterior one-third to one-quarter of the neurocranium. Commensurate with its large cranial capacity (Table 8.9), it has virtually no postorbital constriction. There appear to be no parietal foramina present. Detailed scoring results for the coronal and sagittal sutures are presented in Table 8.10. In oblique superior view, the distinction of the superciliary arch from both the frontal squama and the laterally positioned supraorbital trigone on either side by the obliquely flattened supraorbital sulcus is particularly apparent.

There is a surface irregularity on the frontal to the left of midline ca. 24 mm anterior to bregma; its maximum length is ca. 12 mm, and maximum width ca. 9 mm. There is another small surface irregularity on the right parietal ca. 27 mm to the right of midline and ca. 50 mm posterior to bregma; it is ca. 7 mm in maximum diameter. Finally, ca. 20 mm to the right of the sagittal suture near the most posterior aspect in superior view is an additional small surface irregularity. Each of these is discussed in Chapter 19.

Dolní Věstonice 16

Preservation

The Dolní Věstonice 16 neurocranium consists of the frontal bone, left and right parietal bones, occipital bone, left and right temporal bones, and small fragments of the sphenoid bone. The neurocranium is damaged and missing elements, mainly in the basicranium. The bones of the calvarium are reconstructed on the right side in the area of the inferior margin of the parietal bone, the squamal portion of the right temporal bone, and on the right side of the nuchal plane. The left and right temporal fossae are mostly missing and reconstructed. The splanchnocranium preserves both left and right maxillae, trace remnants of the left nasal bone, and the left and right zygomatic bones. Both maxillae are damaged in the infraorbital region.

The neurocranium and splanchnocranium are joined through the frontal and zygomatic bones via a 12 mm connection on the left side and a 13 mm connection on the right side. There is also a join between the frontal bone and the right maxilla, although here the connection is partly reconstructed, and its original extent is difficult to determine. Reliable contacts also exist between the zygomatic and temporal bones on both left and right sides. Nonetheless, articulation of the mandibular condyles with the glenoid fossae produces a slightly posteriorly positioned maxillary bite point, indicating that the hinging of the splanchnocranium at the frontozygomatic points and preserved frontonasal point might be slightly off. This could be accompanied by reconstructed contact points between the posterior maxillae and the basicranium that are slightly too short.

There is a degree of asymmetry in the orbital and midfacial regions, evident in both anterior and lateral views, that involves postmortem damage and may also include antemortem alterations. The left glenoid fossa, tympanic region, and mastoid process are all slightly more posteriorly placed relative to the right side by at least 10 mm, although the mediolateral distances relative to midline are equidistant.

Postdepositional neurocranial asymmetry is evident in superior view, and it affects both halves. The left side is medially inflected in the region near the lateral coronal suture, and the right side, which is more fragmentary and has been reconstructed, is laterally expanded. The combination of the two produces a long axis that tends to curve toward the left side relative to the supraorbital and frontal bone midline alignment. There is also asymmetry in the anteroposterior location for each parietal tuber, the right side tuber being 10 mm or more anteriorly positioned relative to the left side, consistent with the asymmetry observed in the basicranium.

Anterior View

Frontal Region

The frontal bone of Dolní Věstonice 16 (Figure 8.66) is one of the most complete and undistorted bones in the cranium. There is minimal root mark etching, but the surface detail is very well preserved and virtually all external features are visible. There is postmortem damage along the medial portions of the coronal suture and infilling with reconstructive filler. However, the lateral portions, especially on the left side, are visible. There is some frontal keeling in midline that is apparent from an oblique (inferofrontal) view; it is weakly to moderately expressed, but it is especially evident in the area of metopion. It is similar to Dolní Věstonice 13 for this trait, but the areas lateral to either side of the keeling are slightly more depressed for Dolní Věstonice 16. There is no evidence of a metopic suture, but there is a trace presence of a supranasal suture (Hauser & DeStefano, 1989: plate VI, h), as well as a possible right-side frontal groove (Table 8.3; see below). Approximately 24 mm lateral to metopion on either side are very weakly expressed frontal eminences.

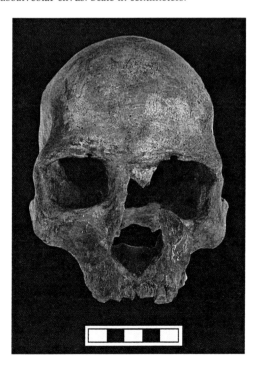

Figure 8.66 Anterior view of the Dolní Věstonice 16 cranium. Note the unilateral presence of a straight, anteroposteriorly oriented groove on the frontal bone superior to the right orbital margin and just medial to the right temporal line that is either a minor healed injury or an epigenetic trait. Note also the altered left infraorbital region, lower piriform aperture and subnasoalveolar clivus. Scale in centimeters.

An antemortem healed lesion is located ca. 30 mm anterior to bregma, nearly in midline. The maximum extent of the healed area is 17.5 mm in anteroposterior length, ca. 12.5 mm in mediolateral width, and 2 mm in depth. Approximately 40 mm superior to the right orbital margin on the frontal bone and 19.3 mm medial to the right temporal line, there is a straight anteroposteriorly oriented groove that is 14.5 mm in length. It is located in the general area in which frontal grooves marking the passage of branches of the supraorbital nerve posteriorly into the scalp are found in varying frequencies. There is no such groove on the left side, although the trait can be unilateral in expression (Hauser & DeStefano, 1989). In addition, much of the frontal bone exhibits a very fine surface porosity that is particularly evident in the supraorbital region but also extends to much of the frontal squama and the parietals in midline. All of these features are discussed in greater detail in Chapter 19.

The least frontal breadth of Dolní Věstonice 16 (97 mm) is the smallest of the Dolní Věstonice and Pavlov males, as is its maximum frontal breadth (118 mm). In the shape relationship between these two variables, it is similar to all of the Pavlovian sample except Dolní Věstonice 14, who shows a relatively larger least frontal breadth. As noted above, the relationship between these two variables is very similar in the European earlier Upper Paleolithic samples (Figure 8.2), but the Qafzeh-Skhul and Neandertal samples have relatively greater least frontal breadth values.

Supraorbital Region

The supraorbital region of Dolní Věstonice 16 shows a distinct separation of medial, midorbital and lateral components. The glabellar region is pronounced in midline, and this robust midline projection extends to the medial half of both right and left supraorbital areas. The superciliary arches each extend superiorly and laterally for ca. 24 mm prior to turning inferiorly and then fading out onto the midorbital regions. The total length of this medial component (encompassing glabella) for both sides combined is ca. 75 mm. The supraorbital trigones are both ca. 6.5 mm thick, and they are distinctly separated from the medial superciliary arch by the obliquely flattened supraorbital sulcus, which is ca. 5 mm thick on either side. The supraorbital sulcus is distinctly separated both mediolaterally and superoinferiorly from the superciliary arch, and it is visible as a distinct entity for virtually the entire length of the supraorbital margin. The supraorbital trigone is inflated and widened anteroposteriorly, and it has a rugose raised anterolateral surface; they correspond most closely to Lahr's (1996) grade 4. This separation of medial, midorbital, and lateral components of the supraorbital region, as with other Pavlovian and earlier Upper Paleolithic humans, separates in Dolní Věstonice 16 from the pattern found in Neandertals (Table 8.2).

Orbital Region

There is a degree of asymmetry in the left and right orbital regions in Dolní Věstonice 16. The right orbit has normal anatomy and is square-shaped, whereas the left orbit, altered in its inferior border (detailed below), shows a more rounded configuration. Nonetheless, even with these factors, Dolní Věstonice 16 shares with the other Dolní Věstonice and Pavlov individuals and other modern specimens orbits that are lower relative to orbital breadth than Neandertals (Figure 8.4; Tables 8.1 and 8.2). Interorbital breadth in Dolní Věstonice 16 is similar to that in Dolní Věstonice 13, and both are slightly smaller than in Dolní Věstonice 15, although relative to bifrontal breadth (Figure 8.3); interorbital breadth in all of them, along with Dolní Věstonice 3, are similarly proportioned.

The superior orbital margins in Dolní Věstonice 16 are rounded rather than sharp (Chapter 7). Both superior orbital margins show bilateral, medium-sized notches on their medial aspects (Hauser & DeStefano, 1989: plate VIII, most similar to d). Internally, the superior orbital roof is intact only for less than 20 mm in both left and right sides. Although there is some filler in both orbits, virtually nothing is left of the medial, lateral, inferior, and posterior internal orbital walls. The inferolateral borders of both orbits are robust in that they are rounded and thick rather than sharp. However, the borders are also raised in relation to the small amount of orbital floor that is present, and they therefore do not attain the most robust grade; they correspond most closely to Lahr's (1996) grade 2. Just inferior to the inferolateral corner of the orbit on the zygomatic bone, there is a small raised zygomaxillary ridge that corresponds to a grade 2 degree of robusticity.

The lower orbital margin on the right side is fairly intact. The distance between zygoorbitale points can be approximated by taking into account the altered anatomy of the left infraorbital border. The breadth across these points relative to facial height (Figure 8.5) in Dolní Věstonice 16 is similar to the other Dolní Věstonice males, and indeed all of the comparative samples show the same relationship. The inferomedial and medial orbital rims are present on the right orbit; however, the lacrimal bone is missing. The left lower orbital rim has been markedly altered. Whether this is due to healed trauma or some combination of the postmortem deformation and subsequent reconstructive efforts is not grossly obvious (Chapter 19). Nonetheless, several aspects of the infraorbital and zygomatic region, as well as mandibular occlusion and tooth wear, provide important clues. These are detailed in the following sections.

Infraorbital Region

The infraorbital region on both sides is damaged in the precise area of the infraorbital foramen and has been filled and reconstructed. The degree of reconstruction is greater in this area than for some of the other Dolní Věstonice individuals and requires a somewhat greater degree of interpretation in order to infer the topography of the infraorbital region. As reconstructed, the right infraorbital region shows a canine fossa, whereas the left shows a flatter infraorbital region and also a different orientation of the infraorbital plane. The latter has been interpreted as showing similarity to, and genetic continuity with, the Neandertals, given their usual inflated, or "puffy," infraorbital configuration (Wolpoff, 1999). For a variety of reasons discussed below, the right-side infraorbital region is probably the more accurate of the two.

About 9 mm of maxillary bone is retained medial to the zygomaxillary suture on the right side, and this medial aspect of the maxilla is oriented in the coronal rather than parasagittal plane, as in the other Dolní Věstonice individuals. Moreover, the frontal process of the right maxilla is everted such that it is oriented predominantly in a parasagittal plane. Therefore, in the transverse plane, the right-side infraorbital region shows the typically modern two-plane configuration (Table 8.2). The infraorbital process of the right zygomatic bone projects anteriorly over the depressed infraorbital region just lateral to the damaged and filled area, creating a horizontal shelf that accentuates the infraorbital depression in a horizontal transect. These features are virtually always found in association with a canine fossa, and it is reasonable to infer that Dolní Věstonice 16's right infraorbital area also had a canine fossa.

The left infraorbital region is misaligned because of the unusual lower border of the orbit, and aspects of both the nasal region and dentognathic features described below. Therefore, prior comments on its "puffy midface" (Wolpoff, 1999) are probably based on pathologically or otherwise altered anatomy.

Nasal Region

The frontal process of the right maxilla has been reconstructed with filler on its external aspect; however, there is a fair amount of actual bone underlying it on its internal aspect. The right nasomaxillary suture has been reconstructed to some degree, but based on internal bone the result is probably close to the original. The more complete right frontal process shows a fair degree of parasagittal eversion and a resulting relatively deep alpha point between the inferior nasomaxillary suture location and zygoorbitale. Even so, this is not as pronounced in Dolní Věstonice 16 as in some other Dolní Věstonice specimens, especially Dolní Věstonice 15 (see lateral view). The orientation of the nasomaxillary suture is inferiorly, rather than horizontally, inflected.

Nasal breadth for Dolní Věstonice 16 (28 mm) is similar to that for Dolní Věstonice 14, and both lie between the highest and lowest values for Dolní Věstonice males, toward the upper end of the Dolní Věstonice sample distribution (Table 8.4; Figure 8.7; Sládek et al., 2000: table 26 with corrected value noted here). Relative to nasal height, this places it, with Dolní Věstonice 14, below most Neandertals and Qafzeh-Skhul individuals in relative nasal breadth. Nasion is superiorly set relative to the superior nasomaxillary points. There appears to be present a small portion of the most superior part of the left nasal bone; there may also be a small slip of nasal bone running along the right nasomaxillary suture, although fill is obscuring the details. The left maxilla is missing the entire frontal process. The lateral margin of the piriform aperture is mostly complete on the right side, and the left side is also mostly complete along its margin.

The lateral walls of the nasal aperture are sharp on both sides. There is no evidence of the inner crest/outer margin separation. The internal lateral walls of the nasal fossa are complete on the right side from the piriform aperture margin posteriorly to the general area bounding the nasolacrimal groove, although the groove itself is either not present or obstructed by filler. The ethmoidal crest is difficult to discern. The more inferiorly located conchal crests are well preserved on both right and left sides. The right-side conchal crest shows a predominantly horizontally positioned anterior component and an obliquely positioned posterior component and the left side shows a predominantly oblique orientation. The conchal crests are not visible in anterior view. A modest turbinal crest is located on either side and contributes to a degree of mediolateral thickening in this area that is often associated with turbinal crest development in extant and fossil *Homo*.

Evaluation of the narial margin at the junction of the internal nasal floor and subnasoalveolar clivus is complicated in this specimen by misalignment of the maxillae and an altered anterior nasal spine configuration. Relative to the right maxilla, the left is relatively inferiorly positioned at the midline by ca. 4 mm, as well as more anteriorly displaced by ca. 4 mm. As discussed in detail in Chapter 19, these and other nasopalatal asymmetries (described below) are due to a growth abnormality (specifically a cleft palate) or, less likely, a healed trauma to the maxilla. Given this, interpretation of the narial margin pattern is complicated (Figure 8.67).

On both sides there appears to be partial fusion of the spinal and turbinal crests, resulting in the formation of a narrow and relatively shallow *fossa intranasalis*, whose point of fusion is relatively far medially from the ante-

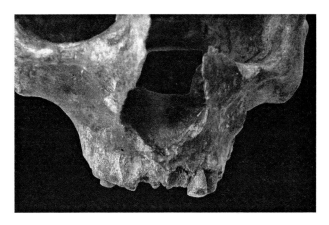

Figure 8.67 Lower border of the piriform aperture in Dolní Věstonice 16. Note the altered left infraorbital region, lower piriform aperture, and subnasoalveolar clivus.

rior nasal spine and whose maximum separation is 3.7 mm and relatively far posteriorly. In addition, there appears to be a very lightly delineated separate lateral crest that swings inferiorly onto the clivus (maximum separation 6.5 mm on the left, 4.5 mm on the right) and then superiorly toward the deformed anterior nasal spine. This configuration can be coded as either a category 3 or 6, following De Villiers (1968), depending on the weighting of various components of the scoring system. It has been previously scored as a category 3 (Franciscus, 1995, 2003b). However, it may be more reliably scored as category 6. Whichever score is appropriate, it is the fused turbinal and spinal crests on either side, and not the lateral crests, that demarcate the internal nasal floor from the subnasoalveolar clivus. The anterior nasal spine is well developed in a mediolateral direction with what might have been some spatulate development laterally. Its anterior prominence is difficult to determine; its superior aspect appears markedly projecting (category 4 or marked; Franciscus, 1995); however, inferiorly it is slight to moderately developed (category 2 or 3) because of the remodeled configuration.

The internal nasal floor is not preserved along the midline (perhaps because of the cleft palate possibility). The lateral portions are preserved and show that the floor would have been level.

Both maxillae retain enough internal anatomy to show both maxillary sinuses through breaks in the lateral walls of the internal nasal fossa. These areas are heavily filled with consolidant and cannot be evaluated.

Subnasoalveolar Clivus

The subnasoalveolar clivus in Dolní Věstonice 16 is considerably altered, as already noted to some degree above. Its height has been relatively shortened because of marked alveolar alteration. The degree to which clivus height may have also been affected by the developmental abnormalities or trauma is unknown. From this perspective, the difference in the vertical positioning of the alveolar apices for the left and right medial incisors can be seen. There is a very small fenestration ca. 1 mm in diameter on the clivus ca. 13 mm above the lateral incisor that is probably postmortem damage.

Left Lateral View

Frontal Region

In left lateral view (Figure 8.68) the frontal bone of the Dolní Věstonice 16 shows a relatively vertical frontal squama up to metopion, where the inclination then becomes much more angled toward bregma. In this, it is most similar to Dolní Věstonice 13–15, with Dolní Věstonice 3 showing higher frontal subtense relative to the chord and Pavlov 1 showing the relatively lowest value (Figure 8.10; Sládek et al., 2000: table 21). It therefore is characteristic of all modern human samples and contrasts with the lower values found among the Neandertals (Figure 8.10; Table 8.1). The glabellar region is markedly projecting, with both a supraglabellar inflection and an infraglabellar notch. The supraglabellar inflection in Dolní Věstonice 16 is not the pronounced supraglabellar sulcus found in Neandertals and other premodern *Homo*, but it is similar to that found in other European Upper Paleolithic specimens (see anterior view description above). The degree of infraglabellar notch development in Dolní Věstonice 16 most closely approximates Lahr's (1996) category 3, and it is similar to the pattern found in all of the Pavlovian individuals, other Upper Paleolithic individuals, and to a large degree the Neandertal sample. Metric comparison for this feature shows Dolní Věstonice 16 to have the most pronounced infraglabellar notch development among the Pavlovian individuals (Figure 8.11).

The distinct separation of medial, midorbital, and lateral components, described above in anterior view, can be further visualized in lateral view. The anterior portion of the temporal line on the frontal bone is well marked and rugose. There is a distinct lateral bulging area just posterior and inferior to the temporal line that is sometimes present as normal variation. The healed lesion anterior to the coronal suture described in anterior view can be seen in lateral view; its full extent is more apparent in this view than in anterior view.

Parietal Region

In left lateral view, the vertex of the cranium is located 24.6 mm posterior to bregma. There is a relatively even

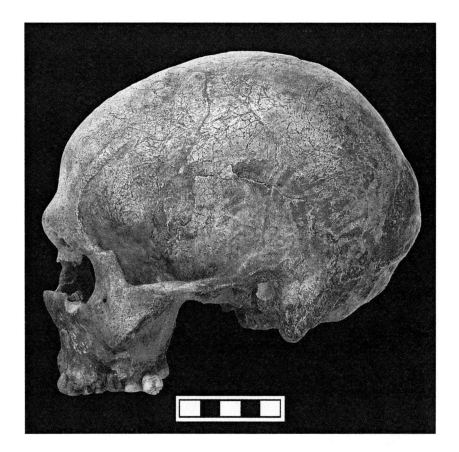

Figure 8.68 Lateral (left) view of the Dolní Věstonice 16 cranium. Note the altered infraorbital region. Scale in centimeters.

curvature between the midline superior neurocranial profile and the posterior parietal plane, as opposed to the more angled configuration in Dolní Věstonice 13 and 15. Dolní Věstonice 16 is most similar to Dolní Věstonice 3, 11, and 15, with a somewhat higher relative value for the bregma-lambda subtense than that of Dolní Věstonice 14 and, especially, Pavlov 1 but well below that of Dolní Věstonice 13. Together, the Dolní Věstonice and Pavlov sample and other modern humans show relatively higher values for both the chord and the subtense compared to Neandertals but with similar proportions (Figure 8.12; Table 8.1; Sládek et al., 2000: table 21).

The left parietal is relatively intact except for a large break posterior to the coronal suture, although there appears to be some postmortem medially directed deformation in that part of the parietal approaching the pterion region. The rugose anterior portion of the temporal line found on the frontal fades and becomes less marked on its continuation along the anterior and posterior parietal. It is difficult to discern the distinction between the superior and inferior temporal lines that usually become most separated in their superoposterior parietal portions.

Pterion Region

The left pterion region of Dolní Věstonice 16 is damaged and has been heavily reconstructed. Although elements of all four bones are present, no useful morphological information can be gleaned from pterion.

Temporal Region

The temporal bone is largely intact except along the most peripheral aspects of the squamosal suture. The fit of the temporal bone onto the parietal is very good, and the connection between the temporal and zygomatic bones along the zygomatic arch is also good, with very little filler along the zygomaticotemporal suture. The temporal bone is robust, although the suprameatal crest in particular is not as robust as on some of the other Dolní Věstonice individuals and does not extend as far beyond porion posterosuperiorly to meet up with the posterior aspect of the temporal line on the parietal, as in other individuals. A small, crest-type suprameatal spine with moderate depression is present (no exact analogue is shown in Hauser & DeStefano, 1989; Table 8.6). Superficially, there is a squamomastoid suture that appears completely closed along its length. However, it is more likely that this feature is instead the limit of the sternocleidomastoideus attachment.

There is some damage to the mastoid process that has been filled in, but a fair amount of morphology can be determined. The mastoid process is large (29.5 mm in oblique length from porion relative to the Frankfurt Horizontal line; 16.1 mm in oblique mediolateral width; 27.9 mm in anteroposterior width from anterior extent to mastoid foramen orthogonal to the Frankfurt Horizontal). No tubercle is present on the left mastoid process anterior or otherwise. The mastoid process extends well below the plane of what is preserved of the juxtamastoid eminence and shows no lateral flattening. In addition to its large size, the mastoid process is rugose, especially in its inferoposterior portion. In all of these features, the mastoid process of Dolní Věstonice 16 is modern rather than Neandertal-like (Table 8.5). The left mastoid process also shows one large and a possible second, medium-sized foramen, which are both filled in with matrix. If the second is a foramen, then their locations are identical to those found on the right side (Table 8.6).

Occipital Region

In left lateral view, Dolní Věstonice 16 shows occipital bun development. Inferiorly, there is marked projection of the external occipital protuberance and the superior nuchal line, creating a prominent inflexion relative to the nuchal plane. The superior projection near lambda, although not as prominent as the inferior relief, is still prominent, and the overall bun development is more marked than that found in Dolní Věstonice 13 and 14, somewhat less marked than in Dolní Věstonice 15 and Pavlov 1, and comparable to that in Dolní Věstonice 2, 3, and 11/12. The occipital chord and subtense configuration in Dolní Věstonice 16 is similar to that in the other Dolní Věstonice and Pavlov males, with moderately large values for each compared to the considerably smaller respective values for Dolní Věstonice 3. In relative terms, Dolní Věstonice 16 has a slightly higher lambda-opisthion subtense. Note that three out of four Neandertals fall in the modern distribution (Figure 8.13; Table 8.4; Sládek et al., 2000: table 21), a pattern also evident in Neandertal versus early modern human immature occipital bones (Trinkaus, 2002a).

Zygomatic Region

The left zygomatic bone is well preserved. There is some damage along the frontozygomatic suture, and the join contains some filler. The join to the maxilla along the zygomaxillary suture is more problematical and involves a fair amount of filler. The connection between the zygomatic and temporal bones along the zygomatic arch is complete and relatively undistorted, with very little filler along the zygomaticotemporal suture. The anterolateral aspect of the zygomatic bone shows some rugosity, but there is only a rather small or trace development of the zygomaxillary ridge just inferior to the inferolateral corner of the orbit, which corresponds to a grade 2 degree of robusticity (Lahr, 1996). There is a single medium- to large-sized zygomaticofacial foramen on the body (Hauser & DeStefano, 1989: figure 35, most similar to a), very similar to that in Dolní Věstonice 15 in placement and size (Table 8.6). There is also strong development of a marginal tubercle that is 41.1 mm in maximum anteroposterior length (Hauser & DeStefano, 1989: figure 36, most similar to d). The origin of masseter extends up onto the lateral surface of the bone.

There is a very straight, long line running obliquely along the lateral face of the frontal process of the zygomatic bone that courses from the marginal tubercle to the lateral orbital margin and then deflects slightly inferiorly along the lower orbital margin; its total length measures 28.5 mm. It is very different from the other surface defects such as root mark etching in its linearity. It appears to have matrix embedded in it, arguing against preparation damage, and it could be interpreted as a cut mark. However, it is a completely isolated mark and therefore does not seem consistent with cut mark damage. Moreover, the burial of Dolní Věstonice 16 and its in situ position (Chapter 4) argues against any intentional postmortem manipulation of the remains.

Orbital Region

In left lateral view, the lateral margin of the orbit is posteriorly positioned relative to nasion and other midfacial points, indicating some degree of midfacial projection. In this feature it is similar to the other Dolní Věstonice males, although not as pronounced as that seen in Neandertals on average (Figure 8.14; Table 8.2). The medial infraorbital border and medial border of the orbit are missing or too damaged to assess on this side.

Infraorbital Region

The left infraorbital region has suffered damage and is heavily reconstructed. There is also the issue of the misalignment of the two maxillae, as described above. As presently reconstructed, the infraorbital region appears to be inflated and to lack the canine fossa depression evident on the other Dolní Věstonice individuals and most modern humans, and thus to show the Neandertal condition. It also appears to have a Neandertal-like parasagittal alignment rather than the two-plane configuration described for all other Dolní Věstonice individuals. However, based on the more intact right side and close scrutiny of the morphology, this appearance of infraorbital inflation and parasagittal alignment is due entirely to the reconstruction (see anterior view and right-side lateral view descriptions).

Enough of the zygomatic process medial to the zygomaxillary suture is present to show the oblique configuration of the lateral infraorbital plane, such that the inferior margin of the zygomatic process of the maxilla is posteriorly positioned relative to the inferior orbital rim (a feature of modern humans), rather than the vertical orientation associated with maxillary inflation on Neandertals. An inflated infraorbital region with an oblique lateral infraorbital alignment does not seem reconcilable and therefore further suggests an erroneous reconstruction on this side and/or a secondary consequence of a pathological alteration. The infraorbital process of the zygomatic bone projects anteriorly over the depressed portion of the lateral maxilla, creating a horizontal shelf over the preserved bone.

Nasal Region

In the nasal region, there is a break along the frontal process of the left maxilla ca. 7.5 mm above the level of the lower orbital margin, with a resultant 18.2 mm gap between this point and what remains of the bone near the nasofrontal suture. This area and the adjacent region lateral to the piriform aperture is too damaged to assess the degree of frontal process eversion and the course of the nasomaxillary suture. Only the tip of the inferior nasomaxillary suture area is preserved on this side. The lateral margin of the piriform aperture is largely complete on the left side, albeit damaged at several places along its margin. The anterior nasal spine shows very little projection from this side, although this is due to some degree to pathological alteration.

Subnasoalveolar Clivus and Alveolar Region

The anterior root of the zygomatic process of the maxilla is located at the anterior edge of the first molar rather than the more posterior placement found in Neandertals (Table 8.2). There is a very large periapical defect above the roots for the first molar (Chapter 19), and marked alveolar remodeling is evident from this perspective. To some extent, the splanchnocranium appears to be too orthognathous. This is evident to some degree when the mandibular condyle is articulated into the glenoid fossa, which produces a slightly posteriorly positioned maxillary bite point. As mentioned above, the hinging of the splanchnocranium at the frontozygomatic points and preserved frontonasal point might be slightly off, and this could be accompanied by reconstructed contact points between the posterior maxillae and the basicranium that are slightly too short. One problem with this interpretation, however, is that the zygomatic arches, as presently reconstructed, will not tolerate too much lengthening, especially the left side, where the zygomaticotemporal sutural reconstruction seems accurate.

Comparisons of the relative projection of the subnasoalveoar clivus (Figures 8.16 and 8.17) indicate that

Dolní Věstonice 16 generally falls in the range of the Dolní Věstonice and earlier Upper Paleolithic scatter, although given the alterations to the clivus detailed above, such comparisons should be seen as approximate.

Overall Size and Proportion in Lateral View

Both the height and the length of the neurocranium in Dolní Věstonice 16 are larger than those in the other Pavlovian males for which these measurements are available (Figure 8.18; Sládek et al., 2000: tables 18 and 19). Overall facial height and facial length in Dolní Věstonice 16 are the largest among the Dolní Věstonice individuals for which reliable measurements are available (Figures 8.19 and 8.20; Sládek et al., 2000: tables 22 and 24). In those cases in which Dolní Věstonice 13 exceeds Dolní Věstonice 16 (e.g., basion-prosthion length), it is likely that reconstructed facial hafting has exaggerated the value in the former. In a wider comparative context, Dolní Věstonice 16 lies toward the upper end of the range for early modern humans in general for these measures. Moreover, the fact that Dolní Věstonice 16 probably had reduced facial height and length because of a combination of degenerative and other factors, makes this comparison a conservative one. Neandertals show larger values for facial height and length.

Right Lateral View

Gross aspects of neurocranial and splanchnocranial morphology that are evident from either side are covered in the preceding section. The following additional features are revealed in right lateral view (Figure 8.69).

Frontal Region

The anterior portion of the right temporal line on the frontal bone is well marked, although not as rugose as that found on the left side. The groove (or possible lesion), found in the general area where frontal grooves are often found and described in anterior view, is apparent from this view of the frontal region.

Parietal Region

The right parietal has been heavily reconstructed from twelve pieces with a substantial amount of filler, and some of the asymmetry evident in superior view (see below) can be accounted for by the reconstruction. The anterior portion of the temporal line found on the frontal fades and becomes less marked on its continuation along the anterior and posterior parietal. It is difficult to discern

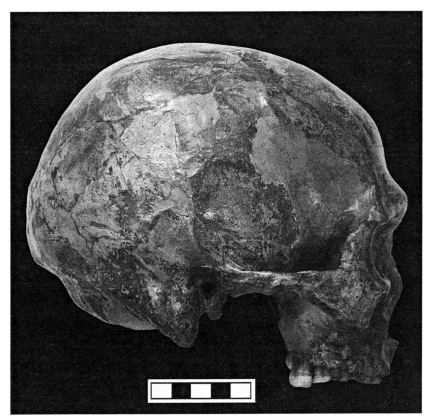

Figure 8.69 Lateral (right) view of the Dolní Věstonice 16 cranium. Note the unilateral presence of a straight, anteroposteriorly oriented groove on the frontal bone superior to the right orbital margin and just medial to the right temporal line that is either a minor healed injury or an epigenetic trait. Scale in centimeters.

the distinction between the superior and inferior temporal lines.

Pterion Region

The right pterion region has been completely filled in and sculpted by a large amount of filler. No observations are possible.

Temporal Region

The squamous of the right temporal bone has been reconstructed from about five pieces. The inferior portion, including the mastoid process, tympanic process, and zygomatic arch, is well preserved. Despite the reconstruction, the fit of the right temporal onto the parietal is reasonable except for some lateral bulging, described in superior view (see below), and the connection between the temporal and zygomatic bones along the zygomatic arch is also good, with little filler along the zygomaticotemporal suture.

The right temporal bone is robust, although the suprameatal crest is not as robust as on some of the other Dolní Věstonice individuals and does not extend as far beyond porion posterosuperiorly to meet up with the posterior aspect of the temporal line on the parietal, as in other specimens. As with the left side, the right side shows a small, crest-type suprameatal spine (the spine is larger here than on the left side) with moderate depression (there is no exact analogue shown in Hauser & DeStefano, 1989; Table 8.6). As with the left side, the right side shows a superficial appearance of a squamomastoid suture that appears completely closed. However, as with its counterpart, it is more likely that this feature is instead the limit of the sternocleidomastoideus attachment.

The mastoid process is large (31.2 mm in oblique length from porion relative to the Frankfurt Horizontal line; 16.1 mm in oblique mediolateral width; 28.2 mm in anteroposterior width from its anterior extent to the mastoid foramen orthogonal to the Frankfurt Horizontal). Unlike the left side, where there is no anterior mastoid tubercle, the right side does have one; however, it is somewhat diffuse. Moreover, there are two additional tubercles on the mastoid, one positioned more posteriorly and one positioned more inferiorly. The mastoid process extends well below the plane of what is preserved of the juxtamastoid eminence and shows no lateral flattening. The right mastoid process is not as rugose as its left counterpart. In all of these features, with the exception of the anterior mastoid tubercle (which, in any case, is very diffuse and supernumerary), the mastoid process of Dolní Věstonice 16 is modern (Table 8.5). In addition, the right side shows two medium-sized mastoid foramina that are both filled in with matrix. They are ca. 8.5 mm apart, and neither appears to be sutural. Both are confined to the temporal near asterion at the level of the external acoustic meatus (Table 8.6).

Zygomatic Region

The right zygomatic bone is well preserved. There is some damage along the frontozygomatic suture, and the join contains some filler. The join to the maxilla along the zygomaxillary suture also involves a fair amount of filler. The connection between the zygomatic and temporal bones along the zygomatic arch is nearly complete and relatively undistorted, with relatively little filler along the zygomaticotemporal suture. The frontal process of the zygomatic has a postmortem crack horizontally across the bone superior to the zygomatic foramen. The anterolateral aspect of the zygomatic shows some rugosity, but there is only a small or trace development of the zygomaxillary ridge just inferior to the inferolateral corner of the orbit, which corresponds to a grade 2 degree of robusticity (Lahr, 1996). In terms of epigenetic traits from this view (Table 8.6), there is a single medium- to large-sized zygomaticofacial foramen on the body (Hauser & DeStefano, 1989: figure 35, most similar to a), that is very similar to that in Dolní Věstonice 15 in placement and size. There is also medium development of a marginal tubercle that is 41.8 mm in maximum anteroposterior length (figure 36, most similar to c). The origin of masseter extends up onto the lateral surface of the bone.

Orbital Region

The lateral margin of the orbit is posteriorly positioned relative to nasion and other midfacial measures, indicating some degree of midfacial projection. The medial infraorbital border and medial border of the orbit have been reconstructed with filler overlying bone that obscures details.

Infraorbital Region

The right infraorbital region is not as heavily reconstructed as the left side, and it does not show the inflated morphology that is present on the left side (the latter most likely due to inaccurate reconstruction and/or pathology as discussed above). From this view, the two-plane configuration described for other Dolní Věstonice individuals is present, if in a more muted condition. Enough of the zygomatic process medial to the zygomaxillary suture is present to show the oblique configuration of the lateral infraorbital plane, such that the inferior margin of the zygomatic process of the maxilla is posteriorly positioned relative to the inferior orbital rim (a feature of modern humans), rather than the vertical orientation associated with maxillary inflation of Neandertals. The infraorbital process of the zygomatic bone projects anteriorly over the depressed portion of the lateral maxilla of the zygomatic process that remains, creating a horizontal shelf over the preserved bone.

Nasal Region

The right frontal process of the maxilla shows some eversion in lateral view as measured by the depth of the alpha point between the inferior zygomaxillary point and zygoorbitale, although to a lesser degree than in all of the other Dolní Věstonice individuals. In this, it comes closest to the Neandertals of all of the individuals across the modern comparative samples (Figure 8.15). To a degree, this provides some support for Wolpoff's (1999) contention that Dolní Věstonice 16 exhibits some Neandertal facial features (Table 8.2). Nonetheless, given uncertainty about the extent to which maxillary developmental or traumatic alteration occurred in this individual, this contention remains problematic and, even if true, would be only one isolated Neandertal feature in a cranium that shows no other Neandertal autapomorphies.

The nasomaxillary suture is present on the right side with what is possibly a narrow strip of attached nasal bone. The course of the suture is inferiorly rather than horizontally oriented, showing the modern human condition. The lateral margin of the piriform aperture is largely complete on the right side, albeit minutely damaged at several places along its margin. The anterior nasal spine shows substantial projection from this side, although this is mainly due to the projection of its superior aspect rather than its subspinale aspect. The pathological alteration of the nasal spine described above is also evident from this view.

Subnasoalveolar Clivus and Alveolar Region

The anterior root of the zygomatic process of the maxilla on the right side is located at the anterior edge of the first molar, as in most modern humans. As was also evident from left-side view, to some extent, the splanchnocranium appears to be too orthognathous. This is evident to some degree when the mandibular condyle is articulated into the glenoid fossa and produces a slightly posteriorly positioned maxillary bite point. As mentioned above, the hinging of the splanchnocranium at the frontozygomatic points and preserved frontonasal point might be slightly off, and this could be accompanied by reconstructed contact points between the posterior maxillae and the basicranium that are slightly too short. As noted previously, however, one problem with this interpretation is that the zygomatic arch on the right side, as well as on the left side, as presently reconstructed will not tolerate too much lengthening.

Inferior View

Occipital Region

In inferior view (Figure 8.70), it can be seen that the occipital region has been reconstructed from at least seven major pieces and connected with filler in several areas, including large areas on the right and left sides medial and posterior to the mastoid processses and across to the foramen magnum. Moreover, large portions of the right side of the occipital squama are reconstructed with filler.

The alignment of the foramen magnum in the anteroposterior axis is largely correct. A prominent median occipital crest beginning 15.6 mm posterior to the foramen magnum is visible, and its orientation is correct. The occipital crest reaches the damaged and poorly preserved inferior nuchal lines, where it ends abruptly and appears to have been snapped off at this point. The anterior portion of the preserved crest is fairly sharp; the development probably most closely approximated a category 2 or 3 in Lahr (1996) prior to being damaged. A well-developed superior nuchal line is evident for most of its course along

Figure 8.70 Inferior view of the Dolní Věstonice 16 cranium. Although damage and restoration prevent confirmation, note the possible midline patency of the palate posterior to the incisive foramen, perhaps representing the remnant of an incompletely formed palatal midline. Scale in centimeters.

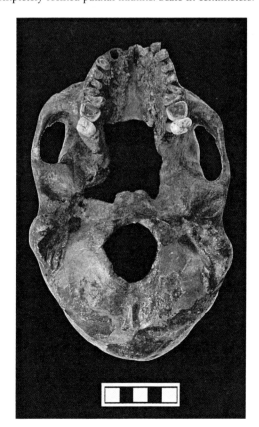

the occipital, although it becomes less rugose, virtually disappearing as it approaches the *tuberculum linearum*. There is a fair degree of asymmetry in the position of the nuchal lines that is probably due to postmortem deformation. The right superior nuchal line is more anteriorly placed, relative to the more posteriorly placed left one (Figures 8.71 and 8.72). This is associated with a distinct posterior bulge of the right occipital in the general area of the supreme nuchal line.

Because of the large amount of reconstruction and filler on the occipital, most of the muscle scarring is not preserved. Portions of attachments for the superior oblique on the right side and the semispinalis capitis on both sides are visible, as are the insertions for rectus capitis posterior minor and major, especially on the left side.

The occipital condyles are well preserved and show minor asymmetry. The maximum length of both condyles is similar (ca. 19 mm). However, the width of the left condyle appears to be expanded (right maximum width: 13.9 mm; left: 16.7 mm); care must be taken not to include the roughened area lateral to the condyle and the attachment for rectus capitis lateralis. This is associated with asymmetries in the first cervical vertebra, although none

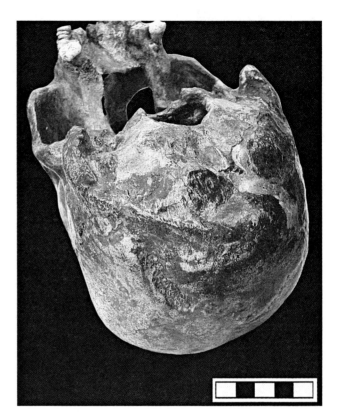

Figure 8.72 Oblique inferior view of the right occipital region of Dolní Věstonice 16. Scale in centimeters.

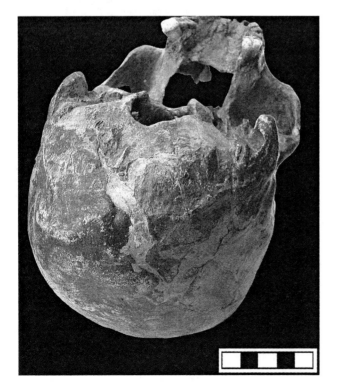

Figure 8.71 Oblique inferior view of the left occipital region of Dolní Věstonice 16. Note the wide, shallow groove coursing along the left superior nuchal line at an oblique angle just lateral to the midline associated with a swelling along the inferior aspect of the groove. It is possible that the groove is a healed injury across the muscle insertion line. Scale in centimeters.

of these include surface alterations indicating degenerative remodeling. The area just lateral to the condyle on the right side in the region of the rectus capitis lateralis insertion is damaged and largely missing. This same area lateral to the left-side occipital condyle retains a small portion but is also largely missing. The hypoglossal canals (both single and unbifurcated: Table 8.8) have been packed with reconstructive filler, and the areas containing the posterior condylar fossae are completely gone and filled.

The breadth of the foramen magnum in Dolní Věstonice 16 (30 mm) is equal to those found in Dolní Věstonice 3 and 14, and all are considerably wider than in Dolní Věstonice 13 (Figure 8.23; Sládek et al., 2000: tables 18 and 19). Foramen magnum length in Dolní Věstonice 16 (36 mm) is nearly as small as that found in Dolní Věstonice 3. Therefore, both Dolní Věstonice 16 and 3 show small and rather rounded foramina magna, Dolní Věstonice 13 shows a relatively small and narrow one, and Dolní Věstonice 14 shows a large and elongated one.

Anterior of the foramen magnum on the inferior basilar surface, a pharyngeal tubercle is weakly present only as a trace discernable by palpation; also, it is expressed more as a transverse ridge rather than a tubercle (Table 8.8). Immediately anterior to the pharyngeal tubercle a shallow pharyngeal fovea with diffuse margins is evident; it is less than 1 mm in depth and because of its shallowness and diffuseness is not scoreable by Hauser & DeStefano (1989) criteria. Lateral to the pharyngeal tubercle, there are modest attachment protuberances for the longus capitis on both sides. Anterior to these, the basilar portion of the occipital is broken away at what appears to be a line close to where the fused sphenooccipital synchondrosis would have been.

Sphenoid Region

Most of the sphenoid is missing. On a small portion of retained right sphenoid, there is a well-preserved, clear single foramen ovale and what might have been a bipartite foramen spinosum. There is no evidence for a foramen of Vesalius (Table 8.8).

Temporal Region

The inferior aspects of the temporal region are well preserved on both sides. The left glenoid fossa, tympanic region, and mastoid process all are slightly more posteriorly placed in the anteroposterior plane relative to the right side by at least 10 mm, and the mediolateral distances relative to midline are equidistant. The right mastoid shows a fairly wide digastric attachment (this area is missing on the left side). The right temporal retains a 10 mm length of styloid process, although its original length cannot be determined. A stylomastoid foramen cannot be seen; it is probably filled in. The left-side styloid process is missing, although its base is present. The areas immediately medial to both styloid processes posterior to the tympanic region are present and include portions of the carotid canals. The posterior glenoid eminences and medial tympanic portions are preserved on both sides. Both glenoid fossae are well preserved and show very little signs of remodeling. The only remodeling occurs bilaterally on the lateral portions of the anterior eminence, where each side has a small, roughly circular, shallow pit with irregular surface area that measures approximately 6.5 mm in diameter (Figure 8.73).

The alignment of the two portions of the zygomatic arch on either side is good, with contact all along each arch. The left zygomatic arch shows a minor medially directed inflection at about midpoint that is most likely postmortem deformation since the entire left side shows some medial inflection that is particularly apparent in superior view (see below). The right zygomatic arch shows a normal curvature along its length. The left infratemporal fossa is positioned more posteriorly along the anteroposterior axis relative to the right by about 10 mm, as is also true for most basicranial features already mentioned. The left infratemporal fossa is longer than the right (right maximum length: 37.3 mm; left: 42.3 mm). From this view, it can also be seen that the angle between the anterior zygomatic arch and the anterolateral aspect of the infraorbital region is more acute on the right side and less angled on the left. It is possible that all of these features of the left zygomatic arch are due to postmortem compression consistent with the left side of the cranium, or it is possible that they are related to the maxillary alterations discussed above. It should be noted that the left infracondylar neck of the mandible (missing its condyle), which does not line up with the left glenoid fossa, is due entirely to a mediolaterally misaligned posterior reconstruction of

Figure 8.73 Glenoid fossae of Dolní Věstonice 16, showing bilaterally localized remodeling on lateral portions of the anterior eminences.

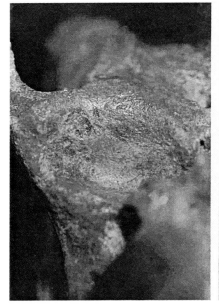

the ramus rather than a mediolateral misalignment of the glenoid fossa. This is evident in the relatively straight inferior border of the right lateral mandibular corpus and ramus compared with the medially inflected inferior ramus border on the left side in inferior view.

The attachment areas for masseter are strongly developed and extend anteriorly to the point where the lateral and anterior aspect of the zygomatic region turns in an acute angle near zygomaxillare; the muscle attachment does not extend onto the inferior maxillary surface. Posteriorly, the muscle scarring seems not to extend beyond the base of the zygomaticotemporal suture on either side. Whereas the anterior vertical wall of the temporal fossa is similar on both sides, the right medial wall is much more medially inflected, although this seems due to reconstruction efforts rather than actual anatomy.

Inferior Palatal Region

The inferior palate is complete in its lateral aspects to the palatomaxillary suture area, although none of the palatine bones is preserved (Figure 11.8). The intermaxillary suture area is reconstructed with filler, as a large V-shaped portion that narrows anteriorly to a point in the area of the incisive foramen. It is about 16.2 mm wide posteriorly. The right palate is slightly inferiorly positioned relative to the left one anteriorly, and the right region is slightly superiorly positioned relative to the left one posteriorly. Moreover, the left palate is slightly anteriorly positioned relative to the right in the transverse plane, and thus the two halves of the incisive fossa are slightly misaligned. It is important to realize, however, that even with this misalignment of the maxilla along the palatal midline, the dentition on both sides falls on the same plane. In other words, the adult dentition came into occlusion and wore down in perfect occlusion with the mandible in an accommodative fashion to the misalignment of the palatal halves. This indicates that the lesion or congenital defect occurred early in the person's life, perhaps well before the adult dentition was in occlusion. It also remains unknown whether the intermaxillary suture would have been complete in this area all along its length or open more posteriorly, as with some forms of cleft palate.

As noted, the palate is deep compared to that of the other Dolní Věstonice and Pavlov individuals, and there is no evidence for midpalatal (intermaxillary) or alveolar tori (Table 8.8). The overall dimensions of the Dolní Věstonice 16 palate fall within the distribution of those in other Dolní Věstonice males (Figure 8.24; Sládek et al., 2000: table 27). It is the largest of the sample in anterior palate breadth, as well as palatal depth. Given the substantial modification of the alveolar process, some of these dimensions may have been larger in their prepathological state. If so, then Dolní Věstonice 16 would have been among the largest of the Dolní Věstonice and Pavlov males in terms of palatal size, as he was for overall craniofacial size.

Inferior Frontal View

The inferior frontal bone has relatively complete orbital plates on only the most anterior aspects of the orbits on both sides. As noted in anterior view, both superior orbital margins show bilateral medium-sized notches on their medial aspects (Hauser & DeStefano, 1989: plate VIII, most similar to d). There is no obvious porosity on the superior internal orbital margins. Medially, the nasofrontal suture and a portion of the frontal nasal spine are present. The ethmoidal notch and all of the ethmoid bone are missing. There is a very large opening into the frontal sinuses that reveals substantial frontal sinus development medially and laterally into all of the superciliary arch region. In this frontal sinus pattern, Dolní Věstonice 16 is very different from the triple burial individuals (see above).

Internal Calvarium View

The large missing area of the sphenoidal region allows assessment of some aspects of the internal calvarium. The frontal crest is present in midline; it is damaged in the immediate area of the posterior sinuses, but its morphology superior to this point is visible. It appears that the crest attenuates and disappears somewhat prematurely in this individual, although precise measurements cannot be made. A series of small and large Pacchionian pits are evident laterally along the midline on both the internal frontal and parietal aspects. The sulcus for the superior sagittal sinus is well preserved. Grooves for the frontal and parietal branches of the middle meningeal vessels are well preserved on the left side, although not on the more heavily damaged and reconstructed right side. Comparison of the position of the internal occipital protuberance with the external occipital protuberance shows that they are not coincident, being separated by at least 10 mm, with the external protuberance more superiorly located.

Posterior View

In posterior view (Figure 8.74), the posterior portion of the sagittal suture is well preserved, and lambda can be determined by using the confluence with the lambdoidal suture. The left lambdoidal suture is preserved for its entire course, whereas the right is complete to about 51 mm of length lateral from lambda; they are more or less symmetrical in their course from lambda toward asterion. Detailed scoring results for the lambdoidal suture are presented in Table 8.10. The left posterior parietal bone

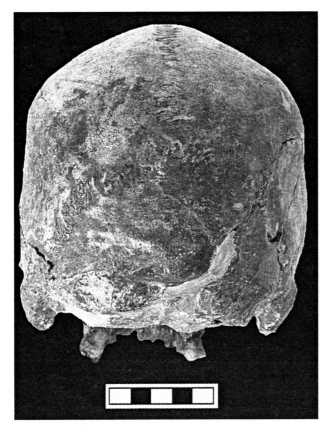

Figure 8.74 Posterior view of the Dolní Věstonice 16 cranium. Scale in centimeters.

is largely intact, whereas the right one has been reconstructed from about nine pieces. Although there is anteroposterior asymmetry of the parietal tuber position, the prominence on either side of the mediolateral plane is fairly symmetrical. There is also a slightly greater posterior projection or bulging in the area between the left lambdoidal suture and the superior nuchal line relative to the right side, probably because of postmortem deformation and reconstructive efforts. The lateral parietal and temporal squama are straight-sided (left side) and straight-sided with a minor lateral bulge near the squamosal suture (right side). Together the parietal tubers and the posterior vertex form the pentagonal configuration that is associated with modern humans. Consistent with the pentagonal shape, the greatest breadth across the parietals is relatively superior.

Maximum cranial breadth for Dolní Věstonice 16 falls between the values for Dolní Věstonice 14 and 13, and its cranial height is the highest among the Pavlovian individuals (Sládek et al., 2000: table 19). This makes Dolní Věstonice 16 rather different from the Qafzeh-Skhul sample and particularly the Neandertals in its neurocranial shape in posterior view (Figure 8.26; Table 8.1). Moderate sagittal keeling is present (category 2, Lahr, 1996).

Inferiorly, the posterior view shows that the mediolateral distance between the foramen magnum/occipital condyle area and the mastoid processes are similar, largely symmetrical, and aligned with the dentition on either side. There is a slight inferior placement of the right condyle relative to the left, but in general the posterior view shows a high degree of mediolateral symmetry. The superior nuchal lines are present on both sides, but they fade out medially as they approach the *tuberculum linareum*. The supreme nuchal line is visible on the left side, coursing nearly 20 mm apart from the superior nuchal lines at its widest extent lateral to the midline and narrowing to eventually converge at the *tuberculum linearum*. The right side, although more damaged and reconstructed, almost certainly followed the same pattern. An 18 mm horizontally roughened, curved line in the midline at the *tuberculum linearum* lies just above the most posterior extent of the occipital crest. This roughened line corresponds to the area that often encompasses an iniac depression (i.e., borders it inferiorly) in the Pavlovian sample (e.g., Pavlov 1 and Dolní Věstonice 14 and 15). As in most of the other Pavlovian specimens (e.g., Dolní Věstonice 14 and 15 and, to a lesser extent, Dolní Věstonice 3 and Pavlov 1), the inferior depression at the *tuberculum linearum* in Dolní Věstonice 16 is matched by an additional distinct depression (the larger of the two) that is more superiorly positioned just below lambda and well above the supreme nuchal line. In Dolní Věstonice 16, the superior depression is large and well defined.

The posterior view also shows the extension of both mastoid processes far below the level of the more medially positioned juxtamastoid eminences.

Superior View

A fair degree of postdepositional neurocranial asymmetry that affects both halves is evident in superior view (Figure 8.75). The left side has been medially inflected in the region near the lateral coronal suture, and the more fragmentary and reconstructed right side has been laterally expanded. The combination of the two produces a long axis that curves toward the left side relative to the supraorbital and frontal bone midline alignment. It is difficult to say which side is less deformed; most likely both show similar levels. The cranium is dolicocephalic, falling toward the more dolicocephalic area of the distribution for individuals across all comparative samples (Figure 8.28). The degree of dolicocephaly is not a function of the deformation.

The greatest width at the parietal tubers is located at the posterior third of the neurocranium. However, there is marked asymmetry in the anteroposterior location for each parietal tuber, the right side tuber being 10 mm or more anteriorly positioned relative to the left. The frontal

Figure 8.75 Superior view of the Dolní Věstonice 16 cranium. Note the antemortem healed lesion located anterior to bregma, nearly in the midline. Scale in centimeters.

tubers are less pronounced, and they show a much smaller degree of anteroposterior asymmetry; the left is slightly more anteriorly positioned than the right. Commensurate with its large cranial capacity (the largest in the Pavlovian sample: Table 8.9), Dolní Věstonice 16 has virtually no postorbital constriction. The coronal suture shows some damage and filler, especially in the bregma region, but is more intact laterally to either side. The coronal suture is well preserved along its length (Table 8.10). As with all of the other Dolní Věstonice individuals, Dolní Věstonice 16 does not appear to have parietal foramina. As Hauser & DeStefano (1989) pointed out, parietal foramina normally transmit emissary veins, according to most anatomy references, that connect the occipital veins with the superior sagittal sinus, as well as anastomoses between the middle meningeal and occipital arteries, although this function has not gone unchallenged. Interestingly, Hauser & DeStefano also note that the heritability estimate for absence of the parietal foramen was significant in an extensive pedigree study by Sjøvold (1984). In oblique superior view, the distinction of the superciliary arch from both the frontal squama and the supraorbital trigone on either side by the obliquely flattened supraorbital sulcus is particularly apparent.

In strict superior view, the occipital region shows a posterior asymmetrical projection or bulging on the left side in the area of the lambdoidal suture. A well-healed antemortem lesion noted in anterior view (Chapter 19) is located ca. 30 mm anterior to bregma, nearly in midline, and its greater depth in the anterior portion of the lesion is evident from superior view.

Dolní Věstonice 17

Dolní Věstonice 17 consists of two small pieces of juvenile neurocranial vault bone (Vlček, 1991; Sládek et al., 2000). They are identifiable as deriving from parietal bones on the basis of the clear meningeal sulci on their endocranial surfaces. They provide no other morphological data of note.

Pavlov 1

Preservation

The cranium of Pavlov 1 consists of a calvarium, a small portion of the basicranium, the left zygomatic bone, and portions of the left and right maxillae. The bones are light ochre to light brown in color, with minute spotting and root etching. The condition of the bone, especially of the calvarium, is good. It preserves the frontal bone, the left and right temporal bones, and most of the occipital bone. There is an additional piece of the occipital (Pavlov 1/3), which is not joined to the calvarium, consisting of the left lateral portion and associated left basilar portion. This piece contains the occipital condyle, the hypoglossal canal, and the condylar fossa with the condylar canal, as well as the anterior margin and left part of the foramen magnum. In terms of the splanchnocranium, Pavlov 1 retains a left zygomatic bone that has been connected to the calotte even though there is no actual bone connection. The left and right maxillae, although relatively complete inferiorly, are missing the superior portions, and there is consequently no direct connection to the rest of the cranium. Besides these, there are several small fragments that cannot be identified or joined.

Articulation of the mandible with the left zygomatic bone at the temporomandibular joint provides a reasonable anatomical position for the mandible. Occlusion of the mandibular and maxillary dentary, with consideration of normal occlusal angle in lateral view, is then possible, which allows some facial morphology and gross dimensions to be estimated (see Vlček, 1997: figures 3 and 32).

Relative to the Dolní Věstonice crania, the Pavlov 1 cranium appears to have suffered less postmortem and reconstructive distortion, despite being disturbed and displaced postdepositionally (Chapter 4), which is evident in all anatomical normae.

Anterior View

Frontal Region

The frontal bone is virtually complete (Figure 8.76). To some degree, the root etching on this bone obscures very fine surface detail; nonetheless, most external features are preserved. A completely open suture is evident at bregma, and there is minimal closure at the midcoronal locations (Table 6.1). There is no evidence of a metopic suture, and there are no frontal grooves on the lateral squamous portions, although for the latter trait the root etching is somewhat more extensive and might be obscuring them (Table 8.3). There is a partially obliterated zigzag-shaped supranasal suture (Hauser & DeStefano, 1989: plate VI, h). There is no sagittal keeling, and there are no frontal eminences.

The frontal squamous appears visually to be narrow and high. The least frontal breadth of Pavlov 1 falls toward the high end of the range for Pavlovian males, exceeded only by Dolní Věstonice 14 (whose value might be affected by postdepositional and reconstructive distortion), and its maximum frontal breadth is the largest among the Dolní Věstonice and Pavlov sample. The relationship between these two variables is very similar in the Pavlovian and European earlier Upper Paleolithic samples (Figure 8.2), with the Qafzeh-Skhul and Neandertal samples showing relatively greater least frontal breadth values.

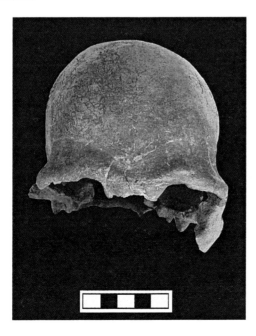

Figure 8.76 Anterior view of the Pavlov 1 calotte. Scale in centimeters.

Supraorbital Region

The supraorbital region of Pavlov 1 shows separation of medial, midorbital, and lateral components, although to a degree that is less pronounced than for the Dolní Věstonice individuals and many other earlier Upper Paleolithic individuals. The glabellar region is pronounced in midline, and this robust midline projection extends to the medial third to half of both right and left supraorbital areas (superciliary arches), each extending superiorly and laterally prior to slightly inflecting inferiorly and then becoming discontinuous for a short distance on the midorbital regions. The lateral-most supraorbital margins, the supraorbital trigones, exceed those of all of the Dolní Věstonice males in superoinferior thickness and anterior projection (and all early modern humans that we have observed). Pavlov 1 is also unusual in that the lateral components appear as massive as the middle components (corresponds most closely to Lahr's, 1996, most robust grade of 4). Moreover, the separation of the lateral components from the medial superciliary arch by the supraorbital sulcus on either side in Pavlov 1 is not as distinct and not as obliquely flattened as in other early modern individuals. Therefore, the "pinched" appearance that is so prominent in the midorbital region of the Dolní Věstonice sample and European Upper Paleolithic specimens in general (Smith & Ranyard, 1980) appears muted in Pavlov 1.

All of this appears to make the supraorbital region in Pavlov 1 more continuous and thus more Neandertal-like (Table 8.2). However, three features must be emphasized in this regard. First the midorbital pinching, however muted on Pavlov 1, is present. Second, the lateral supraorbital margins in the later Neandertals tend to be less massive than the medial margins, even in larger males, whereas in Pavlov 1 they do not thin laterally. Third, the largest earlier Upper Paleolithic early modern human supraorbital regions (e.g., Mladeč 5 and Pavlov 1) show highly angled medial and lateral components in anterior view, even if the midorbital thinning is not as pronounced, whereas Neandertals exhibit more of a gently curving arc across both components. The latter holds even when some midorbital thinning occurs, as, for example, in some of the Vindija Neandertals (Smith & Ranyard, 1980) and Saint-Césaire 1. Pavlov 1 thus extends the range of supraorbital size variation in early modern humans but, in total, appears to fall into a distinctive modern human shape pattern.

Orbital Region

The width of the nasal root in Pavlov 1 appears large, although an accurate measurement for interorbital breadth cannot be obtained. Its estimated bifrontal breadth from the frontomalareorbitale points (111 mm) is larger than that of the Dolní Věstonice males (101 to ca. 107 mm; (Sládek et al., 2000: table 23), and it even exceeds the value of 107.5 mm for the earlier Oase 2.

From what is present on the left side, the orbit was probably more square-shaped than rounded. The superior orbital margins in Pavlov 1 are thick and rounded in cross section, the most extreme, in fact, among the entire Pavlovian sample (Chapter 7). Each superior orbital margin has a very large, blurred medial notch (larger than any shown in Hauser & DeStefano, 1989: plate VIII), the largest in the Pavlovian sample (Table 8.3).

Internally, the superior orbital roof is only intact for a maximum of ca. 18 mm in both sides, and there is no indication of thinning or porosity. The inferolateral border of the left orbit is robust, being rounded rather than sharp, and the border is more or less even in relation to the floor of the orbits, at least to a greater degree than in the Dolní Věstonice individuals, and therefore attains a somewhat more robust score. In this, it corresponds most closely to Lahr's (1996) grade 2 but approaches a grade 3, the most robust score.

Infraorbital Region

The right maxilla of Pavlov 1 (Figure 8.77), consists of a fragment of the alveolar process, spanning the area from the second incisor to the second molar, whose dimensions are ca. 50 mm in mesiodistal length and 15 mm in supero-

Figure 8.77 Anterior view of the Pavlov 1 maxillae. Scale in centimeters.

inferior height. The left maxilla is better preserved, with an alveolar process that is ca. 65 mm in mesiodistal length and ca. 25 mm in maximum superioinferior height and includes ca. 18 mm of the lateral nasal aperture, spanning the area from the first incisor to the second molar. The connection of the left and right maxillae along the intermaxillary suture is not well preserved and has been estimated. The infraorbital region is missing on the right maxilla, but a small portion of it is present on the left side. This area is confined to the most inferomedial portion of the infraorbital region and therefore does not contain what is usually the most concave area of the infraorbital foramen. However, the area that is preserved shows a slight amount of concavity, which appears to be reflecting the parasagittal alignment in the medial portion. This is consistent with the typically modern, two-plane configuration (Table 8.2).

Nasal Region

Pavlov 1 has an estimated nasal breadth of 30 mm, which is among the largest for the Dolní Věstonice and Pavlov sample (Dolní Věstonice 13: 29.2 mm). In this, it also matches the estimated value for Mladeč 8 (30 mm) and is larger than the highest value recorded by Matiegka (1934) for the Předmostí sample (Předmostí 7: 29.0 mm). Nasal height in Pavlov 1 requires a greater degree of estimation because there are no contact points between the maxillae and the calotte. However, articulation of the mandible with the left temporal bone at the temporomandibular joint and occlusion of the mandibular and maxillary dentary, with consideration of normal lateral occlusal angle, facilitate at least a tentative nasal height estimation (see in Vlček, 1997: figures 3 and 32). Based on this, the estimated nasal height of Pavlov 1 (ca. 58 mm) is the largest among the Dolní Věstonice sample, although it scales with its estimated breadth value in a pattern that is very similar to the Dolní Věstonice sample (Figure 8.7; Table 8.4; Sládek et al., 2000: table 26).

The left half of the piriform aperture is complete to nearly midline in the region of the anterior nasal spine and retains most of the lower border of the piriform aperture (Figure 8.78). Although most of the anterior nasal spine is missing, the crest configuration is preserved. There is a single crest consisting of fused lateral, spinal, and turbinal crests (category 1 following De Villiers, 1968; Franciscus, 1995), and there is no evidence for a prenasal fossa. Although the internal nasal floor is incompletely preserved, there is enough to indicate that the floor is completely level with the narial margin and, in combination with its narial margin pattern, presents a total pattern in this area that is different from the derived pattern found in the latest Neandertals.

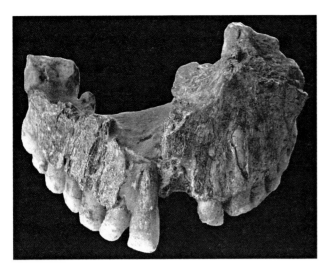

Figure 8.78 Lower border of the piriform aperture of Pavlov 1.

Subnasoalveolar Clivus

The subnasoalveolar clivus in Pavlov 1 appears to be rather vertical, although postmortem damage to the anterior alveolar bone precludes definitive assessment of this and other aspects of anterior clivus form.

Overall Facial Size and Proportion

Overall facial height and breadth in Pavlov 1 appears large. The preservation of the splanchnocranium precludes more detailed metric comparisons.

Left Lateral View

Frontal Region

In left lateral view (Figure 8.79) the frontal bone of Pavlov 1 shows less frontal inclination and projection at metopion relative to the frontal chord than the Dolní Věstonice individuals (Figure 8.10; Table 8.1; Sládek et al., 2000: table 21). Nonetheless, the relationship between the frontal subtense and the chord in Pavlov 1 is more like the modern human configuration than that of the Neandertals, who show a relatively lower subtense value.

The glabellar region is markedly projecting, with both a supraglabellar inflection and an infraglabellar notch. The supraglabellar inflection in Pavlov 1 is not the pronounced supraglabellar sulcus found in late archaic humans, but it is similar to that found in other European Upper Paleolithic specimens. The degree of infraglabellar notch development in Pavlov 1, with allowance for the missing nasal bone articulation with nasion, most closely approximates Lahr's (1996) category 3 and is similar to the pattern found in the Dolní Věstonice males, other Upper Paleolithic individuals, and to a large degree the Neandertal sample. The inflated and anteroposteriorly wide supraorbital trigone can be seen in this view, and from this region, the anterior portion of the temporal line on the frontal bone can be seen as well marked and rugose.

Parietal Region

In left lateral view, the vertex of the cranium is located approximately at bregma. There is less angulation between the midline superior neurocranial profile and the posterior parietal plane in Pavlov 1 than in the Dolní Věstonice individuals, as reflected in the bregma-lambda chord subtense. Pavlov 1 falls on the lower (flatter) portion of the overall distribution, as well as closer to the Neandertal distribution in size (Figure 8.12; Table 8.1; Sládek et al., 2000: table 21). The rugose anterior portion of the temporal line found on the frontal maintains prominence on its continuation along the anterior and posterior parietal. There is an oblique line 60 mm long and less than 5 mm wide that courses across the external left parietal bone. The anterior end is 16.5 mm from the coronal suture and 43.5 mm from the sagittal suture; the posterior end is 67 mm from the coronal suture and 78 mm from the sagittal suture (Chapter 19).

Pterion Region

The junction of the frontal, parietal, temporal, and sphenoid bones in the pterion region is missing.

Temporal Region

The temporal squamous is missing its anterosuperior portions, but the mastoid process, zygomatic process, and suprameatal crest region are all intact, and the fit of the temporal onto the parietal appears sound. There is a 15 mm gap between the elements of the zygomatic arch, but the alignment and filled-in portions appear accurate. The temporal bone is robust, with a well-developed suprameatal crest that continues from the zygomatic process and extends well beyond porion posterosuperiorly to meet up with the posterior aspect of the temporal line on the parietal. There is a small, crest-type suprameatal spine with a small depression that appears to be filled in either with matrix or possibly remodeled bone (there is no exact analogue in Hauser & DeStefano, 1989; Table 8.6). There is a possible trace presence of a squamomastoid suture in the superiormost location (Hauser & DeStefano, 1989: figure 32, closest to b).

The mastoid process is very large in Pavlov 1. Its vertical height at 31 mm surpasses all other Dolní Věstonice

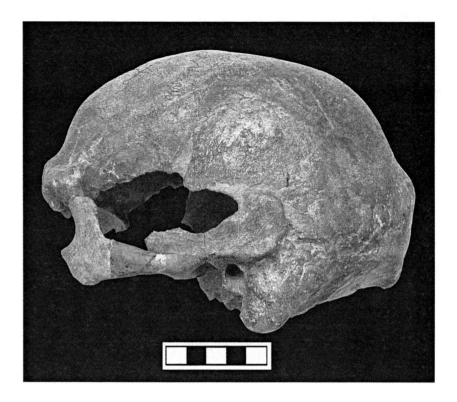

Figure 8.79 Lateral (left) view of the Pavlov 1 calotte. Scale in centimeters.

and Pavlovian individuals (range: 20 to 27 mm; Vlček, 1997: table 9), and it's anteroposterior width (35 mm), although not the largest (Předmostí 3: 40 mm; Dolní Věstonice 14: 37 mm), nonetheless falls at the high end of the total range (23 to 40 mm). A tubercle is present on the mastoid process. It is not anteriorly positioned on the mastoid, as in many Neandertals but in the middle of the process anteroposteriorly, and it is diffuse. The mastoid process extends well below the plane of the juxtamastoid eminence and shows no lateral flattening. In all of these features, the mastoid process of Pavlov 1 is modern (Table 8.5).

In addition to its large size, the mastoid process is rugose. There is one large mastoid foramen that does not appear to be sutural. It is confined to the mastoid process near asterion at the level of the external acoustic meatus, and as in Dolní Věstonice 13–15, it is positioned such that it is more visible in posterior than lateral view.

Occipital Region

In left lateral view, Pavlov 1 shows strong occipital bun development. Inferiorly, there is a pronounced projection of the external occipital protuberance and the superior nuchal line, creating a strong inflection relative to the nuchal plane; in addition, the superior projection near lambda is quite marked. The occipital chord and subtense configuration is similar to those of the Dolní Věstonice males, with moderately large values for each compared to considerably smaller values for Dolní Věstonice 3 (Figure 8.13; Table 8.1; Sládek et al., 2000: table 21).

Zygomatic Region

The left zygomatic bone is well preserved, although its contact points to the frontal and temporal bones are reconstructed with filler. There are two small zygomaticofacial foramina (one on the body and one on the lowest portion of the frontal process) separated by 7.6 mm (Table 8.6). There is also strong development of a marginal tubercle (Hauser & DeStefano, 1989: figure 36, most similar to d).

Orbital Region

The lateral margin of the orbit in Pavlov 1 is posteriorly positioned relative to nasion, although not to the degree found in some of the Dolní Věstonice males.

Infraorbital Region

The small portion of the infraorbital region (Figure 8.80) present on the left side indicates that a trace of the most inferomedial portion of the infraorbital region is present and shows a slight amount of concavity. This appears to indicate the parasagittal alignment in the medial portion that is usually found in modern humans. From this view,

The Cranial Remains

Figure 8.80 Lateral (left) view of the Pavlov 1 maxilla. Scale in centimeters.

the very tip of the anterior root of the zygomatic process of the maxilla appears to be present and is located at the mesial edge of the second molar (Table 8.2).

Nasal Region

As noted above, the anterior nasal spine is missing, and nothing can be gleaned in terms of its size and projection.

Subnasoalveolar Clivus and Alveolar Region

Postmortem damage to the anterior alveolar bone precludes definitive assessment of aspects of anterior clivus form.

Overall Size and Proportion

Overall facial projection and height in Pavlov 1 appears large, based on visual assessment. However, the incomplete preservation of the splanchnocranium precludes metric comparisons.

Right Lateral View

The right side is somewhat less complete than the left, and only a few additional features are noted (Figure 8.81).

Parietal Region

There is a small depressed area ca. 9 mm in diameter on the external posteroinferior aspect of the right parietal, with the center 12 mm from the lambdoid suture. The center of the lesion is ca. 60 mm from lambda. There is also a broad depression on the right parietal 10 to 15 mm in diameter, centered 20 mm from the coronal suture, 54 mm from the sagittal suture, and 26 mm above the temporal line (Chapter 19).

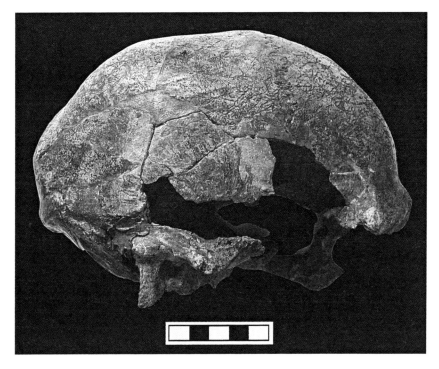

Figure 8.81 Lateral (right) view of the Pavlov 1 calotte. Note the lesion on the posteroinferior right parietal bone near the lambdoid suture and another posterior to the coronal suture above the temporal line. Scale in centimeters.

Temporal Region

In the temporal region, there is a small, crest-type suprameatal spine with a small depression that appears to be filled in either with matrix or possibly remodeled bone (there is no exact analogue in Hauser & DeStefano, 1989; Table 8.6).

Inferior View

Occipital Region

In the inferior view (Figure 8.82) of occipital region, the alignment of the posterior margin of the foramen magnum appears to be largely correct. The breadth of the foramen magnum in Pavlov cannot be measured. However, an estimate of its length (ca. 40 mm) is possible with the articulation of the Pavlov 1/3 piece (not shown in Figure 8.82). This estimate is longer than for all of the Dolní Věstonice individuals except Dolní Věstonice 14 (42 mm; Sládek et al., 2000: tables 18 and 19). Also on Pavlov 1/3, there is a weakly expressed pharyngeal tubercle (Hauser & DeStefano, 1989: see figure 23, a), and a pharyngeal fovea that is oval in shape, with a tub-shaped bottom of shallow to medium depth (Hauser & DeStefano, 1989: p. 138, a in methodology). On the left side, there is no hypoglossal canal bridging (Hauser & DeStefano, 1989: figure 18, corresponding to 1). There are also two weakly developed unfused processes on the preserved left side.

A well-developed median occipital crest (category 3 in Lahr, 1996) posterior to the foramen magnum is visible, and its orientation is in correct alignment. It courses for most of the distance between the foramen magnum and the external occipital protuberance at the junction of the supreme and superior nuchal lines. A well-developed superior nuchal line is evident for its entire course along the occipital, and in midline it joins the *tuberculum linearum*. The lateral-most portions of the superior nuchal lines closest to the posterior mastoid process area are particularly prominent, contrasting with the Neandertal pattern, in which they fade laterally. Although damaged somewhat on the right side, these prominent portions appear to be bilaterally symmetrical (Figures 8.83 and 8.84).

Sphenoid Region

The posterior border of the right foramen spinosum is present.

Temporal Region

The inferior aspects of the temporal region are well preserved on both sides. There is a depressed area on the posterior portion of the articular eminence of the left temporal bone, covering its lateral two-thirds and bordered anteriorly by a vascular groove. There is minimal alteration of the subchondral bone, but there is a distinct raised medial margin (Chapter 19).

Figure 8.82 Inferior view of the Pavlov 1 calotte. Note the depressed area on the posterior shoulder of the articular eminence of the left temporal bone associated with a minor remodeling process; it is probably a secondary effect of dental attrition rather than active degeneration. Scale in centimeters.

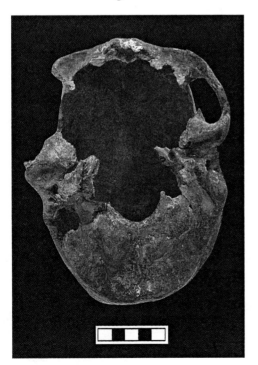

Figure 8.83 Oblique inferior view of the left occipital region of Pavlov 1. Scale in centimeters.

Figure 8.84 Oblique inferior view of the right occipital region of Pavlov 1. Scale in centimeters.

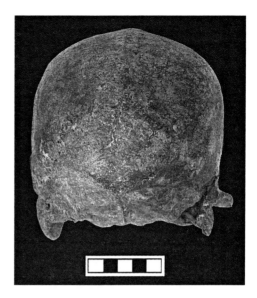

Figure 8.85 Posterior view of the Pavlov 1 calotte. Note the bilateral swelling along the sagittal suture. This is associated with swellings along the external sutures in general. Internal/external radiographs show slight increases in density along the swellings but no other obvious changes. It is unclear whether these are pathological or the result of senile changes in an older individual. Scale in centimeters.

Inferior Palatal Region

The palate is missing substantial portions of bone (Chapter 11; Figure 11.14). The area that is best preserved, the anterolateral portion of the left side, shows absence of palatine tori. Estimated external palate breadth in Pavlov 1 is among the largest of the Dolní Věstonice males, although its length is somewhat shorter (Figure 8.24; Sládek et al., 2000: table 27), and the Pavlovian sample as a whole has smaller palates than the Skhul-Qafzeh and Neandertal samples (Figure 8.24; Table 8.4).

Posterior View

In posterior view (Figure 8.85), significant suture closure is evident at the midlambdoid and lambda locations (see also Table 8.10). The lateral parietal and temporal squama are straight-sided on both sides. Among males, maximum cranial breadth for Pavlov 1 falls between the slightly larger values for Dolní Věstonice 13 and 16 and the slightly smaller values for Dolní Věstonice 14 and 15, whereas its estimated cranial height (ca. 125 mm; Vlček, 1997) is relatively low compared to that for all of these individuals (Sládek et al., 2000: table 19). There is an iniac depression in Pavlov 1, as is the case with all of the other Pavlovian individuals (even in those that are more distinct and pronounced). The depression in Pavlov 1 is not the fully developed, mediolaterally expanded suprainiac fossae associated with a bilaterally protruding occipital torus found in many Neandertals (Table 8.5). The posterior view also shows the extension of both mastoid processes below the level of the more medially positioned juxtamastoid eminence, the usual modern human condition (Table 8.5).

There is a series of swellings on the external surfaces of the calvarium, mostly along the sutures. One of these is a swelling along the lambdoid suture that is slightly distorted by postmortem separation along the suture and reconstruction with filler, but the effect on the pattern is small. There is a posterior displacement of the occipital relative to the parietal bones (also seen in Dolní Věstonice 16), with bony swellings along the medial halves of the two sides of the suture. The right side extends ca. 56 mm from lambda, reaching 25 mm wide, with clear sulci above and below the swelling. The left side extends ca. 60 mm from lambda and reaches 23 mm wide, with sulci on both sides. The swellings are mostly on the occipital bones, but they are also present on the parietal sides of the suture and are discussed in Chapter 19.

Superior View

In superior view (Figure 8.86), the relatively undistorted neurocranium of Pavlov 1 is evident. The cranium is dolicocephalic to a greater degree than that of all of the Dolní Věstonice individuals. It lies at the lower limits of the range for maximum cranial breadth relative to length for all comparative samples (Figure 8.28; Sládek et al., 2000: tables 18 and 19). Its greatest width at the parietal tubers is located at the posterior third of the neurocranium. Commensurate with its large cranial capacity, 1472 cc (Table 8.9), there is no postorbital constriction.

Figure 8.86 Superior view of the Pavlov 1 calotte. Scale in centimeters.

Unlike the Dolní Věstonice individuals, Pavlov 1 has a large parietal foramen present on the right parietal and one very small one on the left side (see discussion of this trait and its genetic affinity implications above in Dolní Věstonice 16). Detailed scoring results for the coronal and sagittal sutures are presented in Table 8.10.

As mentioned above, there is a series of swellings on the external surfaces of the calvarium, mostly along the sutures. In superior view, there is a swelling along the area where the metopic suture would have been located. The swelling starts 41 mm above nasion and gradually expands to a breadth of 48 mm ca. 20 mm anterior of bregma. This swelling is sufficient to create sulci bilaterally along its margins. Along the left coronal suture, there is a very small swelling for ca. 48 mm from bregma. On the right side there is a prominent swelling up to 38 to 40 mm wide anteroposteriorly, extending ca. 38 mm from bregma. It then tapers off ca. 55 mm from bregma along the right coronal suture. Bilateral swelling is evident all along the sagittal suture, especially between 7 mm and 42 mm from bregma, where it reaches a maximum breadth of 24 mm. It then decreases to a more modest breadth of ca. 18 mm until close to lambda. Along the last 29 mm before lambda, there are separate right and left swellings, 11 mm wide on the right and 9 mm wide on the left, with a maximum breadth across the two of ca. 29 mm (these are discussed further in Chapter 19). Finally, in superior view, the distinction of the superciliary arch from both the frontal squama and the laterally positioned and very large supraorbital trigone on either side by the obliquely flattened supraorbital sulcus is particularly apparent.

Pavlov 2

Pavlov 2 consists of a number of elements. Since they entail primarily teeth and their associated alveolar process, they are described in detail in Chapter 11 (Figure 11.16). The largest element is a section of the left maxillary alveolar process, with the second premolar to second molar. The left maxillary segment retains the alveolar process from the distal second premolar alveolus to the third molar alveolus. The associated right side preserves four teeth (first premolar to second molar) and a small fragment of buccal alveolar bone from the distal second premolar to the middle of the first molar. All of these elements are joined together in wax, representing the bony palate.

There is little that can be gleaned from this maxillary piece other than data concerning the teeth and their associated alveoli. The primary feature is a prominent maxillary torus on its lingual side, adjacent to the alveoli of the molars (Figures 11.16 and 19.20). The torus on Pavlov 2 is moderately large but not unusual in its size or location alongside the molars (Chapters 11 and 19). The primary aspect is its presence lingually; such maxillary tori usually occur buccally.

Summary

From a neurocranial perspective, it is apparent that frontal squamous morphology in the Pavlovian sample shows heterogeneity in size dimorphism, with homogeneity in overall shape. The relationship between least and greatest anterior frontal breadth dimensions is very similar in the Pavlovian sample compared to the Qafzeh-Skhul and Neandertal samples, which show relatively greater least frontal breadth values. Variation in frontal squamous shape is evident with the frontal bone of Dolní Věstonice 3 showing a high frontal subtense relative to the chord, Dolní Věstonice 13–15, showing reduced but still relatively steep frontal inclination with marked projection at metopion, and Pavlov 1 showing the lowest value. The high frontal subtense values in the Pavlovian sample overall are also present in other earlier Upper Paleolithic samples, are generally characteristic of recent human samples, and contrast with the lower values among Neandertals. The Pavlovian sample and other modern humans also show relatively higher values for both the bregma-lambda chord and its subtense values compared to Neandertals but with similar proportions. However, there is some variation, with Pavlov 1 falling on the lower (flatter) portion of the overall distribution, as well as closer to the Neandertal distribution in size. It is also clear that the Pavlovian sample and other earlier Upper Paleolithic individuals have cranial heights that are comparable to those of Neandertals, and have cranial lengths that are shorter, resulting in the characteristically modern globular shape.

There is a distinct separation of medial, midorbital, and lateral components of the supraorbital region in all of the Dolní Věstonice individuals that is very distinct from the uniformly rounded pattern characteristic of Neandertals. In this, Pavlov 1 shows a pattern that is somewhat intermediate in overall size and even has a diminution in the degree of separation of the supraorbital elements, although its total pattern is closer to that of modern humans.

There are variable degrees of occipital bun development in the Dolní Věstonice and Pavlov sample, ranging from modest to nonexistent in Dolní Věstonice 13 and 14 to quite marked in Dolní Věstonice 11/12 and 15 and Pavlov 1, and there is overlap in the occipital chord and subtense relationships among the Neandertals and early modern humans. Depressions above inion are found to varying degrees in these Pavlovian individuals. However, even though Neandertal-like suprainiac fossae are present in some recent humans, the depressions in the Pavlovian sample are not the fully developed, mediolaterally expanded suprainiac fossae associated with a bilaterally protruding occipital torus found in most Neandertals, some Middle Pleistocene European humans, and some of the later Pleistocene African specimens.

Foramen magnum size and shape are highly variable in the Pavlovian sample, although some of this variation may be due to postmortem damage and reconstruction. The breadths of the foramina magna in Dolní Věstonice 3, 14, and 16 are all considerably wider than in Dolní Věstonice 13. Foramina magna lengths in Dolní Věstonice 3 and 16 is relatively small and rounded. Dolní Věstonice 13 shows a relatively small and narrow foramen magnum, and Dolní Věstonice 14 shows a large and elongated one.

The mastoid processes in the Pavlovian sample tend to extend well below the plane of the juxtamastoid eminence, show no lateral flattening, and lack a persistent anterior tubercle. In all of these features, the mastoid processes of the sample are modern, rather than Neandertal-like.

From a splanchnocranial perspective, although there are some postmortem and trauma-induced orbital shape asymmetries in the Pavlovian sample, they share with other modern human samples orbital dimensions that are lower relative to breadth, rather than the higher, more rounded orbits found in Neandertals. The lateral margins of the Pavlovian orbits are posteriorly positioned relative to nasion and to other midfacial landmarks, indicating some degree of midfacial projection, although it is not as pronounced as that seen in Neandertals on average. Both the width of the nasal root and the total breadth across the orbits in Pavlov 1 appear exceptionally large, consistent with its exceptional supraorbital development. Even with such levels of individual variation, however, it is interesting to note that the Qafzeh-Skhul and especially Neandertal values for interorbital breadth are not relatively wider than these Pavlovian ones, perhaps reflecting a cold climate response in the upper internal nose.

There is some level of sexual dimorphism in overall absolute facial height and breadth in the Pavlovian sample, with the Dolní Věstonice 3 female generally at the smallest end of the distribution for both measures among the Pavlovian and other earlier Upper Paleolithic samples. She is still comparable to Dolní Věstonice 15 in upper facial height. In contrast, midfacial breadth measurements relative to facial height are similar across the Pavlovian sample, and indeed all of the comparative samples show the same relationship.

Although damaged in much of the Pavlovian sample, enough anatomy is present to establish that their infraorbital regions show the typically modern two-plane configuration that is different from the horizontally flat or convex plane configuration typically found in Neandertals. The one suggestive exception is the configuration in Dolní Věstonice 16, whose more inflated configuration is most likely due to some combination of congenital and postmortem alteration. The extreme degree of infraorbital concavity in Dolní Věstonice 15 and its similar but more muted configuration in the other two members of the triple burial may be additional evidence of a relatively close genetic affiliation among the triple burial individuals. This and the patterning in other epigenetic features documented above (e.g., the absence of parietal foramina in all of the Dolní Věstonice individuals and the similarly shaped unilateral left-side frontal sinus configuration in the triple burial individuals) warrant a more comprehensive analysis of such traits across all later Pleistocene samples, given their potential to contribute to a better understanding of some aspects of their populational dynamics.

The Pavlovian sample shows uniformity in the area between the nasomaxillary suture and inferomedial orbital area, with all being uniformly smooth and lacking the roughened furrow and slight elevation frequently found in both immature and mature archaic *Homo*. They also all share an orientation of the nasomaxillary suture that is inferiorly inflected, as in most modern humans, rather than horizontally configured, as is found in many Neandertals. There is some variation in nasal breadth, with both Dolní

Věstonice 3 and 15 having narrower nasal breadths relative to height, and Dolní Věstonice 13 and Pavlov 1 (along with a few other Upper Paleolithic individuals) having wider than expected nasal breadths. Nonetheless, the Neandertals show larger nasal breadths, given their nasal heights, than the combined earlier Upper Paleolithic modern human sample (with the Qafzeh-Skhul sample intermediate between the two). Moreover, the Dolní Věstonice and Pavlov internal nasal floors are completely level with their narial margins, which in combination with their narial margin patterns presents a total pattern that is different from the derived pattern found in the latest Neandertals.

Finally, the overall dimensions of the Dolní Věstonice 15 palate are relatively small for the Pavlovian sample, and the Pavlovian sample as whole has smaller palates than do the Qafzeh-Skhul and Neandertal samples.

As a result, the morphological pattern seen in the Dolní Věstonice and Pavlov cranial sample, despite abundant postmortem damage and distortion and numerous cases of both minor and major pathological alterations, is one of reasonably robust early modern humans. They share this pattern with other European earlier Upper Paleolithic humans, both those from central European sites such as Brno-Francouzská, Mladeč, and Předmostí and those from across Europe. There is considerable minor variation in aspects of the neurocranium and the upper facial skeleton and some variation in craniofacial robusticity (especially between the three triple burial males and the three isolated males), but the pattern is a consistent one that separates them from the Neandertals but also provides contrasts with recent human samples.

9

The Auditory Ossicles

Petr Lisoněk and Erik Trinkaus

Four auditory ossicles have been preserved from two of the individuals in the triple burial, Dolní Věstonice 14 and 15 (Figures 9.1 and 9.2). The former individual retains a left malleus and incus, both of which are well preserved but have their surfaces covered by thin calcareous encrustations. The malleus of Dolní Věstonice 15 is preserved only from the head to the neck, and the head is damaged. There is a number of surface defects on its right incus, the deepest one lying on the corpus close to the beginning of *crus breve incudis*.

Some standard measurements of the bones are provided in Tables 9.1 and 9.2. These data were originally published in Lisoněk (1992), but they were not included in Sládek et al. (2000) and are therefore repeated here. With them are comparative measurements, as available, for other Late Pleistocene late archaic and early modern human auditory ossicles (from Angel, 1972; Arensburg & Nathan, 1972; Heim, 1982a; Arensburg & Tillier, 1983; Ponce de León & Zollikofer, 1999; Spoor, 2002; see also Tillier, 1999). In

Table 9.1 Measurements of the mallei of Dolní Věstonice 14 and available Neandertal, Qafzeh-Skhul, and Gravettian humans.

	Length	Manubrium Length	Head Width	Width at the *Processus lateralis*	Manubrium Angle
Dolní Věstonice 14	7.9	4.6	2.5	1.6	128°
Middle Paleolithic					
Biache 1	(8.8)	(4.1)	2.7	—	(154°)
La Ferrassie 3	8.3	4.4	2.5	—	150°
Darra-i-Kur 1	8.3	5.0	2.5	—	—
Qafzeh 11	7.3	3.5	2.5	—	155°
Gravettian					
Lagar Velho 1	8.0	4.3	2.3	—	149°
Recent Humans					
Natufians ($N = 11$)	7.7 ± 0.5	4.5 ± 0.4	2.5 ± 0.1	—	143° ± 8.3°
Romans ($N = 16$)	8.1 ± 0.3	4.6 ± 0.3	2.4 ± 0.2	—	140° ± 5.6°
Indians ($N = 30$)	7.8 ± 0.4	4.4 ± 0.5	2.4 ± 0.2	—	137° ± 8.9°
ANOVA P					
Darra-i-Kur as Nean.	0.007	0.186	0.625	—	—
Darra-i-Kur as Q-S	0.752	0.965	0.625	—	—

The values provided for Lagar Velho 1 are the rounded average for the right and left mallei. Fossil data references are in the text; recent human data are from Arensburg et al. (1981). ANOVA P-values are provided across the three samples (Gravettian, Qafzeh-Skhul, and Neandertal), with Darra-i-Kur 1 included in one or the other of the Middle Paleolithic samples. Values in parentheses are estimated.

Table 9.2 Measurements of the incudes of Dolní Věstonice 14 and 15 and available Neandertal, Qafzeh-Skhul, and Gravettian humans.

	Width	Length	Width of Corpus	Bicrural Length
Dolní Věstonice 14	5.5	7.1	3.6	6.6
Dolní Věstonice 15	4.1	6.4	3.4	5.5
Middle Paleolithic				
Biache 1	4.9	7.5	(4.1)	5.9
La Ferrassie 3	5.1	7.2	—	—
Darra-i-Kur 1	4.8	—	—	6.0
Le Moustier 1	4.9	6.8	—	—
Qafzeh 11	5.1	6.8	—	—
Qafzeh 21	4.8	6.4	—	—
Gravettian				
Lagar Velho 1	4.8	—	—	—
Recent Humans				
Natufians ($N = 11$)	5.1 ± 0.2	6.5 ± 0.3	—	—
Romans ($N = 16$)	5.3 ± 0.3	6.6 ± 0.2	—	—
India ($N = 30$)	5.1 ± 0.2	6.4 ± 0.2	—	—
ANOVA P				
Darra-i-Kur as Nean.	0.918	—	—	—
Darra-i-Kur as Q-S	0.894	—	—	—

Fossil data references are in the text; recent human data are from Arensburg et al. (1981). ANOVA P-values are provided across the three samples (Gravettian, Qafzeh-Skhul, and Neandertal), with Darra-i-Kur 1 included in one or the other of the Middle Paleolithic samples. The value in parentheses is estimated.

addition, data on the late Middle Pleistocene Middle Paleolithic Biache 1 incus and malleus (Rougier, 2003) are included in Tables 9.1 and 9.2, even though they are not included in the computation of comparative statistics. Since the ossicles attain their adult size early in development (Scheuer & Black, 2000), it is appropriate to compare those of immature and mature individuals. Comparative data for some of the measurements are provided for three small samples of recent humans, from Arensburg et al. (1981).

The Mallei

The poor state of preservation of the Dolní Věstonice 15 malleus means that most of the observations possible are from the Dolní Věstonice 14 specimen. The preserved portions of the heads of the mallei suggest that the Dolní Věstonice 15 bone was more robustly built than the Dolní Věstonice 14 ossicle, assuming a similar overall length.

The Dolní Věstonice 14 malleus shows considerable angulation between the head and neck relative to the manubrium, which is reflected in a manubrium angle of 128°; this value is lower than those of the three other Late Pleistocene specimens complete enough to provide the angle (Table 9.1), although it remains within the range of variation of the recent human sample. The head exhibits a well-formed articular surface and a marked groove for the insertion of the articular capsule (Figure 9.1).

It is possible to compare the overall dimensions of the Dolní Věstonice 14 malleus to those of four other western Eurasian Late Pleistocene specimens (Table 9.1). It is apparent that four of the Late Pleistocene specimens, including Dolní Věstonice 14, have very similar values in overall length, manubrium length, and head width; only the Qafzeh 11 malleus is markedly smaller. It is principally Qafzeh 11 that contributes to the one low ANOVA P-value in the comparisons, when the Darra-i-Kur 1 specimen is aligned with the Neandertals. However, the overall configuration of the Middle Paleolithic central Asian Darra-i-Kur 1 specimen is closer to those of early modern humans (Angel, 1972), and therefore the more appropriate comparisons are when it is grouped with Qafzeh 11. Except for Qafzeh 11, all of these Late Pleistocene values are close to the means of the recent human samples, as is the earlier Biache 1 specimen. The suggestion of a decrease in ossicle size, especially between Neandertals and early modern humans (Spoor, 2002), is therefore not supported by the available data.

It is also possible to compare manubrium length to total malleus length, which provides an index of 58.2 for Dolní Věstonice 14. The value is exceeded slightly by that of Darra-i-Kur 1 (60.2), and it is above those of Qafzeh 11 (47.9), La Ferrassie 3 (53.0), and Lagar Velho 1 (53.8). An ANOVA analysis including Darra-i-Kur 1 and Qafzeh 11 together provides a $P = 0.926$, indicating variable but nondistinguishing overall proportions. The value of ca. 46.6 for Biache 1 is moderately low.

The Auditory Ossicles

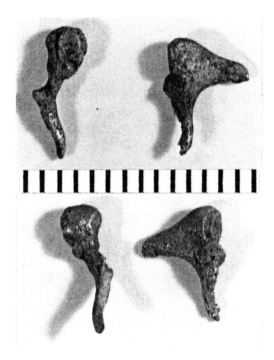

Figure 9.1 Overall views of the Dolní Věstonice 14 left malleus (left) and incus (right). Scale in millimeters.

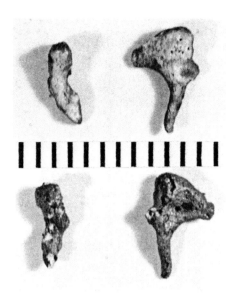

Figure 9.2 Overall views of the Dolní Věstonice 15 right incus and poorly preserved malleus. Scale in millimeters.

The Incudes

Only the *processus lenticularis* of the Dolní Věstonice 14 left incus is missing, which was destroyed antemortem by *otitis media*. Its *crus breve incudis* is contoured on both sides by a concave silhouette, which is not frequently found. A part of the corpus above the *crus breve incudis*, which holds the articular surface and rounds out the groove, is visible (Figure 9.1). This incus has a pronounced channel or a natural incision close to the cusp of *crus breve incudis*, the cause of which has not been satisfactorily clarified (it is probably the impression of a blood vessel). The apex of the *crus longum incudis* without the lenticular process shows the second stage of inflammatory damage caused during the life of the individual.

The Dolní Věstonice 15 right incus is less well preserved (Figure 9.2), with various forms of surface damage. It nonetheless appears to be a more robust bone than the Dolní Věstonice 14 ossicle.

Several dimensions are preserved for these two incudes, but comparative data are limited principally to length and width, with the latter providing the most comparative data. There is a high degree of similarity in size across the Late Pleistocene samples, whether Darra-i-Kur 1 is considered a Neandertal or an early modern human (Table 9.2), and the two Dolní Věstonice specimens bracket the range. Dolní Věstonice 14 has the greatest incus width and is only slightly exceeded in incus length by La Ferrassie 3, whereas Dolní Věstonice 15 has the smallest width and is matched for the smallest length by Qafzeh 21. The index of width to length similarly provides no separation of the samples (ANOVA $P = 0.734$, with Darra-i-Kur 1 aligned with the Qafzeh specimens), and Dolní Věstonice 14 and 15 provide the highest (77.5) and lowest (64.1) values, respectively. The other indices—70.8 for La Ferrassie 3, 71.1 for Darra-i-Kur 1, and 75.0 for Qafzeh 11 and 21—scatter in between. The absolute dimensions and width-to-length index (65.3) of the Biache 1 specimen are similar to those of these Late Pleistocene specimens. All of these values are below the overall mean index of the recent human samples (79.6; $N = 57$), who tend to have similar widths but shorter lengths on average.

Summary

The four auditory ossicles from Dolní Věstonice 14 and 15 support the general pattern of similarities among these bones across the available Late Pleistocene specimens in terms of both size and general proportions. There is some variability in shape, but the differences appear to be as much within a priori defined samples as between those samples. This pattern of within-sample variability is illustrated especially by the contrasts between the ossicles of Dolní Věstonice 14 and 15, who were clearly contemporaneous. When they are combined with the earlier Biache 1 specimen and data from recent humans, there appears to be little discernible trend in auditory ossicle size or proportions through the last 200 millennia.

10

The Mandibular Remains

Robert G. Franciscus, Emanuel Vlček, and Erik Trinkaus

Six associated skeletons from Dolní Věstonice and Pavlov (Dolní Věstonice 3 and 13–16 and Pavlov 1) preserve a largely complete mandible. They are joined by the lateral corpus of Pavlov 3 and a corpus fragment from Pavlov 4. Even though they are proportionately and functionally part of the facial skeleton and therefore fit with the cranial remains described in Chapter 8, they are also described here because of the frequent isolation of mandibles in the human fossil record. In addition, as structural elements subjected to the biomechanical demands of mastication, they have properties that need to be addressed in a manner semi-independent of the crania with which they were originally articulated.

The individual mandibles are described in terms of their preservation and morphology, and then they are compared with respect to a series of metric and discrete traits that have been shown to quantify variability across Middle and Late Pleistocene human mandibles.

Preservation and General Description

Dolní Věstonice 3

The mandible of Dolní Věstonice 3 is largely complete, retaining the corpus and both rami, but it suffered considerable postmortem crushing and bone loss (see Klíma, 1963). Like the cranium (Chapter 8), the mandible was reconstructed from numerous pieces, missing portions were filled with consolidant, and then the assembly was painted with a thick, dark red-brown pigment to resemble ochre. The last aspect of the restoration in particular obscures surface detail to the point where the accuracy of the reconstruction is difficult to ascertain. Moreover, many anatomical surface details are difficult to score and evaluate. Adding to the difficulty of evaluating its morphology is pathological alteration of the left condyle due to trauma to the left side of the face (Chapter 19), which may have produced a host of secondary remodeling alterations to the mandible.

In occlusal view (Figure 19.2), there is a marked misalignment of the lower dental arcade in Dolní Věstonice 3. Relative to the midline, the anterior dentition is oriented obliquely to the left, with each tooth antimere on the left more distally positioned than its right antimere. There is also a slightly greater wear gradient on the left-side molars compared to the right side, but this may well be due to the congenital absence of the left lower third molar, promoting greater occlusal attrition on the left second molar (Chapter 11). The left condylar process was remodeled extensively; the majority of the condyle is gone, with only the remodeled basal aspect retained. The mandibular incisure crest on the left courses to the middle of the remaining condylar surface, unlike its right, nonpathologically altered counterpart, which courses to the most lateral extent (see Chapter 19).

In anterior view, the superior portion of the mental trigone is deflected to the left. The anterior symphysis has a distinct subalveolar concavity, which then descends onto a rounded (in lateral view) tuber symphyseos. As restored, there is no evidence of a right lateral tubercle and only a suggestion of one on the left. It therefore conforms to mentum osseum rank 4 (Table 10.4; Dobson & Trinkaus, 2002). The right incisura is a distinct rounded fossa. The degree of development of the left one is obscured by damage and restoration, but it appears to have resembled the right one.

In right lateral view (Figure 10.1), there is well-developed masseter scarring on the ramus that extends

The Mandibular Remains

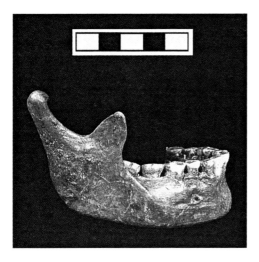

Figure 10.1 Right lateral view of the Dolní Věstonice 3 mandible. View is slightly anterosuperior from *norma lateralis*. Scale in centimeters.

to the second molar. The mental foramen is positioned under the mesial edge of the second premolar. The Dolní Věstonice 3 left medial corpus is damaged and possibly pathological, but the right one presents a rounded mylohyoid ridge and a moderate sulcus below it (Figure 10.2).

In posterior view, there appears to have been continued eruption of the first and second molars, especially on the left side (see Chapters 11 and 19). The right internal gonial aspect exhibits a V-shaped mandibular foramen with a pronounced lingula. The medial pterygoid muscle

Figure 10.2 Medial view of the Dolní Věstonice 3 right ramus, with partial damage of the lingula. Scale in centimeters.

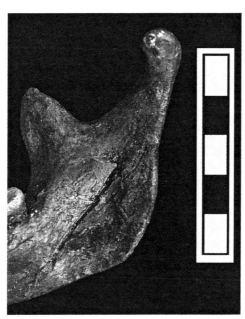

is moderately developed and, unlike the other Pavlovian individuals, exhibits a prominent superior tubercle. The left mandibular foramen is damaged, but from what is retained it appears to have possessed the same antemortem morphology as the right side. The left internal ramal area superficially appears to be postmortem damaged and reconstructed. However, close examination shows that a substantial portion is likely antemortem anatomy that had been remodeled commensurate with the overall pathological alteration of the condyle and temporomandibular joint (Chapters 8 and 19). A remnant of the mylohyoid groove is still present inferiorly, and the area anterior to this groove remnant is remodeled extensively. The superior inner ramal area just inferior to the condyle and coronoid process has also been altered.

From posterior view, it appears that the left condyle had been fractured. There is a clear line that runs around the posterior superior neck below the affected condyle, although given the state of reconstruction, postmortem damage cannot be ruled out. No planum alveolare is present, and there are prominent genioglossal spines.

In left lateral view (Figure 19.4), the left coronoid process is reduced more than the right one. However, there is a broken edge on its current tip, and it remains uncertain to what extent the coronoid tip extended beyond its present level before postmortem damage. The lines of curvature of the preserved adjacent original margins suggest that, at the most, only a couple of millimeters of bone is missing. The mandibular corpus height near the base of the ramus line on the external aspect is smaller on the left side than the right side by 2 mm. Masseter muscle markings on the affected side are still apparent and extend all the way anterior to the mid–second molar.

Dolní Věstonice 13

In occlusal view, the dental arcade is symmetrically oriented with respect to all mandibular anatomy and all aspects of the mandible show near perfect symmetry. The right mandibular condyle has been slightly reconstructed with filler; however, the left condyle is completely preserved. The lateral aspects of the condyles are larger, and they taper toward midline. A greater portion of the condylar mass on each side is situated medial to the superoinferior vertical ramal plane on either side. From this view, it can be seen that the posterior border of each mandibular incisure crest courses to the lateral edge of each condyle. The reconstructed left and completely preserved right coronoid processes are evident from this view, their most superior tips flaring slightly laterally. The greatest breadth of the corpus is located at the midpoint of each second molar.

In anterior view, Dolní Věstonice 13 shows a high degree of symmetry in all aspects. Its anterior symphysis (Figure 10.3) is vertically concave below the alveolar margin, with bilateral anterior buccal depressions and a small vertical ridge separating them. The left buccal depression is slightly deeper than the right one, especially by the first incisor. There is a prominent tuber symphyseos with its peak well above the symphyseal base, such that the infradentale to pogonion distance (22.0 mm) is only 71.7% of the distance to gnathion (30.7 mm). There is then a small sulcus that rounds to gnathion. The lateral tubercles are 9 to 10 mm to each side of the midline near the basal border. The right lateral tubercle is almost completely absent, and the left one below the second incisor to canine interdental septum is very small. It therefore has a mentum osseum rank of 4. The right incisura is consequently essentially absent, and the left one is very shallow (0.2 mm), formed as much by a small horizontal ridge above it as by the minimal lateral tubercle below it.

There is some postmortem alveolar bone loss, especially on the anterior facies of the canine root eminences, and other alveolar aspects are detailed in Chapter 11. The inferior border courses symmetrically from menton on both sides and ends near gonion, with both sides showing mild laterally everted gonial flaring. The left and right anterior ramal margins are oriented vertically along their lengths, and each ends at the tip of the coronoid process with a very slight lateral flare. The posterior portion of the mandibular incisure courses laterally as it approaches each condyle and ends on the most lateral extent of each condyle. The bulk of each condyle is positioned medially with respect to the anteroposterior plane of each ramus. The anteromedial infracondylar area at the top of the neck on the left side shows a moderately rugose pterygoid fovea; this area is reconstructed on the right side. The alignment of the left-side condyle with respect to the transverse plane (which has not been reconstructed) is slightly obliquely angled so that the medial aspect is somewhat more superiorly positioned. In anterior view, the tubercles for the lateral ligament of the temporomandibular joint on the lateral condyle/neck on both sides are visible as minor lateral projections.

In left lateral view, Dolní Věstonice 13 shows the well-developed tuber symphyseos but nearly absent lateral tubercles. There is a diffuse oblique line running from the lateral tubercle up to the mental foramen, which becomes even more diffuse between this point and the anterior ramus border. There is a single mental foramen whose minimum diameter is 4.3 mm. It lies below the root apex of the second premolar. The foramen is filled with matrix, which obscures whether it was divided.

The base of the mandible in left lateral view shows an undulating inferior border that has an inferiorly directed convexity from midline to about the level of the posterior third molar and the vertical portion of the anterior ramal margin, where it becomes concave for a very short distance and then convex again in the gonial region. The angulation between the posterior border of the ramus and the posteroinferior basal border is rounded rather than acute, making gonion a projected osteometric landmark. The root of the anterior ramal margin lies inferior to the second molar on the corpus, and the anterior ramal margin shows a marked anterior concavity at about the level of the occlusal line. The anterior ramal margin covers the posterior one-third to one-half of the third molar, and thus shows the absence of a retromolar space. Posteriorly, the ramal margin shows a slight concavity just inferior to the neck of the mandible.

The mandibular incisure is symmetrical rather than J-shaped, that is, the lowest point is toward the middle of the incisure rather than closer to the condyle (Rak, 1998; Stefan & Trinkaus, 1998a,b), and in strict anatomical position, the tip of the coronoid process lies inferior to the Frankfurt Horizontal plane across the superior-most condylar surface. The left coronoid process is moderate in both width and height, although it has been reconstructed (unlike the other side, which is intact). Muscle scarring for temporalis is not visible on the anterosuperior-most aspect of the coronoid process, although as noted, it is reconstructed to some degree. The insertion area for masseter is large and rugose, showing seven distinct, roughly circular scalloped areas (four inferior and three superior) with mildly developed oblique ridges between the inferior ones. The total area extends anteriorly to the level of the second to third molar interface. The neck below the condyle appears to be as wide as the condyle in lateral view, although this is due to the extension of the mandibular incisure; from medial view the neck appears nar-

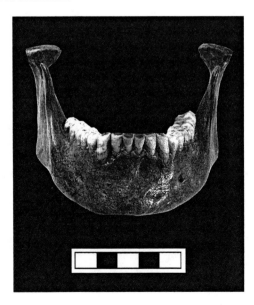

Figure 10.3 Anterior view of the Dolní Věstonice 13 mandible. Scale in centimeters.

rower. There is a small distinct tubercle for the lateral ligament of the temporomandibular joint on the anterolateral condyle/neck interface.

In right lateral view (Figure 10.4), as from the left side, Dolní Věstonice 13 shows the well-developed tuber symphyseos but largely absent lateral tubercle. There is a diffuse oblique line running from the lateral tubercle up to the mental foramen; it is more diffuse than on the left side, but it becomes even more diffuse between this point and the anterior ramus border. There is a single mental foramen below the root apex of the second premolar, whose minimum diameter is 2.8 mm. The foramen is filled with matrix, which obscures whether it was divided. The base of the mandible in left lateral view shows an undulating inferior border configuration identical to that of the left side. The angulation between the posterior border of the ramus and the posteroinferior basal border is more acute than on the left side but still rounded enough to make gonion a projected point. The root of the anterior ramal margin lies inferior to the second molar, and the anterior ramal margin shows a marked anterior concavity at about the level of the occlusal line. The anterior ramal margin covers the posterior half of the third molar in lateral view, and thus there is no retromolar space. Posteriorly, the ramal margin shows a slight concavity just inferior to the neck of the mandible.

The right ramus appears broad relative to total mandibular length, although less so than on the left side. The mandibular incisure is symmetrical, and in anatomical position, the tip of the coronoid process lies inferior to the Frankfurt Horizontal plane across the superior-most condylar surface. The coronoid process is undamaged on this side, and it is moderate in both width and height. There is no visible muscle scarring for temporalis on the anterosuperior-most aspect of the left coronoid process. The insertion area for masseter is large and rugose, showing at least four large scalloped areas, with the superoanterior-most one being the most deeply etched. The total area extends anteriorly to the level of the second to third molar interface. The neck below the condyle appears to be as wide as the condyle in lateral view, although this is due to the extension of the mandibular incisure; from medial view the neck appears narrower. There is a small distinct tubercle for the lateral ligament of the temporomandibular joint on the anterolateral condyle/neck interface.

In posterior view, the medially inflected condyles on both sides are apparent; the right posterior condyle is partly reconstructed with filler. The posterior ramal borders are relatively straight, flaring slightly laterally in the gonial region. Both mandibular foramina are preserved, showing a V-shaped configuration, and both sides show complete mylohyoid bridging. The mylohyoid line is present and well preserved on both sides. On the right it begins at the third molar 4 mm inferoposteriorly to the alveolar bone below the posterior border of the third molar and courses anteroinferiorly to the inferior portion of the sublingual fossa to meet up with the superior border of the digastric fossa. On the left side (Figure 10.5), the mylohyoid line begins at the third molar, 5.8 mm inferoposteriorly to the alveolar bone below the posterior border of the third molar, and courses anteroinferiorly to the inferior portion of the sublingual fossa to meet up with the superior border of the digastric fossa. There is a relatively shallow submandibular fossa inferior to the left mylohyoid line. There are two superior and two inferior genioglossal tubercles. The tubercles encompass an area 10.7 mm high and 6.3 mm wide.

Figure 10.4 Right lateral view of the Dolní Věstonice 13 mandible. Scale in centimeters.

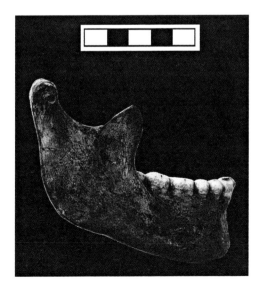

Figure 10.5 Oblique posteromedial view of the Dolní Věstonice 13 left lateral corpus and ramus. Scale in centimeters.

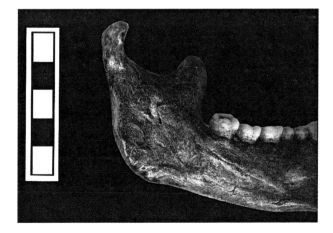

There are two well-defined sublingual fossae extending 14 mm laterally and 11 mm high anteriorly above each mylohyoid line to either side of the genioglossal tubercles. There is a small shallow area in midline below the level of the central incisors roughly 7.8 mm in diameter, with two small foramina at the alveolus to either side of the posterior medial incisors.

In right lateral internal view, the right coronoid process is complete and nonreconstructed. The medial pterygoid insertion is visible but not particularly prominent; there is a modest scallop or sharp edge superiorly, and there are six oblique scallops further inferiorly. There is no hypertrophied superior tubercle. In left lateral internal view, only the coronoid process has been reconstructed partly with filler, and the remaining anatomy is complete and nonreconstructed. The medial pterygoid insertion is visible but not particularly prominent. There appears to be a modest scallop superiorly, and there are four oblique scallops further inferiorly. No enlarged superior tubercle is present.

In basal view, the moderate midline symphyseal thickness is evident. There is a small circular depression anteriorly between the digastric insertions. The insertion areas for the digastrics are well marked and deeply etched; they are larger and deeper medially, and smaller and shallower laterally. Along with those of Dolní Věstonice 14, they are narrower than the fossae of the other specimens (Table 10.1). The lack of development of the lateral tubercles to either side of the midline mental protuberance is also clearly seen here. The submandibular fossa inferior to either mylohyoid line is also evident. The greatest mediolateral width of the corpus at the second molar is apparent from this view, as is the lingula of each mandibular foramen. Also evident is the relative mediolateral thickness of each gonial region. The medial extension of each condyle relative to the vertical ramal margin is also apparent. The high degree of overall symmetry in this mandible is also seen from this view.

Table 10.1 Dimensions of the digastric fossae.

	Right Length	Left Length	Right Depth	Left Depth
Dolní Věstonice 13	18.8	19.9	8.4	8.5
Dolní Věstonice 14	(24.0)	—	(6.4)	(7.0)
Dolní Věstonice 15	19.6	21.9	10.5	10.8
Dolní Věstonice 16	23.7	18.5	10.5	9.3
Pavlov 1	20.1	—	11.5	—

Length is the oblique length from the midline border. Measurements in parentheses are approximate, because of indistinct boundaries of the Dolní Věstonice 14 fossae. The left length of Dolní Věstonice 14 is not included because of the absence of a discernible posterolateral border; the left dimensions for Pavlov 1 are missing because of damage.

Dolní Věstonice 14

In occlusal view, the dental arcade of Dolní Věstonice 14 is fairly symmetrically oriented, and all aspects of the mandible show good symmetry. The left mandibular condyle has been slightly reconstructed with filler posteriorly and the right-side condyle is gone. From this view, it can be seen that the posterior border of the mandibular incisure crest courses to the lateral edge of the condyle. The preserved coronoid processes are nearly vertically oriented. The degree of overlap between the posterior borders of the unerupted third molars on each side with the anterior ramus is evident.

The greatest breadth of the corpus is located at the midpoint of each second molar. Internally, the genioglossal tubercles are evident, although not prominent, in superior view. The tuber symphyseos is evident in midline, although the less prominent lateral tubercles, spaced 28.5 mm apart, are not. The anteroposterior symphyseal thickness is smaller than the other specimens.

In anterior view (Figure 10.6), Dolní Věstonice 14 shows general symmetry, but there are minor asymmetrical aspects, such as the area lateral to each tubercle at the basal border. The anterior symphysis has a concave profile below the incisor crowns, but there is no central vertical ridge and no anterior buccal depressions. However, there are small vertical sulci between the central incisors, left incisors, and left second incisor to canine roots, each 7 to 8 mm high. On the right, there is a large hollow area from the first incisor to the canine roots, ca. 11 mm high and ca. 13.5 mm wide. It is unclear whether it represents a large anterior buccal depression or a resorptive area, but the surface bone appears otherwise normal.

The tuber symphyseos of Dolní Věstonice 14 is also prominent, rounded, and well above the base. Its infradentale to pogonion distance (18.5 mm) is only 60.1% of the distance to gnathion (30.8 mm). Its inferior margin rounds to gnathion with no evidence of a midline sulcus. The lateral tubercles below the second incisors and canines are symmetrical, rounded, slightly rugose, slightly protruding, but well posterior of pogonion. They are positioned ca. 13.5 mm to each side of the midline near the basal border. The incisurae are bilaterally broad, smooth sulci extending toward the mental foramina, 1.2 mm deep on the right and 0.6 mm deep on the left. This combination of features is intermediate between mentum osseum ranks 4 and 5 but close to 4. It should be scored as 4/5.

The inferior border courses symmetrically from menton on both sides and ends near gonion, with both sides showing mildly laterally everted gonial flaring. The anterior ramal margins are oriented mostly vertically along their lengths. The posterior portion of the mandibular incisure courses laterally as it approaches the left condyle,

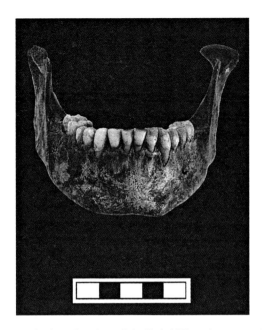

Figure 10.6 Anterior view of the Dolní Věstonice 14 mandible. Scale in centimeters.

and the bulk of the left condyle is positioned medially with respect to the anteroposterior plane of the ramus. The anterolateral infracondylar area at the top of the neck on the left side shows a moderately rugose pterygoid fovea. The alignment of the left-side condyle is completely horizontal with respect to the transverse plane. In anterior view, the tubercle for the lateral ligament of the temporomandibular joint on the lateral condyle/neck on both sides is visible as a larger lateral projection than that for any of the other Pavlovian specimens but considerably smaller than that found in some Neandertals.

In left lateral view (Figure 10.7), Dolní Věstonice 14 shows the well-developed tuber symphyseos and modest lateral tubercles. There is a diffuse oblique line running from the lateral tubercle up to the mental foramen, which then continues up to the anterior ramus border. There is also a line that runs from the tubercle along the basal border of the mandible and superiorly to the anterior ramus margin. There is a single mental foramen whose minimum external diameter is 2.6 mm and is positioned below the root apex of second premolar. The foramen is filled with matrix; it is not known whether it was divided.

The base of the mandible in lateral view shows a straight inferior border to about the level of the distal third molar and the vertical portion of the anterior ramal margin, where it becomes concave for a very short distance and then convex in the gonial region. The angulation between the posterior border of the ramus and the posteroinferior basal border is rounded rather than acute (although more acute than in some of the other Dolní Věstonice and Pavlov individuals). Gonion is reconstructed on this side, and as reconstructed gonion is a projected point. The root of the anterior ramal margin lies inferior to the anterior border of the second molar on the corpus, and the anterior ramal margin shows a marked anterior concavity at about the level of the occlusal line. The anterior ramal margin covers the distal one-third of the unerupted third molar in lateral view and thus shows absence of a retromolar space. Posteriorly, the ramal margin shows a slight concavity just inferior to the neck of the mandible.

The mandibular incisure is slightly asymmetrical or J-shaped; however, the direction of the J is configured in the opposite direction than for a number of the Neandertals. In anatomical position, the tip of the coronoid process lies inferior to the Frankfurt Horizontal plane across the superior-most condylar surface. The coronoid process is moderate in both width and height (although it appears to be wider than taller). Slight muscle scarring for temporalis is visible on the anterosuperior-most aspect of the coronoid process. The insertion area for masseter is large and rugose, showing several distinct scalloped areas with at least three mildly developed oblique ridges. The total area extends anteriorly to the level of the second to third molar interface.

The neck below the condyle appears to be as wide as the condyle in lateral view, although this is due to the extension of the mandibular incisure; from medial view the neck appears narrower. There is a moderately developed distinct tubercle for the lateral ligament of the temporomandibular joint on the anterolateral condyle/neck

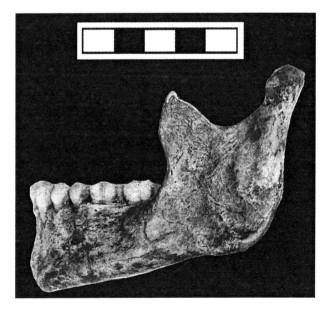

Figure 10.7 Left lateral view of the Dolní Věstonice 14 mandible. Scale in centimeters.

interface that is, 10.0 mm high and extending 1.3 mm beyond the condylar subchondral bone.

In right lateral view, Dolní Věstonice 14 shows the well-developed tuber symphyseos and modestly developed lateral tubercle. The diffuse oblique line running from the lateral tubercle is difficult to make out on this side because of damage and reconstruction. Although reconstructed, there appears to be a single mental foramen on this side whose minimum external diameter is 2.7 mm; it lies below the root apex of the second premolar. The foramen is partially filled with matrix, which obscures whether it was divided.

The base of the mandible in right lateral view shows the same pattern of undulation as the other side (see above). The angulation between the posterior border of the ramus and the posteroinferior basal border is more acute than in some Pavlovian individuals but still rounded enough to make the gonion a projected point. The root of the anterior ramal margin lies inferior to the second molar on the corpus, and the anterior ramal margin shows a marked anterior concavity at about the level of the occlusal line. The anterior ramal margin covers most of the unerupted third molar in lateral view (no retromolar space). Posteriorly, the ramal margin shows a slight concavity inferior to the neck of the mandible at about the occlusal plane.

The mandibular incisure, as in its left counterpart, is slightly asymmetrical but with the J-shape in reverse from that in many Neandertals. In anatomical position, the tip of the right coronoid process would have been inferior to the Frankfurt Horizontal plane across the superior-most condylar surface; this can be determined even though the condyle is missing. The coronoid process is undamaged on this side, and it is moderate in both width and height, but it seems to project anteriorly more than in any other individual in this sample or even on the other side of this specimen. There is no visible muscle scarring for temporalis on the anterosuperior-most aspect of the right coronoid process, although there is a large scallop on the process as a whole. The insertion area for masseter is large and rugose, showing several large scalloped areas; the total area extends anteriorly to the level of the second to third molar interface. On the right side, the condyle is missing, and the neck inferior to the condyle is damaged.

In posterior view, the medially inflected condyle on the left side (partly reconstructed) is apparent. The posterior ramal borders are relatively straight, flaring slightly laterally in the gonial region. Both mandibular foramina are preserved, showing a V-shaped configuration and complete mylohyoid bridging. The mylohyoid line is damaged on the right side. On the left side (Figure 10.8), the mylohyoid line begins posterior and inferior to the unerupted third molar, courses anteroinferiorly to the inferior portion of the sublingual fossa, but does not meet

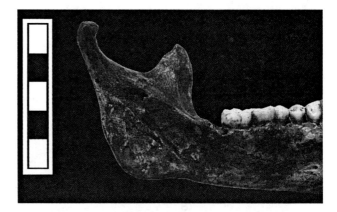

Figure 10.8 Oblique posteromedial view of the Dolní Věstonice 14 left lateral corpus and ramus. The third molar is unerupted. Scale in centimeters.

up with the superior border of the digastric fossa. There is a relatively shallow submandibular fossa inferior to the left mylohyoid line. The genioglossal tubercle area is somewhat amorphous but appears to show a tubercle superiorly, 6.5 mm high and 7.8 mm wide, and a pit inferiorly. There are two poorly defined sublingual fossae (the one on the left is somewhat better defined) anteriorly above each mylohyoid line to either side of the genioglossal tubercle area.

The right internal coronoid process is complete and nonreconstructed; it shows a clear insertion for the temporalis fibers. The medial pterygoid insertion is visible but not particularly prominent. There are modest scallops, and there is no enlarged superior tubercle. The left internal coronoid process is complete. The left medial pterygoid insertion is visible and virtually identical to the right side.

In basal view, the moderate midline symphyseal thickness is evident. The insertion areas for digastrics are shallow and not well marked, especially on the left side, resulting in some uncertainty in their dimensions (Table 10.1). They are fully lingual in position. The modest lateral tubercles to either side of the midline tuber symphyseos are also clearly seen here. The greatest mediolateral width of the corpus at the second molar is apparent from this view. The lingula of each mandibular foramen and the relatively thick mediolateral dimension of each gonial region are evident in this view. The medial extension of the left condyle relative to the vertical ramal margin and the high degree of overall symmetry in this mandible can be seen in this view.

Dolní Věstonice 15

In occlusal view, the dental arcade of Dolní Věstonice 15 is symmetrically oriented, and all aspects of the mandible show near perfect symmetry. The middle aspects of the condyles are larger, and they taper toward medial and lateral endpoints. Virtually all of the condylar mass on each side is situated medial to the superoinferior vertical ramal plane on either side. From this view, it can be seen that the posterior border of each mandibular incisure crest courses to the lateral edge of each condyle. The coronoid processes are largely vertically oriented and show no lateral flaring at the tips. The degree of overlap between the nonocclusal posterior borders of the third molars on each side with the anterior ramus is evident. The greatest breadth of the corpus in Dolní Věstonice 15 is not as wide as in Dolní Věstonice 13, and it is located further distally around the unerupted third molars (ca. 15 mm in breadth on both sides). The tuber symphyseos is evident in midline, although the prominent lateral tubercles (spaced 22.2 mm apart) are not. The moderately thick anteroposterior symphysis is also apparent (in occlusal view but not in anatomical view).

In anterior view (Figure 10.9), there is a large break beginning at midline and running lateroposteriorly to the right down to the inferior border in the area of the mental foramen; the reconstructed alignment nonetheless appears accurate, and other than this, the mandible is well preserved. This view shows a high degree of symmetry in virtually all areas. There is some postmortem alveolar bone loss, especially on the anterior facies of the canine root eminences and left incisor alveoli (empty) eminences.

The superior anterior symphysis of Dolní Věstonice 15 has prominent large, bilaterally present anterior buccal depressions, extending laterally to the mesial sides of

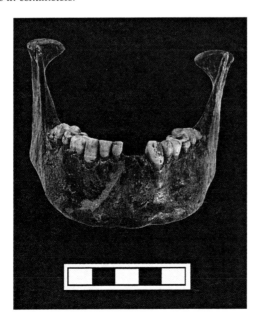

Figure 10.9 Anterior view of the Dolní Věstonice 15 mandible. Scale in centimeters.

each canine root. They are generally symmetrical, but the inferior and lateral margins are more distinct on the right side. The right and left heights are 11.6 and 13.6 mm, respectively, and the breadths are 12.0 and 13.7 mm, respectively. There is a raised rounded area between them.

The prominent and rounded tuber symphyseos is located two-thirds of the distance to gnathion from infradentale (20.0 mm vs. 30.0 mm). Inferiorly there is a clear notch, or short sulcus, from the inferior tuber symphyseos to gnathion. The lateral tubercles are clearly present but also rounded. They are sufficiently prominent in basal view to provide a straight profile to the inferior symphysis but not to form a distinct shelf. They are approximately 11 mm to each side of the midline near the basal border. The right incisure is lost to damage and reconstruction, but the left one is a clear oval fossa with a distinct medial margin, a mediolateral length of 13.5 mm, a height of ca. 8 mm, and a depth of 1.2 mm. These features combine to place the anterior symphysis of Dolní Věstonice 15 between mentum osseum ranks 4 and 5, but close to 5.

The inferior border courses symmetrically from menton on both sides (with allowance for the joined section) and ends near gonion, with the right side showing mildly laterally everted gonial flaring, and the left showing no lateral flaring or gonial inversion (i.e., completely straight). The dental arcade (occlusal dentition) is symmetrical in itself and with respect to all adjacent mandibular anatomy. The anterior ramal margins are oriented vertically along their lengths, with the left side inclined slightly more laterally. The posterior portion of the mandibular incisure courses laterally as it approaches each condyle and ends on the most lateral extent of each condyle. The bulk of each condyle is positioned medially with respect to the anteroposterior plane of each ramus. The anteromedial infracondyle area at the top of the neck on both sides shows a moderately excavated pterygoid fovea, and the left shows a small area of raised remodeled bone in the lateral aspect of the fovea, possibly an ossification of the lateral pterygoid muscle attachment. The alignment of both condyles in the transverse plane appears normal. In anterior view, the tubercles for the lateral ligament of the temporomandibular joint on the lateral condyle/neck on both sides are visible, but they are extremely small and barely projecting.

In left lateral view, Dolní Věstonice 15 shows the well-developed tuber symphyseos and moderate lateral tubercles. There is a very well-defined line that courses from the lateral tubercle along the basal border to the level of the first molar and then superiorly to the anterior ramus border. There is a single mental foramen on this side whose minimum external diameter is 5.0 mm and which lies below the root apex of the second premolar. The foramen is filled with matrix, which obscures whether it was divided.

The base of the mandible in left lateral view shows an undulating inferior border that has an inferiorly directed convexity from midline to about the level of the coronoid process, where it becomes concave for a short distance and then convex again in the gonial region. There is also a slight, and very short, concave undulation just posterior to the lateral tubercle. The angulation between the posterior border of the ramus and the posteroinferior basal border is rounded rather than acute, making gonion a projected point. The root of the anterior ramal margin lies inferior to the second molar on the corpus, and the anterior ramal margin shows a mild anterior concavity superior to the level of the occlusal line. The anterior ramal margin covers the distal one-half of the nonocclusal third molar in lateral view, and thus a retromolar space is absent. Posteriorly, the ramal margin shows a slight concavity just at the occlusal plane.

The mandibular incisure is more symmetrical than J-shaped, and in anatomical position, the tip of the coronoid process lies just inferior to the Frankfurt Horizontal plane across the superior-most condylar surface. The coronoid process is narrow in width and tall in height. Muscle scarring for temporalis is barely visible on the anterosuperior-most aspect of the coronoid process. The insertion area for masseter is large and rugose, although somewhat less rugose than that in Dolní Věstonice 13. It has at least four distinct scalloped areas that are more prominent anteriorly than posteriorly. The total scarring area for masseter extends anteriorly to the level of the second to third molar interface.

The neck below the condyle appears to be as wide as the condyle in lateral view, although this is due to the extension of the mandibular incisure; from medial view the neck appears narrower. There is a very small, barely discernable tubercle for the lateral ligament of the temporomandibular joint on the anterolateral condyle/neck interface.

In right lateral view (Figure 10.10), there is a very well defined line that courses from the left mental tubercle along the basal border to the level of the first molar and then up to the anterior ramus border. There is a single mental foramen on the right side whose minimum external diameter is 6.0 mm, and it lies below the root apex of the second premolar. The foramen is filled with matrix, which obscures whether it was divided.

The base of the mandible in right lateral view shows an undulating inferior border that has inferiorly directed convexity from midline to about the level of the coronoid process (with an uneven break along the join), where it becomes concave for a short distance and then convex again in the gonial region. There is also a slight concave undulation just posterior to the lateral tubercle. The angulation between the posterior border of the ramus and the posteroinferior basal border is more acute than on the left

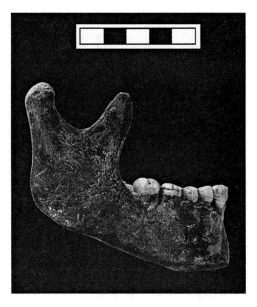

Figure 10.10 Right lateral view of the Dolní Věstonice 15 mandible. Note that the right third molar is partially erupted in an anomalous position (Chapter 11), and it is barely visible above the alveolar process. Scale in centimeters.

side, but it is still rounded enough to make gonion a projected point. The root of the anterior ramal margin lies inferior to the second molar on the corpus, and the anterior ramal margin shows a more marked anterior concavity than the left at about the level of the occlusal plane. The anterior ramal margin covers the posterior one-half of the unerupted third molar in lateral view, and thus there is no retromolar space. If the third molar had erupted normally, rather than against the distal second molar (Chapter 11), even more of the tooth would be covered by the ramal margin. Posteriorly, the ramal margin shows a slight concavity just inferior to the neck of the mandible.

The mandibular incisure is symmetrical, and in anatomical position, the tip of the coronoid process lies inferior to the Frankfurt Horizontal plane across the superior-most condylar surface. The coronoid process is damaged on this side and has been repaired, and muscle scarring for temporalis therefore cannot be assessed; it can be reasonably determined, however, that the process was narrow in width and tall in height. The insertion area for masseter is large and rugose, showing at least four large scalloped areas of equal rugosity. The total maximum area of the scarring extends anteriorly to the level of the second to third molar interface.

The neck below the condyle appears to be as wide as the condyle in lateral view, although this is due to the extension of the mandibular incisure; from medial view, the neck appears narrower. There is a very small tubercle for the lateral ligament of the temporomandibular joint on the anterolateral condyle/neck interface.

In posterior view, the medially inflected condyles on both sides are apparent. The posterior ramal borders are relatively straight, flaring slightly laterally in the gonial region but inflecting medially above the gonial region. Both mandibular foramina are preserved, showing a V-shaped configuration. There is complete mylohyoid bridging on the left side, but incomplete bridging on the right side. The mylohyoid line is present and well preserved on both sides. On the right side, the mylohyoid line begins 9.0 mm inferoposteriorly to the alveolar bone below the posterior border of the nonocclusal third molar and courses anteroinferiorly to the inferior portion of the sublingual fossa to meet up with the superior border of the digastric fossa. On the left side (Figure 10.11), the mylohyoid line begins just distal to the third molar and courses anteroinferiorly to the posterior portion of the sublingual fossa, but it does not meet up with the superior border of the digastric fossa. There is a deeper submandibular fossa inferior to the left mylohyoid line in Dolní Věstonice 15 than in Dolní Věstonice 13. There are two superior genial tubercles and one inferior crest, overall 10.6 mm high and 8.4 mm wide. The sublingual fossae anteriorly above each mylohyoid line to either side of the genioglossal tubercles are not well defined.

The right internal coronoid process is partly reconstructed. The medial pterygoid insertion is visible but not particularly prominent. There are modest scallops, but definitive ridges cannot be seen. No hypertrophied superior tubercle is present. Internally, the left coronoid process is complete and nonreconstructed. The left medial pterygoid insertion is visible and more scalloped than on the right side, still not very prominent, and lacking an enlarged superior tubercle.

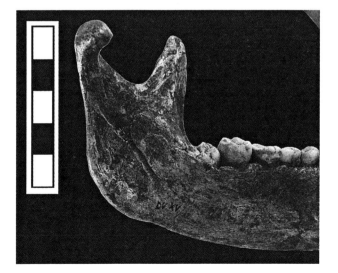

Figure 10.11 Oblique posteromedial view of the Dolní Věstonice 15 left lateral corpus and ramus. The third molar is partially erupted in an anomalous position. Scale in centimeters.

In basal view, the moderate midline symphyseal thickness is evident. There is a prominent tubercle anteriorly between the digastric insertions. The insertion areas for the digastric muscles have a midline peak pointing posteriorly and distinct medial fossae and anterior margins, and they then fade posterosuperiorly and laterally. The moderately prominent lateral tubercles to either side of the midline are clearly seen in basal view, as is the greatest mediolateral width of the corpus at the second molar. The lingula of each mandibular foramen, the medially directed gonial inflection, the medial extension of each condyle relative to the vertical ramal margin, and the high degree of overall symmetry are also seen from this view.

Dolní Věstonice 16

The reconstructed mandible of Dolní Věstonice 16 has several large breaks and large areas of reconstruction, and the left condylar process is completely missing. In occlusal view, the dental arcade is symmetrically oriented, and most aspects of the mandible show general symmetry with each other despite the degree of reconstruction. Principally, the left ramus was positioned too far medially during reassembly. It should be noted that the left infracondylar neck of the mandible (missing its condyle), which does not line up with left glenoid fossa in the cranium when occluded (Chapter 8), is due entirely to a mediolateral misaligned reconstruction of the ramus rather than a misalignment of the glenoid fossa. This is evident in the relatively straight inferior border of the right lateral mandibular corpus and ramus compared with the medially inflected inferior ramus border on the left side in basal view.

Virtually all of the right condylar mass is situated medial to the superoinferior vertical ramal plane. The posterior border of the mandibular incisure crest courses to the lateral edge of the right condyle. The coronoid processes are largely vertically oriented and show slight lateral flaring at the tips. The right coronoid process is deflected more laterally than the left. The degree of overlap between the nonocclusal distal border of the right third molar with the anterior ramus is evident; the left third molar was congenitally absent (Chapter 11). The greatest breadth of the corpus is around the midpoint of each second molar (17.5 mm on both sides). The tuber symphyseos is evident in midline in occlusal view, although the modest lateral tubercles are not. The moderately thick anteroposterior symphyseal thickness is also apparent.

In anterior view (Figure 10.12), Dolní Věstonice 16 shows asymmetry, especially the medially inflected ramal region on the left side that is due to faulty reconstruction. The labial incisor alveoli were altered antemortem

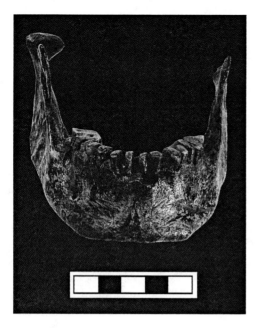

Figure 10.12 Anterior view of the Dolní Věstonice 16 mandible. Scale in centimeters.

(Chapters 11 and 19), but they present a simple concave area above the tuber symphyseos and no evidence of anterior buccal depressions. The tuber symphyseos is a triangular area with no clear median protrusion, which descends into right and left halves as it rounds onto the inferior symphysis. The position of pogonion is consequently lower than in the three previous individuals, being ca. 26.5 mm below an estimated position of infradentale compared to a distance of 34.2 mm to gnathion, or 77.5% of the distance. The inferior tuber symphyseos rounds to gnathion with only a hint of a shallow sulcus. There is a raised rounded inferior edge extending laterally from the tuber symphyseos to the level of the mesial first premolar. The result is lateral tubercles that round laterally and posteriorly and remain minimally prominent. They are spaced ca. 47.2 mm apart near the basal margin. The incisurae are broad with indistinct boundaries and depths of only 1.0 mm (right) and 0.9 mm (left). The mandible therefore has a mentum osseum rank of 4, perhaps tending toward a rank of 4/5.

The inferior border courses symmetrically in the superoinferior plane from menton on both sides (with allowance for the joined section). It ends near gonion, with the right side showing mildly laterally everted gonial flaring and the left showing less lateral flaring, which is due to a faulty reconstruction.

The dental arcade (i.e., the occlusal dentition) is largely symmetrical, and it is symmetrical with respect to adjacent mandibular anatomy. The anterior ramal margins are oriented vertically along their lengths, with both sides inclined slightly more laterally at their superior tips. The posterior portion of the mandibular incisure courses laterally as it approaches the preserved right condyle. The bulk of the condyle is positioned medially with respect to the anteroposterior plane of the ramus. The anteromedial infracondyle area at the top of the neck on the right side shows a moderately excavated pterygoid fovea. The alignment of the condyle with respect to the transverse plane appears normal. In anterior view, the tubercle for the lateral ligament of the temporomandibular joint on the right lateral condyle/neck is visible, but it is extremely small and barely projecting.

In left lateral view, Dolní Věstonice 16 shows a well-developed mentum osseum. There is a well-defined line that courses from the left lateral tubercle along the basal border to the level of the first molar and then up to the anterior ramus border. Although there is some reconstruction to the area, it can be seen that there is a single mental foramen on the left side, whose minimum external diameter is 4.5 mm, that lies below the root of the second premolar. The foramen is filled with matrix, which obscures whether it was divided.

The base of the inferior border on this side is too heavily reconstructed to evaluate, as is also true for the angulation between the posterior ramus and the postero-inferior basal border. The root of the anterior ramal margin lies inferior to the second molar on the corpus, and the anterior ramal margin shows a mild anterior concavity considerably superior to the occlusal plane. It is clear that if the third molar on this side had been present, there would have been no retromolar space, especially prior to the mesial drift of the molars that certainly occurred following their extensive wear; however, even with this drift, a retromolar space would not have been present.

The degree of mandibular incisure symmetricality cannot be assessed on this side because of the missing condyle. The coronoid process is narrower than on the right side and thus appears relatively taller. The insertion area for masseter is large and rugose, showing at least six distinct scalloped areas, as well as oblique ridging. The total maximum area extends anteriorly to the level of the second to third molar interface.

In right lateral view (Figure 10.13), there is a well-defined line that courses from the lateral tubercle along the basal border to the level of the first molar and then up to the anterior ramus border. There is a single mental foramen, whose minimum external diameter is 4.7 mm, that lies below the root apex of the second premolar. The foramen is also filled with matrix.

The base of the mandible in right lateral view shows an undulating inferior border that has an inferiorly directed convexity from midline to about the level of the coronoid process, where it becomes concave for a short distance and then convex again in the gonial region. The angulation between the posterior border of the ramus and

The Mandibular Remains

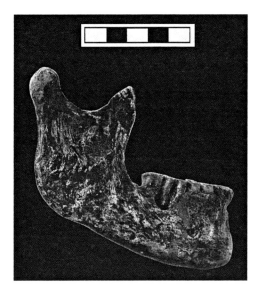

Figure 10.13 Right lateral view of the Dolní Věstonice 16 mandible. Note that the right third molar is partially erupted in an anomalous position, and it is not visible above the alveolar process. Scale in centimeters.

the posteroinferior basal border is more acute than in some Dolní Věstonice specimens but still rounded enough to make gonion a projected point. The root of the anterior ramal margin lies inferior to the second molar on the corpus, and the anterior ramal margin shows a concavity at about the level of the occlusal plane. The anterior ramal margin would have covered the nonocclusal third molar in lateral view. Posteriorly, the ramal margin shows a slight concavity just inferior to the neck of the mandible.

The mandibular incisure is symmetrical, and in anatomical position, the tip of the coronoid process lies inferior to the Frankfurt Horizontal plane across the superior-most condylar surface. There is some muscle scarring for temporalis. The insertion area for masseter is large and rugose, showing at least five large scalloped areas and two or three minor ones with oblique lines. The maximum total area extends anteriorly to the level of the second to third molar interface.

The neck below the condyle appears to be as wide as the condyle in lateral view, although this is due to the extension of the mandibular incisure; from medial view the neck appears narrower. There is a very small tubercle for the lateral ligament of the temporomandibular joint on the anterolateral condyle/neck interface.

In posterior view, the medially inflected condyle on the right side is apparent; its posterior aspect is reconstructed. The posterior ramal borders are relatively straight, flaring slightly laterally in the gonial region on the unreconstructed right side. Both mandibular foramina are preserved, although the right is damaged somewhat. No observations are possible on the left-side because of

damage, but the right side shows a V-shaped configuration and complete mylohyoid bridging.

The mylohyoid line is present and well preserved on both sides. On the right (Figure 10.14), it begins 6.5 mm inferoposteriorly to the alveolar bone below the posterior border of the nonocclusal third molar and courses anteroinferiorly to the inferior portion of the sublingual fossa to meet up with the superior border of the digastric fossa. On the left, it begins well posterior to the area of the absent third molar and courses anteroinferiorly to the posterior portion of the sublingual fossa; it probably met up with the superior border of the digastric fossa, although this area is damaged. There is a relatively shallow submandibular fossa inferior to the left mylohyoid line in Dolní Věstonice 16. There are two superior genioglossal tubercles and one inferior tubercle, 11.5 mm high and 7.2 mm wide, and there is even a single additional tubercle below the latter (an unusual variant). The sublingual fossae anteriorly above each mylohyoid line to either side of the genioglossal tubercles are not well defined.

The right internal coronoid process is well preserved and shows some scarring for temporalis. The medial pterygoid insertion is visible and more prominent than on other Pavlovian specimens. There are several scallops and nonlinear ridges. No enlarged or hypertrophied superior tubercle is evident; the area is slightly damaged but probably complete enough to determine the latter. The left internal coronoid process is complete and nonreconstructed, and it shows a very rugose attachment for temporalis (the most rugose observed in the sample). The medial pterygoid insertion is mostly reconstructed on the left side, but enough is present to indicate that there is no hypertrophied superior tubercle.

In basal view, the moderate midline symphyseal thickness is evident. There is a prominent tubercle anteri-

Figure 10.14 Oblique posteromedial view of the Dolní Věstonice 16 right lateral corpus and ramus, with damage to the mandibular foramen. Scale in centimeters.

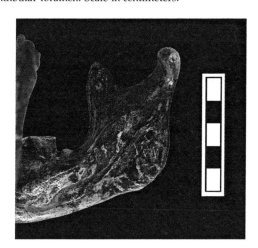

orly between the digastric insertions. The insertion areas of the digastric muscles are rugose and well circumscribed, and they rise to a central peak. The lateral tubercles to either side can be seen from this view. The greatest mediolateral width of the corpus at the second molar, the lingula for the left mandibular foramen, the slightly lateral inflection above each gonial region, and the medial extension of the right condyle relative to the vertical ramal margin are evident in basal view.

Pavlov 1

The Pavlov 1 mandible is relatively complete on the left side. The right side of the mandible is present from midline to just distal of the second molar. In occlusal view, the dental arcade is fairly symmetrically oriented, and all aspects of the retained mandible show near perfect symmetry with each other. The left mandibular condyle is preserved, but the coronoid process is missing. A greater portion of the left condylar mass is not situated medial to the superoinferior vertical ramal plane, as in other specimens, because the neck flares medially on this specimen. From superior view, it can be seen that the posterior border of the mandibular incisure crest courses to the lateral edge of the left condyle. The greatest breadth of the corpus is located at the midpoint of each second molar. The tuber symphyseos is evident in midline in occlusal view. The moderately thick anteroposterior symphyseal thickness is also apparent.

In anterior view (Figure 10.15), Pavlov 1 shows a high degree of symmetry in virtually all aspects except the

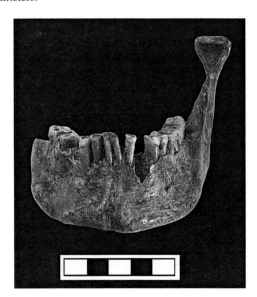

Figure 10.15 Anterior view of the Pavlov 1 mandible. Scale in centimeters.

inferior midline mental eminence. Its symphysis has no anterior buccal depressions, with smooth bone rising to the incisor roots. The tuber symphyseos is vertically tall and rounded, has no central peak, and bifurcates inferiorly. The most prominent points are in fact lateral of the midline just above the inferior margin, but they are not the lateral tubercles. The midline pogonion position, 27.7 mm below infradentale, is 78.0% of the distance to gnathion (35.5 mm), but the more prominent lateral points on the tuber symphyseos are further inferior, 30.5 mm (or 85.9% of the total height) from infradentale. The region of the lateral tubercles is adequately preserved only on the left, and no clear tubercle is present. Despite the lack of lateral tubercles, there are clear incisurae, with depths of 1.2 mm (right) and 1.0 mm (left). It has a mentum osseum rank of 4.

The inferior border courses somewhat nonsymmetrically from menton on the left side, and it ends near gonion (reconstructed to a degree) with no gonial flaring. The left anterior ramal margin is preserved only in the most inferior portion of the left side. The posterior portion of the mandibular incisure on the left side courses laterally as it approaches the condyle and ends on the most lateral extent of the condyle. The bulk of the condyle is positioned medially with respect to the anteroposterior plane of the ramus, but in Pavlov 1, unlike the other Pavlovian specimens, there appears to be a medial shelf emanating from the neck. The entire area along this "shelf" at the top of the neck shows a large but smooth pterygoid fovea. The alignment of the left condyle (which has not been reconstructed) shows it to be horizontal in the transverse plane. In anterior view, the tubercle for the lateral ligament of the temporomandibular joint on the left lateral condyle/neck is visible as a moderately large lateral projection, most similar to that in Dolní Věstonice 14.

In left lateral view (Figure 10.16), Pavlov 1 shows its well-developed tuber symphyseos coursing inferolaterally on the anterior symphysis. There is a diffuse oblique line running from the lateral tubercle up to the mental foramen and then to the anterior ramal margin, as well as a line that runs along the basal border and up to the anterior ramal border; the latter is particularly pronounced. There is a single mental foramen on the left side, whose minimum external diameter is 2.7 mm, that lies below the apex of the second premolar. The foramen is filled with matrix.

The base of the mandible in left lateral view shows an undulating inferior border that has an inferiorly directed convexity from midline to about the level of the distal third molar and the vertical portion of the anterior ramal margin, where it the becomes straight in the gonial region, although this latter part has been reconstructed. The angulation between the posterior border of the ramus and the posteroinferior basal border is rounded but very acute, making gonion an actual rather than a projected

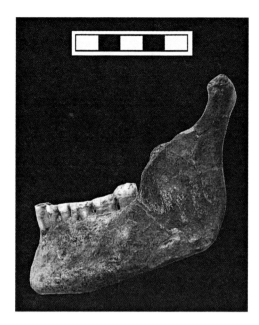

Figure 10.16 Left lateral view of the Pavlov 1 mandible. Scale in centimeters.

point. The root of the anterior ramal margin lies inferior to the second molar on the corpus, and the anterior ramal margin configuration cannot be determined. Posteriorly, the ramal margin shows a slight concavity just inferior to the neck of the mandible.

As preserved, the left side of the Pavlov 1 mandible appears to have a small retromolar space (ca. 2 mm maximum from the top of the third molar). However, the individual experienced mesial drift of his heavily worn teeth (Chapter 11), and the ramus may have been reassembled to the corpus ca. 1 mm too posterior. These developments may have placed the anterior ramal margin on the distal third molar margin when he was fully mature but had experienced little dental attrition. Consequently, Pavlov 1 is given an absent/present designation for a retromolar space.

The insertion area for masseter is large and rugose, showing several distinct, roughly scalloped areas with oblique ridging. The total area extends anteriorly to the level of the second to third molar interface. The neck below the condyle appears to be as wide as the condyle in lateral view, although this is due to the extension of the mandibular incisure; from medial view the neck appears narrower. There is a moderately large distinct tubercle for the lateral ligament of the temporomandibular joint on the anterolateral condyle/neck interface.

In right lateral view, Pavlov 1 shows the mental region as described for the left side. The root of the anterior ramal margin on the right side lies inferior to the second molar. Major damage and reconstruction to this side precludes further observations for this view.

In posterior view, the medially inflected condyle that is apparent on other specimens is not found in the preserved region for Pavlov 1, instead showing a more balanced distribution; the condyle is glued on in Pavlov 1 below the neck but is accurately positioned. The posterior ramal border is very straight. The mandibular foramen is damaged, with the lingula crushed (Figure 10.17). It appears to be open, or V-shaped, but it is probably too heavily reconstructed to evaluate with confidence. The left medial pterygoid insertion is visible but not particularly prominent; many modest scallops are present. There is no hypertrophied superior tubercle.

The mylohyoid line is present and well preserved on the left side and only partly preserved on the right side. On the left side, it begins at the third molar 4.0 mm inferoposteriorly to the alveolar bone inferior to the distal border of the third molar and courses anteroinferiorly to the inferior portion of the poorly defined sublingual fossa. It does not meet up with the superior border of the digastric fossa; it shows instead a curious and unique upward inflection at the second to third molar interface. Posterior to this point, the mylohyoid line is markedly sharp. The right side, although less complete, shows the same basic morphology. There is a relatively deep submandibular fossa inferior to the left mylohyoid line. Instead of separate genioglossal tubercles and/or pits, Pavlov 1 has a single, sharply delimited, common median spine, 11.3 mm high and 6.0 mm wide. The sublingual fossae are very poorly delineated and very shallow.

In basal view, the moderate midline symphyseal thickness quantified above is evident. The insertion areas for the digastric muscles are well marked but shallow, especially on the right side, where they are preserved enough to be measured (Table 10.1). They appear to show

Figure 10.17 Oblique posteromedial view of the Pavlov 1 left lateral corpus and ramus. The mandibular foramen is damaged, and the coronoid process is missing. Scale in centimeters.

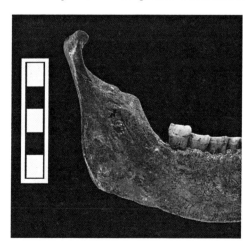

some asymmetry in that the right digastric scar may be larger than the left one, and interestingly, this is associated with a more projecting anterior margin. The greatest mediolateral width of the corpus at the second molar, the balanced position of the remaining condyle relative to the vertical ramal margin, and the high degree of overall symmetry in this mandible are apparent in basal view.

Pavlov 3

Pavlov 3 (Figure 10.18) is a section of lateral mandibular corpus that extends from the second incisor/canine interdental septum to the middle of the second molar alveolus. Buccally below the canine and premolars, it extends down to the level of the mental foramen (16 mm below the alveolar plane), and under the first molar it extends slightly further. Lingually it is preserved to about the same extent. It is possible to determine its corpus breadth at the mental foramen level (ca. 10 mm), which places it with Dolní Věstonice 14 as one of the two narrowest mandibular corpori of the sample (Table 10.3). However, the lack of the inferior half of the mandibular corpus means that it not possible to assess corpus heights or more distal corpus breadths accurately.

The mental foramen, which is preserved on its superior and posterior margins, is below the first premolar, extending under the first to second premolar interdental septum. It is therefore slightly more mesial than those of the other Dolní Věstonice and Pavlov mandibles. It appears to be single, but damage precludes being certain of that observation.

Medially there is a prominent but not rugose mylohyoid line preserved below the second premolar to second molar. If one extends it as a line posteriorly and continues the alveolar plane distally to the third molar region, it is apparent that the strongly sloping mylohyoid line would have passed close to the alveolar plane adjacent to the third molar.

Pavlov 4

Pavlov 4 (Figure 10.19) is a small piece of left lateral corpus, with buccal root alveolus of the third molar, the lateral surface bone adjacent to the second molar roots, and the beginning of the oblique line. The anterior extension of the anterior ramal edge is prominent below the third molar, but its extent cannot be determined.

Comparative Materials and Methods

These Pavlovian mandibles are compared primarily to samples of earlier Upper Paleolithic human mandibles from Europe and the Near East, as well as to earlier samples from the Middle Paleolithic of the same region. The Gravettian remains include specimens from Barma Grande, Caviglione, Cro-Magnon, Fanciulli (Grotte-des-Enfants), Isturitz, Ohalo, Paglicci, Pataud, Sunghir, and Willendorf, plus the earlier Oase mandible. Included with these are metric data on the Předmostí 3 and 4 mandibles (Matiegka, 1934), plus discrete traits derived from photographs of other Předmostí mandibles in Matiegka. The earlier samples consist of the early modern humans from the sites of Qafzeh and Skhul, as well as Neandertal remains from the sites of Amud, Arcy-sur-Cure (Hyène), Banyoles, La Ferrassie, Guattari, Kebara, La Naulette, La Quina, Regourdou, Saint-Césaire, Shanidar, Spy,

Figure 10.18 Lateral view of the Pavlov 3 right mandibular fragment. Scale in centimeters.

Figure 10.19 Lateral view of the Pavlov 4 left mandibular fragment. The posterior break is at the distal third molar. Scale in centimeters.

Subalyuk, Vindija, and Zafarraya. Earlier (late Middle Pleistocene or initial Late Pleistocene) remains from sites such as Aubesier, La Chaise, Krapina, Montmaurin, and Tabun are mentioned only as appropriate, as are the North African late archaic to early modern human remains from Dar-es-Soltane, Nazlet Khater, and Témara. Data derive from Verneau (1906), Lumley (1973), Borgognini Tarli et al. (1980), Wolpoff et al. (1981), García Sánchez (1986), Tillier et al. (1989), Pap et al. (1996), Mallegni et al. (1999), Kozlovskaya and Mednikova (2000), Trinkaus et al. (2003b), J. T. Snyder (personal communication) and Trinkaus (personal observation).

In addition to primary morphological description, these Pavlovian mandibles are assessed with respect to several discrete traits and a limited set of morphometric determinations. The discrete traits are primarily secondary reflections of overall mandibular proportions, which have been useful for assessments of Middle and Late Pleistocene mandibular morphological patterns (e.g., Stefan & Trinkaus, 1998a,b; Rosas, 2001; Dobson & Trinkaus, 2002; Lebel & Trinkaus, 2002b) and probably combine a complex mix of epigenetic traits and developmental and degenerative plasticity (Stefan & Trinkaus, 1998a; Richards et al., 2003). They include mentum osseum rank, mental foramen position, anterior marginal tubercle presence and position, mylohyoid line orientation and position, retromolar space presence, masseter and medial pterygoid surface relief, gonial angle curvature, mandibular notch shape, mediolateral condylar position, mandibular foramen lingular bridging, and superior medial pterygoid tubercle hypertrophy.

The external morphometrics largely follow those of R. Martin (Bräuer, 1988), supplemented by several from Twiesselmann (1973; Table 10.2; Sládek et al., 2000: table 29). These linear and angular measurements are supplemented by cross-sectional assessments of the mandibular symphysis, modeled as a solid beam and oriented relative to the alveolar plane. The resultant cross-sectional geometric parameters, principally second moments of area and their orientations (Table 10.2), were calculated from contours transcribed through the use of polysiloxane dental putty (Cuttersil Putty Plus, Heraeus Kulzer Inc.). Damaged alveolar bone was interpolated by using the preserved margins. The resultant cross sections were projected enlarged onto a Summagraphics III 1812 digitizing tablet; the contours were traced, and the cross-sectional parameters were calculated with a PC DOS version (Eschman, 1992) of SLICE (Nagurka & Hayes, 1980). The resultant values of interest are the anteroposterior and superoinferior second moments of area, the maximum and minimum second moments of area, and the orientation of the maximum second moment of area relative to the alveolar plane (theta). The anatomically oriented second moments of area, as justified in Dobson and Trinkaus (2002), are scaled relative to mandibular length for the anteroposterior second moment of area and relative to dental arcade breadth for the superoinferior second moment of area. For mandible length, the superior length of Twiesselmann (1973) is employed (midsagittal line from the bicondylar midline to infradentale), and dental arcade breadth is taken at the external second molar alveolus.

Given the uncertainty in the degree to which the facial injury of Dolní Věstonice 3 affected mandibular hypertrophy (see above and Chapters 8 and 19), the average of the right and left values (as appropriate) have been used in the metric comparisons, as with all of the other Dolní Věstonice, Pavlov, and comparative specimens. The degree of asymmetry present should have little effect on the results. For example, the minimum ramus breadths, the first/second molar corpus heights, and the mental foramen corpus breadths for Dolní Věstonice 3 are essentially symmetrical and only modestly asymmetrical for the mental foramen corpus heights and molar corpus breadths (Table 10.3). Similar levels of mandibular ramus breadth and corpus diameters asymmetry are common among normal mandibles (Table 10.3); this suggests that the degree of asymmetry of mandibular function was minimal, an interpretation that is supported for the largely symmetrical molar wear once the agenesis of the left third molar is taken into account (Chapters 11 and 19). These comparisons made, it should be stated that the morphology of the superior left ramus is not employed in the following comparative morphological assessments of the Dolní Věstonice 3 mandible.

Overall Mandibular Size and Proportions

The Dolní Věstonice and Pavlov mandibles vary between 96.7 and 109.0 mm in superior length (plus 104.0 for Dolní Věstonice 3; Table 10.2). Their average value of 102.0 falls within 1 standard error of the mean of an earlier Upper Paleolithic sample (102.1 ± 5.4 mm; $N = 11$) and below the mean of a Neandertal sample (110.7 ± 6.2 mm; $N = 19$). The small and variable Qafzeh-Skhul sample provides values of 98.0, 109.0, and 116.5 mm. Even though the ranges of these samples overlap, there is a general shortening of the face between the Middle and Upper Paleolithic, reaching the upper end of recent human ranges of variation in the earlier Upper Paleolithic (Trinkaus, 2003). The Dolní Věstonice and Pavlov remains share in this shift.

Even though the six Dolní Věstonice and Pavlov mandibles provide mandibular lengths and three of them provide accurate bicondylar breadths (with estimation possible for the other three), it is not possible to undertake meaningful breadth/length comparisons given the dearth of sufficiently complete and undistorted Late Pleistocene human mandibles for whom data are available. It

Table 10.2 Mandibular symphyseal cross-sectional geometric parameters and scaling dimensions for the Dolní Věstonice and Pavlov mandibles, with the mandibular symphysis modeled as a solid beam.

	Mandible Superior Length (mm)	Dental Arcade Breadth (mm)	Ant-Post Second Moment of Area (mm⁴)	Sup-Inf Second Moment of Area (mm⁴)	Maximum Second Moment of Area (mm⁴)	Minimum Second Moment of Area (mm⁴)	Theta (°)
Dolní Věstonice 13	100.1	63.5	6,590	22,930	22,961	6,560	92.5
Dolní Věstonice 14	(98.6)	63.6	2,436	13,794	14,148	2,082	100.1
Dolní Věstonice 15	96.7	65.1	4,282	16,405	16,405	4,281	90.1
Dolní Věstonice 16	(109.0)	62.9	7,775	32,768	33,105	7,437	83.4
Pavlov 1	(103.3)	65.2	8,907	37,661	38,725	7,843	100.7

Superior length was calculated from the direct distance between the condylar middle(s) and infradentale and half of the bi-mid-condylar distance (estimated for Dolní Věstonice 14 and 16 and Pavlov 1, because of their ramal damage); their lengths are in parentheses, indicating this bicondylar breadth estimation. Symphyseal cross-sectional data are not available for Dolní Věstonice 3.

is possible, however, to assess the proportions of the ramus breadth and dental arcade breadth versus overall length since many Late Pleistocene mandibles can be oriented close enough to anatomical position to provide a reasonable length determination (and small errors of orientation affect the length only as the cosine of the angle of error).

The scatter of ramus minimum breadth versus mandible superior length (Figure 10.20), provides considerable overlap across the Middle and earlier Upper Paleolithic samples. The high outlier is the Early Upper Paleolithic Oase 1 mandible (Trinkaus et al., 2003b), and it is approached by the Pataud 1 specimen. Not shown but equally broad in relative ramus breadth is the early modern northeast African Nazlet Khater 2 mandible (Crevecoeur & Trinkaus, 2004; Thoma, 1984). The long mandible with the relatively narrow ramus is Qafzeh 9. Interestingly, the Dolní Věstonice and Pavlov mandibles largely span the Late Pleistocene range of variation, with Dolní Věstonice 3 and Pavlov 1 having relatively narrow rami, whereas Dolní Věstonice 13 has a relatively broader one, as noted above.

In contrast to the general, if variable, relationship between ramus breadth and mandible length, dental arcade breadth bears little relationship to overall mandibular length. In addition, as with the palate (Chapter 8), there is little overlap between the Neandertal and early modern human samples, with most of the Neandertals having broader arcades than the early modern humans (only the long La Ferrassie 1 mandible has a slightly narrower dental arcade). This contrast is probably due to their differences in anterior dental crown dimensions (Table 11.6; Stefan & Trinkaus, 1998a), and the same difference in dental arcade breadth emerges during development (Mallegni & Trinkaus, 1997; Trinkaus, 2002b).

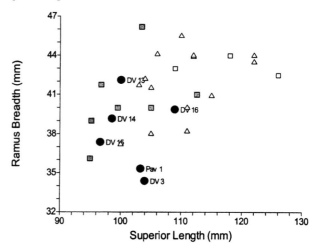
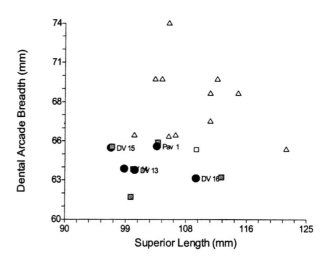

Figure 10.20 Bivariate plots of ramus minimum breadth (above) and dental arcade breadth (below) versus mandible superior length. Solid circles: Dolní Věstonice and Pavlov; gray squares: earlier Upper Paleolithic; open squares: Qafzeh-Skhul; open triangles: Neandertals.

The Mandibular Symphysis

The Anterior Symphysis

As noted for the individual mandibles, each anterior symphyseal region of the Dolní Věstonice and Pavlov mandibles has a uniformly prominent tuber symphyseos, but there is variable development of the lateral tubercles and the associated incisurae. All of them conform to the range of modern human anterior mandibular morphology and in that sense are distinct from the vertical or retreating symphyses, with little development of the mental trigone seen in archaic *Homo* (including the Neandertals).

The predominance of mentum osseum rank 4 (a prominent tuber symphyseos with little or no projection of the lateral tubercles; Dobson & Trinkaus, 2002) or 4/5 in this sample is similar to the pattern seen in other early modern human samples. A mentum osseum rank of 4 is present in 80% of mature earlier Upper Paleolithic mandibles ($N = 15$), and the percentage remains the same with the addition of the ten available immature Aurignacian and Gravettian mandibles ($N = 25$). It is only the mandibles of Cro-Magnon 1 and 3, Lagar Velho 1, and probably Předmostí 14 and 25 (judging from the photographs in Matiegka, 1934) that exhibit a shelflike mentum osseum with truly prominent lateral tubercles. The same general pattern is evident in the Qafzeh-Skhul sample (mature $N = 4$; total $N = 7$), as well as the North African late archaic to early modern human Dar-es-Soltane and Nazlet Khater specimens (Crevecoeur & Trinkaus, 2004; Ferembach, 1976; Thoma, 1984). (Témara 1 appears to have the more archaic mentum osseum rank of 2; Vallois & Roche, 1958).

Symphyseal Orientation

The consistent prominence of the tuber symphyseos of these mandibles produces relatively high anterior symphyseal angles for them. The angles are between the infradentale-pogonion line and the alveolar plane; 90° indicates a vertical symphyseal profile, and angles less than 90° depict retreating symphyseal profiles. The Dolní Věstonice and Pavlov symphyseal angles (Table 10.3; Figure 10.21) are all high, ranging from 94° (Dolní Věstonice 16) to 105° (Dolní Věstonice 14). The only earlier Upper Paleolithic specimen with a value below 90° (Předmostí 3) has a prominent tuber symphyseos but marked alveolar prognathism. The four Qafzeh-Skhul specimens cluster around 90°, which is at the top of the Neandertal range and includes four specimens: Banyoles 1, La Ferrassie 1, Guattari 3, and Vindija 226.

The orientation of the whole symphysis, not just its anterior face, can be assessed by comparing the orientations of their cross-sectional maximum second moments of area relative to the alveolar plane, or theta (Table 10.2; Figure

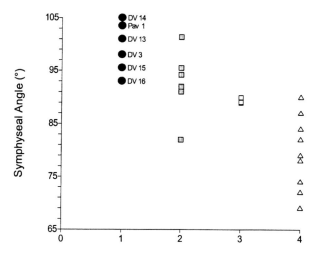

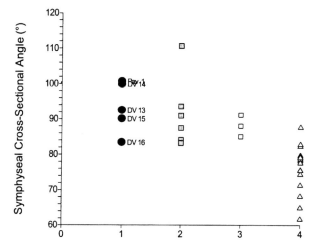

Figure 10.21 Above: distributions of anterior symphyseal (infradentale/pogonion) angles for Dolní Věstonice and Pavlov (1), earlier Upper Paleolithic (2; $N = 6$), Qafzeh-Skhul (3; $N = 3$), and Neandertal (4; $N = 17$) mandibles. Below: distributions of cross-sectional (theta) symphyseal angles for Dolní Věstonice and Pavlov (1), earlier Upper Paleolithic (2; $N = 6$), Qafzeh-Skhul (3; $N = 4$), and Neandertal (4; $N = 13$) mandibles.

10.21). The Dolní Věstonice and Pavlov mandibles again cluster with the other earlier Upper Paleolithic mandibles, are similar to the Qafzeh-Skhul values, and slightly overlap the Neandertal range. The highest Neandertal specimen is Banyoles 1, whereas the very high earlier Upper Paleolithic specimen is Cro-Magnon 1, one of the few earlier Upper Paleolithic mandibles with prominent lateral tubercles forming an inferoanterior shelf across the symphysis.

Symphyseal Cross Sections

In addition to their symphyseal orientations, it is possible to assess the cross-sectional proportions of the mandibular symphysis. This has been done by using external osteometrics (Table 10.3) and minimum versus maximum second moments of area (Table 10.2), the latter more accurately quantifying the distribution of bone in the cross sections. The resultant distributions (Figure 10.22) show a scatter of points for external breadths versus heights, with little difference between the samples. Dolní Věstonice 14 appears to have a moderately thinner symphysis than many of the other specimens. The high outlier is Kebara 2. However, comparisons of the second moments of area provide a tight cluster of Late Pleistocene specimens with two outliers, a relatively thin one (Skhul 4), and a large but otherwise similarly proportioned one (Kebara 2). Therefore, despite the marked shift in symphyseal orientation (Figure 10.21), there is little change in the bone distribution in the symphysis across these samples.

Further, and biomechanically more appropriate, assessment of these symphyses is obtained through the comparison of symphyseal second moments of area to the appropriate mandibular dimensions. When the anteroposterior and superoinferior second moments of area of these symphyses (Table 10.2) are compared to mandible length and arcade breadth, respectively (Figure 10.23), the Dolní Věstonice and Pavlov mandibles fall in the middle of the Late Pleistocene distribution. Particularly with respect to the anteroposterior second moment of area, there are several Neandertal and one Qafzeh-Skhul outliers, but the majority of the specimens fall within a restricted range. As previously documented (Dobson & Trinkaus, 2002), there is little change in human mandibular anterior corpus robusticity in the Late Pleistocene. A similar pattern is evident in their lateral corpori (Trinkaus, 2000c).

The Lateral Corpus

The six more complete mandibles have the mental foramen below the second premolar, whereas Pavlov 3 has it below the first premolar (Table 10.4). In this they are similar to 73.7% ($N = 19$) of the earlier Upper Paleolithic mandibles in having the mental foramen mesial of the second premolar/first molar interdental septum, and among those fourteen with the more anterior mental foramen, all but one have it below the second premolar. The Qafzeh-Skhul sample has a similar pattern, with 66.7% ($N = 6$) of their mandibles having it below the first or second premolar. Among Neandertals, in contrast, 92.6% ($N = 27$) have the mental foramen distal of the second premolar. Since the more mesial position of the foramen is found in long, earlier *Homo* mandibles (Lebel & Trinkaus, 2002b), it is likely that the more posterior relative position of the foramen in the Neandertals is the result of an ancestral facial length (Trinkaus, 2003), shorter dental arcades (Franciscus & Trinkaus, 1995), and the abbrevi-

Table 10.3 Osteometric dimensions of the Dolní Věstonice and Pavlov mandibles.

		Symphyseal Height	Symphyseal Breadth	Anterior Symphyseal Angle	Mental Foramen Height	Mental Foramen Breadth	M1/M2 Height	M1/M2 Breadth	Minimum Ramus Breadth
Dolní Věstonice 3	Right	25.5	15.0	98°	26.1	12.5	23.8	17.7	34.7
	Left				24.2	12.7	23.2	15.5	34.0
Dolní Věstonice 13	Right	31.3	17.0	101°	29.4	12.4	29.1	14.3	41.6
	Left				29.0	12.3	28.4	13.6	42.6
Dolní Věstonice 14	Right	30.3	12.5	105°	29.1	10.0	26.4	13.0	39.1
	Left				28.5	10.0	25.1	13.2	39.2
Dolní Věstonice 15	Right	30.7	15.8	95.5°	28.9	10.5	27.6	12.0	37.2
	Left				27.9	12.0	29.0	11.5	37.5
Dolní Věstonice 16	Right	35.5	16.9	94°	35.6	12.2	(33.0)	14.8	40.7
	Left				37.0	12.8	32.8	15.9	39.0
Pavlov 1	Right	(35.5)	15.9	103.5°	—	—	(33.5)	11.4	—
	Left				36.5	11.3	33.2	11.7	35.3
Pavlov 3	Right	—	—	—	—	(10.0)	—	—	—

Measurements in millimeters and degrees. Measurements in parentheses are estimated because of postmortem damage.

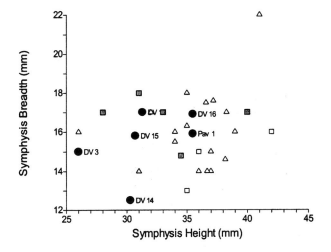

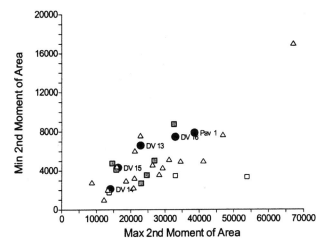

Figure 10.22 Bivariate plots of symphyseal proportions, using external osteometrics (above) and minimum versus maximum second moments of area (below). Solid circles: Dolní Věstonice and Pavlov; gray squares: earlier Upper Paleolithic; open squares: Qafzeh-Skhul; open triangles: Neandertals.

Given the variation in mental foramen position, and also variation in the degree to which the lingual mandibular tori affect medial corpus thickness, the lateral corpus breadth versus height is compared at both the mental foramen position and the first to second molar interdental septum (Table 10.3; Figure 10.24). Slightly different comparative data sets are available for each position. Both comparisons reveal the same pattern: there is little correlation between the variables within samples, and the Neandertals have on average nonsignificantly larger corpus heights (ANOVA $P = 0.089$ and 0.090, respectively) but significantly larger breadths (ANOVA $P = <0.001$ and 0.015, respectively). The difference is particularly noticeable in the mental foramen position comparison. The Dolní Věstonice and Pavlov mandibles all cluster with the other

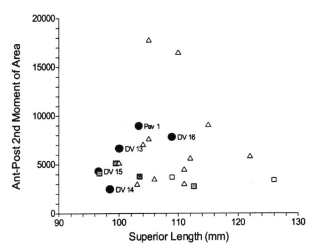

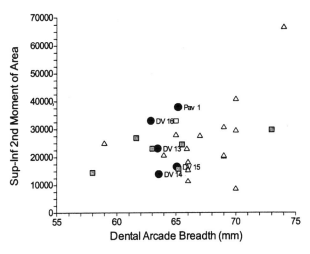

Figure 10.23 Bivariate plots of symphyseal cross-sectional second moments of area versus mandibular dimensions. Solid circles: Dolní Věstonice and Pavlov; gray squares: earlier Upper Paleolithic; open squares: Qafzeh-Skhul; open triangles: Neandertals.

ated inferior alveolar nerve canals shared with recent humans. In either case, the Dolní Věstonice and Pavlov mandibles share the shorter inferior alveolar nerve canal and more orthognathic faces of other early modern humans. All of the Dolní Věstonice and Pavlov mental foramina are, or appear to be, single.

Most of the Dolní Věstonice and Pavlov mandibles lack an anterior marginal tubercle (Table 10.4) since they are present only bilaterally on Dolní Věstonice 13 and unilaterally on Dolní Věstonice 15. The positions of the anterior marginal tubercles in Dolní Věstonice 13 are moderately mesial, whereas the right one on Dolní Věstonice 15 is intermediate between the more distal position found in many Neandertal lineage specimens (Rosas, 2001) and the more mesial position found in other human groups.

Table 10.4 Scoring of mandibular corporeal discrete traits for the Dolní Věstonice and Pavlov mandibles.

	Mentum Osseum Rank	Mental Foramen Position	Anterior Marginal Tubercle Position	Oblique Mylohyoid Line	Mylohyoid Line Close to Third Molar Alveolus	Retromolar Space
Dolní Věstonice 3	4	P2	Absent	Present	Present	Absent
Dolní Věstonice 13	4	P2	P1/P2, P2	Present	Present	Absent
Dolní Věstonice 14	4/5	P2	Absent	Present	Present	Absent
Dolní Věstonice 15	4/5	P2	P2/M1, Absent	Present	Present	Absent
Dolní Věstonice 16	4	P2	Absent	Present	Present	Absent
Pavlov 1	4	P2	Absent	Present	Present	Absent/Present
Pavlov 3	—	P1	—	Present	Present?	—

Bilaterally symmetrical features are listed once; asymmetrical features are listed as "right, left." Features preserved only on one side are listed once; see text and Sládek et al. (2000) for preservation. For the anterior marginal tubercle position, "absent" incidates the absence of the tubercle in otherwise adequately preserved portions of the mandible. The position of the Pavlov 3 mylohyoid line at the third molar is based on linear continuations of the preserved portion of the line and the alveolar plane.

earlier Upper Paleolithic ones, with a greater spread at the molar level than at the mental foramen level.

It is also possible to assess whether the lateral corpus tapers in height as one goes distally along the dental arcade by comparing symphyseal height to molar corpus height (Figure 10.25). With the exception of the La Naulette 1 Neandertal mandible, there is little difference between the samples, with only a slight tapering of the corpus in most of them. The Dolní Věstonice and Pavlov specimens are in the middle of the distribution.

The Medial Corpus

Rosas (2001) has suggested that Neandertals tend to have more obliquely oriented mylohyoid lines than other human groups, and ones that closely approach the third molar alveolar margin. Comparative data are not available for these aspects of this muscle line for recent and early modern humans, but all of the Dolní Věstonice and Pavlov mandibles conform to both of these purported Neandertal patterns (Table 10.4). This applies directly to the six more complete mandibles; the Pavlov 3 mandible preserves the oblique orientation of this line, and linear distal extensions of the mylohyoid line and the molar alveolar plane would place these two lines close to each other as well.

The Lateral Ramus

The Dolní Věstonice mandibles lack retromolar spaces, or a gap between the distal third molar and the anteroinferior ramal margin when viewed in strict *norma lateralis* (Table 10.4). The anterior ramus crosses the second molar in Dolní Věstonice 3; the third molar in Dolní Věstonice 13, 15, and 16; and between the second and third molars in Dolní Věstonice 14. The Pavlov 1 mandible is harder to judge and is given an absent/present designation.

The lateral condylar tubercles remain at or close to the lateral margins of the condyles for all but the left Dolní Věstonice 14 one. Small tubercles are generally characteristic of early modern humans, and they contrast with the large ones seen on some (e.g., Amud 1 and La Ferrassie 1) Neandertals. Associated with these small lateral tubercles, all of the Dolní Věstonice and Pavlov mandibles have the medial positioning of the condyle relative to the mandibular incisure crest, such that the crest meets the condyle in the latter's lateral third or at its lateral margin (Table 10.5). This is the arrangement found in all ($N = 12$) other earlier Upper Paleolithic mandibles, both Qafzeh-Skhul specimens, and ten ($N = 16$) Neandertals. The more lateral positioning of the condyle and consequent more medial "positioning" of the crest on the condyle, found in the highest frequencies in Neandertals and their Middle Pleistocene predecessors (Lebel & Trinkaus, 2002b), is found in high frequencies in recent human juvenile mandibles (Rhoads & Franciscus, 1996; Jabbour et al., 2002) and is probably related to changing mediolateral positions of both the temporomandibular joint and the temporalis muscle acting on the coronoid process. Its distribution, as with most of these mandibular discrete traits (Franciscus & Trinkaus, 1995; Stefan & Trinkaus, 1998a), is therefore likely to be secondary to other, more primary aspects of craniofacial anatomy.

The Medial Ramus

All of the Dolní Věstonice mandibles appear to have the V-shaped pattern or the absence of the horizontal-oval form of a foramen (more appropriately referred to as the

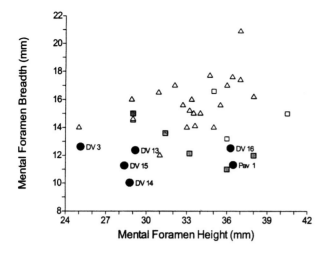

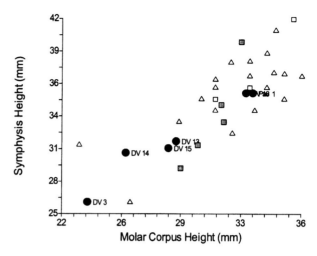

Figure 10.25 Symphyseal height versus lateral corpus height at the mesial intermolar septum. Solid circles: Dolní Věstonice and Pavlov; gray squares: earlier Upper Paleolithic; open squares: Qafzeh-Skhul; open triangles: Neandertals.

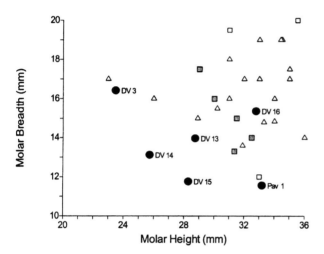

Figure 10.24 Bivariate plots of lateral corpus breadth versus height at the mental foramen (above) and the mesial intermolar septum (below). Right and left values, as available, were averaged before plotting. Solid circles: Dolní Věstonice and Pavlov; gray squares: earlier Upper Paleolithic; open squares: Qafzeh-Skhul; open triangles: Neandertals.

absence of lingular bridging; F. H. Smith, 1978; Jidoi et al., 2001; Richards et al., 2003). This is readily apparent in the Dolní Věstonice 13 to 16 mandibles (Figures 10.5, 10.8, 10.11, and 10.14). The Pavlov 1 left ramus had the lingula crushed into the canal; it appears not to have been bridged, but it cannot be conclusively determined (Figure 10.19). The right side of Dolní Věstonice 3 is not bridged, and the left side, although damaged, appears to have been similar (Figure 10.2).

Of the eight other earlier Upper Paleolithic mandibles that are sufficiently intact to be securely scored (several, such as Brno 2, are too damaged to score), only one exhibits lingular bridging (Oase 1), and it has it only on one side. Předmostí 3 and 4 appear from casts to have lingular bridging, but it is not possible to confirm this observation.

Lingular bridging is absent from the three Qafzeh-Skhul specimens that preserve the area, but it is present on 42.9% ($N = 21$) of the early last glacial Neandertals. It is variably present on earlier Neandertal lineage specimens, from sites such as Aubesier, Krapina, Montmaurin, and Tabun, but it is generally absent from earlier Middle and Early Pleistocene *Homo* (Lebel & Trinkaus, 2002b). It is rare but does occur among recent humans (Jidoi et al., 2001; Richards et al., 2003).

Only the medial pterygoid insertion of Dolní Věstonice 3 exhibits a prominent superior tubercle, the pattern that has been considered to be characteristic of the Neandertals and that contrasts with those of most recent humans. Since it remains uncertain to what extent a prominent superior tubercle reflects an epigenetic character as opposed to a differential development of portions of the medial pterygoid muscle (and its insertions) related indirectly to facial shape and mandibular function (Richards et al., 2003), it is difficult to determine whether the contrasts in this region between Dolní Věstonice 13 to 16 plus Pavlov 1 and most Neandertal mandibles reflects their facial shape differences or some more isolated character. Similarly, it remains unclear whether the more "Neandertal" arrangement of this region in Dolní Věstonice 3 might be secondary to her left zygomatic and mandibular trauma (Chapter 19) and the associated remodeling of the left facial skeleton (Chapter 8).

Summary

The mandibular remains from Dolní Věstonice and Pavlov exhibit a morphological pattern that is fully in

Table 10.5 Scoring of mandibular ramal discrete traits for the Dolní Věstonice and Pavlov mandibles. All features are, or appear to have been, bilaterally symmetrical.

	Masseter Surface Flat	Straight or Inverted Gonial Angle	Mandibular Notch Shape	Condylar Position	Mandibular Foramen Bridging	Medial Pterygoid Surface Deep	Medial Pterygoid Tubercle
Dolní Věstonice 3	Absent	Present	Symmetrical	Medial	Absent	Present	Present
Dolní Věstonice 13	Absent	Absent	Symmetrical	Medial	Absent	Present	Absent
Dolní Věstonice 14	Absent	Absent	Symmetrical	Medial	Absent	Present	Absent
Dolní Věstonice 15	Absent	Present	Symmetrical	Medial	Absent	Present	Absent
Dolní Věstonice 16	Absent	Absent	Symmetrical	Medial	Absent	Present	Absent
Pavlov 1	Present	Present	—	Medial	—	Present	Absent

agreement with their cranial remains, indicating moderately robust early modern humans. This is indicated in their anterior symphyseal morphology and orientation, modest digastric impressions, mesially placed mental foramina (and probably anterior margin tubercles), moderately thin lateral corpori, general absence of retromolar spaces, medial condylar positioning, absence of bridging of the mandibular foramen, and variable but modest development of the medial pterygoid insertion. They also show features considered to be "Neandertal," such as those of the mylohyoid line, but it remains unclear what the distributions of these features are among recent humans. Moreover, these osteometric and discrete traits are probably largely secondary reflections of the overall proportions of their mandibles and the more primary structures within the mandibles and associated crania. They serve, nonetheless, given the taphonomic durability (and paleontological recognizability) of mandibles in the human fossil record, to provide a morphological assessment of the facial anatomy of these Late Pleistocene humans.

11

Dental Morphology, Proportions, and Attrition

Simon W. Hillson

The abundant dental remains from Dolní Věstonice and Pavlov consist of teeth present in the alveoli of the six associated skeletons, in the Pavlov 2 and 3 maxilla and mandible, and in the collections of isolated elements from Dolní Věstonice I and II and Pavlov I (Chapter 5). From Dolní Věstonice I, there are the 31 teeth of Dolní Věstonice 3, plus 16 isolated teeth, 5 of which have been lost. From Dolní Věstonice II, there are 116 teeth from the four associated skeletons, 9 teeth that make up the Dolní Věstonice 36 infant, and 1 isolated tooth. Pavlov I yielded the 27 teeth in Pavlov 1, 11 teeth in Pavlov 2 and 3, and 26 isolated teeth for Pavlov 5 to 28. This provides a total of 237 teeth, of which 232 are currently available for analysis. There are few Paleolithic sites or site complexes that provide as much data on dental remains as do these sites on the Pavlovské Hills.

The teeth and aspects of their morphology, anomalies of position and number, and both normal and unusual patterns of attrition are presented and discussed in this chapter. Dental and alveolar pathological alterations are mentioned in passing here, but they are described in detail in Chapter 19.

Methodological Considerations

The descriptions in this chapter use a number of recording standards to describe different features. The development of different variants of crown morphology are given scores from the Arizona State University Dental Anthropology System (ASUDAS), following the criteria set out by Turner and colleagues (Scott & Turner, 1988; Turner et al., 1991; Hillson, 1996; see Tables 11.1 to 11.3). The state of dental development, where this is important for the interpretation of the specimen, is recorded with the codes of Moorrees and colleagues (Moorrees et al., 1963a,b). For several specimens, root formation is a key piece of evidence (see Chapter 6). Where the root apices of the third molars are fully complete (stage A_c), it is assumed that the individual was at least 18 years of age because a large radiographic study of modern jaws has found that there is a probability of more than 0.85 that this is the case (Mincer et al., 1993). For the deciduous teeth, an important observation is the resorption of the roots before the teeth are exfoliated. The Pavlov material includes many deciduous teeth, in various stages of root resorption, and the tiny remaining vestiges of root in some specimens suggest that they may have been exfoliated before being deposited in the site. Occlusal attrition facets are recorded with the wear stages of B. H. Smith (1984; Table 11.8), and the extent of approximal attrition facets is simply described. In most cases, the approximal facets are confined to the enamel, but where they have penetrated into the underlying dentine this is indicated. One of the unusual features of the Dolní Věstonice and Pavlov dental assemblages is the presence of buccal wear facets on the molars, premolars, and canines of several specimens, a pattern that was first noted for the Pavlovian remains from Předmostí (Matiegka, 1934). Again, most of these are confined to the enamel, and those that expose dentine are pointed out. There is a full discussion of occlusal anomalies and buccal wear at the end of this chapter. Where developmental defects of enamel (enamel hypoplasias) are present, they are mentioned in the descriptions below, but a full discussion of them is given in Chapter 19. The same is true of evidence for dental caries and lesions of the alveolar process.

Table 11.1 Arizona State University Dental Anthropology System (ASUDAS) traits in the sufficiently preserved Dolní Věstonice incisors and canines.

ASUDAS Traits	Dolní Věstonice 13	Dolní Věstonice 14	Dolní Věstonice 15
Shoveling			
Upper right canine		1?	3
Upper right second incisor		1?	3
Upper right first incisor		1?	3
Upper left first incisor	—	1?	3
Upper left second incisor		1?	3
Upper left canine	—	1?	3
Lower right second incisor	1	1	1
Lower right first incisor	1	1	—
Lower left first incisor	1	1	
Lower left second incisor	1	1	
Labial Curvature			
Upper right second incisor		2	4
Upper right first incisor		2	2
Upper left first incisor	2	2	2
Upper left second incisor		2	4
Double Shoveling			
Upper right canine		1	2
Upper right second incisor		0	0
Upper right first incisor		0	0
Upper left first incisor	0	0	0
Upper left second incisor		0	0
Upper left canine	1	1	2
Interruption Groove			
Upper right second incisor		0	0
Upper right 1st incisor		MD	0
Upper left 1st incisor	0	D	0
Upper left second incisor		0	D
Tuberculum Dentale			
Upper right canine		0	5
Upper right second incisor		0?	1
Upper right 1st incisor		0?	3
Upper left 1st incisor	—	0?	3
Upper left second incisor		0?	1
Upper left canine	—	0	5
Mesial Ridge			
Upper right canine		—	0
Upper left canine		—	0
Distal Accessory Ridge			
Upper right canine		—	3
Upper left canine	—	—	2?
Lower right canine	—	—	—
Lower left canine	—	—	1?

The scores are from the ASUDAS system as defined in Turner et al. (1991). Excludes Dolní Věstonice 3 and 16 and Pavlov 1, 5, 22, 23, and 25 because all of their teeth are too worn to score. Also excludes Dolní Věstonice 36 because the crown is insufficiently developed. Symbols: blank: tooth missing; —: too worn to score; ?: score uncertain because of wear or postmortem changes.

Table 11.2 Arizona State University Dental Anthropology System traits in the sufficiently preserved Dolní Věstonice and Pavlov upper molars.

	Dolní Věstonice 3	Dolní Věstonice 13	Dolní Věstonice 14	Dolní Věstonice 15	Dolní Věstonice 16	Dolní Věstonice 36	Pavlov 2	Pavlov 20	Pavlov 28
Upper Molar Cusp 4									
Upper right third molar	—	1–	—	2	1				
Upper right second molar	—	3	1	3	—		3		
Upper right first molar	—	4	4	4	—		—		
Upper left first molar	—	4	4	4	—	4	—		
Upper left second molar	—	3	1	1	—		3		
Upper left third molar	1	1–	1	3	1			4	
Upper Molar Cusp 5									
Upper right third molar	—	5	4/5	0	0				
Upper right second molar	—	2	4	2	—		0		
Upper right first molar	—	0	5	2	—	—			
Upper left first molar	—	0	3	2	—	0	—		
Upper left second molar	—	2	4	2	—		0		
Upper left third molar	0	5	4/5	0	0			1	
Upper Molar Carabelli Trait									
Upper right second molar	—	0	0	0	—		—		
Upper right first molar	—	0	0	2	—		—		
Upper left first molar	—	0	0	7	—	5	—		
Upper left second molar	—	0	0	0	—		—	0	0

The scores are from the ASUDAS system as defined in Turner et al. (1991). Symbols: blank: tooth missing; —: too worn to score; ?: score uncertain because of wear or postmortem changes. The Pavlov 1 teeth are too worn to score.

Dental studies of modern humans would normally number the two permanent premolars first and second, and this convention is followed here. In vertebrate paleontology (sometimes including paleoanthropology), however, they are widely labeled permanent third and fourth premolars, and that is indicated here in parentheses. The same is true of deciduous first and second molars, which in vertebrate paleontology (and sometimes paleoanthropology) are labeled deciduous third and fourth premolars.

It should be noted that during the cleaning, reconstruction, and consolidation of the Dolní Věstonice and Pavlov remains, missing or damaged portions of the crania and mandibles were frequently restored with filler material, most of which can only be removed mechanically. In the process, a couple of teeth were poorly positioned or replaced in incorrect positions. It has not been possible or especially desirable, given the fragility of the remaining bone, to correct these errors on the original specimens. Any such difficulties are explained in the descriptions of the individual dentitions and associated alveolar processes.

Dolní Věstonice 3

Dolní Věstonice 3 retains the bulk of the maxilla and mandible (Figure 11.1), with some large areas reconstructed with a dark molding material. All of the teeth are present in situ, with the exception of the lower left third molar, which appears to have been missing congenitally. All of the teeth and jaws are heavily covered in lacquer, so that it is difficult to see the details of their preservation, but the occlusal surfaces of most of the teeth seem to be marked by a fine porosity, with some cracking.

The right maxilla is broken away in the molar region, and this part of the alveolar process seems to be formed entirely from molding material, which holds the teeth in position. A section of alveolar process has also been lost around the left second incisor, canine, and first premolar. Smaller pieces of the buccal plate of the alveolar process have been lost at the cervix of the left first incisor, first molar, and third molar.

The mandible has been considerably damaged in places, with the whole of the buccal plate of the alveolar

Table 11.3. Arizona State University Dental Anthropology System traits in the sufficiently preserved Dolní Věstonice and Pavlov lower molars and premolars.

	Dolní Věstonice 3	Dolní Věstonice 13	Dolní Věstonice 14	Dolní Věstonice 15	Dolní Věstonice 31	Dolní Věstonice 36	Dolní Věstonice 37	Dolní Věstonice 38	Pavlov 1	Pavlov 3	Pavlov 28
Lower Molar Cusp 5											
Lower right third molar	?	—	—	5				—	—	—	
Lower right second molar	—	0	0	4	0				—	—	
Lower right first molar	—	3	4	5		2			—	—	
Lower left first molar	—	3	4	5		2			—	—	
Lower left second molar	—	0	0	4			0		—	—	
Lower left third molar	—	—	—	5					4		0
Lower Molar Cusp 6											
Lower right third molar	0	—	—	2	0			—	—	—	
Lower right second molar	—	0	0	0		5			—	—	
Lower right first molar	—	0	0	0		5			—	—	
Lower left first molar	—	0	0	0			0		—	—	
Lower left second molar	—	0	0	0					—	—	
Lower left third molar	—	—	—	2	0				0		0
Lower Molar Protostylid											
Lower right third molar	0	6	—	7	0			—	—	—	
Lower right second molar	—	0	0	0		0			—	—	
Lower right first molar	—	0	0	0		0			—	—	
Lower left first molar	—	0	0	0			0		—	—	
Lower left second molar	—	0	0	0					—	—	
Lower left third molar	—	0	—	0					0		0
Lower Molar Fissure Pattern											
Lower right third molar	—	—	—	+				—	—	—	
Lower right second molar	—	+	+	X	Y			+	—	—	
Lower right first molar	—	—	Y	Y		X			—	—	
Lower left first molar	—	—	Y	Y		X			—	—	
Lower left second molar	—	+	+	Y			X		—	—	
Lower left third molar	—	—	—	+					X		Y
Lower Premolar Lingual Cusps											
Lower right second premolar	—	3?	0	0					—	—	
Lower right first premolar	—	5	6	—					—	—	
Lower left first premolar	—	5	6	—					—	—	
Lower right second premolar	—	3?	0	0					—	—	

The scores are from the ASUDAS system as defined in Turner et al. (1991). Symbols: blank: tooth missing; —: too worn or inaccessible to score; ?: score uncertain because of wear or postmortem changes. The Dolní Věstonice 16 teeth are too worn or inaccessible to score. For the lower molar fissure patterns: the '+' pattern is when the mesiobuccal (protoconid), mesiolingual (metaconid), centrobuccal (hypoconid) and distolingual (entoconid) cusps all meet equally together in the central fossa of the occlusal surface; the 'Y' pattern is when the mesiolingual (metaconid) and centrobuccal (hypoconid) cusps meet together in the central fossa, separating the mesiobuccal (protoconid) and distolingual (entoconid) cusps; and the 'X' pattern is when the mesiobuccal (protoconid) and distolingual (entoconid) cusps meet together in the central fossa, separating the mesiolingual (metaconid) and centrobuccal (hypoconid) cusps.

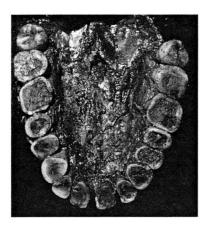 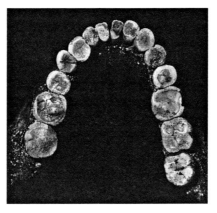

Figure 11.1 Dolní Věstonice 3, upper (left) and lower (right) dentitions, occlusal view.

process and mandibular body replaced by molding material in the area of the right molars (although the lingual side seems to be intact in this region). There seems also to be a narrow crack, repaired with adhesive, running down the symphyseal region between the left first and second incisors. In addition, substantial pieces of the buccal plate of the alveolar process have been lost over the canines, incisors, and left first premolar. The left canine and first premolar have been transposed during reconstruction and glued into positions they would not have occupied during life. The left mandibular condyle shows signs of antemortem alteration (Chapters 10 and 19). Most of it is missing, and the remainder has a flattened, mushroom-like form. The corresponding left mandibular fossa of the skull is also flattened and shallow, so Dolní Věstonice 3 must have accommodated this defect over many years.

Upper Molars

All three upper molars are present on both sides. They are robust and well-worn teeth, but they are among the smaller upper molars in either the Dolní Věstonice or the Pavlov assemblages. Their relative sizes follow the expected pattern for modern humans, in which the second molars are generally largest or equal in size to the first molars and the third molars are smallest. This is accompanied by a progressive reduction from first to third molars in the hypocone (cusp 4), leading to a change in occlusal outline from trapezoidal to triangular.

First Molars

Both upper first molars are worn to the stage where the occlusal surface is entirely composed of dentine, with a rim of enamel around the edge (Smith stage 7), and there are substantial approximal wear facets. There is far too much wear to gain much of an impression of the original crown morphology, but the outline of the occlusal wear facet is square, suggesting that there was a pronounced hypocone (cusp 4). The left first molar has the mesiolingual corner of its enamel rim worn away. It also has a prominent buccal wear facet, penetrating through the enamel to expose a spot of dentine on the crown side.

Second Molars

The upper right second molar is less worn occlusally (Smith stage 4) than the left one (Smith stage 5), but both teeth have marked approximal wear facets. As with the first molars, there is too much wear to see the morphology of the crown, but the outline of the occlusal wear facet is more triangular, suggesting that the hypocone had been reduced relative to the first molars (although still present).

Third Molars

The upper right third molar is only moderately worn (Smith stage 3), and the left one has almost no wear (Smith stage 1; see lower third molars). In both cases, it is still possible to see the pattern of the cusps in the occlusal surface, and the left third molar has a trace hypocone (ASUDAS score 1) whereas the right has none. The irregular fissure pattern of the left third molar has a prominent stained cavity in its central fossa. This may represent a carious lesion, but the coating of lacquer makes it difficult to be sure (see Chapter 19).

Upper Premolars

Both first and second (third and fourth) premolars have been reduced by occlusal wear to rims of enamel with a lower central dentine area (Smith stage 6), but it is still possible to see their original two-cusped form from the hourglass shape of the outline. They have substantial approximal wear facets. The left first and second premolars have buccal wear facets, corresponding to that on the first molar.

Upper Canines and Incisors

Both upper canines have also been reduced by wear to an enamel rim and central dentine area (Smith stage 6), and the outlines show them to have been large teeth with a prominent lingual tuberculum. The upper incisors are even more worn, in that they have lost the enamel rims on their lingual sides (Smith stage 7). There is a slight lingual tilt to the occlusal facet, but because the main axis of the roots was inclined in the jaw, the occlusal plane itself would have been relatively flat. The upper incisors as a whole have become so worn that they do not contact one another, and there are gaps between them with no approximal wear facets. The upper left second incisor has a diagonal scratch across the dentine of its occlusal facet. There is too much wear on the canines and incisors to allow them to be allocated ASUDAS scores for morphology.

Lower Molars

All three lower molars are present in the right side of the dentition, but only the first and second molars are present on the left side since the lower left third molar appears to have been congenitally absent. Heavy wear makes it difficult to review their morphology, particularly of the first molars, but they are not notably smaller than other lower molars in the assemblage.

First Molars

Both first molars have been reduced by occlusal attrition to a raised enamel rim and concave area of dentine in the wear facet (Smith stage 6). There are prominent approximal wear facets.

Second Molars

The lower right second molar is much less worn than the first (Smith stage 4). Some traces of the occlusal fissure pattern remain, and it is possible to see that it had a regular four-cusped crown form. The tooth has both a mesial and a distal approximal wear facet. The lower left second molar is much more heavily worn than the right (Smith stage 6), so the original form cannot be made out, but the occlusal outline is rather rounder. It does not fit the socket well, and the roots are relatively close together for a second molar. There is a possibility that it does not belong to this individual at all, but the reconstruction of the highly fragmented body of the mandible at this point (Klíma, 1963) may explain the displacement. In any case, the high occlusal wear score in the left as opposed to the right second molar is matched in the upper jaw. There is only a mesial approximal wear facet on this tooth with no distal facet, as would be expected if the right third molar were congenitally absent. On the root at the cervix on the mesial side, there is some pitting along the cement enamel junction that appears to be postmortem damage.

Third Molar

There is no lower left third molar and only a smooth bony surface to the alveolar process. It was not possible to take radiographs at the time of examination, but it seems likely that the tooth was congenitally missing. Not only is there no distal approximal facet on the neighboring second molar, but the opposing third molar in the upper jaw has almost no wear (see above). The evidence therefore indicates that the tooth was never present.

The lower right third molar is a large five-cusped form with moderate occlusal wear (Smith stage 3). It is not possible to see the pattern of the fissures, at least partly because there is dark staining in some areas of the fissure system. The fissures look slightly enlarged, and it is possible that the staining represents the early stages of occlusal caries (see Chapter 19). There is a clear mesial approximal wear facet.

Lower Premolars

All of the lower premolars are reduced by occlusal attrition to raised enamel rims with a concave central dentine area in their occlusal wear facets (Smith stage 6). The right premolars have been transposed during reconstruction and glued into sockets they would not have occupied during life. The lower left premolars have clear buccal wear facets, somewhat angled toward the occlusal surface and presumably matching those on the upper left premolars. There are, however, no matching facets on the lower left first molar or canine.

Lower Canines and Incisors

All the lower canines and incisors have been reduced by wear to raised enamel rims with a central dentine area (Smith stage 6 for the canines and stage 7 for the incisors). They show a strong wear gradient from canine to second incisor to first incisor. In spite of the heavy wear, all of the teeth are still in contact and have approximal attrition facets. In the left second incisor these facets have penetrated through the enamel to expose the dentine on both the mesial and distal sides. There is too much wear on the canines and incisors to allow them to be allocated ASUDAS scores for morphology.

Dolní Věstonice 9

Dolní Věstonice 9 is an associated group of three teeth. All of them are either broken postmortem or heavily worn, so identifications are difficult.

Permanent Right Lower Molar (Probably Third)

Most of the enamel has broken away from the crown of the right lower molar (DV 9/1), leaving the dentine exposed, with just vestiges of enamel remaining at the cervix. Two roots are present, completely formed except for the apex tip of the distal root. The mesial root is strongly marked, with a groove down its mesial side. The outline of the cervix indicates that the crown was irregular, suggesting a third molar.

Permanent Left Lower Molar (Probably Third)

Two roots are present in the left lower molar (DV 9/2). They are very similar, but opposite, in form to those of 9/1. Most of the enamel has been lost from the crown, except for the mesiolingual corner, which shows a mesial approximal wear facet and part of the occlusal attrition facet, which seems to have been worn to Smith stage 4.

Permanent Upper First Incisor (Probably Left)

Much enamel has been lost from the crown of the upper first incisor (DV 9/3), except for part of the buccal side. The remains of the occlusal attrition facet show wear to Smith stage 7. The root is fully formed, and its swollen, irregular appearance indicates hypercementosis.

Dolní Věstonice 10

Dolní Věstonice 10 is an isolated deciduous upper canine, perhaps the left (number A17086 DV 1948). It is well preserved, and the crown shows occlusal attrition to Smith stage 5, with the facet sloping to lingual. There is a very marked buccal wear facet. A little over half the root is missing, exposing the root canal in an oblique surface. This seems to be broken, but it may be resorption of the root; it is difficult to be sure.

Dolní Věstonice 13

Dolní Věstonice 13 has a full maxilla and mandible, and most of the teeth are preserved in situ; the missing teeth involve the permanent upper right first and second premolars, canine, first and second incisors, and the permanent upper left second incisor and second premolar. The open sockets of these teeth show that they must have been present when the individual died, but they are now missing. All of the surviving teeth are well preserved, with intact crown surfaces, although there is some cracking of the enamel in most crowns. There is also a slight roughening and staining of the surface of all of the teeth, especially the upper left second and third molars, the buccal side of the upper right first and second molars, and the buccal surfaces of all lower teeth. All of this has the appearance of postmortem change.

The alveolar process of the left maxilla (Figure 11.2), containing the sockets for the upper left second and third molars, seems originally to have been broken away, and it has been glued back into position with part of the lingual bone missing. The position of these teeth appears to be slightly out of the expected alignment, with the upper left third molar reconstructed in a position that implies that it was more fully erupted than the right third molar. In all other respects the teeth match exactly, with the same amount of slight occlusal wear, so the reconstructed position is probably not quite as it would have been during life. The right maxilla was originally broken at the center of the upper first molar, with the loss of a considerable amount of bone, so a section of alveolar process containing the second and third molars with their sockets has been reconstructed in position with a dark-colored molding material. Both maxillae have been damaged between the first incisor sockets along the line of the intermaxillary suture, and this also has been reconstructed with the same material.

In the lower jaw, all of the teeth are apparently still in their original positions (Figure 11.3). There is no evidence of repair to the mandible, except at the right

Figure 11.2 Dolní Věstonice 13, upper dentition, with occlusal view in the center and left and right buccal views.

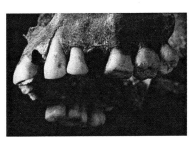

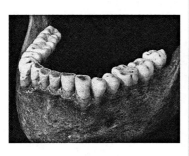
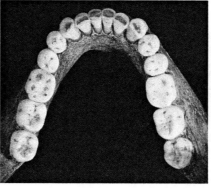

Figure 11.3 Dolní Věstonice 13, lower dentition, with occlusal view in the center and left and right buccal views.

condyle and left coronoid process. The only evidence of postmortem damage is a slight loss of the buccal plate of the alveolar process around the cervix of the right canine and second incisor, the left canine, and the left first molar. The mandible does not articulate well with the skull, and the teeth are difficult to bring into occlusion, but this is not surprising because of the extensive reconstruction around the upper jaw and face.

Upper Molars

All three upper molars are present on both sides. They are large teeth (although not the largest in the Dolní Věstonice assemblage), with robust crowns and bulging, well-developed cusps, and they show the expected modern human progression of decreasing hypocone (cusp 4) size from first to second to third molars and conversely increasing metaconule (cusp 5).

First Molars

Both first molars match well in morphology, with all four main cusps well developed. The distolingual cusp (hypocone or cusp 4) is large (ASUDAS score 4). The distal marginal ridge is crossed by a single groove, and there is no sign of a metaconule (ASUDAS score 0). Occlusal wear (Smith stage 3) obscures the features of the mesial marginal ridge, and there are substantial approximal wear facets to distal and mesial. The right first molar has a buccal wear facet (see below).

Second Molars

The left and right second molars match well in morphology, and the three cusps of the main trigon are well developed (mesiobuccal, mesiolingual, and distobuccal). The distolingual cusp (hypocone or cusp 4) is smaller but still well marked (ASUDAS score 3). A metaconule (cusp 5; ASUDAS score 2) is developed on the distal marginal ridges of both teeth, although it is somewhat obscured by wear. Occlusal wear is not pronounced (Smith stage 2), and there are modest approximal wear facets to mesial. There are no clear distal approximal facets. The right second molar has a buccal wear facet (see below). The cervical third of the crown side is marked by a poorly defined furrow-form hypoplastic defect (Chapter 19).

Third Molars

The main trigon cusps are well developed in both third molars, but there is an irregular pattern of fissures, which makes them seem less well defined. The most prominent feature of the crown is therefore the marginal ridges, which join with the cusps to make an oval rim. The distolingual cusp (hypocone or cusp 4) is almost absent (less than ASUDAS score 1)—almost a bulge at the base of the crown. By contrast, the metaconule is strongly developed (cusp 5; ASUDAS score 5), particularly on the left third molar. Occlusal wear involves just a slight polishing of the enamel of the cusp tips (Smith score 0), and there are no approximal facets, either to mesial or to distal, showing that neither had yet been in occlusion. The right third molar is not yet fully erupted, remaining partially in its crypt well below the occlusal plane. The left third molar has been reconstructed almost in the occlusal plane, but in life it was probably closer in its state of eruption to that of the right molar. Postmortem damage to the jaw has exposed the roots of the left molar, which are fully complete (Moorrees et al. stage A_c).

Upper Premolars

The right first (third) and the left second (fourth) upper premolars are missing, but it is possible to infer the crown morphology from their antimeres and the root morphology from the open sockets.

First (Third) Premolars

The left first premolar crown is a typical two-cusped form, with the lingual cusp strongly skewed to mesial. The mesial developmental groove is not well developed, although occlusal wear (Smith stage 5) is heavy enough to obscure it. There are moderate approximal attrition facets to mesial and distal. The open socket for the right first premolar shows that the root was strongly divided into two (ASUDAS score 2). The left first premolar crown shows a furrow-form hypoplastic defect in the cervical third of the crown side (Chapter 19).

Second (Fourth) Premolars

The right second premolar crown is also a typical two-cusped form. Occlusal attrition is much less advanced, with only spots of dentine exposed (Smith stage 3) and small approximal attrition facets to mesial and distal. The tooth also shows a buccal wear facet (see below). The open socket of the left second premolar shows that the tip of the root was divided into two (ASUDAS score 2).

Upper Canine

The crown of the upper left canine is robust, with a prominent lingual tuberculum and "buttress." Occlusal wear is heavy, with a broad, lozenge-shaped area of exposed dentine (Smith stage 5), and the obliquely formed attrition facet curves down to lingual. The mesial and distal approximal facets are well developed. There is a furrow-form hypoplastic defect in the cervical third of the crown side and a trace of supragingival calculus on the labial surface (Chapter 19). There is too much wear on the surviving canine and incisor to allow them to be allocated most of the ASUDAS scores for morphology (Table 11.1).

Upper First Incisor

Like the canine, the upper first incisor has a robust crown with a prominent lingual tuberculum. Occlusal wear is similarly heavy (Smith stage 5), with an irregular area of exposed dentine and a line of secondary dentine at the center of the attrition facet, which slopes strongly to lingual and distal. The mesial and distal approximal facets are strongly developed.

Lower Molars

All three lower molars are present on both sides, and as with the upper molars, they are robust teeth with strongly bulging crowns. The pattern of cusps is the classic progression for lower molars of modern humans, from five regular main cusps in the first molar to four in the second and an irregular five in the third.

First Molars

Both left and right first molars are five-cusped, with a moderately developed distobuccal cusp (hypoconulid; cusp 5; ASUDAS score 3). The teeth are too worn to make out the fissure patterns. There is a prominent buccal pit that is stained dark in both teeth, but this does not look like a carious lesion (Chapter 19). Occlusal wear has exposed dots of dentine under all five cusps (Smith stage 3), and there are approximal wear facets mesially and distally (especially mesially).

Second Molars

Both left and right second molars are four-cusped and rectangular in occlusal outline. Their fissure patterns take the + (ASUDAS) form. Like the first molars, they have buccal pits, but these are not as pronounced. The teeth are only slightly worn, with slight polishing of the enamel cusp tips (Smith stage 2) and approximal wear facets on the mesial side only. There is a furrow-form hypoplastic defect on the cervical third of the crown side (Chapter 19).

Third Molars

Both third molars have an irregular five-cusped form, which is the most common arrangement in modern humans. This gives the occlusal outline a somewhat rounded form. The fissures are too irregular to make out their pattern. The right third molar has a well-developed protostylid (up to ASUDAS score 6), and a buccal pit is present in both teeth but not strongly developed. The cusp tips are virtually unworn, and there is little evidence of a proper mesial approximal wear facet. The left third molar is fully erupted and, in fact, protrudes slightly above the occlusal plane. The right third molar is not quite in the occlusal plane and is therefore not completely erupted, although the root apices are fully complete (Moorrees et al. stage A_c).

Lower Premolars

First (Third) Premolars

Both lower first premolar crowns are the usual form for modern humans, with a large buccal cusp and much smaller distolingual cusp connected to it by a prominent distal marginal ridge and by an even smaller mesiolingual cusp and mesial marginal ridge (ASUDAS score 5). The mesiolingual cusp is particularly small, and there is a shallow groove separating it from the distolingual cusp.

The tooth crowns show slight occlusal wear, with a dot of dentine exposed at the buccal cusp (Smith stage 3) and small approximal attrition facets mesially and distally. A furrow-form hypopolastic defect is apparent in the cervical third of the crown side (Chapter 19).

Second (Fourth) Premolars

Both lower second premolar crowns similarly take the most common form for modern humans, with a large buccal cusp and two smaller lingual cusps. The mesiolingual cusp is the larger and distolingual the smaller. The fissures have the H pattern that is commonly associated with second premolars having two lingual cusps. Occlusal wear is slight (Smith stage 2), and the small approximal attrition facets expose only enamel. A furrow-form hypopolastic defect is apparent in the cervical third of the crown side, matching that in the first premolar. The left second premolar shows a slight rotation out of normal occlusal relationship (see below).

Lower Canines

Both lower canines have tall crowns with marked occlusal wear (Smith stage 4) in a facet that slopes strongly to distal and buccal. The approximal attrition facets are well developed. There is a furrow-form hypoplastic defect (Chapter 19) in the cervical third to central third part of the crown side.

Lower Incisors

All of the lower incisors are present, with modestly developed crowns, a prominent lingual tuberculum, and only very slightly developed marginal ridges (Table 11.1). The second incisor is heavily worn occlusally (Smith stage 4), but the first incisor is more heavily worn still (Smith stage 5). In both cases, the attrition facet is angled out to labial, and this is particularly noticeable in the first incisors. There are slight occlusal irregularities in the incisor dental arcade at both second incisors. The approximal wear facets are well developed on mesial and distal sides of all lower incisors. There is a small stain on the lingual crown surface of the right first incisor, but this does not appear to be a carious lesion.

Dolní Věstonice 14

This specimen has a full maxilla and mandible, with all of the teeth preserved in situ. The enamel surfaces of most of the teeth are marked by a diffuse roughening and opacity, which gives an etched appearance and probably results from postmortem alterations. In the upper teeth, this is particularly pronounced on the incisor crowns and on the buccal surfaces of the right canine, premolars, and first and second molars. On the lower teeth, the buccal and occlusal surfaces of the left crowns are particularly affected.

The alveolar process of the maxillae (Figure 11.4) is relatively intact, with most damage occurring around the third molars. The supporting bone around the upper right third molar is missing, and the tooth has been glued into the mesial side of its socket against the second molar. Much of the supporting bone around the left third molar is missing as well, together with the lingual plate of the alveolar process at the second molar. The third molar itself has been glued in approximate position on the end of the tooth row. In several areas, a section of the buccal plate of the alveolar process has broken away to expose the cervical region of the root—at the right first and second molars, premolars and canine, both first incisors, and left first molar. Over the root of the left first premolar, a small section of buccal plate has broken away to create a dehiscence-like exposure of the root.

The body of the mandible was broken at the position of the right second premolar, with much of the bone around this tooth lost (Figure 11.5). The break has been reconstructed with molding material. There is also some

Figure 11.4 Dolní Věstonice 14, upper dentition, with occlusal view in the center and left and right buccal views.

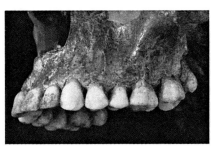
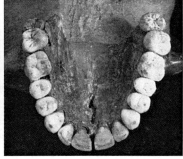
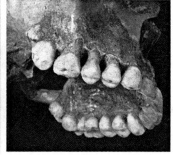

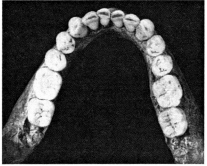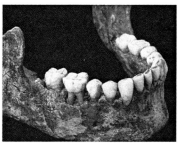

Figure 11.5 Dolní Věstonice 14, lower dentition, with occlusal view in the center and left and right buccal views.

reconstruction around the roots of the molars on this side, particularly on the lingual plate of the mandibular body. The second premolar and the first and second molars are all slightly displaced out of their sockets as a result of the reconstruction. There are two other complete breaks through the mandibular body on the right side, one at the second premolar and one just to distal of the third molar. Little bone was lost at these breaks, and they have been glued together to leave a fine line marking their position. Small pieces of the buccal plate have also broken away around the cervix of the right first premolar, left second premolar, both lower canines, and the left first incisor.

Upper Molars

The molars are all large, robust teeth, with well-developed cusps and bulging crown sides.

First Molars

Both left and right first molars have four well-developed main cusps, including a large distolingual cusp (hypocone or cusp 4; ASUDAS score 4). Both also have a distinct metaconule (cusp 5), which is larger in the right first molar (ASUDAS score 5) than the left (ASUDAS score 3). Occlusal wear is evident in the polish shown at the tips of the cusps, which exposes some small dots of dentine (Smith score 3), and there are modest mesial and distal approximal facets. There is a large, flat buccal wear facet on the right first molar, covering most of the crown side below the mesiobuccal cusp and strongly angled toward the occlusal surface of the tooth (see below).

Second Molars

The second molars have three large cusps in the main trigon and a greatly reduced distolingual cusp (hypocone or cusp 4; ASUDAS score 1). The left second molar also has a prominent metaconule (cusp 5; ASUDAS score 4). Occlusal wear is apparent as slight polishing of the enamel at the tips of the cusps (Smith score 2), and there are modest mesial approximal wear facets but no distal wear facets.

Third Molars

The positions of both third molars have been reconstructed, and whereas the right third molar is low in relation to its neighboring second molar, the left tooth has its occlusal surface almost in the occlusal plane. It is unlikely that this represents the position in life. Both third molars have a very irregular three-cusped form, with a stongly reduced distolingual cusp (hypocone or cusp 4) and a prominent metaconule (cusp 5; ASUDAS score 4/5). The fissure pattern is also very irregular. There is no evidence of wear on the occlusal surfaces or in the mesial approximal area. It is thus clear that neither tooth had erupted fully into the occlusal plane. It is not possible to see the root apices of either tooth because they are buried in wax, nor is it possible to see them in the radiographs.

Upper Premolars

First (Third) Premolars

Both of the upper first premolar crowns have the typical two-cusped form for modern humans, but the mesial developmental groove is not well developed and the lingual cusp is not as reduced as it is in some first premolars (it is almost as large as the buccal cusp). Bulges on the buccal plate of the alveolar process over the root of the left first premolar suggest that this tooth has two buccal roots, and this is confirmed by the radiograph (ASUDAS score 3). The buccal plate has broken away (apparently postmortem) over the more distal of this pair of buccal roots to make a "fenestration" that exposes the root. Occlusal wear in both left and right first premolars exposes dots of dentine (Smith score 3), and there are small approximal wear facets to mesial and distal. There is also a small buccal wear facet on the right first premolar, which is oriented in the same plane as the facet on the first molar.

Second (Fourth) Premolars

The second premolars have a two-cusped form, and they are similar in occlusal outline to the first premolars. Occlusal wear has resulted only in a slight polishing of the cusp tips (Smith score 2), and there are small mesial and distal approximal wear facets. A small buccal wear facet is present on the right second premolar, but this does not line up with the facets on the first molar, first premolar, and canine on this side, suggesting that the second premolar might have been reconstructed slightly out of position. The distal approximal wear of this tooth does not exactly fit the mesial facet of the first molar.

Upper Canines

The upper canines have large, bulging tooth crowns, but although the lingual buttress is quite prominent, the marginal ridges on the lingual surfaces are not very strongly developed (Table 11.1). Occlusal attrition has exposed a dot of dentine at the cusp tips (Smith score 3), there are small approximal facets, and a small buccal wear facet is present on the right canine. This facet lines up with the buccal facets of the first premolar and first molar.

Upper Incisors

The upper incisors are particularly badly marked by surface roughness and staining on their crowns, so little can be seen of the labial surfaces, which makes it difficult to assign ASUDAS scores (Table 11.1). They are, however, large and robust teeth, even though the marginal ridges on the lingual side are not strongly developed (this may be due to abrasion). The right first incisor has a prominent lingual tuberculum, with sharply defined interruption grooves running down its mesial and distal sides (Table 11.1). Similarly, the left first incisor has a distal interruption groove. The right second incisor has a strongly asymmetrical form, more so than the left second incisor. Occlusal wear has exposed a line of dentine in the attrition facets, which slope quite strongly to lingual in all the upper incisors. The first incisors are more worn (Smith score 4) than the second incisors (Smith score 3), and there are modest approximal facets in the enamel around the contact points.

Lower Molars

All three lower molars are present on both sides, but the third molars were still in the process of eruption at the time of death. The rest of the molars seem to have been in normal occlusion at the time of death, except for the right second molar, which looks overerupted in relation to the first molar, and this, in turn, seems overerupted in relation to the second premolar. The mesial approximal wear facet on the first molar is positioned higher than the occlusal surface of the second premolar, which demonstrates that this occlusal anomaly occurred postmortem. It seems to be the result of the reconstruction of the break in the mandible at this point.

First Molars

Left and right first molars match well in morphology, although the details of occlusal morphology are obscured in the left first molar by the diffuse damage to its surface. Both are five-cusped, with the distobuccal cusp rating an ASUDAS score of 4, and the fissure pattern is Y (ASUDAS) in the right first molar. Occlusal attrition has exposed dots of dentine under the cusps in both teeth (Smith score 3), and there are small mesial and distal approximal facets. The right first molar has a large, flat buccal attrition facet involving most of the crown side under the mesiobuccal and centrobuccal cusps (see below). Dental radiographs show that the root apices are fully closed (Moorrees et al. stage A_c).

Second Molars

The left and right second molars match well in morphology, and although the left second molar is damaged, this does not obscure much of the detail. Both are four-cusped teeth with a strongly rectangular occlusal outline and a + (ASUDAS) fissure pattern. The fissure between the mesiobuccal and distobuccal cusps shows an area of dark staining, which has an appearance of dental caries, but this requires careful discussion. There is only a slight polishing of the tips of the cusps, without exposing any dentine (Smith score 2), and there is a modest mesial approximal facet. The teeth are not in occlusion with the third molars, and there is no distal approximal facet. Dental radiographs show that the apices are fully closed (Moorrees et al. stage A_c).

Third Molars

The occlusal surfaces of the third molars are largely hidden because the teeth are still within their bony crypts below the alveolar crest. They appear, however, to have an irregular four-cusped form. Dental radiographs show that both teeth have fully formed roots, but the apices are not closed (Moorrees et al. stage R_c).

Lower Premolars

First (Third) Premolars

Both lower first premolars are similar in form, with a very small mesiolingual cusp and a sharply defined mesiolingual groove dividing it from the much larger distolingual cusp (ASUDAS score 6). This gives the occlusal outline a marked asymmetry. Slight polishing of the main buccal cusp has exposed a tiny dot of dentine (Smith score 2), and there are small approximal wear facets to mesial and distal. There is a prominent, flat buccal wear facet on the right first premolar, covering most of the crown side (see below). The right first premolar is slightly rotated out of normal occlusion. A furrow-form hypoplastic defect is present in the midthird of the crown side in the left first premolar, but it is less easy to see in the right tooth because of the wear (Chapter 19).

Second (Fourth) Premolars

Both second premolars have only one lingual cusp, and a separate distolingual cusp cannot really be made out (ASUDAS score 0). There is a U-form fissure pattern in both teeth. Slight polishing of the buccal cusp tip has exposed a small dot of dentine in the occlusal surface (Smith score 2), and there is a large, flat buccal wear facet covering most of the crown side of the right second premolar. The right second premolar is also quite strongly rotated out of normal occlusion, and this has affected the position of the small approximal wear facets (see below).

Lower Canines

The lower canines are moderately robust teeth, with a well-developed lingual tuberculum and lingual buttress. There is a furrow-form hypoplastic defect in the midthird of the left canine crown side. A large diamond-shaped area of dentine is exposed in the occlusal attrition facet (Smith score 3) that slopes to distal, and there are small approximal attrition facets mesial and distal. A buccal wear facet is present in the right canine, cutting away the most prominent part of the crown side. There is some polishing of the lingual surface

Lower Incisors

Both lower incisors have a prominent lingual tuberculum and poorly marked marginal ridges (Table 11.1). There is a furrow-form hypoplastic defect in the cervical third of the left second incisor crown side (Chapter 19). The occlusal attrition facet exposes a line of dentine in each tooth, but this is larger in the first incisors (Smith score 4) than in the second incisors (Smith score 3). The approximal wear facets are small and expose only enamel, but their position is slightly displaced because of some minor irregularities in the tooth row. There is some polishing of the lingual surfaces.

Dolní Věstonice 15

All of the teeth of Dolní Věstonice 15 are preserved in situ except for the permanent lower left first and second incisors. The open sockets of these teeth are well preserved, and the teeth must have been lost after death. All the surviving teeth are in good condition, with little in the way of surface deposits or alteration to the enamel of the crown. There is a little roughening and staining on the upper molars and some cracking of the enamel in most teeth.

Several parts of the maxillary bones were damaged and have been reconstructed (Figure 11.6). A large area of alveolar process around the right first incisor was broken away and has been reconstructed with wax. The bone around the right second and third molars is missing, and the second molar is glued into position against the first, with only a small amount of reconstruction around it. The left third molar has been glued against the right second

Figure 11.6 Dolní Věstonice 15, upper dentition, with occlusal view in the center and left and right buccal views.

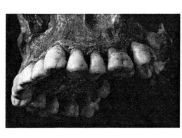
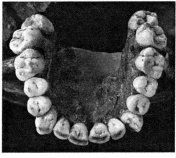
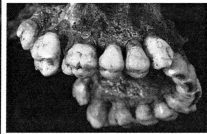

molar. In the same way, the section of alveolar process around the position of the upper left third molar has been lost, and the right third molar has been glued to the left second molar.

Several repaired breaks can be seen on the mandible, and the most prominent starts between the sockets of the lower first incisors at the midline, crossing diagonally to the right and down to the lower border of the mandible (Figure 11.7). A quantity of bone must have been missing from the fractured surfaces because molding material has been used to fill the gap, but this break has not exposed the roots of the teeth. There are other breaks through the body of the mandible at the mesiobuccal root of the lower left first molar and between the first molar and second premolar. These have been directly glued together. Cracks mark the lingual side of the mandibular body on the right side. There is also damage to the buccal plate around the roots of the lower left first and second molars, and this has been repaired with molding material. Examination of the radiographs suggests that all of the teeth are in their correct positions, even near the breaks, with the exception of the upper third molars, where not enough bone remains to be sure about their position.

Upper Molars

All three molars are present on the left and right sides. They are large teeth (among the largest in the Dolní Věstonice and Pavlov assemblages; Table 11.5) with robust crowns and, in general, follow the expected progression of hypocone reduction from first to second and then to third molars, although there is a difference between left and right sides. The third molars have been transposed during reconstruction so that the left tooth is attached to the right maxilla and vice versa, and the hard filler used prevents repositioning to their correct sides. Therefore, even though these teeth are incorrectly placed in the photographs, they are reported below in their original, correct orientations.

First Molars

The left and right first molars match well in crown morphology, each with four well-developed cusps. The distolingual cusp (hypocone or cusp 4) is large (ASUDAS score 4), and there is a small metaconule (ASUDAS score 2), partially obscured by wear in both teeth. The right first molar bears a large Carabelli cusp (ASUDAS score 7), and the left first molar has the Carabelli pit variant (ASUDAS score 2). The occlusal wear (Smith stage 3/4) is sufficiently advanced to obscure the features of the marginal ridges, particularly the mesial marginal ridge, but only the mesial approximal wear facets are well developed. There is a prominent buccal wear facet on the right first molar, affecting the full height of the crown side. Another prominent feature—visible even on the right first molar, with its buccal wear, but particularly clearly defined on the left first molar—is a very marked developmental defect (Chapter 19). This takes the form of a sharp groove at about one-half crown height, running into the buccal pit to give a T shape to the defect.

Second Molars

In both second molars, the main trigon of three cusps is well developed, but the distolingual cusp (hypocone or cusp 4) is reduced—more so on the left second molar (ASUDAS score 1) than on the right second molar (ASUDAS score 3). There is a small metaconule (ASUDAS score 2) on both tooth crowns. The occlusal wear is modest (Smith stage 2), and the approximal wear facets are not well developed. The right second molar has a buccal wear facet on the mesial end of the crown side, matching that in the first molar.

Third Molars

The third molars have been glued back in transposed positions in a way that suggests that they were only par-

Figure 11.7 Dolní Věstonice 15, lower dentition, with occlusal view in the center and left and right buccal views.

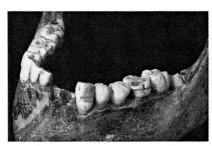
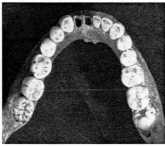
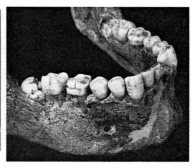

tially erupted, with their occlusal surfaces well below the occlusal plane of the second molars. They have presumably been reconstructed in this position because of the partially erupted lower left third molar (almost in the occlusal plane), which would normally be expected to be in advance of the upper molar eruption. Their almost unworn state (Smith stage 1) and lack of any clear mesial approximal facet suggests that they had at least not been long in the occlusal plane if they had, in fact, reached it at all, but their reconstructed position is not necessarily the position they were in during life. Their root apices are exposed and seem to be fully closed (Moorrees et al. stage A_c).

The third molar crowns show reduction of both the distobuccal and distolingual cusps, but to a variable extent on the left and right sides. The distolingual cusp (hypocone or cusp 4) on the left third molar is larger (ASUDAS score 3) than on the right third molar (ASUDAS score 2). The distobuccal cusp (metacone or cusp 3) is conversely larger on the right third molar (ASUDAS score 3) than on the left (ASUDAS score 2).

Upper Premolars

First (Third) Premolars

The upper first premolars show the typical two-cusped form, and because the lingual cusp is not much smaller than the buccal cusp, the occlusal outline is rather rounded. The lingual cusp is also strongly skewed to mesial. The mesial developmental groove is not really visible in the left first premolar because of wear, but it is not, in any case, strongly developed in the right first premolar. Occlusal wear has exposed an area of dentine underneath the buccal cusp in both teeth, but the left first premolar is more worn (Smith stage 3/4) than the right (Smith stage 3). Approximal facets are strongly developed in both teeth, particularly between the left first premolar and second premolar. The right first premolar has a prominent buccal wear facet, occupying most of that side of the crown.

Second (Fourth) Premolars

The upper second premolars again have a typical two-cusped form, with the expected oval occlusal outline. Occlusal wear is slightly heavier in the left second premolar than the right, but both have a Smith stage 3 score. The approximal wear facets are well developed, particularly on the left side. The right second premolar has a prominent buccal wear facet, including almost the full height of the crown, matching those in the first premolar and first molar.

Upper Canines

Both upper canines are rotated in the jaws. The right canine is more rotated than the left, so both the mesial and the distal approximal wear facets are out of position (see below). In the left canine, the mesial approximal facet is still at the expected contact point, whereas the distal facet is not. These rotations have also affected the occlusal wear facets, particularly in the right canine, although the left canine is slightly more heavily worn. In both teeth, an area of dentine is exposed to give a Smith stage 3/4. The crowns are not very tall for upper canines, but the mesial and distal crown sides bulge out quite strongly above the cement enamel junction, and the teeth have a prominent cingulum and lingual buttress. In addition, there are notably prominent furrows outlining the central cusp on the buccal side, which give a score for double shoveling (Table 11.1). The tuberculum dentale bulges up strongly as a separate cusp. There are strong abrasions on the buccal crown sides of both teeth, and the right canine bears a buccal wear facet, although this is confined to a relatively small area (see below). Both teeth show an area of roughening in the occlusal third of the buccal crown side, which probably represents a defect of enamel hypoplasia (Chapter 19), although this is difficult to see because of the abrasion.

Upper Incisors

First Incisors

There is a slight irregularity of occlusion between the left and right upper first incisors, with the mesial contact point of the left incisor slightly to labial of the right incisor, but this may be due to the reconstruction of the skull (see above). The crowns of the first incisors are robust, with bulging sides, and they have a strong tuberculum, with broad marginal ridges (Table 11.1). The right first incisor has a triple tuberculum dentale extension, whereas the left first incisor has a double tuberculum extension. In both teeth, a rectangular area of dentine is exposed by occlusal wear to give Smith stage 4, and the approximal wear facets are well developed. The labial surfaces of the crowns are quite strongly abraded, so it is not possible to see any details of dental enamel defects.

Second Incisors

The occlusal outline of the upper second incisors is relatively short mesiodistally, and the incisal edge is even shorter. Nevertheless, the marginal ridges are strongly developed, and the area of the tuberculum bulges strongly to lingual (Table 11.1). There are tuberculum furrows in both teeth, confined to the crown in the right second incisor, but with a deeper furrow passing down onto the root as well in the left second incisor. A line of dentine is exposed along the incisal edge of both teeth to give Smith stage 3 for occlusal wear. The approximal wear facets are well marked, and the labial surfaces of the crowns are quite strongly abraded.

Lower Molars

The lower molars are all robust teeth with bulging crowns. As far as it is possible to tell, they are preserved in situ. Their form is somewhat unusual for modern humans in that all three on both sides have five main cusps, and the third molars have six.

First Molars

The five prominent main cusps in the first molars are arranged in a Y (ASUDAS) pattern, with a large distobuccal cusp (hypoconulid or cusp 5; ASUDAS score 5). Occlusal wear has exposed dentine under most cusps to give a Smith stage 4, and there is a prominent mesial approximal wear facet, whereas the distal approximal facet is less clearly developed. A small buccal wear facet is present on the right first molar at the mesial end of the buccal crown side (see below). As with the upper first molars, there is a prominent developmental defect of enamel hypoplasia, taking the form of a sharp groove at about one-half crown height and connecting with the buccal groove to give a T-shaped feature (Chapter 19). This is clearly defined on both sides, and it is filled with material that looks like dental calculus, although it could be postmortem sediment. The buccal pit of the left first molar is marked with a dark stain, which may represent a lesion of dental caries, but it may also represent some postmortem change (Chapter 19).

Second Molars

There are again five main cusps in the second molars, although the distobuccal cusp (hypoconulid or cusp 5) is smaller (ASUDAS score 4) than it is in the first molars. The fissure pattern for the left second molar is Y (ASUDAS), whereas it is X for the right second molar. There is a prominent buccal pit in both teeth, especially in the left second molar, and in this tooth there seems to be a slight stain in the buccal pit, which may represent a carious lesion, although it may also be a deposit. The cusps are much less worn than they are in the first molars, with just a slight polishing to give Smith stage 2. The approximal wear facets are not prominent. There is a furrow-form defect of enamel hypoplasia close to the cervix of the crown in both teeth.

Third Molars

These are particularly large teeth for third molars, and they both have six well-developed cusps. The fissure pattern is + (ASUDAS) in both crowns, the distobuccal cusp is large (ASUDAS score 5), and the cusp 6 is small (ASUDAS score 2). There is also a prominent protostylid (ASUDAS score 7) on the right third molar. The buccal pit of this tooth has a dark stain, which again may be a lesion of dental caries (Chapter 19). Neither third molar shows much wear, although the mesial part of the occlusal surface in the left third molar shows a slight polishing (Smith stage 1). This tooth is still in the process of eruption, but it is almost in the occlusal plane. The right third molar is still partially contained within the jaw because it has erupted to mesial, with its occlusal surface against the distal side of the second molar. It could never have erupted into the normal occlusal position (see below).

Lower Premolars

First (Third) Premolars

The crowns of the lower first premolars have a typical modern human form, with a large buccal cusp and smaller lingual cusp. The expected mesiolingual groove is weakly developed, and it is not really visible because of wear. A small dot of dentine is exposed under the buccal cusp to give a Smith wear stage 2, and there are prominent approximal wear facets. The right first premolar has a prominent buccal wear facet (see below). There is some evidence in both teeth for a hypoplastic defect near the cervix, although this is difficult to see. In the radiographs of the mandibular body on the left side, there is the clear outline of a supernumerary tooth below the first premolar (see below). This seems to have a simple conical crown and a single root, which is still in the process of formation.

Second (Fourth) Premolars

Both second premolars also have one main buccal cusp and a single smaller lingual cusp (ASUDAS score 0), giving their occlusal outline a more rounded form than is usual for recent humans. This is a somewhat unusual variant in anatomically modern humans, and it is noteworthy that the Dolní Věstonice 14 second premolars also show it. Both of Dolní Věstonice 15's second premolars are rotated in the jaw—the right second premolar only slightly rotated and the left more markedly so (see below). They show slight occlusal wear (Smith stage 2) and prominent approximal wear facets. The right second premolar has a small buccal wear facet, although its position on the crown is affected by the rotation of this tooth.

Lower Canines

The lower canines have tall crowns and prominent marginal ridges but not very pronounced lingual buttresses. Dentine is exposed under their main cusp to give Smith stage 3 for wear, and there are prominent approximal wear facets. There are prominent furrow-form hypoplastic defects in the cervical half of the crown side (Chapter 19).

Lower Incisors

Only the right first and second lower incisors are preserved. They have small crowns and modest marginal ridges. Both are moderately well worn, although the first incisor is more worn (Smith stage 4) than the second (Smith stage 3), with moderate approximal wear facets.

Dolní Věstonice 16

The dentition of Dolní Věstonice 16 lacks only the lower left third molar, both upper first incisors, and the upper right second incisor. There is no sign of the third molar at the surface or on radiograph, and it seems likely that this tooth was missing congenitally; there is no distal approximal facet on the lower left second molar and no occlusal wear on the upper third molar. The other missing teeth have left vacant sockets, with no trace of remodeling, and therefore appear to have been lost after death. In general, the teeth are well preserved, with little evidence of postmortem alteration. There are, however, some areas of chipping and erosion.

Much of the alveolar process of both maxillae is present, although the upper parts of these bones have been heavily reconstructed with molding material (Figure 11.8). On the right side, the maxilla was originally broken at the first molar, so the second and third molars are held in an isolated fragment of the buccal portion of the alveolar process, which has been reconstructed against the first molar. There is little original bone on the lingual side of these teeth. Bone has also broken away around the roots of the left third molar, but this tooth is still held in its socket. There is an oblique break in the maxillae between the first incisors. The left portion is displaced slightly upward relative to the right so that the two halves of the dentition are slightly out of alignment (see Chapter 19). Parts of the buccal plate of the alveolar process are missing over both first incisors, left second incisor, and left canine roots.

The mandible is preserved in its entirety, but there are substantial breaks that have been repaired (Figure 11.9). At the lower right first molar there is a break of this type, with a large amount of missing bone that has been reconstructed with dark-colored material. There is also cracking on the lingual side of the mandibular body below the right third molar. A small piece of the buccal wall of the alveolar process has broken away at the cervix of the right canine. On the left side of the mandible, a large break runs obliquely from between the second incisor and canine, passing close to the mental foramen and down to the lower border of the mandibular body below the molars. All of the missing bone in this area has been reconstructed with molding material. There is also a finer break (repaired with glue) running up from the mental foramen to the medial root of the left first molar. The lower left canine, first premolar, and second premolar do not fit well into their sockets, and this seems to be due to the way in which they have been reconstructed between the buccal and lingual plates. For the most part, the lingual side of the left alveolar process is relatively intact, but there is some reconstruction behind the position that should have been occupied by the lower left third molar.

Upper Molars

All of the upper molars except the third molars are very worn, so that little can be said about their occlusal morphology. It is also not possible to make comparisons with other dentitions in terms of size.

Figure 11.8 Dolní Věstonice 16, upper dentition, with occlusal view in the center and left and right buccal views.

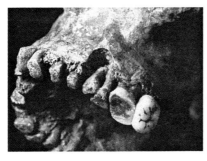
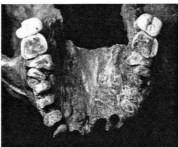

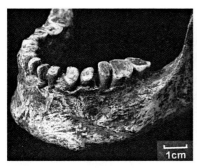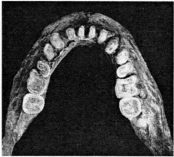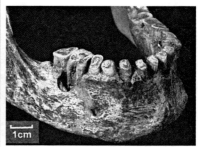

Figure 11.9 Dolní Věstonice 16, lower dentition, with occlusal view in the center and left and right buccal views.

First Molars

On both sides, the first molars have been worn down to root remnants, comprising a separate mesiobuccal root element and combined lingual/distobuccal element (Smith stage greater than 8). The entirely dentine occlusal surface is rounded buccal to lingual, and darker secondary dentine is exposed at the centers of the two root remnants, with no evidence of pulp chamber or root canal exposure. In the alveolar process around the mesiobuccal root of the left first molar there is a periapical cavity, which probably represents a granuloma (Chapter 19).

Second Molars

Both second molars have been worn to expose a broad expanse of the softer dentine in the occlusal wear facet, with a complete enamel rim (Smith stage 7). The rim is lower on the mesial side and, particularly on the left molar, has suffered a little chipping at this point. There is secondary dentine at the center of the wear facet. On both sides, the mesial enamel rim is tightly pressed up against the distobuccal/lingual root element of the first molars, and there is an approximal wear facet. Another approximal facet is present on the distal side. Although nothing can be seen of the original crown form, the square occlusal outline suggests that the hypocone (cusp 4) was reasonably well developed.

Third Molars

Neither third molar is very worn—just a slight polishing to the enamel of the cusps (Smith stage 1)—and this is not surprising because of the lack of opposing third molars in the lower jaw. The cusp and fissure pattern of the crown is irregular, but in addition to the three main cusps of the trigon, the distolingual cusp (hypocone or cusp 4) seems to be present, if only very slightly developed. The occlusal outline is relatively short mesial to distal. On the lingual side of the left third molar crown there is a slight pit, which resembles the pit form of the Carabelli trait, although this is extremely rare in third molars. Also on this tooth is a particularly deep buccal pit.

Upper Premolars and Canines

The remaining teeth in the upper jaw, premolars and canines, are also heavily worn to root remnants (Smith stage greater than 8). Secondary dentine is exposed in each of the rounded dentine occlusal wear facets. None of the teeth meets to mesial or distal, so there are no approximal facets. The alveolar process around them is reconstructed, and it is difficult to record its morphology, but where the interdental plates of bone are exposed, they are porotic and irregularly broken down. This may represent periodontal inflammation (Chapter 19). The upper right first premolar is irregularly placed and does not fit well into its socket; this may be a result of reconstruction.

Upper Incisors

All of the upper incisors are missing, except for a root remnant glued into the left second incisor socket. It looks out of place in terms of its cross section and the contours of the wear facet, and it is not necessarily in its original place. It is more like a lower incisor in the form of its root. The buccal plate of the alveolar process has broken away over the first incisors, and the form of the exposed sockets suggests that bone has been resorbed to accommodate the bulging roots of teeth with hypercementosis. This is a common condition in populations with heavily worn teeth (Chapter 19).

Lower Molars

The lower molars are also very worn, although not to the extent of the upper molars. This may in part be an effect of the difference in the form of the root trunk and the fork of the roots.

First Molars

Both first molars are worn to Smith stage 7, with a rim of enamel remaining at the lingual side of the wear facet but not at the buccal side (the facet slopes markedly to buccal). The roots are still united at the root trunk, with a broad area of dentine exposed between. The darker secondary dentine is visible at the center, and in the right first molar the base of the pulp chamber is exposed in the wear facet. The radiograph shows that the root canals, narrowed to small channels by secondary dentine deposition on their sides, communicate with the occlusal surface. In the alveolar process around the distal root of this tooth there is a periapical cavity, probably a granuloma (Chapter 19), exposed through an opening in the buccal plate. In both left and right first molars, there is an approximal wear facet on the distal side, involving both enamel and dentine, but since each tooth does not meet the second premolar there is no facet to mesial.

Second Molars

The second molars are also worn to Smith stage 7, but the enamel rim is just about complete all around the circumference of the crown. Nevertheless, it is worn quite thin on the distal side. Secondary dentine is exposed in the smooth dentine surface of the occlusal wear facet. There is an approximal wear facet to mesial but not to distal in either tooth.

Third Molars

The right third molar is present, but it is almost completely contained within the bone of the alveolar process, with what should have been the distal side of the crown exposed (Chapter 19). The occlusal surface is pressed against the apical half of the distal root of the second molar. With the heavy wear seen in this dentition, it is likely that there has been considerable continued eruption of the teeth in occlusion, so the third molar was presumably in a higher position relative to the second molar originally. In the radiograph, the roots of this tooth are fully complete at the apex (Moorees et al. score A_c).

There is no sign of the left third molar, and the surface of the alveolar process is perfectly smooth, with no deformity which might suggest bone loss and an apparently normal height of the body at this point. In addition, there is no evidence for a previous tooth socket in the radiograph, although there are variations in density that suggest that there may be some reconstruction or postmortem change not visible from the surface (and there certainly is reconstruction at the gonial angle on the lingual side). Near the crest of the alveolar process, however, the normal texture of trabecular bone is present, and there is little sign of a socket passing through this. The neighboring second molar has no distal approximal wear facet, either on the crown or the exposed part of the root. In addition, the corresponding upper left third molar has very little wear, which suggests that it did not occlude with a lower tooth. All this suggests that the left lower third molar was congenitally absent.

Lower Premolars and Canines

All of the lower premolars and canines are strongly worn, but the second (fourth) premolars are less so than the others. They still have a thin rim of enamel on the lingual side of the polished dentine wear facet, which slopes down to buccal, and thus are scored as Smith stage 7. The first (third) premolars and canines have only root remnants remaining, with dentine wear facets curving from lingual to buccal (Smith stage 8). None of the teeth is in contact with its neighbors, so there are no approximal wear facets. The right canine has a trenchlike loss of the alveolar bone lining the socket, which may possibly represent antemortem bone loss—perhaps a deformity that could be due to periodontal disease (Chapter 19).

Lower Incisors

All of the lower incisors are reduced to short sections of root, with dentine wear facets curving from lingual to labial (Smith stage 8). None of them is in contact with its neighbors, and so there are no approximal facets. There is considerable loss of the buccal bone plate around their roots (and to some extent the lingual plate too), and this probably represents remodeling of the mandible during life. There are also two large cavities around their root apices, exposed by the loss of the buccal plate. One includes the apices of the left second and first incisor, and the other has the apices of the right incisors. The teeth have been glued into their estimated positions (there is not a lot of evidence for this other than the occlusal plane in the other teeth, and the lower left first incisor looks slightly out of place), and there is surplus glue in the cavities, which makes it difficult to see them properly. They seem to have smooth walls and are thus likely to be granulomata exposed by the remodeling of the alveolar process, rather than abscesses. They still, however, represent periapical inflammation and would require the pulp to be exposed to infection (Chapter 19). This presumably occurred through cracking at the side of the secondary dentine in the root canal because there is no sign of caries.

Dolní Věstonice 27

Dolní Věstonice 27 is a deciduous upper left first molar (third premolar). The crown is cracked along a mesial-distal line, and it has been glued together. It is well worn, with Smith stage 6 for occlusal attrition, a large distal approximal facet, and a small mesial facet. There is also a very marked buccal wear facet, occupying the whole side of the crown, including the tubercle of Zuckerkandl. The root has been resorbed close to the cervical edge of the crown.

Dolní Věstonice 31

Dolní Věstonice 31 is an isolated permanent lower right third molar (Figure 11.10). Its cusps are slightly polished to give a Smith stage 2 state of wear. There is a large mesial approximal wear facet and no distal facet (confirming that this is most likely a third molar). There are four well-defined and regular main cusps with a Y fissure pattern (ASUDAS). The crown is marked all around with brown stain, associated with fine porosity on the buccal side, both apparently the result of postmortem changes. The apices of the roots are all fully formed, so it is likely that the individual was at least 18 years of age. The two main roots are well defined, although pressed closely together, and the mesial root has a strong mesial groove running down it. On the lingual side, there is an additional root, branching at the root trunk from the mesial root and running along its full length, tightly pressed against it. Additional roots on the lower molars are well known, but one in this position is highly unusual (Hillson, 1996).

Dolní Věstonice 32

Dolní Věstonice 32 (Figure 11.11) is an isolated permanent lower molar from the left side with five irregular main cusps and two well-separated roots. The enamel on the mesiolingual corner has fractured away, and this, to-

Figure 11.10 Dolní Věstonice 31, permanent lower right third molar, lingual view (left), occlusal view (center), and buccal view (right). The lingual view shows the additional root.

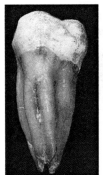

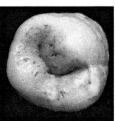
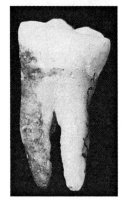

Figure 11.11 Dolní Věstonice 32 (left), a lower third molar (?), in occlusal and buccal views. Dolní Věstonice 33 (right), possibly a supernumerary molar.

gether with the wear, has obscured the fissure pattern. Occlusal wear has reached Smith stage 2, and there is a prominent mesial approximal wear facet but no distal facet. It was therefore clearly in occlusion, but it had no neighboring tooth in contact on its distal side. Thus, in spite of a crown and root form that could well be a second molar or even a first molar, it is more likely to be a third molar. Both roots have fully formed apices, implying that the individual was at least 18 years of age.

The mesial root is considerably larger than the distal root, and the strong groove running down its mesial side almost divides it into two. It is quite common for the root canals of the mesial root to be divided in this way. Both the crown and the root are stained brown, with fine porosity and roughening to the roots and cervical region of the crown, especially on the buccal side. This is presumably due to postmortem change.

Dolní Věstonice 33

Dolní Věstonice 33 is a difficult specimen to identify (Figure 11.11). It is a molariform tooth, whose crown suggests that it comes from the upper left quadrant of the dentition. There is no sign of dental wear of any kind, to either the occlusal or the approximal surfaces. In addition, there is no sign of surface abrasion, and even with the naked eye, it is possible to make out perikymata on the crown surface, together with a furrow-form defect of enamel hypoplasia at the cervix of the crown (Chapter 19).

On the basis of this evidence, it is unlikely that the tooth ever erupted into the mouth.

The occlusal outline is small (8.1 mm mesiodistal diameter; 7.6 mm buccolingual diameter) and circular to triangular in shape. There are three prominent cusps, one of which is connected to the two others by ridges. One of these ridges is relatively straight, and the other is widely curving. The two cusps that are not joined together by a ridge are separated by a fissure that runs into a fossa in the center of the occlusal surface. It is this outline of ridges and fissures that suggests the orientation of the tooth, with the curved ridge being the mesial marginal ridge and the straight ridge representing the oblique ridge of an upper molar trigon. The root has multiple elements, all fused together, and its open apex shows that it was still in the process of development at the time of death. Two grooves run down opposite sides of the root, dividing it into two main elements, one of which is bulkier than the other. There is some chipping of enamel along the cement enamel junction.

There are two possibilities for the identification of this tooth. One is that it is an abnormally small third molar. The other is that it is a supernumerary tooth from the molar row. When these occur, they are usually located at the distal end of the molar row, and they are often described as a fourth molar. The specimen described here is similar to museum specimens of molar supernumeraries. The possibility of it being a reduced third molar should, however, not be ruled out since third molars show a wide range of variation in form.

Similarly, small third molars are rare but known from the Late Pleistocene. One late archaic human, the Amud 1 Neandertal, has an in situ upper third molar (6.8 mm mesiodistal diameter; 7.7 mm buccolingual diameter; Sakura, 1970), which is slightly smaller than Dolní Věstonice 33. A Gravettian specimen, the Paglicci 36 maxilla, possesses a fully formed crown of an unerupted upper third molar in situ of a similar size (7.6 mm mesiodistal; 8.2 mm buccolingual; Mallegni & Palma di Cesnola, 1994). In addition, a more recent Late Upper Paleolithic specimen from Bruniquel–La Faye (Bruniquel 24) exhibits a diminutive lower third molar (7.7 mm mesiodistal; 7.8 mm buccolingual) in situ (Genet-Varcin & Miquel, 1967). Without having the Dolní Věstonice 33 tooth still in situ, resolution of whether it is a small third molar or a supernumerary tooth is difficult.

Dolní Věstonice 36

Dolní Věstonice 36 consists of seven isolated teeth believed to represent one juvenile individual (Figure 11.12) since they were found together (identified among the faunal remains by D. L. West and E. Trinkaus), and the state of development for all of the teeth fits well with a single child a little over 1 year in age, in terms of reference studies of living children. These teeth, which were still in the process of development at the time of death, are delicate, and the enamel is starting to flake away in places.

Deciduous Upper Right First Incisor

The tooth (DV 36/1) has the long incisal edge and low crown, which defines a deciduous upper first incisor. The lingual tuberculum and marginal ridges are notably

Figure 11.12 Dolní Věstonice 36. On the left, upper row from left to right: deciduous upper right first incisor, upper left first molar and second molar, and permanent upper left first molar. Lower row from left to right: deciduous lower left first incisor and second molar and permanent lower left first molar. On the right, the same molars are turned on their sides to show the developing edges of the crowns.

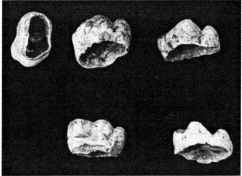

prominent. There is a slight amount of polishing to the incisal edge, which suggests that the tooth may just have been emerging into the mouth at the time of death. The crown is complete and the open (still forming) edge of the root is some 2 mm from the cement enamel junction, possibly around one-third to one-half completed (Moorrees et al. stage $R_{1/2}$). Part of the developing edge of the root is broken away in places, and there are some porous areas of chalky appearance on the crown, which probably represent postmortem change.

Deciduous Upper Left First Molar (Third Premolar)

The deciduous upper left first molar (DV 36/2) consists of a complete crown, with some vestiges of root formation. The broken edges imply that a longer length of root had in fact been formed at the time of death (perhaps almost Moorrees et al. stage $R_{1/4}$). The unworn crown has a large tubercle of Zuckerkandl at the mesial end of its buccal side and two prominent main cusps. The delicate edge of the developing root is broken away in places.

Deciduous Upper Left Second Molar (Fourth Premolar)

In this deciduous upper left second molar (DV 36/3), the unworn crown is complete, with vestiges of root formation around the cervix (Moorrees et al. stage R_i), although it appears that the edge has been considerably damaged. In several places, the enamel edge stands up higher than the dentine, whereas in life there is always a narrow zone of dentine in advance of the enamel. The crown has four prominent main cusps, including a large mesiolingual cusp (hypocone or cusp 4; ASUDAS score 4). There is also a modest development of the Carabelli trait (ASUDAS score 2) and a supernumerary cusplet on the mesiolingual side of the mesiobuccal cusp.

Permanent Upper Left First Molar

This permanent upper left first molar (DV 36/4) consists of a crown, which was still in the process of formation at the point of death (a little over Moorrees et al. stage $C_{3/4}$). The occlusal area and most of the sides were completed. There are four main cusps, with a large hypocone (ASUDAS score 4), a large cusp of Carabelli (ASUDAS score 5), and a supernumerary cusplet on the mesiobuccal cusp (see above). Permanent first molars often follow deciduous second molars in their form, and the similarities between them provide additional evidence that DV36/4 and DV36/3 belong to the same individual.

Deciduous Lower Left First Incisor

This deciduous tooth (DV 36/5) is identified as a lower incisor by its short incisal edge and the modest development of its tuberculum and marginal ridges, and as the first incisor by its relatively symmetrical crown. All three mamelons are clearly defined and seem little worn, and the open root is about half completed or slightly more (Moorrees et al. stage $R_{1/2}$). Some of the developing edge of the root has fractured away, as has some of the thin enamel over the tuberculum (deciduous teeth have thin enamel, which is always vulnerable in archeological specimens).

Deciduous Lower Left Second Molar (Fourth Premolar)

The specimen, a deciduous lower left second molar (DV 36/6), consists principally of the completed crown and just a fringe of root formation at the cervix (Moorrees et al. stage R_i). There are five well-developed cusps, and the fissure pattern is Y (ASUDAS). A furrow marks the buccal crown side at about half crown height. This is probably not a developmental defect of enamel because it is not seen on any of the other teeth. It may possibly be the pit form of the protostylid variant, but it is not well enough defined to be sure. A small part of the developing edge of the root has fractured away.

Permanent Lower Left First Molar

The crown of this permanent lower left first molar (DV 36/7) was almost complete at the point of death (a little over Moorrees et al. stage $C_{3/4}$). It has five main cusps, but the distobuccal cusp (hypoconulid or cusp 5) is rather small (ASUDAS score 2). There is, however, a large cusp 6 (ASUDAS score 5), and the fissure pattern is X (ASUDAS). The chalky, somewhat rough appearance of the crown surface is typical for a crown in the early stages of development in archeological specimens.

Permanent Lower Right First Molar

This partially completed crown, a permanent lower right first molar (DV 36/8), is an exact mirror image of specimen DV36/7.

Permanent Upper Right First Incisor

The crown of this permanent upper right first incisor (DV 36/9) is about one-third complete, somewhat less than Moorrees et al. stage $C_{1/2}$. The mammelons are not marked along the incisal edge, and it is possible that they had not been fully formed at the time of death. The surface texture of the developing crown is chalky, typical of developing crowns in archeological specimens.

Dolní Věstonice 37

Dolní Věstonice 37 is an isolated permanent lower left second molar (Figure 11.13). The crown has a rounded rectangular occlusal outline, four well-developed cusps, an X (ASUDAS) fissure pattern, and a prominent buccal pit. There is a broad furrow-form defect of enamel hypoplasia in the cervical third of the crown side. Dental wear has produced a slight polish to the enamel of the cusps, giving a Smith stage 2 occlusal wear score, and there are well-defined approximal wear facets to mesial and distal. The two roots are closely pressed together and fused. Their apices are fully formed (Moorrees et al. stage A_c), implying that this individual was at least 18 years of age at death. The surface of the root shows a slight roughening, particularly on its buccal surface, which is probably due to postmortem changes.

Dolní Věstonice 38

Dolní Věstonice 38 is an isolated permanent lower right third molar (Figure 11.13). The enamel edges of the crown have broken away around three sides of its circumference, but the prominent cross shape of fissures, meeting in the middle of the occlusal surface, suggests that it had four main cusps and a + (ASUDAS) fissure pattern. What remains of the crown suggests that the tooth was well worn at the time of death, perhaps Smith stage 6. The two large roots are closely pressed together, with fully completed apices, confirming that the individual was mature. In their apical half, both roots are irregularly bulging from the effects of hypercementosis. This suggests that the tooth comes from an older adult and matches the evidence of wear on the crown.

Pavlov 1

The alveolar processes of both left and right maxillae are present in Pavlov 1 (Figure 11.14). They are fractured and have been restored with large areas of molding material. The right maxillary fragment includes all of the teeth, from the permanent second incisor to the second molar, with the third molar glued to the distal end. The left maxillary fragment also contains the teeth from the second incisor to the second molar, but there is no sign of the third molar in its fractured distal end. It is better preserved than the fragment from the right, and it contains part of the socket for the first incisor (which is missing). A strip of the buccal plate has been lost over the canine, and there is damage around the third molar position. The gap between the right and left maxillary fragments has been filled with molding material, within which the right first incisor is held in approximate position. All permanent teeth therefore survive in this heavily reconstructed specimen, with the exception of the upper third molars and the upper left first incisor. Most show some damage to the buccal surfaces of the roots, and there are remnants of what appears to be dental calculus along the cervical parts of the crowns (Chapter 19).

The mandible is missing the right ramus (Figure 11.15), which was broken away at the right third molar, for which part of the socket is still present. Other breaks have been reconstructed. One of these lies between the right first and second premolars and runs through the body just to anterior of the mental foramen. The gap has been filled with molding material. There is also damage around the symphyseal region, where the buccal plate of the alveolar process over the roots of the incisors and left canine has been reconstructed with molding material. Another break lies just behind the left third molar and has again been glued together. All permanent teeth are present except for the right third molar, right second premolar, and left canine. These seem to have been lost postmortem because they have left open sockets and there are approximal wear facets on the neighboring teeth. Most of the surviving teeth, as with the upper dentition, show

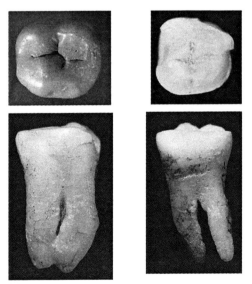

Figure 11.13 Dolní Věstonice 37 (left): a permanent lower left second molar, in occlusal and buccal views. Dolní Věstonice 38 (right): a permanent lower right third molar, in occlusal and buccal views.

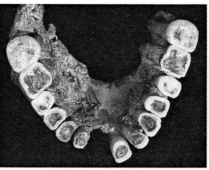

Figure 11.14 Pavlov 1, upper dentition, with occlusal view in the center and left and right buccal views.

some damage to the exposed root surfaces together with remnants of dental calculus on the crowns.

Upper Molars

Third molars were originally present, from the evidence of second molar approximal facets, but they have not been preserved. First and second molars are similar in the size of their occlusal areas, but these show a reduction in the hypocone from first to second.

First Molars

The crowns of the first molars are very worn (Smith stage 7). There is a very similar pattern of dentine exposure in the occlusal wear facets on both sides, with an outline suggesting that the distolingual cusp (hypocone or cusp 4) was large. The approximal wear facets at the mesial and distal ends of the crowns are also large, and each tooth has a prominent buccal wear facet. In the left molar, buccal wear is so severe that dentine has been exposed in the mesial part of this wear facet, which is angled slightly toward the occlusal surface.

Second Molars

The crowns of these teeth are rather less worn (Smith stage 5), but the relative sizes of the cusps can still not be seen directly. The shape of the occlusal outline, however, suggests that the distolingual cusp was reduced relative to that of the first molar. Both left and right second molars have large mesial approximal wear facets, plus smaller and less well defined distal approximal wear facets. In the left second molar, the distal edge of the enamel rim of the occlusal wear facet is beveled in a way that matches the facet on the lower left third molar (see below). Each tooth has a prominent buccal wear facet.

Upper Premolars

First (Third) Premolars

Both of these upper premolars are reduced to an oval-shaped enamel rim, with a broad area of dentine exposed in the occlusal wear facet (Smith stage 6). There are large mesial and distal approximal wear facets. The right first premolar has the buccal wear facet confined

Figure 11.15 Pavlov 1, lower dentition, with occlusal view in the center and left and right buccal views.

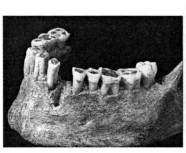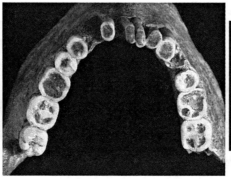

to the enamel, but the left first premolar has such severe buccal wear that it has breached the enamel rim to expose the dentine. As with the first molar and second premolar, this buccal wear facet is somewhat angled toward occlusal.

Second (Fourth) Premolars

The crowns of these second premolars are less worn than the first premolars, with some enamel left in the center of the occlusal wear facet, in addition to a rim of enamel (Smith stage 5). There are large mesial and distal approximal wear facets, and each tooth has a buccal wear facet. As with the first premolar, the left second premolar has a buccal wear facet angled toward occlusal and cutting through the enamel rim to expose the dentine.

Upper Canines

Both upper canines are reduced by wear to an oval enamel rim with a central area of dentine in the occlusal wear facet (Smith stage 6). In the right canine, this wear facet is strongly angled to lingual. In the left canine, the wear is slightly heavier, and the wear facet is more horizontal, although part of it has been fractured away. The right canine has both a mesial and a distal approximal wear facet, whereas the left canine has only a distal facet that is well defined. Both teeth have the buccal wear facet, and this cuts through the enamel rim and exposes the dentine in the left canine.

Upper Incisors

Both first and second upper incisors from the right side are heavily worn down to segments of their roots (Smith stage 7), and as with the right canine, the occlusal wear facet is strongly angled to lingual. The left first incisor is missing, but the surviving right second incisor seems to have been worn to Smith stage 7 with a rather more horizontal occlusal wear facet, although part of it has fractured away postmortem. There is only a small distal facet on the right second incisor because the teeth are so worn that they did not meet their neighbors.

Lower Molars

First Molars

The crowns of the lower first molars are reduced to a rim of enamel, with a broad area of dentine exposed in the occlusal wear facet. The left first molar is slightly more worn (Smith stage 7) than the right one (Smith stage 6). Both teeth are too worn to discern the pattern of the cusps, but the occlusal outline suggests that they had a prominent distobuccal cusp (hypoconulid or cusp 5). In the left first molar, the relatively flat occlusal wear facet is simply angled to buccal, but in the right first molar, it curves strongly down into the mesiobuccal corner of the crown. In this corner, the enamel rim is missing, and it seems likely that there was a fracture during life that has since been smoothed over by attrition. There are large approximal wear facets mesial and distal, and the left first molar has a sharply defined buccal wear facet with such severe attrition that dentine has been exposed on the crown side at the center of the facet. Also, in this tooth, the occlusal wear facet is marked with two broad grooves that run across it from lingual to buccal.

Second Molars

The crowns of the second molars are also well worn (Smith stage 4), with regular occlusal wear facets oriented horizontally. Four areas of dentine are exposed in each tooth, suggesting that all four cusps were well developed. There is also a large mesial approximal wear facet and a smaller distal facet on each tooth. In addition, the left second molar has a small buccal wear facet with rounded margins that is much less well defined than the facet in the first molar.

Third Molar

The lower left third molar stands up proud, with its occlusal surface higher than the occlusal plane of the neighboring second and first molars. Neither a surface examination nor the radiograph gives clear evidence of the cause, although reconstruction of the break in the mandibular body nearby may be involved and adhesive can be seen around the tooth. The wear, however, suggests that it may have stood at least a little proud during life. The occlusal surface itself is only slightly worn, with a little polishing of the cusp tips producing a Smith stage 2 score. There is, however, a beveled facet on the mesial end that angles down toward the occlusal surface of the second molar. This could only have been produced if the third molar occlusal surface was higher than that of the second molar in life, and it matches a similar facet on the upper second molar (see above). The radiograph shows that the root apices were fully formed (Moorrees et al. stage A_c). The crown has five cusps, with the distobuccal cusp (hypoconulid, cusp 5) being large (ASUDAS score 4). The fissure pattern is X (ASUDAS). Deposits of dental calculus are preserved in the fissures and fossae, and there is a prominent, slightly stained cavity in the central fossa. This resembles a carious lesion, although it is possible that it is the result of postmortem change (Chapter 19).

Lower Premolars

First (Third) Premolar

The lower left first premolar has been reduced to a circular enamel rim, with dentine exposed in the occlusal wear facet inside (Smith stage 6). This facet is marked, as in the left first molar, by a broad groove running across it from lingual to buccal. There are also prominent mesial and distal approximal wear facets and a marked buccal wear facet.

Second (Fourth) Premolars

Both lower left and right second premolars have a squarer outline than the first and are less worn (Smith stage 5). A broad groove marks the occlusal wear facet of the left second premolar, as in the first. They also have prominent mesial and distal approximal wear facets, and there is a large buccal wear facet on the left second premolar.

Lower Canine

Only the lower right canine survives. It has been reduced to an oval enamel rim with a central area of dentine in the occlusal wear facet (Smith stage 6), which is angled slightly to buccal. There are also mesial and distal approximal wear facets. The remaining crown sides show a markedly bulging cingulum to buccal and lingual. There is a furrow-form enamel defect on the crown.

Lower Incisors

The position of the lower first incisors in the reconstructed symphyseal region of the mandible looks slightly anomalous. Both the first and second right incisors are heavily worn, down to segments of root, with only a small area of enamel rim surviving (Smith stage 7). The left first incisor is more heavily worn (Smith stage 8), but the left second incisor still has a complete enamel rim around the central area of dentine in its occlusal wear facet (Smith stage 5). There is an oblique groove running across the occlusal facet in this tooth from the mesiolingual corner to the distobuccal corner, roughly parallel with the grooves in the occlusal facets of the premolars and and first molar on this side. Only the second incisors have approximal wear facets on their distal sides because the first incisors were too worn to make contact with their neighbors.

Pavlov 2

Pavlov 2 comprises a number of fragments, which have been joined together with adhesive and wax (Figure 11.16). The largest single fragment is a piece of the alveolar process of the left maxilla, with a permanent upper left second premolar, first molar, and second molar. The right side of the fossil originally had five elements joined together (Vlček, 1997)—upper right first premolar, second premolar, first molar, and second molar—together with a tooth that turned out on reexamination to be an upper left third molar. This last tooth did not fit with the left maxilla fragment either, and so it was removed from the specimen and is reported separately here as Pavlov 20. All the remaining elements are joined together in a wax form of the palate.

The bony part of the left maxillary segment consists of the posterior part of the alveolar process, from the distal wall of the socket for the second premolar to the socket for the third molar, but it is missing the posterior extremity. The third molar was lost postmortem because the socket is still present, but as it is filled with matrix, it is not possible to reconstruct details of the root system. The radiograph shows that all of the remaining teeth have fully complete roots (Moorrees et al. stage A_c).

The alveolar process has a prominent maxillary torus on its lingual side, behind the sockets of the molars. These bony bulges along the lingual margin of the alveolar crest are quite commonly found in recent and Late Pleistocene modern and archaic humans, although they are much

Figure 11.16 Pavlov 2, upper dentition, with occlusal view in the center and left and right buccal views.

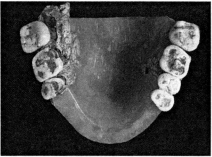

more common in the mandible of recent humans. The most frequent site is the molar region. It has often been suggested that there is a large inherited component in their formation, but there is little good evidence for this. There is variation in the frequency of tori between populations (Campbell, 1925; Moorrees, 1957; Brothwell, 1981; Trinkaus, 1983b), but although the mandibular tori are much better known than the maxillary ones, even with them there is no clear geographical pattern. Another possibility is that the tori may result from remodeling of the alveolar process in response to the forces acting on the teeth as they wear and erupt continuously through life. There is, however, again no clear relationship with heavy wear. One potential problem is variation in the standards for recording them because tori vary from tiny excrescences, which many anthropologists might not regard as an anomaly, up to large formations such as that seen in Pavlov 2.

After the reexamination of Pavlov 2, the right maxilla fragment and associated teeth now contain five elements: right first premolar, second premolar, first molar, second molar, and a fragment of alveolar process lying within the roots of the first molar. These accompany the left maxillary section, with its three teeth.

Upper Molars

First Molars

The upper left first molar is very worn, with four prominent cusps reduced to large areas of dentine in the wear facet (Smith stage 5). The distolingual corner of enamel has been chipped away, and the sharp edges of the fractures suggest that this occurred postmortem. Mesial and distal approximal wear facets are well defined, and there is a less well-defined buccal wear facet. This is most clearly apparent when the tooth is viewed in occlusal outline because the edges of the facet are somewhat rounded.

The upper right first molar shows heavy occlusal wear (Smith stage 5), with four well-developed cusps reduced to large areas of dentine in the wear facet. The pattern of dentine exposure is similar to that of the left first molar, and the size and proportions of the occlusal outline are also similar. There is a large mesial approximal wear facet and a smaller one to distal, as well as a large buccal wear facet with rather rounded margins—best seen from occlusal view. On the distobuccal corner of the cervix is a small deposit of dental calculus. Also at the distal cervix is the trace of a poorly defined furrow-form defect of enamel hypoplasia (Chapter 19).

The associated fragment of the alveolar process contains the complete mesiobuccal root socket of the first molar, together with the mesial walls of the distobuccal and lingual root sockets, and the distal wall of the socket for the second premolar.

Second Molars

The crown of the left second molar has four cusps, including a modestly developed distolingual cusp (hypocone or cusp 4; ASUDAS score 3). There is only a moderate amount of occlusal wear (Smith stage 3), with small dots of exposed dentine. There are well-defined mesial and distal approximal wear facets. In addition, there is a less well-defined buccal wear facet, which, as with the first molar, is easier to see when viewed in occlusal outline. It does not line up well with the buccal wear facet in the first molar, but the reconstruction appears to have left the teeth slightly out of position relative to one another. A small furrow-form defect of enamel hypoplasia runs along just above the cervix on the buccal side of the crown (Chapter 19).

The right second molar also has four cusps, including a small distolingual cusp (hypocone or cusp 4; ASUDAS score 3). There is moderate occlusal wear (Smith stage 3) and well-defined mesial and distal approximal wear facets. In addition, there is a large buccal wear facet with rounded margins. It matches well with the left second molar of Pavlov 2. The mesial approximal facet fits reasonably well with the neighboring first molar, but the tooth is fixed with wax, with its buccal root apices exposed. The roots are fully formed (Moorrees et al. stage A_c). There is a large supragingival calculus deposit along the cervical margin on the buccal side of the crown.

Upper Premolars

First (Third) Premolar

The well-worn crown of the upper right first premolar has two cusps, with the lingual cusp considerably smaller than the buccal one, giving a rather triangular occlusal outline. The occlusal wear facet has large areas of exposed dentine under the cusps (Smith stage 5), and there are prominent mesial and distal approximal wear facets. There is a small flattened area on the buccal side, which may be a buccal wear facet, but it is poorly defined.

Second (Fourth) Premolars

The crown of the left tooth has the typical modern human form for a worn upper second premolar, with two main cusps exposing large areas of dentine in the wear facet (Smith stage 4). There are prominent mesial and distal approximal wear facets. Enamel on the distobuccal and mesiolingual corners has fractured away, and the sharp nature of the break, together with the paler color of the fractured surfaces, suggests that this occurred postmortem. The wear and the fractures make it difficult to confirm the identification of side morphologically, but the distal approximal facet fits well against the first molar.

This crown on the right side also has the distinctive two cusps of the upper premolar, but they are more equal in size. The tooth is less worn than the first premolar (Smith stage 4), with smaller patches of exposed dentine. There are well-defined mesial and distal approximal wear facets. A poorly defined buccal wear facet can be made out when the tooth is seen in occlusal view. It does not line up with the facet on the first molar, but this may be due to reconstruction. There is a small remnant of what appears to be supragingival dental calculus on the mesiobuccal corner of the crown. The mesiolingual corner of enamel has broken away, and the sharp nature of the fracture and light color of the underlying dentine suggest that it occurred postmortem.

Pavlov 3

Pavlov 3 consists of a fragment of mandibular body from the right side, containing four teeth and the socket for another missing tooth, originally joined by a wax arch to an isolated left third molar (Figure 11.17). The principal element of Pavlov 3 is the right mandibular body fragment and associated teeth. The isolated third molar, upon reconsideration, may well not derive from the same mandible, although it is not possible to confirm whether it did. Consequently, the left third molar has been separated from the wax reconstruction and given the number of Pavlov 28 (see below).

The principal, and remaining, portion of Pavlov 3 includes the alveolar process of the mandibular body, fractured at about half body height and including the mental foramen. It runs from just distal of the position for the second incisor to the mesial root socket of the second molar. In situ are the permanent right canine, both premolars, and the first molar. The canine socket is widely spaced from both second incisor and first premolar, with two large diastemata. This is an unusual position for such diastemata in humans, but they are clearly developmental in origin, rather than arising through loss of teeth, because the appropriate teeth are in the right positions in the jaw and there are no signs of the bone remodeling that would accompany tooth loss. The roots of all teeth are exposed by more than the 2 or 3 mm expected of a young adult. This is presumably due to continuous eruption, as the surface of the bone is smooth and shows no signs of the alveolar bone loss, which would have been characteristic of periodontal disease.

In the radiograph of the mandible, only the first molar roots have a clear lamina dura. The premolars have an irregular mix of densities around their sockets, which do not match the normal trabecular structure of the mandible in this region. This is undoubtedly related to the inferior corpus damage and filler associated with consolidation of the specimen.

All of the teeth can be seen, by direct observation or radiographs, to have complete roots (Moorrees et al. stage A_c).

Lower First Molar

The crown of the lower first molar is well worn (Smith stage 5), but it had five cusps because each is marked by a separate area of exposed dentine. There is a large distal approximal wear facet, but the mesiobuccal corner of the crown has been broken away. The sides of this fracture are sharp, and it probably occurred postmortem. It exposes the pulp chamber inside, with a lining of secondary dentine clearly visible in the roof.

Lower Second (Fourth) Premolar

The lower second premolar protrudes out of its socket, so that it rises above the occlusal plane of the other teeth. It is glued into its socket, and an excess of adhesive may cause the anomaly. As with the first premolar, the crown is small and has a markedly circular occlusal outline, which suggests that it probably had only a single lingual cusp. There is slight occlusal wear (Smith stage 3) and small mesial and distal approximal wear facets. As with

Figure 11.17 Pavlov 3, permanent lower right dentition, with buccal view to the left and lingual view to the right.

the first premolar, there are a number of fine furrow-form defects of enamel hypoplasia at the cervix.

The anomalous position of the tooth may well be postmortem and in part related to its reassembly, especially given the internal damage to the premolar sockets evident radiographically. It is possible that this second premolar does not belong to Pavlov 3, but it is generally the correct size and morphology, and in these aspects it resembles the first premolar. It is also worn occlusally to a similar degree, and each tooth has a minor hypoplastic furrow defect in a similar position near the cervix. It is therefore assumed that these teeth are appropriately associated in Pavlov 3, but the high position of the second premolar crown should not be taken as indicating its antemortem position.

Lower First (Third) Premolar

The crown of the lower first premolar is small. It is very worn (Smith stage 7), with the occlusal wear facet sloping very strongly to lingual, exposing a large area of dentine down the root. The slope is so marked that it may represent the site of a crown fracture during life. In addition, there is a small distal approximal wear facet, but none to mesial. At the cervix are a number of fine furrow-form defects of enamel hypoplasia.

Lower Canine

This lower tooth is small for a canine, with a low, narrow crown (Table 11.6) and a weakly developed lingual tuberculum. The crown is very worn (Smith stage 6), and the wear facet exposes a large area of dentine while still retaining a rim of enamel. The wear facet slopes down very strongly to distal and lingual, and it is divided into two planes by an angle at its center. There are no approximal wear facets. The buccal side of the crown is marked by several fine furrow-form defects of enamel hypoplasia (Chapter 19).

Pavlov 5

Pavlov 5 consists of two lower first incisors from one individual (Figure 11.18), originally separate but matched during reexamination: the permanent lower right first incisor (tube number 17; site number 519156) and the permanent lower left first incisor (tube number 20; site number 425656).

Both teeth have a typical lower incisor form, with a short crown (mesiodistal diameter) relative to breadth (buccolingual diameter) and prominent tuberculum but poorly defined marginal ridges. The proportions of the crowns and lengths of the roots are identical. The apices

Figure 11.18 Pavlov 5, permanent lower left and right first incisors, with occlusal view to the left and mesial views to the right.

of the roots are fully developed (Moorrees et al. stage A_c). Both are worn to the point that a broad band of dentine is exposed within a rim of enamel (Smith stage 5), and the occlusal wear facet slopes slightly to lingual. In both teeth there are also large mesial and distal approximal wear facets, and the mesial facets fit closely together. It is on this basis that the two isolated specimens have been matched.

Pavlov 6

Pavlov 6 consists of two isolated deciduous molars (premolars), which appear to come from one individual (Figure 11.19), and they were matched during reexamination: a deciduous upper left first molar (third premolar; tube number 5; site number 737654) and a deciduous upper left second molar (fourth premolar; tube number 6; site number 693656).

Figure 11.19 Pavlov 6, deciduous upper left first and second molars, in occlusal view. The buccal wear facet runs along the upper side of the photograph.

The first deciduous molar has the two-cusp form of crown, worn to Smith stage 6, with only a little patch of enamel from the deepest part of the fissures surviving in the largely dentine occlusal wear facet. There is a rim of enamel around the circumference. A modest tubercle of Zuckerkandl is preserved in the mesiobuccal corner of the crown side, but it may have been modified by wear. There are prominent mesial and distal approximal wear facets, and there is a sharply defined buccal wear facet in which the enamel of the crown side is worn away to expose the underlying dentine. The roots have been resorbed until only remnants are left around the base of the crown.

The second deciduous molar has four well-defined cusps, including a large distolingual cusp (hypocone or cusp 4). It is worn to Smith stage 5, with dots and larger areas of dentine exposed in the occlusal wear facet. There is a prominent distal approximal wear facet, but the mesial facet is less clear, partly because some of the enamel has broken away (probably postmortem) in the mesiobuccal corner. In spite of this, however, the remaining part of the facet fits closely against the distal facet of the first molar. In addition, there is a small buccal wear facet that lines up with the first molar buccal facet when the teeth are placed in position beside one another. In the second molar, it is confined to the mesial part of the buccal crown side. As with the first molar, the roots have been resorbed until only a small remnant remains. This is part of the normal process by which deciduous teeth are exfoliated, and the teeth of Pavlov 6 have reached the point where they may very well have been shed from the mouth before being deposited on the site (Chapter 6).

Pavlov 7

Pavlov 7 (Figure 11.20) is a deciduous lower left second molar (fourth premolar; tube number 1; site number 363256). It has what appears to be a six-cusped form of crown, but it is not possible to make out the fissure pattern because of wear. Part of the distal end of the crown has fractured away, probably postmortem because of the sharpness of the break. Dots of dentine are exposed under the cusps in the occlusal wear facet (Smith stage 4/5), and there are clear mesial and distal approximal wear facets. The roots are resorbed to small remnants at the base of the crown, and they are at the point where the tooth could have been exfoliated before being deposited on the site (Chapter 6).

This specimen may belong to the same individual as Pavlov 8, and both Pavlov 7 and 8 may also belong to the same individual as Pavlov 6.

Pavlov 8

Pavlov 8 (Figure 11.20) is a deciduous lower right second molar (fourth premolar; tube number 3; site number 613356). This tooth has a crown with six cusps, but the fissure pattern is obscured by wear. The cusps are worn to expose small dots and other areas of dentine (Smith stage 4/5) in the occlusal wear facet, and there are large mesial and distal approximal facets. The roots are resorbed to small remnants, and the tooth may have been shed before deposition on the site.

This tooth is very similar in size, morphology, and wear pattern to Pavlov 7, and it may belong to the same individual. However, because the two specimens do not represent neighboring teeth, this cannot be confirmed by matching the approximal wear facets. It has therefore been left as a separate, isolated specimen. In the same way, both Pavlov 8 and 7 may also belong to the same individual as Pavlov 6, but this cannot be confirmed.

Pavlov 9

Pavlov 9 (Figure 11.20) is an additional deciduous lower right second molar (fourth premolar; tube number 2; site number 737554). It has a five-cusped crown. It is worn, exposing dots of dentine in the occlusal surface (Smith

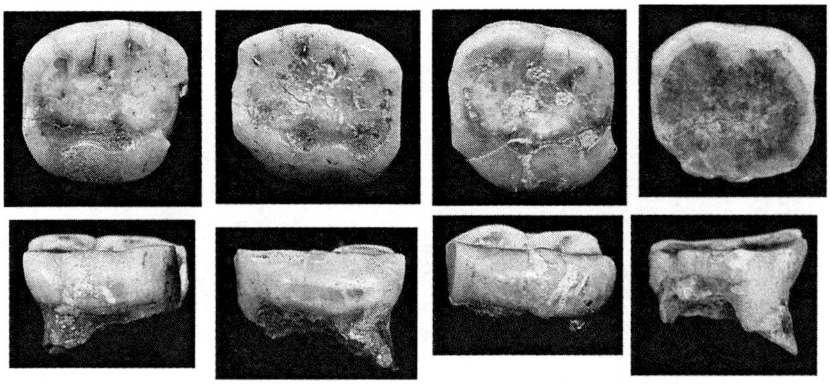

Figure 11.20 Deciduous lower second molars. From left to right in occlusal and buccal views; Pavlov 7 (left), Pavlov 8 (right), Pavlov 9 (right) and Pavlov 10 (right).

stage 4/5), although there is some postmortem damage that obscures the detail. There are also mesial and distal approximal facets. The roots are resorbed to the point at which only small remnants are left at the base of the crown (Chapter 6). The occlusal outline, details of cusps, and form of the wear facet are different from Pavlov 7 and 8, so Pavlov 9 is maintained as a separate specimen.

Pavlov 10

Pavlov 10 (Figure 11.20) is yet another deciduous lower right second molar (fourth premolar; tube number 7; no site number). This tooth has a much more worn crown than Pavlov 7, 8, or 9, with a broad area of dentine inside a rim of enamel in its occlusal wear facet (Smith stage 7). From the patches of darker secondary dentine marking the pulp horns, it is possible to see that the tooth was four-cusped originally. Large approximal wear facets mark the mesial and distal ends. There is a small deposit of calculus on the crown side. The roots are only partly resorbed (Chapter 6).

Pavlov 11

Pavlov 11, a deciduous upper right first molar (third premolar; tube number 4; site number 757754), has a two-cusped form of crown, with a modestly developed tubercle of Zuckerkandl on the mesiobuccal corner of the crown side (Figure 11.21). It is large for a deciduous upper first molar. Wear has exposed two broad areas of dentine in the occlusal wear facet (Smith stage 4/5), and there are both mesial and distal approximal wear facets, with the mesial facet exposing dentine. The roots are re-

Figure 11.21 Pavlov 11 (left), deciduous upper right first molar in occlusal and mesial views. Pavlov 12 (right) deciduous upper right second molar in occlusal and mesial views.

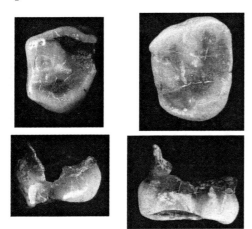

sorbed to small remnants (Chapter 6). For a possible association, see Pavlov 12.

Pavlov 12

Pavlov 12, a deciduous upper right second molar (fourth premolar; tube number 12; site number Pavlov 61), is a heavily worn tooth crown (Figure 11.21) with an irregular area of dentine exposed in its occlusal wear facet and an enamel rim around it (Smith stage 5). The dots of secondary dentine exposed at the horns of the pulp chamber suggest that there were originally four main cusps. There is a large distal and a small mesial approximal wear facet. The roots are resorbed to small remnants (Chapter 6).

The mesial approximal facet of Pavlov 12 fits against the distal facet of Pavlov 11, but they are very differently sized facets, and the fit is not a good one. In addition, Pavlov 12 has rather greater occlusal wear than Pavlov 11, and from the eruption timing of first and second deciduous molars, it would normally be expected that the relative state of wear should be the other way around. For this reason, they have been maintained as separate specimens.

Pavlov 13

This heavily worn specimen (Smith stage 6) is difficult to identify (Figure 11.22). It is most likely a deciduous upper right (?) canine (tube number 8; site number 427956). The occlusal wear facet exposes a diamond-shaped area of dentine with an oval rim of enamel around it, and this is characteristic of canines. The wide flare of the distal and mesial sides confirms it as a deciduous upper canine, but the orientation must be estimated from the asymmetry of the prominent tuberculum. Prominent approximal wear facets mark the mesial and distal ends of the crown. There is virtually nothing left of the root because of resorption, although it can be seen to have had a rounded triangular section, which is again characteristic of deciduous upper canines. It would have been very close to the point of exfoliation or perhaps actually shed (Chapter 6).

Pavlov 14

Pavlov 14 (Figure 11.22), a deciduous upper left canine (tube number 9; site number 203856), is also heavily worn (Smith stage 6). The occlusal outline is a different shape than that of Pavlov 13, with a form of exposed dentine that suggests the tuberculum and marginal ridges were less prominent. It was clearly a smaller tooth originally, even before the effects of wear. There is a large mesial and

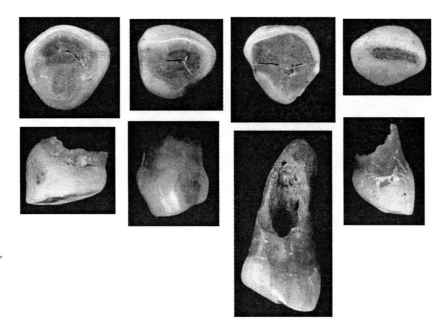

Figure 11.22 From left to right: Pavlov 13, deciduous upper right (?) canine, occlusal (top) and mesial (bottom) views. Pavlov 14 deciduous upper left canine, occlusal (top) and distal (bottom) views. Pavlov 15, deciduous upper left (?) canine, occlusal (top) and mesial (bottom) views. Pavlov 16, deciduous upper right second incisor, occlusal (top) and distal (bottom) views.

a small distal approximal wear facet. The root is resorbed, although not quite so much as in Pavlov 13, and the age implications are more or less the same for the two teeth (Chapter 6).

Pavlov 15

Pavlov 15, a deciduous upper left (?) canine (tube number 21; site number 345856), has advanced occlusal wear (Smith stage 5 and exposure of the pulp), but the root, despite its short length, has only been partially resorbed (Figure 11.22). There are large mesial and distal approximal wear facets. A deciduous tooth with this worn a crown would normally also have advanced root resorption (Chapter 6). It therefore seems likely that Pavlov 15 was retained in the permanent dentition, possibly because the following permanent canine was either impacted in the jaw or was congenitally absent (see below). See also Pavlov 19, which has similar implications.

Pavlov 16

Pavlov 16 is a deciduous upper right second incisor (in tube number 10; site number 932156), with the typical long incisal edge and low, flaring crown of an upper deciduous incisor, but not to the extent expected in a first incisor (Figure 11.22). It is also strongly asymmetrical, which is another particular feature of second incisors. A strip of dentine is exposed in the occlusal wear facet to give Smith stage 4. The approximal wear facets are not clearly developed. The root is about half resorbed, sug-

gesting that this tooth was not exfoliated before the child died (Chapter 6).

Pavlov 17

The crown occlusal outline of Pavlov 17, a deciduous lower right second incisor (tube number 11; site number 377056), is relatively broad labiolingually compared to its mesiodistal length (Figure 11.23). This makes it more likely to be a lower incisor than an upper incisor, and the

Figure 11.23 From left to right: Pavlov 17, deciduous lower right second incisor, occlusal (top) and lingual views (bottom). Pavlov 18, deciduous lower left second incisor, occlusal (top) and mesial (bottom) views. Pavlov 19, deciduous upper left first incisor, occlusal (top) and buccal (bottom) views.

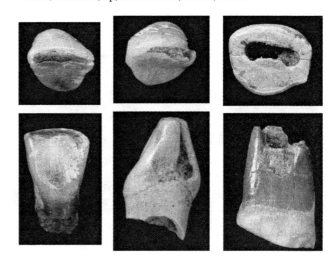

asymmetry of the crown identifies it as a second incisor and gives the side of the dentition that it comes from. A thin line of dentine is exposed on the occlusal wear facet (Smith stage 3). There is a clear distal approximal wear facet, but the mesial facet is not marked. In addition, there is a polished facet on the lingual side of the crown. This suggests either an occlusal anomaly, movement of the tooth with the periodic loosening and tightening up that accompanies the shedding of deciduous teeth, or perhaps some cultural use of the teeth. The root is about half resorbed, which as with Pavlov 16, implies that it had not been shed before being deposited on the site (Chapter 6).

Pavlov 18

Pavlov 18 (Figure 11.23) is a deciduous lower left second incisor (numbered only Pavlov 1956; identified by V. Sládek). A line of dentine is exposed in the occlusal wear facet (Smith stage 4), and there are mesial and distal approximal wear facets. The root is about half resorbed and probably had not been shed before arriving on the site (Chapter 6). In this, it is similar to Pavlov 17, although this tooth does not match Pavlov 17 in either morphology or wear state.

Pavlov 19

Pavlov 19 (Figure 11.23) is a deciduous upper left first incisor (tube number 16; site number 363356). It has advanced occlusal wear (Smith stage 6), with a broad area of exposed dentine and a fully open pulp chamber. There are also large mesial and distal approximal wear facets, both exposing dentine. The root, however, seems not to have been resorbed. As with Pavlov 15, it seems likely that this deciduous tooth was retained among the teeth of the permanent dentition (see above). It is possible that both Pavlov 15 and 19 came from one individual, but it is rare for more than one deciduous tooth to be retained in this way.

Pavlov 20

Pavlov 20 (Figure 11.24) is a permanent upper left third molar, originally fixed with wax to the Pavlov 2 fragment but separated during reexamination (see above). It has a large crown with bulging sides and three broad cusps in the main trigon and a small distolingual cusp (hypocone or cusp 4). There is very little occlusal wear (Smith stage 1), but a large mesial approximal wear facet is present. As expected for a third molar, there is no distal facet, and both the buccal and distal crown sides are coated with

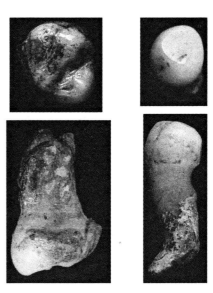

Figure 11.24 Pavlov 20 (left), permanent upper left third molar, occlusal and distal views. The additional root is visible on the left side of the main distobuccal root. Pavlov 21 (right), perhaps a supernumerary tooth, occlusal and side views. The approximal groove is visible in outline on the right side.

what appears to be dental calculus. The roots are short and bulbous and relatively well separated, with completed apices (Moorrees et al. stage A_c), implying that the individual was at least 18 years of age. There is a small additional supernumerary distolingual root (this is rather a rare location for such a root in modern human teeth). All of the roots have an irregularly bulging surface characteristic of hypercementosis, which implies that the individual was a relatively mature adult in spite of the modest wear.

Pavlov 21

Pavlov 21 is another difficult tooth to identify, but it probably represents a permanent lower first premolar or perhaps a supernumerary tooth (Figure 11.24). This specimen has a small crown, with an almost circular outline in occlusal view. It appears to have had one large cusp and one smaller cusp, separated by a poorly defined fossa. On one side of this fossa is another, smaller depression, which separates off a small ridge. The crown side is taller under the larger cusp and lower under the smaller. The root is relatively long, with an oval cross section, and it follows a gently curving path, with a sharp reverse angle in its apical one-third. The surface of the root is rough and irregular, suggesting the presence of hypercementosis.

There is a single facet on the occlusal surface, exposing a small dot of dentine in the larger of the two cusps (Smith stage 3), and it is rather sharply inclined. There is also a single approximal facet on the side of the crown below the lower of the two cusps. This facet appears to be double, with a smooth central ridge separating it into two halves. This implies that the tooth was pressed against two other teeth at their contact points, rather than just against one tooth crown. At the cervical margin underneath the approximal facet there is a transverse groove, containing scratches that run along its length of a type usually termed "interproximal grooving" (see below).

Identification of this specimen is difficult. It clearly erupted and reached occlusion, as shown by the occlusal wear facet. The form of the approximal wear facet could be taken to imply that it lay to one side of the tooth row, pressed against the contact point of two other teeth. The transverse groove at the cervix implies that something was drawn through at this point, either for cleaning or as part of the processing of some material or foodstuff. In the form of the crown and root, Pavlov 21 is most similar to a lower premolar. It is also possible that it is a supernumerary tooth (see below), the form of which may range from a simple conical crown with a single short root, to a complex crown with several cusps and fossae, but still only one root, to a premolar-like tooth or an upper molarlike tooth. It might also be an anomalous (very reduced) third molar. Pavlov 21 does not, however, have any distinctively molariform features. The root is not even partially divided into two or three elements, which would be expected even in the most reduced third molars, and it is rather long. With the wide range of variation seen in third molar form, however, this identification cannot be ruled out. The presence of a single approximal wear facet might be taken as supporting evidence of its identification as a third molar, but its double nature argues against it.

Pavlov 22

A permanent upper left (?) second incisor (tube number 14) represents Pavlov 22 (Figure 11.25). The crown of this tooth is robust, but it is heavily and obliquely worn. The occlusal wear facet exposes a broad area of dentine within a rim of enamel (Smith stage 5/6), and it is strongly angled to distal. There is a marked furrow-form defect of enamel hypopolasia in the cervical third of the crown (Chapter 19). The apex of the root is fully formed (Moorrees et al. stage A_c) and has a slightly irregular surface suggestive of hypercementosis.

Pavlov 23

Pavlov 23 (Figure 11.25) is a permanent lower left first incisor (tube number 15; site number i592256/m84). About half the crown height is worn away, exposing a broad band of dentine in the occlusal wear facet (Smith stage 5). There are prominent mesial and distal approximal wear facets and a polished area, almost a facet, on the lingual side. The cervical half of the crown is marked with a furrow-form enamel hypoplasia defect (Chapter 19). The apex of the root is fully formed (Moorrees et al. stage A_c).

Pavlov 24

Pavlov 24 (Figure 11.25) is a permanent lower left first incisor (tube number 18; site number 514356). The apex of this tooth is fractured away, and there is damage to the root at the cervix and to the crown on the labial side.

Pavlov 25

Pavlov 25 (Figure 11.25) is a permanent lower right first incisor (tube number 19; site number 641456). The crown is worn to Smith stage 5/6, with a broad area of exposed dentine in the occlusal wear facet. There are also prominent mesial and distal approximal wear facets. The apex of the root is fully complete (Moorrees et al. stage A_c).

The distinctive feature of this tooth, however, is that the apical three-quarters of the root has been thinned by

Figure 11.25 From left to right: Pavlov 22, a permanent upper left (?) second incisor, lingual view. Pavlov 23, a permanent lower left first incisor, lingual view. Pavlov 24, a permanent lower left first incisor, lingual view. Pavlov 25, a permanent lower right first incisor, lingual and mesial views. The polishing of the root and the drilled hole are visible.

polishing on both mesial and distal sides to produce a roughly parallel-sided wafer some 1 mm in thickness. In the center of this, a circular hole a little over 1 mm in diameter has been drilled. It has been carefully carried out because the sides of the hole are relatively straight, and it is an even diameter throughout its depth. The crown, on the other hand, does not show any evidence of modification. Drilled animal teeth are not uncommon in European Upper Paleolithic sites, including Pavlov I and the Dolní Věstonice sites, and they appear to have been used as amulets and necklaces, but it is less common for them to be human teeth. Several of these have been noted for western European earlier Upper Paleolithic sites (Henry-Gambier et al., 2004), and more directly relevant, a now lost perforated upper central incisor (Dolní Věstonice 8) was discovered by K. Absolon at Dolní Věstonice I (Vlček, 1991).

Pavlov 26

Two lower incisor roots, held in a fragment of the mandibular alveolar process (tube number 22; site number 374053), make up Pavlov 26. The crowns of these teeth are missing, but it is possible to see that the roots are from permanent teeth and that their apices are fully complete.

Pavlov 27

This specimen is represented only by a small fragment of enamel and dentine from a permanent (relatively thick enamel) cheek tooth (tube number 13; site number Pavlov 61).

Pavlov 28

This permanent lower left third molar (Figure 11.26) was originally associated with the Pavlov 3 mandible, but from the opposite side of the dental arcade (Vlček, 1997). Since it is not possible to confirm its association with the Pavlov 3 right mandible and teeth, it has been separated and given the number of Pavlov 28.

The crown of this tooth has four main cusps and a Y fissure pattern (ASUDAS). It also has only slight occlusal wear (Smith stage 2), with just slight polishing of the cusps and a large mesial approximal wear facet but not a distal facet. The large, bulbous roots are pressed closely together. Their apices are fully completed (Moorrees et al. stage A_c). They have the rough, irregular, and bulging surface characteristic of hypercementosis, which suggests that the tooth comes from a relatively mature individual in spite of the slight occlusal wear.

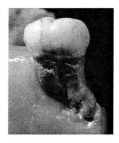

Figure 11.26 Pavlov 28, a permanent lower left third molar, in occlusal and buccal views. This tooth was originally associated with Pavlov 3, and it is shown in the wax mount of that specimen.

Morphology and Relative Size

The dentitions from Dolní Věstonice and Pavlov include large and robust teeth (Tables 11.4 to 11.7). The Dolní Věstonice 3 and Pavlov 1, 2, and 3 teeth are somewhat smaller than those of Dolní Věstonice 13–15, but there is not a large difference between them. All of the teeth overlap in size with the range of more recent modern humans, but the means are high in relation to most recent populations, with the exception of some aboriginal Australian groups (Kieser, 1990). The premolars and molars are similar in size to the other Late Pleistocene samples, which do not differ greatly among themselves, even though the Neandertals have generally higher values for the upper second incisors and the lower first and second incisors. The pattern of crown sizes through each individual dentition is typical of modern humans (Kieser, 1990). Although the Dolní Věstonice and Pavlov teeth are in general robust, particularly Dolní Věstonice 14 and 15, they have moderate or variable scores on the ASUDAS system for those features that add mass to the crown (Tables 11.1 to 11.3). Examples are shoveling, tuberculum dentale, Carabelli's trait, hypocone, and lower molar cusp 6.

Metric comparisons of the deciduous teeth are not presented because of the very small available samples, especially of earlier Upper Paleolithic deciduous teeth. However, there was little change in human deciduous dental crown diameters through the Late Pleistocene and Holocene (Hillson & Trinkaus, 2002), and the Dolní Věstonice and Pavlov teeth are not exceptions to this pattern.

Occlusion and Malocclusion

"Occlusion" can be defined as the way in which teeth fit together, both upper against lower and between neighbors in the same jaw. The relationship between teeth that do not fit together in the expected way is

Table 11.4 Comparative labiolingual crown diameters (mm) for the Dolní Věstonice and Pavlov anterior maxillary teeth.

	I^1	I^2	C^1
Dolní Věstonice 3	—/—	—/—	9.0/9.0
Dolní Věstonice 13	7.3		—/9.8
Dolní Věstonice 14	8.2/8.1	7.6/7.3	10.4/9.8
Dolní Věstonice 15	7.6/7.7	8.0/7.2	—/9.8
Pavlov 22		6.3	
Earlier Upper Paleolithic	7.6 ± 0.5 (18)	7.1 ± 0.5 (18)	9.1 ± 1.0 (21)
Qafzeh-Skhul	8.2 ± 0.5 (12)	7.5 ± 0.6 (11)	9.2 ± 0.8 (10)
Neandertals	8.2 ± 0.5 (32)	8.0 ± 0.7 (34)	9.5 ± 0.7 (30)
Recent Europeans	7.1 ± 0.5 (98)	6.2 ± 0.5 (103)	8.3 ± 0.6 (108)

Right and left values are provided as preserved for the Dolní Věstonice and Pavlov teeth. Summary statistics for the comparative samples average right and left values by individual as available prior to the computation of the mean ± standard deviation (N). Recent European data are from Twiesselmann and Brabant (1967). Complete dental metrics for the Dolní Věstonice and Pavlov remains are in Sládek et al. (2000). —: tooth insufficiently preserved or erupted to measure accurately.

known as "malocclusion." Many archeological and fossil skulls are distorted, so it is often difficult to be sure of the relationships between upper and lower jaws, but the Dolní Věstonice and Pavlov assemblages have many examples of malocclusions within one jaw, and in a number of cases, it is possible to be certain that this would have had an impact on its relationship with the opposite jaw. In the description below, the terms used are those of the Fédération Dentaire Internationale (Baume et al., 1970) system for measuring occlusal traits. One of the difficulties about applying the usual dentists' concepts of occlusion to fossil humans is the very heavy wear. Wear has the effect of reducing occlusal anomalies because continuous eruption of the teeth throughout life allows the dentition to adjust its position as the crowns are reduced in size. The bone is remodeled around the tooth roots in response to the forces acting on the dentition, so malocclusions are reduced. In addition, the cusps, incisal edges, and contact points of the crowns are removed at an early stage, to be replaced by broad attrition facets, so the usual benchmarks for defining malocclusion are simply not there any more. In this way, any minor malocclusions that might have been present in Dolní Věstonice 3 and 16 or in Pavlov 1 when they were younger would leave little trace by the time they died.

Table 11.5 Comparative buccolingual crown diameters (mm) for the Dolní Věstonice and Pavlov posterior maxillary teeth.

	P^1	P^2	M^1	M^2	M^3
Dolní Věstonice 3	9.3/9.3	9.1/9.0	11.6/11.0	11.7/11.9	10.8/11.0
Dolní Věstonice 13	10.6	10.3	12.3/12.5	12.9/13.1	—/12.2
Dolní Věstonice 14	10.3/10.6	10.6/10.3	12.5/12.7	13.0/12.6	—/12.3
Dolní Věstonice 15	9.6/9.6	9.9/9.9	12.1/12.1	13.6/13.3	13.5/12.9
Dolní Věstonice 16	—/—	—/—	—/—	—/—	12.0/13.0
Pavlov 1	—/—	—/—	—/—	12.8/12.7	
Pavlov 2	9.1	8.8/—	11.2/—	—/11.7	
Pavlov 20					12.0
Earlier Upper Paleolithic	9.8 ± 0.6 (22)	10.1 ± 0.8 (20)	12.2 ± 0.8 (32)	12.4 ± 0.9 (26)	11.5 ± 1.2 (21)
Qafzeh-Skhul	10.4 ± 0.4 (9)	10.2 ± 0.8 (10)	12.2 ± 0.7 (18)	12.0 ± 0.7 (9)	11.7 ± 0.7 (6)
Neandertals	10.3 ± 0.7 (28)	10.0 ± 0.7 (22)	12.0 ± 0.8 (34)	12.3 ± 1.0 (25)	11.9 ± 1.4 (29)
Recent Europeans	8.6 ± 0.6 (108)	8.8 ± 0.6 (106)	11.2 ± 0.5 (104)	10.7 ± 0.7 (102)	10.1 ± 0.9 (89)

See footnote in Table 11.4 for an explanation.

Table 11.6 Comparative labiolingual crown diameters (mm) for the Dolní Věstonice and Pavlov anterior mandibular teeth.

	I_1	I_2	C_1
Dolní Věstonice 3	6.1/6.3	6.8/6.6	8.1/8.3
Dolní Věstonice 13	6.7/6.7	7.5/7.6	9.2/9.2
Dolní Věstonice 14	6.4/6.4	7.0/7.2	8.9/8.7
Dolní Věstonice 15	6.1	6.8	9.2/9.0
Pavlov 3			7.8
Pavlov 5	5.5/5.8		
Pavlov 23	5.9		
Pavlov 25	5.4		
Earlier Upper Paleolithic	6.4 ± 0.5 (17)	6.9 ± 0.5 (22)	8.7 ± 0.8 (19)
Qafzeh-Skhul	6.6 ± 0.6 (11)	7.1 ± 0.6 (10)	8.3 ± 0.8 (10)
Neandertals	7.1 ± 0.7 (20)	7.7 ± 0.5 (28)	8.9 ± 0.7 (33)
Recent Europeans	6.0 ± 0.4 (102)	6.3 ± 0.4 (107)	7.8 ± 0.5 (109)

See footnote in Table 11.4 for an explanation.

Irregularities, Rotations, and Diastemata in the Tooth Row

The term "irregularity" is used to describe a failure of the teeth to line up in the normal curve of the dental arcade. This is seen in Dolní Věstonice 13–15. Dolní Věstonice 13 shows irregularities at both left and right lower second incisors (Figure 11.3). The distal end of the right second incisor crown is slightly to lingual of its expected contact with the canine, while the mesial end is in its normal position against the first incisor. The whole of the left second incisor is displaced to lingual relative to the first incisor and canine. In the case of Dolní Věstonice 14 (Figure 11.5), the distal end of the lower right first incisor is slightly to labial of its normal contact with the canine, and the mesial end is slightly to lingual of its normal position against the first incisor. In Dolní Věstonice 15, whose dentition shows several anomalies of occlusion, there is an irregularity between the upper first incisors (Figure 11.27). The distal end of the right incisor is slightly to lingual of its normal contact with the left incisor.

"Rotation" implies that, whereas the central axis of the tooth remains in the line of the dental arcade, the whole crown is rotated so that the mesial and distal con-

Table 11.7 Comparative buccolingual crown diameters (mm) for the Dolní Věstonice and Pavlov posterior mandibular teeth.

	P_1	P_2	M_1	M_2	M_3
Dolní Věstonice 3	8.0/7.8	8.6/8.5	—/11.3	10.9/11.0	10.5/—
Dolní Věstonice 13	8.6/8.6	9.1/9.3	11.1/11.1	11.1/11.0	11.0/10.6
Dolní Věstonice 14	8.6/8.8	9.5/9.3	(11.1)/11.0	11.0/11.0	—/—
Dolní Věstonice 15	8.3/8.5	9.1/9.1	11.4/11.0	11.5/11.9	—/—
Dolní Věstonice 31					11.2
Dolní Věstonice 32					11.3
Dolní Věstonice 37				10.1	
Pavlov 1	—/—	(8.5)		10.9/10.7	9.8
Pavlov 3	—	7.8			
Pavlov 28					11.0
Earlier Upper Paleolithic	8.6 ± 0.5 (15)	8.7 ± 0.6 (14)	11.1 ± 0.7 (34)	10.9 ± 0.8 (27)	10.7 ± 1.0 (14)
Qafzeh-Skhul	8.8 ± 0.5 (8)	8.9 ± 0.6 (8)	11.4 ± 0.6 (15)	10.9 ± 0.7 (10)	10.8 ± 0.7 (7)
Neandertals	9.0 ± 0.7 (33)	8.9 ± 0.8 (34)	10.9 ± 0.6 (48)	11.0 ± 0.7 (35)	11.0 ± 0.8 (42)
Recent Europeans	7.3 ± 0.5 (108)	7.9 ± 0.6 (109)	10.3 ± 0.5 (107)	9.7 ± 0.6 (109)	9.5 ± 0.7 (99)

See the footnote in Table 11.4 for an explanation. Individual values in parentheses are estimated. Individual tooth measurements in parentheses are estimated values.

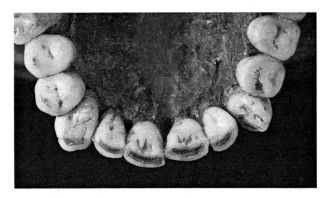

Figure 11.27 Dolní Věstonice 15, upper anterior dentition, in occlusal view. Both canines are rotated (particularly the right), and there is an irregularity between the first incisors.

tact points are out of their normal position against the neighboring teeth. This is seen very slightly in the lower left second premolar of Dolní Věstonice 13 and lower right second premolar of Dolní Věstonice 14. The largest rotations, however, are seen in Dolní Věstonice 15 (Figure 11.27). Both upper canines are strongly rotated, especially the right canine. There is also a strong rotation of the lower left second premolar (associated with the supernumerary tooth; see below and Figure 11.28) and a less pronounced rotation of the lower right second premolar. Similar rotations, especially of premolars, are relatively common among Late Pleistocene human remains. They are present in the lower premolars of the Krapina 58, Malarnaud 1, and La Naulette 1 Neandertals; in the

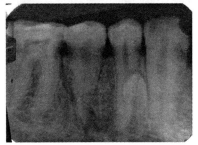

Figure 11.28 The upper image shows the lower left cheek teeth of Dolní Věstonice 15, in occlusal view. The second premolar is rotated. The lower image is a radiograph of the same part of the dentition, showing an unerupted supernumerary tooth under the first premolar.

lower premolars of Qafzeh 7, Qafzeh 11, and Skhul 4 plus an upper premolar of Skhul 4; and in a lower premolar of Pataud 1.

Irregularities are taken to imply a lack of space in the dental arcade. A diastema is the opposite condition, where there is a gap between neighboring teeth. This is seen in Pavlov 3, where the lower right canine is widely spaced from the second incisor and first premolar. There is no evidence of tooth loss or injury, and the bone surface of the alveolar process is smooth.

Anomalies of Eruption

Both Dolní Věstonice 15 and 16 have lower right third molars that have erupted in a way that would never have allowed them to reach normal occlusion (Figure 11.29). In both cases, the occlusal surface of the crown is pressed against the distal side of the neighboring second molar.

This is most marked in Dolní Věstonice 16, where the third molar is almost completely contained within the bone of the jaw, with just its distal crown side exposed through an opening in the crest of the alveolar process. The radiograph shows that the roots are short but fully formed, and the unworn occlusal surface of the crown is against the apices of the roots of the very worn second molar. The third molar was presumably pressed against the second molar crown when Dolní Věstonice 16 was younger. Continuous eruption of the second molar throughout life, with remodeling of the jaw, has apparently left the third molar in the same position, although the socket and roots of the second molar have moved upward. This is a remarkable demonstration of the effects of very heavy wear and continuous eruption upon occlusal anomalies. The case of Dolní Věstonice 16 can be described as an impacted third molar. Strictly speaking, the dentist's term "impaction" implies that the tooth never emerged into the mouth, and it seems likely that the third molar in this case was always covered with soft tissue because it shows no wear even on its distal crown surface.

The case of Dolní Věstonice 15 is not strictly an impaction because at least part of the tooth clearly emerged into the mouth. It is not tilted so strongly toward the second molar as in Dolní Věstonice 16, so the mesial marginal ridge of the occlusal surface is pressed into the cervix on the distal side of the second molar. The roots are fully completed, with the apex closed, as in the other third molars. The distal marginal ridge of the occlusal surface is uppermost, and at least part of the distal crown surface would also have been exposed. There is, however, very little evidence of wear because the third molars had clearly only just erupted.

In both Dolní Věstonice 15 and 16, the right third molars could not have met in normal occlusion, although

Figure 11.29 Malpositioned lower right third molars. On the left is Dolní Věstonice 15, shown, in buccal view above and as a radiograph below. On the right is Dolní Věstonice 16, shown in lingual view above and as a radiograph below (a composite of two images).

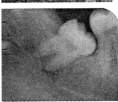
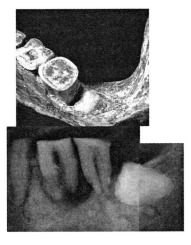

if Dolní Věstonice 15 had lived to be older, wear in the other teeth might eventually have brought upper into contact with lower. If this had occurred, remodeling of the jaw and continuous eruption might in the end have brought the lower right third molar into something approaching a normal position. In Dolní Věstonice 16, however, this did not occur, and the upper right third molar shows very little wear.

Similar impactions are rare in Late Pleistocene humans, known principally in the Cro-Magnon 3 mandibular fragment.

Other anomalies of eruption include the "overeruption" of teeth, in which their occlusal surfaces are high in relation to their neighbors (Figure 11.30). This is seen in the lower left third molar of Dolní Věstonice 13. It is, however, only slightly higher, and as it is clear that the upper third molar was still erupting at the time of death, it is probable that the anomaly would have been adjusted as the teeth came into occlusion. This did not, however, occur in the case of Pavlov 1. Here, the lower left third molar occlusal surface is still higher than the rest of the tooth row, which shows considerable tooth wear. A beveled wear facet slopes down to the second molar occlusal surface, and this is matched by a facet on the upper second molar. The bone to distal of the upper second molar has fractured away and been lost, but it does seem that the upper left third molar was present because there is a small distal approximal attrition facet on the second molar crown. Instead, it seems more likely that the upper third molar had erupted into an anomalous position, which limited its contact with the lower third molar.

Congenitally Missing Teeth

Many living people have one or more of their third molars missing from the end of the tooth row. In some cases, the teeth are impacted and remain hidden inside the jaw, but in many other cases the third molars are not formed at all. This is known as congenital absence, or agenesis. A proportion of the members of most living human populations have at least one third molar congenitally absent, but in some parts of the world well over half the population have this condition. It is known among all Primates (Miles & Grigson, 1990), but it has not frequently been described in fossil hominids. Two individuals from Dolní Věstonice show it: Dolní Věstonice 3 and 16 (Figures 11.1 and 11.31). In both cases, it is the lower left third molar that is missing. The bone surface of the alveolar process

Figure 11.30 Overerupted lower left third molars. On the left is Dolní Věstonice 13 and on the right is Pavlov 1. Both are shown in buccal view. The lower left first molar of Dolní Věstonice 13 shows a deep buccal pit, which is stained. This is a common site for dental caries, but the stain is unlikely to represent a carious lesion in this case. Pavlov 1 shows buccal wear facets on the molars and premolars, although they are not easy to make out in this view (see Figure 11.32). It is, however, possible to see that the buccal facet of the first molar has worn through the enamel, to expose the dentine. In addition, it is possible to make out grooves running across the occlusal wear facets (Figure 11.32).

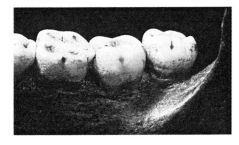
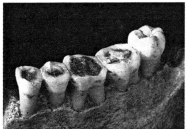

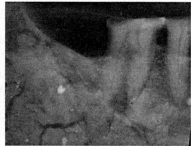

Figure 11.31 Dolní Věstonice 16, congenitally missing lower left third molar. Lingual view of lower left cheek tooth row (left) and radiograph (right) of the second/third molar region.

in that position is smooth, with no sign of injury or other pathology. In the radiograph of Dolní Věstonice 16, there is no sign of the tooth. There is no distal approximal facet on the neighboring lower second molar and little wear on the upper left third molar. All of these features confirm that the lower left third molar was never present in the jaw.

Two isolated specimens may provide indirect evidence of congenitally missing teeth. Pavlov 15 (Figure 11.22) is a deciduous upper canine, and Pavlov 19 (Figure 11.23) is a deciduous upper first incisor. Both show advanced wear, which is not uncommon in deciduous teeth that are about to be exfoliated, but the roots do not show much resorption (see above). Pavlov 19, in particular, has extremely heavy wear, and one plausible explanation is that the tooth was retained in the jaw while all of the other deciduous teeth were replaced by permanent teeth. This is not uncommon, and it is usually due to the congenital absence of the permanent successor, which for Pavlov 15 might be a second incisor and for Pavlov 19 a first incisor. It must also be borne in mind, however, that wear of deciduous teeth is very variable, and the specimens may just represent one extreme of normal variation.

Congenitally missing teeth are rarely documented in the human fossil record, given the simultaneous need for the agenesis of the tooth, the preservation of the alveolar bone, and the radiographic examination of the specimen. The earliest known case of a congenitally absent human tooth is the unilateral agenesis of a lower third molar in the Omo 75–14 *Australopithecus* mandible (Suwa et al., 1996). There is also one case of bilateral first lower incisor agenesis in a Neandertal, the Malarnaud 1 mandible (Heim & Granat, 1995). Among contemporaries of the Dolní Věstonice and Pavlov remains, the now lost Předmostí 3 mandible appears to have had agenesis of one lower third molar, although all three of the other third molars were present (Matiegka, 1934).

Supernumerary Teeth

A supernumerary tooth is one that is additional to those in the normal tooth row. The condition is also known as polygenesis or polydontia. Sometimes, the extra tooth takes on the form of others, and it may even erupt into the tooth row to cause confusion about the identification of the other teeth. Most often, however, it has an anomalous form. Many have a simple conical crown with a single pointed tip and a single root. Others frequently have irregular multiple cusps. If they erupt, supernumerary teeth are often found to lingual of the dental arcade, but often they do not erupt and can only be detected in radiographs. Dental polygenesis is found widely among the Primates, although it is never common (Lavelle & Moore, 1973; Miles & Grigson, 1990). Among living people, 1.9% of dentitions are reported to have at least one supernumerary tooth. They are seen very occasionally in archeological material, but they have rarely been described for fossil hominids (see below). This is almost certainly partly due to the necessity of radiographs for proper study.

The dental radiograph of Dolní Věstonice 15 shows an unerupted supernumerary tooth under the lower left first premolar (Figure 11.28). It appears to have a simple conical crown, with a single root that is still in the process of development. If this root had continued to develop, the tooth might eventually have erupted, although it is difficult to be sure. In the dental arcade above, the first premolar itself is in normal occlusion, but the second premolar is strongly rotated. This may in part be related to the presence of the supernumerary tooth. In living people, 0.7% of dentitions possess a supernumerary tooth in the premolar region (Lavelle & Moore, 1973), so although still rare, this is the most common position for a supernumerary tooth to take.

Two isolated teeth may also be supernumerary, but their morphology may be interpreted in various ways. Dolní Věstonice 33 (Figure 11.11) takes the form of a very small upper molar. It could be a very small third molar, but it might also be a so-called fourth molar. There are supernumerary molars, usually showing clear upper molar morphology, which erupt at the distal end of the molar row. They are very uncommon, but most dental schools possess examples, and personal experience suggests that Dolní Věstonice 33 is very like them. Pavlov 21 (Figure 11.24) takes the form of a lower premolar; it may be an anomalous tooth of that type, but it could also be a supernumerary tooth. In addition, the double approximal facet below the smaller cusp suggests that it may have been either a malpositioned premolar displaced buccally or lingually between two teeth or a similarly positioned supernumerary tooth. As both teeth are isolated specimens, it is difficult to be sure.

The molar region is the rarest position for a supernumerary tooth—0.4% of dentitions in living people (Lavelle & Moore, 1973). One case of a prehistoric fourth molar from a Jomon site in Japan has been published (Suzuki et al., 1995). Supernumerary teeth are largely undocumented among Pleistocene human remains. However, the late Gravettian Pataud 1 skull presents two peg supernumerary teeth, buccally adjacent to the right upper second molar (Legoux, 1975). The smaller of them is within the alveolar process and identified radiographically; the larger one is on the buccal surface of the alveolar process and extended close to the occlusal plane.

Interpretation of Occlusal Anomalies

Malocclusions are unusually frequent in the Dolní Věstonice and Pavlov assemblages. They are not commonly reported in any fossil hominids, including other specimens from Upper Paleolithic contexts, but there is some evidence that they become more common in Mesolithic and Neolithic contexts (Hillson & Fitzgerald, 2003). It is possible that heavy wear reduced minor occlusal anomalies, but there is little evidence that the wear shown in Upper Paleolithic specimens represents a different pattern of attrition to that shown in later material. In any case, even very heavy wear would not reduce the visibility of congenitally missing or impacted third molars, as shown in Dolní Věstonice 3, 15, and 16. It has been suggested that the rise of malocclusions through the Holocene might be explained by the adoption of a softer diet, creating less functional stimulation of the jaws and thus perhaps producing a smaller jaw that was less able to accommodate the teeth (Corruccini, 1991). If this were so, it might also be expected that the teeth of people eating a softer diet would show a different pattern of wear, and as has been observed, this does not seem to be the case. It is possible that anomalies such as third molar agenesis are part of an evolutionary trend toward reduction in the size of the dentition shown by specimens through Upper Paleolithic and Mesolithic contexts. If this is so, then the high rate of third molar agenesis at Dolní Věstonice complicates discussion of the trend. The variation in third molar agenesis between living populations could be taken to suggest an inherited component, and family studies support the heritability of agenesis (Grahnén, 1956; Alt & Türp, 1998). It may quite possibly be that the Dolní Věstonice remains represent a related group, although it must be remembered that Dolní Věstonice 16 was not part of the triple burial, which contained Dolní Věstonice 13–15.

Dental Wear

The pattern of wear in those specimens with more than one permanent tooth in situ is shown in Table 11.8. All of the individuals show a strong gradient of wear, with the heaviest wear at the front of the mouth and the lightest at the back. This is seen particularly strongly in the

Table 11.8 Occlusal attrition scores for Dolní Věstonice and Pavlov associated permanent dentitions, following the system of B. H. Smith (1984).

		Left								Right							
		M 3	M 2	M 1	P 2	P 1	C	I 2	I 1	I 1	I 2	C	P 1	P 2	M 1	M 2	M 3
DV 3	Upper	1	5	7	6	6	5	7	7	7	7	5	6	6	7	4	3
	Lower	—	6	6	6	6	6	7	7	7	7	6	6	6	6	4	3
DV 13	Upper	0	2	3		5	5		5					3	3	2	0
	Lower	0	2	3	2	3	4	4	5	5	4	4	3	2	3	2	0
DV 14	Upper	0	2	3	2	3	3	3	4	4	3	3	3	2	3	2	0
	Lower	0	2	3	2	2	3	3	4	4	3	3	2	2	3	2	0
DV 15	Upper	1	2	3/4	3	3/4	3/4	3	4	4	3	¾	3	3	3/4	2	1
	Lower	1	2	4	2	2	3			4	3	3	2	2	4	2	0
DV 16	Upper	1	7	>8	>8	>8	>8	—				>8	>8	>8	>8	7	1
	Lower	7	7	7	8	8	8	8	8	8	8	8	7	7	7	0[a]	
Pav 1	Upper		5	7	5	6	6	7		7	7	6	6	5	7	5	
	Lower	2	4	7	5	6		5	8	7	7	6		5	6	4	
Pav 2	Upper			4	5	3							5	4	5	3	
Pav 3	Lower											6	7	3	5		

For attrition scores on isolated teeth, see text. Blank: tooth absent; —: crown broken away or congenitally missing.
[a]The tooth is unerupted and in an anomalous position and hence nonocclusal.

less worn dentitions of the younger individuals, but it can also be made out clearly in the heavily worn dentitions of Dolní Věstonice 3 and 16. These gradients are characteristic of hunter-gatherer dentitions (Hinton, 1981; Hillson, 2000, 2001) and would be expected for material from Upper Paleolithic contexts in Europe. In the older individuals, wear has removed the crown and proceeded down the roots until only a vestige is left, particularly in the incisors. As wear progressed, the teeth would have continued to erupt, and eventually the teeth would have become unstable and fallen out. This would be the normal condition throughout most of the history of modern humans. The much less rapid wear of most living people is a very recent phenomenon.

Irregularities of Occlusal Wear

Most of the occlusal wear facets are smooth and evenly curved, usually with a higher rim of the harder enamel around a concave area of the softer dentine. Where the enamel rim is breached, for example, in Dolní Věstonice 16 (Figures 11.8 and 11.9), the facets curve down onto the roots on the side of the heaviest wear. In the lower right first molar of Pavlov 1 (Figure 11.15), the mesiobuccal corner is missing, with a particularly strongly curved attrition facet, which probably represents a fracture of the tooth during life. The lower left first molar of the same individual has two broad grooves marking its occlusal attrition facet, running from lingual to buccal (Figure 11.30 and 11.32). These are not uncommon in archeological material and they are thought to represent the holding of material between the teeth during the processing of food or perhaps the manufacture of artifacts (Milner & Larsen, 1991). Less marked scratches are seen on the occlusal surface of the upper left second incisor in Dolní Věstonice 3.

Buccal Wear Facets

Several individuals from the Dolní Věstonice and Pavlov assemblages show a highly unusual pattern of attrition facets on the buccal surfaces of their molars, premolars, and canines (Tables 11.9 to 11.11). These facets have been described before (Vlček, 1991, 1997) for the permanent teeth of Dolní Věstonice 13–15 and Pavlov 1 and 2. Other examples have been described from Předmostí and Brno 3 (Matiegka, 1934; Vlček, 1997). The present study has added Dolní Věstonice 3 and the deciduous molars of Pavlov 6 and Dolní Věstonice 10. The buccal wear facets (Figures 11.30, 11.32, and 11.33) show a number of characteristics:

1. In permanent teeth, they are found mainly in the first molars, second premolars, and first premolars—sometimes with the addition of the second molars and canines.
2. The facets line up together to make a plane of wear that crosses neighboring teeth.
3. The wear plane is not completely flat but usually shows a slight convex curvature.
4. This plane of wear usually includes both upper and lower teeth, but it is confined to the upper teeth in Dolní Věstonice 13 and on the right side of Pavlov 1.
5. It may be on the left side only (Dolní Věstonice 3), right side only (Dolní Věstonice 13–15), or both sides (Pavlov 1 and 2).
6. Both deciduous and permanent dentitions may be involved.

This pattern of buccal attrition is highly unusual. It seems clear that some hard object was coming into contact with the buccal surfaces of the teeth. The most similar wear facets seen elsewhere in the world are those caused by the wearing of labrets (lip plugs) by the Native Americans of the Northwest Pacific and Arctic coasts of the United States and Canada. The labrets are made of stone, bone, ivory or wood and are worn in incisions in the lips. Most labret facets are found on the lower canines and incisors, sometimes including the first premolars. The lower teeth are thought to be involved because they move relative to the lips, whereas the upper teeth are station-

Figure 11.32 Pavlov 1 buccal wear facets seen in occlusal view. In the upper left cheek tooth row (left), buccal facets on canines, both premolars and both molars line up along a gently curving plane. In the lower left cheek teeth, the facets on both premolars and molars are aligned on a similar plane. It is also possible to make out grooves running from buccal to lingual across the occlusal attrition facets of the second incisor, premolars, and first molar.

Figure 11.33 Dolní Věstonice 14 buccal wear facets. On the left is a buccal view of the lower right cheek tooth row, showing facets well developed on the first molar, second premolar and canine. On the right is an occlusal view of the same teeth, showing facets on first molar, both premolars, and canine, which are approximately aligned along one curved facet (some of the teeth may be out of position because of the reconstruction of the jaw).

ary. The form of the wear facets is very similar to those of Dolní Věstonice and Pavlov, with a broad plane of wear that includes several teeth. It does, however, seem unlikely that a large cheek plug could cause the Dolní Věstonice and Pavlov buccal wear. First, it would need to be a very large stone, bone, or wood plug, necessitating a large incision in the cheek. Second, the wear facets are seen in the Pavlov 6 deciduous teeth (Figure 11.19). The labrets of the American Northwest were started as small plugs, inserted in small incisions. They were gradually increased in size, so the largest labrets would tend to be in adults. Third, the wear facets are seen mostly in both upper and lower teeth or just in the upper teeth. This is unlike the pattern seen in recent labret facets, which seem to be related to the movement of the lower teeth. It seems more likely that the hard object was held loosely inside the cheek. It has been suggested (Vlček, 1997) that pebbles might have been held in the cheek to promote salivation and help alleviate the effects of thirst. This is possible, but a range of possibilities could be considered.

The cheeks might, for example, have been used to hold an implement or in the processing of a food or material.

Approximal (Interproximal) Grooving

The anomalous tooth Pavlov 21 (Figure 11.24) has a groove running transversely below its approximal wear facet, just below the cement enamel junction at the neck of the tooth (Figure 11.34). Scratches in the groove run along its length, showing that it resulted from the passage of some object through the gap between this tooth and its neighbor. Such grooves have been recognized at a number of sites (Ubelaker et al., 1969; Formicola, 1991; Milner & Larsen, 1991; Bermudez de Castro et al., 1997; Ungar et al., 2001; Lebel & Trinkaus, 2002b). Various interpretations are offered, including the use of toothpicks or threads for cleaning teeth or the processing of fibers by pulling them between the teeth (Brown & Molnar, 1990; Ungar et al., 2001; Hlusko, 2003). If Pavlov 21 was in an anomalous position (see above), it might have created a

Table 11.9 Locations of buccal attrition facets on the Dolní Věstonice permanent teeth.

		Left								Right							
		M3	M2	M1	P2	P1	C	I2	I1	I1	I2	C	P1	P2	M1	M2	M3
DV 3	Upper	Abs	Abs	Pres	Pres	Pres	Abs	Abs	Abs	Abs	Abs	Abs	Abs	Abs	Abs	Abs	Abs
	Lower		Abs	Abs	Pres	Pres	Abs	Abs	Abs	Abs	Abs	Abs	Abs	Abs	Abs	Abs	Abs
DV 13	Upper	Abs	Abs	Abs	Abs	Abs	Abs		Abs				Pres	Pres	Pres	—	
	Lower	Abs	Abs	Abs	Abs	Abs	Abs	Abs	Abs	Abs	Abs	Abs	Abs	Abs	Abs	Abs	Abs
DV 14	Upper	—	Abs	Abs	Abs	Abs	Abs	Abs	Abs	Abs	Abs	Pres	Pres	Pres	Pres	Abs	—
	Lower	—	Abs	Abs	Abs	Abs	Abs	Abs	Abs	Abs	Abs	Pres	Pres	Pres	Pres	Abs	—
DV 15	Upper	—	Abs	Abs	Abs	Abs	Abs	Abs	Abs	Abs	Abs	Pres	Pres	Pres	Pres	Pres	—
	Lower	—	Abs	Abs	Abs	Abs	Abs			Abs	Abs	Abs	Pres	Pres	Pres	Abs	—
DV 31	Lower																Abs
DV 32	Lower	Abs															
DV 37	Lower		Abs														

Key to symbols: blank: tooth absent; —: tooth present but insufficient (from wear, damage, or insufficient eruption) to show whether faceting was present; Abs: tooth present with buccal surfaces but no buccal attrition facet; Pres: buccal facet present. Isolated teeth that are insufficiently preserved are not listed.

Table 11.10 Locations of buccal attrition facets on the Pavlov permanent teeth.

		Left								Right							
		M3	M2	M1	P2	P1	C	I2	I1	I1	I2	C	P1	P2	M1	M2	M3
Pav 1	Upper		Pres	Pres	Pres	Pres	Pres	Abs		Abs	Abs	Pres	Pres	Pres	Pres	Pres	
	Lower	Abs	Pres	Pres	Pres	Pres		Abs	Abs	Abs	Abs	Abs		—	Abs	Abs	
Pav 2	Upper		Pres	Pres									Pres?	Pres	Pres	Pres	
Pav 3	Lower											Abs	Abs	Abs	—		
Pav 5	Lower							Abs	Abs								
Pav 20	Upper	Abs															
Pav 22	Upper							Abs									
Pav 23	Lower								Abs								
Pav 25	Lower									Abs							
Pav 28	Lower	Abs															

Key to symbols: blank: tooth absent; —: tooth present but insufficient (from wear, damage, or insufficient eruption) to show whether facetting was present; Abs: tooth present with buccal surfaces but no buccal attrition facet; Pres: buccal facet present. Isolated teeth that are insufficiently preserved are not listed.

food trap that required cleaning with a toothpick or a fiber.

Exfoliated Deciduous Teeth

The Dolní Věstonice and especially Pavlov dental samples are unusual for the large number of exfoliated deciduous tooth crowns. It is possible that many similarly lost teeth exist in Paleolithic dental samples but have not been so identified (but see Alciati et al., 1997; Trinkaus et al., 2000a). This suggests that the immature individuals of these Pavlovian populations spent considerable time at these sites. Moreover, the association of the Pavlov 6 upper first and second deciduous molars, and the possible association of the Pavlov 7 and 8 lower second molars with each other and with those of Pavlov 6, might indicate more long-term and/or frequent occupation of these sites. Reliable data do not exist on the timing and variability of timing of the exfoliation of deciduous teeth, and limited personal experience suggests that it is quite variable. But there is likely to have been some lapse of time between the losses of the Pavlov 6 deciduous molars, and more if Pavlov 7 and 8 derive from the same individual, reinforcing the interpretation (Chapter 2) that these sites on the Pavlovské Hills were a repeated focus of habitation by these Pavlovian human populations.

Summary

Dolní Věstonice and Pavlov provide an important and large Early Upper Paleolithic dental assemblage, both from the permanent and deciduous dentitions. One of the par-

Table 11.11 Locations of buccal attrition facets on Dolní Věstonice and Pavlov deciduous teeth. Isolated teeth that are insufficiently preserved are not listed.

		Left				Right			
		dm2	dm1	dc	di2	di2	dc	dm1	dm2
DV 10	Upper		Pres						
DV 27	Upper				Abs?				
Pav 6	Upper	Pres	Pres						
Pav 7	Lower	Abs							
Pav 8	Lower								Abs
Pav 9	Lower								Abs
Pav 11	Upper							Abs	
Pav 12	Upper								Abs
Pav 14	Upper			Abs					
Pav 16	Upper						Abs		
Pav 17	Lower						Abs		
Pav 18	Lower					Abs			

Dental Morphology, Proportions, and Attrition

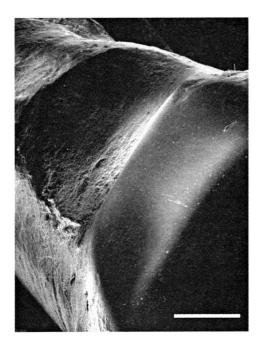

Figure 11.34 Pavlov 21, approximal groove. The image is of an epoxy resin replica (Epotek 301), using a Hitachi S-570 scanning electron mode and the Everhart-Thornley detector. The scale bar represents 1 mm. The crown of the tooth is at the lower right, with the root running to the upper left. On the crown is the approximal wear facet, which has two depressions, suggesting that Pavlov 21 might have been in contact with two other teeth. Above it is the deep approximal groove, with scratches running from side to side across the root.

ticularly useful aspects is the presence of a group of young adults—Dolní Věstonice 13–15—with relatively unworn teeth, which allow the crown morphology to be seen clearly. There is slight variation in dental development among them, relative to skeletal development (Chapter 6). Dolní Věstonice 14 represents an individual a few years younger than Dolní Věstonice 13 and 15, but there is nonetheless a small difference in wear, which gives an idea of the rate at which attrition must have proceeded even in late teenage to young adult life. It is also possible to contrast the dental condition of these younger individuals with the middle-aged adults Dolní Věstonice 3 and 16 and Pavlov 1. These dentitions show heavy attrition, with most of the crowns lost, especially in anterior teeth. In the oldest adult, Dolní Věstonice 16, some of the teeth had been reduced to small root stumps, which with a little more wear and compensating eruption would have become loose and lost. There is evidence of inflammatory conditions in the alveolar process of Dolní Věstonice 16, but these seem to relate primarily to heavy wear and fracturing of teeth. As with the other dentitions (Chapter 19), there is little evidence of dental caries or advanced periodontal disease, which in modern human populations would be a major factor in tooth loss among older adults. This matches closely with the condition of recent hunting, fishing, and gathering people in Greenland, northern North America, Australia, and southern Africa. The Dolní Věstonice and Pavlov human remains therefore show a remarkable range of dental development, from dentitions that were almost fully established to those that were in midterm use and to one that was approaching the point at which teeth had started to wear out.

With the exception of Dolní Věstonice 36, most of the deciduous teeth were probably exfoliated and lost from living children. It is interesting to speculate how they got into the site. The two isolated and exfoliated Dolní Věstonice deciduous teeth (Dolní Věstonice 10 and 27) could represent one or two individuals. The absolute minimum number of children represented by the fifteen isolated and exfoliated deciduous teeth from Pavlov would be three because there are three deciduous lower right second molars (Pavlov 7–9). The maximum number of children would be fourteen because two of the teeth match well (Pavlov 6). Different teeth may survive in the mouth of a child for longer than others, so it is difficult to be sure that teeth of apparently mismatching wear actually represent different children. This number of exfoliated teeth is unusual for an archeological site of any age. There is no particular season in which human deciduous teeth are exfoliated, and in any case there is a variety of different tooth types, which are shed at a range of ages. Perhaps the exfoliated teeth were kept and carried about before being deposited on the sites. It may simply be, however, that the sites were occupied by enough people, over a sufficiently long period, to incorporate many shed teeth among the refuse.

Another factor in the importance of the Dolní Věstonice and Pavlov dental assemblage is the number of occlusal anomalies, including congenitally missing third molars in both Dolní Věstonice 3 and 16, impacted third molars in Dolní Věstonice 15 and 16, and a supernumerary tooth in Dolní Věstonice 15. There are earlier examples of occlusal anomalies in the human fossil record, but these are very few, and Dolní Věstonice is the earliest site at which they are common. It also stands out among other Upper Paleolithic assemblages, even though the morphology of the teeth otherwise fits well within the range of this material.

One final unique feature is the presence of buccal wear facets on no fewer than eight individuals from Dolní Věstonice and Pavlov. They involve substantial loss of enamel (and even some dentine), affecting both upper and lower teeth, left and right, and children as well as adults. Nothing quite like them is seen in any other archeological assemblage, and it clearly involved some unusual behavior or activity.

12

Body Proportions

Trenton W. Holliday

Over the last two decades, the study of the body proportions of Late Pleistocene humans has taken on new significance. Specifically, because of their observed relationship with climate among recent humans (Newman, 1953; Roberts, 1978) and the fact that, whereas they exhibit some phenotypic plasticity, there nonetheless appears to be a strong genetic component (Tanner et al., 1982: 419–420; Katzmarzyk & Leonard, 1998; but see Bogin & Rios, 2003), body proportions have been used as phylogenetic markers to study evolutionarily short-term events, such as the emergence of anatomically modern humans in Europe (Trinkaus, 1981; Ruff, 1994; Holliday, 1997a, 2000a,b). However, with few exceptions (e.g., relative femoral head diameter), the study of body proportions requires relatively complete associated skeletons—a rarity in the human fossil record. Yet the Gravettian human skeletons from Dolní Věstonice and Pavlov include the remains of six individuals (Dolní Věstonice 3 and 13–16, and Pavlov 1) who provide data from more than one limb bone, allowing the examination of body shape.

Materials and Methods

In order to place the body proportions of the Dolní Věstonice and Pavlov sample in its appropriate anatomical context, the fossils are first compared to a large sample of recent humans from Europe and sub-Saharan Africa (maximum N = 710; see Holliday, 1995 for details). They come from Bosnia, Central African Republic, Congo, Czech Republic, England, France, Germany, Kenya, Scotland, South Africa, Sudan, Uganda, and former colonial French West Africa. The samples range in date from ca. 3500 B.P. to the first half of the twentieth century. It should be noted that in Holliday (1995) the 3500-year-old sample from Kerma, Sudan, was grouped with North Africans for statistical analyses. In the current analysis, they are classified as sub-Saharan Africans, which has had the effect of lowering the sub-Saharan African intralimb and limb/trunk index means. However, given that the boundary between North and sub-Saharan Africa is an arbitrary one and that the sample in question comes from a site just south of the Nile's third cataract, the sample's assignment to either North or sub-Saharan African could be justified. For clarity, other North African samples have been excluded from the present analysis; for the majority of limb and body proportions, they are intermediate between Europeans and sub-Saharan Africans.

The Late Pleistocene fossils to which the Dolní Věstonice and Pavlov skeletons are compared include six European Neandertals (La Chapelle-aux-Saints 1, La Ferrassie 1 and 2, Neandertal 1, Regourdou 1, and Spy 2), and twenty Middle Upper Paleolithic (Gravettian; i.e., 20,000 to 28,000 B.P.) skeletons [Arene Candide IP, Baousso da Torre 1; Caviglione 1; Barma Grande 2; Cro-Magnon 3; Fanciulli (Grotte des Enfants) 4–6; Paglicci 25; Pataud 5; Paviland 1; La Rochette 1; Předmostí 1, 3, 4, 9, 10, and 14; and Veneri (Parabita) 1 and 2]. These data were taken on the original specimens, with the exceptions of Předmostí (Matiegka, 1938) and Baousso da Torre and Barma del Caviglione (Trinkaus, 1981).

Bivariate relationships that reflect differences in body shape are elucidated via reduced major axis (RMA) slopes and fossil percentage deviations (i.e., "prediction errors," following R. J. Smith, 1980) from these slopes. In addition, several indices represent three anatomical "complexes": (1) intralimb proportions, (2) relative body linearity, and (3) limb-trunk proportions. The Dolní Věstonice and

Pavlov indices are compared to those of the recent human samples by using a *t* test in which a single individual is compared to a sample (Sokal & Rohlf, 1981: 231).

The intralimb proportion indices are the brachial (radius length/humerus length) and crural (tibia length/femur length) indices. These indices are a general reflection of distal limb segment elongation, although the proximal limb segment shows considerable intergroup variability as well (see Holliday & Ruff, 2001). High brachial and crural indices are interpreted as reflective of long radii (forearms) and long tibiae (legs), respectively. Relative body linearity is examined through femoral length regressed on femoral head anteroposterior (A-P) head diameter (A-P femoral head diameter/femur length indices are also calculated), and body breadth (as reflected in bi-iliac breadth) on stature. Ruff (1991, 1993, 1994) has used this measure repeatedly for fossil human body shape. One means by which he has examined patterning in relative bi-iliac breadth is to regress the bi-iliac breadth/stature index on stature itself. In the current analysis, a similar bivariate operation is performed by using a skeletal proxy of stature (see below). All indices are computed preferentially on the long bones from the left side, but the right side is used in cases where the left measurement is not available. Given the relatively low levels (with the exception of the clavicle) of long bone length asymmetry in humans (Trinkaus et al., 1994), this preferential use of one side should not affect the results.

Four limb/trunk proportion indices have been calculated. These are the lengths of the humerus, radius, femur, and tibia divided by skeletal trunk height (the summed dorsal vertebral body heights of T1 through L5 plus sacral ventral length; Franciscus & Holliday, 1992). Skeletal trunk height (or STH) is often predicted from those elements that a particular individual preserves (i.e., the individual's partial trunk height, or PTH), using a least-squares regression based on a complete recent human series. A predicted skeletal trunk height is only used if the standard error of prediction for an individual is less than 3% of the prediction itself. The prediction formulae used to estimate skeletal trunk height for most of the fossil sample are found in Holliday (1995, 1997a). The predicted skeletal trunk heights of those Dolní Věstonice individuals for which sufficient vertebral remains are preserved were estimated by using the regression equations below.

Dolní Věstonice 3 has a poorly preserved vertebral column but nonetheless preserves the neural canal imprint of the dorsal body heights of T2–T4; the actual dorsal body heights of L4 and L5; and the ventral body heights of T9–T12. The summed total of these elements yields a partial trunk height (PTH) of 174.4 mm. When skeletal trunk height is regressed on this PTH for a complete recent human series ($N = 45$), the following predictive equation results:

$$\hat{Y} = 2.237x + 57.128; r = 0.95$$

This equation results in a predicted skeletal trunk height for Dolní Věstonice 3 of 447.2 mm, with an individual standard error of the estimate of 12.2 mm, or 2.7% of the predicted total. Given this result, the predicted skeletal trunk height of Dolní Věstonice 3 was deemed reliable.

Dolní Věstonice 13 has a relatively complete vertebral column, and the dorsal body heights of T7–T10, L3–L5, and sacral ventral length were summed to produce a PTH of 267 mm. When skeletal trunk height is regressed on this PTH for the same complete recent human series, the following predictive equation results:

$$\hat{Y} = 1.6925x + 30.9462; r = 0.97$$

This yields a predicted skeletal trunk height of 482.9 mm, with a standard error of prediction of 8.9 mm, or 1.8% of the predicted measurement. Therefore the estimate of Dolní Věstonice 13's skeletal trunk height was deemed reliable.

Dolní Věstonice 15 also has a relatively complete vertebral column, and its partial trunk height is computed as the summed dorsal body heights of T2, T3, T7, T8, T12, and L1–L3, plus the summed ventral body heights of T1, T10, and T11 and the sacral ventral length (S1–S5). This computation produces a PTH of 332.1. When skeletal trunk height is regressed on this PTH for the same recent human series, the following predictive equation results:

$$\hat{Y} = 1.3962x + 5.4414; r = 0.99$$

This yields a predicted STH of 469.1 mm, with an individual standard error of prediction of 4.9 mm, or 1.0% of the predicted measurement. Thus, the estimate of skeletal trunk height for Dolní Věstonice 15 was also deemed reliable.

Intralimb Proportions

Brachial and crural index summary statistics for the comparative sample and the Dolní Věstonice and Pavlov humans are found in Table 12.1. As a whole, Dolní Věstonice and Pavlov are intermediate between recent Europeans (who have relatively low indices) and recent sub-Saharan Africans (who have relatively high indices). With regard to specific individuals, Dolní Věstonice 3 and 14 and Pavlov 1 are more similar to the Europeans, whereas Dolní Věstonice 13, 15, and 16 are more similar to the Africans. Relative to other Late Pleistocene fossils, the Dolní Věstonice and Pavlov specimens closely resemble other Gravettian humans in their brachial index values, but they are quite distinct from the European Neandertals, who have long been known to be characterized by abbreviated distal limb segments (Coon, 1962; Badoux, 1965). However, a *t* test fails to find any statistically significant differences in brachial indices between the Dolní Věstonice and Pavlov individuals and either the recent sub-Saharan Africans or Europeans. This is perhaps unsurprising since none of their values would be extremely rare in either group. For example, Dolní Věstonice 3's brachial index falls in the eighty-fourth percentile of the recent European females, and the thirty-first percentile of the recent sub-Saharan African females. Dolní Věstonice 13–16 have brachial indices that fall at the seventy-fourth, sixty-first, seventy-sixth, and ninety-fourth percentiles of recent European males, respectively. In the same order, these individuals fall at the twenty-third, eighteenth, twenty-fifth, and fifty-fourth percentiles of the recent sub-Saharan African males. Pavlov 1 has the lowest brachial index of the group; it falls at the fifty-fourth percentile of the recent European males and the fifteenth percentile of recent sub-Saharan African males. A graphical representation of fossil brachial indices is presented in Figure 12.1. Note that there is little overlap in range between the European Neandertals and the Dolní Věstonice and Pavlov individuals: Neandertal 1 has a slightly higher brachial index (76.1) than Pavlov 1 (75.7); in contrast, the entire Dolní Věstonice and Pavlov sample falls within the range of the European Gravettian group.

For the crural index (Table 12.1), the Dolní Věstonice individuals present a more complex pattern. First, with the exception of the pathological Dolní Věstonice 15 (Chapter 19), the Dolní Věstonice crural index values would not be uncommon in either of the recent human samples. For example, Dolní Věstonice 3's crural index falls at the sixty-sixth percentile of recent sub-Saharan African females and the ninety-fourth percentile of recent European females. The crural indices of Dolní Věstonice 13–16 fall at the sixty-ninth, ninth, ninety-fourth, and seventeenth percentiles of recent sub-Saharan African males, and the ninety-eighth, forty-seventh, ninety-ninth and sixtieth percentiles of the recent European males. The pattern that emerges, then, is that Dolní Věstonice 14 and 16 have lower crural indices, and thus fall closer to the recent European mean. In contrast, Dolní Věstonice 3 and 13 have slightly higher crural indices than the recent sub-Saharan African mean, and individuals with their index values are rare in the European sample. The crural index of Dolní Věstonice 15 is higher, but its value is largely a function of the pathological condition of its femora and their reduced lengths. In fact, the only statistically significant difference uncovered by the single-individual-to-sample *t* test is that Dolní Věstonice 15 has a significantly higher crural index than the recent European sample ($P = 0.004$). With regard to the fossils, the combined Gravettian sample has a mean nearly identical to that of the recent sub-Saharan Africans. Note, too, the long-recognized low crural index mean of the European Neandertals. In Figure 12.2 the Dolní Věstonice and Pavlov humans are similar to their Gravettian cohort. Dolní Věstonice 14 falls just beyond the lower range of

Table 12.1 Summary statistics for brachial, crural, and relative femoral head diameter and relative bi-iliac breadth[a] indices for the Dolní Věstonice and Pavlov sample, European Neandertals, Gravettian-associated humans, and recent humans.

	Brachial Index	Crural Index	Relative Femoral Head Diameter	Relative Bi-iliac Breadth
Recent Europeans	75.0 ± 2.5 (391)	82.7 ± 2.4 (436)	10.4 ± 0.6 (233)	21.2 ± 1.2 (103)
Recent sub-Saharan Africans	78.6 ± 2.8 (152)	85.3 ± 2.4 (158)	9.5 ± 0.6 (131)	19.3 ± 1.5 (72)
European Neandertals	73.2 ± 2.5 (5)	78.7 ± 1.6 (4)	12.1 ± 0.5 (5)	23.9
Total Gravettian[b]	77.7 ± 2.0 (20)	85.1 ± 1.8 (18)	10.3 ± 0.6 (19)	20.0 ± 0.9 (6)
Dolní Věstonice 3	76.7	86.0	9.7	—
Dolní Věstonice 13	77.1	86.6	10.9[d]	19.4
Dolní Věstonice 14	76.2	82.5	10.1	—
Dolní Věstonice 15	77.3	89.6[c]	12.4[d]	20.5
Dolní Věstonice 16	79.2	83.5	10.8[d]	—
Pavlov 1	75.7	—	—	—

Mean ± standard deviation (N) provided for samples.
[a] Computed as bi-iliac breadth/(femoral length + tibial length + skeletal trunk height).
[b] Includes all Gravettian-associated skeletons, both Czech and non-Czech, excluding the pathological Dolní Věstonice 15.
[c] Significantly different from the recent Europeans at $P < 0.05$.
[d] Significantly different from the recent sub-Saharan Africans at $P < 0.05$.

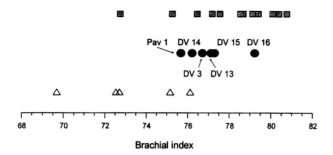

Figure 12.1 Plot of brachial indices of European Neandertals (open triangles), Dolní Věstonice and Pavlov specimens (black circles), and other Gravettian humans (gray squares).

the other Gravettian individuals, whereas the pathological Dolní Věstonice 15 individual falls above their high-end range. There is no overlap between the European Neandertals and any of the Gravettian individuals.

Relative Body Linearity

Femoral head diameter is highly correlated with body mass (Ruff, 1990; Jungers, 1991; Ruff et al., 1991), and femoral length is highly correlated with stature and lower limb length (Konigsberg et al., 1998). Since Bergmann's rule predicts that individuals within a population who live in colder climates tend to exhibit elevated body mass, they should also exhibit large femoral heads. Yet, stature shows no significant relationship with either temperature or latitude (Roberts, 1978). The problem arises when populations adhere to the two most commonly cited ecological rules (Bergmann's and Allen's rules) because they predict different stature outcomes in cold regions (Newman, 1953). Larger body size is expected in members of a widespread population living in cold climates, and stature is one measure of body size. Yet at the same time, Allen's rule predicts that individuals in colder regions should exhibit foreshortened limbs, and foreshortened limbs lead to a reduction in stature. The end result seems to be that adherence to each of these two rules cancels out the other, explaining the zero slope that stature tends to show when regressed on latitude (Ruff, 1994). Nonetheless, empirically it has been seen that cold-climate groups and individuals tend to be characterized by high ratios of femoral head to femoral length (Holliday, unpublished data), and therefore, the ratio is a useful measure of body shape of fossil skeletons, especially since it can be computed from a single bone.

Summary statistics for relative femoral head size (femoral head A-P diameter/femur length) are in Table 12.1. Recent Europeans have high index values, indicating that they possess larger femoral heads relative to the length of their femora than do recent sub-Saharan Africans. Interestingly, the Dolní Věstonice and Pavlov individuals tend to much more closely resemble the recent Europeans than the recent Africans in this regard, with index values that straddle the recent European mean. Despite the fact that Dolní Věstonice 3 appears to lie closer to the sub-Saharan African mean than the European mean, its index value falls at the eighty-sixth percentile of recent sub-Saharan African females and the twenty-fifth percentile of recent European females. The high value for Dolní Věstonice 15 is due to his pathological condition. The t tests indicate that Dolní Věstonice 13, 14, and 16 have significantly higher relative femoral head size indices than recent sub-Saharan Africans (Dolní Věstonice 13 falls outside their range and Dolní Věstonice 14 and 16 fall at the eightieth and ninety-ninth percentile of the recent African males, respectively), and the pathological Dolní Věstonice 15 has a significantly higher relative femoral head size index than either sub-Saharan Africans ($P < 0.001$) or recent Europeans ($P = 0.017$).

In terms of the other fossils, as a whole the Gravettian specimens look more similar to recent Europeans in relative femoral head size than to recent sub-Saharan Africans. Note, in contrast, the extremely high relative femoral head size indices of the European Neandertal sample. These relationships are examined graphically in Figure 12.3. The European Neandertals fall well below both recent human regression lines. With the exception of Dolní Věstonice 15, the Dolní Věstonice sample falls near the recent European RMA slope. Dolní Věstonice 3 and 14 fall between the two recent human slopes, and Dolní Věstonice 13, 15, and 16 fall below the European slope, far from the African RMA line. Likewise, most of the other Gravettian individuals fall about the recent European RMA slope, and only one Gravettian individual (Paglicci 25) falls above the sub-Saharan African line.

The resultant "prediction errors" from the recent human regression lines (Table 12.2) show negative deviations for all of the Dolní Věstonice individuals from the recent sub-Saharan African line. Dolní Věstonice 3 and 14 have small positive deviations from the recent European

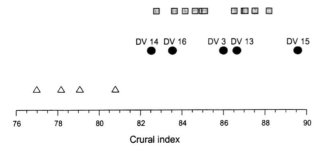

Figure 12.2 Plot of crural indices of European Neandertals (open triangles), Dolní Věstonice and Pavlov specimens (black circles), and other Gravettian humans (gray squares).

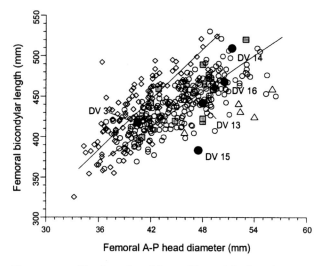

Figure 12.3 Bivariate plot of femoral bicondylar length (mm) regressed on anteroposterior femoral head diameter (mm) for recent sub-Saharan Africans (small diamonds), recent Europeans (small circles), the Dolní Věstonice specimens (black circles), European Neandertals (open triangles), and other Gravettian humans (gray squares). Each recent human group's reduced major axis (RMA) slope is indicated.

line, and Dolní Věstonice 13 and 16 show small negative deviations. Dolní Věstonice 15 falls well below both regression lines because of its pathologically shortened femur. The other Gravettian individuals show a similarly intermediate pattern, and the European Neandertals show the largest negative deviations.

A second measure of overall body linearity is the relative bi-iliac breadth index. In living humans this is calculated as (bi-iliac breadth/stature) × 100. Roberts (1978) and Ruff (1991, 1994) have shown that there is ecogeographic variability in this index. Of course, one problem that plagues researchers studying skeletons is that rarely does one have stature recorded for a skeletonized individual, and for prehistoric skeletons it is always unknown. Thus the issue becomes how best to predict stature—an issue about which there is much debate (Feldesman & Lundy, 1988; Formicola & Franceschi, 1996; Holliday & Ruff, 1997; Konigsberg et al., 1998). For skeletons, Fully's (1956) anatomical method is the most accurate in use (Formicola, 2003), but it is only applicable to very complete skeletons. Rather than predict stature, a proxy variable was chosen—one that represents the lion's share of human stature. This measure is the combined total of skeletal trunk height (STH), femoral bicondylar length, and tibial maximum length. This summed measurement includes all of standing height, except for the intervertebral discs, head, and neck. It has another advantage in that one need not justify which stature prediction formula is the most appropriate for the sample or individuals in question (cf. Holliday & Ruff, 1997).

Note in Table 12.1 that the recent sub-Saharan Africans have lower index values because of their narrow bi-iliac breadths, whereas the Europeans evince higher indices, reflective of their broader trunks. For the fossils, the lone Neandertal (La Chapelle-aux-Saints 1) has the highest value (although it falls within the recent European range). The combined Gravettian sample is intermediate relative to the recent Europeans and Africans, and the two Dolní Věstonice individuals for which the index can be computed (Dolní Věstonice 13 and 15) show different patterns, with Dolní Věstonice 13 falling near the sub-Saharan African mean and Dolní Věstonice 15 falling closer to the European mean. However, t test results fail to discriminate between Dolní Věstonice 13 and 15 and either of the two recent human samples. Dolní Věstonice 13 falls at the seventieth percentile for recent sub-Saharan African males and Dolní Věstonice 15 falls at their eighty-fourth percentile. Dolní Věstonice 13 falls at the European male seventh percentile, and Dolní Věstonice 15 falls at their fortieth percentile.

Figure 12.4 is a scatter plot of relative bi-iliac breadth regressed on the proxy for stature (STH + femoral length

Table 12.2 Percentage deviations (prediction errors or d_{yx}) of Dolní Věstonice, European Neandertal, and other Gravettian femoral lengths from recent European and sub-Saharan African reduced major axis (RMA) slopes of femoral length regressed on femoral anteroposterior head diameter.

	Percentage Deviation from European RMA Slope	Percentage Deviation from sub-Saharan African RMA Slope
Dolní Věstonice 3	+3.9	−2.8
Dolní Věstonice 13	−3.4	−14.9
Dolní Věstonice 14	+5.4	−6.6
Dolní Věstonice 15	−18.2	−31.1
Dolní Věstonice 16	−1.6	−14.1
European Neandertals ($N = 5$)	−9.2 to −18.3	−20.2 to −34.5
Other Gravettian ($N = 13$)	−21.2 to +1.0	−9.1 to +8.7

Dolní Věstonice individuals. Note, too, the extreme position of the La Chapelle-aux-Saints 1 Neandertal.

These patterns are quantified by the use of percentage deviations from the RMA slopes (Table 12.3). The two Dolní Věstonice specimens show intermediate values, although Dolní Věstonice 15 shows only the slightest (+1.9%) deviation from the sub-Saharan African line. Interestingly, the two largest Gravettian individuals (Fanciulli 4 and Barma Grande 2) show higher positive deviations (+20.5% and +21.6%, respectively) from the sub-Saharan African line than does the La Chapelle-aux-Saints Neandertal (+19.7%).

Limb/Trunk Proportions

The humerus/trunk height index (Table 12.4) shows the expected pattern, with recent Europeans evincing low index values (relatively shorter humeri) and recent sub-Saharan Africans having higher index values (relatively longer humeri). The Neandertal mean is almost identical to that of the recent Europeans, and the Gravettian mean is almost identical to that of the recent sub-Saharan African mean. Dolní Věstonice 3 and 13 evince high indices close to the African mean (they fall at the fifty-sixth percentile and fifty-first percentile of recent Africans of the same sex, respectively), whereas the pathological Dolní Věstonice 15 evinces a relatively low index (his value falls at the eleventh percentile of recent African males and the fifty-fourth percentile of recent European males, respectively). However, t tests fail to distinguish any of the Dolní Věstonice individuals from either of the recent human groups; whereas Dolní Věstonice 3 and 13 have high indices, they are not terribly uncommon among Europeans. Dolní Věstonice 3's index value falls at the ninety-sixth percentile of recent European females and Dolní Věstonice 13's index falls at the eighty-ninth percentile of recent European males.

Figure 12.5 is a graphical representation of the humerus length/trunk height index among the fossils. Note

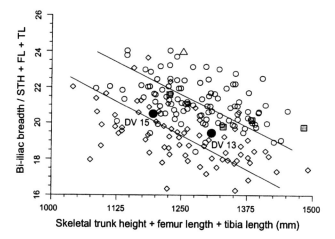

Figure 12.4 Bivariate plot of relative bi-iliac breadth index [bi-iliac breadth/(femur length + tibia length + skeletal trunk height) × 100] regressed on stature "proxy" (femur length + tibia length + skeletal trunk height, in mm) for recent sub-Saharan Africans (small diamonds), recent Europeans (small circles), the Dolní Věstonice specimens (black circles), European Neandertals (open triangles), and other Gravettian humans (gray squares). Each recent human group's reduced major axis (RMA) slope is indicated.

+ tibial length). Although there is overlap between these two groups, note that for any given "stature," the sub-Saharan Africans tend to have relatively narrower pelves than the recent Europeans. With regard to the fossils, Dolní Věstonice 13 and 15 are intermediate between the two recent human groups. Interestingly, whereas Dolní Věstonice 13's relative bi-iliac breadth index value is almost identical to the recent sub-Saharan African mean, it lies between the two regression lines; and whereas the Dolní Věstonice 15 index value is closer to the European mean, in bivariate space the specimen falls closer to the recent sub-Saharan African line than recent European line. The remaining Gravettian individuals fall about the recent European RMA slope; all are farther from the recent sub-Saharan African RMA line than either of the

Table 12.3 Percentage deviations (prediction errors or d_{yx}) of Dolní Věstonice, La Chapelle-aux-Saints 1, and other Gravettian femoral lengths from recent European and sub-Saharan African reduced major axis (RMA) slopes of relative bi-iliac breadth regressed on "stature" (skeletal trunk height + femur length + tibia length).

	Percentage Deviation from European RMA Slope	Percentage Deviation from sub-Saharan African RMA Slope
Dolní Věstonice 13	−7.0	+5.3
Dolní Věstonice 15	−9.5	+1.9
La Chapelle-aux-Saints 1	+9.9	+19.7
Other Gravettian ($N = 4$)	−3.4 to +8.0	+5.3 to +20.5

Table 12.4 Summary statistics for humerus length/trunk height, radius length/trunk height, femur length/trunk height, and tibia length/trunk height indices for the Dolní Věstonice and Pavlov sample, European Neandertals, and Gravettian-associated humans.

	Humerus Length/Trunk Height	Radius Length/Trunk Height	Femur Length/Trunk Height	Tibia Length/Trunk Height
Recent Europeans	63.6 ± 3.4 (124)	47.9 ± 2.8 (123)	88.6 ± 4.4 (123)	73.6 ± 4.3 (124)
Recent sub-Saharan Africans	69.1 ± 4.1 (75)	54.6 ± 4.0 (75)	97.7 ± 6.8 (75)	83.7 ± 6.2 (75)
European Neandertal	62.2, 64.5, 65.1	46.7, 46.9, 42.7	89.1, 89.1	70.4, 71.9
Total Gravettian[a]	69.5 ± 3.6 (8)	54.2 ± 3.1 (8)	95.2 ± 5.1 (8)	82.7 ± 5.3 (7)
Dolní Věstonice 3	69.1	53.0	93.4	80.3
Dolní Věstonice 13	68.8	53.0	91.4	79.2
Dolní Věstonice 15	64.3	49.7	81.6[b]	73.1

[a]Includes all Gravettian-associated skeletons, both Czech and non-Czech, excluding the pathological Dolní Věstonice 15.
[b]Significantly different from the recent sub-Saharan Africans at $P < 0.05$.

that there is little overlap between the European Neandertals and any of the Gravettian individuals. Only the pathological Dolní Věstonice 15 and Paglicci 25 fall within the Neandertal range. Also, aside from Paglicci 25 and Barma Grande 2, Dolní Věstonice 3 and 13 have lower humerus/trunk indices than the other Gravettian individuals.

The summary statistics for the radius length/trunk height index are found in Table 12.4. Here, too, the recent Europeans are characterized by relatively low index values, reflective of relatively shorter radii, and the sub-Saharan Africans are characterized by higher indices, indicative of longer radii. The Neandertals fall very near the recent European mean, and the Gravettian sample mean falls just below the recent sub-Saharan African mean. With the exception of Dolní Věstonice 15, the other Dolní Věstonice individuals appear more "African-like" in their relative radius lengths. The radius length/trunk height index of Dolní Věstonice 3 falls at the forty-fourth percentile of recent sub-Saharan African females and the ninety-eighth percentile of recent European females. Dolní Věstonice 13's and 15's index values fall at the thirty-second and eleventh percentiles of the recent sub-Saharan African males and the ninety-fifth and sixty-sixth percentiles of recent European males, respectively. Despite the fact that the nonpathological specimens appear more "African-like," t tests fail to distinguish any of the Dolní Věstonice individuals from either of the recent human groups.

Figure 12.6 is a graphical representation of the radius length/trunk height indices among the fossils. Once again, the European Neandertals evince the lowest indices, and for this index there is no overlap between the Gravettian sample and the Neandertals—even the pathological Dolní Věstonice 15 individual falls above their range. Dolní Věstonice 15 has the lowest radius/trunk height index of the Gravettian sample, and Dolní Věstonice 3 and 13 fall toward the low end of the Gravettian range; as with the humerus, Paglicci 25 and Barma Grande 2 evince shorter radii than do the remainder of the nonpathological Gravettian sample.

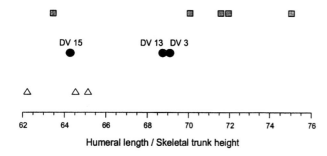

Figure 12.5 Plot of humerus length/skeletal trunk height indices of European Neandertals (open triangles), Dolní Věstonice and Pavlov specimens (black circles), and other Gravettian humans (gray squares).

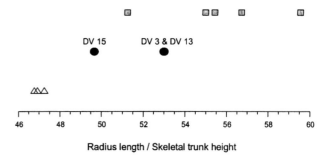

Figure 12.6 Plot of radius length/skeletal trunk height indices of European Neandertals (open triangles), Dolní Věstonice and Pavlov specimens (black circles), and other Gravettian humans (gray squares).

The summary statistics for femur length/trunk height indices (Table 12.4) present a similar picture. The recent Europeans evince lower indices, reflecting relatively shorter femora, whereas the recent sub-Saharan Africans show higher indices as a result of their relatively longer femora. For the fossil groups, the Neandertals fall very near the recent European mean; in contrast, the Gravettian sample falls near the recent sub-Saharan African mean. The Dolní Věstonice humans vary in this regard. Dolní Věstonice 3 and 13 have more "African-like" indices (they fall at the thirty-second percentile and eighteenth percentile of the recent African females and males, respectively), and Dolní Věstonice 15's index is almost 2 standard deviations below the recent European mean. According to t test results none of these three fossils is significantly different from the recent Europeans for this index (Dolní Věstonice 3, 13, and 15 fall at the ninety-first, sixty-eighth, and fourth percentiles of the recent Europeans of the same sex); however, Dolní Věstonice 15 has a significantly lower femur length/trunk height index than the recent sub-Saharan Africans ($P = 0.028$).

The fossil distribution for the femur length/trunk height index is presented graphically in Figure 12.7. The European Neandertal fossils for which this index can be computed (La Chapelle-aux-Saints 1 and La Ferrassie 1) have virtually identical (and low) index values. The Gravettian sample tends to have higher indices, especially Fanciulli 4 and Předmostí 3. An obvious exception to this trend is Dolní Věstonice 15 which because of his pathological femora, has a lower femur length/trunk height index (81.6) than the European Neandertals.

The summary statistics for the tibia length/trunk height index are reported in Table 12.4. As with all the other limb/trunk indices, the recent Europeans evince lower indices, reflective of shorter tibiae, and the recent sub-Saharan Africans have higher indices, reflective of their long tibiae. Among the fossils, the European Neandertals have a lower mean index value than the recent Europeans, which matches the patterning from the crural index, in which their lower limbs evince extreme distal foreshortening. In contrast, the Gravettian mean is very close to the recent sub-Saharan African mean. The three Dolní Věstonice humans for which a tibia length/trunk height index can be computed evince a similar pattern to other limb/trunk indices, with Dolní Věstonice 3 and 13 evincing higher index values (their index values fall at the thirty-second and twenty-sixth percentile, respectively, of the recent sub-Saharan Africans of the same sex), and Dolní Věstonice 15 is characterized by a lower index (his index falls at the sixth and forty-first percentile of sub-Saharan African and European males, respectively). However, according to t tests, none of these specimens' indices is significantly different from either recent human sample (Dolní Věstonice 3 and 13 fall at the ninety-seventh and ninety-third percentiles of the recent European males and females, respectively).

The fossil humans' tibia length/trunk height index is presented graphically in Figure 12.8. All three Dolní Věstonice humans have lower indices than all but one (Barma Grande 2) Gravettian individual, although none of them (including Dolní Věstonice 15) has as low an index value as the European Neandertals.

Summary

The Dolní Věstonice and Pavlov humans, with the sometime exception of the pathological Dolní Věstonice 15, have body proportions similar to those of other Gravettian specimens. Specifically, they are characterized by high brachial and crural indices, indicative of distal limb segment elongation. They also tend to have large femoral heads relative to femoral length, although with the exception of Dolní Věstonice 15, they are not as extreme in this regard as the European Neandertals. In relative bi-iliac breadth, they are intermediate relative to recent

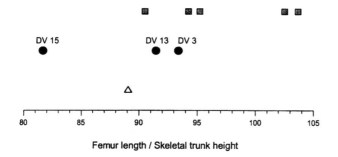

Figure 12.7 Plot of femur length/skeletal trunk height indices of European Neandertals (open triangles), Dolní Věstonice and Pavlov specimens (black circles), and other Gravettian humans (gray squares).

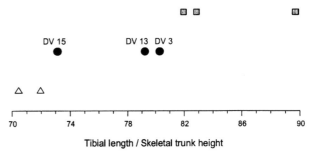

Figure 12.8 Plot of tibia length/skeletal trunk height indices of European Neandertals (open triangles), Dolní Věstonice and Pavlov specimens (black circles), and other Gravettian humans (gray squares).

sub-Saharan Africans and Europeans, although they tend to have slightly narrower pelves than the other Gravettian specimens. In terms of limb/trunk proportions, the specimens for which skeletal trunk height can be predicted (Dolní Věstonice 3, 13, and 15) tend to have different patterns. Dolní Věstonice 3 and 13 more closely resemble other Gravettian specimens, with higher limb/trunk ratios indicative of relatively longer limbs. In contrast, Dolní Věstonice 15 has lower limb/trunk ratios, including the index computed with his seemingly nonpathological tibia (see further discussion in Chapters 13 and 19).

Although most t tests fail to discriminate the individual Dolní Věstonice specimens from either the recent Europeans or recent sub-Saharan Africans, as a whole, in body shape the Gravettian sample (including most of the specimens from Dolní Věstonice and Pavlov) are morphologically closer to the recent Africans than to the recent Europeans. In many cases, recent Europeans of the same sex with index values identical to the Dolní Věstonice and Pavlov individuals are rare indeed. Therefore, the overall pattern that emerges is that the Gravettian humans, despite living in Europe during a glacial period, evince relatively tropically adapted, physiques (Trinkaus, 1981; Ruff, 1994; Holliday, 1997a, 1999). The limb and body proportions of the Dolní Věstonice and (to a lesser degree) Pavlov fossils conform well to this overall pattern.

13

Body Length and Body Mass

Erik Trinkaus

The estimation of overall body length (or stature) and lean body mass from the skeletal remains of Pleistocene humans normally involves several steps of estimation, in which any errors might be (but are not necessarily) compounded. One can therefore query why one would undertake the exercise of estimating these values, rather than merely using the directly measured or minimally estimated dimensions of the appendicular skeleton as indications of body size. The answers are in both the degree of completeness of the remains and the need to have body mass estimations for the appropriate scaling of weight-bearing limb remains.

The Dolní Věstonice 13 to 15 human remains are relatively unusual in preserving at least one of each of the long bones either virtually intact or lacking only portions of the epiphyseal regions, as well as in having complete pelvic remains. In this, they are matched only by several of the Italian Gravettian human remains among their mature contemporaries (e.g., Barma Grande 2, Fanciulli 4 and 5, Ostuni 1, and Paglicci 25); the lost Předmostí 3, 4, 9, 10, and 14 remains; and one Middle Paleolithic specimen (Skhul 4). Among contemporary or older immature remains, only Lagar Velho 1, Sunghir 2 and 3, and KNM-WT 15000 match the completeness of these skeletons.

However, although they preserve sufficiently complete long bones for a reasonable estimation of each appendicular segment length, Dolní Věstonice 3 and 16 lack sufficiently complete pelvic remains for a body breadth measurement. Moreover, in the Pavlov 1 remains only three limb segments are sufficiently intact for reasonable length measurement or estimation and none of the pelvis. In this, these three specimens are more representative of the conditions of preservation of later Pleistocene human remains. For these reasons, in order to systematically compare body size across these samples, it is appropriate to compare both representative long bone lengths and estimations of stature and body mass.

In addition, it has become increasingly documented that assessments of the degree of robusticity (or hypertrophy above the baseline needed for body support) of the weight-bearing limbs require an estimation of body mass. This is due to the principle that the structural rigidity of the anatomical unit under consideration (whether diaphyseal or epiphyseal) is, at a minimum, adapted to support the individual's body mass times the moment arm around which that body mass is operating (Ruff et al., 1993). Moreover, human body form varies ecogeographically, and there were major shifts during the later Pleistocene in the degree of linearity of human populations, as is especially well documented for the northwestern Old World (Trinkaus, 1981; Ruff, 1994; Holliday, 1995, 1997a,b, 1999, 2000a,b; see Chapter 12). As a result, scaling aspects of a long bone solely to its length can seriously distort the perceived degree of hypertrophy of the bone since the length only of a femur, and especially of a tibia of a linear individual with long limbs and a high crural index, will overestimate the baseline load on that limb and suggest a more gracile limb structure, and vice versa for a stocky individual with relatively short limbs. This has been shown in assessments of femoral and tibial diaphyseal strength and of knee biomechanical properties (Ruff et al., 1993, 2000; Trinkaus & Rhoads, 1999; Trinkaus & Ruff, 1999a,b; Trinkaus et al., 1999a; Ruff, 2000b; Trinkaus, 2000c).

Consequently, body lengths (statures) and body masses have been computed for the sufficiently complete Pavlovian human remains and comparative Late Pleistocene samples. For the Pavlovian sample, this is done initially for the Dolní Věstonice 3, 13, 14, and 16 and

Pavlov 1 partial skeletons. It has also been done for the pathological Dolní Věstonice 15 skeleton, but additional considerations are required for it, given its congenitally altered body proportions (Chapter 19). An estimate has also been made for the Dolní Věstonice 35 femoral diaphysis, but solely in order to be able to scale its diaphyseal properties biomechanically.

Comparative Samples

The principal comparative sample for the Pavlovian remains consists of the other Middle Upper Paleolithic (Gravettian) human remains from Europe. This involves both the Pavlovian Předmostí remains (based on the data in Matiegka, 1938) and other remains from sites across Europe. These include specimens from Arene Candide, Baousso da Torre, Barma Grande, Caviglione, Cro-Magnon, Fanciulli (Grotte des Enfants), Paglicci, Pataud, Paviland, La Rochette, Sunghir, Veneri (Parabita), and Willendorf (data from Holliday, 1995; Mallegni et al., 1999, 2000; Kozlovskaya & Mednikova, 2000; Trinkaus, 2000b and personal observation). To these specimens, for the purposes here, have been added the Aurignacian humerus, femur, and two os coxae from Mladeč, providing adequate data (Trinkaus et al., 2005). The Pavlovian Předmostí sample is listed separately and included in the larger European earlier Upper Paleolithic sample. Even though they are smaller and more geographically dispersed, samples of Asian and African earlier Upper Paleolithic specimens have been included. The Asian one includes specimens from Minatogawa, Nahal Ein Gev, and Ohalo, and the African one includes data from Ishango, Kubbaniya, and Nazlet Khater (data from Twiesselmann, 1958; Baba & Endo, 1982; Thoma, 1984; Stewart et al., 1986; Trinkaus, personal observation).

To provide an ancestral background for the Gravettian and especially Pavlovian samples, data are provided for Near Eastern Middle Paleolithic early modern humans from Qafzeh and Skhul (data from McCown & Keith, 1939; Vandermeersch, 1981; Trinkaus, personal observation) and for Near Eastern and European late archaic humans (Neandertals; data from Holliday, 1995; Trinkaus, personal observation).

Body Length Methods

The assessment of relative body length, or stature, involves comparisons of both length measurements of the two principal (and proximal) limb bones, the humerus and the femur, plus estimates of stature (Table 13.1).

The stature estimates were computed by using the formulae provided by Trotter and Gleser (1952) for Euro-Americans and Afro-Americans. In this, the Afro-American formulae are taken to provide a close fit to the relatively linear body proportions of the Qafzeh-Skhul sample, the European earlier Upper Paleolithic humans, and the Near Eastern and African earlier Upper Paleolithic samples. All of them exhibit at least the long, linear leg bones associated with such body proportions and, where documentable, have overall linear body proportions that are closely approximated by the Afro-Americans measured by Trotter and Gleser (see Chapter 12, as well as Holliday, 1997a, 2000b; Trinkaus & Ruff, 1999a,b; Twiesselmann, 1958). The Minatogawa statures were computed by using the Trotter and Gleser Euro-American formulae, as were those of the Neandertals. Given the low crural indices of the Neandertals, in specimens that preserve length only for the tibia (Kiik-Koba 1 and Shanidar 2) their femoral lengths were estimated, based on a sample of Neandertals (FBcL = 1.07 (TMxL) + 69.3; $r^2 = 0.974$; $N = 9$); their statures were then computed by using both their preserved tibial and their estimated femoral lengths. The use of the Euro-American formulae for the Neandertals may slightly underestimate their statures, given the hyperarctic body proportions of these late archaic humans (Trinkaus, 1981; Holliday, 1997b).

It was decided to use only one set of related stature formulae, those of Trotter and Gleser (1952) for documented Euro-Americans and Afro-Americans, fitting them as best as is reasonable to the documented body proportions of the skeletal remains. This was done in order to both account for variation in relative limb and limb segment lengths and provide some consistency across the computed values. The alternative approaches are those of Formicola and colleagues (Formicola & Franceschi, 1996; Formicola & Giannecchini, 1999), in which formulae were developed from associated skeletal remains, and of Vančata (2003), in which a range of formulae pooling data from across the spectrum of recent human populations are utilized and then averaged. The former approach is similar to the one applied here, of using a single set of formulae, except that it does not fully compensate for the variance in limb and limb segment proportions, and its differences with the Trotter and Gleser formulae are small—within documented estimation errors for the original reference samples (Formicola & Giannecchini, 1999). The latter approach also provides minimal differences with alternative techniques (Vančata, 2003), but it does not directly address the issue of body segment length proportions. The issue of body proportions in stature estimation is not new (Holliday & Ruff, 1997), and the use of such generic formulae (or the averaging of a variety of regionally based formulae) can provide significant deviations of the results. At the same time, with all of these estimates, the issue is not whether one is computing the original living statures of these individuals but whether one is providing a consistent set of values for comparisons of the *relative* statures of the individuals in the samples under consideration.

The computation of the statures revealed several fossil cases in which upper limb bones appear relatively short or long compared to the associated leg bones, providing markedly different stature estimates. Since stature is essentially the summed heights of the legs (and feet), vertebral column, and cranium, arm length is not a directly contributing factor. Therefore, stature was computed by two methods, first, by using estimates from as many long bones as provide accurate measures or estimates of their lengths, which were then averaged for the individual (in all cases, right and left long bone lengths, when available, were averaged prior to the computation of statures). This approach maximized sample size. Second, stature was estimated only from the femora and/or tibiae, and the values were averaged if both were available.

For individuals for whom sex can be reasonably assessed, the sex-specific formulae of Trotter and Gleser (1952) were employed. For those individuals whose gender remains uncertain, the male and female values were computed and the results averaged.

Body Mass Methods

The estimation of lean body mass can be done by following two principles. One of them is biomechanical, and the other is geometric.

The first approach to body mass estimation assumes that the articular surfaces of weight-bearing limb articulations respond to the loads placed upon them during development, and that they achieve a size during adolescence that reflects the body mass of the individual at that time. This assumes that the pressure on the articular cartilage and/or the compression on the underlying trabecular bone must remain within a relatively narrow physiological load range, and that subchondral bone growth will respond during development to increases in the habitual load range (Carter & Beaupré, 2001).

This approach has been applied across mammalian species, but it has been applied in particular to the estimation of body mass in recent and fossil humans (e.g., Ruff et al., 1991; McHenry, 1992; Grine et al., 1995; see also Ruff et al., 1997). In particular, it has been applied to femoral head diameters, which scale positively allometrically with body mass (Ruff et al., 1993) and provide reasonably reliable estimates of body mass within recent human samples. In particular, three formulae have been derived to predict body mass. The first formula (from McHenry, 1992) is based on a geographically mixed cadaver and associated skeleton sample of recent humans (BM = 2.239 x FHd − 39.9; $r = 0.98$; $N = 59$). The second is from Grine et al. (1995) and is based on ten sex-specific population means (BM = 2.268 x FHd − 36.5; $r = 0.92$). The third (from Ruff et al., 1991) is derived from radiographic data of living urban North Americans and provides sex-specific formulae of BM = 2.741 x FHd − 54.9 ($r = 0.50$; $N = 41$) for males and BM = 2.426 x FHd − 35.1 ($r = 0.41$; $N = 39$) for females, each then adjusted downward 10% for excess adipose tissue (see Ruff et al., 1997). Each formula has its limitations and its strengths, but together they provide consistent results ($r = .981$; $N = 49$ across the estimates for the three formulae, using the pooled fossil data employed here). As with stature estimation, sex-specific formulae, as available, were employed for specimens with reasonably estimated gender, and both were used and then averaged for specimens of uncertain gender.

The primary limitation of the biomechanical approach to estimating lean body mass is the assumption that the subchondral bone dimensions of the articulation are responding solely to the joint reaction force from weight support. In fact, joint reaction forces several times body mass are frequently generated across the hip joint, and they increase markedly with more vigorous activity levels (Pedersen et al., 1997; Van Den Bogert et al., 1999). Consequently, it remains unclear whether the dimensions of the femoral head are reflecting the baseline loads of body mass or those from body mass plus a significant contribution from activity patterns; this is important because of the elevated robusticity of most immature Pleistocene humans, as reflected in both their femoral and tibial diaphyses and the resultant mature neck-shaft angles (Trinkaus, 1993a; Ruff et al., 1994, 2002; Trinkaus et al., 2002b).

In order to maximize sample sizes, femoral head diameters were used when available from the fossils, but it was predicted from acetabular height when the femur was absent (as in isolated os coxae) or the femoral head was too damaged to measure (recent human sample: FHd = 0.988 x AcetHt − 8.4; $r^2 = 0.896$; $N = 39$). This was applied to Amud 1; Cro-Magnon 4314, 4315, and 4317; Kebara 2; and Mladeč 21 and 22.

As an alternative to biomechanical methods of body mass estimation, Ruff (1991, 1994; Ruff et al., 1997) developed a geometric model. In this approach, the human body is modeled as a cylinder, in which its critical dimensions are its length and diameter. Body core diameter is best approximated by bi-iliac breadth, which is subcutaneous in the living and readily measurable on sufficiently preserved pelvic remains. Stature is known (for the living) or estimated for skeletal remains (as above). The resultant formulae, based on a global sample of sex-specific means, are BM = (0.373 x ST) + (3.033 x BIB) − 82.5 ($r = 0.90$; $N = 31$) for males, and BM = (0.522 x ST) + (1.809 x BIB) − 75.5 ($r = 0.82$; $N = 25$) for females, in which ST is stature and BIB is bi-iliac breadth, and both are measured in centimeters (see Ruff, 1994; Ruff et al., 1997). In addition, it is necessary to convert skeletal bi-iliac breadth to living bi-iliac breadth, using this formula: living BIB = (1.17 x skel BIB) − 3.0 (both in centimeters).

Although based on recent human samples, most are of individuals from nonindustrial, small-scale, nontechnologically aided populations (Ruff, 1994), and therefore they should closely approximate the levels of muscular hypertrophy seen in Pleistocene human populations. Moreover, analysis of modern world-class athletes (Ruff, 2000c) shows that these formulae are reasonably accurate (to within 3% on average) in predicting the weights of individuals with levels of hypertrophy and low body fat that probably approached the more usual condition among Late Pleistocene humans.

In order to estimate body mass by using these geometric formulae, it is necessary to know or estimate both stature and bi-iliac breadth. The estimation of the former is discussed above. The estimation of the latter, when necessary given pelvic preservation, was done in several ways, depending upon the sample. For the European earlier Upper Paleolithic sample, it was predicted by using the least squares regression based on specimens with measures of both bi-iliac breadth and femoral length [BIB = (0.327 x FBcL) + 116.7; $r = 0.812$; $N = 12$]. For the Nahal Ein Gev 1 Near Eastern earlier Upper Paleolithic specimen, the femoropelvic proportions of Ohalo 2 were employed with a slope of 0.237, derived from recent human skeletal samples. The same was done for Minatogawa 1 and 4, using the values from Minatogawa 2 and 3. For Nazlet Khater 1, the mean femoropelvic proportions of three recent Nile Valley samples (separate male and female means) from Holliday (1995) were employed, and the same was done for Ishango A, using the mean proportions of two equatorial African samples. For the Qafzeh-Skhul samples, the proportions of Skhul 4 served as a baseline for the other specimens, as did Kebara 2 for the Neandertals. It is recognized that there is potential error, especially for those samples that are based on a single fossil specimen for reference. However, it is also clear that other methods for estimating overall skeletal proportions would reduce the sample sizes to an uninterpretable minimum. Since any errors are likely to be random, these values are used as a best available approximation.

Body Length Comparisons

The proximal limb segment long bone lengths and the estimated statures for the Dolní Věstonice and Pavlov associated skeletons and comparative samples are in Table 13.1, and they are plotted in Figures 13.1 and 13.2. It is apparent that there was significant variation during

Table 13.1 Comparisons of humeral and femoral lengths and stature estimates for the Pavlovian fossils and comparative samples.

	Humerus Maximum Length (mm)	Femur Bicondylar Length (mm)	Stature: All Bones (cm)	Stature: Leg Bones (cm)
Dolní Věstonice 3	—/(309.0)	—/(417.5)	159.1	158.3
Dolní Věstonice 13	336.0 / 332.0	444.0 / 441.5	169.0	167.8
Dolní Věstonice 14	374.0/—	495.0/(509.0)	180.3	179.0
Dolní Věstonice 16	—/(332.0)	—/(467.5)	170.9	171.4
Pavlov 1	(378.0)/(378.0)	—/(475.0)	178.3	172.3
Předmostí	330.7 ± 17.3 (5)	435.6 ± 31.9 (6)	165.5 ± 8.4 (6)	164.2 ± 9.0 (6)
Earlier Upper Paleolithic: Europe	333.9 ± 25.1 (20)	463.2 ± 35.5 (24)	168.6 ± 7.9 (33)	168.8 ± 8.7 (28)
Earlier Upper Paleolithic: Asia	289.7 ± 31.6 (5)	395.9 ± 39.2 (5)	154.6 ± 8.1 (6)	153.3 ± 10.1 (5)
Earlier Upper Paleolithic: Africa	316.0, 333.0	408.0	161.7, 164.6, 171.1	157.9, 164.6
Middle Paleolithic: Qafzeh-Skhul	330.0, 337.0, 374.0, 379.5	481.1 ± 23.0 (5)	169.9 ± 8.8 (9)	173.6 ± 8.1 (7)
Middle Paleolithic: Neandertals	311.5 ± 13.8 (11)	437.5 ± 25.7 (13)	165.2 ± 5.9 (20)	164.7 ± 7.4 (13)

For the comparative samples, mean ± standard deviation is used for samples ≥5; individual values are used for smaller samples. The Aurignacian specimens from Mladeč are pooled with the Gravettian specimens into a European earlier Upper Paleolithic sample, which includes the Předmostí specimens but not the Dolní Věstonice and Pavlov remains. The Initial Upper Paleolithic Saint Césaire 1 specimen is included with the Middle Paleolithic Neandertals. Right and left values, as available, are provided for the Pavlovian specimens; the averages of right and left values, when available, were used in the summary statistics and stature calculations. Long bone lengths in parentheses are estimated values; see Sládek et al. (2000) for explanations of estimations.

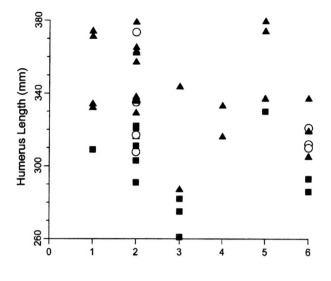

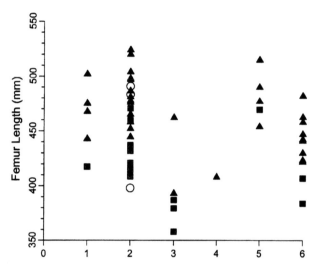

Figure 13.1 Vertical plots of humeral maximum length (above) and femoral bicondylar length (below) for Late Pleistocene human samples. 1: Dolní Věstonice and Pavlov; 2: European Early and Middle Upper Paleolithic; 3: Asian earlier Upper Paleolithic; 4: African earlier Upper Paleolithic; 5: Qafzeh-Skhul; 6: Neandertals. Solid triangles: males; solid squares: females; open circles: sex indeterminate.

the Late Pleistocene in body length. Across the samples, pooling the Pavlovian specimens with the other Early and Middle Upper Paleolithic specimens produces ANOVA P-values < 0.001 for each of the four comparisons (deleting the Neandertal sample places the P-values between 0.001 and 0.002). However, it is apparent that it is principally the mostly very small Asian sample and the relatively high values for the Qafzeh-Skhul sample in the long bone lengths that are producing the differences. In general, despite the small comparative sample sizes, the European earlier Upper Paleolithic specimens are generally taller than their African and Asian counterparts, being matched only by the Qafzeh-Skhul sample. The inclusion or deletion of the upper limb long bones in the stature estimations has little effect except on the Qafzeh-Skhul sample because of the unusually relatively short forearms of a couple of those specimens.

In the context of these data, the female Dolní Věstonice 3 skeleton is similar to the Gravettian females in bone lengths and derived stature estimates. The high male value for the southern Moravian sample is Dolní Věstonice 14, whose stature estimate based on all long bones of ca. 180 cm is exceeded only by those of Barma Grande 2, Fanciulli 4, Mladeč 24, and Qafzeh 8; and its

Figure 13.2 Stature estimates for samples of Late Pleistocene humans, using all available long bones (above) and only lower limb long bones (below). 1: Dolní Věstonice and Pavlov; 2: European Early and Middle Upper Paleolithic; 3: Asian earlier Upper Paleolithic; 4: African earlier Upper Paleolithic; 5: Qafzeh-Skhul; 6: Neandertals. Solid triangles: males; solid squares: females; open circles: sex indeterminate.

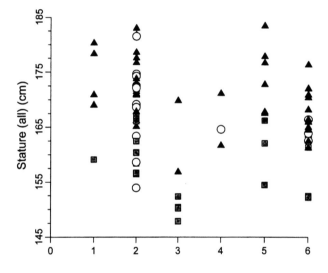

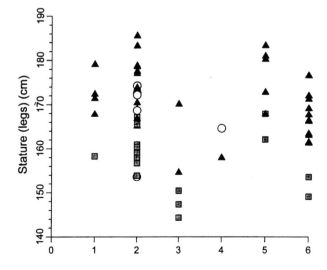

estimate of ca. 179 cm, using leg bones, is only matched or exceeded by Barma Grande 2, Fanciulli 4, Qafzeh 8, Skhul 4 and 5, and Sunghir 1. Among these specimens, Dolní Věstonice 13 is relatively small for a male, but Dolní Věstonice 16 and Pavlov 1 fall in the middle of the Gravettian male range of variation.

Body Mass Comparisons

The comparative values for femoral head diameter and the three body mass estimates are provided in Table 13.2 and graphed in Figures 13.3 and 13.4. The early modern human samples exhibit the same patterning seen in the body length comparisons, with the Asian and African earlier Upper Paleolithic specimens, with the exception of Ohalo 2, being largely below the same-sex values for the European sample. The Qafzeh-Skhul sample is similar to the European earlier Upper Paleolithic one, and the Neandertals, especially the Neandertal males, have relatively higher values. The differences across the samples, with the Pavlovian remains pooled with the European earlier Upper Paleolithic samples, remain significant across the five samples (ANOVA $P = 0.002$ and 0.001 for the femoral head comparisons and <0.001 for the stature/bi-iliac breadth comparisons; these values increase to 0.027, 0.025, <0.001 and 0.003 with the Neandertals removed).

In the body mass comparisons, the Dolní Věstonice and Pavlov male specimens remain close to the European earlier Upper Paleolithic male means. Dolní Věstonice 14 is no longer among the largest individuals, in large part because of its relatively narrow bi-iliac breadth (Chapter 12). Dolní Věstonice 3, despite having its bi-iliac breadth estimated from the more complete Gravettian specimens, has body mass estimates that are moderately low for an earlier Upper Paleolithic female but still well within the range of variation for those females.

The Body Size of Dolní Věstonice 15

Dolní Věstonice 15 suffered from a congenital dysplasia, one that led to differences in proportions of at least the pelvis and femora relative to the remainder of the skeleton, plus a series of other lesions and deformities (Chapters 12, 18, and 19). However, despite these deformities, Dolní Věstonice 15 appears to have remained physically active throughout its life, with normal weight bearing on the lower limbs (Chapter 18). Given these alterations, the stature and body mass of Dolní Věstonice 15 cannot be accurately estimated solely from its long bone lengths since the abnormalities have changed the linear proportions of the individual.

In addition, even though the pelvic remains of Dolní Věstonice 15 indicate that the skeleton is male, the long-term debate concerning its gender (see Chapter 7) means that one should not simply apply male-specific formulae for stature or body mass estimation. Consequently, in the

Table 13.2 Comparisons of body mass indicators and body mass estimates for the Pavlovian fossils and comparative samples.

	Femoral Head Anteroposterior Diameter (mm)	Body Mass from Femoral Head Diameter (kg)	Body Mass from Stature (All) and Bi-Iliac Breadth (kg)	Body Mass from Stature (Legs) and Bi-Iliac Breadth (kg)
Dolní Věstonice 3	40.5/—	54.3	56.2	55.7
Dolní Věstonice 13	47.0/48.0	68.5	65.1	64.7
Dolní Věstonice 14	50.5/51.4	76.5	67.9	67.4
Dolní Věstonice 16	(50.5)/(50.5)	75.5	69.1	68.2
Pavlov 1	—/—	—	71.6	69.4
Předmostí	45.4 ± 3.1 (6)	64.3 ± 7.3 (6)	63.9 ± 6.6	63.1 ± 7.2 (5)
Earlier Upper Paleolithic: Europe	46.9 ± 3.6 (25)	67.6 ± 8.0 (25)	66.5 ± 7.4 (23)	66.6 ± 7.9 (23)
Earlier Upper Paleolithic: Asia	41.1 ± 5.3 (5)	55.1 ± 11.1 (5)	54.4 ± 10.1 (6)	54.9 ± 11.3 (5)
Earlier Upper Paleolithic: Africa	40.5, 48.0	52.2, 69.6	53.4, 55.4	53.4, 54.1
Middle Paleolithic: Qafzeh-Skhul	44.5, 47.2, 47.3, 48.9	63.2, 67.8, 68.0, 71.7	69.9 ± 6.9 (7)	70.4 ± 7.1 (7)
Middle Paleolithic: Neandertals	50.7 ± 3.2 (8)	76.0 ± 7.1 (8)	78.6 ± 7.8 (13)	78.0 ± 8.6 (11)

Samples are arranged as in Table 13.1. Body mass from stature and bi-iliac breadth was computed by using stature estimates from all available long bone lengths (All) and stature estimates only from available leg bone lengths (Legs).

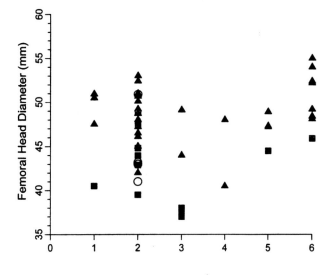

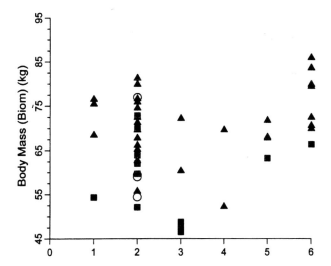

Figure 13.3 Femoral anteroposterior head diameter (above) and body mass estimated from femoral head diameter (below) for samples of Late Pleistocene humans. 1: Dolní Věstonice and Pavlov; 2: European Early and Middle Upper Paleolithic; 3: Asian earlier Upper Paleolithic; 4: African earlier Upper Paleolithic; 5: Qafzeh-Skhul; 6: Neandertals. Solid triangles: males; solid squares: females; open circles: sex indeterminate.

following assessment of body size, whenever sex-specific formulae are available, they are both applied and the results presented but the male-based results are employed for comparisons.

Also, given the nature of the diaphyseal deformities, which affected primarily the right humerus, the left forearm, and the right femur, the lengths from the more normal opposite sides are employed. The lengths of the tibiae, which are normal and highly symmetrical (Chapters 18 and 19), are averaged for stature estimation. The values presented here differ slightly from those in Trinkaus et al. (2001); this is due to remeasurement of the original specimen and do not affect the results substantively.

The brachial index of Dolní Věstonice 15, using the normal left humerus and right radius (79.9), is close to those of other Gravettian humans (77.7 ± 2.0; $N = 20$; see Table 12.1). However, the crural index, using the largely normal left femur and tibia (89.6), is 2.50 standard deviations above the mean of the other Gravettian remains (85.1 ± 1.8; $N = 18$; see Table 12.1), and the crural index of the abnormal right leg is an exceptional 93.7, or 4.78 standard deviations above the comparative mean.

Figure 13.4 Body mass estimates for Late Pleistocene humans from stature and bi-iliac breadth, using all available long bones for stature (above) and only the lower limb long bones for stature (below). 1: Dolní Věstonice and Pavlov; 2: European Early and Middle Upper Paleolithic; 3: Asian earlier Upper Paleolithic; 4: African earlier Upper Paleolithic; 5: Qafzeh-Skhul; 6: Neandertals. Solid triangles: males; solid squares: females; open circles: sex indeterminate.

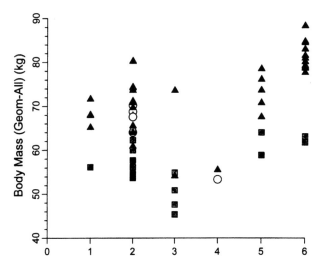

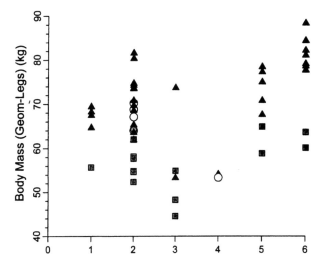

It is also possible to estimate the skeletal trunk height—the sum of T1 to L5 dorsal heights plus sacral (S1 to S5) ventral height—for Dolní Věstonice 15 and twelve other earlier Upper Paleolithic specimens (see Chapter 12). The skeletal trunk height estimate of Dolní Věstonice 15, based on its T1, T10, T11, and sacral ventral heights plus its T2, T3, T7, T8, and T12 to L3 dorsal heights, provides a value of 469 mm. This value is within 1 standard deviation of a Gravettian sample mean (491 ± 34 mm; $N = 12$), even though its left femoral bicondylar length of 383 mm is 2.05 standard deviations below the mean of the Gravettian subsample that provides skeletal trunk height estimates (465 ± 40 mm; $N = 12$). As a result, the indices of femoral or tibial length to skeletal trunk height for Dolní Věstonice 15 (81.6 and 73.1, respectively) are 2.67 and 1.81 standard deviations, respectively, below Gravettian means (95.2 ± 5.1; $N = 8$; and 82.7 ± 5.3; $N = 7$, respectively; see Table 12.4). Relative humerus length (64.3) is less divergent, being 1.44 standard deviations below the Gravettian mean (69.5 ± 3.6; $N = 8$). The limbbone-based formulae therefore should underestimate the original stature of Dolní Věstonice 15.

To provide a more accurate estimate of the stature of Dolní Věstonice 15, first its femoral and tibial lengths were estimated from its skeletal trunk height, using the mean indices of the Middle Upper Paleolithic sample, providing length estimates of 446.0 mm and 388.0 mm, respectively. These were then used to estimate stature as above, providing a male estimate of 168.8 cm and a female estimate of 165.1 cm. Subtracting the difference between the actual and estimated femoral and tibial lengths (6.0 and 3.9 cm, respectively) provides a male estimate of 158.9 cm and a female estimate of 155.2 cm. These estimates are slightly higher than the value (154.3 cm) derived directly from its actual femoral and tibial lengths. Stature was not estimated by using the upper limb bones.

Body mass was estimated geometrically, as above. However, there are difficulties in determining its bi-iliac breadth since postmortem erosion to the sacroiliac surfaces of the sacrum and the os coxae allow a certain amount of movement between the bones when articulated. Consequently, the three bones were articulated, and the reasonable range of bi-iliac breadths was determined. The minimum value (234 mm) is the value that has been previously published (Sládek et al., 2000). The maximum value obtained is 245 mm. Remeasurement did not include the value of 255 mm, employed in Trinkaus et al. (2001), so that value is in error.

Its body mass was computed by using the maximum and minimum bi-iliac breadths and using the gender-specific formulae for both stature and body mass. The results are 50.7 kg and 49.6 kg for male and female formulae, respectively, using the 234 mm bi-iliac breadth; they are 54.6 kg and 51.9 kg for male and female formulae, respectively, using the 245 mm bi-iliac breadth. Averaging the values for the two bi-iliac breadths provides male and female values of 52.7 kg and 50.8 kg, respectively. These values are within the estimation errors of these techniques; however, given the probable male gender of Dolní Věstonice 15 (Chapter 7), the male estimate of 52.7 kg is employed in comparisons using body mass.

Body mass was also estimated by using the femoral head anteroposterior diameters, averaging the results of the three formulae presented above. This approach provides a body mass estimate for Dolní Věstonice 15 of 66.3 kg, using the right femoral head diameter of 46.2 mm, and 68.1 kg, using the left femoral head diameter of 47.5 mm, for an average of 67.2 mm. However, given the unusual hip morphology of Dolní Věstonice 15 (Chapters 18 and 19) and the responsiveness of femoral head size to load levels during development, the mean body mass estimate of 52.7 kg, using stature and bi-iliac breadth, is more likely to be accurate. Increasing the estimate will slightly reduce the perceived robusticity of Dolní Věsonice 15's weight-bearing lower limbs (as assessed in Chapter 18).

The estimated stature of 158.9 cm for Dolní Věstonice 15 is relatively low for a European earlier Upper Paleolithic individual, but it is not the lowest value calculated. It is close to the values for Dolní Věstonice 3 and Předmostí 10, and above those of Předmostí 1 and La Rochette 1. It is also above those of Nahal Ein Gev 1, Minatogawa 1 to 4, and Nazlet Khater 1. If female, therefore, Dolní Věstonice 15 was not unusually short, although it would have been if it were a male. Its estimated body mass of ca. 53 kg, however, is small. The only earlier Upper Paleolithic specimens with smaller values are the insular Minatogawa 2 to 4 females. However, several western Eurasian female specimens are close to it, including Dolní Věstonice 3 (ca. 56 kg), Cro-Magnon 4324 (ca. 55 kg), Nahal Ein Gev 1 (ca. 55 kg), and La Rochette 1 (ca. 54 kg). The body mass estimate for Dolní Věstonice 15 therefore suggests a modest effect of the congenital dysplasia on body size, in the unlikely case that it was female, but a marked reduction in body size in the more likely case that it was male.

The Estimated Body Size of Dolní Věstonice 35

Dolní Věstonice 35 is an isolated proximal femoral diaphysis, which preserves the subtrochanteric area to the midshaft. It is possible to identify the locations of the 80% diaphyseal subtrochanteric section based on the morphology of the proximal diaphysis 191 mm from the distal break and to approximate the position of the midshaft (50%) section at 49 mm from the distal break. Assuming that these locations are accurate, the resultant biomechanical length (proximal neck to distal condyles parallel to the diaphyseal axis) would be 473 mm (142 mm/0.3) Using a least squares regression based on a pooled sample of Pleistocene *Homo* femora (FBcL = 1.008 x FBiomL + 21.3; r^2 = 0.981; N = 30) provides an estimated bicondylar length of 498 mm. This value is clearly approximate, but the morphology of the femur makes it unlikely that it is in error by more than ca. 10 mm. Comparison to the available femoral bicondylar lengths places it close to Dolní Věstonice 14 and several other tall male Middle Upper Paleolithic specimens.

The male Afro-American formula of Trotter and Gleser (1952) provides an estimated stature of 177.3 cm, and the regression formulae for bi-iliac breadth versus femoral length for Gravettian humans furnishes an estimated bi-iliac breadth of 279.5 mm. The male body mass formula using stature and bi-iliac breadth yields an estimated body mass of 73.7 kg. This value is again moderately high for a Middle Upper Paleolithic individual, but not exceptionally so. Given the uncertainty in the original key measurement, femoral biomechanical length, these values should be taken as approximate but unlikely to be seriously in error. They will be used, with appropriate qualifications, only in the necessary scaling of diaphyseal biomechanical properties of the Dolní Věstonice 35 femoral shaft.

Summary

These data therefore indicate that the earlier Upper Paleolithic humans from Europe were generally large for their time period, being matched in stature mainly by the early modern humans from Qafzeh and Skhul and in body mass by the Middle Paleolithic late archaic and early modern humans. In this, the Dolní Věstonice and Pavlov remains are generally similar to the other European Middle Upper Paleolithic remains, with the Dolní Věstonice 14 and 35 males being relatively large, the Dolní Věstonice 3 female being modest in size, and the pathological Dolní Věstonice 15 being exceptionally small for a male. Most of these European earlier Upper Paleolithic individuals are larger than the few Asian and African earlier Upper Paleolithic human remains available.

14

The Vertebral Columns

Trenton W. Holliday

The vertebral column is well represented among the skeletons from Dolní Věstonice and Pavlov. In particular, the individuals from the triple burial (Dolní Věstonice 13–15) preserve the majority of their vertebral elements in various stages of completeness (Sládek et al., 2000; Chapter 5). The purpose of this chapter is twofold. First, detailed morphological description of each individual element of the presacral vertebrae is given (the sacrum is discussed with the pelvis in Chapter 17). Second, osteometric analyses are performed in order to compare the Dolní Věstonice and Pavlov axial skeletal morphology with those of penecontemporary and recent humans. The comparative samples used for these analyses vary considerably, depending on which skeletal elements are being studied, and therefore discussion of the measurements and indices used, as well as the compositions of the comparative samples, will come at the beginning of the osteological analysis.

The axial remains of Dolní Věstonice 16 and Pavlov 1 were separated during excavation and partially reconstructed in the laboratory. The more cranial cervical vertebrae of Dolní Věstonice 3 were separated and cleaned in the late 1940s, when the skeleton was first analyzed (Jelínek, 1954), but the remainder of the vertebrae, all of which were crushed in situ, were left in a block until they were excavated by Trinkaus in the 1990s (Trinkaus & Jelínek, 1997). Their preservation and some uncertainty in the identification of small pieces are due to their in situ condition. The axial skeletons of the triple burial individuals were removed en bloc to the laboratory, cleaned, and reconstructed. However, their vertebrae were subsequently articulated with supporting materials for museum exhibitions (see photos in Vlček, 1991), but they have since been separated. Some of their condition is due to this handling since excavation. Despite the varied histories of these vertebral columns, they are generally well conserved, given their original in situ preservations.

Morphological Description: The Presacral Vertebral Remains

All six of the associated postcranial skeletons from the sites of Dolní Věstonice and Pavlov (Dolní Věstonice 3 and 13–16 and Pavlov 1) preserve at least a few vertebral elements that exhibit a wide range of states of preservation. The vertebral columns of Dolní Věstonice 13 and 15 are largely complete and extremely well preserved. The vertebral column of Dolní Věstonice 14 is relatively complete in terms of preserved elements, but its elements are less well preserved, and some are possibly burn-damaged. Likewise, in terms of elements, the vertebral column of Dolní Věstonice 3 is nearly complete, but much of the column either has been destroyed and/or remains have been embedded in matrix. Finally, Dolní Věstonice 16 and Pavlov I preserve just a few isolated cervical and thoracic elements. Much of this section, and in particular the osteometric analyses, is devoted to the more complete remains, especially those belonging to the individuals from the triple burial, although the less complete specimens are also included in the morphological description and in some of the osteometric analyses.

The Cervical Vertebrae

Among mammals in general and humans in particular, the seven cervical vertebrae exhibit more morphological variation than vertebrae from any other vertebral region (Steele & Bramblett, 1988). In particular, the two cranial cervical vertebrae (C1 and C2) are extensively structurally modified for their roles in the movement and support of the cranium, whereas the more caudal cervical elements tend to be morphologically akin to thoracic vertebrae (Steele & Bramblett, 1988), with the seventh cervical vertebra being the most similar to a thoracic vertebra in its morphology (Aiello & Dean, 1990). The cervical region is the most mobile within the vertebral column since it must accommodate movement of both the head and neck (Levangie & Norkin, 2001). Contributing to this greater mobility are synovial joints, which are unique to the cervical region, called uncovertebral joints. These joints are associated with the cranially projecting uncinate processes so characteristic of the lateral cranial surfaces of the bodies of the cervical vertebrae (and occasionally on T1, as well). In concert these joints facilitate the summed movements typical of the cervical region (Levangie & Norkin, 2001). Despite their small size (which frequently makes them subject to taphonomic destruction), cervical vertebrae are well represented among the Dolní Věstonice and Pavlov skeletons. Dolní Věstonice 3 and 13–16 preserve at least some of all seven cervical elements, and Pavlov 1 preserves the remains of C2.

Atlas

The atlas (C1) of Dolní Věstonice 3 (Figure 14.1) is beautifully preserved. Its lateral masses are relatively small; each has a small ovoid nodule for the transverse ligament. The facet for the dens is ovoid (height: 6.9 mm; breadth: 10.2 mm) and slightly concave. The superior articular facets are large, with slight lipping evident along their margins. Both have double (i.e., anterior and posterior) facets, with the left evincing an unusual morphology in that there is a slight gap between the anterior and poste-

Figure 14.1 Superior view of C1 (left) and C2 (right), Dolní Věstonice 3. Scale in centimeters.

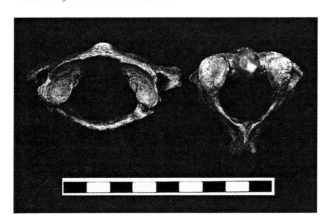

rior portions of its articular surface. In contrast, on the right, the anterior and posterior surfaces have an almost indistinct margin between them. The inferior articular processes are ovoid (right) to kidney-shaped (left) and mediolaterally concave. The left facet is kidney-shaped because of a medial concavity that is evident in inferior view. The transverse processes are relatively big with large, circular transverse foramina. Perhaps the most interesting aspect of C1's morphology is that, rather than having an open groove for the vertebral arteries, each groove is roofed by a bony shelf, with a thicker shelf on the left than on the right (the left shelf measures 6.8 mm in its mediolateral dimension; the right shelf measures 3.3 mm). The vertebral canal evinces the anteriorly constricted shape typical of modern humans.

The atlas of Dolní Věstonice 13 is nearly complete (Figure 14.2). It is missing only a portion of its posterior arch. It has two large transverse processes, although the left one is larger than the right one. Its inferior articular facets are large and almost circular in shape (the antero-

Figure 14.2 Inferior view of the C1 (above) and superior view of the C2 (below) of Dolní Věstonice 13. Scale in centimeters.

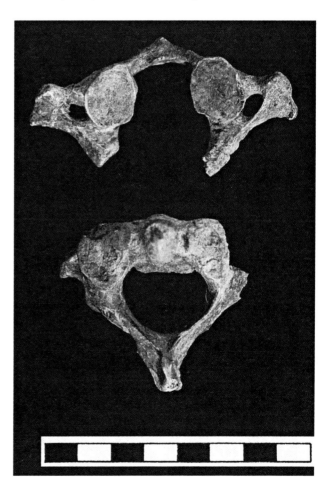

medial to posterolateral axis is slightly longer than the posteromedial to anterolateral axis). The borders of the superior articular facets are well delineated. The anterior tubercle is prominent, and the facet for the dens is circular. Both lateral masses are large, with prominent tubercles for the transverse ligament.

The atlas of Dolní Věstonice 14 is a relatively large bone (Figure 14.3). Its right articular processes are damaged, and both transverse foramina have been destroyed (although enough of the left is present to infer that it was quite large; its parasagittal dimension is 8.8 mm). The anterior and posterior tubercles are not very robust. The relatively shallow left superior articular facet is clearly delimited into two separate (anterior and posterior) facets. The slightly concave facet for the dens is almost circular (breadth: 11.5 mm; height: 9.8 mm), is smooth, and has well-delineated boundaries. The right inferior articular facet is heavily damaged—it preserves only its anterior one-third. The left inferior articular facet is rounded and slightly concave. It has a relatively steep declination toward its lateral side. Its anteroposterior length is 16.0 mm; its transverse breadth is 15.3 mm. Whereas the lateral mass is damaged posteriorly on the right, the tubercles for the transverse ligament are visible on both sides. The left is more clearly delineated and measures

about 6.4 mm in height by 5.2 mm in length. The vertebra is fairly symmetrical, although the right side is a bit more open posteriorly; that is, the posterior arch meets the anterior arch at a more obtuse angle on the right than on the left.

The atlas of Dolní Věstonice 15 is a relatively complete, albeit small bone that is missing its right transverse process (Figures 14.4 and 14.5). Its right side also evinces slight postmortem deformation, such that the right inferior articular facet no longer articulates properly with C2. The left lateral mass is large (the right one is damaged anteriorly). The left transverse process evinces lateral damage, but the left transverse foramen can be measured—height: 6.1 mm; breadth: 5.5 mm. Each of the superior articular facets evinces contiguous, but clearly discrete, anterior and posterior facets. The posterior arch has an undulating inferior margin and is quite robust, with an anteroposteriorly thick (9.9 mm) posterior tubercle. The right inferior articular facet is damaged anteriorly; the left inferior articular facet is almost circular in shape.

Dolní Věstonice 16's C1 is relatively complete (Figure 14.6). However, it is missing a large section of its left lateral mass and articular facets, and there is reconstruction on either side of the posterior tubercle. This latter feature is perhaps the bone's most striking characteristic;

Figure 14.3 Superior view of C1 (above) and C2 (below), Dolní Věstonice 14. Scale in centimeters.

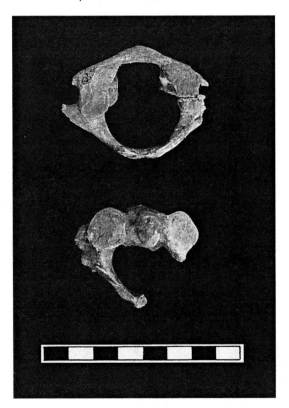

Figure 14.4 Superior view of C1 (above) and C2 (below), Dolní Věstonice 15. Scale in centimeters.

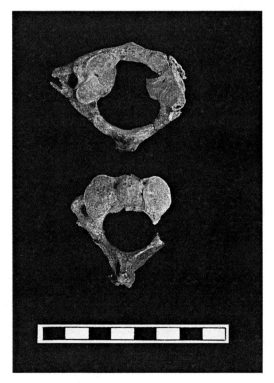

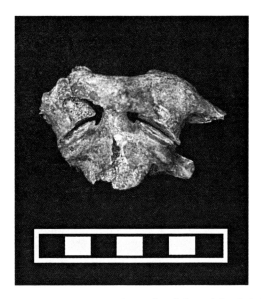

Figure 14.5 Anterior view of articulated C1 and C2, Dolní Věstonice 15. Scale in centimeters.

it is quite impressive in size (height: 13.3 mm; anteroposterior length: 8.2 mm). The anterior tubercle is also large (height: 13.6 mm; anteroposterior length: 7.3 mm), with a large, deep, circular facet for the dens. The lip surrounding the facet for the dens is quite prominent. Both transverse foramina have been broken, but enough of the right is preserved to allow its long axis (parasagittal) measurement: 8.6 mm. The grooves for the vertebral arteries are prominent on both sides. In terms of articular processes, the left is damaged anteriorly, but the right is complete superiorly and nearly complete inferiorly. The superior articular processes are quite deeply concave, and the right facet's anterior and posterior portions are clearly delimited from each other. In contrast, the surface areas of the inferior facets are largely flat. The left evinces heavy damage, and the right is damaged on its medial and anterior surfaces. The lateral mass (preserved only on the right) is large, with the tubercle for the transverse ligament preserved, but broken posteriorly.

Axis

The axis (C2) of Dolní Věstonice 3 has excellent preservation (Figure 14.1). It is in near perfect condition, save only for trivial erosion of the posterior nodules of the spinous process and anterior face of the odontoid process. Both superior articular facets are teardrop-shaped, with a gentle anteroposterior convexity. The left appears to be in a slightly more cranial position than the right. The neural canal is ovoid in shape (breadth: 22.1 mm; height: 15.8 mm). The transverse processes are not particularly robust; the right is blunter and larger than the left, and the transverse foramina within the processes are circular in superior view and ovoid in inferior view. The anterior facet on the dens is ovoid (height: 10.1 mm; breadth: 7.7 mm) and evinces a slight transverse convexity. The laminae are internally to externally thick (5.0 mm left & right) and end in a short, blunt, bifid spinous process that extends more inferiorly on the right.

The axis of Dolní Věstonice 13 is largely complete (Figure 14.2). The caudal surface of its body is anteroposteriorly concave, with a prominent anterior lip that extends more inferiorly than its posterior equivalent. Part of its right transverse process is missing, and a ventral portion of the left superior articular facet has been depressed, postmortem, to roughly 3.5 mm below its more dorsal portion. The left transverse foramen is filled with matrix. The right transverse foramen is missing, as are the anterior tubercle and part of the pedicle. The dens is tall,

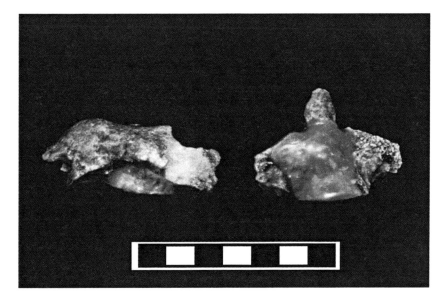

Figure 14.6 Anterior view of C1 (left) and C2 (right), Dolní Věstonice 16. Scale in centimeters.

with a flattened anterior face, evincing a well-delineated facet for C1. The neural canal is almost circular in shape, and the rugose spinous process is minimally bifid. The superior margin of the right lamina has been damaged, as has a small area along the middle posterior margin of the spinous process. Only the anteromedial half of the right superior articular facet is preserved; both inferior articular facets are preserved. The right, which is more rectangular (i.e., less rounded) than the left, is damaged, so part of its lateral inferior face is missing.

The axis of Dolní Věstonice 14 (Figure 14.3) is represented by a nearly complete body (slight damage to its anteroinferior lip), a complete dens, two complete superior articular facets, one complete transverse foramen (left), one partial transverse foramen (right), both pedicles, a partial inferior articular facet (left), and a nearly complete left lamina (the right lamina is missing). The costotransverse bar is preserved on the left, and it is transversely thick (6.6 mm). The spinal canal is almost circular. The small, flat, and ovoid left inferior articular facet evinces lateral damage. The dens is large and prominent. The superior articular facets are slightly convex, circular, and smooth. However, some of the foam intervertebral discs previously glued to the facets remains adhered to them and cannot be removed because of the risk of damage. The left superior articular facet is larger and more ovoid than the more circular right facet. The caudal surface of the body is deeply anteroposteriorly concave. The transverse foramina are large—especially on the left. The small, ovoid, and flat left inferior articular facet is missing its anterolateral quarter.

The axis of Dolní Věstonice 15 is somewhat damaged (Figures 14.4 and 14.5). The dens and body are present, albeit with a large crack running in a sagittal plane just to the right of the dens, and the right half of the inferior lip on the ventral body is missing. The inferior dorsal margin of the body is also eroded, and the dorsal aspect of the inferior ring epiphysis is missing. The cranial margin of the dens evinces light erosion. The right pedicle, along with its associated transverse process, is also missing. The laminae are not particularly robust, but rugose, and give rise to a bifid spinous process. The inferior articular facets are completely destroyed, save a small anterior portion of the left facet. The superior articular facet is ovoid in shape with a convex articular surface.

The axis of Dolní Věstonice 16 (Figure 14.6) is heavily reconstructed. It is difficult to tell if there was any contact between the two halves of the neural arch, since the spinous process and most of the body are reconstructed in wax. Of the portion of the body that is preserved, the inferior left side is quite large and impressive. The dens is also quite large, although the presence of wax makes the evaluation of its height difficult. Both of the ovoid and slightly concave inferior articular facets are heavily damaged. The right facet extends more laterally from the lamina than does the left. It also faces more directly inferiorly, whereas the left faces more inferolaterally. The left superior articular facet has anterior damage and is unusual in being concave, but the right facet is largely complete (it has slight anterior erosion) and shows the more typically convex articular surface. The transverse foramina are ovoid, with mediolateral long axes. The external surface of the lamina is not rugose. Although the neural canal is distorted, its shape was apparently ovoid, with a mediolateral long axis.

The axis of Pavlov 1 is largely complete (Figure 14.7), but it has a fair amount of putty on it, making it difficult to discern real from fabricated bone. The anterior face of the vertebral body is damaged, such that the prominent inferior lip is preserved only to the left of midline. The large right superior articular facet is damaged, and the right inferior articular facet is missing. The morphology of the inferior facet is somewhat unusual in that it is located lateral to the pedicle. The dens is quite large. The left lamina is internally to externally thick (6.4 mm) and is rugose on its external face. The posterior (external) surface of the right lamina evinces heavy damage. The neural canal appears circular in superior view.

Third Cervical Vertebra

The fragmentary third cervical vertebra (C3) of Dolní Věstonice 3 is broken into four separate pieces (Figure 14.8). The first preserves the body and the fragmentary bases of the two pedicles. The body itself is beautifully

Figure 14.7 Superior view of C2, Pavlov 1. Scale in centimeters.

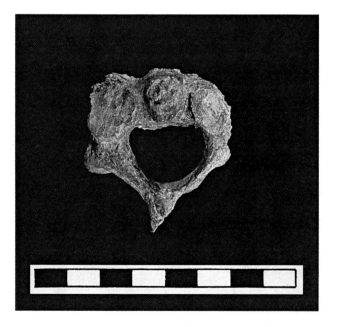

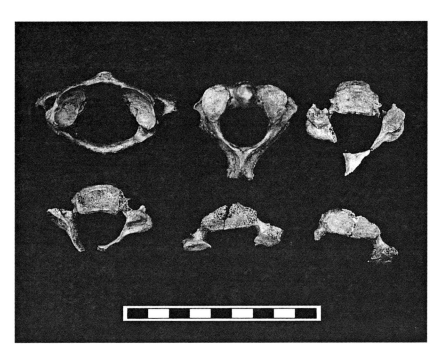

Figure 14.8 Superior view of cervical vertebrae of Dolní Věstonice 3. Top row (left to right): C1, C2, C3; bottom row (left to right): C4, C5, C6. Scale in centimeters.

preserved; only the right uncinate process has damage. The cranial margin of the body is gently concave in the coronal plane and planar in the sagittal plane. The body's caudal surface is deeply dorsoventrally concave and slightly convex in the coronal plane, and the caudal ring epiphyses are fused to the centrum. The second piece includes the left articular facets, along with part of the left transverse process. Also preserved are a small portion of the cranial margin of the left pedicle and an extremely small portion of the inferior margin of the left lamina. The superior articular facet is ovoid and slightly concave. A small portion of its anterolateral face is missing. The inferior articular facet is gently convex and heavily damaged on its lateral side. A third piece retains the right articular processes, including a complete superior articular facet, a virtually complete inferior articular facet, a partial transverse process, and a small portion of the right lamina. The right superior articular facet is ovoid, with matrix adhered along the lateral edge of its relatively flat surface. The right inferior articular facet is circular and, as on the left, evinces a smooth, gentle convexity. The fourth piece preserves the cranial midline of the junction of the two laminae. Internally it has a smooth surface, whereas the external surface has a midline ridge with shallow fossae to either side.

The third cervical vertebra of Dolní Věstonice 13 (Figures 14.9 and 14.10) is largely complete, with a heavily damaged right side. There is a moderately sized uncinate process on the left (none is preserved on the right). There has also been some loss of cortical bone anteromedially along the caudal surface of the body. What little remains of the body's cranial surface suggests it was dorsoven-

trally convex, whereas the caudal surface is dorsoventrally concave. Both anterior tubercles are missing. The right posterior tubercle is large, but the left posterior tubercles are missing. All pedicles and laminae are preserved. The pedicles are thick (left height: 7.4 mm; left breadth: 5.9 mm); the right one is distorted postmortem. The markedly bifid spinous process is short and deviates to the right. The superior articular facets are highly asymmetrical: the right is much smaller than the left. The right facet also evinces a more posteromedial orientation, and it does not protrude as far cranially as the left one. Although the right has suffered loss of cortical bone on its articular surface, its shape is nonetheless conserved—it is ovoid and slightly anteroposteriorly concave. In contrast, the left superior articular facet is broad and circular, and it shows a more directly posterior orientation. In cranial view, the neural arch has a dorsal dome and a flat base formed by the dorsal wall of the body. In inferior view the neural arch is asymmetrical. Specifically, if the dorsal margin of the body faces directly posteriorly, then the spinous process is greatly deviated toward the right. As with the superior facets, the inferior articular facets also differ greatly in size—the right is much smaller than the left. The right also has a planar surface area, whereas the larger left facet is clearly dorsoventrally concave.

The third cervical vertebra of Dolní Věstonice 14 is preserved in two pieces (Figure 14.11). The first includes most of its body, right pedicle, and articular processes. There is a missing midline section of the ventral face of the body about 8.0 mm wide. Posteriorly, there is also a section of body missing from around midline. Both of these areas of missing bone are near a repaired break,

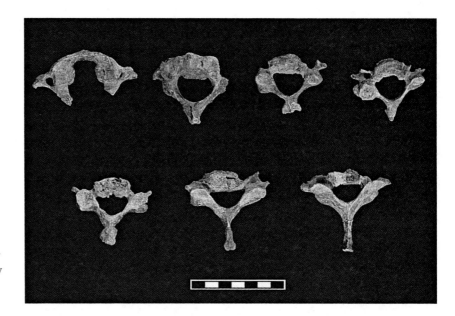

Figure 14.9 Superior view of cervical vertebrae of Dolní Věstonice 13. Top row (left to right): C1, C2, C3, C4; bottom row (left to right): C5, C6, C7. Scale in centimeters.

where the body was previously broken in half. Because of this damage, the superior ring epiphysis is missing at the dorsal midline. A prominent right uncinate process is present; the left is absent. The cranial surface of the body is concave in the coronal plane, although it is decidedly less dorsoventrally concave than is the caudal surface. Both the costotransverse bar and the anterior tubercle are missing. The thick right pedicle is largely circular in cross section, but it evinces slight anterior flattening (height: 6.7

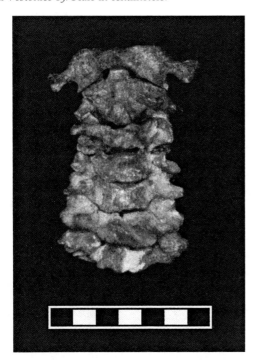

Figure 14.10 Anterior view of articulated cervical vertebrae of Dolní Věstonice 13. Scale in centimeters.

mm; breadth: 5.5 mm). The right superior articular facet is complete except for a small chip missing from its posterolateral corner. It is ovoid in shape and concave in both the sagittal and coronal planes. The right inferior articular facet is largely complete—it is missing a small posteromedial portion that was lost along with the right lamina. It is essentially planar, although there is some suggestion of a mediolateral convexity. The second preserved piece includes the left articular processes, which remain attached to a portion of the left lamina. The left pedicle is not preserved. All of the left inferior articular facet and most of the superior one is present (a portion of its ventral face is missing). The internally to externally thin lamina is smooth on both its internal and external surfaces.

The C3 of Dolní Věstonice 15 is preserved in two pieces (Figure 14.12). The first piece represents the left portion of the body, with a small, inferior portion of the pedicle attached to it. Its anterior surface retains cortical bone only along the lateral one-third or so (ca. 7 mm in width), whereas along the dorsal surface the cortical bone is much better preserved (almost the entire posterior surface is covered in cortical bone). On the caudal surface, cortical bone is found except at the lateral margin. There is no cortical bone on the bone's cranial surface, as it has all eroded away. The cranial half of the pedicle is missing. The second, smaller piece of C3 preserves portions of the articular processes, the root of the anterior process, most of the left lamina, and the spinous process. Only the posterior half of the left inferior articular process is preserved. Its articular surface is ovoid and slightly concave, with a posteromedial to anterolateral long axis. The superior articular process is missing perhaps one-fourth of its caudal portion. Its facet is flat and relatively circular.

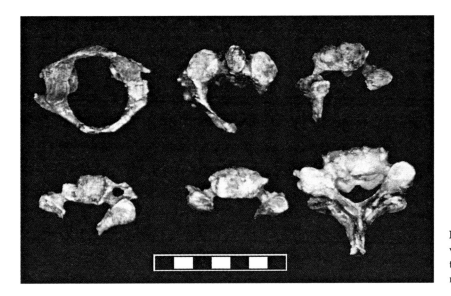

Figure 14.11 Superior view of cervical vertebrae, Dolní Věstonice 14. Top row (left to right): C1, C2, C3; bottom row (left to right): C4, C5, C6/C7. Scale in centimeters.

The spinous process is extremely short and evinces only slight bifurcation.

The third cervical vertebra of Dolní Věstonice 16 preserves most of the body that is glued at midline. The complete right pedicle evinces only trivial damage along its cranial margin, and the right superior and inferior articular processes are nearly complete. The left side of the vertebra is not well preserved, and only a tiny portion of the left pedicle remains. The right pedicle is dorsoventrally thicker than the left (right breadth: 7.0 mm, left: 5.1 mm), although this may be due to the matrix adhered on the ventral aspect of the right pedicle. The vertebral body is not terribly transversely broad, but it is dorsoventrally deep. The caudal surface of the body is slightly dorsoventrally concave, whereas the cranial surface is more flat.

Cortical bone is absent from the ventral and cranial faces of the body. The right inferior articular facet is nearly complete, lacking only its mediocaudal corner, but cortical bone is missing along almost its entire posterior face. The right superior articular facet is largely undamaged; its articular surface is sagittally convex.

Fourth Cervical Vertebra

The fourth cervical vertebra (C4) of Dolní Věstonice 3 is preserved in two pieces (Figure 14.8). The first represents most of the body, a partial left pedicle, the right pedicle, and a small portion of the right transverse process. Also preserved are most of the right articular processes and, posterior to them, much of the right lamina. The laminar

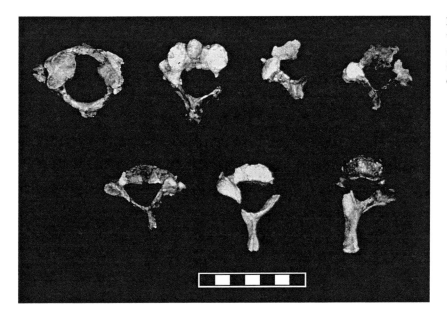

Figure 14.12 Superior view of cervical vertebrae, Dolní Věstonice 15. Top row (left to right): C1, C2, C3, C4; bottom row (left to right): C5, C6, C7. Scale in centimeters.

piece extends slightly beyond midline. The body is transversely wide and craniocaudally short. Its caudal surface is mediolaterally convex and dorsoventrally concave. Its anterior/inferior lip is less pronounced than that of the C3. The cranial surface of the body is dorsoventrally flat and mediolaterally concave. The right uncinate fossa has sustained heavy damage, whereas the left uncinate process evinces a small area of exposed trabecular bone on its ventral face. Ring epiphyses are present, except laterally on the right.

The second piece of C4 includes a very small piece of the left pedicle, most of the left superior articular process, a partial left inferior articular process, and much of the left lamina. This second laminar piece cannot be articulated with the first since it does not extend to the midline. None of the transverse process is preserved on the left, and on the right, portions of it are preserved ventrally and dorsally (the costotransverse bar that would connect the two portions is missing). The right superior articular process is ovoid, with a relatively flat surface, and it is damaged along its superomedial edge. In contrast, the left superior articular facet is not ovoid, but rather, in superior view, it has a slight concavity along its anteromedial edge, giving it an almost teardrop shape. Both inferior articular facets are ovoid and slightly concave. There is also a small (maximum dimension 11.2 mm) isolated piece of a bulbous spinous process that probably belongs to C4, but it may belong to C3. It cannot be articulated with either element.

The fourth cervical vertebra of Dolní Věstonice 13 (Figures 14.9 and 14.10) includes a largely intact body, although its ventral/caudal margin is at present broken up into smaller pieces (held together by glue) and remains separated from the rest of the vertebral body. The body is dorsoventrally narrow and transversely wide. Its entire dorsum is missing, including trabecular bone. The cranial surface of the body is slightly coronally concave but dorsoventrally flat. Its caudal surface is slightly transversely convex but anteroposteriorly concave, with a marked anterior/inferior lip. The right uncinate process is larger than the left. The transverse foramen is complete on the left, although the left anterior tubercle is damaged and the posterior tubercle evinces a loss of cortical bone posterolaterally. The gutter for the nerve is present on the left. On the right, very little of the dorsal part of the transverse process remains, and only a small superomedial piece of its anterior component is preserved. The pedicles are thick, although the right is anteroposteriorly thicker than the left (left height: 7.7 mm, right: 10.0 mm; left breadth: 5.6 mm, right: 7.1 mm). Both superior articular facets are preserved and asymmetrical. The ovoid right facet is planar to slightly concave, and the left one is markedly concave. The left facet is also less vertically oriented than the right one; that is, it lies in a more transverse plane. As with C3, the inferior articular processes follow the superior ones in size, with the left much larger than the right. The left facet is teardrop-shaped, with a paracoronal long axis and posteromedial apex. The smaller right facet is also teardrop- to oval-shaped, with a more parasagittal long axis. Both facets are slightly transversely concave. The laminae are short and internally to externally thin (left laminar height: 10.1 mm, right: 9.9 mm; left thickness: 2.3 mm, right: 2.9 mm), and the right one is longer than the left one. In superior view, the neural canal is arch-shaped, with a flat ventral and convex dorsal margin. The spinous process, which deviates slightly to the left, is not truly bifid but rather short and gracile, and although nearly impossible to quantify, it appears to project more horizontally than does that of the C3.

Dolní Věstonice 14's fourth cervical vertebra (Figure 14.11) is represented by left and right pieces that have been glued together. Together, they preserve most of the body (albeit damaged ventrally), the right anterior tubercle, the costotransverse arch, the right transverse foramen and pedicle, the right articular processes, portions of the right lamina and left pedicle, and the left articular facets. The overall shape of the body is dorsoventrally short and transversely broad. Its inferior surface is almost planar, that is, only slightly concave. The shape of the ventral face of the body is difficult to gauge since a large anterolateral portion is eroded on the left side. However, on the right, the inferior lip is preserved, and it appears less inferiorly projecting than that of the C3. The body's cranial surface is concave, with more marked transverse than dorsoventral concavities. Only the right uncinate process is preserved. It is only moderately expressed, compared to the C3, rising ca. 5.1 mm above the surrounding superior surface. The right anterior tubercle is dorsoventrally thin (1.7 mm) and superoinferiorly tall (5.9 mm), and it forms the anterior boundary of a large transverse foramen—height: 4.9 mm; breadth: 6.1 mm. The gutter for the right spinal nerve is quite well delineated, measuring 6.2 mm in breadth. Only the dorsal wall of the transverse foramen is preserved on the left, yielding a breadth of about 5.2 mm. The right pedicle is robust and only slightly inferiorly declined (it lies in a transverse plane). It is ovoid in cross section, with a slight flattening of its anterior face. The left pedicle is of a similar cross-sectional shape. The left anterior tubercle is represented by an area of missing cortical bone some 4.5 mm high and 4.3 mm wide, located at the anterior junction of the articular processes. The superior articular facets are ovoid and slightly concave. The boundaries of the right one are more discrete than those on the left, which smoothly transition onto the back of the inferior process. The inferior articular process is complete on the right, whereas on the left its posterior one-third is missing. The right inferior articular facet is ovoid and planar, whereas the left appears slightly concave. Both have discrete boundaries separating them from the nonarticular areas. The smooth-surfaced right lamina is about halfway complete; it is cracked, and its cranial margin is displaced into the neural canal, but it remains attached to the pedicle along its caudal margin.

Dolní Věstonice 15 preserves a largely complete C4 (Figure 14.12). Most of the body is present, but there is an area of damage anterolaterally on the left and posterolaterally on the right. There is also much erosion on the cranial margin of the body, so that about half its cortical bone is missing. In anterior view, only the right half of the body has cortical bone running to the base of the right uncinate process (and only the base of this process is preserved). The body's cranial surface is transversely concave, with its dorsal margin less concave than the ventral margin. The right uncinate process would have been marked, but it evinces heavy cranial damage. The inferior/anterior lip is quite prominent, dipping some 4.5 mm below the caudal margin of the body. In inferior view, it is evident that all the cortical bone (as well as some trabecular bone) is missing from the right half. In posterior view, the right lateral third is missing its cortical bone. The middle third is largely complete, although some trabecular bone is exposed. The left lateral third is complete, but it is separated from the middle third by a fissure that runs onto the caudal surface of the body, ending some 4.5 mm from the ventral margin. At midline a short (6.1 mm) posteroinferior segment of the ring epiphysis is preserved. The right pedicle is extremely thick (height: 6.4 mm; breadth: 5.9 mm). There is a swelling at its proximal end, indicating that some of the body is attached, but it cannot be rearticulated to the damaged right side of the body. Both right articular facets are flat and damaged. The lateral two-thirds of the right superior facet remain; there is a clean, superoinferior cut that has removed the remainder of it. More of the inferior articular surface is preserved. It is missing only a small portion of its inferomedial margin. The right lamina is tall (12.0 mm) and is thicker (3.0 mm) than the left one. It is damaged inferoanteriorly, and it is missing a small piece of cortical bone (3.1 mm wide × 1.2 mm high) on its dorsal aspect. The spinous process is short (14.9 mm) and bulbous. Its tip is not truly bifid (although one can make out the two halves of which it is part). On the left, the pedicle, articular facets, lateral wall of the left transverse foramen, and most of the left lamina are preserved. The left pedicle is lacking only its cranial margin, but its dorsoventral diameter (4.1 mm) is considerably smaller than the right (5.9 mm). The superior articular facet is relatively broad and flat, but the inferior articular facet is more concave. As with the pedicles, the left lamina is thinner than the right; its internal-to-external thickness is 1.9 mm.

Dolní Věstonice 16's C4 is represented by two pieces. The first includes the left half of the body, along with a small piece of anterior tubercle, a complete pedicle, virtually complete articular facets, and a small section of the left lamina. The second piece preserves the two right articular facets, as well as the right pedicle and a slight bit of right lamina. Laterally, the body is dorsoventrally shallow, with its dorsoventral dimension increasing toward midline. It is difficult to tell if the cranial surface is concave; it may be slightly transversely concave. The body's caudal surface is dorsoventrally concave, with a moderately projecting inferior lip that shows some signs of osteoarthritis. Specifically, on the lower left, the body is flattened and gives rise to an osteophytic growth that measures 8.5 mm in length and 4.6 mm in breadth. In inferior view, the ring epiphysis is present, although it is damaged laterally because of osteoarthritis. In anterior view, the body preserves three-quarters of its face—it is missing an area from inferolaterally on the right to just to the left of midline along the cranial margin. The root of the anterior process is present. In posterior view, about a third of the posterior face is present—the missing section begins at midline superiorly and tracks down inferolaterally to near the left inferolateral corner. The caudal surface is very smooth—almost polished—again reflective of osteoarthritis. The left pedicle, articular processes, and lamina are still attached to the body, but the right pedicle/articular processes/laminar piece remains separate because of the extensive damage to the right half of the body. The pedicles are thick and circular (left) to ovoid (right) in cross section (left height: 6.2 mm, right: 5.7 mm; left breadth: 5.9 mm, right: 6.7 mm). The ovoid left superior articular facet is concave, with a well-marked inferior border, and it is slightly eroded along its superolateral margin. The ovoid right superior articular facet is also concave, but its inferior border is less discrete than that of the left. Both inferior articular facets are planar and ovoid; the left is missing its lateral one-fourth, and the right is damaged along its lateral and posterior borders. Both laminae are incomplete and heavily damaged.

Fifth Cervical Vertebra

The fifth cervical vertebra (C5) of Dolní Věstonice 3 is in a poor state of preservation (Figure 14.8). The entire ventral face of the body is missing, and the body is split into two pieces (now glued) by a fissure just to the left of midline. Perhaps one-third of the cranial surface remains along the body's dorsal margin, and the uncinate processes have been eroded away. Inferiorly, only the right piece preserves its caudal surface, and in posterior view the left portion of the body is missing its caudal half, although the right side is more complete. The cranial surface appears mildly concave in the coronal plane, and the caudal surface is dorsoventrally concave. Right and left articular processes have been glued to the body. The right side has most of both articular facets preserved, but the ovoid and slightly transversely concave superior facet evinces cranial erosion, and the inferior facet is damaged along its medial margin. The right inferior articular facet is also ovoid and is clearly concave from its cranial to caudal margins. In cross section, the relatively thick and robust right pedicle has a rounded cranial margin and a flattened caudal margin (height: ca. 4.3 mm; breadth: ca. 4.5 mm). The left pedicle is also relatively complete, albeit with some damage along its inferoproximal aspect (height: ca. 4.6 mm; breadth: ca. 4.1 mm). Much of the left superior articular facet is present, although it may have continued inferomedially, which is difficult to evaluate since the entire lamina and inferior articular process are missing.

The fifth cervical vertebra of Dolní Věstonice 13 (Figures 14.9 and 14.10) was in several pieces that have been glued together. The transversely wide and dorsoventrally narrow body had been split into two pieces. The cranial and caudal surfaces of the body are missing most of their cortical bone. It also appears that part of C6's body came off when C5 and C6 were separated (they had been glued together for museum display). The caudal surface appears flat, but it (and the dorsal body) has been extensively damaged. The right uncinate process is marked, but only the base of the left uncinate process remains. The left posterior tubercle is present, but the anterior tubercles are gone, as is the right posterior tubercle. The posterior side of the nerve gutter for C5 is present on the left. Only the dorsal quarter of the gutter is preserved on the right. The pedicles are thick and circular in cross section, but the right cannot be measured because of a crack (left height: 6.9 mm; left breadth: 6.8 mm). The left superior articular facet is much larger than the right, and it is concave, whereas the right one is essentially planar. The ovoid and slightly concave inferior articular facets are more equal in size and form. Both laminae were broken, which resulted in the loss of bone along their cranial margins. The spinous process is relatively large, projects horizontally, and has a large fossa on its inferior aspect. The spine does not appear to be bifid, but there may be a missing tip on the left side where trabecular bone is visible.

The fifth cervical vertebra of Dolní Věstonice 14 (Figure 14.11) is represented by a virtually complete body, two pedicles, right and left articular processes, and a small piece of right lamina. The body, most of which is present, was in two pieces that have been glued together. It is wider than its more cranial counterparts but remains superoinferiorly short. The cranial margin of the body is slightly concave in the coronal plane. The cranial ring epiphysis is worn away ventrally, but it is present along the dorsal and lateral margins. The caudal surface of the body also appears concave in both the coronal and sagittal planes. This appearance is somewhat accentuated by the anterior and posterior inferior lips. The anterior lip is damaged (a small section ca. 4.7 mm in length is missing just to the left of midline). The caudal surface of the body and its associated ring epiphysis are also damaged dorsally near midline. The uncinate processes are heavily eroded, and the right anterior tubercle and both costotransverse bars are missing. On the left, a small (ca. 7.5 mm) portion of the anterior process remains, forming the lateral boundary of the gutter for the ventral ramus of the C5 spinal nerve. Both pedicles are superoinferiorly thick (left height: 7.6 mm, right: 6.4 mm; left breadth: 5.1 mm, right: 5.5 mm), and ovoid in cross section, and have a more or less transverse orientation. The superior articular facets are slightly concave. The right is larger and more ovoid than the left, and it extends more superiorly and laterally. It also lacks a discrete boundary separating it from the nonarticular surface (the left facet has a distinct inferior boundary). The right inferior articular facet is flat and circular and possesses a clearly demarcated articular surface. The left inferior articular facet is missing its posteromedial quarter. Only a tiny portion of the right lamina, which is too small to measure, is present; however, it does appear that the laminae were internally to externally thin.

Dolní Věstonice 15's fifth cervical vertebra comprises three pieces that have been glued together (Figure 14.12). Beginning with the body, most of the left half is preserved, along with a complete pedicle and uncinate process and a partially complete transverse foramen. Cortical bone covers the cranial surface of the body, although there is slight damage along the piece's medial edge. Only the lateral 30% or so of the body's ventral surface retains its cortical bone. The left lateral side preserves cortical bone except near the root of the anterior process. The dorsal face of the vertebra has cortical bone but with slight caudal damage (most of the dorsal aspect of the inferior ring epiphysis is gone). In inferior view, cortical bone is missing dorsally and anteromedially. The cranial surface of the body is largely flat, although it evinces a slight downward slope from its dorsal to ventral margins. In lateral view, the caudal body is clearly concave. This appearance is accentuated by the ring epiphyses, particularly anteriorly. It appears that the inferior/anterior lip would have been as marked as C4's had it been preserved. The left uncinate process is marked—it rises some 3.9 mm above the cranial margin of the body. The large left pedicle originates low on the body and is inferiorly declined in orientation (height: 5.8 mm; breadth: 5.2 mm). The spinous process is small and slightly bifid. The left side of the spine is more bulbous, with damaged cortical bone. The right lamina has a complete cranial margin, although it is damaged along its caudal margin. The left lamina is less complete. It only runs for 14.0 mm along its cranial margin and far less along its caudal margin, which, like the right, is damaged. Thickness cannot be measured on the left lamina, but the right one is 2.8 mm thick near the junction of the two laminae and 2.6 mm thick near its junction with the superior articular facet. Although difficult to evaluate, the inclination of C5's spinous process (as with C3 and C4) seems to be slightly declined, if not horizontal. The right superior articular facet is broad, tall, domed, and relatively flat. The right inferior articular facet is broad and slightly convex. The bone posterior to it is relatively bulbous, with a slight anteroposterior furrow separating it from the superior articular facet. The right pedicle is broad and flat (height: 6.5 mm; breadth: 8.4 mm), and just proximal to the pedicle there is a small cranial piece of the body. This body piece, along with the superior articular facet's medial wall, forms a furrow about 3.9 mm in width, through which the fifth spinal nerve would pass. Inferior to the gutter lies the root of the anterior tubercle.

The fifth cervical vertebra of Dolní Věstonice 16 preserves most of the vertebral body and portions of the left side of the neural canal. In anterior view, only the right uncinate process and most of the right anterior process are missing on the body (along with minor erosion of the anteroinferior cortical bone). In superior view, only the right uncinate process and the region just posterior to it are noticeably missing. In posterior view most of the body is present, except the right posterolateral corner, which is gone. The caudal surface of the body is dorsoventrally concave, with only a moderately projecting inferior lip, and as with C4, this is probably due to osteoarthritis. There are two inferior projections (osteophytes) that rise off of the body's posterior caudal margin. The left is more prominent than the right, but the right may be eroded. The body's cranial face is relatively flat, with a gentle lateral slope up to the left uncinate process. The left uncinate process itself is quite prominent—it rises 4.6 mm above the cranial surface. The left anterior process has only its ventral root preserved. On the caudal surface of the left anterior process is yet another arthritic area (9.7 mm in length x 5.7 mm in breadth)—a flattened pseudoarthrosis of sorts for the uncinate process of C6. The thick and downwardly oriented left pedicle is roughly circular in cross section (height: 6.1 mm; breadth: 6.0 mm). The left superior articular facet is badly damaged along its lateral and superoanterior borders, but with a discrete inferior margin. It was, however, ovoid, with a mediolateral long axis. The ovoid inferior articular facet has a posteromedial to anterolateral long axis and is damaged laterally and posteriorly. Whereas the superior facet appears planar, the inferior facet is clearly concave. The entire posterior side of the inferior articular process is damaged (cortical bone removed), and very little of the lamina is preserved.

Sixth Cervical Vertebra

The sixth cervical vertebra (C6) of Dolní Věstonice 3, like the fifth, is in a poor state of preservation, possessing only a damaged body and small portions of the neural canal (Figure 14.8). Left and right halves of the body have been glued near midline, and the right two-thirds of the body evinces heavy anterior damage. The left uncinate process is preserved in its entirety; the right is damaged anteriorly. The body's cranial surface is relatively flat, except at its lateral margins near the uncinate processes. In contrast, the body's caudal surface is gently concave in both the coronal and sagittal planes. The ring epiphyses are fused onto the centrum and are present throughout, except for two places: a small (16.4 mm long) margin near midline along the ventral face of the cranial surface and dorsally at midline on the caudal surface (ca. 3.0 mm are missing). This latter site is also missing much of its cortical bone. Only the root of the left pedicle remains, and its caudal margin is more complete than its cranial margin. Anterior to this root, the site of the posterior portion of the transverse process evinces heavy damage. On the right, the author glued the pedicle to the body, where there is a good connection between the pieces posteriorly, but much of the body at the anterior pedicle junction is missing. The pedicle is superoinferiorly and mediolaterally thick (height: ca. 5.5 mm; breadth: ca. 5.3 mm). It is somewhat round in cross section, but it is flatter inferiorly with a gently convex superior surface. Although a portion of the right articular processes is preserved, none of the articular surface of the superior articular process has survived. In contrast, much of the inferior process is present. There is an isolated spinous process piece measuring some 23.3 mm in length and 9.3 mm in height that probably belongs to C6. However, one cannot rule out the possibility that it belongs to C5.

The sixth cervical vertebra of Dolní Věstonice 13 is complete (Figures 14.9, 14.10, and 14.13). Its most notable attribute is found on its ventral face—a long, channel-like depression about 7 mm in width running from midline at the cranial margin to the left lateral side of the body's caudal margin. Judging from burial photos, this depression is postmortem, and it is due to the corpus of the mandible being compressed into the ventral face of the vertebra. The anterior portion of the right lateral side of the body also evinces cranial damage. The dorsal half of the body is heavily damaged but glued in place—its cranial face is damaged (with missing cortical bone at posterior midline) from being glued for display. The body is dorsoventrally narrow and transversely wide. The caudal surface of the body is mildly dorsoventrally concave. Anteriorly, only a small portion of the caudal surface's cortical bone is missing just to the left of midline. The cranial surface of the body is missing a slight amount of

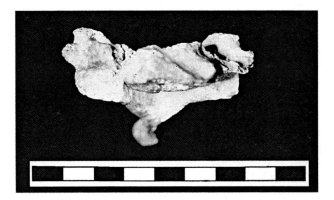

Figure 14.13 Anterior view of C6, Dolní Věstonice 13. Scale in centimeters.

cortical bone on the left. The vertebra does not appear to have had uncinate processes. In superior view, the dorsal margin of the body is almost straight, whereas the ventral body is more convex. Interestingly, there is no transverse foramen on the right side, despite the fact that the entire carotid tubercle is preserved. Nor is there a transverse foramen on the left; rather, only small irregular perforations are present. Therefore, the morphology is unusual in that it does not appear that the vertebral arteries entered the transverse foramina until the C5 level. The nerve gutter on the right is impressive in size (maximum internal diameter = 7.2 mm). It runs 15.7 mm from the vertebral canal. The left gutter is damaged—its lateral extent is missing, as is most of its medial/anterior wall. Both pedicles are preserved, although the right pedicle evinces an imperfect glue join. The superior articular facets are widely spaced from each other, but they are more or less equal in form and size. They are essentially ovoid (with mediolateral long axes) and planar in morphology. In contrast, both inferior facets are slightly superoinferiorly concave, although, like the superior facets, their long axes are mediolateral. The inferior lateral margin of the left inferior articular facet is slightly damaged. The laminae are long (i.e., there is a broad span between left and right articular processes). They are internally to externally thick (left: 2.9 mm, right: 3.1 mm), but not particularly tall (left: 12.8 mm, right: 14.5 mm). The left was broken and has been glued just distal to the articular facets. The vertebral canal is wide, short, and triangular in shape, with a posterior apex. The spinous process is long, with a bulbous tip. It declines slightly (ca. 15°). Its caudal surface is flat, with two lateral ridges barely discernable to either side of midline.

The C6 of Dolní Věstonice 14 is largely complete, but it has been fused postmortem with C7 (Figure 14.11). It may also exhibit postdepositional craniocaudal compression. Its body is much larger than C5's, and it has prominent uncinate processes (although the left is somewhat distorted). When viewed in *norma frontalis*, the body's broad ventral face is inferiorly convex along both its cranial and caudal margins. In superior view, the ventral face of the body is relatively flat, although to some extent, this may be due to postmortem distortion. Also evident in superior view is a fault running just to the left of midline from the dorsal face (missing bone) to just dorsal to the body's ventral face. The costotransverse bars and anterior tubercles are missing. The gutters for the spinal nerves cannot be measured, but they were apparently quite large. The pedicles are thick, although the superior margin of the left is covered with concreted dirt and probably compressed bone. The right pedicle is ovoid in cross section (height: 8.6 mm; breadth: 6.3 mm). Both superior articular facets are ovoid, with mediolateral long axes and discrete margins, but the left one is squatter than the right one, which is more circular in shape. On the right, there is a crack along the inferior medial boundary of the facet that continues down to separate the lamina from the right inferior articular process. On the left, a smaller crack runs from the proximal left lamina to the very inferomedial corner of the left superior articular facet. The laminae are externally to internally thin—their thicknesses (measured at midpoints) are 2.0 mm (left) and 2.5 mm (right). The neural canal is broad and triangular, but the left side may be slightly distorted; it is difficult to know for certain because of the concreted dirt above the pedicle. The spinous process is extremely short (15.3 mm) for a C6, with a small swelling at its tip.

The sixth cervical vertebra of Dolní Věstonice 15 is preserved in two pieces (Figure 14.12). The first preserves most of the body, left pedicle, and articular processes, and the second preserves the right lamina and the spinous process. The body has suffered extensive damage to its ventral face, but it preserves both cranial and caudal ring epiphyses along its dorsal margin. In superior view, the cortical bone on the body's cranial surface appears to have been concave both in the coronal and sagittal planes, but it is heavily damaged along midline. Neither uncinate process is present. The body's caudal surface preserves its cortical bone, and it is coronally concave but only slightly dorsoventrally concave. The entire ventral surface of the body is missing. On the dorsum of the body, a glue join is evident, joining the left third to the right two-thirds. Only the posterior portion of the root of the right pedicle is preserved, but most of the left is preserved; it is quite thick (breadth: 6.2 mm; height: 6.6 mm). On the left there is evidence for the gutter for the sixth spinal nerve, although it is not as well preserved as the one present on C5. Below the gutter, extending anteriorly, is a shelf beyond the pedicle—the anterior tubercle. It is broken anteriorly, with exposed trabecular bone. The left superior articular facet is broad and relatively tall. Inferiorly, its surface becomes gently rounded to form the bulbous posterior portion of the inferior articular process. In other words, there is a gradual change from articular to nonarticular surface. The broad, ovoid, and slightly concave left inferior articular facet has well-delimited boundaries. The inclination of the articular facets appears almost vertical. None of the left lamina is preserved. The only laminar piece that remains is a right partial lamina, which is more complete cranially than caudally. Its external surface is relatively smooth, although there is a fossa (5.6 mm long; 1.9 mm high) just below its superior margin. The lamina is relatively thick—its internal-external thickness is 4.2 mm near the midline. The right laminar piece is also associated with the complete spinous process. The spinous process is not long for a C6 (maximum length: 25.9 mm). It is clearly bifid, with two large, egg-shaped tips, the right one of which extends more inferiorly and posteriorly than does the left one.

The sixth cervical vertebra of Dolní Věstonice 16 is preserved in the form of two heavily damaged pieces that do not articulate with each other but are likely to be two halves of the body of C6. The cranial surface of the body appears to have been flat, whereas caudally it is dorsoventrally concave. Laterally on the left there is osteoarthritis in the form of a bony lip, and the cortical bone on the caudal surface beneath this lip is missing. Cranially, on the dorsal side of the left piece, there is also slight lipping of the margin. The left uncinate process is missing, but the ventral portion of the base of the right is present. The left half also preserves the root of the pedicle and the caudal margin of the root of the anterior process. The left pedicle is roughly circular in cross section (breadth: 6.8 mm; height: 6.7 mm). Posterolaterally on the right, there is osteophytic lipping along the cranial dorsal edge of the body. More of the neural arch is preserved on the right than on the left. The right pedicle points inferiorly, and although complete, it does evince damage. Specifically, glue on the right pedicle precludes taking its anteroposterior measurement, but its height is 7.0 mm. The right superior articular process is all but gone—only a small inferior portion some 9.9 mm wide and 5.9 mm tall remains. The articular surface itself appears to have been concave. The ovoid and concave right inferior articular facet is also heavily damaged, and it evinces a posterolateral osteophytic lip. Its medial boundary is marked by a discrete ridge.

Seventh Cervical Vertebra

Only a fraction of the body of the seventh cervical vertebra (C7) of Dolní Věstonice 3 is intact (Figure 14.14). It is preserved en bloc with several other vertebral and costal elements. Its assignment to C7 is certain, however, since the part that remains has a clear uncinate process and no rib facets. Most of the ventral body is gone, leaving only

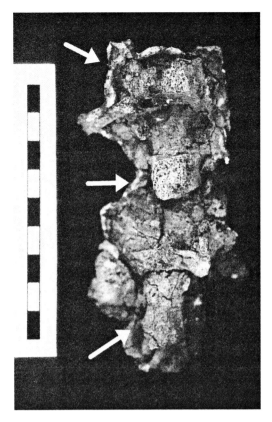

Figure 14.14 Anterior view of lower cervical and upper thoracic vertebrae, Dolní Věstonice 3. Cranial is up. The top arrow indicates the body of C7, the middle arrow indicates the body of T2, and the bottom arrow indicates the natural cast of the neural canal of the upper thoracic vertebrae. Scale in centimeters.

the left front third remaining, and even this portion has sustained lateral damage. The body's caudal face appears to have been concave in both the coronal and sagittal planes. The dorsal margin of the body is the best preserved—here the cranial ring epiphysis is evident. The lateral quarter of the right side of the body is slightly displaced cranially from the left three-quarters. Although this precludes many measurements, the body was clearly transversely wide and relatively tall, especially in comparison to C6.

The seventh cervical vertebra of Dolní Věstonice 13 is fairly complete, although little of its body actually remains (Figures 14.9 and 14.10). The body is preserved in two pieces that are connected by putty. The first piece begins just to the right of midline, and it includes a portion of the right pedicle, but the pedicle is attached to the body by putty. The second piece of the body, which is even smaller, is attached to the left pedicle. It preserves a complete anterior tubercle (unlike on the right, which is missing), but there is no true contact with the right piece. Both pieces evince heavy ventral erosion, but dorsally the right piece preserves up to midline, with full cortical bone and cranial and caudal ring epiphyses present. Although difficult to evaluate, the body was seemingly wide and dorsoventrally narrow. The body's caudal surface is slightly transversely convex—unusual for a C7. The uncinate processes were slight, judging from the left, which is preserved. In superior view, the dorsal face of the body appears to have been relatively flat. The right pedicle is heavily reconstructed with wax and putty. The left pedicle is in better condition, and it is notable in that it is thick both superoinferiorly (9.6 mm) and mediolaterally (9.7 mm). The left anterior tubercle is preserved, but its lateral extension beyond the foramen has been lost. There is a foramen transversum on the left, but since C7 does not typically transmit the vertebral artery, it is quite small: height: 3.2 mm; width: 3.0 mm. The gutter for the C7 spinal nerve is present in its entirety on the left; its origin is present on the right. Both superior articular facets are broad and ovoid with mediolateral long axes. The facets lack discrete margins; rather, they grade into nonarticular surfaces. The two inferior articular facets are of roughly equal size and form. They are both ovoid and slightly concave, with inferomedial to superolateral long axes. The laminae are internally-externally thick (3.4 mm left); 4.2 mm right). There is a crack in the left lamina (as in C6 above and T1 below), and the spinous process was broken off at its base and has been glued to the lamina. The spinous process is long, probably at least 38 mm, although it cannot be measured in its current state. The tip of the process is bulbous but not as wide as that of C6; its width is 8.8 mm and height 8.1 mm. Its declination is approximately 10°. The neural canal is transversely wide, dorsoventrally short, and roughly triangular in shape, with a posterior apex.

Dolní Věstonice 14's C7 has a heavily damaged body compressed into C6 (Figure 14.11). Likewise, the components of the transverse processes have been lost. The neural arch, in contrast, is complete. The body is wide, dorsoventrally narrow, and superoinferiorly short (although the last feature may be due to its being crushed by C6). Only the right third of the vertebral ventral face retains cortical bone; the left two-thirds is eaten away (as is much of the trabecular bone). Dorsally, in the neural canal, cortical bone is present, but caudally, cortical bone is missing from much of the ventral half of the body. The right uncinate process is present, but it remains compressed into the underside of C6. On the left, the posterior wall of the gutter of the C7 spinal nerve is present, but it is damaged at its lateral/distal end. On the right, a gutter is not evident. The pedicles appear anteroposteriorly thick but slightly superoinferiorly compressed (left height: 7.1 mm, right: ca. 6.7 mm; left breadth: 9.0 mm, right: 9.1 mm). The ovoid inferior articular processes have mediolateral long axes. The left is more planar than the right, which is slightly convex. The spinous process, which exhibits a vertebra prominens, is much larger than that of C6 but is short for a C7. The swelling at its tip (which is damaged) measures 8.2 mm in breadth and 8.5 mm in height. The laminae are externally to internally thick: 3.8 mm (left) and 3.9 mm (right). The neural canal is broad and triangular. Because of postmortem deformation, it appears that the right side of the body is encroaching on the canal.

Dolní Věstonice 15's C7 (Figure 14.12) is preserved in two separate pieces—the first is the body; the second includes the right articular processes, lamina, and spinous process. The largely complete, kidney-shaped body of C7 lacks only its right uncinate process, a bit of the fused superior ring epiphysis just to the left of midline (about 7.1 mm), and some of the ventral face below. Its cranial surface is only slightly concave, but its caudal surface is flatter. The left uncinate process is marked, rising 5.1 mm from the surrounding surface. Both anterior tubercles are gone, and the right pedicle is heavily damaged and glued onto the body, and as such its thickness cannot be measured. The left pedicle is complete and is relatively circular in cross section (height: 7.0 mm; breadth: 6.3 mm). The superior articular facet is clearly demarcated from the surrounding nonarticular bone. It is flat and evinces damage to its lateral margin. In anterior view, the bone that forms the ventral face of the superior articular process is eroded, such that only the medial one-third of the process retains its cortical bone. The inferior articular facet is broad, ovoid, and gently concave with clearly demarcated margins. The right lamina is heavily damaged; only its inferior margin is preserved: it is here that there is contact between the lamina and the spinous process. The spinous process is relatively long (33.9 mm) and nearly horizontal in orientation. It evinces a fairly marked vertebra prominens (breadth: 10.4 mm; height: 6.9 mm). The underside of the spinous process is rugose, with lateral ridges and a median sulcus. In posterior view, there appears to be slight clockwise torsion of the spinous process.

The seventh cervical vertebra of Dolní Věstonice 16 is represented solely by a badly eroded left half of the body and pedicle. Only a small portion of cortical bone remains from the anterolateral right side to the posterolateral left side of the cranial surface, and in anterior view only a small (6.7 mm height x 9.6 mm width) section of cortical bone remains on the anterolateral left side. Caudally, most of the left half retains cortical bone, but laterally there is extensive osteoarthritis, with cortical bone erosion (in life) and osteophytic lipping on the far left. In posterior view, only the left lateral third of the body remains covered with cortical bone, and there is osteophytic lipping along the left caudal/dorsal margin. Both the cranial and caudal surfaces of the body appear to have been dorsoventrally concave. The left pedicle is flattened inferiorly and is anteroposteriorly broad (height: 5.6 mm; breadth: 7.5 mm).

The Thoracic Vertebrae

In humans, the twelve thoracic vertebrae are associated with the twelve pairs of costae, and as such are readily identified by the synovial articulations for these ribs that they bear on either their bodies (always) and/or their transverse processes (except for the two caudal ribs). All of the Dolní Věstonice and Pavlov individuals preserve at least one thoracic element, and the members of the triple burial preserve at least part of each of the twelve thoracic elements.

First Thoracic Vertebra

All that remains of the first thoracic vertebra (T1) of Dolní Věstonice 3 (Figure 14.14) are its two pedicles, right inferior articular process, and a portion of the right transverse process, all of which remain embedded in matrix. A natural cast of the neural canal is also preserved in the matrix. The left pedicle appears taller than the right, and it seems likely that some of T1's body remains adhered to it. Both pedicles appear to have been mediolaterally thick with relatively flat inferior margins. Only a small portion of the right transverse process remains. It is characterized by a shallow inferior fossa and is attached to the right pedicle and right inferior articular facet, which remains in articulation with the superior articular facet of T2.

Dolní Věstonice 13's T1 (Figure 14.15) is largely complete but has a badly damaged body and remains fused to T2. Despite this damage, it is apparent that the body was superoinferiorly tall. The entire dorsal half of the body is missing, and thus the pedicles are connected by wax. On the body's cranial face much of the annular ring has been eroded away. The body's caudal surface evinces a gentle coronal convexity, whereas in lateral view the ventral face is flat. Facets for the heads of the first ribs are gone, but the demifacet for the left second rib is present. It is circular and slightly concave, measuring 7.5 mm in both height and length. The pedicles, laminae, transverse, and spinous processes are all complete, although both laminae were broken and glued. The pedicles are mediolaterally thick, and the left pedicle is also superoinferiorly thick. One cannot measure the superoinferior dimension of the right because the T2 superior articular process is crushed up into it (left pedicle height: 9.6 mm; left pedicle width: 9.7 mm; right: 9.5 mm). Both ovoid (mediolateral long axes) superior articular processes are slightly convex, with discrete articular surface boundaries. The right is more inferior in position than the left. The transverse processes are long and thick (left transverse process length: 30.6

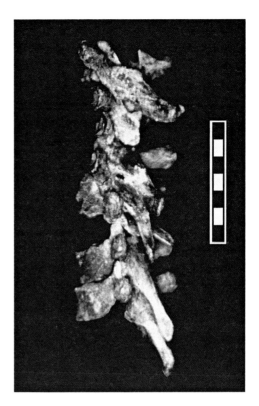

Figure 14.15 Lateral view of articulated T1 through T7, Dolní Věstonice 13. Scale in centimeters.

mm; right: 30.5 mm). Both preserve small ovoid facets for the rib tubercles. The left facet is particularly small, and although damage precludes measuring its height, its width is 5.7 mm. The right facet measures 5.5 mm in height and 7.1 mm in width. The spinous process is slightly longer than that of C7 (40.4 mm vs. 38.4 mm) and has a bulbous tip (width: 8.7 mm; height: 10.8 mm). The declination of the spinous process cannot be accurately measured but is about 10–15°. Neural canal shape is ovoid, with a transverse long axis.

Dolní Věstonice 14's T1 (Figure 14.16) is missing only its left superior articular process. Its body is largely complete, although there is plaster near the right full rib facet. The cranial surface of the body is planar, whereas the caudal surface is undulating, probably because of postmortem distortion. The body, although broad, is not very tall for a T1 and is dorsoventrally narrow. The cranial ring epiphysis is slightly eroded on the left but otherwise is complete. Inferiorly, the annular ring is only noticeable ventrally and dorsally near midline; elsewhere it is eroded. The left full facet for the head of the rib has a clearly demarcated, circular articular surface (breadth: 5.5 mm; height: 5.9 mm). The pedicles are thick and round in cross section (right height: 8.8 mm; right breadth: 8.4 mm). The left pedicle is preserved in two pieces, which cannot be rearticulated well because of postmortem distortion. The first piece preserves

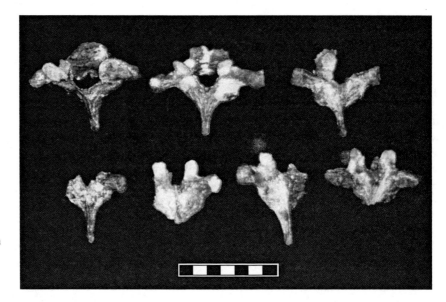

Figure 14.16 Posterior view of T1 through T7, Dolní Věstonice 14. Top row (left to right): T1, T2, T3; bottom row (left to right): T4, T5, T6, T7. Scale in centimeters.

the cranial margin of the pedicle and transverse process. The second piece, attached to the body, includes its bottom half. The superior articular facets are broad and short. The right inferior articular process is well preserved, but the left evinces damage along its inferior margin. The right inferior facet is ovoid, with a superoinferior long axis (it is not as mediolaterally expanded as are the superior facets). The transverse processes are short and more or less horizontal. Both have flat facets for articulation with the rib 1 tubercle: height: 6.2 mm (right) and 5.5 mm (left); breadth: 6.7 mm (left) and 6.3 mm (right). The neural canal is ovoid and dorsoventrally compressed. The spinous process is too declined because of faulty reconstruction. The swelling at its tip measures 8.2 mm in height and 6.4 mm in breadth.

Dolní Věstonice 15's T1 is virtually complete (Figures 14.17 and 14.18). The superoinferiorly tall and dorsoventrally narrow body has sustained posteroinferior damage, especially on the left. In superior view, the body appears kidney-shaped, with a ventrally concave posterior surface and ventrally convex anterior surface. The rib head facets have largely been destroyed, save for the ventral portion of the left one. The region for the left demifacet is missing; on the right the cranial margin is barely visible. The superior articular facets are flat, short, and broad; the inferior articular facets are more circular. The laminae are

Figure 14.17 Anterior view of T1 (top), T2 (left), and T3 (right), Dolní Věstonice 15. Scale in centimeters.

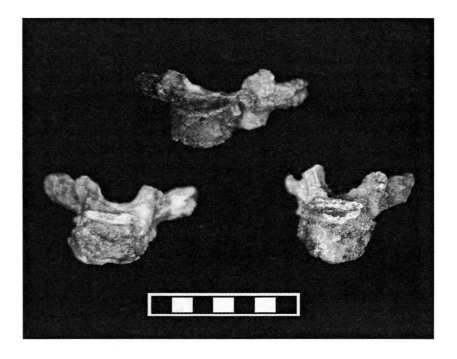

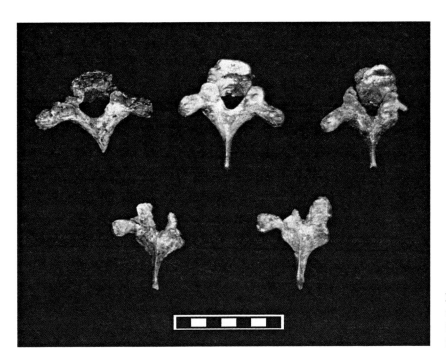

Figure 14.18 Posterior view of T1 through T5, Dolní Věstonice 15. Top row (left to right): T1, T2, T3; bottom row (left to right): T4, T5. Scale in centimeters.

internally to externally thick (left: 7.3 mm, right: 7.2 mm). The transverse processes are horizontal. They appear short but are thick in cross section, with large, circular facets for the rib tubercle. The facet for the left rib tubercle is concave; that on the right is shallower. The spinous process is badly damaged, particularly on its left side, where it has been filled in with putty (the right side also has putty, but less). The spinous process is not preserved to its tip, but nonetheless it appears that it may have deviated to the left. The spinal canal is relatively ovoid in cross section, with greater width than depth.

Dolní Věstonice 16's T1 preserves only the lateral left 25% of the body, with cortical bone present on its cranial, caudal, and lateral margins, as well as on the pedicle and parts of the articular processes and left lamina. Osteoarthritis is evident in the form of lateral osteophytic lipping on both the cranial and caudal faces of the body. The pedicle is circular in cross section (height: 6.2 mm; breadth: 7.2 mm) and somewhat superiorly inclined in orientation. Only a tiny portion of the superior articular facet remains—it is heavily damaged on its lateral, medial, and superior margins. Most of the inferior articular facet appears to be preserved. It is large, ovoid, and deeply concave. Only a minute portion of lamina remains.

Pavlov 1 preserves T1 in the form of a cerviform thoracic vertebra. It is cerviform in that its superior articular facets and body shape are reminiscent of a cervical vertebra, but it is more thoracic-like in the following traits: (1) its left pedicle is thick and rounded; (2) the roots of the transverse processes are thick; (3) the inferior surface of the body is dorsoventrally concave, not planar like C7; and (4) there is no clear nerve gutter, nor is there evidence for a vestigial anterior tubercle. Thus, although it could be a C7, T1 assignment is more likely. All that remains of the body is a small piece attached to the left pedicle; no rib facet or demifacet is visible. The thick pedicle is glued to the lamina; it is slightly ovoid in cross section (height 10.9 mm; breadth 9.7 mm). The superior articular facets are ovoid with mediolateral long axes. The right evinces trivial damage to its cranial margin. Both inferior articular facets were transversely wide but are damaged such that their caudal halves are missing. Only the roots of the transverse processes remain. The laminae are internally to externally thick: 5.2 mm left; 4.9 mm right. Only the root of the spinous process remains, and it does not appear to be significantly declined.

Second Thoracic Vertebra

The second thoracic vertebra (T2) of Dolní Věstonice 3 (Figure 14.14) is represented by a partial body, right partial pedicle and superior articular process, and a left pedicle. All of these elements remain embedded in matrix. There is also a natural cast of the neural canal associated with these remains. The body is missing its entire ventral face, and most of the body to the left of midline is missing. In fact, the only cortical bone remaining is about one-half of the cranial margin and perhaps 20% of the right anterolateral face. The rest of the body is exposed trabecular bone. The cranial face appears both dorsoventrally and transversely concave. The body appears to have been dorsoventrally narrow and relatively superoinferiorly tall.

Atlhough both pedicles are present, most of the left is hidden from view, and the inferior margin of the right is missing. Neither inferior articular process is evident, but the right superior articular process is preserved.

Dolní Věstonice 13's T2 is a badly damaged bone: its right side is crushed into T1, and all that remains of its body are two small pieces attached to the pedicles, although there is seemingly some contact between the two halves posteriorly at midline (albeit trabecular bone only) (Figure 14.15). Since most of the body lacks cortical bone, there is no evidence for rib facets. Both pedicles are distorted. The left is crushed superiorly, and the right is concreted with dirt and bone meal. Only the right inferior articular facet is complete; it is ovoid (superoinferior long axis) and slightly concave, with discrete boundaries. The left is missing its inferior half. The transverse processes have large, ovoid rib facets: breadth 8.8 mm (right); 8.9 mm (left); height 10.6 mm (left). The neural canal is ovoid (transverse long axis) in cross section. The spinous process is missing.

Dolní Věstonice 14 has a largely complete T2, albeit one with a heavily damaged body that was split down the midline and is now filled with putty (Figure 14.16). The body is small in both its dorsoventral and craniocaudal dimensions. There is a large amount of putty along the left superolateral margin of the body, and the entire ventral face of the body is lacking cortical bone. In fact, cortical bone is found in only a few places on the body: (1) laterally on the left (except for the pedicle, which is covered in putty), (2) on the right dorsal aspect, (3) most of the dorsal body (except at midline), and (4) the caudal surface (except its ventral margin). Both the cranial and caudal faces of the vertebrae are planar, and both ring epiphyses are present dorsally. On the left, a small depression 5.1 mm high and 4.5 mm wide corresponds to the articular surface for the head of rib 2. No such facet is preserved on the right. The pedicles are dorsoventrally and superoinferiorly thick, but only the right can be measured since the left is heavily puttied (height: 11.8 mm; breadth: 8.4 mm). Both planar superior articular facets have discrete boundaries. The right is damaged along its inferolateral margin and is more circular, whereas the left is more ovoid (transversely wider). Ovoid and planar left and right inferior articular facets are present but are obscured to some extent by plaster. The junction of pedicles and laminae are heavily covered with putty, so there may be a problem with the orientation of the neural canal (it appears ovoid in cross section, with a mediolateral long axis). The transverse processes are long, straight, and superiorly inclined. The left facet for the rib tubercle is deep, concave, and circular (height: 6.1 mm; breadth: 6.0 mm). The right facet is more planar and superoinferiorly elongated (height: 8.9 mm; breadth: 4.9 mm). Its location is more lateral than the left, which is on the anterior face of the tip. The spinous process is damaged at its tip but was long and spearlike and seemingly greatly declined—although this latter feature is almost certainly due to erroneous reconstruction.

Dolní Věstonice 15's T2 is largely complete (Figures 14.17 and 14.18), although parts of it have been reconstructed. It has a broad body with superoinferiorly thick pedicles, long transverse processes, and a long, robust spinous process. The body is tall and, like the T1, dorsoventrally narrow. It has sustained loss of cortical bone along its ventral face, and there is a postmortem furrow about 7.0 mm wide and 13.4 mm long, running posterolaterally to anteromedially along its caudal surface. The inferior ring epiphysis is missing from near ventral midline to the left posterolateral region. Part of a faux intervertebral disc remains on both the caudal and cranial faces of the body. The left facet for the head of the second rib is covered in plaster, but the right demifacet extends to the cranial border (breadth: 9.4 mm; height: 6.7 mm). Only a small portion of the right caudal demifacet remains; its left counterpart is missing. The superior articular facets are planar and circular. The transverse processes are long (left length: 28.0 mm; right: 28.8 mm). The articulation for the right rib tubercle is not apparent, but there is a flattened area near the tip of the process. The facet on the left transverse process is somewhat covered by plaster, but it is concave and wide but not particularly high. Most of the laminae and the inferior spinous process are plastered. The spinous process is bladelike. It deviates to the right, particularly along its caudal aspect.

Pavlov 1's T2 is probably represented by a dorsal neural arch piece, 45.5 mm wide, that preserves two laminae, a partial spinous process, a small caudal portion of the right pedicle, and two inferior articular facets that are missing their bottom halves. The inferior articular facets were ovoid (superoinferior long axis), with the right somewhat narrower and more superoinferiorly concave than the left. The laminae are somewhat internally to externally thick: 4.8 mm left; 5.2 mm right.

Third Thoracic Vertebra

For Dolní Věstonice 3, there is one small piece of bone visible within the matrix that could be the left pedicle and transverse process of the third thoracic vertebra (T3) although it could also be the left third rib. The body is missing, but there is a natural cast of the spinal canal (Figure 14.14).

Dolní Věstonice 13's T3 (Figure 14.15) is relatively incomplete and remains adhered to T4. On the left, a small posterolateral portion of the body, pedicle, inferior articular process (attached to T4), and lamina are preserved. The spinous process remains a separate piece. On the right, the transverse process and lamina are all that remain. The preserved portion of the body is virtually all trabecular bone. Only the left pedicle remains, and it has T4's superior articular process crushed up into it. Despite this distortion, it is apparent that the pedicle was tall—at least 9.6 mm (probably greater). Its breadth is not as impressive (7.6 mm). Only the left superior articular facet is preserved. It has suffered some damage, making it difficult to judge articular versus nonarticular areas. It is tall and ovoid, with a superoinferior long axis. The transverse processes are short with broad distal ends. Facets for the rib tubercles are large, circular, and concave: breadth: 10.2 mm (right); 12.2 mm (left); height: 9.3 mm (right); 12.4 mm (left). The tip of the left transverse process is damaged just dorsal to the facet. The spinous process is long, and its tip is tall and narrow (height: 12.3 mm; breadth: 4.6 mm). The neural canal was ovoid, with a transverse long axis.

Dolní Věstonice 14's T3 (Figure 14.16) is represented by two pieces that form most of the neural arch—including the left pedicle and superior articular facet, both laminae, a largely complete left transverse process, a broken right transverse process, and a spinous process. The first piece includes the left pedicle with a small portion of the body attached and the left superior articular facet. The pedicle is tall (10.6 mm) and mediolaterally narrow (5.2 mm). The superior articular facet is planar. The second, dorsal, piece includes the laminae, inferior articular facets, and transverse processes. The inferior articular facets are slightly concave, and the right is wider than the left, which is tall and narrow. The left transverse process is long (29.5 mm); at its tip there is a large, concave facet for articulation with the tubercle of rib 3, measuring 9.3 mm in height and ca. 7.4 mm in width (its medial margin is covered in putty). The tip of the right transverse process is gone. The spinous process is long (35.7 mm), superiorly convex, narrow, and bladelike, with a modest swelling at its tip (tip breadth: 5.1 mm; height: 9.3 mm).

The third thoracic vertebra of Dolní Věstonice 15 (Figures 14.17 and 14.18) includes a fragmented body, pedicles, superior and inferior articular processes, laminae, most of a spinous process, and a left transverse process. The tall and dorsoventrally shallow body has gentle ventral convexity in superior view. In right lateral view, most of the cortical bone caudal and ventral to the pedicle is gone. There is also some trabecular bone visible anteriorly near midline. In *norma frontalis* this area is visible (as are others), and in *norma lateralis* on the left much of the cortical bone is gone. In inferior view as well, there is wax from the display preparation, with areas of exposed trabecular bone. In superior view, trabecular bone is visible along the left lateral ventral margin, along with residual faux intervertebral disc. The cranial ring epiphysis is also heavily eroded. The right pedicle is attached to the body. It is damaged inferiorly but has a complete cranial margin. The left pedicle is largely complete, except along its internal margin. It is not attached to the body because the left section of the body is missing. However, on the left pedicle a portion of the facet for the head of the third rib is present (whereas on the right it is missing). The superior articular processes are tall and narrow. The right evinces slight anterolateral damage. Only the left transverse process is present. It is long and horizontal, with a concave, ovoid (breadth: 11.1 mm; height: 7.9 mm) facet for articulation with the tubercle of the third rib. The tip of the spinous process is missing, but the spinous process is rectangular and steeply declined (54°).

Fourth Thoracic Vertebra

The fourth thoracic vertebra (T4) of Dolní Věstonice 3 is represented solely by a very fragmentary body, still in matrix (Figure 14.13), that has been rotated so that its caudal surface is facing laterally (left) and its cranial surface is facing medially (right). Most of the body is missing or is represented by trabecular bone. Only small areas of the bone's superoanterior margin and caudal surface preserve cortical bone.

Dolní Věstonice 13's T4 (Figure 14.15) is the first in the thoracic series since T1 to preserve a large portion of its body. Only the left side of the body preserves cortical bone, and even here the bone is somewhat crushed and distorted, with a fault running anterosuperiorly to posteroinferiorly on the left. A small remnant of the demifacet for the left rib 5 is preserved. Only the left pedicle remains, and it has been crushed up into T3, precluding its measurement. Its associated superior articular facet also remains adhered to T3. The left transverse process is somewhat superiorly inclined. The right transverse process, however, is filled with wax near its origin; no true contact between it and the lamina is evident. Rib facets on the transverse processes are not large: height: 10.0 mm (right), 9.2 mm (left); breadth: 10.1 mm (right), 8.9 mm (left). The left lamina is complete, and most of the right lamina is preserved. They unite to form a long, spearlike spinous process with a tapered end and modest swelling. It appears that the neural canal was ovoid (dorsoventrally narrow), but it cannot be measured (the right pedicle is missing and the inferior portion of the dorsal body is damaged). The right inferior articular process is also missing, but the left is complete. It is planar and ovoid, with a superoinferior long axis.

Dolní Věstonice 14's T4 is connected by wax and glue to T5, T6, and T7 (Figure 14.16). It is represented by a dorsal piece that includes the laminae, the inferior articular facets, the right transverse process, and the spinous process. The left inferior articular facet is much wider and taller than its counterpart on the right. Both are planar, but the left is more rectangular in shape and the right is more ovoid (with a superoinferior long axis). The right transverse process is 27.2 mm long and superiorly inclined, and at its tip there is an ovoid, concave articular facet for the tubercle of rib 4 (breadth: 9.2 mm; height: 8.2 mm). The spinous process is long (40.0 mm) and spear-like, with a cranial central ridge and a slightly expanded tip that appears to deviate slightly to the left (tip height: 4.6 mm; breadth: 6.0 mm).

All that remains of he Dolní Věstonice 15's T4 (Figure 14.18) is a partial left pedicle (which includes a slight amount of bone from the body), a left superior articular facet, a left transverse process, two laminae, two inferior articular facets, and most of a spinous process. The pedicle's caudal surface is eroded. On the right, there is a partial superior articular process and a complete inferior articular process. The left superior articular facet evinces superolateral damage. The two inferior articular facets are asymmetrical, with the right more inferiorly placed than the left. The left transverse process is somewhat superiorly inclined; its length is 27.4 mm. The concave circular facet for the rib tubercle is present: height: 7.0 mm, breadth: 7.4 mm. Most of the spinous process is preserved (perhaps 5 mm of its length is missing). It is less bladelike and more triangular in cross section than the more cranial elements, with a central ridge and sloping sides. It seems to deviate slightly to the right.

Fifth Thoracic Vertebra

Very little of Dolní Věstonice 3's fifth thoracic vertebra (T5) remains. The most identifiable piece is the right inferior articular process. Its facet is ovoid (superoinferior long axis) and concave (height: 11.2 mm; breadth: 5.5 mm) and has sustained damage along its lateral and inferior margins. As with the other thoracic vertebrae, there is a natural cast of the neural canal and a small (5.1 mm x 2.9 mm) piece of bone from the left superolateral portion of the body.

All that remains of the Dolní Věstonice 13's T5 (Figure 14.15) is a left transverse process that is glued to its lamina, a spinous process that is missing about 20 mm of its distal end, and most of the right lamina. The transverse process has sustained damage along its inferomedial margin. Its facet is ovoid and slightly concave, with an inferomedial to superolateral long axis (height: 7.3 mm). The left inferior articular facet is planar and ovoid, with well-defined margins and a superoinferior long axis. The right inferior articular facet is missing about one-third of its area because of a break beginning near its lateral midpoint and descending inferomedially to near the inferomedial corner of the facet.

Dolní Věstonice 14's T5 is represented by a fairly complete neural arch; very little of the body is preserved (Figure 14.16). Unfortunately, it remains sandwiched between T4 and T6, and thus its morphology is difficult to assess. Only small pieces of the body are adhered to the pedicles. The left side preserves more than the right and includes part of the facet for the head of the fifth rib, which measures 10.5 mm in breadth and was at least 6.2 mm high (it may have continued slightly beyond what is preserved). The pedicles are tall, but the right seems mediolaterally thinner than the left, probably because of reconstruction, as there is a large amount of putty on the internal side of the right pedicle (left height: 11.4 mm, right: 12.4 mm; left breadth: 4.6 mm). All of the articular facets are present. The tall superior articular facets have discrete boundaries to their articular surfaces and are oriented in a paracoronal plane. The left facet is more cranial in position relative to the laminae than is the right, which does not extend as far superiorly above its lamina. The right inferior articular facet is conserved in its entirety; the left preserves its cranial half. The distal end of the relatively short right transverse process is repaired with fill. No spinous process is preserved.

All that remains of the Dolní Věstonice 15's T5 are two partial laminae with inferior articular processes, left transverse process, right pedicle with a small portion of the body attached, right superior articular facet, a tiny portion of the left superior articular process (sans facet), and a virtually complete spinous process (Figure 14.18). The right pedicle is tall and mediolaterally narrow: height: 11.3 mm; breadth: 5.5 mm. The right superior articular facet is planar and tombstone-shaped, and it has trivial damage along its superomedial margin. The inferior articular facets are ovoid and planar; the right is damaged along its caudal margin. There is a fissure between the left transverse process and lamina. Most of the left transverse process is preserved; its length is about 27.3 mm. At its distal end it has a barely notable depression for the rib facet, with an approximate height of ca. 6.5 mm and a breadth of ca. 5.9 mm. There is also a small area of exposed trabecular bone at the inferior juncture of the left lamina and the spinous process, corresponding to a break between them. The spinous process is triangular in cross section and relatively robust—it is transversely thick throughout and possesses an expanded tip (tip height: 15.3 mm; breadth: 4.5 mm). Along its superior margin it is smooth on the left, but on its right it is concave, with a long, shallow sulcus measuring 6.2 mm by 19.8 mm.

Sixth Thoracic Vertebra

The Dolní Věstonice 3's sixth thoracic vertebra (T6) is present in at least three pieces—one that has been removed from matrix and at least two others that remain embedded in matrix. The embedded pieces include the right superior articular process and pedicle. Only a small section of it is visible, as it is largely surrounded by matrix. Inferior to this piece, but also embedded in matrix, is a section of lamina, along with two inferior articular facets. A relatively narrow (ca. 16 mm) area of the laminar internal surface is visible. Whereas the left inferior articular facet remains attached to the main laminar piece, it appears that the right inferior articular facet broke off postmortem. The latter has a horizontal crack separating its cranial two-thirds from its caudal one-third. The left facet has an area of cortical bone loss in its mediocranial margin. Both facets are slightly concave and ovoid with superoinferior long axes. As both are damaged along their lateral margins, all measurements are estimates—facet height: ca. 10.9 mm (right) and ca. 10.5 mm (left); facet breadth: ca. 7.7 mm (right) and ca. 6.2 mm (left). The piece that has been removed from matrix warrants further discussion. It measures some 18.4 mm in depth and 24.3 mm in height and includes portions of the bodies of T6 and T7 that have been fused together postmortem. The midsection of the body is present, albeit in a badly eroded state. The largest area (15.3 mm in height x 11.4 mm in breadth) of cortical bone is found on the body's dorsal surface, where the two nutrient foramina are preserved. On the anterior face an area of cortical bone is preserved, with a matrix-filled gap separating the T6 and T7 bodies. The ventral face of T6 has been compressed dorsally into the body. The right side of the body has also been badly crushed, and the body's caudal face reveals only trabecular bone. The cranial surface of T6 has a small area of preserved cortical bone, and dorsal body height is probably reliable: 15.3 mm.

The sixth thoracic vertebra of Dolní Věstonice 13 (Figure 14.15) is the most complete thoracic vertebra thus far discussed for this individual. Most of the body is present—on the right it is eroded posteroinferiorly, and on the left it is eroded superolaterally. The pedicles, laminae, articular processes, and the spinous process are all complete—although there is glue connecting the left transverse process to the neural arch, and the spinous process and inferior articular facets are glued on as well. The vertebral body is not sagittally deep and appears to have been only slightly longer than it is wide. The ventral face of the body evinces a gentle superoinferior concavity. The sides and dorsal face of the body are similarly concave, but the cranial and caudal surfaces are relatively flat. The annular rings appear to have been fused (their absence along the dorsal margin is probably due to taphonomic erosion), and there is no evidence of osteoarthritis. Both demifacets for the heads of the sixth ribs are present: the left facet is damaged, but the right one has a distinct border and is slightly convex and oval, with a craniodorsal to caudoventral long axis (height: 6.4 mm; breadth: 6.0 mm). On the caudal left side of the body is the area for the demifacet for rib 7, but its cortical bone is missing, precluding its measurement. On the right, the entire area for the demifacet is gone. The pedicles are tall and relatively narrow: height: 11.9 mm (left), 12.8 mm (right); breadth: 5.7 mm (left), 5.9 mm (right). Both superior articular facets are nearly circular and planar, although the left is wider than the right. The inferior articular facets are ovoid (with superoinferior long axes) and are subtly mediolaterally concave. The left is glued and somewhat larger than the right. The transverse processes project horizontally. Both preserve ovoid facets for articulation with the tubercles of the sixth ribs. The right facet is larger and more distinct than the left: height: 9.3 mm (right); (ca. 7.5 mm (left); breadth: 9.3 mm (right) 8.3 mm (left). The vertebral canal appears almost circular in cross section. The spinous process is long, with a spearlike tip, and is greatly declined (83°). The moderate swelling at its tip measures 7.5 mm in breadth and 6.4 mm in height.

Dolní Věstonice 14's T6 (Figure 14.16) is represented by its posterior portion, including the left pedicle and transverse process, both laminae, and the spinous process. Only the slightest trace of body remains attached to the left pedicle, although because of the wax reconstruction one cannot determine if a rib facet is present. The pedicle itself is tall and relatively narrow (height: 11.5 mm; breadth: 8.4 mm). Only the left superior articular process is preserved. It is planar, tall, and relatively thin (height: 8.9 mm; breadth: 7.7 mm). Its long axis is oriented slightly off vertical (inferomedial to superolateral). Both laminae are preserved, along with their associated inferior articular facets. However, neither of the two facets is well preserved. The right is broken in half, and the left is partially covered in plaster. The relatively short left transverse process is inclined slightly above horizontal. On the anterior face of its tip is a flat circular area with nondemarcated boundaries for articulation with the tubercle of rib 6 (breadth: 8.4 mm; height: 9.3 mm). The spinous process was broken but has been reglued. It is long (45.6 mm) and bladelike, with a modest swelling at its tip that is slightly damaged. It is also bent in its midsection so that its distal end is deviated to the left. Along its superior margin it evinces a central ridge, but there are no associated sulci to either side.

Dolní Věstonice 15's T6 (Figure 14.19) is largely complete, although a fair amount of plaster is present. Among the preserved elements are a body, all articular facets, pedicles and laminae, and a spinous process. Both transverse processes are missing. The body appears to have had a distinct wedge, so its dorsal height is much greater than its ventral height. To some extent, this may be a function of preservation and/or reconstruction, but it was probably true of the original as well. Unfortunately, because of damage to the specimen, one cannot compute its wedging angle (see osteometric analyses). The ventral face of the vertebra is relatively straight superoinferiorly. However, laterally, when viewed in *norma frontalis*, the body is more superoinferiorly concave. The concavity

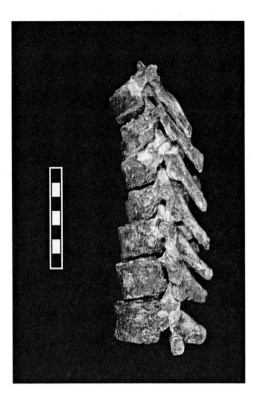

Figure 14.19 Lateral view of articulated T6 through T12, Dolní Věstonice 15. Scale in centimeters.

appears to be more extreme on the right, but this could be due to the fact that the right side preserves more of the body's posterior aspect. In inferior view, the body is ventrally convex and dorsally planar. Its dorsoventral dimension is seemingly small. However, in many areas, the body is heavily reconstructed. The caudal surface is the one body surface on which neither wax nor plaster was used. Elsewhere, their use was liberal. Neither of the demifacets for the sixth rib heads is preserved. The demifacet for rib 7 is preserved only on the right. It is likely that neither pedicle is correctly articulated with the body. Both are tall and narrow (left height: 10.6 mm, right: 11.2 mm; left breadth: 4.8 mm, right: 5.7 mm). The superior articular facets are flat, tall, and narrow. The right is damaged, but the left is well preserved. The inferior articular facets are relatively planar and circular. The neural canal is ovoid in cross section (mediolateral long axis). The spinous process is long (50.6 mm), needle-like, and triangular in cross section. It has a tall central ridge running down its back, which has formed a shallow fossa on the right side; the left side is flatter. Its tip is only moderately swollen: breadth: 5.0 mm; height: 10.7 mm. Although it appears to be greatly inclined (ca. 90°), it does not appear that the pedicles have been properly articulated with the body, and thus this measure is unreliable.

Seventh Thoracic Vertebra

Dolní Věstonice 3's seventh thoracic vertebra (T7) remains embedded in matrix along with T8 through T12 (in anatomical series); T7 itself is represented solely by a relatively indistinct mass of bone from the vertebral body, measuring about 22.5 mm in transverse dimension. On its ventral face it is heavily encrusted, and in posterior view it appears that the laminae have been compressed into the posterior portion of the body.

The seventh thoracic vertebra of Dolní Věstonice 13 is complete (Figure 14.15). Its body is almost circular in transverse cross section; that is, it is only slightly longer than wide. Dorsal body height is greater than ventral body height because of normal thoracic curvature. Laterally, the body is superoinferiorly concave, whereas the cranial and caudal surfaces are flat. The ventral face of the body evinces only the slightest concavity in lateral view. The annular rings are fused onto the centrum; the cranial one is damaged only in one small location—about 5.0 mm are missing from the left side. There is no demifacet on the left side for the left seventh rib: it has been abraded away. The demifacet for the right seventh rib is clearly present, and when the rib is articulated with the vertebra, it meets both the demifacet and the transverse facet. In contrast, when the left seventh rib is articulated with the left transverse process, its head fails to make contact with the vertebral body. The demifacet for rib 7 on the right is large, ovoid, and concave, with a posteroinferior to superoanterior long axis (breadth: 12.5 mm; height: 7.8 mm). The body is also damaged in the region of the demifacet for the left eighth rib; this section of the body is crushed into the left pedicle and superior articular process of T8. However, the large (breadth 9.7 mm; height 7.1 mm), slightly concave, and semicircular demifacet for rib 8 on the right side flares out away from the body. Its pedicles are tall and mediolaterally thin, although a small portion of the cranial margin of the left may have been sheared off: (left height: ca. 12.7 mm, right: 13.5 mm; left breadth: 6.6 mm, right: 6.8 mm). Whereas the left superior articular facet is more triangular, the right is squarer. Both are planar and roughly the same size, and both are oriented in a paracoronal plane. The right is missing cortical bone from its center. It has a discrete inferior border; the left facet has damage along its inferior margin. The inferior articular facets are planar and ovoid, with superoinferior long axes; the left is wider than the right. The vertebral canal is relatively small and nearly circular. The transverse processes are rounded, with the right longer and slightly more cranially oriented than the left: (right length: 29.2 mm, left: 29.5 mm). The rib facets on the transverse processes are also oval and only slightly concave (left height: 8.5 mm, right: 7.2 mm; left breadth:

7.5 mm, right: 8.9 mm). The spinous process has been glued to the laminae. It is long (50.8 mm), spearlike, and declined (ca. 74°), and it has no prominent central ridge.

Dolní Věstonice 14's T7 (Figure 14.16) is represented by a posterior piece lacking only its spinous process. Each pedicle has a small piece of body attached. On the right, there is a depression on the dorsolateral body that corresponds to the articulation of the head of the seventh rib (breadth 6.7 mm; height 4.2 mm). No evidence of an articular facet is seen on the corresponding area on the left. The right pedicle evinces putty along its caudal margin and is in a more cranial position than the right, presumably because of postmortem distortion. The pedicles are tall and narrow (left height: 12.1 mm; left breadth: 4.7 mm, right: 4.5 mm). The transverse processes are somewhat asymmetrical. The left is longer than the right, and although both are slightly inclined above the horizontal, the left is more cranially inclined. Also, in the transverse plane, the long axis of the left runs about 7° dorsal to the coronal plane, whereas the right is about 22° dorsal to it because of postmortem distortion. Both preserve small, flat facets for the rib tubercles (right height: 5.6 mm, left: 4.5 mm; right breadth: 6.8 mm, left: 5.1 mm). The superior articular facets are small; the right is broader than the left. Only the broad and planar right inferior articular facet is present. It has putty along its medial border and may be broken off inferiorly.

Dolní Věstonice 15's T7 (Figure 14.19) is largely complete, but its body is severely damaged and the transverse processes are missing. A large chunk of the anterosuperior body is missing—it almost appears to be discovery damage—as if the front upper quarter of the body had been chopped off. The T8 demifacets are also eroded away. In inferior view, the body is ventrally convex and somewhat dorsoventrally elongated. In anterior view, its lateral margins are slightly concave superoinferiorly. In superior view, its dorsal margin is deeply concave, yielding an almost circular neural canal. Both facets for rib 7 are damaged and encroach onto the cranial margin of the body and pedicle bases. Although damage along the facets' dorsal margins makes measurements unwise, the left is particularly large (the preserved portion is 9.5 mm in height), whereas the right is 8.3 mm in height. The demifacets for the eighth ribs are large, particularly in their superoinferior dimensions (left height: 15.7 mm; right breadth: 9.4 mm, left: 9.2 mm). The pedicles are tall and relatively narrow (left height: 10.8 mm, right: 10.6 mm; left breadth: 5.8 mm, right: 6.4 mm). The superior articular facets are planar, tall, and narrow. The right is in a considerably more cranial position than the left—perhaps reflective of pathology. The inferior articular facets are ovoid and planar. The spinous process is relatively short (49.3 mm) and spearlike, with an expanded tip. It is quite declined, about 86°. The rounded expansion of its tip measures 11.6 mm (height) and 4.7 mm (breadth). It also exhibits a gentle central ridge along its dorsum, with more expression of a lateral fossa on the right than on the left.

Eighth Thoracic Vertebra

Dolní Věstonice 3's eighth thoracic vertebra (T8) is represented by a badly crushed body with ventral concretions and dorsal display material adhered to it. The right side has exposed trabecular bone over its entire surface (except dorsally, where it is covered in matrix), whereas the left side retains concreted matrix. Ventrally there is a large hole (ca. 9.5 mm in height and breadth) worn into the body, and although the cranial surface is visible, it, too, is heavily eroded. The body appears to exhibit ventral wedging, but this could be due to postmortem taphonomic processes. Red ochre is present along the left lateral side of the body.

Dolní Věstonice 13's T8 is largely complete (Figures 14.20 and 14.21), although its left transverse process is badly damaged and the dorsal wall of the vertebral canal is glued to the pedicles. The caudal body flares laterally, making its caudal surface much larger than its cranial surface. The body is slightly dorsoventrally elongated, with a gentle superoinferior concavity laterally and ventrally. As is typical for an eighth thoracic vertebra, there

Figure 14.20 Superior view of T8 through T12, Dolní Věstonice 13. Top row (left to right): T8, T9; bottom row (left to right): T10, T11, T12. Scale in centimeters.

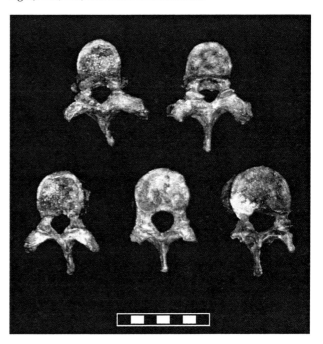

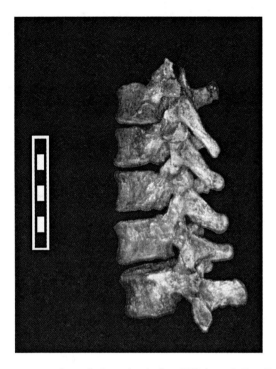

Figure 14.21 Lateral view of articulated T8 through T12, Dolní Věstonice 13. Scale in centimeters.

is little wedging; dorsal and ventral body heights are roughly equal. It, like all the other vertebrae thus far examined for this specimen, is free from degenerative changes. Perhaps the most striking features on the body are its large, flaring inferior (T9) demifacets. Both are circular and slightly concave—height: 10.0 mm (left), 8.9 mm (right); breadth: 10.2 mm (left), 9.4 mm (right). The right cranial (T8) demifacet has a crack through it and is shallow, with poorly defined margins, and thus it cannot be accurately measured. In contrast, although the left T8 demifacet is also shallow, its margins are more discrete (height: 6.7 mm; breadth: 11.6 mm). The pedicles are tall and mediolaterally narrow, but neither can be measured. The left retains a portion of the T7 body crushed into it; the right evinces a large gap filled with glue. Both superior articular facets are ovoid and planar. The inferior articular facets are ovoid and slightly concave, with discrete margins and superoinferior long axes. The long right transverse process is oriented superoposteriorly; it has been glued to the lamina. At its tip a planar to slightly convex articular facet for the rib tubercle is found. It is small and oval-shaped with discrete borders (height: 6.4 mm; breadth: 8.5 mm). The spinous process is short (45.1 mm) and deviates to the right, probably because of postmortem distortion. It is spearlike, with a modest swelling at its tip that measures 4.7 mm in breadth and 6.4 mm in height.

The Dolní Věstonice 14's T8, T9, and T10 are stuck together more or less as they were in situ (Figures 14.22 and 14.23). The bodies of all three vertebrae are barely discernable from the surrounding matrix, and they may have been fire-damaged (Svoboda, personal communication); T8 has a craniocaudally short, dorsoventrally long, and transversely narrow body. No wedging is evident. On the left, a portion of the facet for the head of the eighth rib is visible. It appears to have been ovoid, with a height of 7.1 mm and a breadth ca 9.9 mm. The demifacet for T9 is clearly visible on the left inferior body, and it is circular (ca. 7 mm in diameter). The matching demifacet is barely discernable on the right. The pedicles are tall, but the superior articular facets that rest upon them are relatively gracile. Only the right transverse process is present, and it is heavily reconstructed. It was, however, short, globular, and somewhat horizontal in inclination. The spinal canal is ovoid, and the spinous process is short (35.4 mm) and inferiorly declined (64°).

The Dolní Věstonice 15's T8 is almost complete—only the left transverse process and a posterolateral portion of the body on the caudal right side are missing (Figure 14.19). In addition, many areas have been liberally reconstructed with plaster. The body is rather cylindrical in its form. In superior view, its sides and ventral face are gently convex, and the body is not dorsoventrally elongated. It appears that the only place where the superior ring epiphysis remains is posterolaterally on the left; the inferior ring epiphysis is missing at the posterior midline. The ventral face of the body is superoinferiorly straight when viewed in *norma lateralis*. In contrast, its sides are superoinferiorly concave when viewed in *norma frontalis*. There is also a

Figure 14.22 Anterior view of fused (postmortem) T8 through T10, Dolní Věstonice 14. Scale in centimeters.

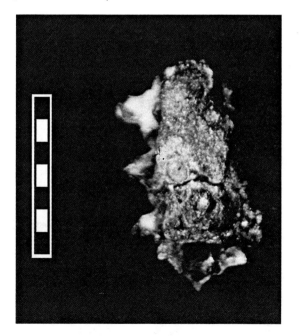

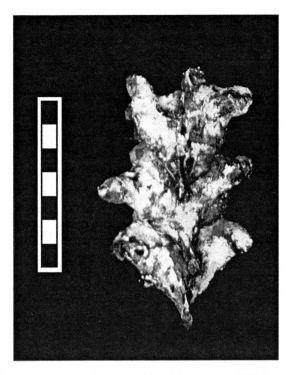

Figure 14.23 Posterior view of fused (postmortem) T8 through T10, Dolní Věstonice 14. Scale in centimeters.

very subtle (but nonpathological; see osteometric analyses) wedge shape to the vertebra; its dorsal body height exceeds that of ventral body height. The area for articulation of the heads of the eighth ribs is covered in plaster on both sides. On the left, however, the large demifacet for rib 9 is present (height: 6.8 mm; breadth: 8.1 mm). It is slightly triangular in shape, with a superior apex. The roots of both pedicles are damaged; there is plaster near their junctions with the body. The pedicles are tall and narrow (left height: 12.1 mm, right: 11.1 mm, left breadth: 6.5 mm, right: 6.4 mm). All articular processes are complete. The superior articular facets are large, with the left markedly wider than the right. The left also has a more rounded shape, whereas the right is more ovoid, with an inferomedial to superolateral long axis. Both inferior articular facets are ovoid and planar. The right evinces damage along its lateral margin. The short (25.1 mm) right transverse process is inclined superiorly but is badly damaged, with plaster evident on its anterior and superior faces. The facet for the rib tubercle is not visible because of slight inferoanterior erosion near its tip. The neural canal is slightly ovoid, with a mediolateral long axis. The spinous process is long (47.6 mm), narrow, spearlike, and greatly declined (ca. 82°). Its dorsal margin is smooth, without a clearly definable central ridge or associated sulci. Its tip is only modestly expanded (height: 6.9 mm; breadth: 5.8 mm). Although more symmetrical than many of the more cranial elements, it still deviates slightly to the right.

Ninth Thoracic Vertebra

The body form of Dolní Věstonice 3's ninth thoracic vertebra (T9) is evident in a heavily eroded piece that remains sandwiched between the damaged bodies of T8 and T10. Only tiny pieces of cortical bone are present on the left lateral aspect of the body; elsewhere it is made up of trabecular bone and matrix. Ventral body height is estimated at about 15.5 mm. Superolateral to the right side of the body is a small (8.7 mm x 7.7 mm) separate piece that may be the right pedicle. Red ochre is present along the entire left lateral aspect of the body.

Dolní Věstonice 13's T9 (Figures 14.20 and 14.21) exhibits only trivial damage to the left transverse process (anteriorly and posteriorly) and right transverse process (anteriorly). The left transverse process was glued to its lamina, and the laminae themselves are glued to the pedicles. The body is the most elongated yet for the specimen and is apparently free from degenerative changes. It is relatively short and is slightly shorter on the right than on the left. Although the cranial and caudal surfaces are flat, the ventral, dorsal, and lateral walls of the body evince gentle superoinferior concavities. Dorsal and ventral body heights are roughly equivalent. Both cranial and caudal faces of the body are in superb shape; the fused ring epiphyses are perfectly preserved. There is, however, slight (just 4.6 mm across) damage to the cranial margin on the left posterolateral quadrant. The cranial (rib 9) demifacets on the body are found at the junction of the pedicle and body. The left is large and well demarcated. It is concave, ovoid, and blends into the cranial surface of the body (height: 6.3 mm; breadth: 10.7 mm). The right is less well delineated, but this is due to postmortem destruction of the facet's caudal margin. The left inferior demifacet is nothing more than a small circular bump (breadth: 4.2 mm; height: 4.3 mm); its counterpart on the right has been largely lost. The pedicles are tall and relatively robust—height: 11.6 mm (left), 11.4 mm (right); breadth: 7.9 mm (left), 8.0 mm (right). The superior articular facets are large and planar, and perhaps one-fifth of the left's area is missing superomedially. Both are tall, with rounded superior margins and a relatively narrow interfacet distance. The inferior articular facets are ovoid, with superoinferior long axes. Both have articular surfaces that are well delineated from their nonarticular portions, although the boundaries of the left are more discrete than those of the right (the medial border of the right grades into nonarticular surface, but on the left there is a sharp medial break). The left inferior articular process also extends more caudally than does the right. The neural canal is ovoid, with a transverse long axis. The transverse processes are short and broad in cross section. The facets on their tips for the rib tubercles are large, with erosion evident along the cranial and dorsal margins of the left (right breadth: 12.0 mm; right height: 9.5 mm, left: 10.6 mm). The short but spearlike spinous process is thick in cross section. It has a tubercle at its tip (tip height: 7.3 mm; breadth: 5.1 mm) and does not have a superior ridge. Its declination is relatively low (45°), and it is deviated to the left (the opposite of the elements above and below it).

Dolní Věstonice 14's T9 (Figures 14.22 and 14.23) is in the middle of three conjoined lower thoracic elements with badly preserved (perhaps burnt) bodies. Its body appears wider than that of T8, but measurements of the body cannot be taken because of the enormous amount of matrix still attached. There is some evidence for the demifacet for the left tenth rib, but it is badly eroded. The right pedicle is badly damaged, but the pedicles are tall (left height: 13.1 mm). On the body and intruding onto the left pedicle is a damaged circular facet for articulation with the head of the ninth rib (height: 5.4 mm; breadth: 4.7 mm). The superior articular facets are preserved. Measurements are not possible on the right, but because of damage to the left inferior articular facet of T8, the left superior articular facet of T9 is visible and broad (ca. 9.9 mm). The slightly declined transverse processes have relatively deep, ovoid (particularly on the left) rib facets (right height: 9.4 mm, left: 10.5 mm; right breadth: 7.8 mm, left: 6.6 mm). The spinous process is relatively short, deviates to the right, and has a small swelling at its base (tip height: 6.4 mm; breadth: ca. 3.9 mm). It is spearlike in its morphology, with a declination of 45°.

Dolní Věstonice 15's T9 is a badly damaged vertebra (Figure 14.19). The entire right anterior one-third of the body has been cleaved off and is missing, so only the left dorsal half and right dorsal fifth of the body remain. Even these preserved parts evince heavy damage—there is posterolateral damage to both sides of the caudal body. Despite this damage, the neural canal remains unaffected. Thus, slightly more than half the body remains, mostly to the left of midline and in the dorsal two quadrants of the body. Faux intervertebral disc material adheres to the body along both its cranial and caudal surfaces. The body is superoinferiorly concave on the preserved left side. The body also flares, such that its cranial surface is much smaller than its caudal surface. The facet for the head of the ninth rib is present. It is a large and double facet; that is, there is a rounded posterior articular portion and then, anteriorly across a slight furrow, lies another articular portion (facet height: 8.3 mm; breadth: 9.6 mm). The left demifacet for rib 10 is present; it is very slight indeed—only a bump measuring about 4.9 mm in height by 5.7 mm in breadth. It may have sustained dorsal damage and thus may have been larger. The superior ring epiphysis is present only on the left, from the left dorsal margin ventrally to the coronal midpoint of the left half. The inferior ring epiphysis is missing on the left as well. Despite the presence of putty at the external surface of the junction of the body with the right pedicle, it is apparent that both pedicles are articulated with the body in the appropriate position because the internal surface of the right pedicle reveals good contact. Measurements of the pedicles cannot be taken, and there is an area of bone growth (callus?) along the superior margin of the right pedicle. All articular facets are present. As was the case in T8, the left superior articular facet is larger than the right. The inferior articular processes are broad, and whereas the right evinces inferior damage, the two inferior facets are equal to each other in size. The neural canal is ovoid, with a mediolateral long axis. The relatively steeply inclined right transverse process is preserved; only the root of the left transverse process remains. The right transverse process is short but thick in cross section (breadth: ca 12.2 mm; height: 12.1 mm). Its entire ventral margin is covered in putty, and as a result the facet for the tubercle of the ninth rib is not preserved. The spinous process is short (44.4 mm), thin, spearlike, and declined (ca. 68°). It deviates slightly to the right, and its tip is somewhat expanded (height: 8.1 mm; breadth: 5.1 mm). It is smoothly convex along its superior margin, with no evidence of lateral sulci.

Tenth Thoracic Vertebra

The Dolní Věstonice 3's tenth thoracic vertebra (T10) remains sandwiched in matrix between the badly damaged bodies of T9 and T11, and as with T9, this vertebra is represented only by the form of its ventral body, which lacks cortical bone. In fact, T10 is in worse condition than T9. The best-preserved area of the vertebra is a small (7.4 mm in depth x 13.7 mm in width) section of the body's caudal surface. Ventral body height is estimated at about 17.5 mm. In addition, a small (7.2 mm in height x 5.3 mm in depth) piece lateral to the body is apparently a portion of the right pedicle. There is a small amount of red ochre on the left superolateral aspect of the body.

The relatively small T10 of Dolní Věstonice 13 is complete (Figures 14.20 and 14.21), showing only minor damage. Ventrally, there is a small (6.6 mm in height x 12.7 mm in width) area of missing bone at midline just above the caudal margin, and there is a small (ca. 5.0 mm) area of missing cortical bone on the right side, along the cranial margin. The body's dorsal height is markedly smaller than its ventral height, which would seem to indicate that the vertebral column is beginning to move out of its thoracic and into its lumbar curvature. If this is the case, then it is occurring at a more cranial level than is typical (see osteometric analyses). The body is also quite broad relative to its dorsoventral dimension, and its dorsal surface and sides are deeply concave superoinferiorly. Its ventral surface evinces a lower degree of curvature than the sides, and the right side of the body is slightly shorter superoinferiorly than the left. The cranial and caudal surfaces are planar. The area surrounding what would have been the body's facet for the head of the left tenth rib is missing. On the right, there is a glue join through the facet, which is ovoid and has an interesting morphology: on the body it is wide and short (breadth: 10.8 mm; height: 7.1 mm), but there is a small section (ca. 4.5 mm in breadth) of the facet that actually extends onto the superior articular facet. The pedicles were glued to the body. They are tall and wide in cross section, although glue joins prevent quantifying this. Both superior articular facets are present, but the right's superomedial border has been eroded. Also, whereas the left is a simple, tombstone-shaped facet that is oriented with a vertical long axis, the right facet's long axis is oblique—inferomedial to superolateral—and the facet itself is more ovoid as well. The inferior articular facets also remain, but the inferior one-fourth of the right facet is missing. The left is complete and is largely ovoid, albeit with an irregular lateral edge (cranially to caudally, it is laterally convex, then concave, then convex again). The left inferior articular process has a more discrete medial boundary than the right. On the left, there is a distinct bony ridge separating the articular versus nonarticular surfaces, whereas on the right the articular surface grades into the nonarticular surface. The articular portion of the left facet is cranial to the nonarticular surface. The neural canal is ovoid with a mediolateral long axis. The transverse processes are short and blunt, with superoinferiorly expanded tips. They have flat superior surfaces, point posteriorly, and are not very steeply inclined. Whereas the facet for the rib tubercle is visible on the left (located ca. 5.7 mm from its tip), there is no visible tubercle on the right. The spinous process is somewhat spearlike, but it is slightly shorter and blunter than that of the T9 and has a clearer tubercle (breadth: 6.0 mm; height: 10.0 mm). It is minimally declined (24°), and deviates slightly to the right. Whether this latter phenomenon is due to postmortem deformation cannot be ascertained.

Dolní Věstonice 14's T10 (Figures 14.22 and 14.23) is the caudal of the three heavily damaged (burned?) thoracic vertebrae that are adhered to each other, and its body is in the worst shape of the three. On the left, a piece of cortical bone 15.6 mm in height by 16.3 mm in breadth has been displaced laterally away from the body. The caudal surface of the body is almost completely destroyed, as is most of the body's spinal canal margin. No evidence of either rib head facet remains. Although body shape is difficult to discern, it was apparently broad. The pedicles are in reasonably good shape and are tall (left height: 14.9 mm, right: 14.6 mm). Both superior articular facets are short and broad. The inferior articular facets are damaged caudally—the right more so than the left. The left transverse process is broken, so only its root remains. The right transverse process is present: its estimated length is ca. 26.5 mm, and it has a small but damaged rib facet on its superior anterior margin. The neural canal was ovoid (mediolateral long axis). The tip of the spinous process is missing a small, but unknown, amount of its length. Although it cannot be quantified, the spinous process appears to be steeply declined for a T10, short, blunt, and deviated slightly to the right.

Dolní Věstonice 15's T10 is relatively complete (Figure 14.19), although it has suffered extensive damage to the following areas: the body's right cranial surface—except for the ventral right side, the cranial margin of the right transverse process, and lamina—and the body's dorsal surface within the neural canal. It also lacks a left transverse process, and the right superior articular process is missing. Much trabecular bone is exposed on the body's caudal surface as well. The body is large and bulky, transversely broader than dorsoventrally deep, and it appears to gain height toward its ventral face. There is a gentle superoinferior concavity to the lateral faces of the body, but ventrally it presents a straight vertical face. The cranial and caudal surfaces are flat, and both evince faux intervertebral disc substance—more so on the cranial than caudal surfaces (only a small amount remains on the anterior portion of the inferior ring epiphysis). The superior ring epiphysis is missing at dorsal midline and along the right side about 10.0 mm dorsal to the ventral midline. The inferior ring epiphysis is largely preserved—only a small posterolateral portion is missing from the left. The facets for the head of rib 10 are not well defined; the right is more obvious than the left, which is barely discernable. The right facet is smooth, flat, and in a relatively inferior position, but its cranial margin is damaged. Its breadth is 9.1 mm, and its height was at least 8.1 mm (there may have been more of the facet superiorly that has been lost). The pedicles appear thick in all regards. The right pedicle is heavily damaged along its cranial margin, and there is a glue join between it and the body. The left pedicle is in better shape; it is only slightly damaged along its caudal margin. The only measure that can be reliably taken is left pedicle breadth: 8.2 mm. The left superior articular process is short and ovoid with a transverse long axis (which is somewhat unusual). The inferior articular facets are flat and ovoid; the left is damaged but appears to have been larger than the right. Only the right transverse process is preserved; it is short and minimally inclined, and near its distal end it is extremely thick in cross section, both dorsoventrally and especially superoinferiorly (transverse process breadth: 10.0 mm; height: 15.4 mm). It has a clearly delineated, slightly convex superoanteriorly facing facet for the tubercle of rib 10 (facet height: 7.9 mm; facet breadth: 8.4 mm). The spearlike spinous process is short, moderately declined (ca. 56°), and deviates slightly to the right. It is almost circular in cross section, with a smooth dorsal surface and a moderate nodule at its tip that measures 7.7 mm in height and 5.6 mm in breadth.

Eleventh Thoracic Vertebra

The remains of Dolní Věstonice 3's eleventh thoracic vertebra (T11) embedded in matrix between the damaged bodies of T10 and T12. All that remains of its body are small pieces of cancellous bone and one small (7.5 mm height x 6.1 mm breadth) left piece of the superolateral body. However, three key pieces of the vertebra are present on the right. The first represents most of the proximal pedicle; it was apparently tall (its height is estimated at 10.1 mm) and fairly wide in cross section (its breadth is estimated at 6.1 mm). Dorsal to it is a piece that includes a distal portion of the pedicle and the superior articular facet. On the distal pedicle is an ovoid fossa (length 4.5 mm x height 2.5 mm) with an inferomedial to superolateral long axis that appears to have been the articulation point for the head of the right eleventh rib. The third piece is caudal to the second and includes an ovoid-shaped inferior articular facet that is heavily damaged along its inferior and medial margins. It appears that the articular surface of this facet was oriented in a paracoronal plane, indicating it was not lumbariform in its morphology.

Dolní Věstonice 13's T11 is in a nearly perfect state of preservation (Figures 14.20 and 14.21). Of its major structures, only the distal one-third of the left transverse process is missing. Its body is transversely wider than dorsoventrally deep, and as with T10, ventral body height is greater than dorsal body height (although this latter measurement cannot be taken because of missing cortical bone). The lateral faces of the body evince deeply superoinferiorly concave sides, and the ventral and dorsal faces are also markedly concave. Some cortical bone is missing off of the cranial and caudal surfaces dorsally at midline, and the ring epiphyses are also missing in these areas. On the body and running onto the pedicles are two large, ovoid (anteroposterior long axis), concave facets for the eleventh ribs. The right is more concave than the left (left breadth: 15.1 mm, right: 15.0 mm; left height: 9.1 mm, right 8.9 mm). The pedicles themselves are quite robust: both are thick and tall in cross section (left height: 16.1 mm, right: 16.4 mm; left breadth: 11.4 mm, right: 10.4 mm). The superior articular facets are ovoid and are roughly of the same size and inclination. They are slightly mediolaterally convex and appear to be shorter than those of the more cranial thoracic elements. The inferior articular processes are dorsoventrally thick (right: 5.4 mm), left: 5.7 mm). The inferior articular facets found on them are ovoid (superoinferior long axes) and are much thinner mediolaterally than more cranial thoracic elements. Both facets have well-defined margins, but the left has a more clearly defined medial margin than does the right. The neural canal is slightly triangular, with a posterior apex and a mediolateral long axis. The transverse processes are present, but the left is missing its distal one-third. Both are short and horizontally and dorsally oriented. The spinous process is short but spearlike, with minimal declination (8°). Along its length, it evinces an inferior crest that separates two inferior fossae.

Dolní Věstonice 14's T11 is preserved in two pieces (Figure 14.24). The first is a neural arch piece, including the pedicles, laminae, all articular and transverse processes, and the spinous process. The second includes the badly damaged body that remains adhered to the body of T12. In spite of its damaged state, the body looks to have been dorsoventrally shallow and transversely broad with a tall ventral face. The entire dorsal margin of the body is missing, and a large cavity 26 mm wide and 18 mm tall is evident posteriorly, where cortical bone and most of the trabecular bone has been eaten away. At ventral midline, the cortical bone is also missing, leaving a large (22.9 mm x 7.9 mm) area gouged out. The entire left side of the body is similarly eaten away—only trabecular bone is visible. In contrast, on the right, a small section of cortical bone remains on the anterolateral one-third of the ventral face. It measures 19.2 mm across its base and 6 mm across its apex. This area suggests a gentle superoinferior concavity on the lateral wall of the verte-

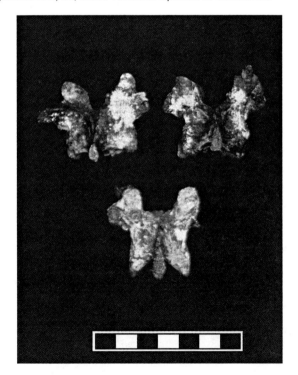

Figure 14.24 Posterior view of (top row, left to right) T11, T12; (bottom row) L1, Dolní Vostonice 14. Scale in centimeters.

bral body. In superior view, perhaps one-half of its original area retains cortical bone, and a small section of the superior ring epiphysis is preserved on the ventral left side. On the right, it is present anterolaterally as well but is distorted. There is no reliable contact between the body and the pedicles. The pedicles are tall, and the right is broader in cross section than the left (left height: 15.1 mm, right: 14.9 mm; left breadth: 7.0 mm, right 8.0 mm). Only the right pedicle evinces evidence of a rib facet: it is a smooth, convex ovoid area 9.2 mm in breadth by 6.8 mm in height. The superior articular facets are planar and ovoid, with the right showing medial damage. The left superior articular facet is complete—it is broad and short with an oblique (inferomedial to superolateral) long axis. All of the inferior and superior articular facets have discrete margins separating articular from nonarticular areas. However, this cannot be assessed on the cranial margin of the right inferior articular process, which is covered in putty.

Perhaps the most interesting aspect of this vertebra involves the morphology of its inferior articular facets; T11 is one of this specimen's transitional vertebrae with regard to articular facet morphology. Specifically, the right inferior articular facet is clearly lumbariform in its morphology, both in terms of its shape and orientation (see osteometric analyses). Its orientation is parasagittal, like those of a lumbar vertebra, and in this regard it differs from the left facet (a fact reflected in the superior facet angles of T12). The left facet is more thoraciform in its morphology. It is ovoid with a superoinferior long axis, and it has an orientation that is best described as paracoronal, whereas the right facet is more convex and more parasagittal in its orientation. It is wider than the left one as well.

The neural canal appears to have been somewhat circular in cross section, although this is difficult to assess without the preservation of the dorsal body. The right transverse process has been heavily reconstructed with putty. In contrast, the left is complete, except for an area of missing cortical bone along its inferior margin. It is relatively short (26.1 mm), blunt, and gracile in its superoinferior cross section. Although it cannot be quantified, the spinous process does not appear to have been declined. It is short (28.5 mm) with a relatively blunt tip.

Dolní Věstonice 15's T11 (Figure 14.19) remains in two separate pieces—a heavily eroded body and a damaged neural arch. Neither pedicle is present, but the laminae, damaged inferior articular facets, and spinous process are present. The left superior articular facet, and right transverse process are also relatively complete. Its large, broad body is dorsally incomplete and cannot be articulated with its posterior arch because of missing pedicles. The size of the body increases caudally; the cranial surface is considerably smaller than the caudal surface. The caudal half of the body is more complete than the cranial half, and the entire dorsal one-fourth of the body is missing. The ring epiphyses are fused and present (at least in some areas) both cranially and caudally. For example, the superior ring epiphysis remains along the ventrum of the body, but just dorsal to it, the cortical bone is gone, and there is a deep invagination due to missing trabecular bone as well. The ring epiphysis also retains some of the faux intervertebral disc material adhered to it. Much of this same material is adhered to the caudal body as well. The body is markedly concave on the sides, and dorsal body height would have been greater than ventral height. In *norma lateralis*, one can see that the ventral face of the body is also slightly superoinferiorly concave. Although they are not preserved, the pedicles were large, judging from the size of the laminar root of the right pedicle. The right superior articular process is heavily damaged; the left articular process is complete. The facet on the left is relatively small but tall and thin. Both superior articular processes are anteroposteriorly thick, and there is a deep V-shape between them down to the root of the spinous process. The inferior facets are large but heavily damaged and are oriented in a paracoronal plane. Only the root of the left transverse process remains; the damaged right transverse process is inclined at a fairly high angle and also is markedly posterior in its orientation. The tip (ca. 5 mm?) of the spinous process is missing, but it was apparently a very short and fairly steeply declined spinous process.

Twelfth Thoracic Vertebra

Dolní Věstonice 3's twelfth thoracic vertebra (T12) remains embedded in matrix caudal to T11. The vertebra itself is quite fragmentary, and its position in the matrix is shifted in such a way that its caudal surface is facing laterally to the left. It is represented by a heavily damaged body, the remnants of the two laminae and the right pedicle. The only portion of cortical bone remaining on the body is the left lateral portion of the caudal surface. The remainder of the body's caudal surface is heavily eroded, revealing a large (25.8 mm wide x 18.3 mm deep) pit lined with trabecular bone. At its deepest point, the pit appears to terminate just caudal to the cranial surface of the vertebra, but this cannot be verified because the bone remains embedded in matrix. The right pedicle is tall but narrow (height: 12.3 mm; breadth: ca. 3.0 mm). The inferior margin of the pedicle is clearly visible and is rounded in cross section. Posteroinferiorly on the pedicle is a small, circular (ca. 3.4 mm in height; 3.5 mm in breadth) tubercle representing the articulation point for the right twelfth rib. It is cracked along its dorsal margin. The laminae are extremely distorted, and neither of the inferior articular facets is preserved. The laminae surround a natural cast of the spinal canal that measures ca. 12.8 mm in height and ca. 16.9 mm in breadth.

Dolní Věstonice 13's T12 is largely complete (Figures 14.20 and 14.21), and it is noticeably larger than T11. In superior view the body is kidney-shaped. Along the caudal surface a wedge of bone measuring 12.4 mm along its base is missing from the dorsal margin. A section of cortical bone ca. 12.1 mm in breadth is also missing from the extreme lower left dorsal corner. In fact, most of the left posterior aspect of the body is missing, and plaster makes up the entire left superolateral area to which the left pedicle is bound. The body evinces a gentle superoinferior concavity on its lateral and ventral faces. Both the superior and inferior disc surfaces are planar. Vertebral body height reduces from the dorsal lateral margins ventrally and toward the median plane. Thus, dorsal height, had it been preserved, would have been much greater than ventral height, but this feature is clearly nonpathological. Both pedicles have been damaged—the left more so than the right, which is largely complete. As a result, they are heavily reconstructed, prohibiting their measurement; however, they were much taller than they were wide. The facet for the twelfth rib is missing (see below) from the left pedicle, but it is present, albeit heavily reconstructed, on the right. The right facet appears to have been quite large (its height is 9.3 mm) and circular. Close examination of the left pedicle suggests that there may not have been a twelfth rib on the left side, which may explain why none was recovered (see Chapter 15). Specifically, on the right pedicle, the rib facet is located directly beneath the superior articular process. The corresponding region is preserved on the left, but there is no facet. Both superior articular facets are small, planar, and ovoid. The right has a rift filled with putty through its midsection. However, if it is accurately reconstructed, then the right superior articular facet was both taller and wider than the left. The inferior articular facets are large, though only the left is complete, as the right is missing approximately one-fourth of its caudal area. Both facets are slightly anteroposteriorly convex. As is most common, T12 is the transitional vertebra in terms of the orientation of its articular facets. Its superior facets lie in a paracoronal plane; its inferior facets lie in a parasagittal plane. The left mammillary process is broken off, but the right is present and is quite large (length: 16.0 mm; maximum anteroposterior diameter: 8.3 mm). The laminae are almost completely vertical. Vertebral canal height cannot be measured, but the canal width of 20.5 mm is suggestive of a mediolateral long axis for its cross section. The left transverse process is damaged, but the right is preserved. Both are extremely small—the right's maximum diameter (superoinferior) is only 6.1 mm, and their lengths are 15.3 mm (left) and 16.3 mm (right). The right transverse process is even smaller than the small accessory process found posterior to it (the accessory process was lost on the left). The spinous process is rectangular and blunt, with a fairly prominent tubercle at its tip (tip height: 15.8 mm; breadth: 6.6 mm). It has little declination (15°) and exhibits two caudal sulci separated by a central ridge.

Dolní Věstonice 14's T12 (Figure 14.24) is preserved in two separate pieces—the ventral portion of the body that remains adhered to T11 and a neural arch piece with only trivial damage. The body was transversely wider than dorsoventrally deep. In posterior view, the body retains some cortical bone in the neural canal; the largest section is a piece just to the left of midline, measuring 15.2 mm in height and 10.0 mm in breadth. To either side and below this piece, however, great expanses of both cortical and trabecular bone are missing, such that there is no good contact between the body and the pedicles. In *norma lateralis* on the right, cortical bone is present except for the dorsal one-fourth, which is eroded away. In *norma lateralis* on the left, cortical bone is only visible near the ventral face. In anterior view, cortical bone is present throughout, except just to the left of midline, where it has been gouged out of an area some 21.1 mm high and 10.6 mm wide. In inferior view, the right side is better preserved than the left in terms of cortical bone: only the dorsal one-fifth to one-fourth of the cortical bone is missing on the right; on the left, most of it is gone, except for the ventromedial one-fourth. Concreted onto its caudal surface at midline is a section of matrix corresponding to the intervertebral disc. Inferior to this matrix is a small amount of bone, probably from the cranial margin of L1 body. Laterally, the body is gently superoinferiorly concave, but ventrally it presents an almost flat surface that is gently transversely convex.

Both pedicles are superoinferiorly thick and somewhat gracile in breadth. The left is more complete than the right; the right is damaged along its superior margin, and putty is evident on its medial face (left height: 16.9 mm; left breadth: 7.4 mm, right: ca. 6.5 mm). On the left pedicle, a small (height: 4.6 mm; breadth: 4.8 mm), circular rib facet is preserved. Its center is located 9.8 mm from the cranial margin of the pedicle, 20.7 mm from the dorsal point on the mammillary process, and 8.8 mm from the inferior margin of the pedicle. Is the comparable area of the right pedicle preserved? According to the above measurements, if the right rib originated at the same spot as did the left, then there is no facet for it. Therefore, given the fact that no right rib 12 was recovered for Dolní Věstonice 14, it is possible that, as with Dolní Věstonice 13, there was congenital absence of a twelfth rib. In terms of superior articular facet morphology, the left and the right facets are strikingly asymmetrical (reflective of the difference in the inferior articular facets of T11). Neural canal shape was probably triangular at this level, with a dorsal apex. The right superior articular facet is in a parasagittal orientation with a markedly large mammillary-accessory process. The left side's mammillary process, in contrast, is smaller, not so much in its dorsoventral dimension but in its superoinferior dimension. The left facet is more thoraciform. As mentioned above, it is oriented in a paracoronal plane, and its articular surface is more planar. In contrast, the right facet is more lumbariform. It lies in a more parasagittal orientation, and its surface is more concave. The inferior articular facets are also asymmetrical. The left facet is larger and more inferiorly projecting than the right, but both have discrete, ridgelike boundaries that separate the facet proper from the nonarticular bony surfaces. The nonarticular portion of the right inferior articular process is much wider than that of the left. For example, the distance between the spinous process and the lateral point on the articular surface of each inferior articular facet reveals this asymmetry: 17.8 mm (left) and 19.4 mm (right). What is unusual about the T12 inferior articular facets is that both are in a *paracoronal* (thoraciform) orientation. The blunt spinous process is damaged at its tip. It does not appear to have been declined, although this cannot be quantified. It has central crests both superiorly and inferiorly, deviates slightly to the left, and is relatively short (its estimated length is 27.6 mm).

Dolní Věstonice 15's T12 (Figure 14.19) is represented by an apparently complete vertebra, albeit with areas of fill throughout. The body is large and transversely wider than dorsoventrally deep. In superior view, it is kidney-shaped, with a gentle internal contour (i.e., a gently concave dorsal margin). It also evinces a shift in body height, so the dorsum is much higher than the ventrum. This is particularly evident in lateral view, as the cranial surface is steeply declined toward the ventral. In other words, the body is wedged but not pathologically (in fact, its degree of wedging is almost identical to the recent human mean; see osteometric analyses). When viewed laterally and anteriorly, the sides of the body are greatly concave superoinferiorly. Ventrally, cortical bone is missing to the left of midline, and there is a small area (10.5 mm x 7.5 mm) on the left superolateral superior margin where trabecular bone is visible. Inferiorly, trabeculae are visible over much of the surface. The dorsal body in the neural canal is heavily reconstructed with putty. There is also a large amount of putty on the right where the pedicle meets the body; in fact, the right pedicle appears to be composed primarily of putty. As a result, the rib facet on the right side is covered (except for its cranial margin) by putty. The left pedicle is more complete, but there is a lip on the left pedicle that runs onto the cranial margin of the body, which could be evidence of osteoarthritis. No rib facet is visible on the left side. Both pedicles are extremely large in cross section, especially in their superoinferior dimension. Measurements are somewhat problematic because of the putty, but some can be taken—pedicle height: 17.8 mm (left), 19.3 mm (right); pedicle breadth: 9.6 mm (left). Both superior articular processes are tall and broad, but the right is larger than the left. The medial borders of the superior articular processes form a great V down to the root of the spinous process. With regard to its inferior articular facets, T12 is transitional, as the inferior articular facets face laterally (i.e., are oriented in a parasagittal plane). The neural canal is much wider than it is deep and is somewhat triangular in cross section, with its apex at the junction of the two laminae. Both transverse processes are heavily damaged. The left is gone; its root is now covered with plaster. Approximately half of the right transverse process remains. The spinous process is short (28.6 mm), blunt, and lumbariform. It is moderately declined (32°) and deviates slightly to the right. It has a large superoinferior diameter, with a slight tubercle at its tip (tip height: 13.8 mm; breadth: 5.9 mm). It evinces a central dorsal crest with two lateral sulci.

The Lumbar Vertebrae

The lumbar vertebrae are the largest of the true vertebrae, whose corpora, or bodies, are easily distinguished from other vertebral elements because of their large size and (usually) their lack of rib facets. The great size of lumbar vertebrae is due to their obvious role in weight transmission; they support the weight of the head, neck, upper limbs, and most of the trunk (White & Folkens, 2000). Among Dolní Věstonice and Pavlov humans, only Dolní Věstonice 3 and the three members of the triple burial preserve lumbar elements.

First Lumbar Vertebra

Dolní Věstonice 3's first lumbar vertebra (L1) is represented by a poorly preserved neural arch piece that remains embedded in a chunk of matrix with the other four lumbar vertebrae. The lower four lumbar vertebrae are more or less in anatomical position, but L1 is rotated so that its cranial surface faces caudally. Preserved on the piece are a large partial right pedicle, a partial left lamina, two extremely fragmentary inferior articular processes, and a badly damaged portion of the right transverse process. The right pedicle is quite robust and relatively square in cross section. Its cross section is impressive in both height and breadth (height: ca. 12.2 mm; breadth: 11.0 mm), and its external surface is smooth and relatively flat. The transverse process is preserved for 14.9 mm of its length. It is relatively thin and spearlike in morphology, with a superoinferior cross-sectional diameter of about 6.1 mm and an anteroposterior cross-sectional diameter of about 3.1 mm.

Dolní Věstonice 13's first lumbar vertebra is largely complete (Figures 14.25 to 14.28), although it is missing

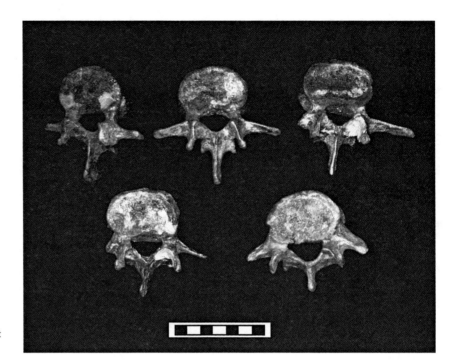

Figure 14.25 Superior view of lumbar vertebrae, Dolní Věstonice 13. Top row (left to right): L1, L2, L3; bottom row (left to right): L4, L5. Scale in centimeters.

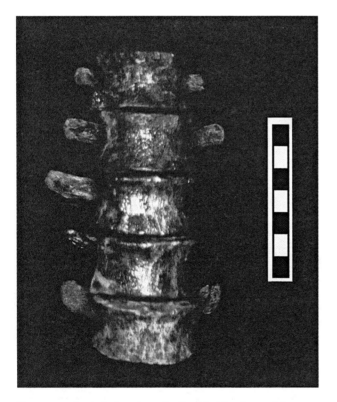

Figure 14.26 Anterior view of articulated lumbar vertebrae, Dolní Věstonice 13. Scale in centimeters.

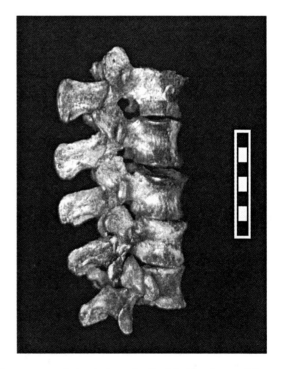

Figure 14.27 Lateral view of articulated lumbar vertebrae, Dolní Věstonice 13. Scale in centimeters.

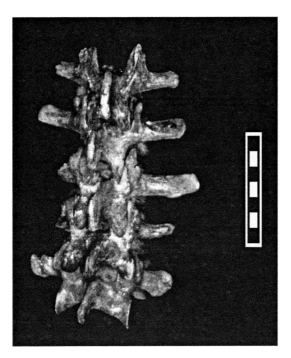

Figure 14.28 Posterior view of articulated lumbar vertebrae, Dolní Věstonice 13. Scale in centimeters.

its left transverse process, and areas of the body are heavily reconstructed. In particular, the left posterolateral aspect of the body is gone, as is much of the dorsal body. On the right side of the body, two sections of bone are missing. The first section is superolateral, located some 22 mm ventral to the dorsal margin of the body. The second damaged area is inferolateral, about 15 mm ventral to the dorsal margin. Dorsally, the right inferolateral corner is attached to the other pieces by wax and putty. Thus, although there appears to be good contact, wax covers much of the articulation between the right pedicle and the body. The join of the left pedicle to the body has been (reasonably) estimated. The body itself is large and broad. It is deeply superoinferiorly concave on all sides, although its cranial surface is slightly transversely convex. In contrast, the caudal surface is slightly transversely concave. Were it preserved, dorsal body height would have been greater (albeit not much greater) than ventral body height.

The pedicles are tall but not particularly wide. Both are damaged inferiorly, the left more so than the right, precluding measurement of their height (left breadth: 7.7 mm, right: 7.1 mm). Both superior articular facets are damaged. They are concave and oriented in a parasagittal plane; the left is more dorsoventrally expanded than the right, but the right has a greater craniocaudal dimension. Both inferior articular facets are large and mediolaterally thick, but the ventral one-third or so of the left one is missing. They are teardrop-shaped, convex, and oriented in a parasagittal plane. The mammillary processes are large, particularly the right one, which rises some 19.4 mm above the midline of the pedicle (the left rises only 17.7 mm). The accessory process is broken off on the right, but it is relatively large on the left. Only about half of the left transverse process remains. The right transverse process is complete; it is rectangular in shape and horizontal in orientation, but it is neither particularly long nor robust. Its distal end evinces damage to its superior margin; its cross-sectional measurements are height: about 10.4 mm; breadth: 4.1 mm. The neural canal is ovoid in cross section, with a mediolateral long axis. In lateral view the large spinous process appears trapezoidal, and it exhibits a large posterior tubercle (at its root it measures 21.7 mm tall, but at its tip its height is 26.6 mm). It evinces only minimal declination of about 15°. Inferiorly, the spinous process has a central crest 1.2 mm in thickness that separates two small fossae.

Dolní Věstonice 14's L1 is lacking its body, and as such it is represented solely by a partial neural arch (Figure 14.24). It is the most cranial vertebra of the specimen to have a lumbarlike spinous process. Only the very root of the right pedicle remains, precluding any measurement. The left pedicle, on the other hand, is tall and narrow (height: 15.9 mm; breadth: 7.5 mm) and has a small section of the body attached to it. Its external surface evinces a central fossa bounded by superior and inferior ridges that lead to the left transverse process. Both superior articular facets are widely spaced from each other and have discrete boundaries (although the boundaries of the right are more marked). The right facet is ovoid (superoinferior long axis), as is the left, but the right is larger. With regard to articular facet morphology, L1 is the third transitional vertebra for the Dolní Věstonice 14 individual. Specifically, the superior articular processes are short, broad, and thoracic-like, and they are in a paracoronal plane, whereas the inferior facets are in a parasagittal orientation. Also, the superior articular processes are not associated with prominent mammillary processes. If these processes were present (the root of one is visible on the left, so it appears that they were), then they were minimal and not associated with the superior articular facets. The inferior articular facets are teardrop-shaped, with discrete boundaries, especially medially, where each evinces a ridge. They are parasagittally oriented and are slightly dorsoventrally convex. The right extends more laterally than the left. The left transverse process is broken; only 17.4 mm of its length remains. Even less of the right transverse process remains—just 11 mm of its length. The spinous process is blunt and evinces central ridges both superiorly and inferiorly. Its tip measures 16.2 mm in height.

Dolní Věstonice 15's L1 is largely complete (Figures 14.29 and 14.30); the body and spinous process have trivial damage, both transverse processes are missing, and the

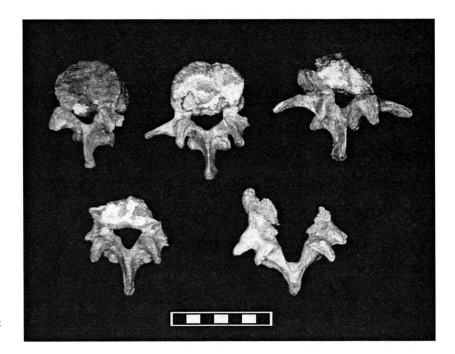

Figure 14.29 Superior view of lumbar vertebrae, Dolní Věstonice 15. Top row (left to right): L1, L2, L3; bottom row (left to right): L4, L5. Scale in centimeters.

inferior margin of the left pedicle is heavily eroded. The body is quite broad, and it is much smaller in its dorsoventral than transverse dimension. It is unusual for an L1 in that its vertebral body height decreases greatly from the dorsum of the body to its ventral margin (the more typical pattern is a slight decrease from dorsal to ventral height), and although no pathology is readily apparent, recall that this individual has an array of pathological conditions (see Chapter 17). The body is also markedly superoinferiorly concave on its lateral and ventral faces. In superior view, its dorsal concavity is quite gentle, giving the neural canal a wide ovoid shape. The caudal aspect of the body is both transversely and dorsoventrally more expanded than its cranial aspect. On both sides the posterolateral margins of the body are extensively damaged, although the right is less damaged than the left one.

The right posterolateral portion of the body preserves a small part of the pedicle. Although this pedicle is joined to the body, distal to its root it has a glued break. The articulation of the body to the neural arch is imperfect, not because of damage to the right pedicle, but because the left pedicle articulates in too cranial a position on the body, apparently because of postmortem distortion. Judging from the right side, the pedicles were tall and mediolaterally broad. No cross-sectional measurement of either pedicle is possible because the right is glued and the inferior portion of the left is missing. The superior articular facets are large and ovoid, with superoinferior long axes, and they are parasagittal in orientation. The parasagittally oriented inferior articular facets are quite large and nearly complete. The left is smaller than the right, and the right one is characterized by a larger (9.8 mm in height x 6.3 mm in breadth) swelling posterosuperior to its articular surface. The facet on the right is more circular than that on the left, which is ovoid, with a superoinferior long axis. The spinous process is short (30.0 mm) and convex along its inferior and superior margins. It evinces minor damage along its superior margin at its tip.

Second Lumbar Vertebra

Dolní Věstonice 3's second lumbar vertebra (L2) is represented solely by a fragmentary piece of the body, with a maximum width of 39.2 mm and a maximum height of 23.1 mm. There is virtually no remaining cortical bone, and the trabecular bone is heavily damaged. Its left side is filled with matrix, and its right side is trabecular bone only and is missing its cranial half. It is separated from L3 below by an intervening space roughly the width of an intervertebral disc.

Dolní Věstonice 13's second lumbar vertebra (Figures 14.25 to 14.28) is largely complete, but there are small areas of missing bone on the body. The pedicles have been glued to the body, and the inferior articular processes have been glued to their superior counterparts. In superior view its body is kidney-shaped and even more transversely expanded than that of L1. It, too, evinces a smooth superoinferior concavity on its sides and ventral face. Both the cranial and caudal surfaces of the body are generally planar, but the caudal surface of the body is larger than the cranial one. There are several areas of missing cortical bone on the body, which include small portions

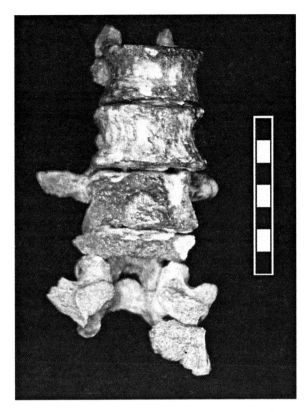

Figure 14.30 Anterior view of articulated lumbar vertebrae, Dolní Věstonice 15. Scale in centimeters.

of the right lateral cranial and caudal margins, and there is putty on the right anterolateral cranial face. However, on its caudal surface, much of the area within the annular rings was hollowed out and subsequently filled with wax. The dorsal wall of the body is badly damaged. The annular rings are missing at dorsal midline. Specifically, a stretch of the superior annular ring about 38.3 mm in length is missing, and a 26.9 mm stretch is missing from the inferior ring epiphysis. Putty is visible in the dorsal wall of the body, and the pedicles are glued to the body. There is also trivial damage to the cortical bone along the ventral midline near both the cranial and caudal surfaces.

The pedicles are tall but not particularly thick (left pedicle height: 16.1 mm, right: 17.5 mm; right pedicle breadth: 7.4 mm, left: 9.4 mm). The mammillary processes and accessory processes combine to create huge superior articular processes with maximum (dorsoventral) diameters of 20.6 mm (left) and 19.9 mm (right). The superior articular facets are tall, ovoid, and concave. The piece that includes both of the inferior articular processes is glued onto the base of the superior articular process and transverse processes. The inferior processes are long and relatively robust (although they are the most gracile of the lumbar elements). The convex and teardrop-shaped inferior articular facets are oriented in a parasagittal plane.

The vertebra also has long, fairly robust (albeit narrow in dorsoventral cross section), and slightly inclined transverse processes. The height at the tips is 11.5 mm (right and left), and the anteroposterior thicknesses at the tips are 2.5 mm (right) and 2.7 mm (left). The neural canal is wide and dorsoventrally narrow. It is somewhat triangular in cross section because of an apex at the internal junction of the two laminae. The spinous process is as long as that of L1 (its length is 39.1 mm), but its craniocaudal dimension is not so great. It is declined 22°, and it has an inferior ridge 1.4 mm thick with sulci to either side. It manifests a large tubercle at its tip, albeit not as large as L1's (tip height: 21.8 mm; breadth: 5.7 mm).

Dolní Věstonice 14's L2 (Figures 14.31 and 14.32) is represented by two pieces that cannot be articulated because of damage: the first preserves most of the neural arch, and the second preserves a portion of the inferior body, the caudal surface of which remains adhered to L3. Few landmarks on the inferior body are discernable, but it is apparent that the piece preserves the caudal ventral margin near midline, and it continues posteriorly for 25.2 mm, probably to a point just anterior to the neural canal. It also appears that the ventral face of the vertebra was superoinferiorly straight, as opposed to concave. As with the T11 and T12 bodies, the L2 body is separated from the L3 body by a layer of concreted matrix corresponding to the position of the intervertebral disc. This wedge of matrix ranges in thickness from 6.5 mm ventrally to 9.5 mm dorsally.

The pedicles retain small amounts of the body at their proximal ends. Both are tall with transversely expanded cross sections (left height: 14.7 mm, right: 14.5 mm; left breadth: 8.3 mm, right: 8.5 mm). Both superior articular facets are concave and parasagittally oriented. The right

Figure 14.31 Lateral view of L2 (left) and L3 (right), Dolní Věstonice 14. Scale in centimeters.

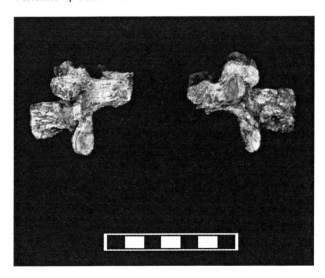

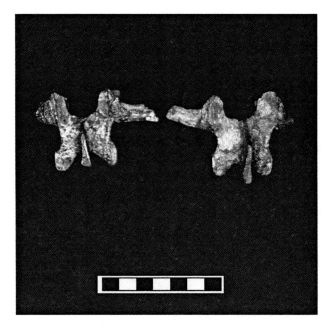

Figure 14.32 Posterior view of L2 (left) and L3 (right), Dolní Věstonice 14. Scale in centimeters.

has suffered slight damage along its anterosuperior margin—a small (6.0 mm x 2.3 mm) section of cortical bone is missing. Both mammillary and accessory processes are impressive, but the right is larger and extends more posteriorly than does the left; their dorsoventral dimensions are 21.4 mm (right) and 18.6 mm (left). The inferior articular processes are thick, teardrop-shaped, and parasagittal in orientation. The right is slightly convex, and the left is planar to slightly concave and slightly larger. The right articular surface area is separated by a ridge from its nonarticular portion; at its widest the ridge measures 2.5 mm. In posterior view the laminae evince several small fossae. The vertebral foramen appears to have been triangular, with its apex at the internal junction of the laminae. The left transverse process is broken off; all that remains is a missing area of cortical bone 11.7 mm in height and 8.9 mm in breadth. The right transverse process is narrow and bladelike; its ventral face is flat to slightly convex, and it is more dorsally convex. It was at least 35 mm long (and likely longer), somewhat dorsally oriented, and rectangular (its tip is missing). The spinous process is blunt and short (31.7 mm), with an expanded tip and a caudal, but not cranial, central ridge (tip height: 19.1 mm; breadth: 5.4 mm).

The Dolní Věstonice 15's L2 is represented by two pieces: the first is the dorsal portion of a vertebra (left partial pedicle, two laminae and associated articular processes, left transverse process, and spinous process; (Figures 14.28 and 14.29). The second piece is a vertebral body that preserves all but its cranial posterolateral portions (i.e., where the pedicles attach). The body is transversely broad and dorsoventrally shallow, with deep lateral curves and a shallow ventral curve. Its base is broader than its crown, and there is slight wedging (though not as much as was encountered in L1). It is also taller on its left than on its right. The cranial margin, rather than being straight, dips slightly at midline. In contrast, the caudal surface (which has plaster on its left side) is planar.

On the body's ventral face, about 12 mm to the right of midline and about halfway between the cranial and caudal surfaces, is a hole too large and too ventral to be a nutrient foramen. Another, larger hole is evident in the caudal surface dorsally on the left and appears to be a Schmorl's node (Chapter 19).

Only the left pedicle is preserved; the right presents only part of its inferior margin. Judging from the left, the pedicles were thick (particularly superoinferiorly)—height: 16.3 mm (left); breadth: 10.6 mm (left). The mammillary and accessory processes are large. On the right, the accessory process has a small, lateral projection that distinguishes it from the mammillary process. On the left, they cannot be distinguished (left dorsoventral dimension of mammillary process: 18.0 mm, right: 20.3 mm). The right superior articular process is larger and its facet projects more posteriorly than the left. The slope between these two processes is U- (not V-) shaped. The inferior articular processes are mediolaterally thick, but the right inferior articular process and its associated facet are larger than their counterparts on the left. Also, the right articular process is closer to the spinous process than is the left one. The right transverse process is missing, but the left one is horizontally oriented and relatively robust. Its length is 31.7 mm, its breadth at the tip is 3.7 mm, and its height at the tip is 7.9 mm. In anterior view it appears rectangular, but in superior view it thins toward its tip. The rectangular spinous process is short, broad, and blunt, and it has an inferior crest with associated sulci (tip height: 21.4 mm; breadth: 7.6 mm). It is deviated to the right and appears as if it had been twisted counterclockwise from behind.

Third Lumbar Vertebra

Dolní Věstonice 3's third lumbar vertebra (L3) is represented by a very fragmentary body 40.6 mm in breadth that remains sandwiched between L2 and L4. The left side is missing its ventral one-half to two-thirds, exposing trabecular bone. Dorsal to this damaged area it retains cortical bone, but here, too, it is crushed. A small (20.7 mm mediolateral x 8.7 mm anteroposterior x 7.5 mm superoinferior) left anterolateral body piece is separated from the remainder of the body and is adhered to the cranial surface of the L4. Finally, a small (31.1 mm anteroposterior x 9.9 mm mediolateral) portion of the cranial margin is present as well.

Dolní Věstonice 13's L3 is largely complete; its body has extensive dorsal damage and trivial erosion on its right side (Figures 14.25 to 14.28). The right pedicle is attached to the body by wax, and the left transverse and right accessory processes are broken. The body is large and transversely expanded but short, with deep superoinferior concavities laterally (especially on the right, which is more markedly concave than the left). The concavity is also present ventrally, but it is not as pronounced as laterally. In superior view the curves are gentle, with a very slight posterior convexity. The cranial and caudal surfaces of the body are in superb shape; they are slightly transversely concave, and the caudal surface is larger than the cranial.

The pedicles do not appear to be as tall as more cranial elements. They are, however, wide (left height: 15.0 mm, right: 14.5 mm; left breadth: 10.7 mm, right: 8.7 mm). The left lamina has a glue join. The neural canal is broad but not dorsoventrally deep, and it is triangular in shape (with a dorsal apex). It is a smaller vertebral foramen relative to more cranial elements, but it looks particularly small when compared to the size of the vertebral body. The mammillary and superior articular processes together make up a large superior articular process; the two processes, though merged, have a clear sulcus between them. Maximum anteroposterior diameter of the combined processes is 21.0 mm (right) and 20.5 mm (left), and their combined mediolateral breadth is 8.5 mm (right) and 6.0 mm (left). Both superior articular facets are concave and oriented in a parasagittal plane. The inferior articular processes (especially the right) are extremely thick. The facets on them are convex and teardrop-shaped; the right is larger than the left, and it extends more inferiorly as well. Yet perhaps the most striking feature of L3 is its enormous right transverse process (the left is broken off at its root). It is glued but essentially complete, and it measures 45.2 mm in length. The height of its distal end is 10.6 mm; the breadth of its distal end is 4.1 mm. The spinous process is fairly long (38.1 mm) and blunt, and the inferior margin of its tip is missing. It is only somewhat declined (21°) and is smaller than the spinous processes of either L1 or L2. It evinces an inferior central ridge separating two lateral sulci.

Dolní Věstonice 14's L3 is represented by a partial body that remains attached to L2 and a neural arch that is missing its right transverse process (Figures 14.31 and 14.32). The body is quite eroded and distorted. It is represented by a smaller cranial and larger caudal piece that have been glued together by trabecular bone, not by good bony contact. The caudal piece preserves from the posterolateral right side, which evinces cortical bone, anteriorly to the ventral face of the bone, which also preserves cortical bone. The cranial piece is badly damaged. Its cranial surface remains attached to the inferior disc surface of L2. The cranial margin preserves cortical bone; elsewhere it presents trabecular bone. The pedicles are short, broad, and tall (left height: 13.6 mm, right: 14.5 mm; left breadth: 10.2 mm, right: 10.7 mm). A small piece of superolateral body is attached to each pedicle, but neither piece makes good contact with the body piece discussed above.

The superior articular processes are concave and oriented parasagittally. Large mammillary and accessory processes are associated with the superior articular processes. The mammillary process of the right is larger and extends more posteriorly than that of the left (right anteroposterior diameter: 22.4 mm, left: 20.0 mm). Two large, teardrop-shaped inferior articular processes are also manifest, but the left one is larger than the right one. The right facet is convex, whereas the left is slightly concave. The two facets are oriented in a parasagittal plane (although the right more so than the left). Both have discrete boundaries, and medially they are separated by a furrow from the swollen nonarticular portion of the inferior articular process. The neural canal was wide—almost ovoid but with a slight peak at the internal junction of the two laminae. The left transverse process (the right is missing) is long, bladelike, slightly inclined, and rectangular in anterior view. It curves dorsally (i.e., it is ventrally convex and dorsally slightly concave), and its anterior face is more superoinferiorly planar than its superoinferiorly convex posterior face. Its tip is broken off, but most of its length is preserved (preserved length is 35.6 mm). The spinous process is the largest of the five lumbars; it is rectangular, tall, and blunt (tip height: 23.9 mm; breadth: 5.3 mm). It possesses an inferior central ridge with a sulcus to each side. Its cranial half is deviated slightly to the right.

Dolní Věstonice 15's L3 is represented by a complete neural arch and roughly half of the body (Figures 14.29 and 14.30). Ventrally, all that remains of the body is a small section of cortical bone and inferior ring epiphysis near midline. The section of cortical bone measures about 19 mm in width and 15 mm in height. The remainder of the ventral body (including trabecular bone) is missing. The entire dorsum of the body is present (there is some plaster at midline on the dorsal face and a bit more anterolaterally on the left caudal margin). The cranial portion of the body is complete dorsally, and it continues until about 11 mm anterior to the root of the left pedicle. Where present, the ring epiphyses are fused to the centrum. Caudally, the epiphysis runs for 18.6 mm at midline ventrally and is also present for the entire dorsal half of the body. Cranially, the ring epiphysis is present along the dorsal margin (but putty is present), and also for 7 mm along the left side. There is no superior ring epiphysis ventral to the right pedicle. In *norma frontalis*, it is evident that the vertebra is superoinferiorly concave on the left. This cannot be assessed ventrally, where cortical bone remains only caudally at midline. On the right, the small bit of cortical bone present is also suggestive of a superoinferiorly concave body. The dorsum of the body has only the slightest hint of a superoinferior concavity—it is almost straight.

Both pedicles are mediolaterally thick and tall (left height: 14.8 mm, right: 15.1 mm; left breadth: 11.9 mm, right: 12.2 mm). The mammillary and accessory processes are also large (left anteroposterior diameter: 21.8 mm, right: 19.6 mm). The superior articular processes are large and dorsoventrally expanded. Whereas the inferior articular processes are large, it is apparent in inferior view that the inferior articular facets are asymmetrical. Specifically, the left inferior articular process is closer to the body than is the right. It is as if the caudal margins of the two inferior articular processes had been twisted counterclockwise. This asymmetry is much less obvious from above since, although the caudal body is closer to the left inferior facet, the body's cranial margin is closer to the right superior articular facet. There is also abnormal torsion along the neural arch that surrounds a very small neural canal.

The transverse processes are long (right: 37.5 mm, left: ca. 38.5 mm) and spear-shaped. Both are damaged along their superolateral margins. The tip of the left one looks complete, but it is filled in with putty ventrally. In superior view, both processes curve posteriorly (i.e., are anteriorly convex), and both are anteroposteriorly thin as one moves distally along the process. The spinous process is longer than that of L2, and it is tall (height at its tip: 19.2 mm) and blunt. It, too, has a thick (3.3 mm) ridge separating inferior sulci, with the right sulcus more prominent than the left.

Fourth Lumbar Vertebra

The fourth lumbar vertebra (L4) of Dolní Věstonice 3 remains embedded in matrix between the third and fifth lumbar vertebrae. It is badly damaged and present only in tiny pieces, most of which lack cortical bone and some of which are separated from the rest of the body. The preserved maximum width of the body is ca. 28.8 mm, and its preserved body height is 26.1 mm.

The Dolní Věstonice 13's L4 is almost complete (Figures 14.25 to 14.28); its body is missing a chunk of its left posteroinferior margin, and the left transverse process is missing. The body is very broad, with putty liberally applied to many areas. In its right posterolateral corner there is an area about 18.3 mm dorsoventrally by 13.2 mm mediolaterally that is filled with putty. This damage affects only the annular ring for 1.9 mm along the dorsal margin and 7.1 mm on the dorsal right side. In inferior view one notes that because of the missing left posterolateral piece, 12.5 mm of annular ring is missing; the putty visible elsewhere on the rest of the caudal surface is minor fill. For example, anterolaterally on the right, there is putty along the annular ring for about 14.2 mm, beginning about 10.8 mm from ventral midline. Traces of cortical bone from the annular ring can be seen beneath the putty. The caudal disc surface is both transversely and dorsoventrally concave. In lateral view, the body evinces deep superoinferior concavities on its sides, the left of which is particularly deep. Ventrally, there is only a shallow concavity. The body's cranial surface is also mediolaterally concave, and ventral body height slightly exceeds dorsal body height. Although there is little difference in body height between the left and right sides along the dorsal margin, ventrally the right side of the body is visibly shorter than the left.

The robust pedicles are circular in cross section; that is, their mediolateral breadths rival their superoinferior heights: height: 14.1 mm (left); 13.7 mm (right); breadth: 12.0 mm (left); 10.5 mm (right). The large, concave, superior articular facets are oriented in a parasagittal plane and are taller than they are wide. The mammillary and superior articular processes combine to yield a large process (right dorsoventral diameter: 19.4 mm, left: 18.2 mm), but combined they are not as mediolaterally thick as those of L3: 4.9 mm (right) and 7.0 mm (left). On both sides there is a sulcus separating the superior articular and mammillary processes, but the right one is more conspicuous (right accessory thickness: 4.3 mm, left: 4.4 mm). The mediolaterally expanded inferior articular processes are widely spaced from each other. Their facets are teardrop-shaped, convex, and oriented in a parasagittal plane. The neural canal is triangular in shape, with its base at the dorsal margin of the body and its apex at the internal junction of the laminae. The gracile, spearlike right transverse process is slightly inclined (only the base of the left is present). The spinous process is long (33.3 mm), somewhat declined (35°), and evinces an inferior central ridge that separates two sulci.

The Dolní Věstonice 14 L4 is represented by a caudal part of its body and a neural arch with damage to its transverse processes. The caudal body of L4 is heavily eroded in its midsection, such that it appears to be a ring of bone. However, there is enough preserved to get an accurate sense of caudal body width and to determine that whereas the vertebral body is superoinferiorly concave on its sides, ventrally it is relatively flat. The body is also transversely expanded and relatively shallow in its dorsoventral dimension. It is difficult to get a sense of its superoinferior dimension, as very little of the cranial margin is preserved (only a small anterolateral section is present on the right side). However, it is almost certain that ventral body height would have exceeded dorsal body height.

Both pedicles are proximodistally short and mediolaterally broad in cross section. Although tall, they are not as high as those of more cranial elements such as T12, L1, or L2 (left height: 14.1 mm, right: 14.2 mm; left breadth: 13.3 mm, right: 12.9 mm). Both superior articular facets are oriented in a parasagittal plane, although the right is oriented more parasagittally than is the left. Both have clearly delineated articular verses nonarticular surfaces, with furrows separating the articular from nonarticular portion of the process. The right facet is more concave than the left, which is only moderately concave—almost planar. Inferior to each superior articular facet is a fossa, the right of which is deeper than the left (right fossa height: 6.3 mm, left: 5.4 mm; right fossa width: 4.4 mm, left: 5.2 mm). The mammillary processes are relatively modest in size (right dorsoventral dimension: 17.5 mm, left: 19.2 mm). The inferior articular processes are broadly spaced from each other; they are tall, teardrop-shaped, and convex. The neural canal would have been triangular in shape—a broad-based triangle with its apex at the internal junction of the laminae. Much of the right transverse process is preserved; it appears that the tip (perhaps 20% or less of its total length) is broken off. Its preserved length is 30.1 mm. All that remains of the left transverse process is a hole at its base measuring 9.8 mm in breadth and 10.3 mm in height. The spinous process is blunt and somewhat declined (25°). Its tip is slightly damaged, and the inferior surface of its tip is deviated to the right. It evinces a prominent inferior central ridge, 0.9 mm thick at its midsection, that separates two fossae.

Dolní Věstonice 15's L4 is represented by a superoposterior body and neural arch that is lacking only its transverse processes (Figures 14.29 and 14.30). None of the body is preserved caudal to the level of the pedicles, and none is present ventral to the coronal plane that would divide the body into equal dorsal and ventral halves. Thus, only a tiny section of the mediolaterally expanded body remains. The cranial surface of the body has a fair amount of faux intervertebral disc glued to it. In superior view, the ventral wall of the neural canal is gently concave.

The pedicles are thicker than they are tall and rectangular in cross section. Both superior articular processes are large, although the right appears to have more of a posterior extension than does the left. Their articular surfaces are concave. The associated mammillary processes are relatively large—anteroposterior diameter: 19.7 mm (right) and 19.5 mm (left). The inferior articular processes are bulbous, and their articular surfaces are somewhat convex. The left inferior articular facet is larger than the right. The spinal canal is small and triangular, with a dorsal apex. The spinous process is long (34.6 mm), is slightly declined (26°), and evinces a slight swelling at its tip. Its inferior surface has some plaster on it, but it is split by a central ridge that separates two fossae, the left of which is larger than the right. It appears that an epiphyseal line is present along the posterior right side of the spinous process. Finally, the spinous process appears to deviate toward the right halfway along its course before returning to the midline. The superior portion is more deviated than is evident from an inferior view (i.e., the inferior half of the spinous process appears more symmetrical).

Fifth Lumbar Vertebra

Relative to the other lumbar vertebrae of Dolní Věstonice 3, which also remains in matrix, the fifth (L5) is in somewhat better shape, although it, too, is battered. It is represented solely by its badly damaged body, the ventral surface of which is 37.1 mm in breadth and 22.3 mm in height. Posteriorly, what little bone that can be seen indicates that the dorsal aspect of the body was much shorter than its ventral aspect. A second piece of L5, a section of cancellous bone, remains adhered to the superolateral margin of the first sacral segment.

Dolní Věstonice 13's fifth lumbar vertebra (Figures 14.25 to 14.28) is largely complete, aside from a missing section of body. In superior view, the body is ovoid in shape: it is mediolaterally broad, with a short dorsal body height and a tall ventral body height. A wedge of bone (13.8 mm anteroposteriorly by 13.2 mm mediolaterally) is missing from the left dorsal corner of the body next to the pedicle. The superior ring epiphysis is missing at midline. The inferior ring epiphysis is missing just to the left of midline for about 26.6 mm, near the ventral margin of the left pedicle. It is also eroded inferior to the right pedicle for about 7.9 mm. Dorsally, there is a crack through the body about 4 mm to the right of the left pedicle. Laterally the body is somewhat superoinferiorly concave, but ventrally it is gently concave. The caudal surface of the body is transversely and dorsoventrally concave, whereas the cranial surface is planar.

The pedicles are extremely broad in cross section (left pedicle height: ca. 12.5 mm, right: 11.7 mm; left breadth: ca. 19.2 mm, right: 20.8 mm) because the transverse processes come off the junction of the pedicle with the body, adding to their width. The left pedicle has a break in it that is associated with the missing section of body. The root of the left transverse process is also broken at the same location, and it has another break about 8 mm from its rounded tip. Both of the superior articular facets are concave, but the right is larger than the left. The ventral portions of their faces lie in a paracoronal plane, whereas the dorsal aspects of their faces are parasagittal. The mammillary processes are not distinguishable from the superior articular processes (right anteroposterior diameter: 21.0 mm, left: 19.7 mm). The right lamina is glued to the bottom of the superior articular process. The inferior articular processes are large, thick, and widely spaced. The convex facets themselves are oriented in a parasagittal plane, which is unusual for an L5. The vertebral canal is large and triangular. The spinous process is short (26.7 mm), not too blunt, and only moderately declined (19°). Its tip measures 12.2 mm in height and 5.9 mm in breadth. It evinces an inferior central crest, but no fossae are evident to either side.

Dolní Věstonice 14's L5 is largely complete; its body has some damage, as does its right superior articular facet and the right mammillary process. The right transverse process is missing. There is also trivial damage to the left inferior articular facet, left transverse process, and spinous process. The body of L5 is heavily damaged anteromedially. Specifically, it has suffered a gash on its ventral face that runs from its right superolateral side inferomedially down through and slightly beyond midline. Putty fills some of this gash at its caudal portion. In superior view, putty is also evident just to the left of midline, running dorsoventrally from the dorsal margin to just shy of the ventral margin. In inferior view, the body is complete, except along its left anterolateral margin. Putty is evident posteroinferiorly, as well as in the vertebral canal. As is typical for an L5, the body is much shorter dorsally than ventrally. Laterally, it is only very slightly superoinferiorly concave, and although difficult to assess, in lateral view its ventral face appears to have been flatter. The body is ovoid in superior view, and it is much more transversely broad than the other lumbar vertebrae.

The pedicles are large, especially if measured with the transverse processes, which share their roots—pedicle height: 20.4 mm (left), 19.8 mm (right); breadth: 12.0 mm (left), 12.3 mm (right). Both superior articular facets are concave and are oriented nearly halfway between the sagittal and coronal planes. The right is missing its inferolateral portion: perhaps one-fourth of its original size is gone. The right mammillary process is broken (perhaps 40% remains), but the left is large (dorsoventral diameter: 20.7 mm). The inferior articular processes are very widely spaced, with the right smaller and in a more inferior position than the left. Both inferior processes evince complete facets, although the left has putty on its cranial margin. As expected for a fifth lumbar vertebra, the mediolaterally concave facets are oriented in a paracoronal plane. The vertebral foramen is triangular in cross section. The right transverse process is represented solely by a gaping hole at its origin some 22.0 mm in breadth and 11.8 mm in height. The left transverse process is rounded and spearlike, although its tip is damaged. It is somewhat superiorly and dorsally inclined. Its cross section is rounded ventrally and superiorly, whereas it is dorsoventrally concave caudally. Its dorsal face is flat, with a fossa some 11.9 mm long and 3.8 mm high. The spinous process is largely complete; a small portion of its tip is damaged. It has a very simple morphology (flat and rectangular); there is no cresting either superiorly or inferiorly.

Dolní Věstonice 15's L5 is relatively poorly preserved; it is represented by its neural arch and three small pieces of the body (Figures 14.29 and 14.30). Aside from small pieces of the body that remain attached to the cranial margins of the pedicles, only one other tiny portion of the body can confidently be assigned to L5— a piece that is glued inferomedially to the left transverse process. This piece includes the caudal margin of the body, as well as a small section of the left lateral face. It indicates that the body was extremely superoinferiorly concave, at least laterally, and that the dorsal wall of the body was extremely short (i.e., exhibited the nonpathological dorsal wedging common in L5). The pedicles are relatively broad, short, and tall in cross section (left height: 11.4 mm, right: 11.6 mm; left breadth: 16.3 mm, right: 16.2 mm). The mammillary/superior articular processes are large—dorsoventral diameter: 19.6 mm (right), ca. 18.5 mm (left). The superior articular facets are large and only slightly concave. The left is more circular, whereas the right is more ovoid, with an anteroposterior long axis. They are spaced widely apart, but the inferior articular facets are more widely spaced still. The inferior articular facets are planar and lie in a paracoronal plane. The left is larger than the right. In cross section, the neural canal is thimble-shaped, with its rounded end dorsal; it is extremely dorsoventrally elongated. The right transverse process is spearlike in morphology and horizontal in inclination. The left transverse process is small and spearlike, although its tip is damaged along its dorsal margin. The spinous process is small and beaklike.

Osteometric Analyses: The Presacral Vertebral Remains

The human vertebral column has undergone extensive modification for bipedal locomotion, and thus it is an anatomical area of interest to paleoanthropologists. However, much of this structural modification occurred relatively early in human evolution (Latimer & Ward, 1993), and because of strong stabilizing selection has remained relatively stable for well over 1600 millennia. Yet despite its conservative nature, metric and nonmetric differences in vertebral morphology do exist among Late Pleistocene humans (Ogilvie et al., 1998; Brown et al., 2002) and are worthy of exploration. Vertebrae from paleontological/archeological contexts are relatively rare (Brain, 1981) because of their small size and their composition (they have a high proportion of trabecular bone). If for no other reason, then, the fact that the Dolní Věstonice and Pavlov sample preserves such a large number of vertebrae is significant.

Materials and Methods

A number of standard and nonstandard osteometric measurements for the Dolní Věstonice and Pavlov vertebrae were reported in Sládek et al. (2000), and more nonstandard metrics have been given in the morphological description above. These data serve as a baseline for comparison of the Dolní Věstonice and Pavlov vertebrae to those of recent humans, but also to penecontemporary humans. In particular, the Dolní Věstonice and Pavlov vertebrae are compared to those humans Gravettian-associated, as well as to the European Neandertals. Unfortunately, given the relative rarity of fossil vertebrae and the atypical nature of many of the vertebral measurements used, there are few penecontemporary humans to which the Dolní Věstonice and Pavlov remains can be compared. Data on Paglicci 25 and Arene Candide IP are included in some of the analyses, as are data on the Regourdou 1 Neandertal. Also, data for the Předmostí specimens from Matiegka (1938) are used in some of the analyses.

In terms of recent humans, the analyses primarily employ data from the literature. Specifically, means and standard deviations for several measurements of the thoracic and lumbar vertebrae were taken from Latimer and Ward (1993), and individual measurements of the vertebral canal were taken from MacLarnon (1993). The former sample is from the Hamann-Todd Collection at the Cleveland Museum of Natural History, and it includes a maximum of thirty adult individuals who possessed twelve thoracic and five lumbar vertebrae. Not all of these individuals preserve all of the measurements used, and thus smaller sample sizes occur for most of the analyses and are noted as such. The latter sample comprises just seven individuals (two males and five females) from the Spitalfields collection at the Natural History Museum in London. These data are supplemented by data on recent humans (Amerindians and Euro-Americans; total $N = 174$) from the Maxwell Museum of Anthropology at the University of New Mexico.

The measurements and indices to be investigated include the following: (1) degree of vertebral wedging of thoracic and lumbar vertebrae; (2) interfacet distances of thoracic and lumbar vertebrae (and indices thereof); (3) thoracic and lumbar superior articular facet orientation; (4) spinous process angle of thoracic and lumbar vertebrae; (5) vertebral costal facet angles of T1–T9; and (6) vertebral canal height and breadth of cervical, thoracic, and lumbar vertebrae. All of the above measurements for the Dolní Věstonice and Pavlov individuals are in Sládek et al. (2000). The quantitative methodology by which these measurements and indices for recent humans are compared to the Dolní Věstonice and Pavlov sample includes standard univariate and bivariate statistical methods such as t tests (including both the two-sample comparison and the single observation vs. a sample of varieties) and percentage deviations from expectation [(observed measure – predicted measure)/observed measure; R. J. Smith, 1980]. This formula has been modified to read (fossil value—recent human mean value)/fossil value (following Franciscus & Churchill, 2002).

Degree of Vertebral Body Wedging

In lateral view, the adult human vertebral column evinces four normal curves. The cervical and lumbar regions are ventrally convex, whereas the thoracic and sacral regions are ventrally concave. These latter two curves are referred to as primary curvatures because they are present in neonates (Aiello & Dean, 1990). In contrast, the cervical and lumbar curvatures are developmental, and they are due to holding the head upright (cervical) and, for the lumbar region, the upright posture associated with bipedality (Aiello & Dean, 1990). These secondary curvatures aid in bringing the body's center of gravity over the feet during bipedal locomotion. Although much of these curvatures are due to differences in the dorsal versus ventral height of the intervertebral discs, they are also evident in the bodies of vertebrae. Vertebrae that are part of a ventrally convex portion of the vertebral column tend to have dorsally superoinferiorly compressed and ventrally superoinferiorly expanded bodies; the opposite is true of vertebrae from a region with a ventrally concave curvature. Vertebrae that evince a reduced dorsal or ventral height are referred to as wedged. Latimer and Ward (1993: 271) use the following formula to estimate the degree of wedging in thoracic and lumbar vertebrae:

$$a° = 2 \tan^{-1} (Y/X)$$

where Y = (vertebral dorsal body height – ventral body height)/2, and X = vertebral body's cranial dorsoventral diameter. This method of quantifying the degree of wedging is arguably preferable to computing a simple ratio of dorsal-to-ventral body height since it takes into account the body's dorsoventral dimension (Latimer & Ward, 1993).

Twelve vertebrae from Dolní Věstonice 13 and 15 are complete enough that their degree of wedging can be computed, and in Table 14.1 these specimens are compared to the recent human means reported by Latimer and Ward (1993) by using a t test in which a single individual is compared to a sample (Sokal & Rohlf, 1981: 231). Negative index values indicate those vertebrae that are dorsally wedged, that is, have a shorter dorsal than ventral body height. Positive values, in contrast, are those individuals who evince ventral wedging, that is, have a shorter ventral than dorsal body height. The t test results

Table 14.1 Comparison of wedging angle (in degrees) of Dolní Věstonice thoracic and lumbar vertebrae versus that of a sample of recent humans.

	Vertebra	Degree of Wedging[a]	Recent Human Mean ($N = 20$)[b]	Recent Human s.d.	t_s[c]	P
Dolní Věstonice 13	T7	5.19	3.73	1.41	1.011	0.325
	T8	3.75	3.23	1.53	0.332	0.744
	T9	−0.21	1.67	1.98	−0.927	0.366
	T10	−2.39	1.90	1.58	−2.650	0.015*
	L3	−1.80	−3.11	2.06	0.621	0.542
	L4	−1.22	−3.72	2.42	1.008	0.326
	L5	−8.93	−7.93	2.12	−0.460	0.651
Dolní Věstonice 15	T2	3.58	−0.19	2.27	1.621	0.122
	T3	6.48	1.03	1.99	2.673	0.015*
	T8	2.19	3.23	1.53	−0.663	0.515
	T12	4.55	4.38	1.97	0.084	0.934
	L1	8.42	3.00	2.39	2.213	0.039*

*Significantly different at $P < .05$.
[a]Degree of wedging is calculated as in Latimer and Ward (1993:271).
[b]Recent human data are from Latimer and Ward (1993).
[c]The t-statistic was computed by using a modified version of Student's t test (Sokal & Rohlf, 1981: 231).

indicate that three of the vertebrae statistically exhibit significantly different ($P < 0.05$) degrees of wedging from those of their recent human counterparts. First, the tenth thoracic vertebra of Dolní Věstonice 13 is dorsally wedged, which is unusual for a thoracic vertebra. In fact, the simple ratio of dorsal-to-ventral body height for this specimen (94.6) falls below the third percentile of a recent human sample (Amerindians and Euro-Americans; $N = 77$). Unfortunately, because of damage to the dorsal margins of the body, T11 dorsal height could not be measured for Dolní Věstonice 13, but it, too, had a shorter dorsal than ventral body height, suggesting an unusual lower thoracic lordosis for this individual. However, all of the lumbar vertebrae of Dolní Věstonice 13 appear to be normal, so it does not appear that the specimen had an exaggerated lumbar lordosis.

The second unusually wedged vertebrae belong to the pathological Dolní Věstonice 15. Specifically, T3 has a higher dorsal than ventral body height, which is in the expected direction, but the magnitude to which the vertebra is ventrally wedged is unusual—its ratio value of 110.7 falls at the ninetieth percentile of the same sample of recent humans ($N = 73$). Its T6 also appears to be ventrally wedged (see morphological description), but neither dorsal nor ventral body height was preserved for this element. The third unusually wedged vertebra of the Dolní Věstonice and Pavlov sample is the first lumbar vertebra of Dolní Věstonice 15. It evinces a much higher dorsal than ventral body height. Among recent humans, there is generally only a moderate (ca. 3°) amount of ventral wedging; at this level the vertebral column is expected to begin the transition to an anteriorly convex state. The simple ratio of dorsal-to-ventral body height corroborates the wedging angle result; Dolní Věstonice 15's L1 value (119.2) falls just under the ninetieth percentile of the above recent human sample ($N = 78$).

Interfacet Distance and Relative Interfacet Distance Index

Unlike nonhuman hominoids, whose interfacet distances (i.e., those distances between the articular processes) in the lumbar region either exhibit stasis or decrease as one moves caudally down the lumbar spine, human interfacet distances increase as one moves caudally through the lumbar elements (Aiello & Dean, 1990; Latimer & Ward, 1993). Two nonmutually exclusive explanations are offered for this uniquely human pattern. The first is that humans have wider sacra than our African ape relatives, and in order to compensate for this, the breadth between the left and right articular facets on human lumbar vertebrae must increase as one moves caudally (Aiello & Dean, 1990). The second is that only by increasing the breadth between the left and right lumbar articular processes are the vertebrae able to overlap, or imbricate (Latimer & Ward, 1993). This imbrication is critical because it allows for the normal human lumbar lordosis without having these articular processes impinging on, and/or potentially damaging, intervening structures such as the laminae or pars interarticularis (Latimer & Ward, 1993).

Interfacet distance (or superior interfacet distance; Sládek et al., 2000) is measured as the straight-line distance between the midpoints of the left and right superior articular facets. Figure 14.33 shows the intervertebral

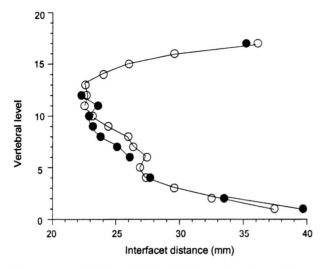

Figure 14.33 Profiles of interfacet distances (mm) along the thoracolumbar column (after Latimer and Ward, 1993) for Dolní Věstonice 13 (black circles) and recent human means (open circles).

facet distance profiles for the mean of a sample of recent humans ($N = 20$) and Dolní Věstonice 13. For this and similar plots, vertebral level 1 is the fifth lumbar vertebra, and vertebral level 17 is the first thoracic vertebra. Thus, as in the body, the more cranial vertebrae are at the top of the figure, and the more caudal elements are at the bottom. Only means are shown for the recent humans. Note that although interfacet distance is great in the upper thoracic and lower lumbar regions, it reduces in the mid- to lower thoracic region. The overall profile of Dolní Věstonice 13 looks very similar to recent humans, an observation borne out by the t test results presented in Table 14.2. These univariate tests indicate that none of Dolní Věstonice 13's interfacet distance percentage deviations from recent humans (which range in magnitude from +1.2% to -9.2%) is significantly different from recent humans.

Figure 14.34 shows the same recent human interfacet distance profile, this time compared to Dolní Věstonice 14. This specimen also shows great similarity to the recent human sample in the upper thoracic and lower lumbar regions, but there appear to be some differences in the lumbar and lower thoracic region, a supposition borne out to some degree by the t test results in Table 14.2. Note that the eleventh thoracic vertebra has a significantly wider (+11.8%) interfacet distance than expected. The third lumbar vertebra's interfacet distance is also much larger (+14.3%) than the recent human mean, but according to the t test, it is not statistically not significantly greater. The other interfacet distances closely mirror those of the recent human sample, evincing percentage deviations ranging in magnitude from −0.5% to +9.3%.

Figure 14.35 shows the recent human interfacet distance profile with values for both Dolní Věstonice 15 and Pavlov 1 (the latter of which only preserves this measurement for T1) superimposed. Dolní Věstonice 15's profile closely approximates the recent human pattern, an observation that is corroborated by the t test results in Table 14.2. None of Dolní Věstonice 15's interfacet distance percentage deviations (which range in magnitude from −0.9% to +9.3%) is significantly different from the recent human sample.

Although interfacet distance shows the expected caudal increase in the lumbar vertebral column, to some extent this could be a function of the fact that vertebrae become larger as one moves caudally (Latimer & Ward, 1993). One means by which to correct for this phenomenon is to create an index of a particular vertebral element's interfacet distance relative to its own body's transverse diameter (Latimer & Ward, 1993). Figure 14.36 shows the recent human mean profile ($N = 20$) for this index with values for Dolní Věstonice 13 superimposed. Note that among recent humans, the index steadily increases caudally throughout the thoracic column but then decreases in the lumbar region, indicative of the fact that even correcting for these vertebral elements' larger sizes their interfacet distances are increasing at a still greater rate. Although his values are generally greater than the recent human means, Dolní Věstonice 13's profile is similar in shape to that of recent humans, and it appears that for T10, T12, and L2, Dolní Věstonice 13 has considerably larger interfacet indices than the recent human mean. The t test results in Table 14.2 indicate that only two of these vertebrae have statistically significantly different indices (at $P < 0.01$) from recent humans. First, T10 has a considerably higher index (+15.6%) than the recent human sample. The second vertebra, T12, also has a considerably higher index (+19.1%) than the recent human sample. This suggests that at least for these two vertebrae, Dolní Věstonice 13 has a relatively smaller interfacet distance than expected.

The interfacet index values for Dolní Věstonice 14 and 15 are superimposed over the same recent human profile in Figure 14.37. All of the Dolní Věstonice 14 and 15 index values fall very close to the recent human profile, an observation corroborated by the t test results in Table 14.2. The greatest deviation (+13.1%) is for the T1 of Dolní Věstonice 14; most of the other deviation values fall within the 1–3% range (essentially within measurement error).

Superior Articular Facet Angle

Latimer and Ward (1993: 280) measured the superior articular facet angle as the angle a line perpendicular to the plane tangential to the centroid of the facet makes with

Table 14.2 Percentage deviations of Dolní Věstonice and Pavlov thoracic and lumbar superior interfacet distances and vertebral body breadth and superior interfacet distance indices from recent human means.

	Vertebra	Percentage Deviation, Interfacet Distance ($N = 30$)	Percentage Deviation, Body Width/Interfacet Distance Index ($N = 20$)
Dolní Věstonice 13	T1	−2.6	
	T6	−1.8	
	T7	+4.5	+12
	T8	−1.2	+3.0
	T9	−5.2	+9.3
	T10	−9.2	+15.6†
	T11	−5.1	+10.9
	T12	−5.2	+19.1†
	L2	+1.2	+11.6
	L4	+3.0	+7.8
	L5	+5.7	
Dolní Věstonice 14	T1	+9.3	+13.1
	T2	+1.7	−3.2
	T5	−0.5	
	T7	+6.8	
	T11	+11.8*	
	T12	+7.9	
	L1	+6.6	
	L2	+8.4	
	L3	+14.3	
	L4	+2.1	
	L5	+5.2	−0.5
Dolní Věstonice 15	T1	+1.6	
	T2	−2.7	
	T3	−0.9	
	T6	−8.1	
	T7	−4.9	
	T8	−1.2	+5.5
	T9	+2.8	
	L1	+6.3	+2.9
	L2	+9.3	+1.8
	L3	−5.3	
	L4	−4.9	
	L5	−2.3	
Pavlov 1	T1	−2.0	

*Significantly different at $P < 0.05$; †significantly different at $P < 0.01$; significance determined by using modified Student's t test (Sokal & Rohlf, 1981: 231).

the median plane. In most thoracic vertebrae, where the articular surfaces face posterolaterally (i.e., the facets themselves are oriented in a paracoronal plane), the normal is anteromedial, and the angle is computed as a negative angle. In contrast, for lumbar vertebrae, where the articular surface of the superior articular facets tend to face posteromedially (i.e., the facets themselves are oriented in a parasagittal plane), the normal is anterolateral, and the angle is computed as a positive angle. In humans, this angle tends to decrease in the lumbar column (i.e., the lumbar superior articular facets become more coronally oriented as one moves caudally down the column). As with interfacet distance, this has much to do with the normal human lumbar curvature since vertical compressive forces will tend to anteriorly displace the lower lumbar vertebrae; facets that are more coronally oriented help resist this displacement (Latimer & Ward, 1993). The lumbosacral junction also plays a role, particularly with re-

The Vertebral Columns

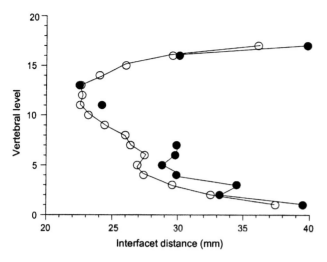

Figure 14.34 Profiles of interfacet distances (mm) along the thoracolumbar column (after Latimer & Ward, 1993) for Dolní Věstonice 14 (black circles) and recent human means (open circles)

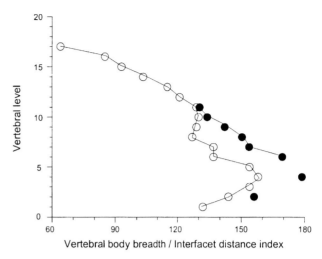

Figure 14.36 Profiles of vertebral body breadth/interfacet distance indices along the thoracolumbar column (after Latimer & Ward, 1993) for Dolní Věstonice 13 (black circles) and recent human means (open circles).

gard to the inferior articular facets. Specifically, the ventrally facing inferior articular facets of L5 help to prevent it from sliding off of the steeply declined cranial surface of the sacrum (Aiello & Dean, 1990).

Figure 14.38 shows the mean superior articular facet angle profile for the recent humans ($N = 15$) with the values for Dolní Věstonice 13 superimposed. The right and left sides were averaged for Dolní Věstonice 13, and when only one side was present, it was used. Differences between the profiles are apparent in the thoracic region, but overall the profiles are similar, especially with regard to the transition between T12 and L1, for which the two profiles are virtually identical. Table 14.3 presents t test results that quantify the differences in superior articular facet angle between recent humans and Dolní Věstonice 13 for the thoracic vertebrae, and Table 14.4 presents t test results for the lumbar vertebrae. Almost all of the thoracic vertebrae (with the exception of T10 and T12) have statistically significantly greater negative indices, indicating that the superior articular facets face more laterally than

Figure 14.35 Profiles of interfacet distances (mm) along the thoracolumbar column (after Latimer & Ward, 1993) for Dolní Věstonice 15 and Pavlov 1 (black circles) and recent human means (open circles).

Figure 14.37 Profiles of vertebral body breadth/interfacet distance indices along the thoracolumbar column (after Latimer & Ward, 1993) for Dolní Věstonice 14 and 15 (black circles) and recent human means (open circles).

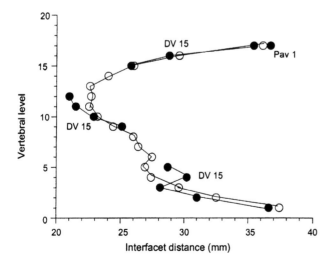

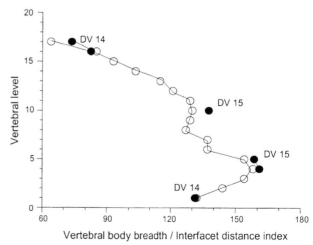

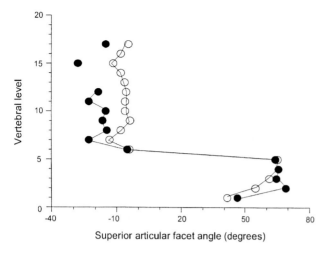

Figure 14.38 Profiles of superior articular facet orientation (in degrees) for Dolní Věstonice 13 (black circles) and recent human means (open circles).

expected. Of the lumbar vertebrae, only L4 is significantly different from the recent human sample mean; it evinces a more medially facing superior articular facet than expected. The functional significance (if any) of this finding is uncertain, and it is possible that measurement error, lack of sufficient variability in the small recent human sample, and/or postmortem distortion are playing a role in this result.

Dolní Věstonice 14's superior articular facet angle profile is superimposed over the recent human mean profile in Figure 14.39, and the differences in facet orientation between this specimen and recent humans are immediately apparent. Note in particular the radical change in orientation from T11 to T12 to L1. Recall that each of these three vertebrae in Dolní Věstonice 14 could be considered transitional with regard to articular facet morphology. Specifically, the right inferior articular facet of T11 is lumbarlike in shape and orientation, whereas the same facet on the left is more thoracic-like in morphology. The superior articular facets of T12 mirror this pattern, with the left more thoraciform in its morphology, and the right more lumbariform. The inferior articular facets of T12 present a paracoronal, thoracic-like orientation. Thus, the superior articular facets of the first lumbar vertebra have thoracic-like superior articular facets to meet the T12 facets, but the inferior articular facets are more lumbarlike in morphology and orientation. These observations explain the unique profile presented for this specimen in Figure 14.7.

The t test results in Tables 14.3 and 14.4 indicate that most of Dolní Věstonice 14's thoracic superior articular facets are oriented in directions significantly different from those of the recent human sample; only T5 is similar to recent humans. In contrast, with the exception of the unusual first lumbar vertebra, the superior articular facets of the remaining lumbar vertebrae are oriented as expected.

Dolní Věstonice 15 presents a profile in Figure 14.40 that is similar in many respects to the recent human sample, especially with regard to the transition between T12 and L1. However, there are some striking differences in the thoracic column, and these differences are illuminated in Table 14.3. Note, too, that the superior articular facet angle of the first thoracic vertebra of Pavlov 1 is significantly different from the recent human sample. Table 14.3 reveals no significant differences in lumbar facet orientation between Dolní Věstonice 15 and recent humans.

Spinous Process Angle

Spinous process angle is defined by Latimer and Ward (1993) as that angle that the spinous process makes relative to the plane of the body's cranial surface. They note that there are two regions along the vertebral column where in recent humans the spines tend to project more horizontally: the upper thoracic region and the lumbar region. In the mid-thoracic region, the spines are more inferiorly oriented, but this, they argue, is due to the high degree of imbrication exhibited in this region.

Figure 14.41 shows the mean recent human thoracolumbar spinous process angle profile (N = 20) with the preserved values for Dolní Věstonice 13 superimposed. The recent human profile shows the increased steep inclination of the spinous processes in the mid-thoracic region, with more horizontal processes evident in the lumbar and upper thoracic regions. What is interesting is that except in the lower thoracic region, Dolní Věstonice 13 tends to have a higher spinous process angle (i.e., a more steeply declined spinous process) than expected. The t test results presented in Table 14.3 and Table 14.4 suggest that only some of these differences are statistically significant, however; specifically, the angles for T6, T7, T11, and L4 are significantly different from those of the recent human sample.

Figure 14.42 shows a similar pattern for Dolní Věstonice 14 and 15, which with one exception (T9 of Dolní Věstonice 14) have more steeply declined spinous processes than the recent human mean. However, as presented in Tables 14.3 and 14.4, only three of the thirteen differences are actually statistically significant (T7, T8, and T9 of Dolní Věstonice 15), all of which are more steeply inclined than expected.

Vertebral Costal Facets Angle

The angle made between the median plane and a line drawn between the two vertebral costal facets (i.e., body and transverse process) is referred to as the vertebral costal facets angle by Latimer and Ward (1993). This angle

Table 14.3 Results of modified Student's t tests for the thoracic vertebrae of Dolní Věstonice and Pavlov versus recent human means for superior articular facet angles, spinous process angles, and costal facet angles.

	Vertebra	Superior Articular Facet Angle ($N = 15$)		Spinous Process Angle ($N = 20$)		Costal Facet Angle ($N = 10$)	
		t_s	P	t_s	P	t_s	P
DV 13	T1	−4.686	<0.001	—	—	—	—
	T3	−3.648	0.003	—	—	—	—
	T6	−6.993	<0.001	2.901	0.009	1.313	0.222
	T7	−6.480	<0.001	2.176	0.042	1.778	0.109
	T8	−12.449	<0.001	—	—	2.286	0.048
	T9	−3.551	0.003	−0.567	0.577	−0.774	0.459
	T10	−1.683	0.115	−1.894	0.074	−0.592	0.568
	T11	−3.227	0.006	−3.310	0.004	—	—
	T12	−0.335	0.743	−0.545	0.592	—	—
DV 14	T1	−8.701	<0.001	—	—	3.342	0.009
	T2	—	—	0.726	0.477	—	—
	T5	−1.826	0.089	—	—	—	—
	T6	−6.724	<0.001	—	—	—	—
	T7	−0.381	0.709	—	—	—	—
	T8	−6.224	<0.001	1.067	0.299	2.058	0.070
	T9	6.020	<0.001	−0.567	0.577	2.128	0.062
	T11	−2.898	0.012	—	—	—	—
	T12	13.066	<0.001	—	—	—	—
DV 15	T1	−0.892	0.387	—	—	0.891	0.396
	T2	1.021	0.324	0.515	0.612	0.121	0.906
	T3	−1.879	0.081	1.687	0.108	0.195	0.850
	T4	−3.828	0.002	—	—	—	—
	T5	0.058	>0.950	—	—	—	—
	T6	−0.269	0.792	—	—	—	—
	T7	−1.715	0.108	4.096	<0.001	—	—
	T8	−11.066	<0.001	4.215	<0.001	—	—
	T9	−3.273	0.006	2.398	0.027	—	—
	T10	−5.178	<0.001	−0.332	0.744	—	—
	T11	−2.898	0.012	—	—	—	—
	T12	−1.508	0.154	1.307	0.207	—	—
Pavlov 1	T1	−5.131	<0.001	—	—	—	—

is greater in humans than in African apes because of the posterior expansion of the human ribcage. Figure 14.43 shows the recent human profile ($N = 10$; open circles) with the values for Dolní Věstonice 13–15 added. For the fossils, left and right angles were averaged; in those specimens that preserve only one side, that side is used in the analysis. Note the close match of Dolní Věstonice 15's upper thoracic angles with those of recent humans. In contrast, both Dolní Věstonice 13 and 14 tend to fall to either side of the recent human mean profile. The t test results in Table 14.3, however, indicate that only two of these angles are statistically significantly different from the recent human mean—T8 of Dolní Věstonice 13 and T1 of Dolní Věstonice 14, both of which have much more posteriorly oriented transverse processes. It seems unlikely that the T8 transverse process is more posterior in orientation in order to accommodate a posteriorly deeper thorax since the respiratory area of the specimen's eighth rib is not significantly greater than that of recent humans (see costal skeleton osteometric analyses in Chapter 15). It could be that the vertebra has been reconstructed incorrectly since the right transverse process has been glued to the lamina. In fact, the most parsimonious explanation for the orientation of this transverse process (and that of Dolní Věstonice 14's T1) is postmortem distortion.

Vertebral Canal Height and Breadth

All tetrapods show cross-sectional expansion of the spinal cord (and, hence, vertebral canal) in the caudal cer-

Table 14.4 Results of modified Student's t tests for lumbar vertebrae of Dolní Věstonice versus recent human superior articular facet angles and spinous process angles.

	Vertebra	Superior Articular Facet Angle ($N = 15$)		Spinous Process Angle ($N = 20$)	
		t_s	P	t_s	P
Dolní Věstonice 13	L1	−0.146	0.886	−0.904	0.377
	L2	0.000	>0.999	1.700	0.105
	L3	0.534	0.602	0.727	0.476
	L4	2.201	0.045	2.952	0.008
	L5	0.645	0.530	−0.104	0.918
Dolní Věstonice 14	L1	−8.525	<0.001	—	—
	L2	0.072	0.944	—	—
	L3	−1.219	0.243	—	—
	L4	0.472	0.645	1.245	0.228
	L5	−1.160	0.265	0.705	0.489
Dolní Věstonice 15	L1	−0.878	0.395	1.000	0.330
	L2	−0.931	0.368	—	—
	L3	−0.634	0.536	1.765	0.094
	L4	−1.100	0.290	1.416	0.173
	L5	−0.451	0.659	—	—

vical and cranial lumbar regions corresponding to an increase in the number of neurons for the enervation of the fore- and hindlimbs, respectively (Giffin, 1990, 1992). Figure 14.44 (after MacLarnon, 1993) shows the profiles of vertebral canal dorsoventral height and transverse breadth for recent humans with the values for Dolní Věstonice 13 superimposed. Note that in recent humans both the cervical and lumbar expansions are primarily an expansion in canal breadth, not canal height. Note, too, that the canal increases in breadth caudal to L2, which is below the level at which the spinal cord ends. At this level the neural canal is filled with the nerves that make up the cauda equina, and there is no obvious physiological need for a greater amount of neural canal space. As Latimer and Ward (1993: 278) note, the best explanation for this phenomenon is that it reflects the increased interfacet distance of the lower lumbar vertebrae discussed above.

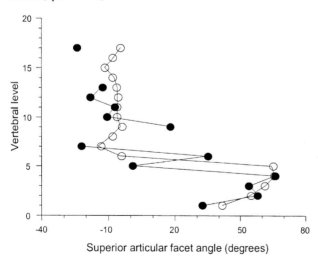

Figure 14.39 Profiles of superior articular facet orientation (in degrees) for Dolní Věstonice 14 (black circles) and recent human means (open circles).

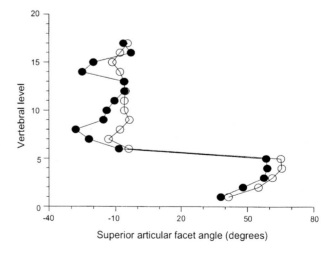

Figure 14.40 Profiles of superior articular facet orientation (in degrees) for Dolní Věstonice 15 (black circles) and recent human means (open circles).

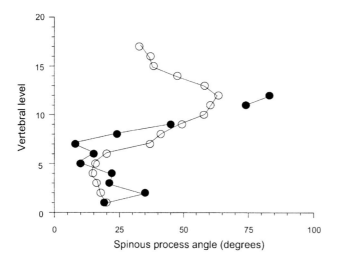

Figure 14.41 Profiles of spinous process angles (in degrees) for Dolní Věstonice 13 (black circles) and recent human means (open circles).

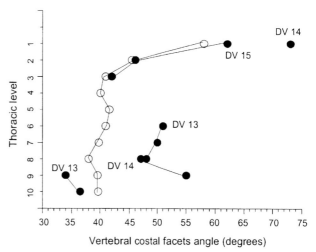

Figure 14.43 Profiles of vertebral costal facet angles (in degrees) for Dolní Věstonice 13, 14, and 15 (black circles) and recent human means (open circles).

The vertebral canal breadths and heights of Dolní Věstonice 13's presacral vertebrae are similar to those of the recent human means, although canal breadths of the lower thoracic canal fall below the range of the small sample ($N = 7$) of recent humans, as do the canal heights of C6, C7, and L2–L5. The vertebral canal profile of Dolní Věstonice 14 is superimposed over the recent human one in Figure 14.45. For this specimen, too, canal height is lower than the recent human sample range in the lower cervical and upper thoracic regions. However, the most striking difference between Dolní Věstonice 14 and the recent human sample is the specimen's enormous C1 canal height, which exceeds its own canal breadth by 10%. Figure 14.46 shows Dolní Věstonice 15's profile superimposed over the recent human one. There are some similarities, but as with Dolní Věstonice 14, the canal heights of upper thoracic vertebrae fall below the range of the small recent human sample, as do the canal heights of T7–T9 and canal breadths of T6–T9.

A simple means by which to consolidate the above relationships is to create a vertebral canal cranial height and cranial breadth index. Figure 14.47 shows the profiles for this index for recent humans, Dolní Věstonice 13–15,

Figure 14.42 Profiles of spinous process angles (in degrees) for Dolní Věstonice 14 and 15 (black circles) and recent human means (open circles).

Figure 14.44 Profiles (after MacLarnon, 1993) of vertebral canal dorsoventral height (triangles) and transverse breadth (diamonds) for Dolní Věstonice 13 (gray symbols) and recent human means (open symbols).

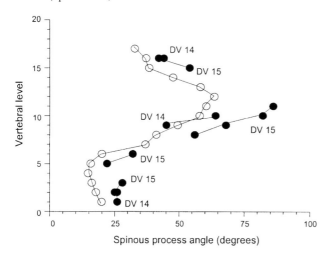

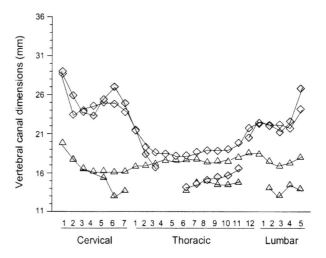

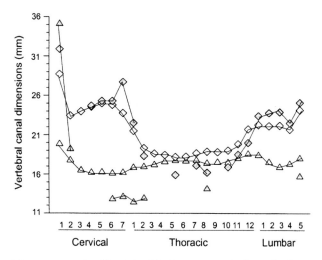

Figure 14.45 Profiles (after MacLarnon, 1993) of vertebral canal dorsoventral height (triangles) and transverse breadth (diamonds) for Dolní Věstonice 14 (gray symbols) and recent human means (open symbols).

one European Neandertal (Regourdou 1), and two Gravettian-associated specimens from Italy (Arene Candide IP and Paglicci 25). Note that among the recent humans, low indices mark the cervical region, which is reflective of the increase in canal breadth found here. Then the index increases greatly in the thoracic (and especially midthoracic) region, where the vertebral canal tends to be circular to slightly transversely expanded. The index again falls in the lumbar region (reflecting the increase in canal breadth typical of this region). Note that with a few exceptions, the fossil samples tend to have lower indices for all presacral vertebrae. Notable exceptions include the

Figure 14.46 Profiles (after MacLarnon, 1993) of vertebral canal dorsoventral height (triangles) and transverse breadth (diamonds) for Dolní Věstonice 15 (gray symbols) and recent human means (open symbols).

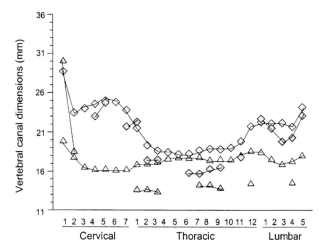

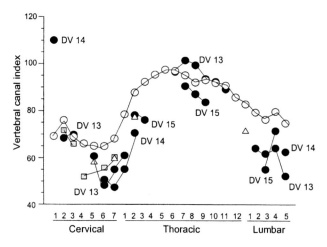

Figure 14.47 Profiles of vertebral canal index (vertebral canal height/vertebral canal breadth x 100) for Dolní Věstonice 13, 14, and 15 (black circles), European Neandertal (Régourdou 1; open triangles), non-Czech Gravettian specimens (Paglicci 25 and Arene Candide IP; gray squares), and recent human means (open circles). A low index value reflects a dorsoventrally short spinal canal.

very large C1 of Dolní Věstonice 14, C3 of Paglicci 25, and the mid- to lower thoracic region of Dolní Věstonice 13, which tends to fall just above the recent human mean profile. Recall that this last specimen's vertebral canal breadths and heights were smaller than those of the recent human sample.

Figure 14.48 shows the same recent human and Dolní Věstonice profiles; however, included this time are data from the Předmostí males (Předmostí 3, 9, and 14), who preserve the two vertebral canal dimensions. Note that

Figure 14.48 Profiles of vertebral canal index [(vertebral canal height/vertebral canal breadth) x 100] for Dolní Věstonice 13, 14, and 15 (black circles), Předmostí Gravettian males (Předmostí 3, 9, and 14; gray squares), and recent human means (open circles). A low index value reflects a dorsoventrally short spinal canal.

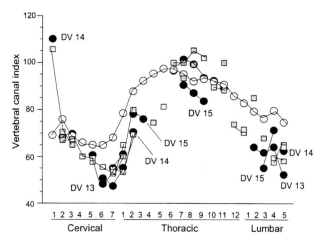

Předmostí 3 has a C1 vertebral canal index that is almost as high as that of Dolní Věstonice 14. In general, the Předmostí males are similar to the Dolní Věstonice and Italian Gravettian samples in that they tend to have lower on average cervical and lumbar canal height and breadth indices than recent humans, which is reflective of more transversely expanded vertebral canals in these regions.

Figure 14.49 shows the Předmostí females, whose pattern is quite similar to that of the males from the site. Most of their cervical and lumbar vertebrae evince lower canal height and breadth indices than the recent human means. Notable exceptions include the C1 of Předmostí 4 and 10, which fall close to the index value for Dolní Věstonice 14 (and Předmostí 10 slightly exceeds his value).

But are any of the above differences statistically significant? Table 14.5 presents the results of t tests comparing the combined Gravettian sample (i.e., Dolní Věstonice, Arene Candide, Paglicci, Pavlov, and Předmostí) with the recent human sample. Given the small sample sizes of

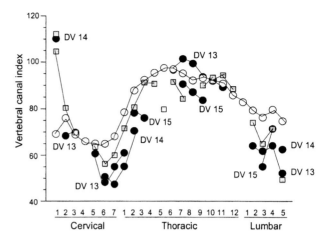

Figure 14.49 Profiles of vertebral canal index (vertebral canal height/vertebral canal breadth x 100) for Dolní Věstonice 13, 14, and 15 (black circles), Předmostí Gravettian females (Předmostí 4 and 10; gray squares), and recent human means (open circles). A low index value reflects a dorsoventrally short spinal canal.

Table 14.5 The t test results of comparisons of vertebral canal index [(canal height/canal breadth) x 100] between Gravettian humans from the sites of Arene Candide, Dolní Věstonice, Paglicci, Pavlov, and Předmostí versus recent humans ($N = 7$).

Vertebral Element	Gravettian Mean	s.d. (N)	Recent Human Mean	s.d. (N)	t_s	p
C1	108.1	3.63 (4)	69.1	5.25 (7)	−13.027	<0.001†
C2	71.0	4.87 (6)	75.9	5.63 (7)	1.663	0.125
C3	67.6	2.18 (5)	68.8	3.62 (7)	0.694	0.504
C4	56.0	5.70 (2)	66.0	4.97 (7)	2.454	0.044
C5	60.3	2.49 (4)	65.0	4.16 (7)	2.031	0.073
C6	53.2	3.65 (5)	64.9	2.82 (7)	6.249	0.001†
C7	54.7	4.47 (7)	68.1	3.15 (7)	6.466	<0.001†
T1	62.2	6.89 (7)	78.4	4.74 (7)	5.134	<0.001†
T2	75.7	5.35 (5)	87.7	3.20 (7)	4.888	<0.001*
T3	83.6	10.73 (2)	92.2	3.66 (7)	2.040	0.081
T4	82.6	11.37 (2)	95.3	5.40 (7)	2.409	0.047
T5	80.5	1.06 (2)	97.4	5.23 (7)	4.338	0.003
T6	96.0	4.31 (3)	96.9	4.77 (7)	0.282	0.785
T7	93.9	7.03 (5)	95.0	3.13 (7)	0.389	0.705
T8	95.6	8.16 (5)	92.0	3.41 (7)	−1.056	0.317
T9	92.3	7.66 (4)	93.1	3.04 (7)	0.237	0.818
T10	91.8	1.58 (4)	91.8	1.99 (7)	0.013	0.990
T11	92.9	5.42 (4)	90.7	1.41 (7)	−1.062	0.316
T12	81.2	10.31 (2)	85.7	5.23 (6)	0.872	0.417
L1	71.0	1.02 (2)	82.8	0.02 (7)	3.153	0.016
L2	74.4	10.45 (3)	79.3	4.26 (7)	1.121	0.295
L3	62.5	5.54 (4)	76.3	4.51 (7)	4.524	0.001*
L4	64.8	6.47 (5)	79.7	4.05 (7)	4.922	<0.001*
L5	58.6	6.56 (6)	74.8	7.29 (7)	4.160	0.002

*Significantly different at family-wide α-level of 0.05; †Significantly different at family-wide α-level of 0.01; the significance was determined by using the sequential Bonferroni technique (Rice, 1989).

both the recent humans and the fossils, statistical power $(1-\beta)$ is low, and thus the expectation is that even if real differences exist, we may not be able to detect them. However, because these are multiple comparisons between the same samples, in order to avoid making a type I (α) error, a sequential Bonferroni correction factor (Rice, 1989) was applied in order to arrive at a stricter family-wide α-level. Thus, the finding of eight statistically significant differences is conservative, and the differences are in all likelihood real ones.

Perhaps the clearest difference is that between the Gravettian sample and the recent humans in the canal index for C1, where the Gravettian mean index of 108.1 is over 7 standard deviations above the recent human mean. For this vertebral element, the difference is not so much due to transverse expansion of the canal as to height values evinced by most of the Gravettian individuals in the large canal. The Gravettian sample also evinces lower indices in the caudal cervical and cranial thoracic vertebrae (C6–T2). This is the region of the cervical spinal cord expansion, and the index appears to be the result of spinal canals that are both transversely wider and, in many cases, dorsoventrally narrower than those of recent humans. Finally, the caudal lumbar vertebrae (L3–L5) of the Gravettian sample evince lower canal height and breadth indices than recent humans; here, too, the difference is due to a combination of expanded transverse canal dimensions and slightly contracted dorsoventral canal dimensions. The functional significance of these differences, if any, is unclear. It seems most likely that this is a reflection of the greater body size and/or brain size (Ruff et al., 1997) of the Upper Paleolithic sample relative to recent humans.

Summary

The vertebral columns of the Dolní Věstonice and Pavlov sample are in most aspects similar to those of recent humans. This is not surprising, given the strong stabilizing selection under which the vertebral column, with its close ties to the sensitive but extremely important spinal cord, is placed. However, in other aspects of vertebral anatomy, these humans more closely resemble their penecontemporaries than they do recent, industrialized humans. This, too, is not surprising, given the time differences involved, as well as differences in activity levels between Late Pleistocene hunter-gatherers and those who live in industrialized societies.

For example, the presacral vertebrae of Dolní Věstonice and Pavlov although clearly anatomically modern in morphology, nonetheless show some interesting differences from those of recent humans. Many of these differences are probably due to postmortem distortion of the remains, except for the atypical wedging of the lower thoracic vertebrae of Dolní Věstonice 13 and the unusual lower thoracic and upper lumbar articular facet morphology of Dolní Věstonice 14. However, the most striking differences between the Dolní Věstonice and Pavlov sample (and indeed Gravettian-associated humans as a whole) and recent humans are shape differences in the vertebral canal. Specifically, the vertebral canal of Gravettian-associated humans tends to be transversely broader than those of recent humans in the lower cervical, upper thoracic, and lower lumbar regions, and the vertebral canal of C1 (or atlas) of Gravettian-associated humans tends to evince a pronounced dorsoventral expansion.

15

The Costal Skeletons

Trenton W. Holliday

The costal skeleton [costae (ribs) and sternebrae] is reasonably well represented among the skeletons from Dolní Věstonice and Pavlov. It is best preserved in the triple burial individuals: Dolní Věstonice 13–15, each of which preserves at least one (and more often both) rib(s) from each thoracic level (Sládek et al., 2000; Chapter 5). In this chapter, detailed morphological description of each individual costal element is given, working craniocaudally through the ribs and ending with the sternebrae. Following these descriptions, osteometric analyses are undertaken. These analyses compare the Dolní Věstonice and Pavlov costal morphology with that of recent and penecontemporary humans. Discussion of the measurements and indices used in these analyses, as well as the compositions of the comparative samples, comes at the beginning of each analytical section. As mentioned in Chapter 14, the less complete axial remains of Dolní Věstonice 16 and Pavlov 1 were separated during excavation and partially reconstructed in the laboratory. In contrast, the costal skeletons of the triple burial individuals were removed en bloc to the laboratory, cleaned, and then reconstructed.

Morphological Description: The Costae and Sternebrae

Ribs and sternebrae are rarely preserved in paleontological/archeological contexts, for a variety of reasons. In cases where carnivores have access to bodies, ribs may be chewed and even swallowed by some predators. Rapid burial of remains (either intentional or fortuitous) increases the likelihood that ribs will be preserved, but even in these cases ribs are often broken and fragmentary because of their long, curved shape and relatively thin cortical bone. Similarly, the bones that make up the sternum are not particularly dense, and because of their high trabecular bone composition they, too, may not survive taphonomically. Given the wide range of their state of preservation, then, the skeletons from Dolní Věstonice and Pavlov vary greatly in the number of costae and sternebrae preserved. Pavlov 1 has only a single identifiable rib fragment associated with it. Dolní Věstonice 3 has only two isolated rib pieces and a crushed rib cage that remains embedded in matrix. Several elements of the rib cage of Dolní Věstonice 16 are preserved, but all are in a very fragmentary state and most cannot be identified by number. The best-preserved ribs from these sites are those from the triple burial. These three individuals, Dolní Věstonice 13–15, preserve largely complete rib cages in excellent condition, and Dolní Věstonice 13 and 15 also preserve sternal elements. It is the bones of these three individuals to which most of this section is devoted, and only those ribs that can be identified by number will be discussed here (Figures 15.1 to 15.14).

Costae (Ribs)

First Ribs

The neck and partial head of the first right rib of Dolní Věstonice 3 is one of two rib fragments that have been removed from the matrix in which most of her axial skeleton still remains. The head is badly damaged, so most of its cancellous bone is visible—only a trace of cortical bone remains on the proximal end of the head. The internal margin of the neck is sharp, whereas its external margin is more convex. It is broken just proximal to the tubercle; its preserved length is 26.2 mm.

The right first rib of Dolní Věstonice 13 is missing the head, but its sternal end is complete. The superoproximal portion of its tubercle is also missing (there are trabeculae exposed where it has been eroded away), and there is fill on its posterolateral margin. The groove for the subclavian vein is broad (17.3 mm on its ventral margin) and shallow. The scalene tubercle is relatively flat and broad (ca. 13.7 mm proximodistally × 8.7 mm dorsoventrally]. The area of attachment for the middle scalene muscle is relatively concave, with discrete margins along all its proximal edge, where it blends into the tubercular region. The attachment area for subclavius is proximodistally elongated (28.2 mm × 8.1 mm), and it extends all the way to the distal end of the shaft. The inferior surface of the rib is relatively smooth and featureless. The sternal end of the rib has a beveled cranial surface, and it is superoinferiorly and, to a lesser extent, internally to externally concave.

The left first rib of Dolní Věstonice 13 (Figure 15.1) also has damage to its head; only the head's dorsal aspect remains, although a small (6.2 mm proximodistally × 7.2 mm superoinferiorly) section of the cortical bone from the articular portion of the head is preserved. The distal 25% of the rib (ca. 10 mm in length) is also missing. Although measurements of the right rib's neck are unreliable because of its missing head, it is nonetheless obvious that the neck of the left rib is much longer than its right counterpart. The left first rib is also less robust than its right counterpart, and the grooves for the subclavian vein and artery are shallower. The dimensions of the scalene tubercle are 6.7 mm dorsoventrally and 10.9 mm proximo-

distally. The margins surrounding the attachment for the middle scalene muscle are less discrete on this rib than they are on the right. As on the right first rib, the area of attachment for the middle scalene muscle is gently concave; it measures 32.1 mm proximodistally and 9.9 mm internoexternally. The tubercle, whose breadth is about 12.0 mm, forms a more open angle with the neck than did the right rib. Only the proximal 16.5 mm of the subclavius attachment is preserved. Whereas no costal groove is evident on the right rib, the left has a distinct costal groove about 6.0 mm in width running along the ventral face of its caudal surface. Aside from the groove, the caudal surface of the rib is relatively convex, whereas its cranial surface is more planar than that of the right, which exhibits more relief.

Dolní Věstonice 14's right first rib is nearly complete (Figure 15.2). In overall shape, the rib evinces a very sharp angle distal to which its corpus is quite straight. The sternal end has some trivial damage on its caudal/external margin, as does the neck. Proximally, the epiphysis of the head is missing, and the subchrondral area distal to the missing epiphysis is damaged, with a superoinferior postmortem cut 3.2 mm wide at its caudal end and 2.0 mm wide at its cranial end. The rib is not very robust; in fact the sulci for the subclavian vein and artery are not distinct. The scalene tubercle is manifest as a slight ridge along the internal margin of the corpus—here there is a thickening of the shaft about 37 mm from the sternal end. The scalene tubercle blends in with a slightly rugose area that serves as the attachment for the subclavius (the boundaries of this muscle's attachment are, however, indistinct). There is also a depression on the cranial margin of the proximal shaft for the middle scalene muscle

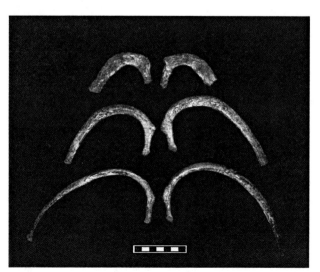

Figure 15.1 Superior view, ribs 1–3, Dolní Věstonice 13. Scale in centimeters.

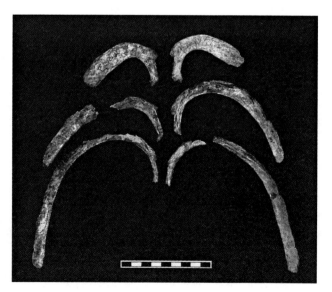

Figure 15.2 Superior view, ribs 1–3, Dolní Věstonice 14. Scale in centimeters.

that measures 14.7 mm in proximodistal length and 9.4 mm in internal-external breadth. On the posterior/superior margin of the neck is another small sulcus 6.3 mm long and 2.1 mm wide that corresponds to the attachment of the medial costotransverse ligament. The caudal surface of the rib is smooth and relatively featureless.

The left first rib of Dolní Věstonice 14 is not as complete as the right. Its distal end is badly damaged, especially along its cranial margin. However, in terms of length, most of the rib is preserved, at least caudally. There is also some damage to the superior margin of the tubercle and the posterior margin of the head. In overall shape, the left rib, like the right, has a straight corpus distal to the angle. The epiphysis of the head is missing, revealing the beveling associated with an immature metaphysis. The neck is stouter than that of the right, and its external margin is flat. The neck evinces a shallow sulcus measuring 12.7 mm in length and 3.9 mm in height that corresponds to the attachment of the medial costotransverse ligament. As with the left first rib, this rib is not robust, nor are the sulci for the subclavian artery and vein discernable. The attachment for the middle scalene muscle is manifest as a cluster of three shallow, ovoid depressions that in total measure 15.0 mm proximodistally by 8.1 mm internoexternally. Both the caudal and cranial surfaces of the rib (aside from the above depression) are relatively smooth and featureless, aside from scattered root marks on the cranial surface.

The Dolní Věstonice 15 right first rib (Figure 15.3) is missing perhaps 15 mm of its sternal end, and it has two large areas of fill on its distal shaft; otherwise, it is complete. The posterior margin of the neck evinces a sulcus about 79 mm in length and 2.8 mm in height located just proximal to the tubercle that corresponds to the attachment of the medial costotransverse ligament. The tubercle shows the loss of cortical bone along its posterior margin, but it is quite large. Its articular portion is slightly superoinferiorly convex. The shaft is relatively straight and for a first rib is largely featureless. The sulci for the subclavian artery and vein are present but are not particularly marked. In part this may be due to damage; the sulcus for the vein has putty on it. Likewise, the tubercle for the anterior scalene muscle is not particularly marked, but it, too, is filled with putty along its distal edge. The area for the attachment of the middle scalene muscle is not marked either. The entire caudal surface of the rib is smooth and featureless. The inferior table extends about 11.6 mm distal to the superior table, which has been lost, exposing trabecular bone.

The left first rib of Dolní Věstonice 15 shows damage to its anterior (external) sternal end (with trabecular bone exposed), as well as to the caudal surface of the head and neck; otherwise it is complete. The posterior margin of the neck has a sulcus for the medial costotransverse

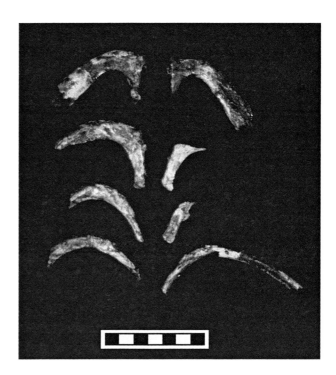

Figure 15.3 Superior view, ribs 1–4, Dolní Věstonice 15. Scale in centimeters.

ligament, but matrix adhers to the posterior neck, obscuring the view of this feature. The tubercle of this rib is massive, and it projects cranially and laterally to a seemingly extreme degree. Like its counterpart on the right, the shaft of this rib is quite straight. Sulci for both the subclavian artery and vein are clearly visible. Just distal and lateral to the groove for the subclavian vein is a raised area that cannot be measured because of damage along its anterior margin that corresponds to the origin of the subclavian muscle. In contrast, the area of attachment for the middle scalene muscle is not delineated; rather there is only a shallow depression with indistinct boundaries. Internal and distal to this shallow depression, however, is the raised scalene tubercle, the point of attachment for the anterior scalene muscle; and distal and lateral to this tubercle is a rugose area corresponding to the attachment of the tendinous slip of one of the origins of serratus anterior. None of these muscle attachment areas is marked. The caudal surface of the rib is smooth and featureless, and overall the rib is quite gracile.

Second Ribs

The Dolní Věstonice 13 right second rib is largely complete (Figure 15.1). All that is missing is about 20 mm of the sternal end, which is eroded away, and there are minor areas of fill on the cranial margin of the neck and angle. The head evinces minor damage to its proximal surface. The neck is both internally to externally broad and superoinferiorly tall (height: 6.8 mm; breadth: 7.5 mm). The articular tubercle is marked. The attachment site for the serratus anterior muscle is also marked, but it is not possible to discern the attachment of the first versus second digitations of this muscle. The areas of attachment for the posterior scalene muscle and the serratus anterior are separated by a shallow sulcus. The posterior scalene muscle attachment is about 41 mm in length, and the attachment for serratus anterior is around 30 mm in length. Caudally, there is a shallow costal groove just distal to the tubercle that measures about 5 mm in breath and 49.6 mm in length. Distal to this point the caudal surface is smooth and featureless.

The left second rib of Dolní Věstonice 13 is virtually complete, with only trivial damage to the cranial margin of its head and sternal end. There is also a small (3.9 mm in height × 5.0 mm in proximodistal breadth) area of missing cortical bone on the posterior margin of the neck. The rib has a prominent articular tubercle, the nonarticular portion of which is particularly rugose. The rib also evinces a marked crest along its cranial margin running some 48.0 mm in length that corresponds to the attachment of the posterior scalene muscle. The area of attachment for the first two digitations of serratus anterior is also rugose and measures about 28 mm proximodistally by 13 mm dorsoventrally. As was true for its right counterpart, the area of each digitation's attachment cannot be discerned; rather it is a combined attachment. Distally, the remainder of the cranial surface is smooth, flat, and featureless, and beyond the articular tubercle the shaft is relatively rectangular in cross section. The caudal surface of the rib is also relatively smooth, although there is a mild costal groove proximally, running to about one-third of the shaft's length.

The Dolní Věstonice 14 right second rib (Figure 15.2) is represented by two pieces that cannot be joined. It is possible that there was contact between the proximal and distal pieces along the rib's internal border; however, the contact is not certain, and thus the pieces remain separate. The proximal piece includes a complete tubercle, a neck that is damaged or missing its inferior proximal half, and a small anterior and cranial portion of the head. The piece extends distally from the tubercle about 48 mm. The neck is superoinferiorly short, and its proximal cranial portion preserves the upward slope to the head. Its external margin is flattened, with slight ridges marking its superior and inferior boundaries; it measures 5.3 mm in height and was at least 17.9 mm in length, although it could have been slightly longer (part of the neck is missing). The articular portion of the tubercle has discrete boundaries and is superoinferiorly convex. The cranial margin of the neck is smooth and featureless, but as one moves distally along the corpus, the cranial surface becomes more rugose, and the area of attachment of the iliocostalis thoracis, while incompletely preserved, is quite rugose. The inferior surface of the corpus, which extends farther than the superior surface, is relatively smooth, flat, and featureless.

The distal piece preserves most of the remainder of the rib. The piece is broken at a point just proximal to its distal end, and when paired up with the left rib it appears that it is missing its distal 12 mm. The distal piece has a repaired break (with putty) about 42.4 mm from its preserved proximal end. Proximally, the distal piece evinces a sulcus for the intercostal muscles that runs along the internal margin of its cranial surface. It measures 5.5 mm in internal-external width, and it was at least 10.8 mm in proximodistal length (its proximal end is missing). Just distal to the sulcus, also on the internal margin, is a tubercle corresponding to the attachment of the first digitation of the serratus anterior muscle that measures 4.4 mm in proximodistal length by 5.6 mm in internal to external breadth. Distal to this tubercle, the remainder of the cranial surface of the rib is smooth and featureless. The external surface of the rib is damaged along its entire course, particularly along its distal end. The entire inferior surface of the distal shaft is smooth and featureless as well, and it evinces a sharp internal margin.

The left second rib of Dolní Věstonice 14 is largely complete; it is missing its head and a small section of its distal end, and much of its corpus has been repaired with black fill. The neck is internally to externally broad and superoinferiorly thicker than its counterpart on the right. Most of the posterior margin of the neck is marked by a large sulcus, corresponding to the attachment of the medial costotransverse ligament, that measures 8.1 mm by 2.3 mm. Unlike the right second rib, it is difficult to distinguish the articular portion of the tubercle from its nonarticular portion. Distal to the tuberosity along the internal margin of the corpus is a long sulcus, at least 51 mm in length and about 6.3 mm in breadth, that corresponds to the attachment of the intercostal muscles. Its length is difficult to determine because the sulcus blends in with another sulcus near midshaft that corresponds to the attachment of the first digitation of the serratus anterior muscle. On the external margin of the rib just distal to midshaft is the rugose area of attachment for the second digitation of serratus anterior. Distal to this tuberosity, the rib is smooth along its cranial margin. Its entire inferior face is largely smooth and featureless. At least a portion of the internal surface of the sternal end is preserved. There are, however, two repaired breaks distal to the rugosity. The first is located 40.1 mm from the preserved distal end; the second, which unlike the first evinces fill, is located about 14.2 mm from the sternal end.

Perhaps half of the Dolní Věstonice 15 right second rib remains—it preserves from a damaged head and neck to near midshaft (Figure 15.3). The head and neck are badly damaged along their caudal margins. The posterior neck evinces a sulcus 3.1 mm in height and at least 10.3 mm in length, corresponding to the attachment of the medial costotransverse ligament. The prominent tubercle faces cranially and has a damaged caudal margin. The inferior margin of the shaft is flat and featureless. In contrast, the superior surface of the shaft has sulci that correspond to the attachment sites of numerous muscles. The most proximal sulcus is an oblong one, measuring 14 mm in proximodistal length and 5 mm in internal/external breadth, that corresponds to the attachment of the posterior scalene muscle. This sulcus is separated from a smaller, more distal ovoid sulcus, which corresponds to the attachment of the first digitation of the serratus anterior muscle. It measures 8.1 mm in length by 4.0 mm in breadth. Distal to it is a large sulcus that is incomplete distally and internally because of damage. This sulcus measured at least 7.7 mm in internal/external breadth and corresponds to the attachment of the second digitation of serratus anterior.

The left second rib of Dolní Věstonice 15 is represented solely by a partial neck and a small internal piece of the shaft distal to the damaged (and mostly missing) tubercle. The posterior surface of the neck evinces an oval-shaped sulcus 9.8 mm in length and 3.2 mm in height, corresponding to the attachment of the medial costotransverse ligament. The cranial margin of the neck is flat, whereas the inferior margin is convex.

Third Ribs

The Dolní Věstonice 13 right third rib (Figure 15.1) looks complete, but there is an area near midshaft about 14 mm in length that has been reconstructed with putty, and it is uncertain if the bone beneath it is continuous. Although most of the internally to externally narrow neck appears to be present, the head is missing; and although the sternal end is present, it evinces extensive damage along its ventral and superior margins. Cranially, there is a relatively broad (ca. 5.2 mm) and long (ca. 58.4 mm) sulcus for the intercostal muscles that begins roughly 46 mm distal to the articular tubercle, and it extends just distal to the reconstructed section near midshaft. Along the rib's caudal margin, the costal groove is present proximally, ending about 74 mm distal to the articular tubercle. The articular tubercle has discrete boundaries, and its ovoid articular facet is slightly superoinferiorly convex and slightly proximodistally concave. The distal shaft is smooth and featureless along both its internal and external margins.

The left third rib of Dolní Věstonice 13 is largely complete; only the cranial margin of the head and the cranial and ventral faces of the distal end are eroded. Also, near midline there is a small (27.2 mm in length) section of the caudal margin of the rib that has been reconstructed with putty. The crest on the cranial neck for the attachment of the superior costotransverse ligament is damaged. The ovoid articular facet is slightly superoinferiorly convex and relatively proximodistally flat and is separated from the nonarticular portion by a sulcus measuring around 3 mm in breadth. The nonarticular portion of the tubercle is rugose and relatively large (ca. 19 mm dorsoventrally by 9 mm superoinferiorly). Just proximal to the articular portion of the tubercle on the rib's external surface lies a small (9.3 mm proximodistally x 2.9 mm superoinferiorly) sulcus near the rib's cranial margin that corresponds to the attachment of the medial costotransverse ligament. On the cranial margin directly superior to the articular portion of the articular tubercle lies a small sulcus corresponding to the attachment of the lateral costotransverse ligament, and distal and cranial to it lies a long sulcus for the intercostal muscles that continues to near midshaft and varies in breadth from 3.5 mm to 5.5 mm. Distal to this sulcus the remainder of the rib's cranial margin is smooth to slightly sharp. Along its caudal edge, the costal groove is evident from the proximal end to near midshaft; the remainder of the caudal and internal surfaces are smooth and featureless.

The Dolní Věstonice 14 right third rib (Figure 15.2) is heavily damaged along its external and inferior margins, but nonetheless most of its length is preserved in a single piece. Although it preserves most of its neck, its head is missing; matrix embedded in the neck gives the impression of an articular surface, but none is preserved. The neck evinces a matrix-filled sulcus on its external surface that corresponds to the attachment of the medial costotransverse ligament. The cranial margin of the neck is damaged, but distal to the neck it is undamaged, and the crest corresponding to the attachment of the superior costotransverse ligament is evident. The caudal margin of the neck is sharp, albeit not as sharp as the corresponding portion of the left rib. The articular portion of the tubercle is ovoid with discrete boundaries. The nonarticular portion evinces heavy damage that extends beyond the angle to near midshaft, where the cortical bone is missing, leaving fill and trabecular bone exposed. Along the proximal cranial margin is a flat sulcus about 78.7 mm in length that corresponds to the attachment of the intercostal muscles. The sulcus flattens out near midshaft; the remainder of the external surface is smooth, convex, and featureless. The caudal margin of the shaft is badly damaged for most of the rib's length, but a costal groove is evident on two small areas preserved near midshaft.

The left third rib of Dolní Věstonice 14 is represented by two pieces. The first is a small, proximal piece about 56 mm in length that preserves the tubercle, part of the neck, and the iliocostalis region. The second piece (which does not articulate with the first) is about 156 mm in length and preserves most of the shaft to near its sternal end. The neck evinces superior and inferior crests. The superior crest runs along the central/cranial margin and corresponds to the attachment of the superior costotransverse ligament. It is damaged proximally, where trabecular bone is exposed. Inferior to this crest, on the external margin of the neck, is a sulcus, about 10.1 mm in length and 2.9 mm in height, that corresponds to the attachment of the medial costotransverse ligament. The relatively large articular facet is ovoid (almost circular) in shape, with a proximodistal long axis. It is superoinferiorly convex, evinces trivial damage to its articular surface, and is missing its lateral margin. The inferior costal point is not preserved, but the iliocostalis region is robust. The caudal margin of the distal piece evinces heavy damage, with patches of black fill throughout. Nonetheless, a costal groove is evident along the proximal 106.9 mm of the piece. It begins as a high groove that gradually decreases in height as the rib corpus becomes more bladelike. There is also a sulcus running along the distal piece's cranial margin at its proximal end, corresponding to the attachment of the intercostal muscles, that is surrounded by internal and external ridges. At its widest point it measures 5.2 mm and is at least 54.8 mm in length. Distal to this sulcus, the cranial surface of the rib becomes smooth and convex. Perhaps the most striking feature of this rib is its minimal declination.

All that remains of the Dolní Věstonice 15 right third rib (Figure 15.3) is a small proximal piece that includes the neck, articular tubercle, and a heavily curved portion of the shaft proximal to the angle. The neck is superoinferiorly tall and internally to externally narrow. Both its superior and inferior margins are convex, but the inferior margin is internally to externally narrower than the superior margin. The neck has relatively heavy damage along its proximal external aspect, but two small sulci (one superior and one inferior) are evident there that correspond to the attachment of the medial costotransverse ligament. The more superior sulcus measures 2.0 mm in height and minimally 8.3 mm in length. It could be longer; its proximal end is missing. The inferior sulcus measures 1.7 mm in height and 7.1 mm in length. A third sulcus lies directly superior to the articular facet. It measures 8.7 mm in length and 3.6 mm in height and corresponds to the attachment of the lateral costotransverse ligament. The articular facet is large, ovoid, and proximodistally convex. It has minor damage along its inferoproximal margin, but its margins are otherwise discrete. The rugose nonarticular portion of the tubercle is heavily damaged, with fill and exposed trabecular bone evident laterally and inferiorly. The inferior surface of the shaft is smooth and featureless. The inferior costal point is not preserved; hence the value for tubercle/iliocostal line distance reported in Sládek et al. (2000) is erroneous.

All that is preserved of the left third rib of Dolní Věstonice 15 is a small proximal piece that includes the neck and about 6 mm of shaft beyond the articular portion of the tubercle. The neck is superoinferiorly tall and internally to externally narrow (the superior margin is wider than the inferior). Most of the neck is preserved since it expands proximally to form the head, but none of the head's articular surface is preserved. The internal surface of the neck is smooth and featureless. As was observed on the right third rib, two horizontal sulci, one superior and one inferior, are evident on the external surface of the neck. These sulci correspond to the attachment of the medial costotransverse ligament. The superior one measures 9.0 mm in length and 2.6 mm in height; the inferior one is 6.4 mm long and 2.0 mm in height. There is a well-delineated ovoid (proximodistal long axis) articular tubercle with trivial damage along its inferior margin. Its articular surface is superoinferiorly convex.

Fourth Ribs

The Dolní Věstonice 13 right fourth rib (Figure 15.4) is largely complete, although its head is damaged superiorly, and the superior margin of the neck has exposed trabeculae. Much of the inferior half of the articular portion of the articular tubercle is missing (articular facet breadth: 9.4 mm), and a major portion (ca. 48.7 mm proximodistally) of the proximal shaft beginning proximally at the inferior iliocostalis point is reconstructed. The iliocostalis line is well defined. Only a small, inferior portion of the sternal end is preserved. On the cranial margin of the neck is a small (10.5 mm in length by 2.7 mm in breadth) sulcus for the attachment of the superior costotransverse ligament. Distal to this sulcus is a second sulcus along the cranial margin measuring 14.5 mm in length and 2.5 mm in breadth in the region of the tubercle for the attachment of the lateral costotransverse ligament, but distal to this tubercle there is a large area of fill on the cranial margin distal to the iliocostalis line. A long sulcus along the cranial margin that corresponds to the attachment of the intercostal muscles emerges from this area of fill, but it flattens out near midshaft. Beyond midshaft the cranial margin is flat, but at almost two-thirds of its length it becomes superiorly convex and evinces a gradual internal to external thinning toward its distal end. A costal groove is present along the entire caudal margin, but it is not marked.

The left fourth rib of Dolní Věstonice 13 is largely complete, with only trivial damage to the superior margin of the head and inferior margin of its distal end. There are two glue joins on the rib. The first is just proximal to midshaft and has small areas of fill on both its superior and inferior margins. The second is about 30 mm from

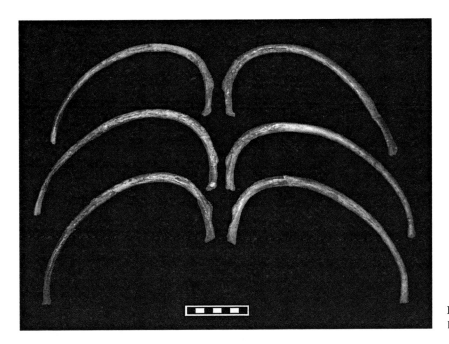

Figure 15.4 Superior view, ribs 4–6, Dolní Věstonice 13. Scale in centimeters.

the distal end. In posterior view the cranial margin of the neck has a triangular crest for the attachment of the superior costotransverse ligament. The rib is internally to externally thick until about one-third of its length, at which point it narrows. The articular facet lacks discrete boundaries, and just distal to it the nonarticular portion of the tubercle is missing cortical bone. The nonarticular portion of the tubercle is large and rugose, much more so than its right counterpart. On the cranial margin superior to the facet, a small (7.8 mm in length × 2.8 mm in breadth) sulcus for the attachment of the lateral costotransverse ligament is present. Just distal to the tubercle along the cranial margin a broad, flat sulcus corresponding to the attachment of the intercostal muscles is evident. It runs for about 80 mm, to a point just proximal to midshaft, before it narrows and the cranial margin becomes smooth, convex, and relatively featureless. There is a costal groove evident along the caudal surface of the rib, but the groove becomes hard to discern near midshaft, and it is essentially absent from the sternal end.

The Dolní Věstonice 14 right fourth rib (Figure 15.5) is preserved in two separate pieces. The first piece, 72.2 mm in length, preserves only a tiny portion of the neck, most of the tubercle, and a small internal-inferior portion of the proximal shaft. The second piece, 164 mm in length, preserves the distal end of the rib, perhaps to within 10 mm of the sternal end. The neck is represented solely by three thin prongs of cortical bone, the largest of which is posterior. The articular facet is missing its inferior proximal quarter. The nonarticular portion of the tubercle and attachment for iliocostalis thoracis is rugose. The inferior surface of the tubercular region exhibits a flattened, sulcuslike area 3.6 mm in width. The distal piece has a badly damaged external shaft at its preserved proximal end. Proximally, the distal piece's cranial surface is broad and convex, but it becomes internally to externally thinner, with sharper margins, near midshaft. There is a slightly raised area on the external surface of the rib, about 40 mm from its distal end that corresponds to one of the origins of the serratus anterior. The shaft's caudal margin is damaged for roughly three-fifths of its length, but a costal groove is evident throughout; it is superoinferiorly deeper proximally (5.6 mm) than distally (2.5 mm).

The left fourth rib of Dolní Věstonice 14 is represented for certainty solely by a small proximal piece, 62 mm in length, which preserves most of the neck, tubercle, and a small portion of the shaft. A small superomedial portion of the head's articular surface is also preserved and has the beveling consistent with the young age at death of the specimen. The neck is tall (9.5 mm) and relatively thin (5.7 mm). Its cranial and caudal margins are both convex, but the caudal edge is sharper. The sharp caudal crest of the neck continues onto the tubercle as the internal rim of the corpus. The superoinferiorly convex articular facet is large and ovoid (proximodistal long axis) with discrete boundaries. The nonarticular portion of the tubercle is rugose, but the inferior costal point is not preserved. The cranial margin of the proximal shaft is broadly convex.

Dolní Věstonice 15's right fourth rib (Figure 15.3) is represented by a small, curved, proximal piece that includes a partial head and neck, tubercle, and a small

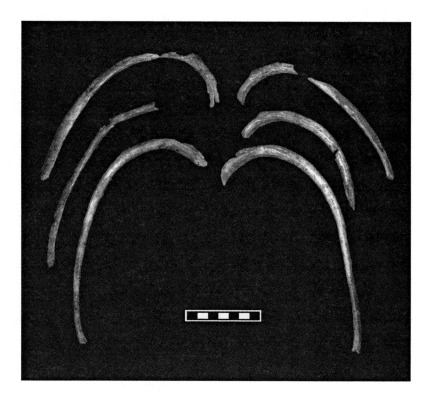

Figure 15.5 Superior view, ribs 4–6, Dolní Věstonice 14. Scale in centimeters.

segment of the shaft distal to the iliocostalis line. The only portion of the articular surface of the head that remains is its superior/internal quadrant, measuring 6.3 mm in height and 6.0 mm in breadth. The articular facet is ovoid, and although its distal margin is well delineated, the inferior margin is damaged and the proximal and superior margins are not discrete. Its surface is slightly proximodistally concave and superoinferiorly convex. The relatively long neck is superoinferiorly tall and internally to externally narrow. Proximally, its external surface is heavily damaged and repaired with fill. The cranial margin of the neck is smooth and convex, but just distal to it lies the crest for the superior costotransverse ligament. Distal to this crest, the cranial margin of the proximal shaft is internally to externally broad and relatively flat. No ridges or other muscle markings are observed along its cranial margin. In contrast, the external ridge of the corpus is inferiorly marked (although it is repaired with fill for part of its course), and it would have formed the costal groove more distally had the distal shaft been preserved.

The left fourth rib of Dolní Věstonice 15 is represented by two pieces that almost join. In fact, there may be a small area of contact between them along their superoexternal faces, but this is uncertain, and thus they remain separate. The first piece preserves most of the neck, articular tubercle, and a small segment of proximal shaft distal to the tubercle. The second piece preserves a small segment of the proximal shaft. The neck of this rib is tall, long, and internally to externally narrow. Its external surface evinces a sulcus with indiscrete margins that corresponds to the attachment of the medial costotransverse ligament. The cranial margin of the neck has exposed trabecular bone, but the thin crest corresponding to the attachment of the superior costotransverse ligament is present. Distally this crest fades into a smoothly convex cranial surface. The margins of the ovoid articular facet are well delineated, except for the cranial margin, which blends into the shaft. The facet is slightly proximodistally concave and superoinferiorly convex. Just superior and distal to the articular facet is another, ovoid tubercle measuring about 8.4 mm (long axis; superoproximal to inferodistal) by 6.1 mm (short axis; superodistal to inferoproximal) that corresponds to the attachment of the lateral costotransverse ligament. Within the tubercle is a shallow groove, and there is trivial damage to the tubercle's inferior margin. The shaft's caudal margin is smooth and internally to externally broadens distal to the articular tubercle. At the distal end of the proximal piece is an inferior external ridge that will form the costal groove distally. The second, distal, piece is characterized by a broad and convex cranial margin evincing fill along its proximal aspect. The inferior margin is damaged proximally; distally, it evinces a subtle costal groove.

Fifth Ribs

The Dolní Věstonice 13 right fifth rib is largely complete (Figure 15.4), and it evinces a greater cross-sectional diameter than its more cranial counterparts. The rib is internally to externally thick for much of its course, only thinning out for the last quarter of its running length. It also has a marked articular tubercle, as well as a prominent costal groove. The costal groove itself is especially deep proximal to midshaft. The head is largely intact, although its articular surface is slightly abraded. There are two small areas of fill along the cranial margin and superoposterior face of the neck. The neck exhibits a long, low crest that corresponds to the attachment of the superior costotransverse ligament. Part of the sternal end is present, but it, too, is damaged, particularly along its cranial margin—and in fact, much of the distal end is reconstructed. In addition, the caudal projection corresponding to the lateral extent of the iliocostalis thoracis has sustained damage. Superiorly and proximally, a sulcus for the intercostal muscles, with a breadth of about 5.4 mm, runs to near the angle. Distal to this point, the rib's cranial margin becomes flat, and beyond midshaft it becomes gently convex. Along the posterior face of the rib, just superior to the articular tubercle, is a deep sulcus, measuring 2.3 mm in height and 19.8 mm in length, for the attachment for the levator costae brevis. On the rib's external surface, about 50.0 mm from the sternal end, there is a flattened area corresponding to an attachment of the serratus anterior muscle.

The left fifth rib of Dolní Věstonice 13 lacks its head, has damage to the caudal surface of its sternal end and evinces a broad area of fill near one-third of its running length. It is particularly thick in cross section along the first one-third of its running length, after which it becomes quite tall and narrow. The articular portion of the articular tubercle is separated from the nonarticular portion by a shallow sulcus. Two sulci are evident along the cranial margin of the neck: one along the internal margin, the other along the external margin. They are separated from each other by a small crest for the attachment of the superior costotransverse ligament. The internal sulcus measures about 6.1 mm in length and 1.6 mm in breadth and, as with the crest, is part of the attachment of the superior costotransverse ligament. The second sulcus is found along the cranial margin just superior to the nonarticular portion of the tubercle. It measures about 24.0 mm in length and 3.3 mm in breadth and is for the attachment of the lateral costotransverse ligament. The ridge that separates the two sulci continues distoexternally, where it becomes an external ridge that is bounding a broad (ca. 4.3 mm) sulcus that runs to about one-third of the rib's running length, at which point the rib is heavily reconstructed. Beyond about 15 mm of reconstruction, the cranial margin of the rib appears smooth, convex, and featureless, aside from a few root marks. The caudal surface mirrors the cranial surface, in the sense that proximally, there is a broad, flat sulcus about 5.7 mm in breadth, separated by a prominent external ridge and a small, almost nonexistent internal ridge. The external ridge begins to demarcate the costal groove as one moves distally (the costal groove disappears at about two-thirds of the running length).

The Dolní Věstonice 14 right fifth rib (Figure 15.5) is represented solely by a piece of shaft 149 mm long from near midshaft. The piece has a repaired break about 31.5 mm from its preserved proximal end. The caudal margin of the piece is intact, but the superior margin is badly damaged along most of its course; for example, the distal 76.9 mm of the cranial surface is missing, and there is fill evident for 16.2 mm of the proximal end. Its proximal end is internally to externally thick, but about 74 mm distal to the preserved proximal end, it becomes thin and bladelike. A costal groove is present throughout. The groove is surrounded by internal and external ridges proximally, and it measures 3.9 mm in width. In contrast, distally, its margins are not so distinct.

The piece begins to thin out at its presumed midshaft, and just distal to its presumed midshaft the corpus appears to deviate inferiorly, reflective of the S-shaped curvature characteristic of human lower vertebrosternal and upper vertebrochondral ribs. There are few anatomical descriptions of this phenomenon in which these ribs make a slight cranial rise (or at minimum maintain a horizontal orientation) near midshaft before declining again toward the sternal end. The best description of this anatomical feature is given by Franciscus and Churchill (2002: 320) in describing the right eighth rib of Shanidar 3: held in anatomical position and seen in lateral view, the rib describes a sigmoid curve, descending from the vertebral articulation to the posterior angle, remaining relatively horizontal from the angle to midshaft, and then descending again towards the sternal end.

The left fifth rib of Dolní Věstonice 14 is represented by what were originally two pieces that have been glued together. The bone is preserved from the neck to just beyond midshaft (preserved length is 131 mm). There is also a possible third distal shaft piece (although this piece may belong to the left fourth rib; see below). Proximally, the rib is quite thick and robust. The neck is internally to externally thick; its superior surface is broad and convex, and its inferior surface is sharp, with a thin internal ridge. The tubercular region is robust. The inferiorly positioned articular facet is large and ovoid with a proximodistal long axis. The shaft beyond the tubercle is smooth and relatively featureless. On its cranial surface, there is a long sulcus, bounded by external and internal ridges, that corresponds to the attachment of the intercostal muscles. The sulcus is about 5 mm in width and at least 58 mm in length (it probably continued distally, but the section just beyond the sulcus is missing because of a break 59.6 mm from the distal end). Distal to the break, the cranial surface of the rib is flat, and then it rapidly becomes more convex near midshaft. Along the rib's inferior margin is a marked costal groove. Distally, the costal groove becomes superoinferiorly taller and shallower. Its breadth is 4.5 mm at its proximal end; its maximum breadth distally is 5.9 mm. Another piece of rib shaft, measuring 126 mm in length, is most likely the distal end of the left fifth rib since the coloration of its proximal end matches the coloration at the distal end of the piece discussed above. However, there is no good contact between the pieces, and thus it cannot be positively assigned. It could therefore belong to the left fourth rib. Its entire cranial margin evinces either putty or is broken, and the caudal margin is equally poorly preserved. There is a costal groove throughout the piece, although proximally it is more prominent.

The Dolní Věstonice 15 right fifth rib (Figure 15.6) preserves its proximal third, including much of its head, the entire neck, articular tubercle, angle, and a small segment of shaft distal to it. The head is tall and relatively narrow. It is missing the inferior one-fourth of its articular surface; the portion that remains measures 9.3 mm in height and 9.7 mm in breadth. The neck is tall and thin. On its caudal external surface is an ovoid (11.6 mm proximodistally × 4.3 mm superoinferiorly) fossa for the attachment of the medial costotransverse ligament. The cranial margin of the neck evinces a crest for the attachment of the superior costotransverse ligament that runs along the cranial margin of the corpus to a point just distal to the tubercle. The large, ovoid articular facet has an indiscrete cranial margin that blends into the nonarticular portion. The facet evinces slight damage to its distal and inferior margins, and it is slightly proximodistally concave. The nonarticular portion of the tubercle and iliocostal region evince rugose attachments for the iliocostalis thoracis and levatores costarum (longus and brevis). The inferior costal point is evident but slightly damaged. The caudal surface of the corpus has a marked external ridge, which distally will form the costal groove. The cranial margin of the corpus distal to the superior crest is broad and relatively flat. However, distal to a repaired break (about 50.9 mm from the articular tubercle), two subtle ridges (external and internal) for the intercostal muscles are visible on the otherwise flat and featureless superior surface. The preserved distal end is missing its cranial surface for about 18.3 mm, but it is evident that the corpus is gradually reducing in breadth.

Only a small proximal portion of the left fifth rib of Dolní Věstonice 15 is preserved, but it nonetheless includes most of the neck, the articular tubercle, the angle, and a short (40 mm) segment of shaft distal to the inferior costal point. The neck is relatively tall and narrow and evinces a fossa on its caudal external face, for the attachment of the medial costotransverse ligament, measuring 9.1 mm in length and 3.2 mm in height. There is a central ridge along the inferior margin of the neck, and a crest corresponding to the attachment of the superior costotransverse ligament is found along its cranial margin, which ends just proximal to the articular tubercle. The articular tubercle is large, ovoid, and slightly proximodistally concave. It has minor cortical bone loss on its inferior proximal surface, and its superior and distal margins blend into the nonarticular portion of the tubercle. In contrast, its posterior border is discrete and projects from the shaft. The iliocostal and nonarticular region of the tubercle is rugose, and the angle of the rib is marked. The inferior costal point is damaged, as is the inferior margin beginning some 14.3 mm distal to it. The cranial surface of the shaft is smoothly convex, whereas the inferior surface evinces an external ridge that begins just distal to the tubercle.

Figure 15.6 Superior view, ribs 5–7, Dolní Věstonice 15. Scale in centimeters.

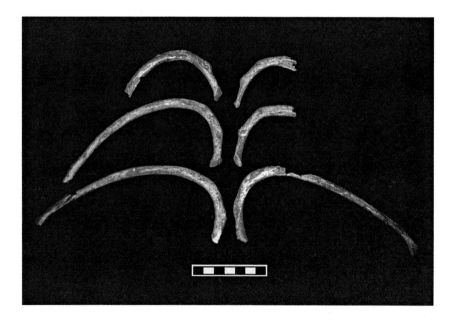

Sixth Ribs

The Dolní Věstonice 13 right sixth rib is complete (Figure 15.4), with an area of fill just proximal to midshaft. It is thick in cross section for about two-thirds of its running length, after which point it internally to externally thins. It has a prominent costal groove for most of its length (the groove is much less marked for the last one-fourth of its running length). The head is missing a small (5 mm in height × 3.5 mm in breadth) area of cortical bone on the superoposterior quadrant of its articular surface. Just distal to this damaged area, the cranial margin of the neck has a small (3.2 mm in height × 5.1 mm in length) area of exposed trabecular bone. Inferiorly, the posterior margin of the neck evinces a sulcus measuring 7.7 mm in length and 3.4 mm in height for the medial costotransverse ligament. There is a slight groove separating the articular and nonarticular portions of the tubercle. The articular portion of the tubercle is ovoid with discrete margins, slightly proximodistally concave, and slightly superoinferiorly convex, with trivial damage along its inferoproximal margin. Cranial to its nonarticular portion is a sulcus measuring 33.5 mm proximodistally and 3.7 mm superoinferiorly for the lateral costotransverse ligament. The cranial margin of the proximal two-thirds of the shaft is flat, with only a slight hint of ridges along its external and internal margins. The distal one-third of the rib has a more convex cranial margin. The inferior margin of the rib has trivial damage throughout its course, and near midshaft it is largely putty for about 39 mm of its running length. There is slight postmortem damage along the cranial margin just proximal to the complete sternal end.

The left sixth rib of Dolní Věstonice 13 is complete, exhibiting only minor erosion along the cranial margin of its sternal end. It is very thick in cross section and maintains this thickness for nearly two-thirds of its running length. The rib shows strong declination and marked torsion along its long axis. Its head is intact, and the neck reveals only trivial damage along its superior and posterior margins. Two sulci are evident on the neck's posterior face. The first, measuring 6.0 mm in length and 2.6 mm in height, is ovoid and is found along the caudal aspect of the posterior neck. It is for the medial costotransverse ligament. Just superior to the nonarticular portion of the tubercle is the second sulcus, which actually begins more proximally on the neck and measures 14.3 mm in length and 2.5 mm in height. It is for the lateral costotransverse ligament and is bounded internally by a sharp ridge. The prominent articular facet is ovoid and slightly proximodistally concave and superoinferiorly convex. It has trivial damage to its inferior and proximal margins. The proximal shaft distal to the tubercle is somewhat circular in cross section (although its cranial margin is relatively flat). A break with wax fill is evident about 75 mm distal to the tubercle, and just distal to this is a small area of fill (9.8 mm proximodistally × 3.1 mm internoexternally) along the cranial margin. Near midshaft along the cranial margin is a pair of ridges, one internal and one external, for the intercostal muscles, which converge to form a single ridge located at a point that is close to two-thirds of its running length. Distal to this point (ca. 63.5 mm from the sternal end) is an area of fill some 14.5 mm in length. Likewise, the distal 24.1 mm of the rib's cranial surface is fill. The caudal surface of the rib forms a broad base proximally, but the base gradually narrows distally. A costal groove is evident but becomes less marked after three-fourths of running length.

The Dolní Věstonice 14 right sixth rib (Figure 15.5) is not as robust as the left one, but it preserves much of the neck and tubercle and most of the shaft (it probably ends just a few millimeters shy of the sternal end). There are three repaired breaks along the corpus, as well as areas of dark fill along the shaft. The proximal third of the rib is internally to externally broad, and it evinces a deep posterior curvature. The distal two-thirds of the rib is quite straight and more bladelike in morphology. The neck is tall and narrow, with crests on its cranial and caudal margins. On its external margin a portion of the sulcus for the medial costotransverse ligament is preserved. The caudal neck evinces relatively heavy damage. The neck's superior crest, which is for the superior costotransverse ligament, extends for another 22 mm distal to the tubercle, at which point it flattens out. The ovoid articular facet is preserved, but it is heavily abraded and is characterized by an inferoproximal to superodistal long axis. Because of the abrasion, it cannot be determined if its articular surface was convex or flat. The inferior costal point is preserved, but in general, the nonarticular portion of the tubercle evinces extensive external damage. Beyond the tubercle, the superior surface of the shaft is broad and flat, only becoming more convex around midshaft. A costal groove extends from the angle to just proximal (12 mm) to the sternal end. At its widest (near the angle), it is 6.8 mm in breadth, and at its narrowest (its distal end), it is 3.4 mm in breadth. Along the cranial internal margin at the distal end is a sulcus, 3.8 mm in breadth and at least 17.8 mm in length, that corresponds to the distal attachment of the external intercostal muscle. The shaft of the rib evinces the classic S-shape, with moderate declination and marked torsion along its long axis.

The left sixth rib of Dolní Věstonice 14 is almost complete. It preserves a small portion of its head all the way to a point just proximal to its sternal end. The preserved part of the head's articular surface is a small internal/inferior section; elsewhere the cortical bone is missing. The neck is tall and internally to externally narrow. Its cranial margin is damaged, but at its terminus the crest for the superior costotransverse ligament is preserved.

The margin of the inferior neck is sharp. The articular portion of the tubercle is prominent, and although matrix remains adhered to it, it is ovoid with a relatively flat surface area. The nonarticular portion of the tubercle is heavily eroded. Its inferior margin, in particular, is heavily damaged, especially in the region near the inferior costal point. Distal to the tubercle, the inferior external ridge is either repaired with fill or is missing. The proximal shaft is broad and cranially flat. Near midshaft the rib becomes internally to externally thinner, and despite a large amount of fill, it is apparent that the rib becomes superiorly convex (it is also very straight in this region). Along its damaged inferior margin, it is nonetheless apparent that a shallow costal groove was present along its distal aspect. As with the right sixth rib, on the distal cranial surface there is a sulcus that emerges from the simple convexity more proximal to it that is at least 27 mm long and 2 mm wide. It corresponds to the distal attachment of the external intercostal muscle. The rib evinces relatively low declination, and an S-shape is not apparent in lateral view. This latter feature is likely a result of the damage along the inferior margin of the corpus. There is, however, a high degree of torsion along the shaft's long axis.

The Dolní Věstonice 15 right sixth rib (Figure 15.6) is well preserved, and it includes most of the head, all of its neck, and perhaps three-fourths of its running length beyond the tubercle. Much of the head's double articular facet remains. The superior facet (for T5) is larger, and all of it appears to be preserved (there is damage along the external face of the neck, with exposed trabecular bone). Its height is 9.2 mm; its breadth is 8.0 mm. The smaller inferior facet (for T6) is 6.1 mm in height and breadth and is damaged along its external margin. The neck is proximodistally short and craniocaudally tall and narrow, with a flat inferior surface. The crest associated with the superior costotransverse ligament is present along its cranial margin, fading into the shaft 45.2 mm from the proximal end. The neck's external surface evinces a sulcus measuring 9.0 mm in (proximodistal) length and 2.6 mm in (craniocaudal) height for the medial costotransverse ligament. Although its position is evident, the articular facet has been eroded away. Just distal to the missing facet is a break through the corpus repaired with fill. Distal to the break, the iliocostal area is rugose, with a marked inferior costal point. The cranial margin of the tubercular region evinces a flat superior sulcus that measures 23.9 mm in length and 6.5 mm in breadth for levator costae longus. The cranial margin distal to this sulcus is broad and convex, and it gradually narrows toward the distal end. Inferiorly, the costal groove is wide proximally and gradually narrows as the rib's cross-sectional diameter decreases. The external shaft is smooth and featureless. Only the inferior portion of the shaft is present for the distal 40.6 mm; the cranial shaft is missing and fill is evident. In lateral view, the rib evinces the classic S-shape of human ribs, and torsion is evident along its long axis.

Only a proximal segment of the left sixth rib of Dolní Věstonice 15, with neck, tubercle, angle, and a small segment of shaft, is preserved. None of the head's articular surface is present, but most of the neck is preserved and it is tall and narrow. The neck evinces the crest for the superior costotransverse ligament along its cranial margin. Just distal to the tubercle the ridge fades into two subtle ridges, one internal and one external, for the intercostal muscles. The large superoinferiorly convex and proximodistally flat ovoid articular tubercle has discrete boundaries and protrudes caudally from the shaft. The iliocostal region is rugose, and the inferior costal point is damaged and repaired with fill. The proximal shaft is broad with a smoothly convex superior surface. The caudal surface shows the beginnings of a costal groove, but the lateral margin is heavily damaged.

Seventh Ribs

The Dolní Věstonice 13 right seventh rib (Figure 15.7) is thick, robust, and complete, with only minor areas of fill along its cranial margin, each associated with breaks. The first is located just distal to the angle, and the second is found about 80.6 mm from the sternal end. The large articular surface of the head is intact, although there are two small areas of damage at the junction of the head and neck. The first is along the cranioexternal margin and measures 2.2 mm in height by 5.3 mm in length. The second is found along the inferior margin of the junction; it measures 5.9 mm in height by 3.4 mm in length. The articular portion of the tubercle is circular, and its articular surface is flat to slightly proximodistally concave. The cranial margin of the facet blends into the nonarticular portion of the tubercle, which itself is large and rugose. The external margin of the neck evinces two sulci. The first, near the caudal margin of the neck, is for the medial costotransverse ligament, and it measures 8.4 mm in length by 2.7 mm in height. The second, larger (20.2 mm in length × 3.5 mm in height) and more cranially positioned sulcus, extends the length of the neck. It is associated with a marked crest, and these two features are for the superior costotransverse ligament. The cranial external face of the nonarticular portion of the tubercle has a lateral costotransverse ligament sulcus measuring 10.4 mm in length by 3.5 mm in height. On the cranial margin just distal to the articular facet is a flat sulcus, measuring 10.8 mm in length by 2.6 mm in breadth, for the levator costae longus muscle. The proximal cranial margin of the corpus does not evince a sulcus for the intercostal muscles, as was seen in the fifth and sixth ribs, nor is it as flat as more cranial ribs, exhibiting a more smoothly

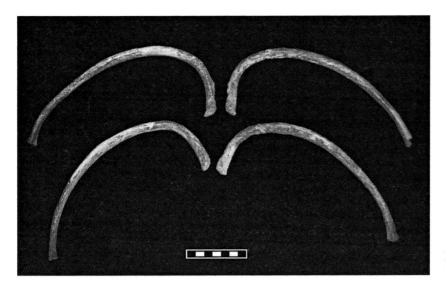

Figure 15.7 Superior view, ribs 7 and 8, Dolní Věstonice 13. Scale in centimeters.

rounded cranial margin. However, as with more cranial ribs, near midshaft the rib changes shape from a more rounded cross section to a narrower, taller one. On the caudal margin is a prominent costal groove for the proximal two-thirds of the shaft. In lateral view this rib is the first for this individual to possess a sinuous shape, with a rise near midshaft. The rib also has a relatively high degree of torsion along its long axis.

The sternal end of the left seventh rib of Dolní Věstonice 13 is heavily reconstructed with fill, and there is a repaired break just distal to the angle (ca. 67 mm distal to the articular tubercle); otherwise the rib is complete. The rib has a relatively thick corpus that gradually tapers distally so that it becomes internally to externally narrower. The head, with its large articular surfaces for T6 and T7, is perfectly preserved. There is, however, a small (5.1 mm in length x 2.4 mm in breadth) area along the cranial margin of the neck that is damaged. Beyond this damaged area is the crest for the superior costotransverse ligament. It begins along the external margin, but it runs internally in the tubercular region. This crest ultimately flattens out about 53 mm distal to the head. Inferior to this crest along the external margin of the tubercle region is a sulcus, measuring 16.8 mm in length by 4.2 mm in height, for the lateral costotransverse ligament. The articular portion of the tubercle is teardrop-shaped, with its peak located on its caudal margin. The area surrounding the tubercle is quite robust, with morphology similar to that of rib 6. The cranial margin beyond the tubercle is relatively smooth and rounded, although from roughly one-half to two-thirds running length along the shaft, there is a suggestion of internal and external ridges for the intercostal muscles. Inferiorly, there is a marked costal groove, which narrows as the corpus of the rib narrows. About 35 to 70 mm from the distal end, the caudal margin is flat, not convex as were those of the more cranial ribs. On the distal external surface, from about 42 mm proximal to the sternal end, is a flattened area for one of the tendinous origins of serratus anterior. In lateral view the rib evinces the subtle S-shape along its course, and a high degree of torsion is evident along its long axis.

The Dolní Věstonice 14 right seventh rib (Figure 15.8) is somewhat complete; it is missing its head, part of its neck, and a portion of its distal end. There is also an area of fill evident along the cranial margin near midshaft. The preserved portion of the neck is damaged inferiorly and internally, but it does preserve the crest for the superior costotransverse ligament. The crest runs proximally from the neck's external margin (where it is damaged) toward the internal margin (where it is marked) near the tubercle and then moves to midline. It fades into a smooth, flat, cranial margin about 45.7 mm distal to the preserved proximal end. Along the inferior surface of the neck and tubercular region is an internal ridge that fades midway between the tubercle and the angle. There is also an inferior external ridge, however, which despite erosion is quite prominent and only fades at nearly two-thirds of its running length (ca. 39.9 mm proximal to the preserved distal end). It forms part of the costal groove, which is proximally broad (ca. 5 mm), but narrows distally (ca 2.7 mm). The nearly circular and flat articular facet is large, with discrete boundaries. The nonarticular portion of the tubercle is relatively small, but the inferior costal point is clearly visible. In superior view the rib is proximally relatively broad but gradually tapers to a more bladelike morphology. This is associated with a shift from a relatively flat cranial surface to a more convex one in the angle region. The distal 15.3 mm of the cranial shaft has evidence of an internal sulcus for the external intercostal muscle. The rib does have the classic S-shape of human

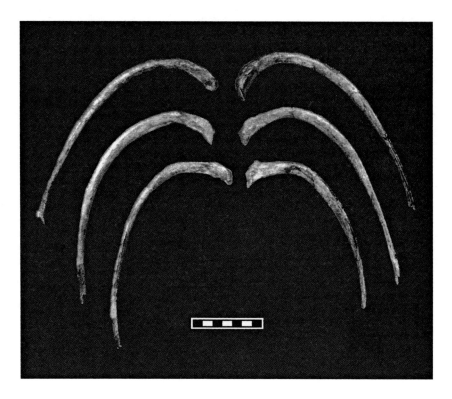

Figure 15.8 Superior view, ribs 7–9, Dolní Věstonice 14. Scale in centimeters.

ribs, although the rise of the shaft at midshaft is not as marked as in its left counterpart. The rib appears to have a relatively steep declination, but it does not appear to have as much torsion along its long axis as might be expected.

The left seventh rib of Dolní Věstonice 14 is largely complete, despite absence of the head and perhaps 20 mm of its distal end. There are areas of fill along much of the rib's cranial margin, and the distal one-fourth of the rib (ca. 83.8 mm) is missing its superior surface. The cranial margin of the neck and the crest for the superior costotransverse ligament are damaged proximally, but distally, in the tubercular region, the crest moves to the rib's internal margin, where it fades into the shaft's convex cranial surface. Along the inferior surface of the proximal rib (neck and tubercle) is a sharp internal crest that forms the internal border of a sulcus roughly 3 mm in width in the tubercular region and 5.4 mm in width (its widest point) about 46.3 mm distal to the tubercle. The articular facet has a somewhat convex surface and erosion along its distal and cranial margins. Distal to the tubercle, it is the inferior external ridge that is prominent. This bounds the costal groove, although the external crest flattens 111.7 mm (straight-line distance) from the center of the articular facet. The costal groove is flat, with very slight internal and external ridges. The distal end of the rib is beginning to expand in its internal-external dimension, suggesting that most of the rib's length is preserved. The rib appears to exhibit a relatively steep declination and an S-shape (with a rise at midshaft), and it is characterized by a fair degree of torsion along its long axis.

The Dolní Věstonice 15 right seventh rib (Figure 15.6) preserves a partial neck, articular tubercle, angle, and probably over three-fourths of the length of the shaft. Much of the neck's proximal, cranial, and external margins evince fill. The caudal margin of the neck is flat. The neck is tall and has the crest for the superior costotransverse ligament along the undamaged portion of its cranial margin. The crest itself begins proximally in an external position, but distally it snakes its way internally, ending 22.8 mm distal to the midpoint of the articular tubercle. There it blends into a broadly convex cranial margin that gradually tapers in its internal-external dimension. The articular portion of the tubercle is quite large, with an indiscrete cranial border that blends into the tubercle's nonarticular portion. The facet itself is ovoid and is proximodistally flat and superoinferiorly convex. The rib appears to have had a very slight posterior angle. Aside from the area just proximal to the angle, which is rugose, the external surface of the rib is smooth and featureless. A broad (13.6 mm in length × 12.6 mm in height) area of the external face of the mid-distal shaft is missing. The inferior margin of the corpus has both an internal ridge that gradually becomes smooth distally (forming the costal groove) and an external ridge on which the articular facet is situated, and which ultimately becomes the quite prominent inferior costal point and lateral/external margin of the costal groove. The rib evinces only a subtle S-shape but has a high degree of torsion along its long axis.

The left seventh rib of Dolní Věstonice 15 is represented by a proximal piece that preserves a portion of the head; most of the neck, tubercle, and angle; and, beyond the angle, about 36 mm of the shaft. The caudal two-thirds of the head's articular surface for T7 is preserved. The neck is tall and narrow with a flat caudal margin. Along the neck's external surface is an ovoid fossa that is 5.2 mm in height and at least 11.7 mm in length (it may have continued proximally) for the medial costotransverse ligament. The neck's cranial margin has the crest for the superior costotransverse ligament. This crest begins on the external margin and then moves to the internal margin near the proximal border of the articular facet. The crest then fades into the smoothly convex upper surface just distal to the distal margin of the articular tubercle. On the cranial margin directly superior to the articular tubercle is a fossa, about 9.4 mm in length and about 3.3 mm in breadth, for the levator costae brevis muscle. The articular facet itself is quite large, with an indistinct cranial boundary. It is proximodistally flat and superoinferiorly convex, and it has fill along the caudal aspect of its distal margin. The caudal and external surface of the tubercular region exhibits two crests. The first runs cranially to the articular facet, where it fades into the iliocostalis region. The second continues as a sharp line internally to the facet, which becomes blunt and smooth some 11.2 mm distal to the facet's distal border. There it forms the internal border of the costal groove; the external border is more or less contiguous with the articular tubercle. Although damaged along most of its course (especially at the inferior costal point), it is apparent that the external crest was quite marked at the angle.

Eighth Ribs

The Dolní Věstonice 13 right eighth rib is essentially complete (Figure 15.7). It does, however, have two repaired breaks distal to the angle and proximal to midshaft, a large area of fill along its cranial margin at midshaft, and an area of exposed trabecular bone along the cranial margin of the neck. It is not as thick in cross section as more cranial ribs, and externally to internally it narrows much earlier (ca. 20 mm distal to the angle) than do its more cranial counterparts. The large double facet for T7 and T8 is preserved, with one small area on the cranial facet evincing cortical bone loss. The thin crest for the superior costotransverse ligament is damaged proximally, fading into a convex cranial surface about 57 mm from the head/neck margin. Below the crest along the external surface of the rib in the tubercular region is a sulcus, measuring 28.9 mm in length by 3.4 mm in height, for the lateral costotransverse ligament. Inferiorly, there is a prominent costal groove up to approximately three-fourths running length. Proximal to the angle, however, the costal groove's form is more like that of a sulcus, with ridges bounding its internal and (particularly) external margins. The rib's superoinferior dimension decreases distal to midshaft, from 17.3 mm at the distal end of the broad area of fill to only 11.8 mm about 60 mm from the sternal end. Its superoinferior height then increases so that at its sternal end it measures 14.8 mm. In lateral view the rib has the classic S-shape of human ribs, and it exhibits a greater degree of torsion along its long axis than do its more cranial counterparts.

The left eighth rib of Dolní Věstonice 13 is complete, with repaired breaks just distal to the inferior costal point and about 23.4 mm from its sternal end, as well as trivial damage along its cranial margin near the sternal end. The large head, with its double facet for T7 and T8, is preserved in its entirety; the facet for T8 dwarfs that for T7. The crest corresponding to the attachment of the superior costotransverse ligament follows a sinuous path, moving from external to midline to internal as one moves distally from the head to just beyond the tubercle (it ends about 52 mm distal to the head/neck junction). The articular portion of the tubercle has a discrete distal boundary, but its proximal margin blends into the surrounding bone. The nonarticular portion of the tubercle is robust and rugose. On the external margin of the neck is a small sulcus, measuring 6.9 mm in length by 3.7 mm in height, for the medial costotransverse ligament. There is also a broad sulcus about 7.4 mm in length located just cranial to the nonarticular portion of the tubercle for the lateral costotransverse ligament. Distal to the tubercle the cranial margin of the rib corpus is smooth and convex. Its internal surface has a clearly demarcated costal groove, which is present to about 80% of its running length (51 mm proximal of its sternal end), after which point the caudal surface becomes relatively flat. Its corpus is not as internally to externally thick as rib 7, but as one moves distally along the shaft and it tapers externally to internally, the shaft's cranial-caudal dimension rapidly increases. The sternal end is perfectly preserved; it is marked by large cranial-caudal and narrow external-internal dimensions. In lateral view the rib evinces the classic S-shape and is also characterized by a high degree of torsion along its long axis.

The Dolní Věstonice 14 right eighth rib (Figure 15.8) has a heavily damaged neck and damaged caudal margin, and it is missing its distal 20–30 mm. There are large areas of fill on the non-articular portion of the tubercle and distal to midshaft; otherwise, the rib is complete. At the base of the neck a small (3.3 mm in height × 2.7 mm in breadth) portion of the articular surface of the head remains. Although the cranial half of the neck is missing, just distal to the neck the crest for the superior costotransverse ligament is visible at midline; it fades into the cranial surface about 31.5 mm distal to the centroid of the tubercle. The external surface of the neck evinces a portion of the sulcus for the medial costotransverse ligament. Inferiorly, the distal portion of the neck that is preserved is relatively flat, with a slight internal crest. Beyond the tubercle, the internal crest's edge becomes blunt and forms part of the costal groove. The costal groove all but disappears distal to midshaft, but this is probably due in part to damage along the caudal margin (the distal 32 mm is present, however, and no costal groove is evident). The ovoid articular tubercle is large and flat with discrete margins. There is a broad area of fill just distal to the facet. The corpus of the rib is moderately broad proximally, but it becomes more bladelike at the angle. Its cranial surface is convex throughout, becoming sharper as it nears the distal end. There is a repaired break about 14.6 mm proximal to the preserved distal end. Beyond this break, the rib is blackened, as if it had been burned; this could therefore be associated with the apparent burning of the eighth through tenth thoracic vertebrae of this specimen (see Chapter 14). The rib has much torsion along its long axis and shows the S-shape characteristic of human ribs, evincing a marked rise at midshaft.

The left eighth rib of Dolní Věstonice 14 is mostly present. It is missing the distal 34 mm of its superior margin, but the shape of the distal end suggests that most of the rib's length is preserved. A small (10.9 mm in height × 8.1 mm in breadth) portion of the articular surface of the head is also present. The external surface of the neck preserves portions of two sulci for the medial costotransverse ligament. The cranial sulcus for the superior costotransverse ligament is damaged throughout its course along the neck, but it reemerges in the tubercular region, first along the rib's posterior aspect, then moving to the internal margin, and ultimately blending into the superior surface of the shaft about 30 mm beyond the center of the articular tubercle. The inferior neck has external and internal ridges that are about 4.1 mm apart. The internal ridge is much more prominent than the external, yet the external ridge increases greatly in size distal to the articular tubercle (although it is badly damaged in the angle region). Distally, it bounds the costal groove. The groove's tallest point is about 30.1 mm distal to the center of the articular tubercle, where it is 5.3 mm in height. Distally, it is much shorter. At midshaft, it is only 2.4 mm tall, and at the distal end no ridge is visible at all, only a flat surface 3 mm in diameter. The corpus of the rib is internally to externally thinner than the seventh ribs. The entire rib is essentially bladelike. Its cranial surface is convex and sharp. The rib is steeply declined, evinces the classic S-shape of human ribs, and shows a high degree of torsion along its long axis such that its corpus is turned outward in its midsection (i.e., its cranial surface is located more externally, whereas the caudal surface is located more internally).

The Dolní Věstonice 15 right eighth rib (Figure 15.9) is largely complete, preserving a partial head; most of the neck, tubercle, and angle; and about 80% of its shaft. Some of the articular surface of the head is preserved, mostly along its internal margin. The neck is tall. Its external surface has an ovoid sulcus, measuring 7.6 mm in proximodistal length and 3.7 mm in superoinferior height, for the medial costotransverse ligament. A damaged crest for the superior costotransverse ligament is present along the neck's cranial margin. It fades into a flat sulcus 45.1 mm distal to the proximal end. This sulcus, for the levator costae longus muscle, measures 26.7 mm in length and 4.9 mm in breadth. Beyond this sulcus, the cranial margin of the rib is smoothly convex. It is, however, only nominally thick proximally, and it quickly becomes more bladelike near one-third of its running length. The rib is

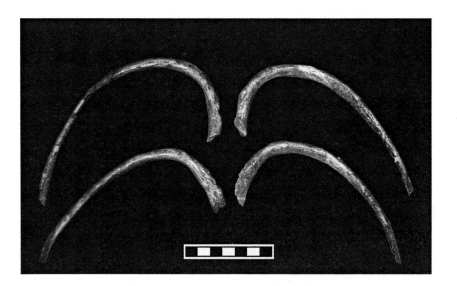

Figure 15.9 Superior view, ribs 8 and 9, Dolní Věstonice 15. Scale in centimeters.

internally to externally thin near midshaft, after which it again expands. None of the margins of the articular tubercle is well delineated, and as a result it is difficult to distinguish the articular from nonarticular portions of the tubercle. The rib's external surface is rugose in the iliocostal region but smooth and featureless thereafter. There are two notable breaks in the shaft. The first is located about 41.8 mm from the proximal end, just distal to the flat cranial sulcus; it has been repaired with fill. The second is located near midshaft, about 52.9 mm distal to the previous break. Here a large portion of the shaft is missing and has been replaced with wax. Along the inferior margin, the tubercular region evinces an internal ridge that smoothes out to form a broad costal groove about 22 mm distal to the midpoint of the articular tubercle. The costal groove is present along those portions of the inferior margin that are undamaged. The preserved distal end of the rib is superoinferiorly expanding, suggesting that most of the rib is preserved. The shaft has a clear S-shape, with its cranial point just proximal to midshaft. It is moderately declined and has a high degree of torsion along its long axis.

The left eighth rib of Dolní Věstonice 15 preserves a portion of its head, all of the neck, and about three-fourths of its shaft. Trabecular bone is exposed for most of the rib's proximal end, yet the inferior one-third of the articular surface of the head is preserved; it measures 7.1 mm in height and 8.7 mm in breadth. The neck is tall and proximodistally short. The neck's superior crest for the superior costotransverse ligament begins along the external margin and then snakes internally at the distal margin of the articular tubercle. It then disappears into the cranial margin of the rib about 24.3 mm distal to the midpoint of the articular tubercle. Along the external surface of the neck is a fossa for the medial costotransverse ligament measuring 7.9 mm in length by 4 mm in height. The inferior surface of the neck evinces two crests—one internally and one externally. The external crest blends into the cranial margin of the articular facet, whereas the internal crest becomes less sharp, forming the costal groove about 19.2 mm distal to the midpoint of the articular tubercle. The costal groove remains broad to a point about 33.5 mm distal to the angle, after which the caudal margin of the shaft is heavily damaged. It reemerges at a point some 95 mm distal to the angle, where it is narrower (ca. 4.4 mm) but quite prominent. The articular portion of the tubercle is large and ovoid. Its surface is slightly proximodistally concave and superoinferiorly convex. Its boundaries are indistinct, especially superodistally, where it blends into the nonarticular portion of the tubercle. As with the right eighth rib, there is a flat ovoid sulcus just distal to the superior crest for the levator costae brevis muscle. It measures about 32.4 mm in length and 4.6 mm in breadth. Beyond this sulcus the cranial margin of the rib is smoothly convex. The corpus of the rib is not internally to externally thick. In lateral view it has the classic S-shape with a midshaft peak, and the rib has a high degree of torsion along its long axis.

Ninth Ribs

The Dolní Věstonice 13 right ninth rib is almost complete (Figure 15.10). There is a break just proximal to the inferior costal point (ca. 62.5 mm from the head/neck margin), and there are two areas where the bone is eroded, revealing trabecular bone. The first is along the cranial margin of the neck, and the second is a small portion of the cranial margin of the sternal end. The large double

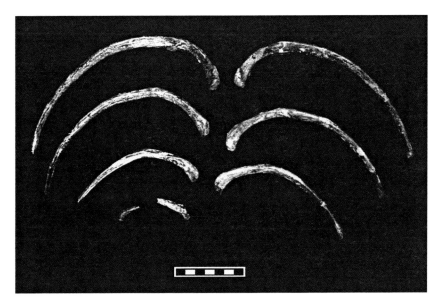

Figure 15.10 Superior view, ribs 9–12, Dolní Věstonice 13. Scale in centimeters.

facet for T8 and T9 is perfectly preserved; the demifacet for T8 is much larger than that for T9. The small, round, and slightly convex articular facet has discrete boundaries except along its superoproximal margin, where it blends into the surrounding bone. Just distal to the facet is a small, round depression, measuring 3.9 mm in height by 4.1 mm in length, for the medial costotransverse ligament. The crest associated with the attachment of the superior costotransverse ligament is damaged along the neck, but it is prominent along the internal cranial margin in the tubercular region. The crest also forms the internal boundary of a sulcus located between the tubercle and the angle, measuring 29.4 mm in length by 4.4 mm in breadth, that corresponds to the attachment of the lateral costotransverse ligament. The remainder of the cranial surface of the rib's corpus is smoothly convex and relatively featureless. The rib is relatively gracile, and like rib 8 it begins to internally to externally flatten just proximal to the angle. The rib has a very prominent inferior costal point. There is a sulcus proximal to this point, with external and internal ridges bounding it. Distal to this point is a marked costal groove—up to about two-thirds running length (ca. 68.4 mm from the sternal end); it is bounded by a single ridge externally, but subsequent to this point, there are two ridges that bound a central, shallow sulcus for the intercostal muscles. The external surface of the rib is quite smooth, and it is only mildly rugose proximal to the angle. In lateral view it is evident that the rib's superoinferior dimension decreases rapidly distal to midshaft, only to increase slightly at the sternal end (the sternal end thickens more in an external/internal dimension than a superoinferior one). In medial view the sternal end is ovoid in shape and wider than it is tall, with a relatively deep pit associated with the attachment of the costal cartilage. The rib also evinces steep declination distal to the angle, albeit with a slight rise near midshaft. It also shows a relatively high degree of torsion along its long axis.

The left ninth rib of Dolní Věstonice 13 is incomplete, as it is missing about 15 mm of its distal shaft. There is a break with fill near the inferior costal point (ca. 69 mm from the head/neck margin), and the cranial margin of the neck has fill as well. The large head, with its double facets for T8 and T9, is perfectly preserved. The external surface of the neck has a shallow fossa, measuring some 7.6 mm in length and 2.6 mm in height, for the medial costotransverse ligament. The articular portion of the tubercle is larger than its counterpart on the right. It is ovoid, has a slight superoinferior convexity, and has discrete boundaries. The thin crest for the superior costotransverse ligament is damaged on the neck, but it emerges in the tubercular region, first externally but then rapidly shifting to the rib's internal margin. The area for the insertion of iliocostalis thoracis and levatores costarum longus and brevis (i.e., much of the nonarticular portion of the tubercle) is quite large and rugose. The inferior costal point is quite prominent, as is the costal groove. The costal groove becomes superoinferiorly deeper and externally/internally thinner distal to midshaft. This is associated with a change in the shape of the corpus, which tapers quickly so that its distal end is both superoinferiorly short and dorsoventrally narrow. Distally (ca. 65 mm proximal to the sternal end), the costal groove becomes a sulcus bounded by internal and external ridges. The cranial margin of the corpus is convex throughout much of its length, although in the area from the tubercle to the angle there is a narrow, flat sulcus on the superior/lateral border, measuring about 57 mm in length and 4 mm in breadth, for the intercostal muscles. In lateral view the rib has a steep declination with a slight rise at midshaft (as was seen in the right ninth rib). There is also considerable torsion along the rib's long axis.

The Dolní Věstonice 14 right ninth rib (Figure 15.8) preserves a portion of its head, and despite heavy concentrations of fill, all but a small distal portion is present. Its neck is proximodistally short and craniocaudally tall and narrow. On its cranial margin the crest for the superior costotransverse ligament is present; it fades into the corpus about 34 mm distal to the midpoint of the articular tubercle. Distal to this point, the cranial surface of the rib is smooth and convex. The articular facet is extremely large and prominent. It is ovoid in shape, with an inferoproximal to superodistal long axis and a slightly superoinferiorly convex articular surface. Superoproximal to the facet is a large ovoid pit (5.6 mm in height × 9.1 mm in width) for the medial costotransverse ligament. Inferiorly, a crest that appears just proximal to the articular tubercle on the internal surface becomes more rounded about 25.6 mm distal to the center of the facet, where it forms the internal boundary of the costal groove. An external ridge (which despite being broken, is quite prominent) forms the external boundary of the costal groove. The external crest is damaged and diminishes in size distal to midshaft. The costal groove measures 4.7 mm at its widest point (ca. 49.2 mm distal to the center of the articular tubercle). Distal to this point, the groove is much narrower; at approximately four-fifths of running length it is 2 mm wide. The rib is characterized by a sharply bent angle. The entire corpus of the rib is bladelike; nowhere along its course can this rib be said to be internally to externally thick. The rib also declines and decreases greatly in superoinferior diameter as one moves distally. In lateral view there is a marked rise at midshaft (i.e., the classic S-shape). The rib also has a high degree of torsion along its long axis.

The left ninth rib of Dolní Věstonice 14 is narrow in cross section, and it is largely complete; it lacks most of its head and a short segment of its distal end. There are also areas of dark fill, particularly in the angle and midshaft regions. At its proximal end, a small (4.2 mm in height × 6.5 mm in breadth) inferior portion of the articular surface of the head is preserved; the remainder of the proximal end of the rib is exposed trabecular bone. The neck is proximodistally short and craniocaudally tall. It is missing much of its proximal and cranial aspect. Along the neck's distal cranial margin is the crest for the superior costotransverse ligament. As with the right ninth rib, there is a pit superoproximal to the articular facet for a portion of the attachment of the medial costotransverse ligament. This pit is smaller than its counterpart on the right, measuring only 4.8 mm in width and 3.1 mm in height. The articular facet is large, ovoid, and flat, but it is smaller than that on the right ninth rib. It has trivial damage to its distal end; elsewhere, its borders are discrete. The nonarticular portion of the tubercle evinces a rugose attachment for iliocostalis thoracis. The cranial margin of the corpus of the rib is badly damaged, especially distally, but where it is preserved it has a smoothly convex shape. Inferiorly, there is a marked external crest, and a costal groove is present except for the most distal 49 mm of the shaft. The inferior external crest is slightly damaged along its entire course. The corpus decreases greatly in superoinferior diameter at about three-fourths of the running length. There is a sharp angulation of the rib at the angle, and the rib declines greatly. The corpus has a slight rise at midshaft and a fair degree of torsion along its long axis.

Perhaps three-fourths of the Dolní Věstonice 15 right ninth rib is preserved (Figure 15.9), including a partial head, largely complete neck, tubercle, and much of the shaft. Only a very small (5.0 mm in height and 7.6 mm in breadth) inferior portion of the articular surface of the head is preserved. The neck is craniocaudally tall, proximodistally short, and narrow. There are two small sulci on the neck's external surface. The first is just proximal to the articular facet. Measuring 7.6 mm in length by 3.2 mm in height, it is for the medial costotransverse ligament. The second is associated with the superior crest of the superior costotransverse ligament. This crest blends into the cranial margin just distal to the articular tubercle. Distal to the superior crest is a flat, ovoid sulcus, about 22.4 mm in length and 5.0 mm in breadth, that is for the levator costae brevis muscle. Distal to this sulcus the cranial margin of the rib's shaft is smoothly convex, although it internally to externally thins at about one-third its running length. The ovoid tubercle is large, with relatively well-demarcated boundaries. It is proximodistally concave and has minor damage along its superodistal margin. The inferior margin of the proximal shaft evinces an internal and external crest. The internal crest flattens about 21 mm distal to the center of the articular facet, where it forms part of the costal groove. The costal groove is present along the rib's entire course, narrowing gradually with the rib's superoinferior dimension. There is minor damage to the external inferior crest, beginning 19 mm distal to the center of the articular tubercle. The inferior costal point is marginally projecting. Much of the distal shaft evinces fill along its external and cranial margins. The rib does show the classic S-shape of human ribs, with a midshaft rise, and it is strongly declined with a fair degree of torsion along its long axis.

Perhaps half of the left ninth rib of Dolní Věstonice 15 is preserved, including a portion of the head and neck, tubercle, angle, and proximal shaft to a point just distal to midshaft. Much of the caudal margin of the rib, including the inferior costal point, has been repaired with fill. A small (6.4 mm in height × 8.4 mm in breadth) inferior/external portion of the articular surface of the head is preserved. The neck is proximodistally short, craniocaudally tall, and narrow. Its external surface evinces a shallow, ovoid pit just proximal to the articular tubercle, measuring 7.2 mm in length and 3.6 mm in height, for the medial costotransverse ligament. Likewise, the cranial margin of the neck evinces the crest for the superior costotransverse ligament. It blends into a smooth, convex, superior surface at the distal margin of the articular tubercle. The caudal neck has an internal groove that flattens about 17 mm distal to the center of the articular tubercle, where it forms a broad costal groove that gradually narrows along its course. The large, circular articular tubercle is slightly concave; its articular surface faces inferiorly. There is a small (4.1 mm in height × 4.1 mm in length) area of exposed trabecular bone just cranial to it on the markedly rugose nonarticular portion of the tubercle. As with the right ninth rib, the angle of this rib appears deep. The cranial margin of the corpus remains smoothly convex along its course, and the rib considerably narrows in internal to external dimension distal to the angle. Likewise, the rib corpus's superoinferior diameter decreases gradually, too, as did its counterpart on the right. The distal 33 mm of the cranial margin is missing, as are small areas of the external table's cortical bone distal to midshaft. Aside from the iliocostal area, most of the external surface is smooth and featureless. The rib is characterized by the classic S-shape of human ribs, and a high degree of torsion is evident along its long axis.

Tenth Ribs

The Dolní Věstonice 13 right tenth rib is almost complete (Figure 15.10), with only minor erosion along the cranial margin of the neck and very slight wear on the lateral/external margin of the sternal end. There is also a repaired break in the shaft about 59 mm distal to the head/neck margin. Along the cranial margin of the neck and tubercular region is a large, posteriorly projecting crest for the superior costotransverse ligament. Along the inferior, posterior margin of the proximal rib is a small ridge associated with a shallow depression, which is probably the articular portion of the tubercle (no articular facet was noted, however, on the right transverse process of T10; see Chapter 14). The iliocostalis region of the rib is quite rugose, and the iliocostalis line and inferior costal point are clearly visible. Along the rib's cranial margin are two shallow sulci associated with the intercostal muscles. The first runs along the cranial margin, beginning above the tubercle (ca. 29 mm from the head/neck junction to ca. 67 mm distal to the head/neck junction, about one-fourth of the running length). The second sulcus begins about 59 mm distal to the distal margin of the head and ends about 80 mm distal to the head/neck margin. Distal to this sulcus, the cranial margin of the rib is smooth, convex, and featureless. Along the inferior margin, there is a marked costal groove from the angle to roughly four-fifths of its running length, at which point it becomes more of a sulcus on the internal, distal aspect of the rib, with a slight ridge that appears to bisect the rib into superior and inferior halves. The rib has a high degree of declination; however, torsion along its long axis, although present, is minimal.

The left tenth rib of Dolní Věstonice 13 is missing perhaps its distal 20 mm of length. It is otherwise complete. It also has four repaired breaks. In proximal view the head of the rib is teardrop-shaped, with its peak at the cranial margin. Along the cranial margin of the neck and tubercular region is a short, thin crest for the superior costotransverse ligament. Just inferior to this crest along the external surface of the rib is a sulcus for the medial costotransverse ligament. The sulcus is bounded superiorly by the superior crest and inferiorly by a smooth ridge of bone, the articular portion of the tubercle. Along the cranial margin distal to the superior crest is a shallow sulcus, about 3.6 mm in breadth, for the intercostal muscles. It runs for 41.9 mm along the cranial margin, beginning at the tubercle. Distal to this point, the cranial margin of the rib is smooth, convex, and featureless. Aside from the rugosity in the iliocostal, tubercle, and neck regions, the external surface of the rib is relatively smooth and featureless. The caudal surface exhibits a proximal sulcus, which runs along the neck for about 41 mm. This sulcus, roughly 3.2 mm in breadth, is bounded by internal and external ridges for the intercostal muscles. There is also a prominent costal groove distal to the inferior costal point (although it is almost gone by the preserved distal end). The rib evinces a strong amount of declination; however, little torsion is evident along its long axis.

Dolní Věstonice 14's right tenth rib (Figure 15.11) preserves most of the bone from its damaged neck to perhaps just 20–30 mm shy of the distal end. There are broad areas of fill just distal to the angle. None of the head's articular surface is preserved. The external surface of the neck and tubercle (the boundary between these two features is not evident since the articular facet cannot be discerned) evinces little relief. There is, however, a sulcus on the external surface, measuring about 4.9 mm in height by 11.1 mm in length, for the lateral costotransverse ligament. Proximal and superior to the sulcus is an area that retains adhered matrix; perhaps under the matrix lies the articular facet. The neck and tubercular region are relatively cylindrical (i.e., circular in cross section), but distal to the angle the corpus becomes decidedly more bladelike. The rib's cranial surface is convex throughout its course. The rib also decreases greatly in superoinferior dimension distal to midshaft, and it is apparently steeply declined. There is a sharp angulation at the angle, but

Figure 15.11 Superior view, ribs 10–12, Dolní Věstonice 14. Scale in centimeters.

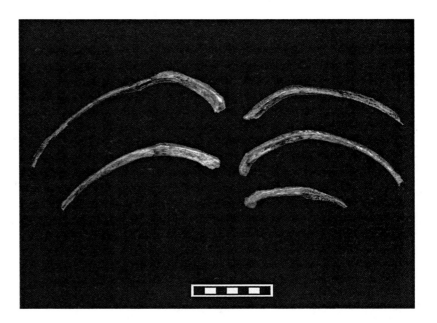

unfortunately, about 24 mm distal to the angle, there is a postmortem break that has resulted in the external table being crushed inward. This distorts the shaft inward, making calculations of within-rib curvature area unreliable. The costal groove is prominent throughout, but it is particularly marked for the first 97 mm distal to the angle, where the external inferior ridge is quite large. The distal end of the rib is missing its cranial half; trabecular bone is exposed for the distal 51.5 mm. The rib is steeply declined, but it does not appear to be characterized by a high degree of torsion.

The left tenth rib of Dolní Věstonice 14 is less complete than the right. It preserves from a point just distal to the articular tubercle to perhaps two-thirds of running length. Proximally, a portion of an external sulcus for the lateral costotransverse ligament, measuring at least 6.2 mm in length and 3.1 mm in height, is preserved. There is a central ridge for the superior costotransverse ligament along the proximal 34.3 mm of the rib. Distal to this point, the cranial margin of the rib becomes smooth, convex, and relatively featureless. The rib is also circular in cross section for the proximal 29 mm of its course; distal to this point it becomes more spearlike in morphology (i.e., it internally to externally narrows). In the area proximal to the angle, the costal groove is relatively small, but distally the groove is quite marked, with a large (5.8 mm near midshaft) external inferior ridge. A large (15.5 mm in length and 8 mm in height) area of fill is evident on the external table near the angle. As with its more complete counterpart, the left tenth rib has marked angulation at its angle.

The Dolní Věstonice 15 right tenth rib (Figure 15.12) is nearly complete; its entire head, neck, and tubercle and about 80% of the shaft are preserved. The teardrop-shaped articular surface of the head has a single articular facet for T10. It is deeply concave, or pitlike, with its deepest point located inferiorly. The neck is proximodistally short and, relative to the corpus, is quite tall. On its external surface, just proximal to the articular facet, is a small pit, measuring 3.1 mm in height and 2.3 mm in length, for the medial costotransverse ligament. The cranial margin of the neck is marked by a slightly damaged crest for the superior costotransverse ligament. This crest begins proximally along the external margin, but then it moves internally at the midpoint of the articular tubercle. It blends into the smooth superior surface 12 mm distal to the articular tubercle midpoint. The inferior neck has two crests, neither of which is marked. The first (external) crest forms the external margin of the articular tubercle, and the second (internal) crest blends into the shaft to form the costal groove about 16 mm distal to the articular tubercle midpoint. The articular portion of the tubercle is large, almost circular, and inferiorly facing. It is gently proximodistally concave and has damage along its cranial/external margin. The rib corpus is superoinferiorly short (with a slight expansion near midshaft) and internally to externally narrow throughout, but it becomes even more narrow distal to the angle. A costal groove is present throughout but narrows considerably. Its external ridge, though damaged, is not particularly large. The rib has a relatively modest degree of curvature at its posterior angle. It appears to have been greatly declined but shows only minimal torsion along its long axis.

The left tenth rib of Dolní Věstonice 15 is fairly complete; it is missing its head, but it includes a portion of the neck and tubercle and about 75% of its shaft. The inferior neck is all but destroyed. However, its cranial surface evinces a small crest for the superior costotransverse ligament. The portion of the crest that is preserved is located along the neck's external margin, and it blends into

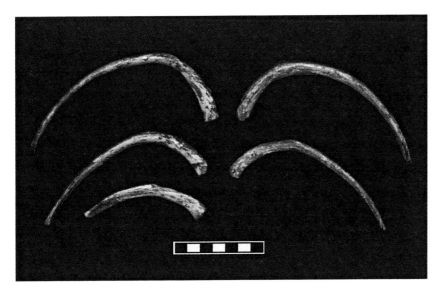

Figure 15.12 Superior view, ribs 10–12, Dolní Věstonice 15. Scale in centimeters.

a smoothly convex cranial surface 8.5 mm distal to the midpoint of the articular tubercle. The articular facet is large, circular, and inferiorly facing, with a slightly internally to externally convex surface. Proximally, the inferior margin of the rib evinces two crests, one internal and one external. The external crest is largest at the angle, decreasing in size both proximally and distally. The internal crest blends into the shaft about 25 mm distal to the midpoint of the articular tubercle. The costal groove that is associated with these crests is present, but subtle, throughout the rib's course. The shaft of the rib is craniocaudally short throughout its course, although just distal to the angle it is somewhat expanded. Its cranial surface is smooth and convex and is relatively internally/externally thick proximal to the angle. It gradually thins thereafter. The rib has a steep declination and a clear rise at midshaft. The iliocostal region is rugose. Except for this, both the external and internal faces of the shaft are smooth and relatively featureless. The curvature of the angle is slight.

Eleventh Ribs

The Dolní Věstonice 13 right eleventh rib is only missing about its distal 10 mm; otherwise it is complete (Figure 15.10). An area 9.3 mm in length by 4.0 mm in height on the external surface just distal to the tubercle is excavated and is for the lateral costotransverse ligament. Perhaps the most striking feature of the rib is a broad, flat sulcus on its cranial margin, approximately 47 mm in length and 5.5 mm in width. Located along the proximal end of the rib, it begins about 16 mm distal to the distal margin of the head and is for the intercostal muscles. Distally, it gives rise to a single ridge that runs for about 17.8 mm; the remainder of the cranial margin is relatively sharp and featureless. Inferiorly, the rib has a rugose iliocostal region that extends about 10 mm distal to the cranial sulcus described above. Although the angle of the rib is quite evident, the inferior costal point is not as marked as that of the left eleventh rib. A costal groove is present but not particularly marked. The rib is surprisingly round in cross section for an eleventh rib (see osteometric analyses). It shows steady declination but very little torsion along its long axis.

The left eleventh rib of Dolní Věstonice 13 is missing only what appears to have been a very small portion of its distal end, although it does have four repaired breaks, one of which has a large amount of putty associated with it. There is trivial damage to the cranial margin of its neck as well. The rib's proximal end is complete and is characterized by a large head that is still somewhat encrusted with matrix. On the external surface of the neck about 6.9 mm distal to the head/neck margin is a large pitted area (7.4 mm in length × 2.4 mm in height) that corresponds to the rib tubercle. The area of attachment of iliocostalis thoracis is rugose, with a marked inferior costal point. Running along the cranial margin of the neck and tubercular region is a sulcus about 45 mm in length and 4 mm in breadth for the intercostal muscles. This sulcus is bounded by almost indistinct internal and external ridges. Inferiorly, the costal groove begins about 35 mm distal to the head, and it continues distally for about 100 mm prior to blending into a smooth, convex, inferior margin about 14.9 mm proximal to the preserved distal end. The rib has steady declination but very little torsion.

Most of Dolní Věstonice 14's right eleventh rib is preserved (Figure 15.11); only the head and a small portion of the distal end are missing. Since eleventh rib tubercles are non-articular, it is difficult to discern the neck from the tubercular region of this bone. The angle can be discerned, however. There is a repaired break in the rib, just distal to the angle (ca. 61 mm from the proximal end), where the distal piece is displaced slightly inwardly. Proximal to this break, the cranial margin of the rib is convex and relatively broad; distal to it, the rib narrows considerably. The superior margin does, however, remain convex. In contrast, the inferior surface is either sharp (the proximal 10.1 mm of the neck/tubercle region) or flat (the remainder of the bone). Distal to the tubercle, the flat inferior surface has two small ridges—one internal and one external. The external ridge is the larger of the two. The sulcus between these ridges (for the last intercostal muscle) varies in breadth from 2 mm (23 mm from the preserved proximal end) to about 4.3 mm at the preserved distal end. The external surface of the rib is smooth, except for the area of the angle, where it is quite rugose. The rib evinces a gentle curvature.

The left eleventh rib of Dolní Věstonice 14 is almost complete. Most of its head is present, and it appears to preserve some of its distal end. The rib is not robust but rather externally to internally thin and bladelike. It evinces steep declination and is gently curved in a transverse plane. The cranial margin of the head and neck (as with the right eleventh rib, it is difficult to discern the neck from the tubercular region) is heavily damaged, with missing cortical bone along the proximal 18 mm of running length. The superior surface is somewhat flat for its proximal 50.1 mm (about one-third of its running length); thereafter, it becomes more convex. The inferior margin is featureless for its proximal 45 mm. Distal to this point, however, a costal groove is evident. The maximum height of the costal groove is 5.4 mm, and its height diminishes distally. The costal groove measures 4.0 mm in height about 32 mm from the distal end, and distal to that point it seems to disappear altogether—although inferior damage makes this impossible to say with certainty. The distal end of the rib is preserved, although it has trivial damage to its inferior margin, and the distal 16.7 mm of running length along the cranial margin is also damaged, exposing trabecular bone.

The Dolní Věstonice 15 right eleventh rib is essentially complete (Figure 15.12). There is, however, a break 68.4 mm from the preserved distal end that has been repaired with fill. The distal end is damaged, but it is unlikely that more than 1 mm of distal shaft is missing. The head of the rib, which has trivial damage to its inferior margin, is internally to externally broad along its inferior margin and more narrow superiorly. The neck tapers in superoinferior dimension toward its distal end. The external surface of the neck evinces a pit measuring 4.4 mm in height and 4 mm in length for the medial costotransverse ligament. The iliocostal region is not completely smooth but is only mildly rugose. The inferior costal point is clearly visible. The costal groove is almost nonexistent. Proximally, it is more apparent than distally, but it is subtle throughout. The proximal one-third of the corpus is relatively internally to externally thick. Distal to that, the shaft is much thinner and more bladelike, with slight internal and external ridges along its cranial margin for the intercostal muscles. Where the cranial margin is undamaged it is smoothly convex. The rib as a whole is steeply declined and only moderately curved in the transverse plane (although its posterior angle is quite marked).

The left eleventh rib of Dolní Věstonice 15 has a complete proximal end and is missing perhaps the last 5 to 7 mm of its distal end. There are two breaks repaired with fill. The first is at midshaft (74.6 mm from the proximal end); the second is 21.8 mm distal to the first. In proximal view its head is kidney-shaped, with its ventrum anteriorly convex and its dorsum slightly anteriorly concave. The external surface of the neck evinces a relatively indistinct tubercle about 15.6 mm distal to the head for the medial costotransverse ligament. The neck's cranial surface is convex; but distal to the neck, the cranial margin evinces a flattened area 37.5 mm in length and about 6 mm in breadth. Distal to this flattened area, the remainder of the cranial surface is convex. The corpus of the rib is internally to externally thicker proximal to the angle, after which it is thin and bladelike. The iliocostal region is rugose, and the iliocostalis line is clearly marked along the rib's external face. Distal to the iliocostalis line, both external and internal faces are smooth and featureless, except for the costal groove, which is quite distinct along the rib's internal face. In fact, the groove is much more marked on this rib than on the right one. The rib's superoinferior dimension is greatest (15.1 mm) some 26 mm distal to midshaft (it is reduced both proximal and distal to this point). The rib declines extensively and is only moderately curved in a transverse plane.

The Dolní Věstonice 16 left eleventh rib is the only rib fragment of Dolní Věstonice 16 that can be assigned to both number and side. This rib is represented by a preserved piece of proximal shaft 90.6 mm in length, and it is preserved from a point about 32.2 mm proximal to the angle to a point 68.3 mm distal to it. Proximal to the angle, the external surface of the rib evinces the rugosity associated with the attachment of iliocostalis thoracis. Also proximal to the angle, both the inferior and superior surfaces of the rib are relatively flat—the superior surface is wider (4.3 mm) than the inferior surface (2.6 mm). Distal to the angle, the piece is bladelike, with a smooth, convex, superior surface and an inferior surface with only the slightest hint of a costal groove. The rib curves more inwardly at the angle and is characterized by a steep declination.

Twelfth Ribs

The Dolní Věstonice 13 right twelfth rib (Figure 15.10) is represented by a proximal piece about 39 mm in length that includes the head and a distal piece about 26 mm in length that includes the distal end. These two pieces cannot be joined, and thus they are separated by missing bone of an indeterminate length. There is trivial damage to the caudal/internal margin of the head and slight damage to the external margin of the process located above the articular portion of the head. The dorsal surface of the proximal piece is rugose, and the articular surface of the head evinces a small pit (1.4 mm in height × 1.8 mm in breadth) that is seemingly premortem yet not appear to be pathological. The proximal piece also manifests a clear sulcus inferiorly, just distal to the head, about 18 mm long and 3 mm wide, for the quadratus lumborum muscle. This sulcus is bounded by a marked dorsal crest and a less marked ventral crest. Along the rib's cranial margin, a smaller, flattened sulcus is found for the intercostal muscles. It begins about 12 mm distal to the articular surface of the head and runs for approximately 15 mm. It is roughly 2 mm wide and is bounded by very small crests. The distal piece has rugosities dorsally and evinces a strong anterior curvature. It has a relatively deep caudal sulcus that measures about 12.5 mm by 3.5 mm and corresponds to the insertion of the levator costae longus muscle. There is also a much shallower sulcus running the length of the piece ventrally, as well as a ridge that runs along the cranial margin of the piece that corresponds to the attachments of the intercostal muscles.

The Dolní Věstonice 14 left twelfth rib is almost complete (Figure 15.11). It preserves most of its head, as well as its sternal end. The head has an ovoid cranial facet and a thinner, nonarticular portion on its caudal aspect. There is a marked angle between the neck and shaft. The proximal shaft is parallel-sided (i.e., its superior and inferior margins are parallel) for its proximal 37.0 mm of running length. Distal to this point (which is just distal to midshaft) the shaft swells out, expanding both superiorly and, especially, inferiorly. For the distal 15 mm of its running length, however, the shaft again reduces its superoinferior dimension (e.g., midshaft height: 8.5 mm; swelling, maximum height: 13.0 mm; distal end height: 6.9 mm). The maximum swelling is the location of a repaired break 63.6 mm from the proximal end and 31.2 mm from the distal end. The cranial margin of the rib has minor damage to either side of this break. The inferior surface of the shaft evinces three sulci for the quadratus lumborum. The first is proximal to midshaft and measures 20.1 mm in length by 2.6 mm in breadth. The second spans midshaft and measures 21.2 mm in length by 3.6 mm in breadth, and the third, which is found near the maximum swelling point of the shaft, measures 9.9 mm in length by 4.0 mm in breadth. The distal end has minor erosion.

The Dolní Věstonice 15 right twelfth rib is virtually complete (Figure 15.12). It is missing perhaps 1 to 2 mm of its distal end. The rib is characterized by a superoinferiorly tall and externally to internally narrow head, with cranial and caudal proximal projections. In proximal view, the head is ovoid, with a slight depression along its lateral border. There is a subtle costal groove proximal to midshaft. Also near midshaft, the internal to external dimension of the shaft reduces. Along the cranial surface of the rib, beginning 15.2 mm distal to the head, there is a flattened area about 4 mm in breadth. Distal to this area, the cranial surface is a single, thin convexity (as is the entire caudal margin). The internal surface of the distal rib exhibits a shallow depression that was 7.7 mm in height and at least 23.5 mm in length.

Sternebrae

For Dolní Věstonice 13, a complete manubrium and mesosternum are preserved (Figure 15.13). The manubrium is relatively robust. It is wide, with a prominent jugular notch and large clavicular notches. There is no evidence of osteoarthritis. The manubrium evinces some areas of fill, including along the posterior margin of the jugular notch. There are also broad areas of fill on both its anterior and posterior surfaces. Cortical bone is missing from the right posterior portion of its base. Cortical bone loss is also evident inferior to both the right and left first costal notches. There is relatively heavy damage cranial to the first costal notch on the right, where a superoinferiorly long (ca. 23.5 mm) area just inferior to the right clavicular notch (corresponding to the site of the synchondrosis between the right first rib and the manubrium) has been abraded, so that trabecular bone is visible. The manubrium is thick and robust, becoming dorsoventrally thicker toward its base. There is slight asymmetry evident in the manubrium. First, the left clavicular notch appears to be in a more anterior position than does the right (i.e., it appears to project more from the anterior surface than

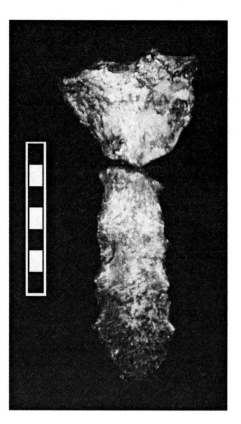

Figure 15.13 Anterior view, sternum, Dolní Věstonice 13. Scale in centimeters.

does the right). Second, when the manubrium is articulated with the mesosternum, the two bones appear to articulate asymmetrically, with more of the articular surface present on the right. What results is a small right costal notch and a large left costal notch for the right and left second ribs, respectively. This apparent asymmetry is somewhat inexplicable, as there is little evidence of asymmetry in the second ribs, and it may be that more bone is missing on the inferior left margin of the mesosternum than there appears to be (putty is evident here). Vlček (1991) maintained that a significant portion of bone was missing from the inferior manubrium. However, cortical bone is visible on the left side of the inferior manubrium, which seems to indicate that the asymmetry is not an artifact of differential preservation (and see below).

The mesosternum of Dolní Věstonice 13 is also beautifully preserved, and it is missing only a small amount of cortical bone superoposteriorly (primarily on the right side) and also anterolaterally on the right, between the third and fourth costal notches. It is relatively long and wide and slightly asymmetrical—the left costal cartilages articulate in a slightly more inferior position than do those on the right. Regarding the potential asymmetry of the second costal notches mentioned above, the cranial margin of the mesosternum itself is shaped such that the second sternal notch is larger on the left than on the right—again, consistent with asymmetry, and not with its being a preservation artifact.

Dolní Věstonice 15 preserves its second sternebral segment, that is, the cranial portion of the mesosternum (Figure 15.14). The piece is dorsoventrally thick (9.5 mm at midline; 11.3 mm at left lateral margin). Its cranial margin, or sternal angle, is badly damaged, and only a small area of the articular surface for articulation with the costal cartilage of the left second rib is preserved. Caudally on the left, there is a large facet for articulation with the costal cartilage of the third rib (height: 9.0 mm; breadth: 7.3 mm). The cranial surface is transversely narrower than the caudal surface, primarily because of postmortem damage, and the ventral surface preserves more cortical bone than does the dorsal surface, which is missing a large right craniolateral section of its cortical bone. The segment is slightly longer than the second sternebral segment of Dolní Věstonice 13 (30.8 mm vs. 27.8 mm).

It should be noted that a piece of what appears to be a mesosternal segment is preserved with the Pavlov 1 remains and has been described as such (Vlček, 1997). However, its overall dimensions would make it an exceptionally large human sternal segment, especially in relation to those preserved for Dolní Věstonice 13 and 15. It is therefore unlikely to derive from the Pavlov 1 remains and undoubtedly became mixed with them during the postdepositional disturbance of the remains (see Chapter 4).

Osteometric Analyses: The Costae

Ribs remain relatively understudied in human paleontology since so few of them survive the taphonomic processes

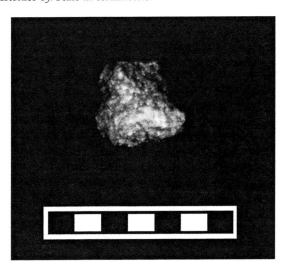

Figure 15.14 Anterior view, second sternebral segment, Dolní Věstonice 15. Scale in centimeters.

responsible for the creation of the fossil record. Even when ribs do survive, if the rib cage is incomplete, questions arise as to which rib fragment represents which element, confounding the ability to compare the rib(s). Added to this problem is the fact that ribs are relatively complex structures because of their three-dimensional anatomy. For this reason there has been little agreement on how best to quantify this anatomy (cf. Arensburg, 1991, vs. Jellema et al., 1993, vs. Franciscus & Churchill, 2002).

Materials and Methods

The following analyses follow the terminology and measurement definitions of Franciscus and Churchill (2002). This facilitates a ready comparison of the Dolní Věstonice sample to Franciscus and Churchill's recent human comparative sample (Euro-American males; $N = 20$), as well as to the fossil humans for which they present data, including the European Late Upper Paleolithic-associated Obercassel 1 and 2, the Gravettian Cro-Magnon 4320, and the western Asian Shanidar 3 and Kebara 2 Neandertals. The data collected on the Dolní Věstonice ribs as well as descriptions of these measurements, are in Sládek et al. (2000). Most of the following analyses include only data from the triple burial individuals, that is, Dolní Věstonice 13–15. No data are reported for the isolated right rib of the Pavlov 1 specimen, since this fragment could not be assigned a number, nor were data presented on Dolní Věstonice 3, whose preserved ribs (with the exception of a few isolated pieces) remain crushed in matrix. Only the left eleventh rib of Dolní Věstonice 16 is included in the analyses because it was the only rib of that individual that could be assigned to a number with confidence.

Franciscus and Churchill (2002) used a vertebral proxy for thoracic height to which they scaled some rib measurements. Since the primary focus of their analysis was Shanidar 3, they used the summed dorsal vertebral body heights for T11 through L5 because these elements are preserved for that specimen (Trinkaus, 1983b). In order to compare Dolní Věstonice 13 to their sample, it was necessary to predict the specimen's T11 to L5 summed dorsal height since Dolní Věstonice 13 does not preserve the dorsal heights of all seven elements. It does, however, preserve the ventral heights of T11 through L2 and the dorsal heights of L3 to L5, the sum total of which is 174.7 mm. Therefore, for a recent human sample that preserves all ventral and dorsal body height measurements (combined Euro-Americans and Amerindians; $N = 54$), a formula was computed to predict the summed dorsal body heights from the combination of ventral and dorsal elements preserved in Dolní Věstonice 13. The resulting predictive equation,

$$\hat{Y} = 0.9442 \,(174.7 \text{ mm}) + 18.7545 \;(r = 0.95)$$

produced a T11 to L5 estimate of 183.7 mm for Dolní Věstonice 13, with an individual standard error of prediction of 4.6 mm, about 2.5% of the predicted measure. Thus, the predicted T11 to L5 measurement was deemed acceptably accurate.

Finally, the quantitative methods used to compare the above measurements and indices of the Dolní Věstonice to recent humans and other Late Pleistocene fossils include standard univariate and bivariate statistical methods such as t tests and percentage deviations: (observed measure − predicted measure)/observed measure (R. J. Smith, 1980). As was the case for the vertebral analyses (Chapter 14), here this formula reads (fossil value − recent human mean value)/fossil value (following Franciscus & Churchill, 2002).

Rib Measurements

Table 15.1 presents the percentage deviations of the Dolní Věstonice 13 ribs from the recent Euro-American male sample for all rib measurements and indices. Dolní Věstonice 13's deviations range from 0% for third rib articular tubercle height to +38.7% for eleventh rib midshaft minimum/maximum diameter index. A modified version of the t test for comparing a single individual to a sample (Sokal & Rohlf, 1981) indicates that only five of the ninety-six comparisons are statistically significant, all of them in a positive direction from the recent human sample. These include the minimum/maximum shaft diameter index at midshaft for ribs 2 and 11, midshaft maximum diameter for rib 5, midshaft minimum diameter for rib 9, and minimum/maximum shaft diameter at midshaft for rib 11.

Table 15.2 presents the percentage deviations from the recent Euro-American male means for those rib measurements preserved on Dolní Věstonice 14. This individual has deviations that range from 0% for the minimum shaft diameter at the angle of rib 7 and the midshaft maximum diameter of rib 11 to the midshaft maximum diameter of rib 12, which is 48.2% smaller than expected. The t tests indicate that only eight of the seventy-three comparisons (or ca. 11%) are statistically significantly different. These are the tubercle-iliocostal line distance and posterior angle chord for ribs 8 and 9, rib 9's minimum shaft diameter at the angle and at midshaft, and the minimum/maximum shaft index at midshaft for ribs 9 and 12. Of these, only two are positive deviations (angle minimum diameter of rib 9 and midshaft index of rib 12); the remaining measurements are smaller than expected.

The percentage deviations of Dolní Věstonice 15's rib measurements from the recent Euro-American male means (Table 15.3) range in magnitude from 0% for the eighth rib's minimum shaft diameter at the angle to 57.1% smaller than the recent males for the posterior angle chord of rib 4. As with the other specimens, however, only a minority of these comparisons (fifteen of eighty, or ca. 19%) are statistically significantly different, and of these significant differences, all but four are in the negative direction (i.e., the measurement for Dolní Věstonice 15 is smaller than expected). This is a higher percentage of significant differences than was evident for the other members of the triple burial, and one for which two non-mutually exclusive hypotheses are offered. The first possibility is that the specimen's small size is responsible for

Table 15.1 Percentage deviations[a] of Dolní Věstonice 13's rib measurements and indices from recent Euro-American male means.

Measurement	Rib 2	Rib 3	Rib 4	Rib 5	Rib 6	Rib 7	Rib 8	Rib 9	Rib 10	Rib 11
Tubercle/iliocostal Line Distance		−2.6	−8.3	−12.1	−5.6	−4.9	−4.3	−1.6	+0.9	
Posterior angle Chord		−3.5	−9.2	−9.6	−6.1	−6.1	−6.1	−3.3	−2.5	
Maximum shaft Diameter at angle	−1.6	+6.8	+4.2	−11.4	−3.0	+3.9	+4.9	+12.9	−7.6	−16.2
Minimum shaft Diameter at angle	+8.8	+3.7	+8.7	+9.1	+9.8	+13.6	+11.3	+20.2	+20.5	+23.1
Articular tubercle Height		0	+10.7	+4.4	+6.3	+15.5	+13.8	+20.0	−21.3	
Articular tubercle Breadth		+18.3	+12.8	−15.5	+8.1	+7.9	+8.2	−7.7	−55.6	
Midshaft maximum Diameter	−9.7	+7.1	−3.9	+23.2*	+15.6	+21.1	+16.1	+3.6	−1.9	−18.7
Midshaft minimum Diameter	+30.1	+17.1	+13.8	+7.3	+11.4	+9.9	+17.1	+14.3*	+22.1	+28.8
Neck length						−14.4				
Neck S-I diameter						−14.5				
Tuberculo/ventral Chord						−4.5				
Tuberculo/ventral Subtense						+14.2				
Shaft index at angle	+9.4	−4.5	+1.7	+17.6	+11.5	+8.8	+4.5	+6.3	+24.9	+33.0†
Shaft index at midshaft	+36.5†	+8.0	+16.0	−21.1	−6.1	−16.5	−1.1	+10.2	+23.0	+38.7*

*Significantly different at P < 0.05; †significantly different at P < 0.01; significance was determined by using the modified Student's t test (Sokal & Rohlf, 1981: 231).

[a]Percentage deviation is calculated as follows: (value for Dolní Věstonice 13 − Euro-American male mean)/value for Dolní Věstonice 13] × 100 (following Franciscus & Churchill, 2002).

Table 15.2 Percentage deviations[a] of Dolní Věstonice 14's rib measurements and indices from recent Euro-American male means.

Measurement	Rib 2	Rib 3	Rib 4	Rib 5	Rib 6	Rib 7	Rib 8	Rib 9	Rib 10	Rib 11	Rib 12
Tubercle-iliocostal Line distance				−4.9	−19.4	−21.2	−29.4*	−34.9*	−21.0	−10.0	
Posterior angle chord				−1.6	−16.3	−18.9	−27.6*	−38.6*		−7.2	
Maximum shaft Diameter at angle				−30.9	−7.8	0.0				−4.0	
Minimum shaft Diameter at angle				−8.4	−3.4	+6.3	+5.5	+18.6*	+17.6	+11.8	
Articular tubercle Height		−18.6	+4.3	−18.2	+6.3	+17.8	+3.4	+18.9			
Articular tubercle Breadth		+3.8	+18.0	+12.8	+2.5	+28.6	+21.2	+20.5			
Midshaft maximum Diameter	−1.3		+1.5	+3.7	−0.8	−6.6	+3.8	+5.4	+7.6	0.0	−48.2
Midshaft minimum Diameter	+19.4	−9.4	−6.2	−28.8	+1.3	+15.1	−6.8	−36.4*	−6.0	+14.8	+18.0
Shaft index at angle				+16.3	+3.1	+5.0				+14.1	
Shaft index at midshaft	+20.7		−9.0	−34.2	+1.0	+18.8	−13.6	−45.5*	−15.6	+13.0	+43.6*

*Significantly different at $P < 0.05$; significantly different at $P < 0.01$; significance was determined by using the modified Student's t test (Sokal & Rohlf, 1981: 231).
[a]Percentage deviation is calculated as follows: (value for Dolní Věstonice 14 − Euro-American male mean) / value for Dolní Věstonice 14) x 100 (following Franciscus & Churchill, 2002).

these differences. The second possibility is that the pathological condition of Dolní Věstonice 15 individual is playing a role in these differences. In order to investigate these and related issues, the next section presents a series of graphical representations of those data for which significant differences between the Dolní Věstonice specimens and recent humans exist.

Minimum/Maximum Shaft Diameter Indices

Univariate testing reveals that several of the significant differences between the Dolní Věstonice and recent humans were manifest in the form of shaft thickness indices, or the ratios between the minimum and maximum shaft breadths taken at both the angle and at midshaft. Figure 15.15 shows the minimum/maximum shaft diameter index profile (as taken at the angle) for recent human males (means only), Dolní Věstonice 13, and Obercassel 1 and 2. A high index value is indicative of a rib with a more circular cross section, and a low index value is reflective of a rib that has a more oblong, or ovoid, cross section (i.e., one with a high maximum dimension and/or a low minimum dimension). As shown on the graph, among recent humans, at the angle, ribs 3–5 tend to be the most circular ribs in cross section, and ribs 9–11 tend to be the flattest (rib 1 would also fall into this category but was not included in the analysis).

Dolní Věstonice 13 shows a similar profile to that of recent human males, albeit with steeper ascent and descent. Rib 5 shows a particularly marked index, although one that is not significantly different from the Euro-American males. Rib 10 also evinces a much higher index than expected, although it, too, is not significantly different from the recent human males. However, the index value for rib 11 is 33% larger than expected, and it is significantly rounder in cross section than the recent

Table 15.3 Percentage deviations[a] of Dolní Věstonice 15 rib measurements from recent Euro-American male means.

Measurement	Rib 2	Rib 3	Rib 4	Rib 5	Rib 6	Rib 7	Rib 8	Rib 9	Rib 10	Rib 11	Rib 12
Tubercle/iliocostal Line distance			−56.6*	−47.8*	−37.5*	−33.4	−41.2†	−22.7	−28.4	−37.4	
Posterior angle chord		−56.7*	−57.1*	−48.1*	−41.6†	−34.8	−44.0†	−27.0	−32.3	−38.9*	
Maximum shaft diameter at angle	+17.0*	+11.3	+10.2	+6.6	−0.7	+9.7	+1.9		−5.4	+3.0	+9.0
Minimum shaft diameter at angle	+13.1	−9.9	−10.5	−11.1	−10.8	−2.3	0.0	+2.5	+4.1	+11.8	+12.7
Articular tubercle height		−6.1	−11.7	−3.2	+6.3	+31.8*	+5.1	+22.1	+24.0*		
Articular tubercle breadth	+6.2	−13.9	+11.8	+28.8	+35.8	+14.1*	+18.6	+12.5			
Midshaft maximum diameter						−5.8		+3.6	+0.6	−2.4	−6.8
Midshaft minimum diameter						+2.7	−46.5*	−13.2	−15.2	+5.5	+14.6
Shaft index at angle	−5.6	−25.2	−27.0	−20.2	−11.3	−14.9	−4.4		+7.6	+7.9	+3.2
Shaft index at midshaft						+6.2		−18.6	−16.8	+5.8	+18.4

*Significantly different at $P < 0.05$; †significantly different at $P < 0.01$; significance was determined by using the modified Student's t test (Sokal & Rohlf, 1981:231).
[a]Percentage deviation is calculated as follows: (value for Dolní Věstonice 15 − Euro-American male mean) / value for Dolní Věstonice 15) x 100 (following Franciscus & Churchill, 2002).

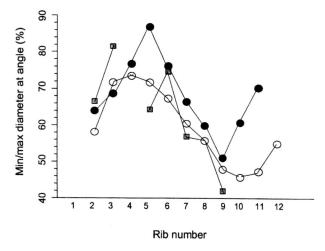

Figure 15.15 Profiles of shaft shape indices (at the angle) for Dolní Věstonice 13 (black circles), Upper Paleolithic (Obercassel 1 and 2; gray squares), and recent Euro-American males (open circles). A high index value is indicative of a rib with a more circular cross section.

males at $P < 0.01$. Also note that Dolní Věstonice 13 has higher indices than the later Obercassel specimens, except for the second and third ribs.

Figure 15.16 shows the same recent human male and Obercassel profiles with the profiles for Dolní Věstonice 14 and 15 and the single eleventh rib value for Dolní Věstonice 16 superimposed. Dolní Věstonice 14 evinces a higher than expected index for ribs 2 and 5, although neither of these is statistically significantly different, according to the t tests. Dolní Věstonice 14's index values for the second, sixth, seventh, and eleventh ribs are quite

close to expectation, as is Dolní Věstonice 16's eleventh rib. What is somewhat surprising are the index values for Dolní Věstonice 15, who at the angle has consistently narrower rib cross sections—except for the caudal four ribs that approximate the recent human male pattern.

Figure 15.17 presents the minimum/maximum shaft diameter index profiles taken at midshaft for Euro-American males (means only), Dolní Věstonice 13 and 16, European Neandertals (La Chapelle-aux-Saints 1, La Ferrassie 1, and Neandertal 1) and Gravettian humans (Cro-Magnon 4320). The sixth ribs of the Euro-American males have the highest indices (i.e., are rounder in cross section) at midshaft; the index decreases from the sixth rib both cranially and caudally, albeit with a slight rise for rib 11. The profile of Cro-Magnon 4320 shows a very similar pattern, although its index values for ribs 6 and 7 are considerably higher than those of the Euro-Americans. The Neandertals show a slightly more caudal peak at rib 7. Dolní Věstonice 16 evinces an eleventh rib midshaft minimum/maximum index that is almost identical to that of the Euro-Americans. Dolní Věstonice 13 shows an unusual pattern, with the second and eleventh ribs much more circular in cross section than expected. Recall that for both of these indices, Dolní Věstonice 13 was significantly different from the Euro-American males.

Dolní Věstonice 14 and 15 are superimposed on the Euro-American male mean, Cro-Magnon, and Neandertal profiles in Figure 15.18. At midshaft, Dolní Věstonice 15 closely resembles recent humans for the minimum/maximum shaft diameter index. Dolní Věstonice 14 has a profile similar to that of the Euro-American males, with two key exceptions: it has a much flatter ninth and much

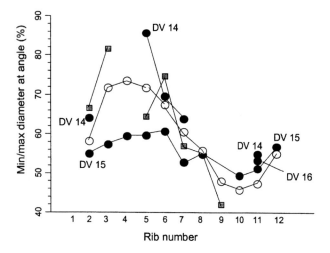

Figure 15.16 Profiles of shaft shape indices (at the angle) for Dolní Věstonice 14, 15, and 16 (black circles), Upper Paleolithic (Obercassel 1 and 2; gray squares), and recent Euro-American males (open circles). A high index value is indicative of a rib with a more circular cross section.

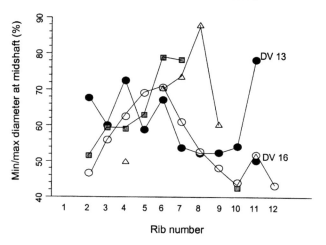

Figure 15.17 Profiles of shaft shape indices (at midshaft) for Dolní Věstonice 13 and 16 (black circles), European Neandertals (La Chapelle-aux-Saints 1, La Ferrassie 1, and Neandertal 1; triangles), Gravettian (Cro-Magnon 4320; gray squares), and recent Euro-American males (open circles). A high index value is indicative of a rib with a more circular cross section.

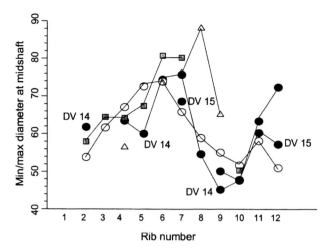 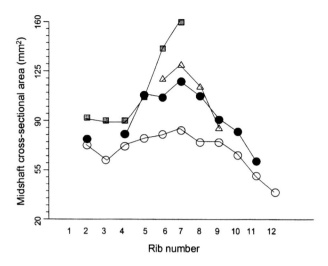

Figure 15.18 Profiles of shaft shape indices (at midshaft) for Dolní Věstonice 14 and 15 (black circles), European Neandertals (La Chapelle-aux-Saints 1, La Ferrassie 1, and Neandertal 1; triangles), Gravettian (Cro-Magnon 4320; gray squares), and recent Euro-American males (open circles). A high index value is indicative of a rib with a more circular cross section.

Figure 15.19 Midshaft cross-sectional area (mm²) profiles for Dolní Věstonice 13 (black circles), European Neandertals (La Chapelle-aux-Saints 1, La Ferrassie 1, and Neandertal 1; triangles), Gravettian (Cro-Magnon 4320; gray squares), and recent Euro-American males (open circles).

rounder twelfth rib than expected. Recall that t tests distinguished this individual from the recent humans for both of these indices at $P < 0.05$.

Midshaft Cross-Sectional Area

As is evident above, many of the differences between the Dolní Věstonice sample and recent humans are differences in shaft thickness indices, but some shaft thickness measurements themselves also evinced significant differences. It may therefore be informative to investigate these shaft dimensions further by using them to compute rib shaft cross-sectional areas. To accomplish this, the cross-sectional area at midshaft was predicted from maximum and minimum shaft diameters by using the ellipse formula. Figure 15.19 shows the resulting profiles for the midshaft cross-sectional area for Euro-American males (means only) Dolní Věstonice 13, European Neandertals (La Chapelle-aux-Saints 1, La Ferrassie 1, and Neandertal 1), and Cro-Magnon 4320. For all of the ribs, the Late Pleistocene humans have greater cross-sectional areas than the recent human males. Also note that Cro-Magnon 4320's sixth and seventh rib values exceed those of the Neandertals. The rib cross-sectional area peaks at the seventh rib for all groups, with steeper profiles evident for the Pleistocene than for recent humans. Dolní Věstonice 13 has a similar profile to the Neandertals and Cro-Magnon 4320, although the last specimen far exceeds Dolní Věstonice 13 in the cross-sectional area of the sixth and seventh ribs.

Figure 15.20 presents the same cross-sectional area profiles as before, except that Dolní Věstonice 14 (black circles) is now superimposed on them. This specimen more closely approximates the recent human pattern than does Dolní Věstonice 13. It only exceeds the recent human mean for the seventh rib, which as expected evinces the greatest cross-sectional area. The cross-sectional areas of ribs 5 and 9 are somewhat smaller than expected.

The midshaft cross-sectional area values for Dolní Věstonice 15 and 16 are superimposed over the recent humans, Neandertals, and Cro-Magnon 4320 in Figure 15.21. Dolní Věstonice 15's profile is almost identical to

Figure 15.20 Midshaft cross-sectional area (mm²) profiles for Dolní Věstonice 14 (black circles), European Neandertals (La Chapelle-aux-Saints 1, La Ferrassie 1, and Neandertal 1; triangles), Gravettian (Cro-Magnon 4320; gray squares), and recent Euro-American males (open circles).

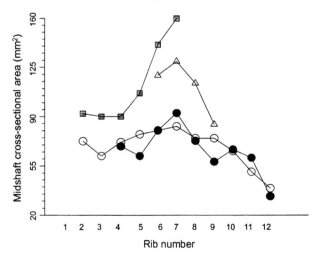

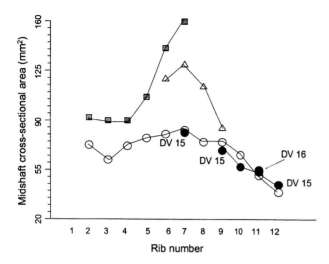

Figure 15.21 Midshaft cross-sectional area (mm²) profiles for Dolní Věstonice 15 and 16 (black circles), European Neandertals (La Chapelle-aux-Saints 1, La Ferrassie 1, and Neandertal 1; triangles), Gravettian (Cro-Magnon 4320; gray squares), and recent Euro-American males (open circles).

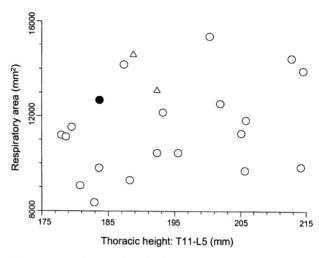

Figure 15.22 Scatter plot of eighth rib tuberculoventral chord (mm) versus summed dorsal vertebral body heights (mm) for T11 to L5 in Dolní Věstonice 13 (solid circle), Neandertals (Kebara 2 and Shanidar 3; triangles), and recent Euro-American males (open circles). The relationship among the recent Europeans is not significant at the 0.05 level.

that of the recent human males. However, given the small size of this individual (Chapter 4), perhaps this finding is indicative of ribs that are actually larger in cross-sectional area than expected for such a diminutive male. The eleventh rib of Dolní Věstonice 16 also falls right on the Euro-American male mean for midshaft cross-sectional area.

Evaluation of Respiratory Area

Franciscus and Churchill (2002) were able to reconstruct the respiratory area of the Shanidar 3 Neandertal's eighth rib by using two measurements that in concert can be used to capture the area within the rib's primary curvature: the tuberculoventral chord and subtense. These measurements are available for the eighth rib of Dolní Věstonice 13, facilitating its comparison to Shanidar 3, Kebara 2, and the Euro-American comparative sample, in order to evaluate the specimen's thoracic shape and volume.

The tuberculoventral chord is defined as the straight-line distance from the centroid of the rib's articular tubercle to the ventral margin of the sternal end (McCown & Keith, 1939). As a measurement of dorsoventral thoracic depth, it has a distinct advantage over the straight-line distance between the head and sternal end of the rib in that it more accurately captures the rib's anteroposterior dimension since the posterior curvature so characteristic of human ribs brings the tubercle to a position well posterior to the head.

Figure 15.22 is a bivariate plot of eighth rib tuberculoventral chord on the thoracic height proxy discussed earlier (i.e., summed dorsal vertebral heights of T11 through L5) for the Euro-American males (open circles), Kebara 2 and Shanidar 3 Neandertals (triangles), and Dolní Věstonice 13 (black circle). Note that the relationship between the X and Y variables is not significant for the Euro-American sample and that although the Neandertals evince relatively long tuberculoventral chords (reflective of their dorsoventrally deep chests), they are exceeded by some Euro-American males in this measure. As a result, the two Neandertals fall within the recent human scatter. Dolní Věstonice 13, in contrast, has a relatively short tuberculoventral chord, indicative of an anteroposteriorly shallow chest, but he, too, lies well within the Euro-American male scatter.

The tuberculoventral subtense is measured as the straight-line, perpendicular distance from the tuberculoventral chord to the lateral margin of the rib (McCown & Keith, 1939). As such it is a measure of transverse thoracic breadth. Figure 15.23 is a scatter plot of tuberculoventral subtense on the summed dorsal body heights of T11 to L5. As for the tuberculoventral chord, the relationship between these variables in the Euro-American sample is not significant. Unlike his tuberculoventral chord measurement, Dolní Věstonice 13 evinces a relatively large tuberculoventral subtense, especially relative to thoracic height. This suggests that Dolní Věstonice 13 is characterized by a somewhat broader thorax than Euro-American males on average. However, this specimen and the two western Asian Neandertals (Kebara 2 upper and Shanidar 3 lower) fall within the Euro-American male scatter, albeit at the high end (i.e., relatively wide thoraces) of the recent male range.

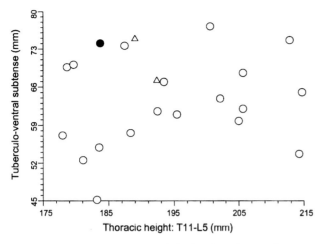

Figure 15.23 Scatter plot of eighth rib tuberculoventral subtense (mm) versus summed dorsal vertebral body heights (mm) for T11 to L5 in Dolní Věstonice 13 (solid circle), Neandertals (Kebara 2 and Shanidar 3; triangles), and recent Euro-American males (open circles). The relationship among the recent Europeans is not significant at the 0.05 level.

Using the dorsoventral rib measurement (tuberculoventral chord) and the mediolateral rib measurement (tuberculoventral subtense) as the height and width of half an ellipse, respectively, one can estimate the area within the primary curvature of the rib, referred to as respiratory area by Franciscus and Churchill (2002). Figure 15.24 is a bivariate plot of the respiratory area of rib 8 on thoracic height for Dolní Věstonice 13, Kebara 2 upper, and Shanidar 3 lower and the Euro-American males. As before, the relationship between these two

Figure 15.24 Scatter plot of eighth rib respiratory area (mm²) versus summed dorsal vertebral body heights (mm) for T11 to L5 in Dolní Věstonice 13 (solid circle), Neandertals (Kebara 2 and Shanidar 3; triangles), and recent Euro-American males (open circles). The relationship among the recent Europeans is not significant at the 0.05 level.

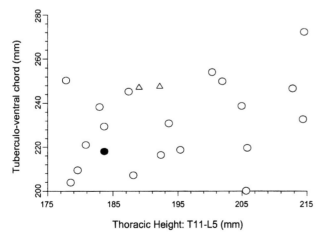

variables is not significant among the recent Europeans. Dolní Věstonice 13 falls at the high end of the respiratory area for Euro-American males (but well within their range). Likewise, Shanidar 3 and especially Kebara 2 have large respiratory areas, particularly relative to thoracic height. However, one Euro-American male exceeds both their values for respiratory area (and three more exceed Shanidar 3 alone), and the two Neandertals appear to fall within the recent human scatter.

Osteometric Analyses: Sternebrae

Sternal elements are even rarer than ribs in the human fossil record, but like ribs they may provide significant clues into thoracic shape as related to either ecogeographical patterning (Bergmann's Rule) or respiratory physiology (high-altitude adaptation or increased vital capacity due to activity levels). As recently discussed by Churchill (1998) and Franciscus and Churchill (2002), it has long been argued that Neandertals were characterized by voluminous, barrel-shaped chests. The data presented by Franciscus and Churchill for Shanidar 3 and Kebara 2 (and discussed above) suggest that these deep chests are more likely because of cold adaptation (Bergmann's Rule) than elevated activity levels and high vital capacity. However, both Shanidar 3 and Kebara 2 are western Asian Neandertals, and thus they may not show as high a degree of thoracic expansion as would European Neandertals.

Franciscus (1989) also posited that if Neandertals did, in fact, have such large thoracic cavities, then this should be reflected in the size of their sternebrae, and in particular in the length of their mesosterna. Indeed, he found that the one European Neandertal (Regourdou 1) appeared to have a longer mesosternum than recent humans or western Asian Neandertals. Could, then, either Dolní Věstonice 13 or 15 be characterized by a Neandertal-like mesosternum? In order to answer this question, we now examine data presented in Franciscus (1989). As he notes, sternebral segments of Neandertals are rare, but at least four individuals (Kebara 2, Regourdou 1, Shanidar 4 or 6, and Tabun 1) have partially preserved mesosterna. Kebara 2 has a perfectly preserved second sternebral segment (i.e., the cranial segment of the mesosternum); the other segments of the mesosternum are present but badly crushed. Regourdou 1 has complete second and third sternebral segments. Shanidar 4 or 6 (the association of the sternum to the individual is uncertain) preserves its second sternebral segment, and Tabun 1 preserves portions of its second, third, and fifth sternebral segments preserved (Franciscus, 1989).

Franciscus (1989) used a combined sample of Amerindian and Euro-American mesosterna ($N = 31$ for

Table 15.4 Mesosternal measurements (in mm) for Dolní Věstonice 13 compared to recent humans (mean ± standard deviation) and Neandertals (after Franciscus, 1989).

	Second Sternebral Length	Third Sternebral Length	Second + Third Sternebral Length	Total Mesosternal Length
Dolní Věstonice 13	27.8	27.2	55.0	100.1
Dolní Věstonice 15	30.8	—	—	—
Kebara 2	32.1	—	—	(105.0)
Regourdou 1	38.0	32.0	70.0	(117.0)
Shanidar 4 or 6	32.9	—	—	—
Tabun 1	32.0	29.0	61.0	(103.3)
Recent Euro-Americans	32.2 ± 3.7	26.7 ± 3.9	58.9 ± 6.9	100.9 ± 12.5
Recent Amerindians	32.4 ± 4.8	26.8 ± 4.0	59.2 ± 8.1	98.4 ± 14.8

Data for recent humans are from Franciscus (1989), as are the data for Regourdou and Tabun 1; data for Shanidar 4/6 are from Trinkaus (1983b); data for Dolní Věstonice 13 and Kebara 2 are measurements I have taken). Parentheses indicate estimated measurements.

Amerindians; N = 88 for Euro-Americans) to create a regression to predict total mesosternal length from partially preserved sternebrae. He found that whereas one could not accurately predict total mesosternal length when only the second sternebral segment is present, one can accurately predict total mesosternal length from the combined height of the second and third sternebral segments. Using the resulting regression formula, he was able to predict total mesosternal length for Regourdou 1 and Tabun 1. To his data I add my assessment of the mesosternal length of Kebara 2 of about 105 mm.

Table 15.4 presents these data, and an interesting pattern emerges. First, it appears that neither of the Dolní Věstonice specimens nor the western Asian Neandertals is characterized by particularly long mesosterna. All of these fossil humans have mesosternal segments that fall well within 1 standard deviation of the recent human means. However, Regourdou 1, the only European Neandertal to preserve its mesosternum, appears to have a longer one than any of the other fossil humans (Franciscus, 1989). Thus, it may be the case that only European Neandertals are characterized by an extremely barrel-shaped chest. The two Dolní Věstonice specimens are not Neandertal-like in this regard; their sterna fall well within the recent human range of variation.

Summary

There are few differences between the Dolní Věstonice sample and recent humans in costal or sternebral morphology. For some rib measurements, Dolní Věstonice 15 is smaller than recent human males, but this may be a reflection of size or a secondary result of the specimen's pathological condition. Only the left eleventh rib of Dolní Věstonice 16 is identifiable as to number, and it consistently falls on or near the Euro-American male mean. At midshaft, Dolní Věstonice 14 has a more circular cross section for its twelfth rib and a flatter cross section for its ninth rib than do recent humans. Dolní Věstonice 13 has a greater midshaft cross-sectional area than recent humans, and it is more similar to other Late Pleistocene specimens in this regard. Finally, in terms of respiratory area, Dolní Věstonice 13's apparently dorsoventrally narrow and mediolaterally wide chest combine to create a respiratory area that is at the high end of the Euro-American range.

16

The Upper Limb Remains

Erik Trinkaus

The human remains from Dolní Věstonice and Pavlov retain an exceptional number of well-preserved upper limb remains, from the claviculosternal articulation to the terminal hand phalanges (Sládek et al., 2000; Chapter 5). With only minor damage to some of the epiphyses and thinner areas of the scapulae, the bones to the distal radii and ulnae are essentially complete for both arms of Dolní Věstonice 13 and 15, and they preserve variable numbers of hand remains (Chapter 5; Table 16.5). The shoulder and arm bones are only slightly less complete for Dolní Věstonice 14, with damage mostly to the clavicular ends, the proximal humeri, and the distal forearms (which was exacerbated by the lack of fusion of the distal radial and ulnar epiphyses). The arm remains of Dolní Věstonice 16 and then Dolní Věstonice 3 have progressively greater damage to the epiphyses, but their arm bones, plus the shoulder remains of Dolní Věstonice 16, are sufficiently intact to permit reasonable estimation of their original interarticular lengths. The hand remains of Dolní Věstonice 13–15 are variably preserved and incomplete, but Dolní Věstonice 16 preserves an almost complete set of its right hand bones and a number of its left hand elements. There are also scattered elements of both hands of Dolní Věstonice 3. The Pavlov 1 arm bones are less complete, having been displaced postmortem (Chapter 4). However, by combining the right and left sides of Pavlov 1, it is possible to extract a major amount of information from its shoulder and arm remains; only overall clavicular and scapular dimensions remain largely unknown. These associated skeletons are joined by isolated arm and hand bones of Dolní Věstonice 34, 41, 50, and 53.

It is therefore possible to assess both the overall proportions and detailed aspects of the morphology of these Pavlovian upper limb remains. The relative lengths of the long bones are considered in Chapter 12, and they provide background for the considerations here.

Materials and Methods

The assessment of these Pavlovian upper limb remains involve both detailed morphological descriptions of the bones and morphometric comparisons of them to relevant samples of Late Pleistocene human remains. The principal comparative sample is of Gravettian human remains from Europe, which derive from the sites of Arene Candide, Baousso da Torre, Barma Grande, Caviglione, Cro-Magnon, Fanciulli (Grotte-des-Enfants), Fontéchevade, Paglicci, Pair-non-Pair, Pataud, Paviland, Předmostí, La Rochette, and Sunghir. Even though they are geographically, archeologically, and chronologically close to the Dolní Věstonice and Pavlov sample (Svoboda et al., 1996; Chapter 2), the Předmostí human remains are included as part of the greater Gravettian sample since the focus here is on the comparative framework of the remains from the Pavlovské Hills. To these Gravettian remains are added the slightly earlier Aurignacian remains from Mladeč. The more distant comparisons, for considerations of temporal trends through the Late Pleistocene, are to the Middle Paleolithic samples of late archaic and early modern humans. The former consists of the European and Near Eastern Neandertals and includes specimens from Amud, Bisitun, La Chapelle-aux-Saints, La Ferrassie, Kebara, Neandertal, La Quina, Regourdou, Shanidar, and Spy, plus the initial Upper Paleolithic (Châtelperronian) Saint-Césaire specimen. The latter are the Near Eastern remains from Qafzeh and Skhul. The earlier, terminal Middle to Initial Late Pleistocene remains from Krapina and Tabun are only referred to in the context of a more ancestral pattern. Data are from primary descriptions of the remains (e.g., Verneau, 1906; Matiegka, 1938; McCown & Keith, 1939; Endo & Kimura, 1970; Sergi et al., 1974; Vandermeersch, 1981, 1991; Heim, 1982b; Trinkaus, 1983b, 2000b; Formicola, 1990; Vandermeersch & Trinkaus, 1995; Churchill & Formicola, 1997; Mallegni et al., 1999; Kozlovskaya & Mednikova, 2000; Trinkaus et al., 2005), supplemented by personal analysis of many of the original specimens and with additional data from Churchill (1994), Holliday (1995), and J. T. Snyder (personal communication).

The lengths of the long bones serve both for overall proportions and for the size scaling of the upper limb remains. There has been some discussion about the appropriate manner in which to scale upper limb mechanical properties, namely, whether bone (or beam) length alone should be used or whether it should be combined with a body mass estimate (Ruff et al., 1993; Trinkaus & Churchill, 1999; Churchill & Smith, 2000; Ruff, 2000b). Although body mass does seem to have an effect in some comparisons, when mechanical equivalency is expected the human upper limb is not primarily weight bearing. In particular, although all clinically normal modern humans are capable of supporting and displacing their body masses effectively on their legs, many medically normal individuals are not capable of supporting their body weight on both of their arms in more than a passive suspensory position and are incapable of even supporting body weight in suspension on one arm (Trinkaus, personal experience). Consequently, the mechanical properties of the Pavlovian and other Late Pleistocene human arm bones will be scaled only to long bone lengths, as an approximation of mechanical beam length.

In addition, the primary use of the human upper limb is for burden carrying and, especially, the manipulation of objects and technology. Therefore, the human upper limb is used primarily as part of an open kinematic chain (Levangie & Norkin, 2001). Given this, the relevant load arms for muscles, against which their power arms should be scaled for considerations of mechanical advantages, are the lengths of the associated limb segments. These limb segment lengths are closely approximated by long bone lengths for most of the arm.

For many of the Pavlovian arm bones, length was simply measured on the well-preserved bones (especially for Dolní Věstonice 13 and 15), with minor estimation of damaged epiphyseal regions in some cases (especially the bones of Dolní Věstonice 14 and 16). The Dolní Věstonice 3 arm bones all have damaged epiphyseal regions, and their lengths were estimated principally through the use of least squares regressions based on recent human samples between the preserved (approximately intermetaphyseal) lengths and standard interarticular lengths (Trinkaus & Jelínek, 1997; Sládek et al., 2000). The Pavlov 1 right radius is largely complete, and it provides lengths for its less complete ulnae through least squares regressions based on recent human paired forearm bones. Its right humerus lacks the distal articulations, and its left humerus lacks the proximal epiphysis; by aligning the multiple, shared, preserved landmarks, it was possible to estimate a combined length for the two bones.

Several of these length estimations assume that asymmetries in long bone lengths will be small, as has been documented for recent human samples (less than ca. 2% of length; Trinkaus et al., 1994; Čuk et al., 2001), even though the right/left within-sample differences sometimes reach statistical significance (Sakaue, 1997; Čuk et al., 2001). The same appears to apply to most articular dimensions (Trinkaus et al., 1994; Sakaue, 1997). However, aspects of upper limb bones that respond plastically to the mechanical demands placed upon them, especially the diaphyses, frequently show marked asymmetry in recent and Late Pleistocene human remains (Trinkaus et al., 1994; Churchill & Formicola, 1997; Sakaue, 1997; Trinkaus & Churchill, 1999). Consequently, long bone lengths are assumed to be symmetrical unless otherwise documented for individuals; given the dearth of fossils that preserve both sides of these bones with intact lengths, this becomes a necessity in any case. The same is applied to some articulations. However, diaphyseal measures, including diameters, circumferences, cross-sectional parameters, and muscle attachment dimensions, are considered separately (with a few exceptions because of limited comparative sample sizes) for the right and left sides.

The majority of the morphometric comparisons employ standard osteometrics (Senut, 1981; Bräuer, 1988), and they are listed for the Dolní Věstonice and Pavlov human remains in Sládek et al. (2000), which also lists cross-sectional geometric measurements for the clavicles, humeri, radii, and ulnae. Of these, meaningful comparative samples are available only for the humerus (Churchill, 1994; Trinkaus & Churchill, 1999). Moreover, given the complex mechanical constraints on the clavicular diaphysis, with multiple oblique muscle attachments on it, plus its role as a strut between the glenohumeral joint and the sternum, the mechanics of the clavicular diaphysis are currently unclear. Similarly, although the forearm as a unit acts as a beam, with principally longitudinal muscle forces impinging upon it, the variable positions of the radius and ulna through pronation and supination—plus the mechanical effects of the interosseous membrane—mean that bending and torsional stresses acting on either bone cannot be assumed to be straightforward. For these reasons, even though cross-sectional measures are provided in Sládek et al. (2000), they will not be considered in this context.

The cross-sectional parameters are measured at standard percentages of interarticular (biomechanical) bone length, with 0% being distal (Ruff & Hayes, 1983a; Ruff et al., 1993; Trinkaus, 1997b). They were calculated from cross sections taken perpendicular to the diaphyseal axis. Each section was reconstructed from the external (subperiosteal) contour; transferred through the use of polysiloxane dental putty (Cuttersil Putty Plus, Heraeus Kulzer Inc.); and from the anterior, posterior, medial, and lateral cortical thicknesses, obtained from parallax-corrected measures of biplanar radiographs of the long bone diaphyses. The endosteal contour was then interpolated. The resultant cross sections were projected enlarged onto a Summagraphics III 1812 digitizing tablet, the contours were traced, and the cross-sectional parameters were calculated by using a PC DOS version (Eschman, 1992) of SLICE (Nagurka & Hayes, 1980). The resultant values, as tabulated in Sládek et al. (2000), are total subperiosteal and cortical areas (TA and CA), the anteroposterior and mediolateral second moments of area (I_x and I_y), the maximum and minimum second moments of area (I_{max} and I_{min}), and the polar moment of area ($J = I_{max} + I_{min}$; for further considerations, see Ruff, 2000a).

The various morphometric analyses are occasionally provided as ratios (or indices), but given the mathematical difficulties with ratios, the comparisons are most often presented graphically as bivariate plots. Because of bilateral asymmetry, these bivariate plots are usually paired.

The Clavicles

Substantial portions of both right and left clavicles remain for all six of the associated Pavlovian skeletons. However, they vary in completeness from sections of the proximal curvature for Dolní Věstonice 3 (Figure 16.1), to substantial portions of both clavicles with serious damage to the ends for Pavlov 1 (Figure 16.6), to largely complete clavicles but with erosion of the sternal and acromial ends for Dolní Věstonice 14 and 16 (Figures 16.3 and 16.5), and to essentially complete bones with trivial edge erosion for Dolní Věstonice 13 and 15 (Figures 16.2 and 16.4). Consequently, all except Dolní Věstonice 3 and Pavlov 1 provide clavicular length measures (or reasonable estimates thereof), and all except Dolní Věstonice 3 provide information on several aspects of the surface morphology.

The Sternal Ends

The Dolní Věstonice 13 proximal clavicle has a strong dorsoinferior beak. The right proximal surface is strongly undulating, with a ventrosuperior pit, whereas the left one is gently concave in a dorsoinferior to ventrosuperior

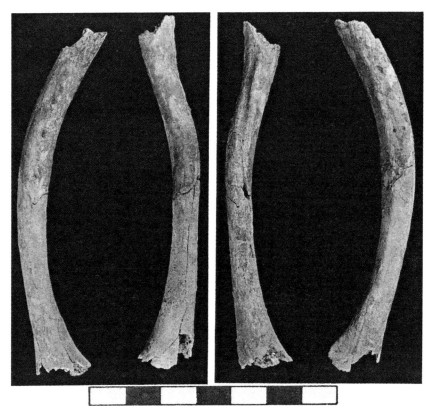

Figure 16.1 Superior/cranial (left) and inferior/caudal (right) views of the Dolní Věstonice 3 clavicles. Scale in centimeters.

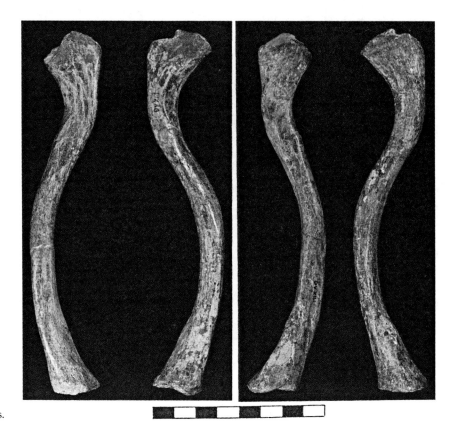

Figure 16.2 Superior/cranial (left) and inferior/caudal (right) views of the Dolní Věstonice 13 clavicles. Scale in centimeters.

direction. The costoclavicular facet is damaged on the right side, but what remains is an undulating rugose area 25 mm long, starting ca. 20 mm from the proximal end. The left one is a raised knob of bone, 17.7 mm long by 8.6 mm wide, located 16.3 mm from the proximal end, with a smooth surface. Neither costoclavicular facet presents a broad area for the attachment of the costoclavicular ligament.

The proximal end of the Dolní Věstonice 14 left clavicle exhibits a strong dorsoinferior beak; the proximal surfaces are both absent. The left costoclavicular facet is only a slightly roughened strip 18.6 mm long and 4.9 mm wide along the inferior crest, with no alteration of the bony contour. The damaged right one is similar but larger, being 27.5 mm long and 6.5 mm wide.

The sternal ends of the Dolní Věstonice 15 clavicles exhibit the flat, irregular, and pitted surfaces of the metaphyses for unfused sternal epiphyses (Chapter 6). These facets are associated with proximal dorsoinferior beaks, which extend primarily dorsally. The right costoclavicular facet is a large rugose area within the general contour of the bone, built up proximally and excavated distally, measuring 19.5 mm long and 12.8 mm wide. The left one is a smooth area barely separated from the surface, slightly rough, and excavated over a length of 19.8 mm and a breadth of 9.0 mm.

The proximal ends of the Dolní Věstonice 16 and Pavlov 1 clavicles are damaged, and there is no evidence of their sternal articular surfaces. Both of the Dolní Věstonice 16 clavicles and the left Pavlov 1 clavicle retain the distal ends of the costoclavicular facets, which are rugose areas moderately built up but still largely within the contours of the proximal shafts.

The Diaphyses

The clavicular diaphysis, following Llunggren (1979), consists of the portion from the sternal end to the conoid tubercle, since the distal end from the conoid tubercle to the acromial facet consists functionally of the claviculoscapular articulation. All of the clavicular diaphyses of the Dolní Věstonice and Pavlov fossils are evenly curved, show variable but minimal to moderate degrees of torsion, and have largely smooth surfaces. The Dolní Věstonice 13 ones are slightly angled along the dorsosuperior margin of the proximal third of the diaphysis, but they are otherwise smooth. The Dolní Věstonice 14 surfaces are evenly rounded, whereas those of Dolní Věstonice 15 are flat to slightly concave on the proximal two-thirds of the ventral surface along the pectoralis major attachment area, especially on the right side. Both of the Dolní Věstonice 16 clavicles are flat ventrally, with the right one being

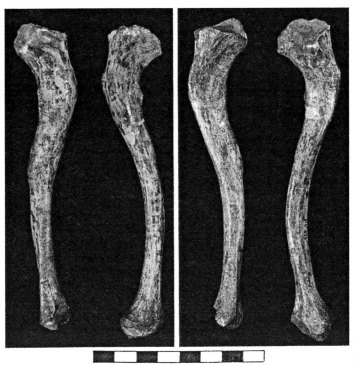

Figure 16.3 Superior/cranial (left) and inferior/caudal (right) views of the Dolní Věstonice 14 clavicles. Scale in centimeters.

rough along the proximal two-thirds of the ventral surface for pectoralis major. Similarly, the Pavlov 1 clavicles are flat to concave along their ventral surfaces. In addition, both Dolní Věstonice 16 and Pavlov 1 show more development of the angles along the dorsal edges of their diaphyses.

The Pavlovian clavicular diaphyses are variable in their curvatures, with Dolní Věstonice 14 having relatively straight clavicles and the others having more markedly curved ones. Comparative data are extremely limited, consisting entirely of Neandertals. Nonetheless, the proximal curvature indices (Olivier, 1951–1956) of the right

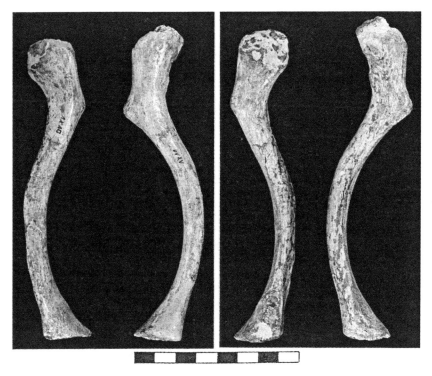

Figure 16.4 Superior/cranial (left) and inferior/caudal (right) views of the Dolní Věstonice 15 clavicles. Scale in centimeters.

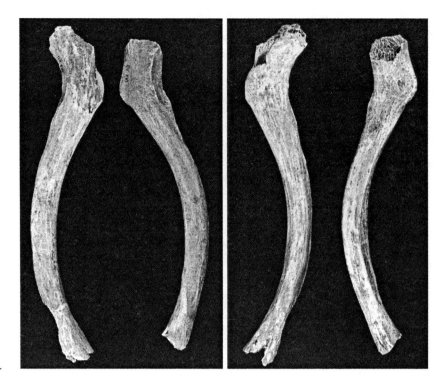

Figure 16.5 Superior/cranial (left) and inferior/caudal (right) views of the Dolní Věstonice 16 clavicles. Scale in centimeters.

clavicles are 15.9, 13.3, and 13.6 for Dolní Věstonice 13, 15, and 16, respectively, which are similar to those of three Neandertals that provide indices, Neandertal 1 (ca. 13.0), La Ferrassie 1 (12.3), and Regourdou 1 (15.0). The left clavicular ones are mostly lower, with three Neandertals (La Ferrassie 1, Kebara 2, and Regourdou 1) providing indices of 11.0, 7.8, and 12.7, respectively, and Dolní Věstonice 13 and 14 having indices of 14.3 and 8.0. However, the left clavicle of Dolní Věstonice 15 is quite curved, with an index of 17.6. The right clavicle of Dolní Věstonice 3 is more curved than its left one (Figure 16.1), even though they are not sufficiently preserved to apply standard metrics. The earlier (terminal Middle Pleistocene) right clavicle of Krapina 142 has an index of 16.8, but other

Figure 16.6 Superior/cranial (left) and inferior/caudal (right) views of the Pavlov 1 clavicles. Scale in centimeters.

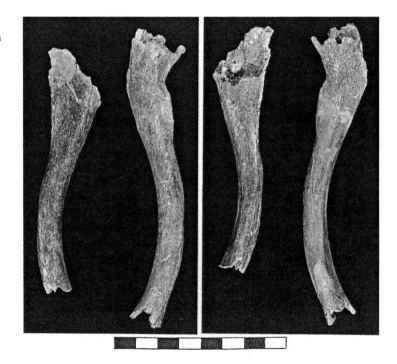

Krapina clavicles (especially Krapina 143 and 153) are considerably straighter. The biological significance of these variations in clavicular curvature remains obscure.

Similarly variable are linear metrics relating to the clavicular mid-diaphyses. Measures of robusticity, taken as the mid-circumference versus the maximum length (for which the most comparative data are available; Figure 16.7), provide right and left scatters in which there is little separation of the Late Pleistocene samples. The four Dolní Věstonice individuals providing both length and circumference, individuals 13 to 16, are all well within the scatters of the comparative values. Diameters providing an assessment of diaphyseal shape, the vertical (craniocaudal) diameter versus the sagittal (ventrodorsal) one (Figure 16.8), also indicate considerable scatter, with little separation of the samples. Most of the individuals have the dorsoventral diameter as the larger one, whereas the opposite holds for all of the Dolní Věstonice right clavicles, but not for Pavlov 1, and for most of the Dolní Věstonice left clavicles (all but Dolní Věstonice 15 and Pavlov 1).

The Acromial Ends

The largely complete distal ends of the Dolní Věstonice 13 clavicles present minimally developed conoid tubercles. In fact, the conoid tubercles are only a slight swelling of bone—rather than distinct tubercles—which can be located principally by the slight angles that they form. The trapezoid lines are small and faint but clear, and there is generally little distal rugosity of the bone for the trapezius attachment or the claviculoacromial articulation. There is a distinctly rugose concave area inside the distal curve for the attachment of deltoideus; it is more excavated on the right side but has more of a raised dorsosuperior edge on the left side. The flat and smooth acromial facet measures 12.9 mm by 8.2 mm.

The Dolní Věstonice 14 conoid tubercles form distinct beaks, especially on the left side, but they are neither rough nor turn inferiorly (Figure 16.3). The trapezoid lines are raised but minimally rugose. The deltoideus attachment areas are well demarcated bilaterally. The right one has an irregular low ridge marking its anterosuperior border, and the left one has a small spine of bone, 5.8 mm long and projecting 1.6 mm ventrally, within the attachment area. It appears to represent a modest enthesopathy, especially given the late adolescent age of Dolní Věstonice 14. The more distal area of the acromial end is moderately rugose, but the distal end proper is absent.

The conoid tubercles of the Dolní Věstonice 15 clavicles are markedly asymmetrical (Figure 16.4). The right one is a large rounded beak, which projects dorsally and slightly inferiorly, producing a slight concavity next to it on the inferior surface. The left one, however, is a distinct

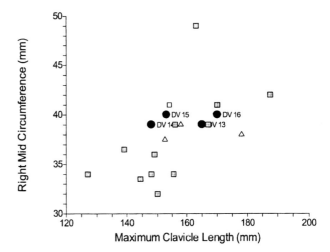

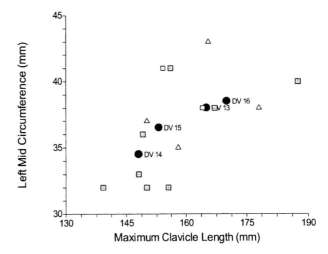

Figure 16.7 Bivariate plots of the right (above) and left (below) clavicular mid-circumference (50% of maximum length) versus maximum length. Solid circles: Dolní Věstonice specimens; gray squares: European Gravettian specimens; open squares: Qafzeh-Skhul specimens; open triangles: Neandertals.

protuberance, continuing the internal (dorsal) curve of the diaphysis, being a rounded right angle at the tubercle and then being separated from the acromial end by an open concavity. The left dorsoventral diameter across the conoid tubercle is 123.3% of the right one, whereas the comparable ratios are 80.0%, 95.7%, and 76.0% for Dolní Věstonice 13, Dolní Věstonice 14, and Pavlov 1, respectively. The level of asymmetry is in fact the same as that for Dolní Věstonice 13 and Pavlov 1, but in the opposite direction. Moreover, for the other individuals, the asymmetry is in the overall sizes of the acromial ends, whereas for Dolní Věstonice 15 it is principally in the relative degrees of projection of the conoid tubercles.

There is also some asymmetry in the Dolní Věstonice 15 trapezoid lines, such that the right one is barely no-

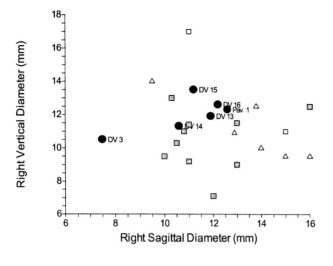

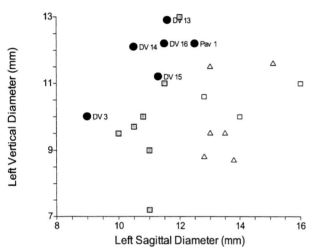

Figure 16.8 Bivariate plots of the right (above) and left (below) clavicular mid (50% of maximum length) superoinferior versus anteroposterior diameters. Solid circles: Dolní Věstonice and Pavlov specimens; gray squares: European Gravettian specimens; open squares: Qafzeh-Skhul specimens; open triangles: Neandertals.

ticeable, just a small round tubercle where it reaches the anterior edge. In contrast, the left one is a raised rugose area, which is small by the conoid tubercle and becomes progressively larger distally. Therefore, there appears to have been significant asymmetry in the hypertrophy of the conoid and trapezoid ligaments in the Dolní Věstonice 15 shoulders. The attachments for deltoideus within the distal curvature also vary. The right one is minimally rugose and largely flat, with none of the concavity seen in the other Dolní Věstonice clavicles. The left one is also not particularly concave, but it has a distinct bony spur within it, 7.7 mm long and 2.3 mm high.

The conoid tubercle is preserved only on the left clavicle of Dolní Věstonice 16. It is a large beak that goes directly dorsally, with a flat inferior surface adjacent to it. The right trapezoid line is indistinct, but the left one is a rugose line, albeit within the plane of the bone. There is a marked bilateral excavation of the distal ventral curvature for deltoideus, with breadths of 7.7 mm on the right and 7.5 mm on the left, smooth surfaces within them, and sharp superior edges.

Both of the conoid tubercles of Pavlov 1 are damaged, but they appear to have remained within the contours of the bones. The left trapezoid line is not distinct as a line but blends in with the generally roughened surface. For the deltoideus attachment area within the distal curve, the left clavicle exhibits a clear and large ventrally directed crest of bone ca. 22 mm long. On the right side there is only a small crest of bone. The concave areas are rugose on the right side and smooth on the left, being 7.4 mm and 6.6 mm wide, respectively.

The Scapulae

Given their fragile supraspinatus and infraspinatus surfaces, few Late Pleistocene scapulae are largely intact. However, all sides of the associated skeletons from Dolní Věstonice and Pavlov preserve significant portions of their scapulae. Those of the triple burial are the most complete (Figures 16.9 to 16.11), but major portions of the other three (albeit separated into pieces for Dolní Věstonice 3 and 16) remain (Figures 16.12 and 16.13). Given this, it is possible to make detailed comparisons of four of their scapular mediolateral dimensions and spinous shapes and of all six of their glenoid fossae and axillary borders.

Relative Scapular Breadth

The morphological lengths, or the distances from the middle of the glenoid fossa to the point at which the spine meets the vertebral border, are available for Dolní Věstonice 13 to 16. Comparative data are scarce for other Late Pleistocene scapulae (Figure 16.14), being available for only four other earlier Upper Paleolithic specimens, two Qafzeh-Skhul fossils, and four Neandertals. In the past (e.g., Trinkaus, 1983b, 1992), scapular length has been compared to humeral length, but it now appears unclear whether the resultant contrasts are influenced principally by scapular dimensions or humeral length. A more limited comparison, but one that should more directly reflect the relative hypertrophy of the scapula and its associated musculature (since the dimensions of the scapula directly affect the moment arms for muscles, such as trapezius and serratus anterior, which rotate it, and the absolute size of the bone is influenced by the degree of development of the rotator cuff muscles), is

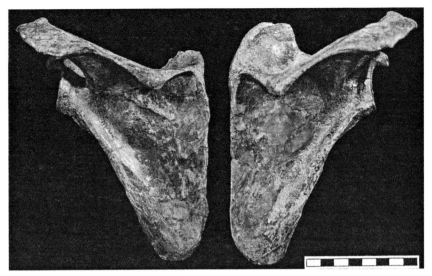

Figure 16.9 Dorsal view of the Dolní Věstonice 13 right and left scapulae. Scale in centimeters.

to compare the scapular length to clavicular length. The resultant distribution places the Neandertal specimens at the top of the distribution, further confirming that they had relatively broad scapulae (see Trinkaus, 1983b). The two Qafzeh-Skhul specimens are relatively low, despite their modest clavicular lengths. The few European earlier Upper Paleolithic specimens, including most of the Dolní Věstonice specimens, fall in between. The one relatively broad Pavlovian scapula is that of Dolní Věstonice 14, whose clavicles (and hence shoulders) were relatively abbreviated. The moderately, but not exceptionally, narrow scapulae are those of Dolní Věstonice 15.

There therefore appears to be a general pattern in which the Neandertals have the relatively broadest scapulae, the Qafzeh-Skhul sample has among the narrowest, and the European early modern humans are intermediate and somewhat variable in this feature.

The Scapular Spine

One of the distinctive features of the Dolní Věstonice scapulae, noted by Vlček (1991) and used by him to emphasize a close kin relationship among these individuals, is the shape of their scapular spines. This is most noticeable on the well-preserved scapulae of Dolní Věstonice 13 (Figure 16.15; see also Figure 16.9). In this feature, the scapular spine makes a straight line from the acromion to about three-quarters of the distance to the vertebral border. At that point, it makes a sharp angle craniomedially, to reach the vertebral border approximately where, craniocaudally, one would expect it to intersect the border.

Figure 16.10 Dorsal view of the Dolní Věstonice 14 right and left scapulae. Scale in centimeters.

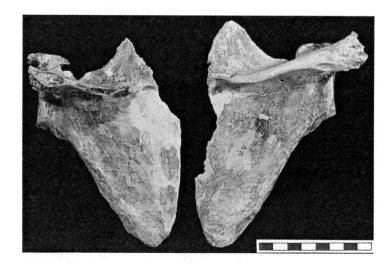

Figure 16.11 Dorsal view of the Dolní Věstonice 15 right and left scapulae. Scale in centimeters.

More specifically, on the Dolní Věstonice 13 scapulae, the base of the spine (where it meets the infraspinatus and supraspinatus surfaces) makes a direct line from the middle of the glenoid fossa to the vertebral border. The dorsal surface of the spine, however, deviates caudomedially from the spine base, at an angle of ca. 25° on the right and ca. 30° on the left. This results in a caudal projection of the more medial portion of the spine over the infraspinatus surface, for ca. 15 mm on the right and ca. 13 mm on the left. Then, in order to correct for the deviation, the medial end of the dorsal spine angles craniomedially for a distance of 48.5 mm on the right and 47.5 mm on the left to reach the vertebral border at the same location as the spine base. Consequently, what appears as the principal feature, the caudal lipping of the more medial spine, is in fact a secondary consequence of the more primary feature, namely, that the dorsal spine angles markedly craniolaterally to caudomedially relative to its ventral base.

The same pattern is evident on the scapulae of Dolní Věstonice 14 to 16 and probably on those of Pavlov 1 (the Dolní Věstonice 3 scapulae are too damaged in that region to provide information). On the Dolní Věstonice 14 right scapula (the left spine is too damaged medially), the same pattern is present but slightly less pronounced than with Dolní Věstonice 13 (Figure 16.10). The angle between the dorsal spine and the ventral base is ca. 15°, and the overhanging lip is 8–9 mm. On the Dolní Věstonice 15 scapulae, the left one is too damaged to determine the extent of the deviation, but the preserved portion indicates its presence (Figure 16.11). The right scapula has had the dorsal spine damaged and partially reconstructed, although the underlying bone is sufficient to indicate a dorsal spine angle of at least 10° to 15°, but the extent of the original overhang is unknown. On Dolní Věstonice 16, the medial half of the left spine is absent, but portions of the right one are preserved to the vertebral border (despite damage and bridging of the break for stabilization, it is clear from the ventral spine that the orientation is correct) (Figure 16.12). Using the relative orientations of the preserved ventral spine base and the lateral portion of the dorsal spine, an angle between them of ca. 20° can

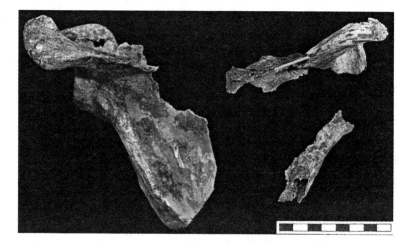

Figure 16.12 Dorsal view of the Dolní Věstonice 16 right and left scapulae. Scale in centimeters.

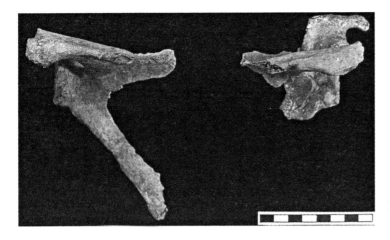

Figure 16.13 Dorsal view of the Pavlov 1 right and left scapulae. Scale in centimeters.

be estimated. The Pavlov 1 scapulae are too incomplete to even estimate the angle since neither one reaches close to the vertebral border (Figure 16.13), but the left one does show a clear angle between the lateral dorsal spine and the more medial ventral spine.

This pattern of the scapular spine, as suggested by Vlček (1991), does seem to unite the Gravettian human remains from the sites on the Pavlovské Hills. However, assessing whether it can be used as a morphological (and hence genetic) marker for these populations, since they may span as much as a millenium in time (Chapter 2), requires assessing whether this pattern is present elsewhere among European Gravettian human populations. Scapulae of three of the approximately contemporaneous Předmostí individuals, Předmostí 3, 4, and 10, are illustrated by Matiegka (1938); all of them show the strong caudal lipping of the medial dorsal spine, but only Předmostí 4, and to a lesser extent Předmostí 10, shows the strong angulation between the ventral and dorsal spine. The only other Gravettian scapula for which data are available, Pataud 5, also exhibits the caudal lipping of the margin, but it does not appear to have the strong dorsal versus ventral angulation. It therefore appears likely, but in need of further data for confirmation, that this pattern is a peculiarity of the Pavlovian populations of central Europe.

Figure 16.15 Craniodorsal view of the Dolní Věstonice 13 scapular spines. Scale in centimeters.

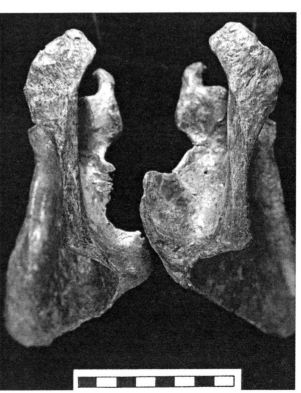

Figure 16.14 Bivariate plot of scapular breadth (morphological length) versus clavicular maximum length. Solid circles: Dolní Věstonice specimens; gray squares: European Gravettian specimens; open squares: Qafzeh-Skhul specimens; open triangles: Neandertals.

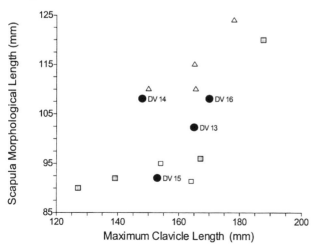

The Glenoid Fossa

All six of the associated skeletons preserve at least one essentially complete glenoid fossa, with the three from the triple burial preserving both sides intact (Figures 16.16 and 16.17). Since the proportions of the glenoid fossa have been a focus of Late Pleistocene human paleontological attention (Vallois, 1928–1946; Churchill & Trinkaus, 1990), these specimens therefore provide significant additional data.

The Dolní Věstonice 3 right glenoid fossa is smooth and evenly concave craniocaudally. Its dorsal margin is evenly convex, but its ventral margin has a distinct caudoventral bulge, which produces a modest notch in the middle of the ventral margin. The Dolní Věstonice 13 glenoid fossae are smooth, have no pits in them, and exhibit a slight ventral notch on the right side and a more pronounced one associated with a caudoventral bulge on the left side. The Dolní Věstonice 14 fossae are similarly smooth, and they are most clearly concave in their caudal portions. The left one has even and straight to convex borders, whereas the right one exhibits a distinct cranioventral notch, associated with a small caudoventral bulge. The glenoid fossae of Dolní Věstonice 15 are generally similar, but they exhibit largely symmetrical, small cranioventral notches on their borders. The right one has experienced localized osteoarthritis in the middle of the surface (Chapter 19), but the subchondral changes have not altered the contours or curvature of the fossa.

Both of the fossae of Dolní Věstonice 16 are present, but only the left one is intact (Figure 16.17). There is only a hint of a cranioventral notch on the left one, produced by a modest caudoventral bulge. Both are gently concave craniocaudally but only slightly dorsoventrally. There is

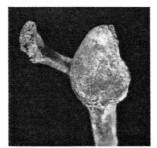

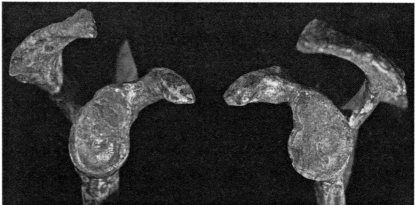

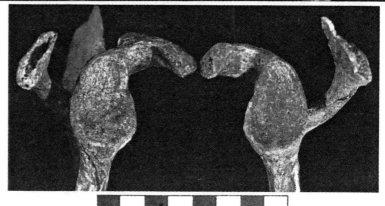

Figure 16.16 Lateral views of the Dolní Věstonice 3 right glenoid fossa (top), the Dolní Věstonice 13 right and left glenoid fossae (middle), and the Dolní Věstonice 14 right and left glenoid fossae (bottom). Scale in centimeters.

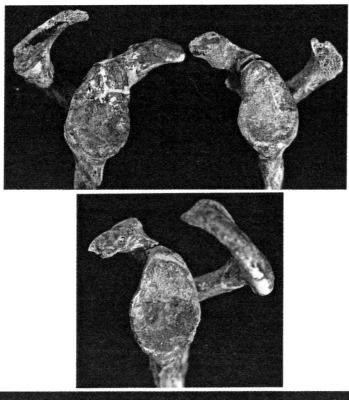

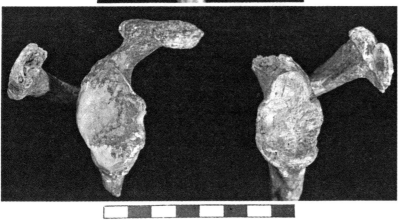

Figure 16.17 Lateral views of the Dolní Věstonice 15 right and left glenoid fossae (top), the Dolní Věstonice 16 left glenoid fossa (middle), and the Pavlov 1 right and left glenoid fossae (bottom). Scale in centimeters.

slight osteophytic lipping on the cranioventral left margin and the craniodorsal right margin, neither one sufficient to alter the contours of the fossae.

The Pavlov 1 glenoid fossae are both largely intact, but damage and restoration of the right dorsal margin means that only the left fossa breadth can be measured. Both of them have undulating and slightly irregular subchondral bone, which has been accentuated slightly by postmortem erosion. The ventral margin of the right fossa has a pair of protuberances, a smaller one caudally and a larger one cranially, that produces an irregular margin of the surface. These appear to be secondary ossifications into the glenoid labrum that extended the dimensions of the fossa. The intact left one has irregular ventral and dorsal margins, resulting in small cranial notches on both sides of the fossa, but it has a more caudal one on the dorsal side (Figure 16.17). Given the older age at death of Pavlov 1 (Chapter 6), it is tempting to attribute these irregularities to age changes, but similar ones are not apparent on the glenoid fossae of Dolní Věstonice 3 and 16. They may be related to habitual glenohumeral loading patterns, but they do not appear to be pathological; there is an irregularity of the inferior medial right humeral head (see below), but it is in a position not likely to come into contact with the protuberances toward the middle of the right ventral glenoid margin.

In the past (e.g., Churchill & Trinkaus, 1990, and references therein), it has been argued that there was a significant shift in relative glenoid breadth with the emergence of modern humans, and this has been related to changes in the habitual degrees of loading of the glenohumeral articulation in the extremes of medial and lateral rotation (Churchill & Trinkaus, 1990). However, those comparisons have been principally between Neandertals and recent humans, and the Neandertal sample has included specimens from Krapina and Tabun now known to be terminal Middle Pleistocene and initial Late Pleistocene in age. Previous assessments have also pooled left and right glenoid fossae to maximize sample size.

To reassess this pattern in light of the middle of the Late Pleistocene, data for Gravettian and Neandertal glenoid fossae height and breadth (there are no sufficiently complete Qafzeh-Skhul glenoid fossae) were plotted for right and left scapulae (Figure 16.18). In the right fossa, three of the Neandertal specimens do indeed have relatively narrow fossae, although La Ferrassie 1 has a higher relative value. Dolní Věstonice 3 has a moderately high relative breadth, although it is exceeded by Arene Candide IP. In the left glenoid fossa, all of the Late Pleistocene specimens follow a relatively narrow distribution, and there is no separation between the Neandertals and the early modern humans. Addition of the late Middle Pleistocene specimens from Krapina and Tabun (Churchill, 1994) continues the same pattern, with the majority of the right late archaic glenoid fossae being relatively narrow but with no real difference in the left side.

The Axillary Border

Probably more than any other feature of the scapula, the form of the crests and sulci along the axillary border has been a focus of Late Pleistocene human paleontology (e.g., Boule, 1911–1913; Vallois, 1928–1946; Dittner, 1976; Trinkaus, 1977; Vandermeersch, 1991; Churchill, 1994, 1996). In this, there is a clear trend from a predominance of dorsal sulci among the Neandertals (including the Upper Paleolithic Saint-Césaire 1) to mostly a bisulcate pattern among Upper Paleolithic individuals to a predominance of the ventral sulcus pattern among recent humans (Table 16.1).

The Dolní Věstonice and Pavlov scapulae present a predominantly bisulcate morphology of this region, but there is some variation between and within individuals, and there is considerable variation in the manner in which the dorsal and ventral buttresses, the dorsal and ventral sulci, and the lateral crest relate to each other (Figure 16.19).

The Dolní Věstonice 3 left axillary border, although damaged, exhibits near its middle a clear dorsal buttress and a sulcus 8 mm wide, bordered by a clear but not angled lateral crest, and then a concave ventral sulcus; this makes it a bisulcate border. The right one, however, presents a contrasting pattern, in that there is little dor-

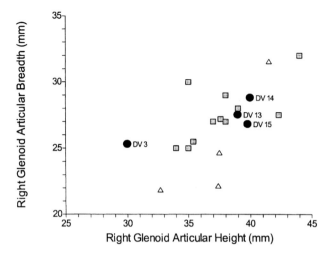

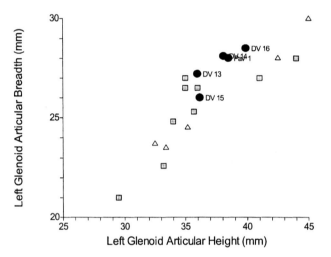

Figure 16.18 Bivariate plots of glenoid articular breadth versus articular height for right (above) and left (below) scapulae. Solid circles: Dolní Věstonice and Pavlov specimens; gray squares: European Gravettian specimens; open triangles: Neandertals.

sal buttress, there is a prominent ventral buttress with a ventral sulcus, and the lateral crest near the craniocaudal middle of the border (where it is broken) is toward the dorsal side but nonetheless is more lateral than dorsal; this makes it intermediate between a clear ventral sulcus and a bisulcate one, and it has been scored as such.

The two axillary borders of Dolní Věstonice 13 show little distinct relief, with the edges largely convex around the lateral margin. On both of them, the dorsal buttress is stronger than the ventral one, and it is more strongly differentiated from the infraspinatus surface than the ventral buttress is from the subscapularis surface. This pattern is more pronounced on the right than on the left. However, there is little development of either dorsal and

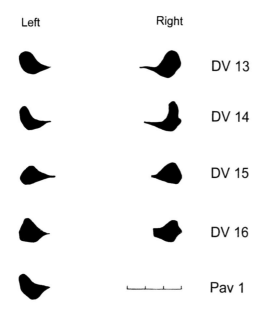

Figure 16.19 Cross sections of the scapular axillary borders of the Dolní Věstonice 13 to 16 and Pavlov 1 left scapulae (from top to bottom on the left) and the Dolní Věstonice 13 to 16 right scapulae (from top to bottom on the right). Medial is toward the midline, and dorsal is above. All were taken when viewed caudally. Mediolateral orientations are approximate because of variable preservation and irregularities of the infraspinatus surfaces. Scale in centimeters.

ventral sulci, and the poorly developed lateral crests remain in the middles of the lateral borders. They are therefore both considered to be bisulcate.

The axillary borders of Dolní Věstonice 14 are asymmetrical, with the right one exhibiting a marked dorsal buttress, a clear dorsal sulcus, little development of the ventral buttress, and no ventral sulcus. The lateral crest is more toward the dorsoventral middle of the lateral face, but it is no more so than are those of several Neandertal scapulae, which are normally considered to have dorsal sulci (Churchill, 1994). It is scored as having a dorsal sulcus. The left one, in contrast, has a larger dorsal buttress, but similar to those of Dolní Věstonice 13, it is generally convex and shows little development of either dorsal or ventral sulci. Since the lateral crest is in the middle of the lateral border, it is bisulcate.

The axillary borders of Dolní Věstonice 15 closely resemble those of Dolní Věstonice 13 in having generally flat to convex contours, a larger dorsal than ventral buttress, little development of sulci, and a lateral crest that is largely in the middle of, although in this case slightly dorsal on, the lateral margin. They are therefore bisulcate.

The Dolní Věstonice 16 right scapula presents a symmetrical cross section, in which there are largely equal dorsal and ventral buttresses and sulci, and a prominent lateral crest is located in the middle of the lateral margin. The left one has greater development of the dorsal buttress and a larger dorsal sulcus, but the lateral crest remains within the middle third of the lateral margin, and there is still a clear ventral buttress and sulcus. They are both classified as bisulcate.

Only the left scapula of Pavlov 1 preserves the axillary border, and it is close to the form of the Dolní Věstonice 13 borders. There is a more prominent dorsal buttress, a more evident ventral sulcus, greater separation of the dorsal buttress from the adjacent scapular surface, and a lateral crest in the middle of the lateral margin. Even though it is asymmetrical in cross section, it is best scored as bisulcate.

These categorizations of the axillary borders of these six Pavlovian individuals provide them with a predominance of the bisulcate pattern, nonetheless accompanied by one (Dolní Věstonice 3 right) that could be classified as ventral and another (Dolní Věstonice 14 right) that is considered to be dorsal. This distribution is generally similar to that of the other European earlier Upper Paleolithic humans (Table 16.1). The Qafzeh-Skhul sample is similar but too small to be reliable. Both of these early modern human samples contrast with the Neandertal sample, but they do so in terms of frequencies rather than presence/absence of the different forms. None of these

Table 16.1 Frequencies of scapular axillary border forms for Neandertals, Qafzeh-Skhul humans, and European earlier Upper Paleolithic humans.

	Ventral Sulcus	Bisulcate	Dorsal Sulcus	N
Dolní Věstonice & Pavlov	4.2%	87.5%	8.3%	6
Earlier Upper Paleolithic	16.7%	66.7%	16.7%	15
Earlier Upper Paleolithic plus Pavlovian	13.1%	72.6%	14.3%	21
Qafzeh-Skhul	0.0%	100%	0.0%	2
Neandertals	0.0%	25.0%	75.0%	10
Recent Europeans	75.8%	23.8%	0.4%	120
Recent Amerindians	86.6%	13.4%	0.0%	119

Fossil data are modified from Churchill (1994) and from personal observation. Recent human data are from Trinkaus (1977).

Late Pleistocene samples approaches the high frequencies of the ventral sulcus pattern seen in later Holocene samples. Despite these clear trends, which suggest a shift to more ventral sulci, or dorsal lateral crests, with decreasing upper limb robusticity, the functional significance of these changes, if any, remains elusive (Churchill & Trinkaus, 1990; Churchill, 1996).

The infraglenoid tubercle is preserved on all but the Pavlov 1 scapulae, and they are generally long and prominent (Table 16.2). Those on Dolní Věstonice 15's scapulae are especially noticeable for their large dimensions, given the otherwise modest dimensions of these bones.

The Supraspinatus and Infraspinatus Surfaces

As with most prehistoric scapulae, the thin bone of the supraspinatus and infraspinatus surfaces and the subscapularis surface is variably preserved on these bones. However, a couple of them are exceptionally intact (Figures 16.9 to 16.13).

The Dolní Věstonice 13 superior margins have a broad concavity, which becomes a distinct notch adjacent to the coracoid processes. The more caudal portions have weak margins for teres major and modest ridges for subscapularis. The Dolní Věstonice 14 superior margin is largely preserved on the left side, and it has a narrower concavity with a small notch next to the coracoid process. Caudally, the muscle markings are generally weak, especially along the lateral margin. The Dolní Věstonice 15 scapulae also have a notch next to the coracoid process, and then the margin curves superiorly and medially. There is a general similarity between these superior margins of the Dolní Věstonice 13–15 scapulae, but they also show distinct individual variations.

The Dolní Věstonice 16 scapulae preserve few of their margins, but the left one has a distinct crest along the cranial half of the ventral axillary margin. It rises on the ventral buttress adjacent to the infraglenoid tubercle and then extends largely parallel to the axillary border, moving gradually onto the subscapularis surface to fade out in the middle of the caudal half of the bone.

The Humeri

The Dolní Věstonice and Pavlov human remains retain a substantial series of humeri. All six of the associated skeletons preserve most of both bones, and these are supplemented by the Dolní Věstonice 41 distal humeral shaft (Figures 16.20 to 16.26). All of the associated humeri have largely complete diaphyses, and the epiphyses of Dolní Věstonice 13 and 15 are virtually complete, with minor restoration. The Dolní Věstonice 14 distal epiphyses are intact, but both of the heads have been secured to the surgical necks with wax, the left one more tenuously than the right. The Dolní Věstonice 16 right humerus lacks the head and the distal epicondyles, but the left one is more complete despite separation of the head from the surgical neck; the head was replaced slightly too distally, and 2 mm have been added to the length measurements to compensate for it. The Pavlov 1 right humerus lacks only the bulk of the distal articulations, whereas the left one is missing all of the proximal end; as mentioned above (see also Sládek et al., 2000), their lengths were estimated by aligning anatomical landmarks on the two bones, assuming length symmetry. The humeri of Dolní Věstonice 3, however, sustained extensive damage to all of the epiphyses, and only the diaphyses provide morphological data. The Dolní Věstonice 41 right humerus preserves only a portion of the distal diaphysis with the proximal end of the supraolecranon sulcus, and it therefore provides data only on distal diaphyseal cross-sectional proportions.

The humeri of Dolní Věstonice 3, 13, 14, 16, and 41 plus Pavlov 1 appear to be normal without pathological alterations. Only some minor changes to the medial head of the Pavlov 1 right humerus could be considered lesions, and they are extremely minor changes with few likely consequences (see below and Chapter 19). However, the right humerus of Dolní Věstonice 15 has undergone serious pathological alterations, which are discussed in detail in Chapter 19. Of concern here is their effect on the bone's overall morphology and biomechanical characteristics. The localized osteoarthritis of the glenohumeral joint is not relevant in this context since it did not alter the shape or position of the humeral head. The distal diaphyseal abnormal curvature (Figure 16.21), however, altered both the distal diaphysis and the orientation of the distal epiphysis. It changed the mediolateral angle of the distal articulations such that the cubital angle, which is a normal 92° on the left humerus, is only 82° on the right one. It does not appear to have altered the bone's torsion; its angle is 21° as opposed to 24° on the left, and both angles are well within normal ranges of variation—in fact

Table 16.2 Development of the scapular infraglenoid tubercle, in millimeters.

	Side	Length	Thickness	Projection
Dolní Věstonice 3	Right	23.5	5.5	3.5
Dolní Věstonice 13	Right	(33.0)	6.3	5.0
	Left	(33.0)	3.9	5.1
Dolní Věstonice 14	Right	24.8	6.1	(3.7)
	Left	24.0	5.0	(4.0)
Dolní Věstonice 15	Right	22.8	8.6	3.1
	Left	25.0	7.0	5.0
Dolní Věstonice 16	Left	32.3	4.7	5.1

Values in parentheses are estimated.

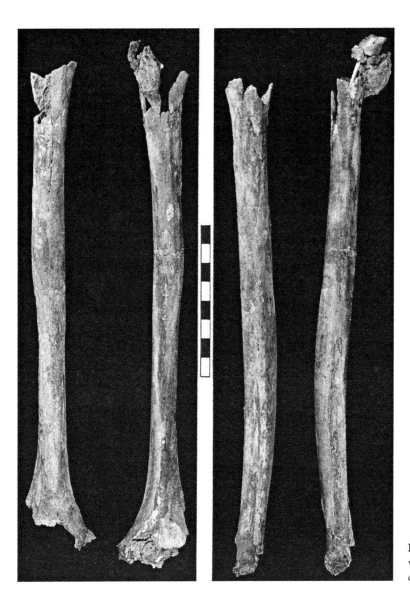

Figure 16.20 Anterior (left) and medial (right) views of the Dolní Věstonice 3 humeri. Scale in centimeters.

between those of Dolní Věstonice 13 (32°) and Dolní Věstonice 14 (9°). This angular displacement can only have changed the trajectory of forces through the distal humerus. However, there is no evidence of osteoarthritis on any of the cubital articulations, indicating that they continued to function in a normal fashion without significantly altered joint reaction force magnitudes or directions. The medial deviation of the distal diaphysis would have increased mediolateral bending of the diaphysis near the angle, and some remodeling of the diaphyseal bone to compensate for it would be expected. However, the bone is externally and radiographically normal, without any indication of altered histology or density (to the extent determinable macroscopically on fossilized bone), indicating that the bone can be analyzed morphologically relative to the other Pavlovian humeri, taking into consideration the immediate effects of the abnormality.

It is also possible that the left humerus of Dolní Věstonice 15 was hypertrophied in response to the implied impaired function of the right arm, or atrophied in conjunction with the alterations of the left radius and ulna (see below and Chapter 19). Those possibilities and any biomechanical consequences of the right humeral deformity are considered below and in Chapter 19.

The Proximal Epiphyses

The proximal epiphyses (heads, tubercles, and bicipital sulci) of the Dolní Věstonice and Pavlov humeri are notable for being unremarkable. The Dolní Věstonice 13 heads are smooth, and the greater and lesser tubercles are prominent but smooth, with distinct facets for the rotator cuff muscle insertions. There is some lipping of the greater tubercles over the lateral bicipital sulci. The heads of the Dolní

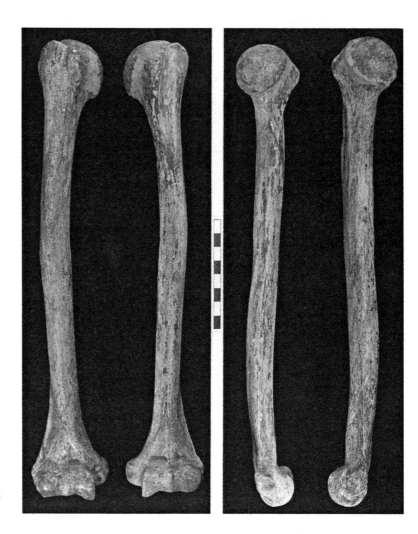

Figure 16.21 Anterior (left) and medial (right) views of the Dolní Věstonice 13 humeri. Scale in centimeters.

Věstonice 14 humeri are evenly rounded despite minor erosion. The Dolní Věstonice 15 proximal epiphyses are notable only for their prominent tubercles; the glenohumeral osteoarthritis is described in Chapter 19. Only a portion of the left head subchondral bone and the distal portions of the tubercles remain on the Dolní Věstonice 16 humeri, and there is some rugosity of the tubercles.

The head and tubercles of the Pavlov 1 right humerus are generally as would be expected for a large male. There is a raised area 5.7 mm wide across 25 mm of the inferior medial humeral head; it appears to be subchondral but nonarticular, since an unusual degree of adduction of the glenohumeral articulation would be necessary to bring it into contact with the inferior glenoid surface. As with the irregularities of the glenoid fossae (see above), it may be related to some minor, but not pathological, alterations of the glenoid labrum.

Diaphyseal Morphology

The Pavlovian human humeral diaphyses provide an abundance of information in terms of external metrics, cross-sectional geometric parameters, and muscle markings (Sládek et al., 2000). Each is described, and then they are evaluated metrically in terms of diaphyseal shape, robusticity, and one muscle marking.

The Dolní Věstonice 3 humeri are notable primarily for the lack of strong relief on their diaphyseal surfaces. Both diaphyses have smooth external surfaces with rounded "crests" and generally convex planes. Only the dorsal supraolecranon surface shows any concavity, and only the lateral supracondylar crest, for about 35 mm proximal of the olecranon fossa on both bones, shows any clear angulation. The deltoid tuberosities are sufficiently elevated to distinguish them from the adjacent subperiosteal bone, but there is little rugosity on them. The insertions for latissimus dorsi and teres major on the proximal diaphyses are barely discernible. The insertions for pectoralis major are clearly marked by modest excavations on both bones, with a firmly delimited porous sulcus for each one.

The Dolní Věstonice 13 humeral diaphyses are similarly very smooth on most of their surfaces. The right one is clearly larger than the left. They exhibit a gentle lateral bowing in anterior view, and a suggestion of an S-curve in medial view. The lateral bowing is not due to the deltoid tuberosity, which is only a minimally raised area on the anterolateral diaphysis. The right deltoid tuberosity is trivially rugose, whereas the left one is polished smooth. The pectoralis major tuberosities are clearly present, but the rugosity of them is minimal (see Figure 16.27). The teres major and latissimus dorsi insertions can be located if one knows the appropriate anatomy, but they completely blend into the anteromedial proximal diaphysis. The lateral supracondylar crests are readily apparent in anterior view, and the medial supracondylar crest forms a distinct angle for ca. 75 mm proximal of the epiphysis.

The diaphyses of the Dolní Věstonice 14 humeri are also very smooth on their surfaces; almost all of the roughness on them is due to postmortem root etching. They are noticeable for the marked lateral convexity of their midshafts, in which the right one approaches an angle between 50% and 60% of length (0% is distal), and the left one has a more even curve. The modest deltoid tuberosities contribute little to the effect. The S-curve in medial view is more apparent than on the Dolní Věstonice 13 humeri. The shafts are also highly asymmetrical in size, but not particularly in morphology. The proximal and midshaft muscle markings are very modest. The pectoralis major insertions are noticeable only as slight rugosities. The teres major and latissimus dorsi tuberosities are barely present, and the deltoid tuberosities consist of slight roughenings of minimal swellings. The lateral supracondylar crests are clearly present, especially on the left, where it forms a small flange of bone. The modest rugosities of the diaphyseal muscle markings, but not necessarily their dimensions, may be due in part to the late adolescent age of the individual.

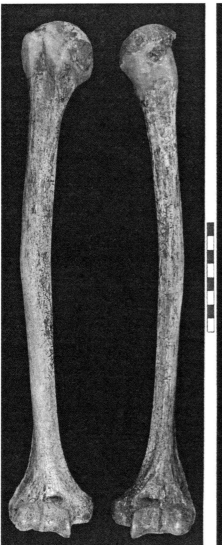

Figure 16.22 Anterior (left) and medial (right) views of the Dolní Věstonice 14 humeri. Scale in centimeters.

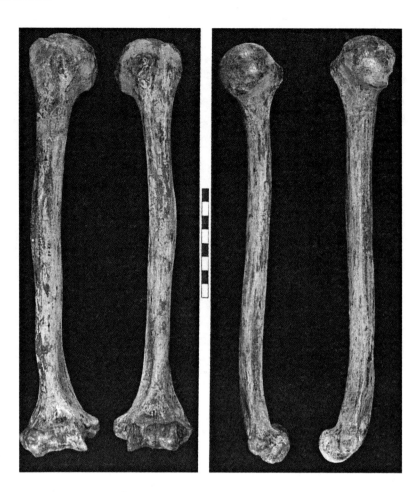

Figure 16.23 Anterior (left) and medial (right) views of the Dolní Věstonice 15 humeri. Scale in centimeters.

The Dolní Věstonice 15 humeri, despite the distal right abnormal curvature, appear as short and stout bones. They have less of the lateral bowing seen in the other two triple burial individuals, and the deltoid tuberosities have more of an effect on the lateral contours. In medial view, the proximal halves are essentially straight, with the normal anterior concavity of the distal diaphysis. The muscle markings are also lightly marked but show some variation. The right pectoralis major tuberosity is mostly within the contour of the bone, whereas the left one is a small, raised, smooth ridge. There is little evidence for the latissimus dorsi and teres major tuberosities. The right deltoid tuberosity is a broad, rounded, and prominent swelling, and the left one is a more distinctive and narrow swelling of bone. The lateral supracondylar crests are modest on both humeri, but the medial ones form small flanges of bone; the right one may be due in part to the diaphyseal bending, but the left one is large as well.

The smooth surfaces of the humeri of Dolní Věstonice 13 to 15 may be attributable in part to their young ages, although this would not apply to Dolní Věstonice 3. Moreover, it cannot explain the very smooth surfaces of the Dolní Věstonice 16 humeri. The last are quite straight, with only a modest lateral convexity near 60% of length and a hint of an S-curve in medial view. The pectoralis major tuberosities are actual discernible tuberosities, especially on the left side, with some rugosity and central grooves, and teres major and latissimus dorsi produced raised rugose areas. The deltoid tuberosities, however, are very modest. The supracondylar crests are strong laterally but weak medially.

The Pavlov 1 humeri are relatively straight in anterior view, with the lateral convexity, in this case being produced by the swelling for the deltoid tuberosities. There is a clear S-curve in medial view, especially on the more complete right bone. The pectoralis major tuberosities are bilaterally distinct grooves, with little raising of the bone above the shaft surface; there is a slight rugosity along the lateral lips of the grooves, especially on the left one (Figure 16.25). The latissimus dorsi and teres major attachments are evident as slight grooves with minimal rugosity. Although the deltoid tuberosities are in general swellings of the proximal part of the midshaft, the apparent muscle insertions consist of thin ridges of clear bony rugosity along the middles of the swellings, almost rough crests rather than tuberosities, especially on the left bone.

The pectoralis major tuberosities of these humeri have largely distinct medial and lateral boundaries, and their breadths can therefore be measured consistently. If one assumes a balance between the antagonist and synergist muscles, its tuberosity should provide a general indication of muscular hypertrophy of the shoulder region. The pectoralis major breadths of the Pavlovian remains have therefore been plotted against humeral length, as an approximation of their load arms (Figure 16.28). The resultant plots for the right and left humeri show the Neandertals falling largely above the early modern humans; the Aurignacian Mladeč 24 humerus is the most gracile of the early modern humans. The Amud 1 and La Ferrassie 1 Neandertals fall among the early modern humans in the left arm distribution. The Pavlovian fossils fall generally among the early modern humans, although Dolní Věstonice 3 and 14 are relatively gracile, Dolní Věstonice 13 and 16 are relatively robust in the right arm, and the latter remains are high in the left arm. In both of these distributions, but especially in the left one, Dolní Věstonice 15 has one of the highest relative values for an early modern human. Given that the size of the pectoralis major tuberosity varies considerably in recent human samples (Churchill & Smith, 2000), it is likely that these similarities and differences reflect mostly activity-related hypertrophy among these Late Pleistocene samples.

It is interesting that the deltoid tuberosities of these humeri are generally modest in their development, in some cases being little more than a slight rugosity over a trivial swelling of the lateral midshaft. In this, they are similar to the earlier Mladeč 24 humerus (Trinkaus et al., 2005). Other Gravettian humeri are variable in the development of the deltoid tuberosity, but some of them are as gracile as these Pavlovian remains (Matiegka, 1938; Trinkaus, 2000b). At the same time, the deltoid attachments on the anterior distal surfaces of the Dolní Věstonice and Pavlov clavicles are consistently well marked with concave facets and/or small bony crests. Since the deltoid muscle fibers both insert onto the humeral diaphysis (the deltoid tuberosity) and blend with fibers of brachialis, it remains uncertain to what extent the development of the deltoid tuberosity per se reflects the hypertrophy of the deltoid muscle as opposed to the proportion

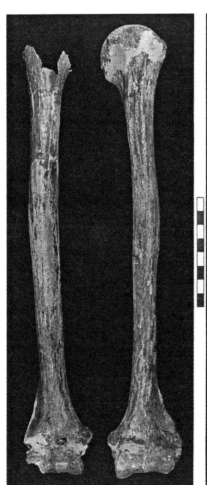 

Figure 16.24 Anterior (left) and medial (right) views of the Dolní Věstonice 16 humeri. Scale in centimeters.

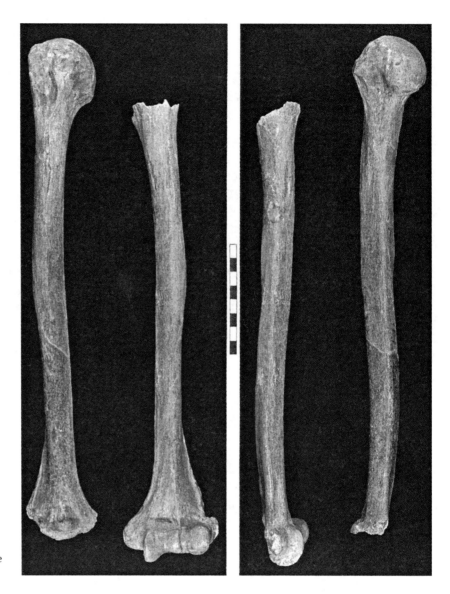

Figure 16.25 Anterior (left) and medial (right) views of the Pavlov 1 humeri. Scale in centimeters.

of its fibers penetrating the periosteum of the lateral humeral diaphysis.

Diaphyseal Asymmetry

All of these Pavlovian humeri are larger on the right side, as are those of all nonpathological late archaic and early modern humans. Only two of them provide direct measurements of the lengths of both humeri, Dolní Věstonice 13 and 15, and the differences are modest. A percentage of the asymmetry of articular length, calculated as the absolute value of the difference of the measures divided by their mean, provides values of 1.51 for the former and 3.12 for the latter. The Dolní Věstonice 15 value is influenced mostly by the abnormal curvature of its distal end. However, both values fall within the range of five Předmostí humeri (0.30, 0.62, 1.60, 1.77, and 5.55), and mean length differences of 1 to 3 mm are common in recent human samples (Trinkaus et al., 1994; Sakaue, 1997; Čuk et al., 2001; Table 19.6).

As a measure of overall diaphyseal rigidity, the polar moment of area was used to assess diaphyseal asymmetry. Using the fourth root of the polar moment of area (since it is in mm^4), one sees that Dolní Věstonice 3 has a low level of asymmetry, similar to the recent human median value; Dolní Věstonice 13 and 14 have rather high values, exceeding the recent human range; and Dolní Věstonice 16 and Pavlov 1 fall near the upper limits of the recent human sample (Table 16.3). Dolní Věstonice 15's humeri, despite the right abnormalities, have a level of asymmetry in the middle of the Pavlovian range, with the abnormal right one larger. These Pavlovian values are

The Upper Limb Remains

Figure 16.26 Posterior (left) and medial (right) views of the Dolní Věstonice 41 right distal humerus. Scale in centimeters.

similar to those of Late Pleistocene humans in general, although one individual (Barma Grande 2; see Churchill & Formicola, 1997) has a high level of asymmetry.

Diaphyseal Shape and Robusticity

The diaphyses of Late Pleistocene humans are characterized generally by a decrease in circularity, or a relative increase of the anterolateral to posteromedial direction, of the middle and proximal diaphysis. This is evident in the external midshaft measures of diaphyseal shape (Figure 16.29), in which most of the Neandertals fall on a line above but parallel to the early modern human one. There is nonetheless some overlap between the distributions—for the smaller size range plus the very large value for Barma Grande 2 in the right humeral comparison and in the larger size range in the left one. All of the Dolní Věstonice specimens are well within the earlier Upper Paleolithic distributions, although Pavlov 1 is in the middle of the overlap between the samples.

This pattern remains in cross-sectional comparisons of the maximum versus minimum second moments of area for midshaft (50%) and midproximal shaft (65%) (Figures 16.30 and 16.31). In the left humerus, all of the Pavlovian specimens are with the majority of the early modern humans. In the right humerus, only Pavlov 1 is above the early modern humans and among the major-

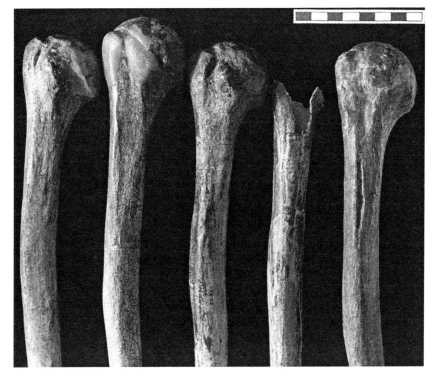

Figure 16.27 Anteromedial views of the Dolní Věstonice 13 to 16 and Pavlov 1 proximal humeri (from left to right), showing the markings for the insertions of the thoracohumeral and scapulohumeral muscles. Scale in centimeters.

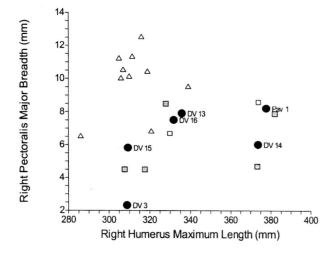

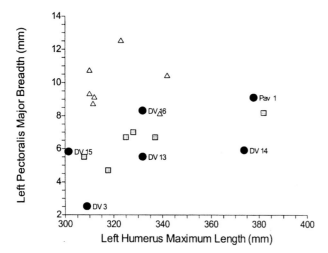

Figure 16.28 Bivariate plots of pectoralis major tuberosity breadth versus maximum length for right (above) and left (below) humeri. Solid circles: Dolní Věstonice and Pavlov specimens; gray squares: European earlier Upper Paleolithic specimens; open squares: Qafzeh-Skhul specimens; open triangles: Neandertals.

ity of the Neandertal sample. The Aurignacian Mladeč 23 and 24 humeri are with the later Gravettian remains.

Percent cortical area, or cortical area relative to total subperiosteal area, is sometimes taken to be a measure of robusticity, but it is best seen as a measure of the endosteal versus subperiosteal depositional and resorptive histories of the diaphysis, rather than strictly a measure of relative strength (Ruff & Hayes, 1983b; Ruff et al., 1994). Comparisons across Late Pleistocene samples for the mid-distal (35%) humerus show similar and consistent patterns across the samples (Figure 16.32). In the right humerus, the moderately high Pavlovian value is for Dolní Věstonice 41, and the low value is for the older Dolní Věstonice 16, perhaps reflecting its older age (Ruff & Hayes, 1983b). In the left humerus, all of the Pavlovian specimens are close to the other distributions, and the two high values are Skhul 4 and Shanidar 1, the latter perhaps influenced by the absence of an effective right arm (Trinkaus, 1983b).

The best measures of diaphyseal robusticity are those that scale diaphyseal properties against appropriate measures of structure size (Ruff et al., 1993), for which beam length is most appropriate for the upper limb (see above). In order to be able to include the maximum sample initially, distal minimum circumference was plotted against humeral length (Figure 16.33), producing a general pattern of a decrease in humeral robusticity between late archaic and early modern humans, despite considerable overlap and a high value for Barma Grande 2. The Dolní

Table 16.3 Asymmetry of the humeral diaphysis, computed as the absolute value of the difference between the two sides divided by the mean of the two values, expressed as a percentage.

	35% Polar Moment of Area
Dolní Věstonice & Pavlov	
Dolní Věstonice 3	2.25
Dolní Věstonice 13	16.07
Dolní Věstonice 14	17.10
Dolní Věstonice 15	13.24
Dolní Věstonice 16	9.55
Pavlov 1	8.26
Gravettian	
Barma Grande 2	24.75
Cro-Magnon 4294	4.90
Fanciulli 4	12.95
Paglicci 25	5.97
Pataud 26.230	0.59
Qafzeh-Skhul	
Skhul 2	3.06
Skhul 5	6.08
Skhul 7	5.49
Neandertals	
La Chapelle-aux-Saints 1	11.20
La Ferrassie 1	10.07
Kebara 2	17.09
Spy 2	4.94
Recent Humans	
Median	2.36
95th Percentile	6.69
Maximum	8.84
N	169

The fourth roots of the polar moments of area are employed to linearize the values. The pooled recent human samples are those used in Trinkaus et al. (1994) and Churchill and Formicola (1997), but the percentage asymmetry is calculated differently.

radius (Chapter 19), and there is no necessary reason to expect one to compensate for the other.

The Distal Epiphysis

There is little of morphological note on the distal epiphyses. Dolní Věstonice 14 and Pavlov 1, and to a lesser extent Dolní Věstonice 16, show strong development of the coronoid and radial head fossae, and Dolní Věstonice 14 appears to have some extra bone growth within them. However, none of the humeri exhibit septal foramina.

The percentage of asymmetry of the distal articular surfaces, for which the distal articular breadth is used here, varies between 0 and almost 9 in the recent human samples.

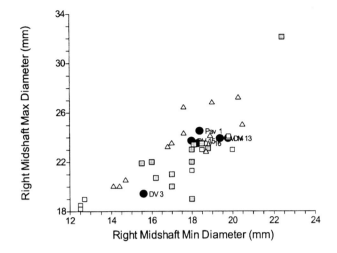

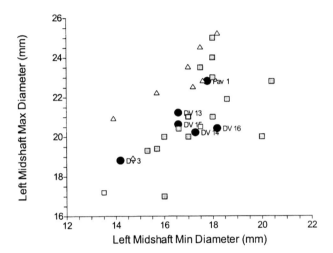

Figure 16.29 Bivariate plot of humeral midshaft maximum versus minimum diameter for right (above) and left (below) humeri. Solid circles: Dolní Věstonice and Pavlov specimens; gray squares: European earlier Upper Paleolithic specimens; open squares: Qafzeh-Skhul specimens; open triangles: Neandertals.

Věstonice and Pavlov specimens are aligned with other early modern humans, although Dolní Věstonice 3 is rather gracile and Dolní Věstonice 13, 15, and 16 are moderately robust.

The same pattern holds for cross-sectional comparisons, using the 35% polar moment of area as a measure of diaphyseal rigidity (Figure 16.34). Interestingly, Dolní Věstonice 15 is among the more robust Gravettian humeri, in both the right and the left arms. It is possible that its right 35% humeral rigidity was increased by the abnormal bending strains from the diaphyseal deformity, but this does not explain the high level of hypertrophy of the left humerus. Both arms sustained deformities, the right one in the humerus and the left one in the ulna and

Figure 16.30 Bivariate plots of humeral midshaft (50%) maximum second moment of area versus minimum second moment of area for right (above) and left (below) humeri. Solid circles: Dolní Věstonice and Pavlov specimens; gray squares: European earlier Upper Paleolithic specimens; open squares: Qafzeh-Skhul specimens; open triangles: Neandertals.

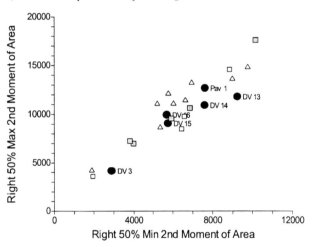

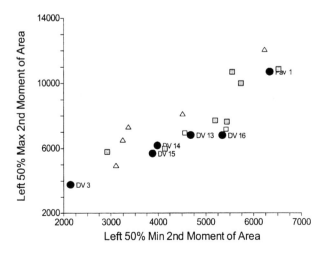

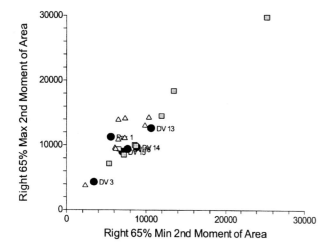

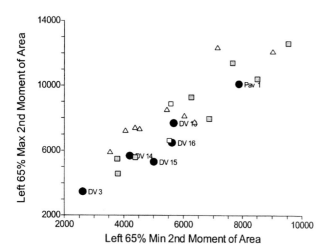

Figure 16.31 Bivariate plots of humeral midproximal shaft (65%) maximum second moment of area versus minimum second moment of area for right (above) and left (below) humeri. Solid circles: Dolní Věstonice and Pavlov specimens; gray squares: European earlier Upper Paleolithic specimens; open squares: Qafzeh-Skhul specimens; open triangles: Neandertals.

All of the Late Pleistocene paired humeri providing both measurements fall within this range, although two of them, Dolní Věstonice 15 and Barma Grande 2, fall outside of the 95% interval of the recent human sample (see Chapter 19).

The Ulnae and Radii

The antebrachial bones of the Pavlovian human remains follow much the same pattern of preservation as their humeri (Figures 16.35 to 16.46). Those of Dolní Věstonice 13 and 15 are essentially complete. The bones of Dolní Věstonice 14 and 16 are largely intact in the cubital region

and in their diaphyses (Dolní Věstonice 16 a little less so than Dolní Věstonice 14), but they lack major portions of their distal epiphyses. Pavlov 1 has one largely complete bone (the right radius), but the other bones were truncated. Dolní Věstonice 3 has the least complete bones, with one bone (the right radius) preserving most of its length, with crushing near and on the epiphyses and the other bones—despite the reassembly of some diaphyseal splinters from the crushed burial (Trinkaus & Jelínek, 1997)—being incomplete and partially distorted. In addition, there is an isolated fragment of a proximal left radius, Dolní Věstonice 50.

Figure 16.32 Bivariate plots of humeral mid-distal shaft (35%) cortical cross-sectional area versus total subperiosteal cross-sectional area for right (above) and left (below) humeri. Solid circles: Dolní Věstonice and Pavlov specimens; gray squares: European earlier Upper Paleolithic specimens; open squares: Qafzeh-Skhul specimens; open triangles: Neandertals.

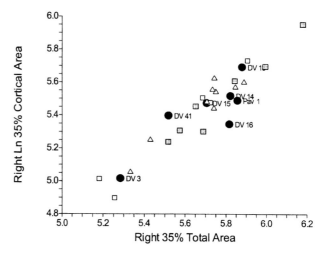

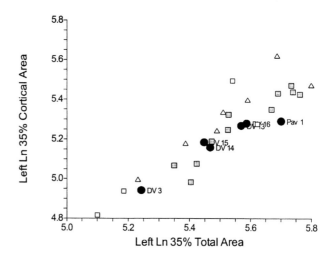

The Upper Limb Remains

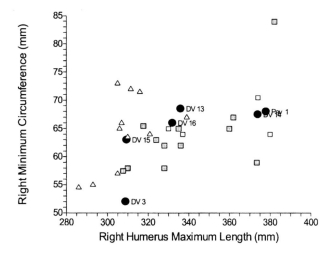

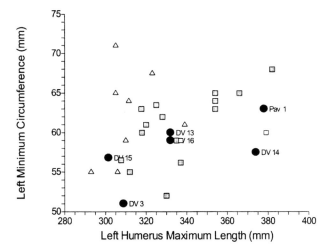

Figure 16.33 Bivariate plots of humeral distal minimum circumference versus maximum length for right (above) and left (below) humeri. Solid circles: Dolní Věstonice and Pavlov specimens; gray squares: European earlier Upper Paleolithic specimens; open squares: Qafzeh-Skhul specimens; open triangles: Neandertals.

The Proximal Ulnar Epiphyses

The proximal epiphyses of the Dolní Věstonice and Pavlov ulnae are generally unremarkable for those of early modern humans. Those of Dolní Věstonice 3 are largely crushed and provide little information. The intact epiphyses of Dolní Vestonice 13 are in line with their proximal diaphyses, angle slightly laterally, and are generally symmetrical (the right side has trivially larger measurements for most dimensions). The triceps brachii attachments on the olecranon processes are moderately rugose, and there are clear fossae for the insertion of brachialis, especially on the right ulna. The supinator crests are modest. In addition, the right ulna has a bony process on the medial half of the midtrochlear notch, in

the area that is normally nonarticular. The angular process is 8.2 mm mediolateral and 3.6 mm proximodistal, and it projects up to 1.7 mm from the adjacent subchondral bone. There is no sign of it affecting either the other aspects of the trochlear notch or the subchondral bone of the right humeral trochlea. The Dolní Věstonice 14 proximal ulnae are also in line with their proximal diaphyses, although they, too, angle laterally relative to the diaphyseal axis. Both ulnae exhibit strong supinator crests, which lip slightly dorsally, but the brachialis tuberosities are poorly marked with little concavity, especially on the left.

The ulnae of Dolní Věstonice 15 present some difficulties in their assessment. The right one articulates with the medially deviated distal right humerus, yet it shows little if any effect of that marked abnormality of the dis-

Figure 16.34 Bivariate plots of humeral mid-distal (35%) polar moment of area versus biomechanical length for right (above) and left (below) humeri. Solid circles: Dolní Věstonice and Pavlov specimens; gray squares: European earlier Upper Paleolithic specimens; open squares: Qafzeh-Skhul specimens; open triangles: Neandertals.

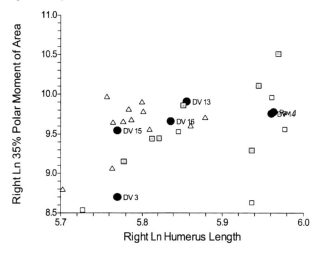

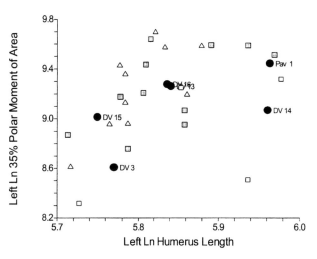

Figure 16.35 Anterior views of the Dolní Věstonice 3 radii and ulnae. Scale in centimeters.

(-1°). Yet, the dimensions of the olecranon and coronoid processes and the trochlear notch, although generally smaller than those of the right side, are in approproximately the same proportions as those of the other Pavlovian ulnae. Nonetheless, the supinator crest is moderately prominent for 30 mm proximodistal on the right bone and clearly present for 27 mm proximodistal on the left bone. The brachialis tuberosities are symmetrically rugose, but there is no change in the distal coronoid contours for them.

The proximal ulnae of Dolní Věstonice 16 and Pavlov 1 are slightly damaged, but they are otherwise unremarkable. The Dolní Věstonice 16 olecranon processes have modest enthesopathies for the triceps brachii tendon (Chapter 19), its supinator crests are moderately large, and the brachialis tuberosities are symmetrically rugose, with slight concavities near their distal ends. The

Figure 16.36 Anterior views of the Dolní Věstonice 13 radii and ulnae. Scale in centimeters.

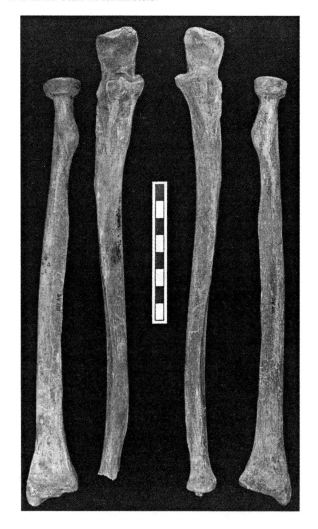

tal humerus. Certainly, its angular orientations, both sagittal and coronal (Sládek et al., 2000), are very close to those of the other Pavlovian ulnae. The left one, although articulating with a normal distal humerus, continued distally onto a pathological diaphysis with what appears to be a healed mid-distal diaphyseal fracture (Figure 16.39; Chapter 19). This alteration was combined with an abnormal curvature of the left radius to produce some alterations of the orientation of the proximal ulna. In particular, it is markedly medially displaced relative to the long axis of the proximal diaphysis, and its coronal trochlear angle, instead of being slightly medial (4° to 10° for the other Pavlovian ulnae, 7° for the Dolní Věstonice 15 right ulna), is minimally laterally oriented

The Upper Limb Remains

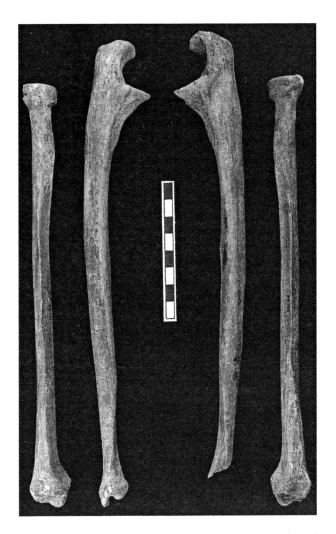

Figure 16.37 Medial views of the Dolní Věstonice 13 radii and ulnae. Scale in centimeters.

Pavlov 1 supinator crests are separated from the radial facets and are moderately well developed. The brachialis tuberosities present a rugose concavity on the right but only a rugose surface on the left.

In addition to these morphological considerations, it is possible to assess the relative heights, or dorsovolar dimensions, of the Dolní Věstonice and Pavlov olecranon and coronoid processes. Middle and Late Pleistocene archaic humans, including both African, western Asian, and European specimens, have largely volar orientations of the trochlear notch, which are produced by similar heights of the two processes that border it (Trinkaus, 1983b; Churchill et al., 1996). This pattern is illustrated by plots of the coronoid height relative to the olecranon height for Late Pleistocene specimens (Figure 16.47), in which almost all of the Neandertal specimens fall below the distributions for early modern humans (the one ex-

ception is the Spy 1 left ulna). In the right ulnar comparison, Dolní Věstonice 13 to 15 fall at the top of the overall distribution, and in the left one Dolní Věstonice 13 and 14 and Pavlov 1 are in the middle of the early modern human distribution.

The Proximal Radial Epiphyses

The Dolní Věstonice 3 right proximal radius is sufficiently complete to allow the estimation of its radial length, and the left distal tuberosity permits assessment of its orientation, but they are otherwise seriously damaged. The complete proximal radii of Dolní Věstonice 13 exhibit very round radial heads and necks and smooth tuberosities, with the maximum medial projection dorsally on the

Figure 16.38 Anterior views of the Dolní Věstonice 14 radii and ulnae. Scale in centimeters.

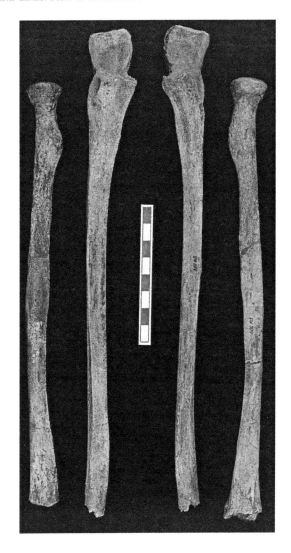

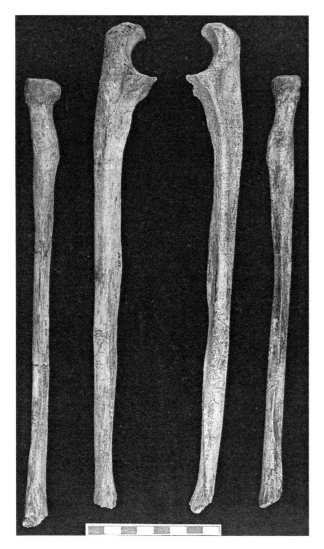

Figure 16.39 Medial views of the Dolní Věstonice 14 radii and ulnae. Scale in centimeters.

tuberosity. The Dolní Věstonice 14 proximal radii provide necks that are flattened anteromedially to posterolaterally and tuberosities that remain smooth.

The Dolní Věstonice 15 proximal radii are asymmetrical, in that the left one is clearly shorter than the right one; the left articular length of 222.0 mm is 6.0 mm shorter than the right one. Of this 6.0 mm difference, 4.3 mm is contained within the head/neck length (proximal head to midtuberosity), resulting in a much shorter neck on the left side. In addition, there is axial rotation of the left radius, which results in the radial tuberosity being more anteriorly oriented (category 1; see below) than the right one. The tuberosities are otherwise smooth with greater projection dorsally. From each one, a low and rounded anterolateral ridge extends distally, especially on the right side. These asymmetries and other deformities of the left radius are discussed in Chapter 19.

The Dolní Věstonice 16 radii have mediolaterally flattened necks. The tuberosities have flat, rugose surfaces with excavated middle portions and raised dorsal margins. The Pavlov 1 proximal radius is preserved principally on the right side, with the left side retaining only the distal tuberosity. The neck is flattened slightly. The tuberosity is a flattened and prominent surface, with an antemortem pit in the middle. There are rounded trabeculae within the pit, which is ca. 5 mm in diameter. The dorsal margin is prominent along the full length of the tuberosity. The Dolní Věstonice 50 proximal radius has a rather round neck and what appears to be a smooth and prominent tuberosity.

Several features of the proximal radii can be evaluated metrically, of which a few of the more salient aspects are considered here. It has been argued (Trinkaus & Churchill, 1988; Trinkaus, 2000c) that there was a decrease in the effectiveness of biceps brachii with the transition to early modern humans, as reflected in both their rela-

Figure 16.40 Anterior views of the Dolní Věstonice 15 radii and ulnae. Scale in centimeters.

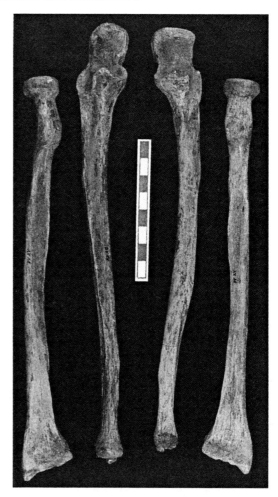

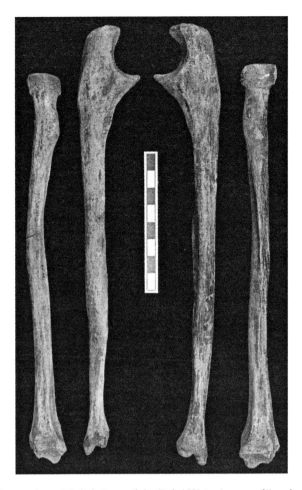

Figure 16.41 Medial views of the Dolní Věstonice 15 radii and ulnae. Scale in centimeters.

suming that the size of the tuberosity reflects biceps brachii tendon size and hence muscle size. The resultant value, which is head/neck length times the product of the tuberosity diameters (Figure 16.49), reduces the differences across the Late Pleistocene samples. A few of the Neandertals retain their high positions relative to the other specimens, but the majority of the specimens, including all of the Pavlovian ones, fall within a relatively modest range of variation (the Dolní Věstonice 15 left radius is not included in this comparison, given its abnormalities).

The other indication of biceps brachii function concerns radial tuberosity orientation, which varies from directly medial to anteromedial. The more medial the orientation of the tuberosity, the greater the range of pronation/supination over which biceps brachii will function efficiently as a supinator (Trinkaus & Churchill,

Figure 16.42 Anterior views of the Dolní Věstonice 16 radii and ulnae. Scale in centimeters.

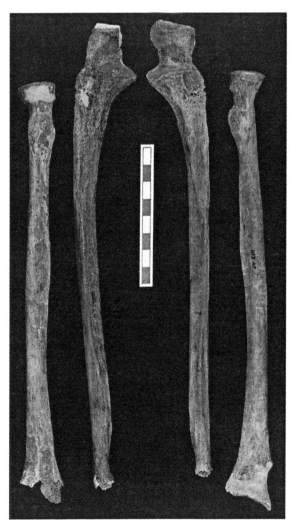

tive radial head/neck lengths and their radial tuberosity orientations. If their head/neck lengths are plotted against radial length, the latter serving as the load arm for cubital flexion and the former as an approximation of the power arm for biceps brachii (Murray et al., 2002), the late archaic humans have generally longer radial head/neck lengths, and the Pavlovian remains fall largely with the other early modern humans (Figure 16.48). The one earlier Upper Paleolithic right specimen with a relatively long head/neck length is Fanciulli 4, and the Pavlovian specimen with the moderately longer head/neck length is Dolní Věstonice 13.

In these data plots, the right and left radii of Dolní Věstonice 15 are included, despite their asymmetry and the deformities of the left bone. In both of them, however, the proportions of head/neck length to articular length remain similar, in the middle of the earlier Upper Paleolithic distributions (Figure 16.48).

However, it is possible to combine the head/neck length with the dimensions of the radial tuberosity, as-

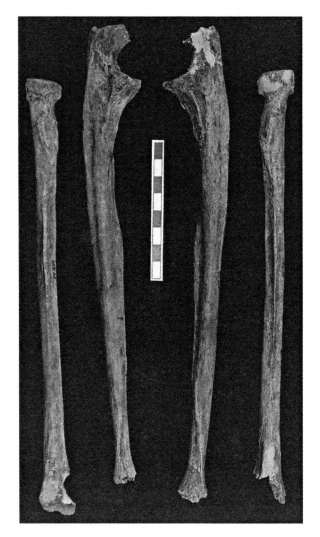

Figure 16.43 Medial views of the Dolní Věstonice 16 radii and ulnae. Scale in centimeters.

Upper Paleolithic specimens are very similar to the recent humans, with their overall frequency of 78.8% for position 2 being very close to the overall frequency of 81.2% for this form among the recent humans and the remainder of their radii exhibiting the more anteriorly oriented tuberosities (position 1). For the individual Pavlovian specimens, Dolní Věstonice 3 (left), 13–15 (right), and 50 exhibit position 2, Pavlov 1 exhibits position 1, and Dolní Věstonice 16 has its crest on the dorsal border of the tuberosity and is scored as position 1/2.

The Ulnar Diaphyses

The Dolní Věstonice 3 ulnar diaphyses, which preserve the proximal two-thirds to three-quarters, are relatively

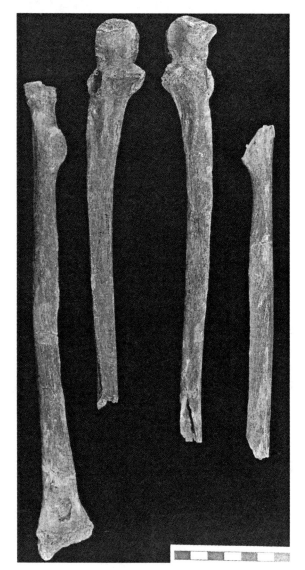

Figure 16.44 Anterior views of the Pavlov 1 radii and ulnae. Scale in centimeters.

1988). To categorize the variation, the position of the radial tuberosity was scored relative to the proximal end of the interosseous crest, such that a line drawn proximally from the crest to the tuberosity would be posterior of the tuberosity (position 1), in the dorsal third of the tuberosity (position 2), or intersecting the middle third of the tuberosity (position 3). Archaic humans predominantly exhibit position 3, with a fully medially oriented tuberosity (Trinkaus & Churchill, 1988), and this is the most common pattern among the Neandertals (Table 16.4). However, a full third of the Neandertal specimens have the tuberosity more anteriorly oriented, and one specimen (La Quina 5) has the tuberosity fully anterior of the interosseous crest. In contrast, among recent humans the strictly medial orientation is rare, and the moderately anteromedial orientation (position 2) is the most common one (Table 16.4). The small Qafzeh-Skhul sample is variable, but the earlier

The Upper Limb Remains

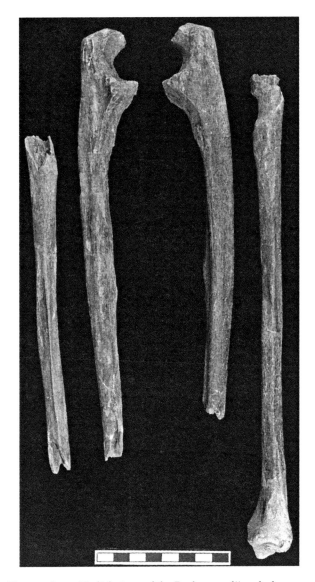

Figure 16.45 Medial views of the Pavlov 1 radii and ulnae. Scale in centimeters.

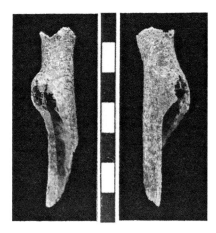

Figure 16.46 Posterior (left) and medial (right) views of the Dolní Věstonice 50 proximal radius. Scale in centimeters.

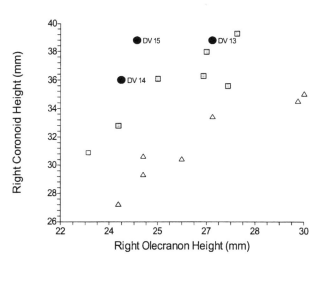

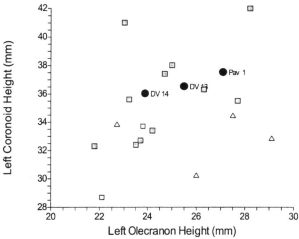

Figure 16.47 Bivariate plots of right and left coronoid height versus olecranon height. Solid circles: Dolní Věstonice and Pavlov specimens; gray squares: European Gravettian specimens; open squares: Qafzeh-Skhul specimens; open triangles: Neandertals.

straight but appear to be curving medially toward the pronator quadratus attachments. Both have distinct dorsal convexities along the proximal portions. The interosseous crests are moderately rugose proximally, producing in addition small sulci alongside of them. The Dolní Věstonice 13 ulnar diaphyses are smooth with angled but blunt interosseous crests and small and rounded pronator quadratus crests. They have a clear S-curve in anterior view with a lateral deviation distal of the pronator quadratus attachment areas. They are gently convex dorsally.

The Dolní Věstonice 14 ulnar shafts are quite straight, with only a hint of an S-curve in anterior view. The interosseous crests are clear and sharp to the level of the pronator quadratus crests, and those pronator quadratus

359

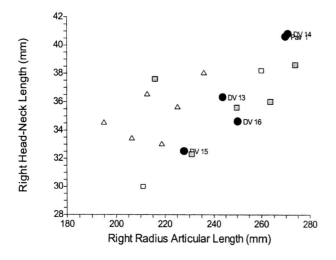

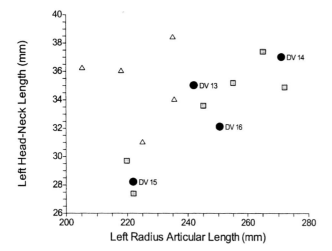

Figure 16.48 Bivariate plots of right and left radial head and neck length versus articular length. Solid circles: Dolní Věstonice and Pavlov specimens; gray squares: European Gravettian specimens; open squares: Qafzeh-Skhul specimens; open triangles: Neandertals.

The Dolní Věstonice 16 ulnar diaphyses are essentially straight in anterior view; they give the impression of a lateral convexity proximally, which is a product of the medial expansion of the proximal shaft for the coronoid process. There is none of the distal curvature associated with the pronator quadratus crest seen in the ulnae of Dolní Věstonice 3 and 13–15. The shafts similarly lack any clear dorsal curvature. The shafts are generally smooth with sharply angled interosseous crests. The pronator quadratus crests are among the most clearly demarcated of these ulnae, but they project minimally from the anteromedial margins of the bones.

The Pavlov 1 ulnae lack their distal thirds, but the preserved portions, especially of the left ulna (Figure 16.44 and 16.45), indicate very straight diaphyses, simi-

Figure 16.49 Bivariate plots of right and left head and neck length times tuberosity area (length times breadth) versus radial articular length. Solid circles: Dolní Věstonice and Pavlov specimens; gray squares: European Gravettian specimens; open squares: Qafzeh-Skhul specimens; open triangles: Neandertals.

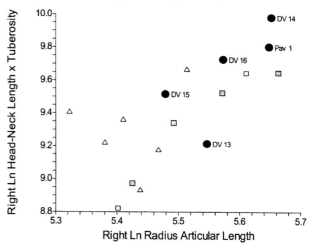

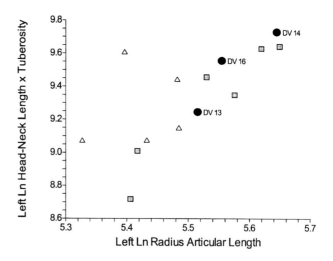

crests are strongly marked on both sides. Dorsally, the ulnae are very straight, once one is distal of the brachialis and supinator level.

The Dolní Věstonice 15 ulnar diaphyseal morphology must be based on the right bone since the left one was altered pathologically. The right shaft has an S-curve in anterior view, which is principally created by a lateral convexity near the distal supinator crest and a medial convexity by the mid pronator quadratus crest. The shaft is straight in between. The left bone is similar to the right one in curvature, despite the obvious alteration from the callus of the healed fracture (Figure 16.40). The right interosseous crest is moderately prominent for ca. 50 mm along midshaft; the left one was altered by the lesion. Both pronator quadratus crests are small, although they can be easily located.

Table 16.4 Frequencies of radial tuberosity position relative to the interosseous crest for Neandertals, Qafzeh-Skhul humans, and European earlier Upper Paleolithic humans.

	Position 1	Position 2	Position 3	N
Dolní Věstonice & Pavlov	21.4%	78.6%	0.0%	7
Earlier Upper Paleolithic	20.0%	80.0%	0.0%	20
Earlier Upper Paleolithic plus Pavlovian	20.4%	79.6%	0.0%	27
Qafzeh-Skhul	25.0%	50.0%	25.0%	4
Neandertals	3.8%	30.8%	65.4%	13
Recent Euro-Americans	27.3%	72.7%	0.0%	44
Recent Afro-Americans	18.4%	77.5%	4.1%	49
Recent Amerindians	10.0%	90.0%	0.0%	40
Recent Aleuts	11.6%	86.1%	2.3%	43

Fossil data are modified from Churchill (1994) and from personal observation. Recent human data are from Churchill (1994).

lar to those of Dolní Věstonice 16. The interosseous crests are sharply angled, but there is little else of note on them.

Although a series of diaphyseal diameters and cross-sectional properties are provided (Sládek et al., 2000), there is generally little of note in their proportions that varies through the Late Pleistocene (Trinkaus, 1983b; Hambücken, 1995). The one feature that appears to contrast between late archaic and early modern humans is the development of the pronator quadratus crest, a feature that is poorly developed in all of the Pavlovian ulnae. Comparisons of the maximum diameter across the crest to the minimum diameter at the same position (Figure 16.50) do provide an average difference between the Neandertals and the early modern humans. The one high outlier for the Qafzeh-Skhul sample is the Skhul 4 right ulna, and the relatively lower values for the Neandertals are La Ferrassie 2 and Regourdou 1 on the right and Shanidar 6 on the left. The absolutely large specimen in both comparisons, but not with an especially large crest, is Barma Grande 2. The Dolní Věstonice values are all comfortably aligned with the few other early modern humans for whom data are available, especially on the left side.

The Radial Diaphyses

The radial diaphyses of Dolní Věstonice 3 are relatively straight in anterior view; despite the damage to the right lateral diaphysis (Figure 16.35), the join between the proximal and distal pieces is clean and the contours accurate. There is a pronounced dorsal convexity to the right shaft, which is also not the result of postmortem distortion; its etiology is not apparent since neither of the preserved right arm bones shows any abnormalities. The shafts have an even tear-dropped cross-sectional shape, and the interosseous crest is prominent on both radii.

The Dolní Věstonice 13 radial diaphyses are very straight in both planes (Figures 16.36 and 16.37). They are

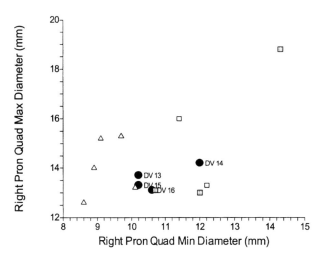

Figure 16.50 Bivariate plots of the maximum versus the minimum diameters at the pronator quadratus tuberosity for right and left ulnae. Solid circles: Dolní Věstonice specimens; gray squares: European Gravettian specimens; open squares: Qafzeh-Skhul specimens; open triangles: Neandertals.

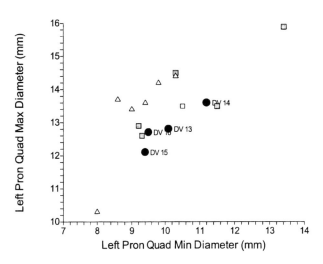

rounded laterally with modest interosseous crests. On the left there is a prominent rounded ridge extending distally from the anterolateral tuberosity, forming an anteromedial concavity that has a rugose area 22 mm long and 5 mm wide within it. The right bone has a smaller ridge and none of the associated rugosity medial of it.

The Dolní Věstonice 14 radial diaphyses are similarly straight, although the right one is slightly more dorsally convex. The interosseous crests are clear but modest, being slightly larger on the right bone. Each radius has a rugose ridge running distal of the anterolateral tuberosity for ca. 48 mm on the right and ca. 39 mm on the left.

The right radial diaphysis of Dolní Věstonice 15 is slightly more laterally curved than those of the other two members of the triple burial, but it is dorsally straight as well. The interosseous crest is prominent, has small sulci running along its dorsal and volar margins, and arcs dorsally through midshaft. The left radial diaphysis presents no evidence of healed abnormalities (traumatic or otherwise), but it presents a curious dorsal convexity that produces an angle of 8° between the proximal and distal portions of the shaft. The center of the angle is ca. 85 mm from the head, or 38% of the distance distally along the shaft. However, in anterior view, the bone exhibits a normal lateral curvature, even slightly less than that of the right radius. The interosseous crest appears normal, as it follows the dorsal curve of the whole diaphysis. Further discussion of the irregularities of the bone are included with its paleopathological analysis.

The Dolní Věstonice 16 radial diaphyses are quite straight laterally and dorsally, and they exhibit clearly delimited, but not particularly projecting, interosseous crests. Each one has a distinct rugosity for the pronator teres muscle near midshaft. The Pavlov 1 radial shafts are also rather straight in both planes. Their interosseous crests are delimited by adjacent small longitudinal sulci, and the right crest is thickened along its medial edge. Yet, they remain minimally projecting. There are also clear pronator teres tuberosities on both lateral midshafts. There are no ridges running distally from the anterolateral radial tuberosities.

The proportions of these radial diaphyses can be assessed in several ways. Of particular interest are the cross-sectional proportions and degrees of lateral curvature. The former, given limited comparative data for full cross sections, is assessed by using external diameters at the maximum interosseous crest development (the mid-proximal shaft; Figure 16.51). Even though details of cross-sectional shape are generally similar across the Late Pleistocene samples, there is a trend for the early modern humans to have less development of the interosseous crest or larger relative anteroposterior diameters. There is some overlap between the samples, and the degree of separation is greater in the right radius. The

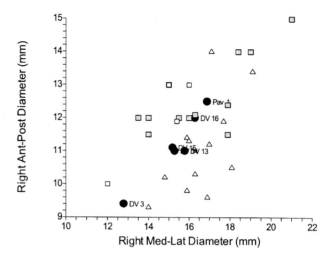

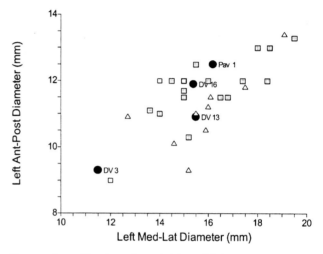

Figure 16.51 Bivariate plots of right and left anteroposterior versus mediolateral diameters at the radial interosseous crest. Solid circles: Dolní Věstonice and Pavlov specimens; gray squares: European earlier Upper Paleolithic specimens; open squares: Qafzeh-Skhul specimens; open triangles: Neandertals.

Dolní Věstonice and Pavlov radii are in the middle of the distributions, albeit generally closer to the other early modern human specimens.

The degree of lateral curvature of the radius appears to be related, in particular, to the size of the power arm for pronator teres during active pronation, and it is therefore unclear whether it should be treated as an absolute value or scaled to the length of the arm (Youm et al., 1979; Swartz, 1990). In either case, it is clear that, on average, the Neandertals have greater radial curvature than do the early modern humans, although a couple of Neandertals (La Ferrassie 2 and Shanidar 6 in the right radius; Shanidar 1 and 6 in the left radius) have relatively straight radii, and one Qafzeh-Skhul specimen (Skhul 7) has a markedly curved radius (Figure 16.52). Especially for the right bone, where the comparative samples are larger, the Dolní Věstonice and Pavlov radii fall with the majority of the other early modern humans and at the bottom of the Neandertal distribution

An indirect reflection of this curvature is the neck/shaft angle. There is a suggestion of a decrease in this angle across the samples from the Neandertals to the earlier Upper Paleolithic (Figure 16.53), but it is not significant. The Dolní Věstonice and Pavlov radii fall comfortably within or at the lower end of the ranges of variation of the other early modern humans.

The Distal Epiphyses

Only the left ulna of Dolní Věstonice 13 and the two ulnae of Dolní Věstonice 15 preserve their distal ulnae intact

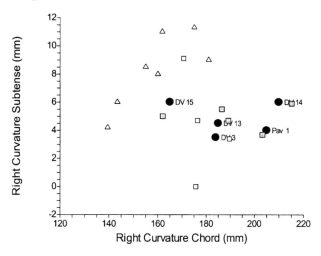

Figure 16.52 Bivariate plots of right and left radial lateral curvature subtense versus chord. Solid circles: Dolní Věstonice and Pavlov specimens; gray squares: European Gravettian specimens; open squares: Qafzeh-Skhul specimens; open triangles: Neandertals.

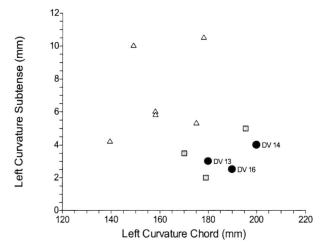

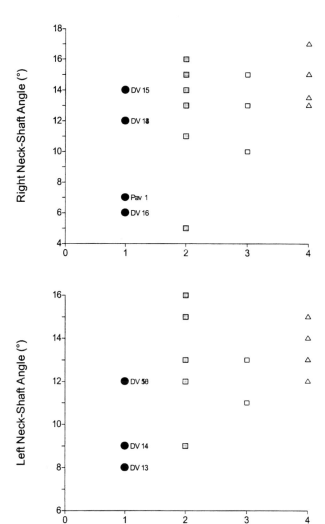

Figure 16.53 Bivariate plots of right and left radial neck/shaft angle. Solid circles: Dolní Věstonice and Pavlov specimens; gray squares: European Gravettian specimens; open squares: Qafzeh-Skhul specimens; open triangles: Neandertals. Note that Dolní Věstonice 13 and 14 have the same right angle (12°) and that Dolní Věstonice 16 and 50 have the same left angle (12°). Similarly, in the Neandertal sample, two right specimens exhibit an angle of 13° and three have an angle of 15°; and in the left radius, three specimens have an angle of 15°, for N = 7 on each side. In the Gravettian sample, there are no duplicates in the right radial angles, but in the left one there are two specimens each with angles of 9°, 13°, 15°, and 16°, for N = 9.

(there is a separate piece of the Dolní Věstonice 14 left head). None of the heads are remarkable, and the styloid processes are well developed and prominent.

The Dolní Věstonice 13 distal radii each has a prominent pair of tendon dorsal tubercles, there are large and blunt styloid processes, and the lunate and scaphoid facets are separated only near the dorsal and volar margins of the carpal facet. The Dolní Věstonce 14 distal radial epiphysis shows clear separation of the lunate and scaphoid facets. On Dolní Věstonice 15 each distal radius has a single dorsal tubercle with a groove on its medial side. These are largely symmetrical, but the right one is slightly larger. The lunate and scaphoid facets are separated only by small notches on the dorsal and volar margins of the carpal facet. The Dolní Věstonice 16 left distal radius lacks its margins, but there is no separation of the lunate and scaphoid facets on the carpal facet. The Pavlov 1 distal right radius has a single dorsal tubercle, a prominent styloid process, and only dorsal and volar separation of the scaphoid and lunate facets.

The Hand Remains

The six associated skeletons from Dolní Věstonice and Pavlov preserve varying numbers of their hand remains, from the thirty-eight known for Dolní Věstonice 16 to the single metacarpal for Pavlov 1 and the isolated phalanges of Dolní Věstonice 34 and 53 (Figures 16.54 to 16.59). In general, as is evident in Table 16.5, Dolní Věstonice 3 and 16 preserve a large number of hand remains, whereas the triple burial individuals preserve far fewer. However, many of the Dolní Věstonice 3 remains are extremely fragmentary, especially of the right hand, which was found folded under the right shoulder and thorax (Trinkaus & Jelínek, 1997), and the remainder of the metacarpals are incomplete. The Dolní Věstonice 16 remains are more intact, but many of the edges are abraded, and some of the

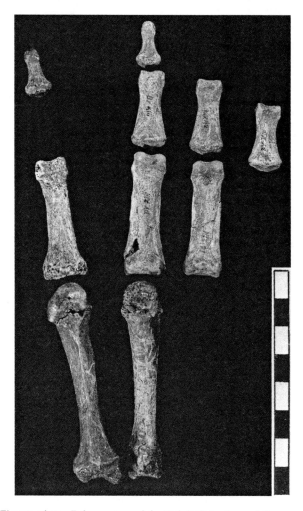

Figure 16.54 Palmar view of the Dolní Věstonice 3 left-hand phalanges. Scale in centimeters.

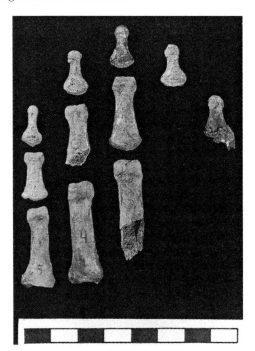

Figure 16.55 Palmar view of the Dolní Věstonice 13 left metacarpals 2 and 3 with associated phalanges. The assignment of the middle phalanges to digit is based on relative size and the absence of pathological lesions; the assignments of the distal phalanges are based on size and should be considered suggestive. Scale in centimeters.

bones are incomplete. The hand bones from the triple burial tend to be generally better preserved individually, especially for Dolní Věstonice 13 and 15, but the sets of bones are less complete. The single bones from Pavlov 1 and from Dolní Věstonice 34 and 53 do little more than indicate their presence.

All of the hand remains, except the partly crushed bones for the Dolní Věstonice 3 right hand, were mixed within each individual, and those of Dolní Věstonice 14 and 15 were partially mixed between them. As a result, carpal, metacarpal, proximal phalangeal, and pollical distal phalanx identification and side were determined on the basis of morphology (despite some difficulties with fragmentary metacarpal shafts). The middle and ulnar distal phalanges were sorted on the basis of size and, to a lesser extent, morphology.

The middle three middle phalanges were originally sorted largely by size (Sládek et al., 2000; Figures 16.54 to 16.58), with the lengths in the order of 3 > 2 > 4 > 5; their lengths should probably be ordered as 3 > 4 > 2 > 5 (the usual recent human pattern), but the lengths of the second and fourth middle phalanges are frequently close enough to make the difference of little consequence. Moreover, the allocation to digit of the middle phalanges of Dolní Věstonice 15 (Figure 16.57) is confirmed by the matching osetoarthritis across the second and fifth proximal interphalangeal articulations; the thus securely identified second middle phalanx is slightly longer than the associated fouth one (29.3 mm vs. 28.5 mm).

For Dolní Věstonice 13 (Figure 16.55), the three preserved middle phalanges have been assigned to the three ulnar digits of the more complete left hand (Sládek et al., 2000). If one assumes that they do indeed derive from the left hand, none of them can derive from the second digit since the distal trochlea of the second proximal phalanx exhibits marked osteoarthritic degeneration, and none of

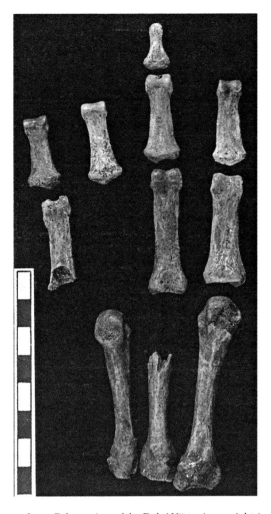

Figure 16.57 Palmar view of the Dolní Věstonice 15 right (or probably right) metacarpals and phalanges. Scale in centimeters.

the middle phalangeal bases exhibits any pathological alterations (Figure 16.55; see Chapter 19).

Among the distal phalanges, the fifth ones could usually be separated according to their smaller distal tuberosities, but the other three, even if all present, rarely presented criteria for their accurate sorting by ray. The digit numbers provided in Sládek et al. (2000) and the positions of the terminal phalanges in Figures 16.54, 16.55, 16.57, and 16.58 generally follow a scheme of 3 > 2 = 4 > 5, but these assignments should be taken as tentative.

The middle and distal phalanges, for all but Dolní Věstonice 3 (her isolated hand remains were, by default, left), were generally assigned to the side that has the more complete set of metacarpals and proximal phalanges. For Dolní Věstonice 16, where, based on symmetry, there are several pairs of middle phalanges, the actual sides were not assigned in Sládek et al. (2000); but right and left are indicated in Table 16.5 since one of each side is present,

Figure 16.56 Palmar views of the Dolní Věstonice 14 right (or probably right by association) metacarpal 2, proximal phalanges, and middle phalanges. Scale in centimeters.

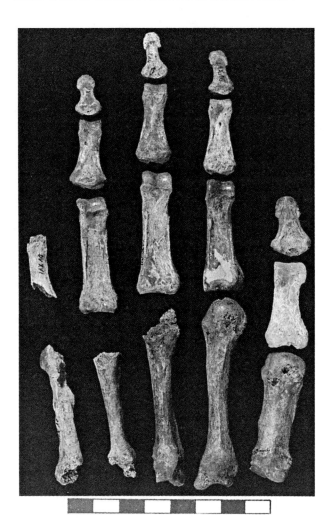

Figure 16.58 Palmar view of the Dolní Věstonice 16 right metacarpals and phalanges. Scale in centimeters.

changes functionally (e.g., Trinkaus & Villemeur, 1991; Niewoehner, 2000) is beyond the scope of this descriptive monograph. The presentation will focus principally on those aspects for which adequate comparative analyses have been completed and frameworks exist for the interpretation of these Pavlovian human hand remains.

The Proximal Carpals

The proximal carpals of the Dolní Věstonice remains are generally unremarkable. The scaphoid bones of Dolní Věstonice 3 and 16 have rather modest tubercles, whereas the one of Dolní Věstonice 14 is large and prominent. The lunate bones have variable separation of the facets for the capitate and hamate bones, with the facets being clearly separated on the Dolní Věstonice 16 lunate bones but not on that of Dolní Věstonice 15. The one pisiform bone, that of Dolní Věstonice 14, is rather small for so large an individual, and the triquetral bones exhibit nothing of note.

The Carpometacarpal Articulations and Associated Tuberosities

The only individual to preserve the first carpometacarpal articulation is Dolní Věstonice 16, for whom the right side is largely complete with both bones. The carpometacarpal surfaces are fully sellar, and they lack the vertical flatness characteristic of many Neandertal and earlier archaic *Homo* first carpometacarpal articulations (Trinkaus, 1983b, 1989; Niewoehner, 2000). However, they are

Figure 16.59 Palmar views of the Dolní Věstonice 34 (left) and 53 (right) middle hand phalanges. Scale in centimeters.

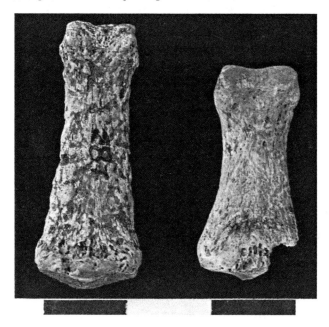

even if it cannot be determined which one of each pair belongs on which side.

The Dolní Věstonice hand remains exhibit little in the form of pathological lesions. When present, these lesions consist of proximal interphalangeal osteoarthritis in Dolní Věstonice 13 and 15, plus bilateral scaphoid-trapezium osteoarthritis in Dolní Věstonice 16. None of these changes appears to have affected more than the immediate articulation, and they are therefore considered in detail in Chapter 19. The congenital dysplasia of Dolní Věstonice 15 had no apparent effects on its hand remains.

It is clear from analyses of Late Pleistocene hand remains that there were a series of significant changes in hand anatomy between late archaic and early modern humans and that some changes continued through the evolution of modern humans (Musgrave, 1970; Trinkaus, 1983b; Trinkaus & Villemeur, 1991; Villemeur, 1994; Niewoehner et al., 1997; Niewoehner, 2000, 2001). However, the level of analysis appropriate for deciphering these

Table 16.5 Summary inventory of the Dolní Věstonice and Pavlov hand remains.

	DV 3	DV 13	DV 14	DV 15	DV 16	Pavlov 1	DV 34 & 53
Scaphoid	L		R	R	RL		
Lunate	RL		L		RL		
Triquetral	RL		R				
Pisiform			R				
Trapezium					R		
Trapezoid	L		R	R	R		
Capitate				R	RL		
Hamate	RL			R	RL		
MC 1	RL				R		
MC 2	RL	RL	R	R	RL		
MC 3	RL	L		RL	RL	L	
MC 4	RL			R	RL		
MC 5	RL				RL		
PP 1	L		R		R		
PP 2	L	L		R	RL		
PP 3	R	L	R	R	RL		
PP 4	R	L	R		RL		
PP 5	R		R	R	?		
MP 2	R		?	R	RL		?
MP 3		?	R	R	RL		
MP 4	R	?		R	RL		?
MP 5	R	?		R			
DP 1	L				R		
DP 2–4	RRR	??		R	????		
DP 5	R						

R: right or probable right bone preserved; L: left or probable left bone preserved; ?: bone of uncertain side preserved. Multiple entries for distal phalanges indicate the number of bones preserved but with uncertainty about the digits represented. For details on preservation, see Sládek et al. (2000).

also largely distinct from those of recent humans, falling among other earlier Upper Paleolithic specimens and outside of the range of variation of recent human samples, principally in their modest degree of palmar beak formation on the first metacarpal base and generally less dorsopalmarly concave/convex surfaces (Niewoehner, 2000). As with interpretations of archaic *Homo* first carpometacarpal articulations, a higher degree of dorsopalmar flattening of the surface suggests an adjustment for higher habitual levels of axial joint reaction force (Trinkaus, 1989), and in this Dolní Věstonice 16 is aligned with other earlier Upper Paleolithic specimens in being intermediate between archaic humans and Holocene people.

The trapezium tubercle, which decreases in size from late archaic to early modern humans (Trinkaus, 1983b), is not preserved on the Dolní Věstonice 16 bone.

The bases of the second metacarpals are variable in their trapezium facets and dorsal tubercles. Dolní Věstonice 13 and 14 have large trapezium facets, whereas that of Dolní Věstonice 16 is relatively small. All three have large dorsoradial tubercles, which are distinctly flat on Dolní Věstonice 15 and 16. There is little significant variation in the trapezoid/metacarpal 2 articulation. The Dolní Věstonice 14 trapezoid has a small palmar crest within the radioulnar rounding of the metacarpal 2 facet, the Dolní Věstonice 16 metacarpal 2 facet has a distinct ridge separating the radial and ulnar sides of the facet, and the Dolní Věstonice 15 facet is evenly rounded radioulnarly.

There is an average change between late archaic and recent humans in the orientation of the capitate facet on the second metacarpal, from a more proximodistal (or parasagittal) orientation to one obliquely oriented to the long axis of the metacarpal (Niewoehner et al., 1997). This is reflected in an angle between the plane of the dorsopalmar middle of the capitate facet and the long axis of the metacarpal 2, taken in the radioulnar plane of the bone. Two recent human samples provide means of about 50° [Amerindians: 44° ± 9°; $N = 39$; Euroamericans: 56° ± 9°; $N = 53$ (Niewoehner et al., 1997)], whereas a Neandertal sample provides a mean of 22° (± 9°, $N = 10$). The values for the Dolní Věstonice remains are all close to the recent human means (Dolní Věstonice 3: 42°; Dolní Věstonice 13: 56° and 48°; Dolní Věstonice 15: 55°; Dolní Věstonice 16: 56° and 66°). One non-European earlier

Upper Paleolithic specimen for which data are available, Ohalo 2, has angles of 59° and 50°. The Dolní Věstonice remains therefore are very similar to recent humans in this feature.

Associated with this orientation of the capitate facet on the second metacarpal is the orientation of the metacarpal 2 facet on the capitate bone. This is quantified by the angle between the metacarpal 2 and 3 facets, in which a higher angle indicates a more parasagittal orientation of the metacarpal 2 facet. The value for Dolní Věstonice 15 of 40° is very close to the means of two recent human samples (Amerindians: 46° ± 7°; N = 41; Euroamericans: 39° ± 9°; N = 53; Niewoehner et al., 1997) and below the higher but variable mean for a small Neandertal sample (59° ± 12°; N = 6); two earlier specimens, Krapina 200 and Tabun 1, have values of 65° and 64°, respectively. The capitate bones of Dolní Věstonice 16 lack a clear angle between the metacarpal 2 and 3 facets, and it has been possible to approximate the angle for the right bone at ca. 25°. Both of these Dolní Věstonice values are below the known range of the Neandertal capitate bones, indicating—along with the capitate facets on their second metacarpal bones—a recent human pattern of capitate/metacarpal 2 articular orientation, and one value is in contrast with the Neandertals (and earlier archaic humans; Lorenzo et al., 1999) despite some overlap between the samples.

Related to these facet orientations is the degree of projection of the styloid process of the third metacarpal bone. The relative proximal projection of the styloid process can be assessed by an index of the styloid projection (maximum length minus articular length) versus the articular length. For the three Dolní Věstonice individuals with sufficiently intact third metacarpals, the values are Dolní Věstonice 3: 3.32; Dolní Věstonice 13: 4.22 and 2.91; Dolní Věstonice 16: 5.81 and 6.40. These values are similar to those of a Neandertal sample (4.94 ± 1.95; N = 8) and the earlier Tabun 1 (3.62), and above those of the Near Eastern Ohalo 2 (1.45 and 0.88) and Qafzeh 9 (1.49) specimens. There is therefore little change in the relative projection of the styloid process across these samples.

Otherwise, the Dolní Věstonice and Pavlov capitate and third metacarpal bases present a series of minor variations. The Dolní Věstonice 13 metacarpal 3 capitate facet has a slight twist and a midulnar projection of the surface. The metacarpal 2 facet is strongly concave up onto the radial side of the styloid process. And the metacarpal 4 facet has two lobes that are joined in the middle. The Dolní Věstonice 15 capitate has a flat metacarpal 3 surface and little rounding of the dorsoradial corner for the styloid process. The associated capitate facet on the metacarpal 3 base is flat, with a small proximal extension for the styloid process. Indeed, the projection of the styloid processes (2.0 and 2.3 mm) is due mostly to the ulnar turning of the base relative to the shaft since the styloid processes project only 0.9 mm from the plane of the facet. There is a large metacarpal 2 facet up onto the radial styloid processes.

The Dolní Věstonice 16 capitate bones are strongly rounded dorsoradially for the styloid processes, and as mentioned above, it is not possible to clearly distinguish between the metacarpal 2 and 3 facets. The capitate facets on the metacarpal 3 bases are undulating, with a midulnar projection; this makes it difficult to measure the projection of the styloid processes relative to the facet, and hence their approximate indication in Sládek et al. (2000). The metacarpal 2 facets are large and concave, and they appear to reduce the sizes of the capitate facets radially. The base of the Pavlov 1 third metacarpal is eroded, which makes it impossible to be certain of the extent of the styloid process or the margins.

The hamate bones vary in their facets for the fourth and fifth metacarpals. Those of Dolní Věstonice 3 have a clear angle between the two facets, and both are concave. The Dolní Věstonice 15 hamate bone has a concave metacarpal 4 facet but a convex metacarpal 5 facet. The two facets on the Dolní Věstonice 16 hamate bones blend in with each other and cannot be clearly separated. Both the fifth metacarpals of Dolní Věstonice 3 and 16 have radioulnarly concave facets, or fully sellar surfaces. The fourth metacarpals of Dolní Věstonice 15 and 16 are notable for the strong ulnar deviations of their bases.

The hamuli of these individuals are generally modest in size, with Dolní Věstonice 3 and 15 having rather small ones and Dolní Věstonice 16 having slightly larger ones. The hamuli of Dolní Věstonice 15 and 16 are largely vertical, whereas those of Dolní Věstonice 3 deviate strongly distally. This modest development of the hamulus is accompanied by a very small abductor tubercle on the ulnar base of the Dolní Věstonice 3 fifth metacarpal bone, although the Dolní Věstonice 16 fifth metacarpals have relatively large ones.

The Pollex

Information on pollical anatomy comes from Dolní Věstonice 3, 14, and 16. Only the last has a sufficiently complete pollex to permit overall assessments, but it was possible during "excavation" of the Dolní Věstonice 3 postcranial remains to measure the lengths of the first and third metacarpals before they descended into small fragments (Trinkaus & Jelínek, 1997).

The relative length of the pollex was assessed by comparing the metacarpal 1 length to that of the metacarpal 3 (Figure 16.60), and it is apparent that there is a fairly regular relationship between these lengths across the Late Pleistocene samples. Only Caviglione 1 and Předmostí 14 among the early modern humans have moderately short

dertals except La Ferrassie 2 are well above the early modern human distribution. As has been noted before (Trinkaus, 1983b; Trinkaus & Villemeur, 1991; Villemeur, 1994), this relative shortening of the proximal phalanx and more marked lengthening of the distal phalanx among the Neandertals enhanced their power grips relative to those of modern humans. The Dolní Věstonice specimens, in both comparisons, fall with the other early modern humans, who are similar to recent humans in these proportions (Trinkaus & Villemeur, 1991).

The Dolní Věstonice 3 distal first metacarpal bone has little more than a line for the opponens pollicis "crest," and the distal tuberosity of the distal phalanx is very small for the size of the bone and is rather smooth. The Dolní Věstonice 14 proximal pollical phalanx has a distinct

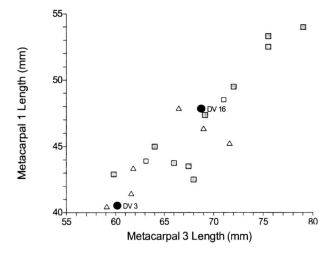
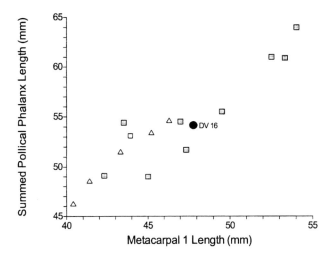

Figure 16.60 Metacarpal 1 versus metacarpal 3 length (above) and summed pollical phalangeal length versus metacarpal 1 length (below). Solid circles: Dolní Věstonice specimens; gray squares: European Gravettian specimens; open squares: Qafzeh-Skhul specimens; open triangles: Neandertals.

Figure 16.61 Bivariate plots of proximal pollical phalangeal length (above) and distal pollical phalangeal length (below) to metacarpal 1 length. Solid circles: Dolní Věstonice specimens; gray squares: European Gravettian specimens; open squares: Qafzeh-Skhul specimens; open triangles: Neandertals.

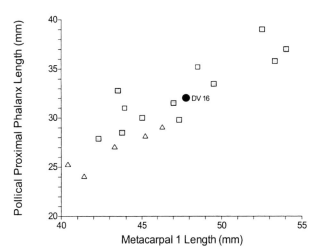
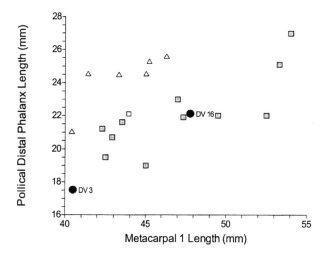

thumb metacarpal bones, and they are joined by Kebara 2. The others, including Dolní Věstonice 3 and 16, fall along a relatively narrow distribution. Similarly, there is little difference between these Late Pleistocene samples in summed pollical phalangeal length to metacarpal 1 length (Figure 16.60), with only Paglicci 25 and Předmostí 10 having slightly lower relative values.

When the individual pollical phalanges are compared to metacarpal 1 length, however, a different pattern emerges (Figure 16.61). In the comparison of proximal phalanx to metacarpal 1 length, the Neandertals are largely along the lower margin of the early modern humans, with only La Ferrassie 2 being close to the latter and Paglicci 25 falling among the former. In the comparison of distal phalanx to metacarpal 1 length, all of the Nean-

swelling on its radial palmar base, adjacent to a notch on the radial side of the palmar edge of the facet for the metacarpal head. The Dolní Věstonice 16 metacarpal 1 has a discrete concave scalloped area with a tubercle at its distal end for the first dorsal interosseous muscle. However, its opponens pollicis crest is minimally projecting on the distal radial shaft. On the head, the radial peak is very small. There is an ulnar palmar extension of the head articular surface, which becomes concave proximally and fits into a modest concavity on the proximal phalangeal ulnar base in full flexion. The distal pollical phalanx has large radial and ulnar tubercles, which produce a maximum base breadth 3.1 mm larger than the proximal articular breadth. The proximal palmar fossa ("flexor pollicis longus pit") is smooth and slightly oval, and it is bordered by a palmar swelling onto the midshaft. There is a prominent peaked crest for the insertion of flexor pollicis longus. The distal ungual tuberosity is narrow and smooth, and it has only a slight serration along its dorsal margin. The spines are small and not separated from the sides of the shaft, with the ulnar one protruding slightly more than the radial one. There is a modest (3°) ulnar deviation of the bone relative to its base. The angle is similar to those of three earlier Upper Paleolithic specimens (Caviglione 1: 2°; Pataud 4: 2°; Pataud 6: 4°) and Qafzeh 8 and 9 (2° and 1°); it is below the range (4° to 11°) of a Neandertal sample (6.3° ± 2.1°; $N = 7$).

The Ulnar Digits

The metacarpal bases are described above with the carpometacarpal articulations. The remainder of the metacarpals and the ulnar digit phalanges present a scattering of features of note.

The Dolní Věstonice 3 metacarpal shafts are notable for their weak to imperceptible markings for the interosseous muscles. This is combined with a complete absence of a crest for the opponens digiti minimi on the fifth metacarpal. The only metacarpal head, that of the right second metacarpal, is damaged and provides little information other than its overall dimensions. The proximal phalanges have weak flexor tendon sheaths along their more distal palmar diaphyses (Figure 16.54). There are slight indications for the insertions of the flexor digitorum superficialis tendons on the middle phalanges and small rugose pits at the bases of the distal phalanges for the flexor digitorum profundus tendons. The terminal phalanges have small tuberosities with poorly developed spines on the ones from the three middle digits and none on the one for the fifth digit.

The metacarpals 2 and 3 of Dolní Věstonice 13 have moderately strong interosseous muscle markings. The metacarpal 2 heads are strongly asymmetrical, with a marked dorsoradial beveling of the subchondral bone. In conjunction with this, each bone has a large palmar radial process, extending 3.0 mm on the right and 4.2 mm on the left (Figure 16.55). Moreover, although the head articular breadths are given (Sládek et al., 2000) as 14.8 and 15.0 mm, without the palmar radial process they would measure only 11.8 mm and 10.8 mm, respectively. The proximal phalanx 2 has a large proximal radial tubercle for the first dorsal interosseous muscle, but the flexor tendon sheath attachments do little more than form angles on the sides of the palmar shaft. The flexor tendon sheath lines are even weaker on the third and fourth proximal phalanges. Similarly, the three middle phalanges have little evidence of the insertions of the flexor digitorum superficialis tendons, and the two distal ones have essentially no indications of the flexor digitorum profundus tendons beyond the capsular attachments. The distal tuberosities are small, are rounded, and extend little beyond the sides of the narrow shafts. Only one of them (the one positioned at the end of the second ray) has a hint of spines, but they do not project beyond the diaphyseal margins.

Dolní Věstonice 14 preserves only a nondescript shaft of a second metacarpal, plus portions of five ulnar phalanges (Figure 16.56). The third proximal phalanx has a distinct proximal radial tubercle, and the three proximal phalanges have clear and sharp, but nonprojecting, flexor tendon sheath ridges. The two middle phalanges have distinct proximal radial and ulnar ridges for the flexor digitorum superficialis insertions but no evidence of tendon attachments on the palmar surfaces of the shafts.

The Dolní Věstonice 15 metacarpal 2 shaft curves slightly radially, and its head has a clear beveling off of the dorsoradial corner. There is a moderately large proximal radial palmar tubercle, producing a maximum articular breadth of 14.2 mm, rather than the 11.3 mm of the head proper. There is little of note on the fourth and third metacarpals. The flexor tendon sheaths increase in prominence from barely present on the second proximal phalanx to distinct but small on the fifth proximal phalanx. The proximal phalanx 2 has a large tubercle for the first dorsal interosseous muscle, and there is a smaller radial tubercle on the proximal phalanx 3. The middle phalanges from the middle three rays are modest, with weak muscle markings, and the one from the little finger has no trace of muscle markings. The one distal phalanx has no separate trace of the flexor digitorum profundus tendon attachment, and the smooth tuberosity is well within the borders of the shaft, lacking clear spines.

The second and third metacarpals of Dolní Věstonice 16 have clear, nonprominent markings for their interosseous muscle attachments, but they are less strongly marked on the fourth and fifth metacarpals. The metacarpal 2 head has the same distinct beveling of the dorsoradial head, and there is a radial palmar extension of the head, which exaggerated the breadth from 12.8 mm for the head proper to an articular breadth of 14.9 mm. The right fifth metacarpal shaft has a definite marking for the opponens digiti minimi muscle, which consists of small tubercles on the ulnar side near midshaft and a small crest on the ulnar palmar shaft about two-thirds of the length distally along the shaft. In this, it is relatively unusual among early modern humans in showing such a marking for this muscle, although about three-quarters of the late archaic human fifth metacarpals exhibit crests for this muscle at least as large as those of Dolní Věstonice 16 (Trinkaus, 1983b). The Dolní Věstonice 16 proximal phalanges stand out in the Pavlovian sample for having strongly marked flexor tendon sheaths, especially on the middle three digits but also on the fifth ray (Figure 16.58).

The Dolní Věstonice 16 middle phalanges exhibit varying levels of rugosity for the flexor digitorum superficialis tendons. The most strongly marked are those on the bone identified as from the second digit, but it is the third-longest middle phalanx and may therefore come from the fourth digit. The next strongest in marking are those on the longest phalanx, from the third ray, followed by those on the shortest phalanx, on the fifth ray. Also, there is some asymmetry in the strengths of the markings between the pairs of middle phalanges, and it is tempting to say that the more strongly marked ones come from the right hand because of the asymmetry with right-side dominance further proximally on the arm. The four terminal phalanges have distinct grooves on their bases for the flexor digitorum profundus tendons, modest shafts, and larger tuberosities than in any of the other Dolní Věstonice terminal hand phalanges. They are also the only ones that extend beyond the sides of the shafts, and three of the four have distinct proximal radial and ulnar spines.

The two isolated middle phalanges, Dolní Věstonice 34 and 53, are eroded and provide little surface detail. They have been assigned to either second or fourth digits according to their overall sizes, but these identifications remain tenuous, given their isolated natures.

Summary

These upper limb remains from the Dolní Věstonice and Pavlov individuals therefore provide a relatively complete picture of their shoulder, arm, and hand skeletal anatomies. The overall picture that they paint is one of moderately gracile bones for a Pleistocene human huntergatherer sample, but one that is largely in the middle of the known range of variation for European earlier Upper Paleolithic humans. Indeed, only the Neandertal Saint-Césaire 1 specimen (Trinkaus et al., 1999c) stands out from these other earlier Upper Paleolithic remains in some aspects of robusticity. It is nonetheless possible to characterize a number of aspects of their upper limb remains in the context of Late Pleistocene late archaic and early modern human remains.

The clavicles are notable for their small costoclavicular facets, variable proximal shaft curvature and conoid tubercles, but consistently well-marked anterodistal deltoid muscle attachments. The last is of interest, given the consistently weak development of the deltoid tuberosity on all of their humeral diaphyses. Their scapulae are relatively narrow, have moderately broad glenoid fossae, and exhibit axillary borders that are variable but mostly have the bisulcate pattern. All of them exhibit a curious angulation of the dorsal scapular spines, which may be (as suggested by Vlček, 1991) a peculiarity of these Pavlovian humans.

Their humeri exhibit smooth, moderately robust diaphyses with rather weak markings for the muscles that frequently leave distinct indications of their attachment sites—pectoralis major, latissimus dorsi, teres major, and deltoideus. The modest breadths of the pectoralis major insertions are characteristic of earlier Upper Paleolithic humans, but they nonetheless remain small in their development. The diaphyses are rounder than those of late archaic humans, and they exhibit variable levels of asymmetry.

In the forearm, the proximal epiphyses possess the anteroproximal orientation of the ulnar trochlear notch and relatively short radial necks characteristic of more recent humans, but the large dimensions of their radial tuberosities seem to counteract the effects of those shorter radial necks. Their ulnar diaphyses have variable development of an S-curve, and their radii consistently exhibit little lateral curvature (also reflected in their low neck/shaft angles). Their radial interosseous crests are generally nonprojecting, as are their ulnar pronator quadratus crests.

The modest development of humeral and antebrachial muscle marking is also apparent in their hand remains, which exhibit variable scaphoid tubercles, a small pisiform bone, small hamuli, minimal opponens pollicis crests, variable opponens digiti minimi crests, variable metacarpal 5 adductor tubercles, mostly very small flexor tendon sheath crests, flexor digitorum superficialis and profundus insertions that vary from invisible to modest, and small, nonprojecting distal phalangeal tuberosities. Only the Dolní Věstonice 14 scaphoid tubercle and the Dolní Věstonice 16 metacarpal 5 adductor tubercle and opponens digiti minimi crest indicate any musculoligamentous hypertrophy, and their hand remains are elsewhere rather gracile. The thumb length proportions of Dolní Věstonice 3 and 16 match those of early and recent modern humans, and the first and fifth carpometacarpal articulations of the same two individuals are fully sellar. The distal pollical phalanx exhibits little ulnar deviation.

The distal tuberosities are notable for their small sizes (all except those of Dolní Věstonice 16 within the margins of the shafts), smooth surfaces, and frequent absence of clear proximal spines.

It is interesting that the upper limb remains of Dolní Věstonice 15, despite the deformities of the right distal humerus and the left forearm, exhibit levels of hypertrophy that are consistently similar to those of the other Pavlovian human remains. There is nothing to indicate any disuse atrophy of these shoulder, arm, and hand bones.

In most of these various features, the Dolní Věstonice and Pavlov remains are less robust than those of the preceding late archaic humans, suggesting a significant decrease in load levels in the upper limb. However, they are also less robust in some features than later Upper Paleolithic Europeans, as well as scattered recent human samples (Churchill, 1994; Churchill & Smith, 2000; Holt et al., 2000), suggesting that these Pavlovian people reaped the benefits of a mechanically efficient Gravettian technology without the added burdens of the intensive hunting and gathering and then agriculture of some later human populations.

17

The Pelvic Remains

Erik Trinkaus

The Dolní Věstonice and Pavlov pelvic remains consist principally of the three largely complete pelves, those of the triple burial individuals. To these are added pieces in varying states of completeness and distortion for Dolní Věstonice 3 and 16 (Figures 7.1–7.4, 7.6, 17.1–17.7). Dolní Věstonice 3 has portions of both ilia, but they were either crushed (on the left side) or reduced to an inferior section (on the right side). The associated sacrum retains only portions of three bodies crushed into the left ilium. The pelvis of Dolní Věstonice 16 is largely present, but it is currently in eight identifiable pieces and innumerable smaller fragments. Its sacrum, as with all of its vertebral remains caudal to T1, was reduced to unidentifiable fragments. None of the pelvis of Pavlov 1 survives, and no isolated human pelvic pieces have been identified from these sites.

The morphological features that relate to sexual diagnosis are presented and evaluated in Chapter 7, and the overall breadths of the more complete pelves as they relate to body proportions are compared in Chapter 12. This section, therefore, addresses the remaining features, primarily those related to detailed morphological aspects of the hip bones (ossa coxae) and sacra by individual. Despite a wealth of morphometric values that can be derived from these remains, especially those of Dolní Věstonice 13 to 15 (Sládek et al., 2000), comparative Late Pleistocene data are scarce and most of the values relate to either sexual diagnosis or overall size. Parturitional issues (per Tague, 1992) are not directly relevant since only Dolní Věstonice 3 is female (Chapter 7) and her pelvic remains are insufficiently preserved to evaluate such reproductive issues.

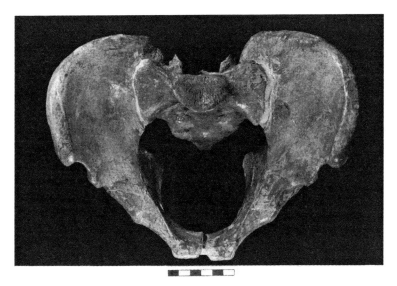

Figure 17.1 Superior view of the Dolní Věstonice 13 articulated pelvis. Scale in centimeters.

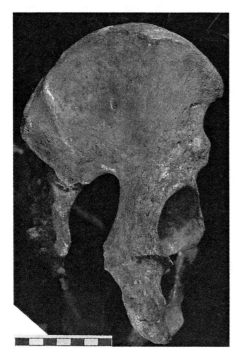

Figure 17.2 Posterolateral view of the Dolní Věstonice 13 right ilium and ischium. Scale in centimeters.

The Acetabulae

The remaining portions of the Dolní Věstonice 3 acetabulae are sufficient only to provide an approximate height for the left side and paleopathological observations on the weight-bearing area of the lunate surface on both sides (see Chapter 19). The Dolní Věstonice 13 acetabulae have smooth lunate surfaces. The notches are approximately square, with corner extensions toward the anterior inferior iliac spine. There is little separation in relief between the lunate and notch surfaces, especially on the ischial portions. Both of the Dolní Věstonice 14 acetabulae are damaged with restoration and matrix, but it is possible to see that the lunate surfaces are even with square notches. The Dolní Věstonice 15 acetabulae similarly have even lunate surfaces with square notches. With both Dolní Věstonice 13 and 15, the right acetabular height is slightly smaller than the left one (1.8 mm for Dolní Věstonice 13; 1.5 mm for Dolní Věstonice 15), but these fit with their slightly smaller right femoral head diameters (1.0 mm smaller for Dolní Věstonice 13; 1.5 mm for Dolní Věstonice 15). The fragments of the Dolní Věstonice 16 acetabulum provide little information.

The Ilia

The left ilium of Dolní Věstonice 3 was extensively crushed, and the right one preserves mostly the inferior portion. At least the lower portion of the right ilium appears to have had a moderate degree of iliac flare. The arcuate lines are evenly rounded, without a sharp angle between the false and true pelves. On the posteroinferior external surface, well preserved on the right but evident on the left, there is a raised and prominent rugose area, almost forming a tubercle, which is 29.5 mm anteroposterior by 18.0 mm superoinferior on the right side; it appears to be a rugose origin for gluteus medius bilaterally.

The Dolní Věstonice 13 ilia are smooth externally, without even a trace of the line for gluteus minimus. The iliac pillar is minimally developed, and there is a small external peak bilaterally at the top of each pillar. The iliac crest is relatively thick anterior of the pillar and then even wider for the anterior superior iliac spine, with a minimal thickness of 12.4 mm and 13.5 mm anterior of the

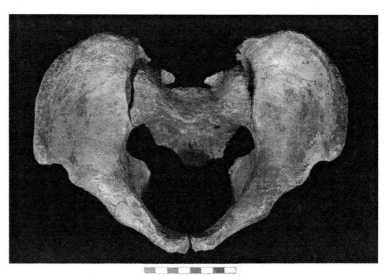

Figure 17.3 Superior view of the Dolní Věstonice 14 articulated pelvis. Scale in centimeters.

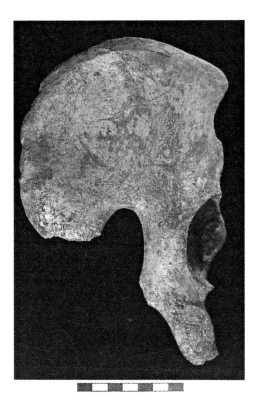

Figure 17.4 Posterolateral view of the Dolní Věstonice 14 right ilium and ischium. Scale in centimeters.

pillar, but then 15.2 mm and 16.5 mm thick at the anterior superior iliac spine on the right and left, respectively. The anterior superior iliac spine thickness is produced by a continuous internal arc combined with an external flaring at the spine. The anterior inferior iliac spine is internally rotated and slightly concave internally, with a clear bulge externally. The inferior edge is almost on the acetabular margin. There is no fossa or rugosity for the reflected head of the rectus femoris muscle. The iliac fossa is smooth, with a palpable depression just lateral of the auricular surface. The right arcuate line is smooth and rounded, and the left one has a slight angle but no clear line. The iliac tuberosities are relatively smooth and thick, with maximum thicknesses of 25.6 mm and 23.0 mm on the right and left sides, respectively.

The external surfaces of the Dolní Věstonice 14 ilia are smooth, with modest swellings for the pillars. There is no apparent thickening of the crest at the pillar, but the iliac crest secondary center of ossification appears to have been only partially fused. The anterior superior iliac spine is moderately thickened—13.3 versus 12.5 mm on the right and 13.5 versus 12.3 mm on the left—between the spine thickness and the minimum thickness just behind it. They turn only slightly externally. The anterior inferior iliac spines meet the acetabular margins, and they have a trace of an external bulge on the left and a small one on the right. The iliac fossae are even concavities, with a slightly greater concavity anterolateral of the auricular surfaces. The arcuate lines are rounded, with a line on the right side. The iliac tuberosities are damaged, but they appear to have mild external rugosity.

The Dolní Věstonice 15 ilia have smooth external surfaces with little evidence of pillars but distinct hollows dorsally just ventral of the iliac tuberosities. The iliac crest tapers anteriorly to minimum thicknesses of 9.3 mm and 8.4 mm. The right one thickens little at the anterior superior iliac spine (10.2 mm), but there is a more marked thickening on the left side (11.3 mm); the thickening of the left anterior superior iliac spine is both internal and external. The anterior inferior iliac spines are curved inward, and there are small shelves between the spines and the acetabular margins. The iliac fossae are evenly concave without the depressions adjacent to the auricular

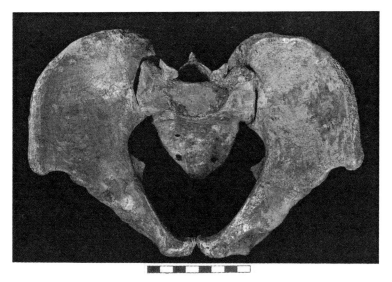

Figure 17.5 Superior view of the Dolní Věstonice 15 articulated pelvis. Scale in centimeters.

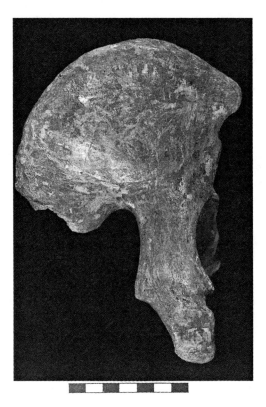

Figure 17.6 Posterolateral view of the Dolní Věstonice 15 right ilium and ischium. Scale in centimeters.

surfaces present on the ilia of Dolní Věstonice 13 and 14. The arcuate lines are rounded.

From the pieces of the Dolní Věstonice 16 ilia it is possible to note that the external gluteal lines are clearly rugose, there are strong lateral peaks on the iliac crests above the iliac pillars, the right anterior superior iliac spine turns externally, and the left arcuate line is rounded.

The Ischia

The Dolní Věstonice 13 ischiae have smooth tuberosities whose edges project minimally. They are strongly rotated laterally. There is no trace of the obturator internus bursa sulcus; it was probably located slightly above the tuberosity. The ischial spines are damaged, but the tips clearly were strongly rotated internally.

The Dolní Věstonice 14 ischial tuberosities are only partially preserved, but they are relatively narrow and are strongly rotated laterally. There is no trace of the obturator internus bursa sulcus; it was probably proximal of the tuberosity. The ischial spines were strongly internally oriented.

Dolní Věstonice 15 similarly exhibits strongly laterally rotated ischial tuberosities and internally oriented ischial spines (the left one is complete). There is a trace of the left obturator internus bursa attachment on the left side by the ischial spine, placing it above the tuberosity. There is a strong transverse crease across each superior tuberosity.

The right and left ischial tuberosities of Dolní Věstonice 16 are also laterally rotated, and the left one exhibits an undulating tuberosity but without sharp furrows or ridges. The obturator internus sulcus did not impinge on the tuberosity.

The Pubic Bones

Dolní Věstonice 13 exhibits short and fat superior pubic rami. There is a swelling above each acetabulum, forming a slightly rugose iliopsoas groove anteromedial of the anterior inferior iliac spine. There is a clear pectineal line up to each pubic tubercle, forming a small crest on the right side. The right crest largely fades out before the tubercle, but the left one forms a raised and rounded ridge that leads into the tubercle. The right tubercle is rounded, extending mostly anteriorly, and then is squared-off anteriorly. The left one is similar. The ischiopubic rami are markedly rotated anteriorly, starting just inferior of the symphysis, especially on the left side. They then continue to the ischial tuberosities along straight lines, but with only moderate thickening. There is an appearance of thickening in medial view, but it is largely a product of the anterolateral rotation of the rami.

The Dolní Věstonice 14 superior pubic rami are short and stout, and the obturator sulci are poorly delineated.

Figure 17.7 Posterolateral view of the Dolní Věstonice 16 left inferior ilium and ischium. Light gray areas are reconstructions of missing portions. Scale in centimeters.

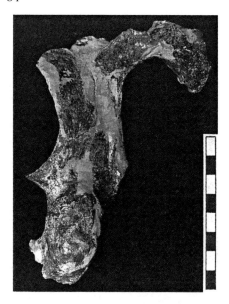

The pectineal crests are more acute angles than crests, and they spread out anteromedially for the pubic tubercles. The tubercles are damaged but appear to extend only ventrally and not laterally. The ischiopubic rami are also rotated anterolaterally, but much less so than those of Dolní Věstonice 13.

The Dolní Věstonice 15 superior pubic rami are very short and stout, with a narrow symphyseal body that tapers inferiorly. There are prominent pubic tubercles, and the pectineal crests are prominent on the dorsal portions of the rami above the acetabulae (especially on the right), tapering down to the surface above the obturator foramina. The ischiopubic rami appear everted, but they are partially damaged.

Little of the Dolní Věstonice 16 pubic bones remains, but there is evidence of a small pectineal crest—more of an acute angle than a crest—above the left acetabulum.

None of the three Dolní Věstonice specimens preserving the pubic region exhibit the mediolateral elongation of the pubic bone and the ventral thinning of the superior ramus that are characteristic of Neandertals (Trinkaus, 1976b, 1984b; Hublin et al., 1998) and appear to be found throughout archaic *Homo* (Arsuaga et al., 1999; Rosenberg et al., 1999). They have the derived morphology and proportions that are particularly characteristic of recent human males.

The Sacra

As mentioned in Chapter 14, the sacra are described here as part of the pelvis, even though they are developmentally part of the axial skeleton. Four of the individuals from Dolní Věstonice provide portions of the sacrum, Dolní Věstonice 3 and 13–15. The first is extremely fragmentary, retains only portions of S1 to S3, and is laterally crushed into the left ilium and the incomplete L4 and L5 vertebral bodies. Only small portions of the ventral S1 to S3 bodies are visible. The other three sacra appear to be largely complete, but they have been restored with filler to varying degrees (Figures 17.8 to 17.10). The Dolní Věstonice 13 sacrum remains in the assembled pelvis for that individual, but the Dolní Věstonice 14 and 15 sacra are separate. Despite the presence of numerous ossa coxae pieces for Dolní Věstonice 16, none of its sacrum has been recognized from among the fragments.

The Dolní Věstonice 13 sacrum presents an even, gradual ventral curve, with a modest ventral concavity. There is a hint of incomplete fusion between the S3 and S4 ventral bodies, but the S1 annular ring is fully fused on. The craniodorsal edge of the sacral canal is 17.4 mm below the level of the sacral plateau. The lower edge of the auricular surface is at the level of the middle S3.

The sacrum of Dolní Věstonice 14 has a ventral curve with a well-developed ventral concavity at the level of the S3. The fusion of the ventral bodies is difficult to discern because of filler along the surface. They were all laterally completely fused, but there is evidence of incomplete fusion between S3 and S4 and between S4 and S5. It is also likely that the two more cranial fusion lines were partly patent. The craniodorsal edge of the sacral canal is 14.6 mm below the plane of the base.

The Dolní Věstonice 15 sacrum appears long, narrow, and strongly curved at its caudal end. This is largely a product of having six vertebral elements, of which the sixth one is equivalent to the first coccygeal vertebra (see below). As a result, its ventral arc is normal in curvature down to the S5, and then it curves distinctly ventrally for the S6/Cx1. This is reflected in ventral subtense to arc indices for Dolní Věstonice 15, in which the index for the cranial five vertebrae is 20.1 but increases to 24.6 for the six vertebral elements. However, neither of these values is exceptional since they fall between those for Dolní Věstonice 13 (18.8) and 14 (24.7). It is difficult to ascertain the degree of ventral body fusion, given restoration with filler, but the S4 to S5 join was probably open and

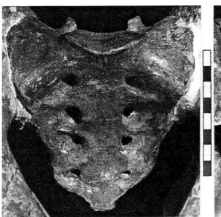 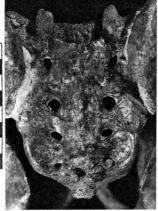

Figure 17.8 Ventral (left) and dorsal (right) views of the Dolní Věstonice 13 sacrum. It is partially obscured by being mounted in the articulated pelvis. Scale in centimeters.

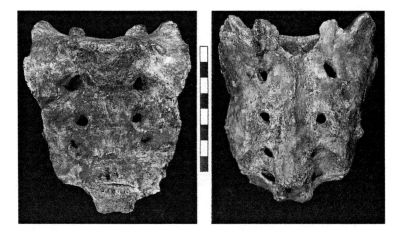

Figure 17.9 Ventral (left) and dorsal (right) views of the Dolní Věstonice 14 sacrum. Scale in centimeters.

the S5 to S6/Cx1 surfaces were separate. The craniodorsal margin of the sacral canal is asymmetrical since it is lower on the left, but the midline is relatively high, being 6.0 mm below the base. The S1 spinous process also deviates to the left. There is also a slight deviation of the caudal two vertebrae to the left, but it is unclear to what extent this might be due to the restoration of the damaged bone.

Sacral Vertebral Number

The number of sacral vertebrae in the Dolní Věstonice 3 sacrum cannot be determined, but the relative proportions of the three cranial ones that are present are in agreement that it originally possessed five vertebrae. The Dolní Věstonice 13 and 14 sacra are each made up of five vertebrae. Since they retain articular surfaces for both the last lumbar and the first coccygeal vertebrae, they indeed had only five sacral vertebrae.

The Dolní Věstonice 15 sacrum, however, exhibits six vertebrae. The most cranial vertebral element is morphologically similar to first sacral vertebrae, but the most caudal one resembles a first coccygeal vertebra fused on to the S5. The presence of twelve thoracic and five lumbar vertebrae in the Dolní Věstonice 15 axial skeleton (Chapter 14) indicates that this expansion of the sacrum is from a caudal, and not a cranial, extension of the bone.

Among recent humans, the majority have five sacral vertebrae, a significant minority have six, and very rare individuals have four or seven. In a pooled Japanese, Euro-American, and Afro-American sample ($N = 967$; Schultz, 1930), the dominant pattern is found in 77.7% of individuals, whereas the pattern seen in Dolní Věstonice 15 occurs in 21.4% of the sample. There is therefore nothing exceptional in either arrangement among recent humans.

Sacral Proportions

The sacra of Dolní Věstonice 13 and 14 have overall breadth/length (anterosuperior breadth to ventral sacral length) proportions similar to those of other Late Pleistocene and recent humans. The resultant indices of 93.8 and 105.2, respectively, are close to those of the two Gravettian specimens that provide both measurements (Cro-Magnon 1: 107.2; Fanciulli 4: 100.0) and the three similarly complete Neandertals (Kebara 2: 107.0; Shanidar 1: ca. 96.7; Shanidar 3: ca. 114.7; Trinkaus, 1983b; Holliday, 1995). All of these values fall within a range of means of recent human samples (Radlauer, 1908).

In contrast, an index of the same sacral measurements for Dolní Věstonice 15 (ca. 84.9) makes it appear exceptionally long and/or narrow. However, if one uses a ventral length only to the caudal S5, the resultant index (ca. 93.4) is within measurement error of the Dolní Věstonice 13 value and well within recent human ranges of variation.

Similar proportional assessments can be made by comparing ventral length to femoral bicondylar length. Using the average of their right and left femoral lengths,

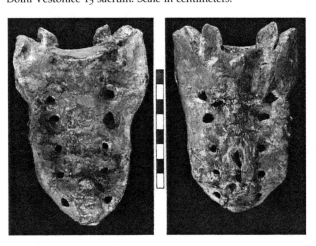

Figure 17.10 Ventral (left) and dorsal (right) views of the Dolní Věstonice 15 sacrum. Scale in centimeters.

these values are 25.3 and 20.9 for Dolní Věstonice 13 and 14. The Gravettian Fanciulli 4 and 5 specimens provide indices of 21.9 and 24.8 (Holliday, 1995), and the Neandertal Shanidar 1 and 3 remains provide estimates of 23.5 and 23.3 (Trinkaus, 1983b). A recent human (Egyptian) sample has similar indices (males: 21.4 ± 1.7; N = 25; females: 22.8 ± 1.9; N = 31; Warren, 1897). Again, the sacral size of Dolní Věstonice 15 is difficult to scale, given both its six elements and its relatively short lower limbs (Chapters 12 and 13). The use of its less affected left femur and the six vertebra sacral ventral height provides an high index of 28.5, whereas an S1 to S5 ventral height reduces the index to 25.9. The latter value, despite the relative abbreviation of its femoral lengths, is close to those of Dolní Věstonice 13 and Fanciulli 5 and between 1.6 (female) and 2.6 (male) standard deviations from the recent human sample.

Finally, the breadths of the Dolní Věstonice sacra can be evaluated relative to their bi-iliac breadths. Dolní Věstonice 13 and 14 yield sacral to bi-iliac breadth indices of 39.8 and 42.5, whereas the range of estimates for the Dolní Věstonice 15 bi-iliac breadth provides a range of indices of 38.0 to 39.7. None of these values is exceptional. One Neandertal (Kebara 2) has an index of 39.0, and a Gravettian sample has a range of 37.3 to 45.6 (41.6 ± 2.8; N = 7; Holliday, 1995).

Consequently, the overall proportions of the Dolní Věstonice 13–15 sacra are similar to those of other Late Pleistocene and recent humans, especially if only the five cranial vertebrae of the Dolní Věstonice 15 sacrum are employed.

The Sacral Hiatus

The sacral hiatus of Dolní Věstonice 13 extends cranially to the lower part of the S4, and it has a large nubbin of bone (distinct from a sacral cornua) 19.3 mm wide and 12.2 mm high, which bridges across the top of the hiatus. Caudally, the processes on the S5 arch inward across the exposed canal. The Dolní Věstonice 14 hiatus is similar, at the same level with a swelling of bone at least 17 mm wide and 12.5 mm high, bridging the cranial end of the hiatus. The S5 processes also strongly curve medially across the exposed canal. The Dolní Věstonice 15 hiatus lacks the prominent nubbin of bone across its cranial end, but it extends more cranially, ending in the middle of the S3.

This variation in the height of the sacral hiatus falls well within the recent human ranges of variation. In samples of Euro-Americans (N = 519), Afro-Americans (N = 694), and Japanese (N = 92; Trotter & Lanier, 1945; Sekiguchi et al., 2004), the majority of the individuals (83.8%, 79.7%, and 84.8%, respectively) have a sacral hiatus that extends at least to the lower part of the S4. A sacral hiatus to at least the middle of the S3 is uncommon but not exceptional in recent humans: 6.2%, 3.9%, and 19.6%, respectively. Comparative Gravettian data are not available. Among Neandertals, all of the specimens (N = 6) have relatively high apices, with three having it around the caudal S3, two by the S2/S3 juncture, and one at the S2 level (Trinkaus, 1983b; Rak, 1991).

Summary

The pelvic remains of the Dolní Věstonice individuals are therefore generally similar to those of recent humans. The Dolní Věstonice 3 and 16 remains tend to have more prominent muscular attachment markings than do those of the triple burial individuals, possibly in part because of their more advanced ages at death. As with the rest of its skeleton, the Dolní Věstonice 15 pelvis presents a suite of unusual features, but most of these are related either to its overall small size or minor aspects that fall within normal ranges of variation (e.g., its six sacral vertebrae and high sacral hiatus). All of the specimens, where sufficiently complete to evaluate it, have overall proportions within expected Late Pleistocene and recent human ranges of variation.

18

The Lower Limb Remains

Erik Trinkaus

The lower limb remains from Dolní Věstonice and Pavlov include most of the long bones for five of the individuals with associated skeletons, the five from Dolní Věstonice, plus major portions of the femora of Pavlov 1. However, pedal remains are less complete, being present in significant numbers for three of the individuals, Dolní Věstonice 3, 15, and 16, and they are eroded and/or incomplete for all of those individuals Of these remains, as with the upper limb remains, those that are retained for Dolní Věstonice 13 and 15 are generally exceptionally well preserved, with primarily minor erosion to the more fragile edges (much of which has been restored with filler).

The bones of Dolní Věstonice 14 are generally more heavily root-etched, and fragile portions of the epiphyses are frequently broken or incomplete. His long bones sustained considerable epiphyseal erosion, but not so much as to prevent reasonable estimation of most the long bone lengths for at least one side. However, ironically, Dolní Věstonice 16 retains the most complete pedal remains, despite erosion to many of their edges.

Whereas the bodies of Dolní Věstonice 13 to 16 were buried in largely extended or semiflexed positions (Chapter 4), Dolní Věstonice 3 was buried in a flexed position. The Dolní Věstonice 3 body became more tightly flexed through decomposition and sediment compaction (Trinkaus & Jelínek, 1997; Chapter 4). As a result, the knees ended up so tightly flexed that the left calcaneus became embedded into the left femoral greater trochanter. The various lower limb epiphyses were all crushed and hence destroyed to varying degrees. Yet, the diaphyses of all of the long bones are largely intact, and it is possible to estimate reasonably their original lengths by using regressions based on the preserved portions of the bones (see tables in Sládek et al., 2000). The pedal remains, surprisingly, are well represented, and even four sesamoid bones are preserved.

The only lower limb remains that survive from the disturbed Pavlov 1 skeleton are the femora from the proximal diaphysis to the condyles, two pieces of tibial condyles, and a small section of fibula. For the femora, it was necessary to estimate the dimensions of the missing trochanters, done by using the dimensions of the large male femora from Dolní Věstonice, in order to determine a reasonable biomechanical length (proximal neck to condyles). From this measurement, estimates of the standard interarticular lengths were calculated (see tables in Sládek et al., 2000).

To these remains are added an isolated proximal half of the femoral diaphysis from Dolní Věstonice I (Dolní Věstonice 35) and ten isolated remains from the Dolní Věstonice II localities, including two portions of (possibly the same) femur (Dolní Věstonice 40 and 43), two fibular pieces (Dolní Věstonice 42 and 48), and six pedal remains (Dolní Věstonice 39, 44, 46, 47, 49, and 52; Trinkaus et al., 1999b, 2000c; Sládek et al., 2000).

These remains therefore permit a series of comparisons regarding aspects of the long bone diaphyses, most of the articulations (including patellae, which are present for five individuals), and to a lesser extent the feet.

Materials and Methods

The considerations of these Pavlovian lower limb remains involve both morphological descriptions of the bones and morphometric comparisons of them to relevant samples of Late Pleistocene human remains. The principal comparative sample is of Gravettian human remains from

Europe, which derives from the sites of Arene Candide, Barma Grande, Caviglione, Cro-Magnon, Fanciulli (Grotte-des-Enfants), Paglicci, Pataud, Paviland, Předmostí, La Rochette, Sunghir, Veneri (Parabita), and Willendorf. Even though they are geographically, chronologically, and archeologically close to the Dolní Věstonice and Pavlov human remains, the Předmostí and Willendorf specimens are included in the larger Gravettian sample; the Předmostí data all derive from Matiegka (1938). To these are added the slightly earlier Aurignacian remains from Fontana Nuova and Mladeč. The more distant comparisons, for considerations of temporal trends through the Late Pleistocene, are to the Middle Paleolithic samples of late archaic and early modern humans. The former consist of the European and Near Eastern Neandertals and include specimens from Amud, La Chapelle-aux-Saints, La Ferrassie, Fond-de-Forêt, Hortus, Kiik-Koba, Neandertal, Pofi, La Quina, Regourdou, Rochers de Villeneuve, Santa Croce (Bisceglie), Shanidar, Spy, Stadel, and Subalyuk, plus the initial Upper Paleolithic (Châtelperronian) Saint-Césaire specimen. The latter are the Near Eastern remains from Qafzeh and Skhul. The earlier, terminal Middle to Initial Late Pleistocene remains from Krapina and Tabun are only referred to in the context of a more ancestral pattern. Data are from primary descriptions of the remains (e.g., Verneau, 1906; Klaatsch & Lustig, 1914; H. Martin, 1923; Matiegka, 1938; McCown & Keith, 1939; Endo & Kimura, 1970; Sergi et al., 1974; Trinkaus, 1975a, 1983b, 2000b; Vandermeersch, 1981; Heim, 1982b; Formicola, 1990; Kunter & Wahl, 1992; Vandermeersch & Trinkaus, 1995; Chilardi et al., 1996; Mallegni et al., 1999, 2000; Kozlovskaya & Mednikova, 2000; Trinkaus et al., 2005), supplemented by personal analysis of many of the original specimens and additional data from Holliday (1995), Holt (1999), R. Macchiarelli (personal communication) and J. T. Snyder (personal communication).

Human lower limbs, in contrast to our upper limbs, are primarily weight bearing. Therefore, considerations of their degrees of robusticity need to establish a baseline above which the bones exhibit hypertrophy from activity levels exceeding that needed for minimal weight bearing and locomotion. These activity levels, or more appropriately the strains they place on the bones, are a consequence of the individual's level, frequency, and duration of locomotion, since bone responds to both the levels of force placed upon them and the durations of those stresses. In other words, bone responds through hypertrophy to withstand both peak loads and fatigue from repetitive loading (Trinkaus et al., 1994; Lieberman & Crompton, 1998; Carter & Beaupré, 2001; R. B. Martin, 2003). In addition, since the lower limbs are weight bearing and since humans habitually carry resources across the landscape (especially during the Upper Paleolithic; see Svoboda et al., 1996, and Trinkaus et al., 2001, for the Pavlovian), the effective weight on the lower limbs equals the individual's body mass plus any habitual loads that the individual might be carrying.

In order to establish a loading baseline above which hypertrophy, or robusticity, of the lower limb bones can be assessed, it is necessary to take into account both body mass and the effective beam length (Ruff et al., 1993). A variety of techniques have been developed for controlling for both body mass and beam length during the past decade (Ruff et al., 1993; Trinkaus, 1997a; Trinkaus & Ruff, 1999a,b; see comments in Trinkaus & Ruff, 2000), all of which are based on the same principal and involve similar levels of estimation. Moreover, these recent techniques incorporate the previously largely unrecognized importance of body mass, as well as beam length, in the scaling of lower limb diaphyses, something that was lacking in previous "robusticity indices" (see R. Martin, 1928; Trinkaus, 1976a). Experience has shown that the best, and most transparent, approach is one that utilizes, as directly as possible, measurements or estimates of both beam length and body mass and then scales lower limb measures to them as appropriate.

Beam length is closely approximated by the biomechanical lengths of the bones in question (interarticular lengths, removing the effects of the femoral head and neck for the femoral diaphysis), and they are so employed. Body mass can be estimated through either lower limb articular dimensions, making biomechanical assumptions, or through a geometric model of the human body, using stature and bi-iliac breadth (Chapter 13). For the reasons given in Chapter 13, the latter technique is employed here, especially since it remains unclear to what extent the lower limb articulations of these Late Pleistocene individuals might have been affected by activity levels and, in the case of the femoral head, differences in pelvic proportions. In particular, the body mass estimates employed are those involving stature estimates from femora and/or tibiae as available plus either measured or estimated bi-iliac breadth, calculated as possible by using sex-specific formulae (Chapter 13). For Dolní Věstonice 15, despite ongoing discussions regarding its gender (Chapter 7), the male estimate is employed; the male and female differences nonetheless are less than 2 kg and therefore well within estimation error.

Since the human lower limb is weight bearing during the stance phase, it acts as a closed kinematic chain (Levangie & Norkin, 2001) between the body core and the point of ground reaction force. For this reason, in assessing articular biomechanics, it is best to estimate the load arm according to these principles. For example, the load arm of body mass at the knee for knee extension during the stance phase is neither femoral nor tibial length nor their sum but, rather, the perpendicular distance from the line of action between the hip and ankle articulations to the knee, assuming a fixed ankle joint (see Trinkaus & Rhoads, 1999 for justification and derivation); the power arm is the distance from the knee axis of rotation to the line of action of quadriceps femoris. Extension at the ankle is simpler since the load arm is from the ankle to the point of ground reaction force (normally the metatarsal heads during propulsion) and the power arm is from the insertion of triceps surae to the same point of contact, the foot acting as a second-class lever (Levangie & Norkin, 2001).

For many of the Pavlovian leg bones, length was simply measured on the well-preserved bones (especially for Dolní Věstonice 13 and 15), with minor estimation of damaged epiphyseal regions in some cases (especially the bones of Dolní Věstonice 14 and 16). The Dolní Věstonice 3 leg bones all have damaged epiphyseal regions, and their lengths were estimated principally through the use of least squares regressions based on recent human samples between the preserved (approximately intermetaphyseal) lengths and standard interarticular lengths (Trinkaus & Jelínek, 1997; Sládek et al., 2000). The Pavlov 1 femora lack their proximal portions, and as described above, the missing trochanteric region was estimated by using the more complete Dolní Věstonice male femora, and the remaining lengths were estimated by using least squares regressions based on Pleistocene and recent human femora.

Several of these length estimations assume that asymmetries in lower limb long bone lengths will be small, as has been documented on average for recent human samples (Macho, 1991; Trinkaus et al., 1994; Čuk et al., 2001). Moreover, asymmetries in articular and diaphyseal dimensions in the lower limb tend to be smaller than those of the upper limb; and although they can reach significance in some samples, they tend to be relatively random with respect to side across multiple samples, despite larger dimensions on the left side in some samples (Ruff & Hayes, 1983b; Macho, 1991; Trinkaus et al. 1994; Anderson & Trinkaus, 1998; Čuk et al., 2001). For this reason, in contrast with the upper limb; the lower limb comparisons pool data, as available, for right and left bones, taking the average of the values per individual prior to the computation of sample statistics or the plotting of points on graphs.

At the same time, it should be noted that the Dolní Věstonice 15 right femur is clearly abnormal in curvature and proximal epiphyseal orientations (Chapter 19). Yet, the pelvis, distal femora, patellae, tibiae, fibulae, and pedal remains exhibit only the trivial levels of asymmetry that are characteristic of normal human lower limbs. The left femur also appears largely normal, even though it is clearly too short for the overall size of the individual, exhibits a possibly (but not necessarily) abnormally lower neck/shaft angle, and might have an exaggerated anterior bowing of the proximal diaphysis. However, since the individual was fully ambulatory for most of its two decades of life, these lower limb bones can be assessed in terms of morphology and biomechanics, much as can the remains of the other Pavlovian individuals. However, considerations of the right femoral abnormalities and possible irregularities in the other leg bones will be taken into account in the discussions of the comparative assessments of its morphology and proportions.

The majority of the morphometric comparisons employs standard osteometrics (Trinkaus, 1983b; Bräuer, 1988), as are listed for the Dolní Věstonice and Pavlov human remains in Sládek et al. (2000). Sládek and colleagues (2000) also list cross-sectional geometric measurements for the femora and tibiae. The cross-sectional parameters were measured at standard percentages of interarticular (biomechanical) bone length, with 0% being distal (Ruff & Hayes, 1983a; Ruff et al., 1993). They were calculated from cross sections taken perpendicular to the diaphyseal axis. Each section was reconstructed from the external (subperiosteal) contour, transferred through the use of polysiloxane dental putty (Cuttersil Putty Plus, Heraeus Kulzer Inc.), and from the anterior, posterior, medial, and lateral cortical thicknesses, obtained from parallax-corrected measures of biplanar radiographs of the diaphyses. The endosteal contour was then interpolated. The resultant cross sections were projected enlarged onto a Summagraphics III 1812 digitizing tablet, the contours were traced, and the cross-sectional parameters were calculated by using a PC DOS version (Eschman, 1992) of SLICE (Nagurka & Hayes, 1980). All cross sections were digitized twice and the results averaged.

The resultant values, as tabulated in Sládek et al. (2000), are total subperiosteal and cortical areas (TA and CA), the anteroposterior and mediolateral second moments of area (I_x and I_y), the maximum and minimum second moments of area (I_{max} and I_{min}), and the polar moment of area ($J = I_{max} + I_{min}$); for further considerations, see Ruff (2000a). The second moments of area are employed to measure relative bending rigidity of the diaphyses, and the polar moment of area is used as an overall measure of both generalized bending rigidity and torsional strength (Ruff, 2000a). In this, it is fully recognized that these measures become increasingly approximate as the cross sections deviate from ellipses. Moreover, the polar moment of area (and the related polar section modulus) indicates resistance to torsional strains principally when the cross sections are close to circular, and its accuracy decreases when the ratio of I_{max} to I_{min} markedly exceeds 1 (Daegling, 2002). Yet, given the complex combinations of axial, bending, and torsional loads upon lower limb diaphyses (Lengsfeld et al., 1996; Taylor et al., 1996; Duda et al., 1997, 1998), it is unlikely that this will seriously affect the interpretation of the polar moment of area as reflecting the overall mechanical rigidities of the diaphyses in question.

The various morphometric analyses are occasionally provided as ratios (or indices), but given the mathematical difficulties with ratios, they are most often presented graphically as bivariate plots.

The Femora

The femoral remains from Dolní Věstonice and Pavlov consist of largely complete femora from the triple burial, slightly less complete bones from Dolní Věstonice 16, diaphyses plus some epiphyseal data from Dolní Věstonice 3, diaphyses plus distal epiphyses of Pavlov 1, and the incomplete diaphyses of Dolní Věstonice 35, 40, and 43 (Figures 18.1 to 18.8). The last two provide only 35% and 65% cross-sectional data and are not considered in detail here. The remainder provide information on proximal and distal epiphyseal morphology and, especially, diaphyseal shape and relative strength.

The Proximal Epiphyses

Femoral heads are present and in their original osteological locations on the Dolní Věstonice 13–15 femora, and they are displaced on the Dolní Věstonice 3 and 16 remains. However, the trochanters are largely intact on the Dolní Věstonice 16 femora, and portions of the left neck and greater trochanter are present on Dolní Věstonice 3.

There is little of note on most of the femoral heads per se. The Dolní Věstonice 13 right fovea capitis is offset dorsally and has a dorsal lip; the fovea measures 13.3 mm proximodistally and 14.2 mm (plus 6.1 mm for the lip)

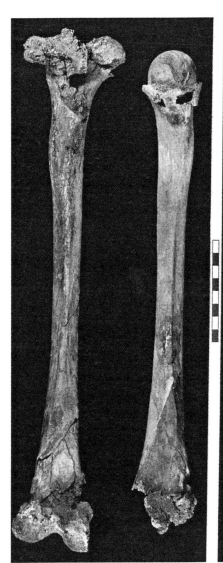
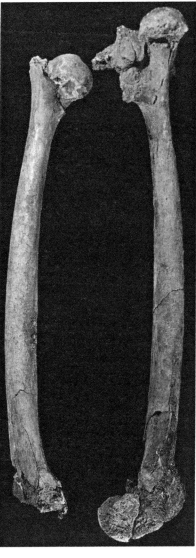

Figure 18.1 Posterior and medial views of the Dolní Věstonice 3 femora. Scale in centimeters.

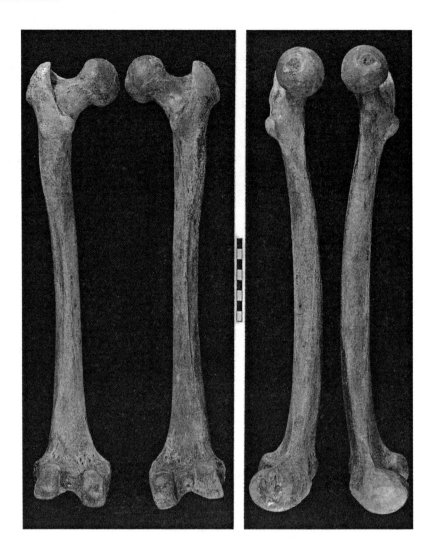

Figure 18.2 Posterior and medial views of the Dolní Věstonice 13 femora. Scale in centimeters.

anteroposteriorly. The left one is damaged but similar. Dolní Věstonice 14 has dorsally offset foveae. The right one is ovoid proximoanterior to distoposterior (24.3 mm × 17.5 mm), and the left one is round (18.0 mm × 17.0 mm). There is an anteroproximal extension of a slightly depressed surface on the right one. The right femoral head of Dolní Věstonice 15 has a broken pit where the fovea capitis was located, plus a low, crescent-shaped depression 8.5 mm from the fovea. The left one is a round rugose area 15 mm in diameter. There is an extra piece of bone protruding medially from the left head (Figure 18.11); it is a fragment of acetabular bone fused on postmortem. The Dolní Věstonice 16 femoral heads are separated and eroded, and they provide little more than difficult-to-orient diameters (hence the parentheses indicating estimation in Sládek et al., 2000).

The left greater trochanter of Dolní Věstonice 3 was impacted postmortem (and postburial) by the left calcaneus, which obscured most of its morphology (Figure 18.1).

The Dolní Věstonice 13 trochanters are very well preserved. There is a prominent tubercle for gluteus minimus on the right side and a smaller one on the left femur. The digital fossae are single and clearly delineated on both sides. The greater trochanters have prominent bulges anteriorly for gluteus medius, and they lip over the neck posteromedially. There is a narrow sulcus just above the medial side of the ridge on each greater trochanter. The lesser trochanters are very projecting, are smooth on their surfaces, and present small medial lips. There is a third trochanter at the proximal ends of each gluteal tuberosity; the right one is very small, only a slight raising of the proximal end of the gluteal tuberosity, whereas the left one is a distinct knob of bone, 18.1 mm proximodistal and 10.7 mm anteroposterior. The spiral line is faint but discernable on the left and clear by the lesser trochanter on the right side.

The greater trochanters of Dolní Věstonice 14 are unremarkable. The right gluteus minimus insertion is a flat and slightly rugose raised area, and the left one is damaged but similar. The digital fossae are large and single. The lesser trochanters are also very prominent, and they lack any medial lipping. The right third trochanter is a large (32 mm × 14 mm) swelling that blends in with the bone surfaces around it, whereas the left one is a smaller distinct nubbin of bone, 14 mm by 11 mm. The spiral lines are slightly raised, smooth on the right, and minimally rugose on the left.

The proximal right femur of Dolní Věstonice 15 has been distorted by the abnormal curvature and twisting of the proximal diaphysis (Figure 18.4). The gluteal minimus and medius markings are clear but modest on the greater trochanter. The lesser trochanters are markedly prominent, with small medial lips. The right one points slightly distally because of the abnormal curvature of the proximal diaphysis. As with the other two triple burial individuals, there is a third trochanter on each femur. The right one is a distinct nubbin of bone, 18.5 mm by 12.0 mm; the left one is larger and more prominent, 24.0 mm by 14.5 mm. The spiral lines are symmetrical, and both of them fade just beyond the lesser trochanters. Both of these proximal femora are especially notable for the largely horizontal orientations of their heads and necks (see neck/shaft angles below); the right neck orientation is sufficiently medial to make the top of the greater trochanter higher than the proximal surface of the head when the bone is placed in anatomical position.

The proximal femora of Dolní Věstonice 16 are less complete, but the region is sufficiently intact on the left femur to present several details. The greater trochanters are incomplete, but there is a distinct lateroanterodistal lipping of the left one. The left lesser trochanter is damaged and appears less prominent than those of the triple burial, but it also has modest medial lipping. The third trochanters are small nubbins of bone, 14.7 mm by 10.5 mm on the right and 19.0 mm by 11.4 mm on the left. Both spiral lines are prominent from the lesser trochanters to the anteromedial corners, but they then become indistinct across the anterior surface. The left femur of Pavlov 1 preserves the distal part of a distinct base of a third trochanter at its proximal break, but the size of the prominence is unknown.

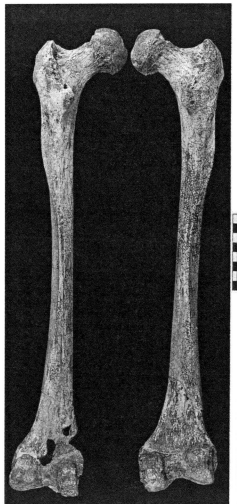

Figure 18.3 Posterior and medial views of the Dolní Věstonice 14 femora. Scale in centimeters.

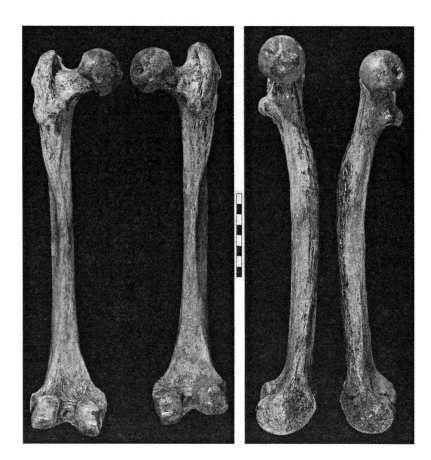

Figure 18.4 Posterior and medial views of the Dolní Věstonice 15 femora. Scale in centimeters.

One of the features that can be ascertained for five of the individuals is their neck/shaft angles (Figure 18.9). The European Gravettian sample (120.9° ± 5.7°; N = 17; 119.9° ± 4.2° without Arene Candide IP) has values close to those of both archaic humans (including the Neandertals; 120.6 ° ± 4.9°, N = 8) and more recent hunter-gatherers (Trinkaus, 1993a; Grine et al., 1995). The one high outlier for the Gravettian sample is the adolescent Arene Candide IP. The Qafzeh-Skhul sample, plus an isolated early modern human from East Africa (KNM-ER 999; Trinkaus, 1993b), is simply anomalous, having values that closely match those of recent urban populations (Trinkaus, 1993a; Anderson & Trinkaus, 1998). All of these angles reflect differential degrees of hip loading—from locomotion and burden carrying—during growth and development, principally during the first decade of life (see Anderson & Trinkaus, 1998, and references therein). The four Dolní Věstonice individuals with normal hip regions have neck/shaft angles that are either in the middle of the Gravettian distribution or, in the case of Dolní Věstonice 3, elevated but not excessively so (her value is matched by a number of later Upper Paleolithic specimens; Trinkaus, 1993a).

The low value is the left femur of Dolní Věstonice 15, whose angle of 113° is below the lowest Gravettian value (Předmostí 3: 114°) but within measurement error of it. It is nonetheless matched by the angle of the Amud 1 Neandertal, and two earlier archaic *Homo* specimens (KNM-WT 15000 and Berg Aukas 1) have slightly lower neck/shaft angles (Walker & Leakey, 1993; Grine et al., 1995). The left angle of Dolní Věstonice 15 is therefore low but by no means exceptional for a Pleistocene foraging population. The abnormal right femur has an angle of 111°, which is low but still within Pleistocene human ranges of variation. Indeed, in a pooled sample of 1350 recent humans (agricultural and urban; see Anderson & Trinkaus, 1998), there are 5 apparently normal specimens (<0.4%) with neck/shaft angles of 110° or 111° but none below 110°.

The Dolní Věstonice 15 right femoral head is nonetheless rotated slightly posteriorly. Its anteversion angle, measured between the dorsal condylar plane and the plane through the middle of the head and greater trochanter (Martin measurement 28) is −5°. This compares with a more normal value of 0° for the left femur, although slightly negative anteversion angles are known for clinically normal recent humans (Bråten et al., 1992). The anteversion angles are 5° and 19° for Dolní Věstonice 13 and 13° and 21° for Dolní Věstonice 14.

In the context of these neck/shaft angle variations, it is possible to compare the relative head and neck lengths of the triple burial femora. Appropriate assessment of femoral head and neck length requires an assessment within the context of the whole hip region (McLeish & Charnley, 1970; Lovejoy et al., 1973; Ruff, 1995), which is beyond the scope of this paleontological description. However, it is possible to assess them by a comparison to femoral head diameter, as an indicator of body mass and overall size. In this, the measure of head and neck length is the trochanteric biomechanical neck length of Lovejoy et al. (1973), which is the distance perpendicular to the shaft axis from the lateral greater trochanter to the tangent to the proximal femoral head; it approximates the moment arm for the gluteus medius abductor muscle with reference to the femur. The resultant distribution (Figure 18.10) shows some scatter but little difference between the samples; the European Neandertals, with their large femoral heads, have slightly shorter relative neck lengths on average, but they are matched by a couple of early modern humans. The addition of both earlier specimens (from Broken Hill, Krapina, and Tabun) and Asian early Upper Paleolithic specimens (from Minatogawa and Ohalo) changes little in the distribution. The Dolní Věstonice 13–15 femora are generally similar to the others, even though they are among those with longer femoral necks. The Dolní Věstonice 15 left femur clusters closely with the Dolní Věstonice 13 femora and among the other Late Pleistocene specimens.

The Proximal Diaphyses

The proximal femoral diaphyses of Late Pleistocene humans shift from a predominantly subcircular form in Middle Paleolithic samples (Neandertals, including the Initial Upper Paleolithic Saint-Césaire 1, and all of the Qafzeh-Skhul humans except Qafzeh 8) to a cross-sectional shape with a lateral expansion in the Early Upper Paleolithic. The lateral expansion, inappropriately referred to for the past century as "platymeric," is produced largely by a prominent swelling or ridge of bone, the gluteal buttress (or lateral buttress), along the

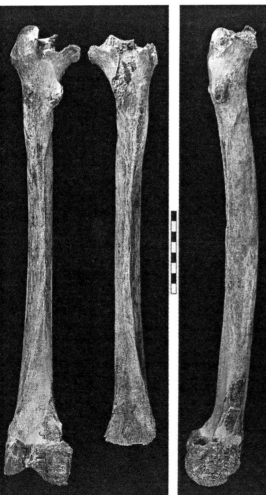

Figure 18.5 Posterior and medial views of the Dolní Věstonice 16 femora. Scale in centimeters.

Figure 18.6 Posterior and medial views of the Dolní Věstonice 35 femur. Scale in centimeters.

lateral diaphysis adjacent to and slightly distal of the lesser trochanter. The gluteal buttress tends to be slightly posterior of directly lateral, usually in line with the varying degrees of anteversion of the femoral head and neck. This lateral expansion is therefore probably an osteological response to tensile strains on the proximal lateral diaphysis from mediolateral bending of the region because of the medial offset of the femoral head (Brekelmans et al., 1972; Rybicki et al., 1972; Aamodt et al., 1997; see also Ruff, 1995), even though it is recognized that muscle forces in general (Taylor et al., 1996; Duda et al., 1997) and tension in the iliotibial tract in particular (Rybicki et al., 1972; Rohlmann et al., 1982) will reduce most of the bending strains. A decrease in the neck/shaft angle can increase these strains (Brekelmans et al., 1972), but with the exceptions of the Qafzeh-Skhul sample, Dolní Věstonice 3 and Arene Candide IP at the high end and Dolní Věstonice 15, Předmostí 3 and Amud 1 at the low end, most of these Late Pleistocene specimens have neck-shaft angles within a relatively narrow range (Figure 18.9).

The proximal diaphyses of the Dolní Věstonice 3 femora exhibit prominent gluteal buttresses, which have modest sulci on both their anterior and posterior margins and produce a clear convexity to the lateral proximal diaphyses (Figure 18.11). The Dolní Věstonice 13 buttresses are less separated from the proximolateral diaphyses, with the right one exhibiting a shallow sulcus along its anterior surface but a flat area on the left side. Those of Dolní Věstonice 14, in contrast, are large flanges, which are sufficiently separated from both the anterior and posterior diaphyseal surfaces by distinct sulci to enable the measurement of their thicknesses (8 mm on the right and 10 mm on the left). They curve around from the anterolateral surface to the posterolateral diaphysis as they approach the linea aspera distally.

The right gluteal buttress of Dolní Věstonice 15 has been altered in orientation by the proximal diaphyseal distortion, but there is a strong swelling that curves posterolaterally as it goes distally to blend with the proximal linea aspera. The more normal left buttress is sufficiently separated from the adjacent diaphysis to have modest concave sulci anteriorly and posteriorly, and it, too, distinctly curves around to the proximal linea aspera. The gluteal buttresses of Dolní Věstonice 16 are clear, and they are distinguished by anterior sulci and posterior concavities accentuated by rugose gluteal tuberosities (see below), but they do not project markedly laterally (Figure 18.12).

The isolated Dolní Věstonice 35 femur, similar to that of Dolní Věstonice 14, has a markedly pronounced gluteal buttress (Figure 18.12). It is sufficiently separated from both the anterior and posterior diaphyseal surfaces by longitudinal sulci to permit measurement of its midthickness (12 mm). In lateral view, it is slightly convex anteriorly, since it curves posteriorly to blend in with the shaft adjacent to the proximal linea aspera. The Dolní Věstonice 43 femur piece preserves a sufficient portion of its gluteal buttress to indicate that it was rounded and lacked a sulcus between it and the anterior diaphyseal surface.

The Pavlov 1 gluteal buttresses are not very projecting, similar to those of Dolní Věstonice 16, but they are distinguished by being thick anteroposteriorly and separated from the adjacent anterior diaphysis by longitudinal sulci, especially on the right.

Archaic *Homo* and some modern humans exhibit a swelling of bone along the proximal medial femoral diaphysis, which appears to represent a structural response to mediolateral bending (Trinkaus, 1976a, 1984a). This swelling is present in modest form in some earlier Upper Paleolithic femora (e.g., those from Cro-Magnon), but it is entirely absent from all of the Dolní Věstonice and Pavlov femora.

The mediolateral expansion of the proximal (or subtrochanteric) diaphysis has been quantified by using traditional perpendicular external diameters at the maximum development of the gluteal buttress to maximize sample

The Lower Limb Remains

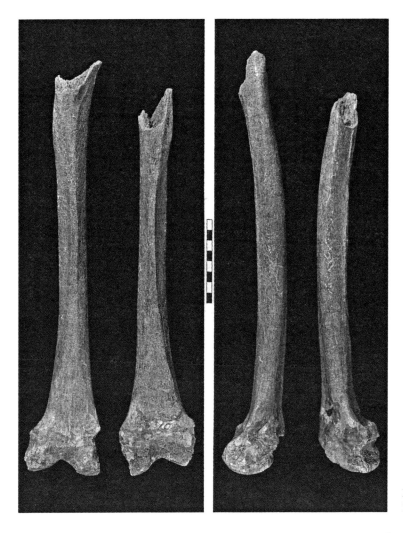

Figure 18.7 Posterior and medial views of the Pavlov 1 femora. Scale in centimeters.

Figure 18.8 Posterior views of the Dolní Věstonice 43 (left) and 40 (right) right femoral diaphyseal sections. Scale in centimeters.

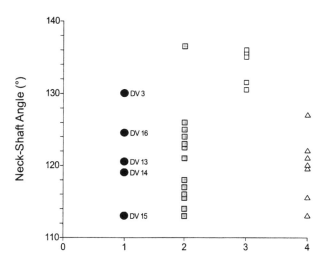

Figure 18.9 Plot of femoral neck/shaft angles for (1) Dolní Věstonice specimens ($N = 5$), (2) Gravettian specimens ($N = 17$), (3) Qafzeh-Skhul humans ($N = 6$), and (4) Neandertals ($N = 8$).

389

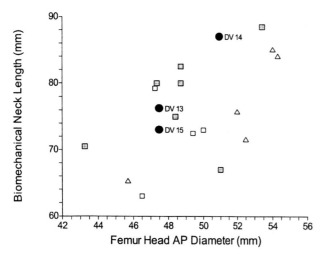

Figure 18.10 Bivariate plot of trochanteric biomechanical neck length versus femoral head diameter. Solid circles: Dolní Věstonice specimens; gray squares: European Gravettian specimens; open squares: Qafzeh-Skhul specimens; open triangles: Neandertals.

size and include the Předmostí femora (the equivalent of the meric index). The resultant distribution (Figure 18.13) places most of the Neandertals and some of the Qafzeh-Skhul femora closer to the subcircular upper portion of the graph, with Skhul 5 being an unusually high outlier. The majority of the earlier Upper Paleolithic femora appears to have more ovoid femora or at least ones with markedly greater mediolateral diameters. The two Pavlovian femora with the rounder proximal diaphyses are the Dolní Věstonice 15 left femur (the smaller one) and the average of the Dolní Věstonice 16 femora, with their moderately projecting gluteal buttresses. The lower value is for Dolní Věstonice 14, with its prominent gluteal buttresses.

Despite these distributions of external diameters, comparison of a smaller sample of femora, for which cross-sectional measures of bone distribution are available (Figure 18.13), provides a possibly contrasting picture. In using anteroposterior versus mediolateral second moments of area, any significant differences between these samples in these perpendicular, mechanically relevant distributions of bone disappear. The Neandertals are in the middle of the early modern human distribution, Skhul 5 is unexceptional, and the Dolní Věstonice and Pavlov femora are well within the overall distribution. The large and low values, indicating more mediolateral reinforcement, are Barma Grande 2 and Qafzeh 8, and the two lowest Pavlovian specimens are Dolní Věstonice 14 and Pavlov 1. It is possible that use of perpendicular second moments of area in which the principal axis is aligned with the plane of anteversion would reinforce the Middle to Upper Paleolithic contrasts evident in the external diameters, but the fragmentary nature of most of these Late Pleistocene femoral diaphyses would reduce the sample sizes to a low level.

The other prominent feature of the femoral proximal diaphysis is the development of the gluteal tuberosity. This rugosity along the posterolateral proximal diaphysis and the medial side of the gluteal buttress (when the latter is present) provides insertion for a significant portion of gluteus maximus, the remainder of which inserts into the iliotibial tract. It has been noted to be markedly developed among the Neandertals (Trinkaus, 1976a), although scaling it to femur length and body mass largely eliminates the difference between the Neandertals and early modern humans (Trinkaus, 2000c).

The gluteal tuberosities of Dolní Věstonice 3 exhibit minimal rugosity with little alteration of the bony contour, with the left bone showing slightly more rugosity than the right one. Those of Dolní Věstonice 13 and 14 similarly present only modestly rugose surfaces. The tuberosities of Dolní Věstonice 15 are clear and moderately rugose, but the right one has been twisted around the shaft in conjunction with the deformity. The Dolní Věstonice 16 tuberosities contrast in being strongly rugose and producing slight concavities bilaterally. The Dolní Věstonice 35 and 43 tuberosities are weakly marked, given the development of the buttress, especially on Dolní Věstonice 35. Pavlov 1,

Figure 18.11 Reconstructed cross sections of the Dolní Věstonice 3, 13, 14, and 15 femora at 80% of length (subtrochanteric level). Right femora are on the left, medial is toward the midline, anterior is above, and all are viewed distally. The endosteal contours are interpolated from parallax-corrected biplanar radiography. Scale in centimeters.

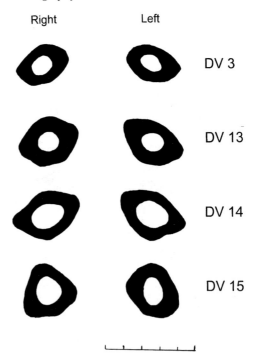

midshaft and the distal third of the diaphysis (Figure 18.15). The Dolní Věstonice 13 pilaster is present but small, with only a slight concavity along its lateral side. The linea aspera is faintly marked and rounded. Dolní Věstonice 14 exhibits clear and relatively prominent pilasters, which are concave laterally along midshaft. The associated linea aspera is minimally marked but rises to an angle laterally. Dolní Věstonice 15 also has modest pilasters, which are broad and have only slight lateral concavities. The linea aspera is wide and smooth on each. Each Pavlov 1 femur is very similar to these specimens, in having a modest pilaster, nonetheless with lateral concavity along midshaft and a linea aspera that is broad and largely smooth (Figure 18.16).

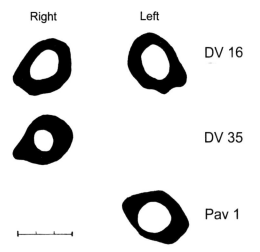

Figure 18.12 Reconstructed cross sections of the Dolní Věstonice 16 and 35 and Pavlov 1 femora at 80% of length (subtrochanteric level). Right femora are on the left, medial is toward the midline, anterior is above, and all are viewed distally. The endosteal contours are interpolated from parallax-corrected biplanar radiography. Scale in centimeters.

Figure 18.13 Bivariate plots of subtrochanteric anteroposterior versus mediolateral diameters (above) and of 80% anteroposterior versus mediolateral second moments of area (below). Solid circles: Dolní Věstonice and Pavlov specimens; gray squares: European earlier Upper Paleolithic specimens; open squares: Qafzeh-Skhul specimens; open triangles: Neandertals.

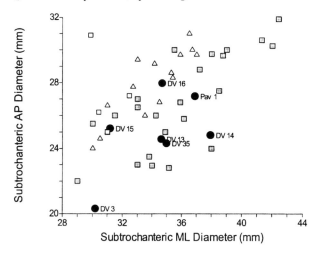

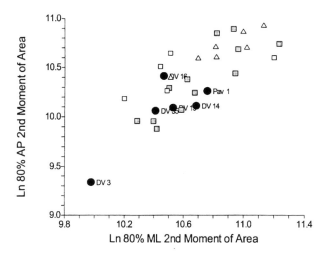

similarly, has only moderately rugose tuberosities, which are slightly concave, especially on the left. None of these gluteal tuberosities exhibits a hypotrochanteric fossa.

A plot of gluteal tuberosity breadth, an estimate of attachment area taken as the maximum breadth of the rugosity itself, against estimated body mass times femur length (to account for both the load arm and the baseline load) removes the differences between most of the Late Pleistocene specimens (Figure 18.14). The higher values are for two Neandertal females (La Ferrassie 2 and Shanidar 6), a small Gravettian isolated femur (Cro-Magnon 4324), and Dolní Věstonice 15. The remainder fall within a narrow cluster, bounded above by La Ferrassie 1 and below by Dolní Věstonice 3. The other Pavlovian specimens are unexceptional, even though Dolní Věstonice 13 is moderately high in the distribution.

The Middle of the Diaphysis

The femoral midshaft has been repeatedly noted to shift from a relatively round structure among late archaic humans to one that is variably anteroposteriorly expanded, principally through the development of a pilaster, among early modern humans (Twiesselmann, 1961; Trinkaus, 1976a; Trinkaus & Ruff, 1999a). There is marked variability in this feature in the Předmostí sample (Matiegka, 1938) and some variability in the Qafzeh-Skhul sample (Trinkaus & Ruff, 1999a), even though most early modern human femora present a clear pilaster.

The Dolní Věstonice 3 femora exhibit modest pilasters, which reach their maximum development between

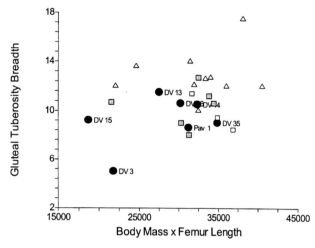

Figure 18.14 Bivariate plot of gluteal tuberosity breadth versus body mass times femoral length. Solid circles: Dolní Věstonice and Pavlov specimens; gray squares: European earlier Upper Paleolithic specimens; open squares: Qafzeh-Skhul specimens; open triangles: Neandertals.

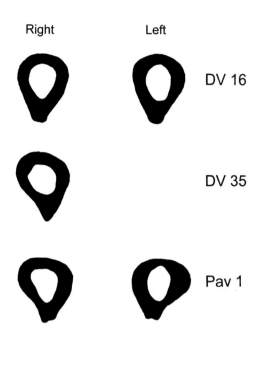

Figure 18.16 Reconstructed cross sections of the Dolní Věstonice 16 and 35 and Pavlov 1 femora at 50% of length (midshaft level). Right femora are on the left, medial is toward the midline, anterior is above, and all are viewed distally. The endosteal contours are interpolated from parallax-corrected biplanar radiography. Scale in centimeters.

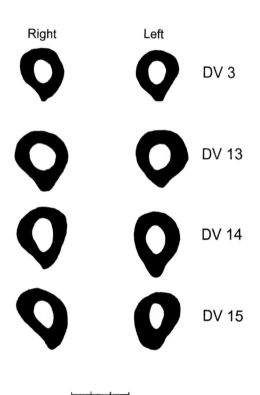

Figure 18.15 Reconstructed cross sections of the Dolní Věstonice 3, 13, 14, and 15 femora at 50% of length (midshaft level). Right femora are on the left, medial is toward the midline, anterior is above, and all are viewed distally. The endosteal contours are interpolated from parallax-corrected biplanar radiography. Scale in centimeters.

The Dolní Věstonice 16 femora have similar pilasters, with minimal lateral concavity, but their lineae aspera are strongly marked and very rugose, with medial lipping of the muscle attachment area through the midshaft region. The Dolní Věstonice 35 pilaster is moderately prominent, with a linea aspera that becomes gradually broader as it goes distally, mostly through the expansion of its medial margin.

The shape of the femoral midshaft can be quantified by comparisons of anteroposterior to mediolateral external diameters (the pilastric index) and anteroposterior to mediolateral second moments of area (Figure 18.17). In the comparison of external diameters, there is some overlap between the samples, but there is a general trend for the early modern humans to have greater relative anteroposterior dimensions. All of the Dolní Věstonice specimens are well within the upper portion of the earlier Upper Paleolithic distribution, and only Pavlov 1 has a moderately low value, overlapping the Neandertal distribution, along with Předmostí 10. In the cross-sectional comparison (which unfortunately cannot include the Předmostí femora), however, there is little overlap be-

The Lower Limb Remains

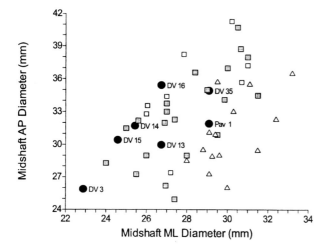

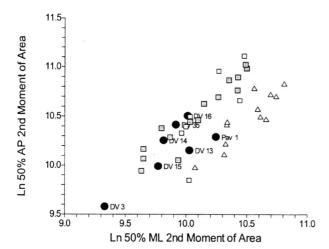

Figure 18.17 Bivariate plots of femoral midshaft anteroposterior versus mediolateral diameters (above) and similarly oriented second moments of area (below). Solid circles: Dolní Věstonice and Pavlov specimens; gray squares: European earlier Upper Paleolithic specimens; open squares: Qafzeh-Skhul specimens; open triangles: Neandertals.

second, which compares external measures of shaft size to femur length, approaches traditional robusticity indices. The third, which assesses the bone's relative bending and torsional strength relative to body mass and bone length, is the closest approximation to comparisons of properly scaled bone strength, or robusticity.

The comparisons of midshaft and subtrochanteric cortical area versus total subperiosteal area (Figure 18.18) show no separation of the Late Pleistocene samples; there are only a couple of outliers. The very low value in the midshaft (50%) comparison is the adolescent Arene Candide IP specimen. The three slightly lower Pavlovian 16 specimens are Dolní Věstonice 18 and 35 and Pavlov 1, and the highest of the Pavlovian specimens is the left

Figure 18.18 Bivariate plots of cross-sectional cortical area versus total subperiosteal area at midshaft (above) and the subtrochanteric level (below). Solid circles: Dolní Věstonice and Pavlov specimens; gray squares: European earlier Upper Paleolithic specimens; open squares: Qafzeh-Skhul specimens; open triangles: Neandertals.

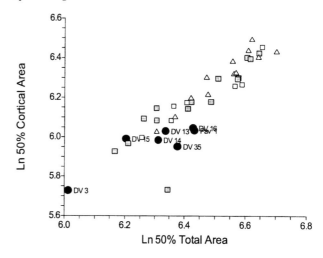

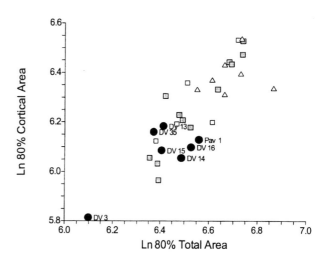

tween the samples, with only Pavlov 1 and Skhul 3 and 6 among the early modern humans along the edge of the Neandertal distribution. The remainder of the earlier Upper Paleolithic specimens (including the Aurignacian Mladeč 27 femur) are well above the Neandertals.

Diaphyseal Robusticity and Curvature

The relative strength of the femoral diaphysis has been approached in a variety of ways, of which three will be considered in detail. The first, a comparison of relative cortical area, is as much an assessment of the individual's life history of endosteal and subperiosteal deposition and resorption as an indication of diaphyseal robusticity. The

femur of Dolní Věstonice 15. In the subtrochanteric (80%) comparison, there is more scatter but little separation of the samples. The one low Neandertal outlier is Spy 2.

To assess relative shaft size by traditional osteometrics, a pseudopolar moment of area was computed by modeling the shaft as a solid ellipse and adding the resultant mediolateral and anteroposterior second moments of area (Figure 18.19). This approach assumes similar degrees of cortical area (which is reasonable), but it will exaggerate the amount of bone present the more a cross section deviates from an ellipse. It should therefore overestimate the strength of the more pilastric early modern human femora. Despite this problem, the Neandertals are near or above the more robust side of the early modern human distribution, joined by Dolní Věstonice 15. However, both the Neandertals and Dolní Věstonice 15 have relatively short femora compared to body mass (Chapters 12 and 13) vis-à-vis the early modern human samples, which increases the perceived level of diaphyseal robusticity. This comparison, therefore, as with all traditional "robusticity indices" (Trinkaus, 1983b; Bräuer, 1988), distorts the data both in their measure of diaphyseal bone quantity and distribution and in their scaling.

To correct for these problems, even though it decreases sample size (from forty-six to thirty-nine) and introduces estimation errors, the polar section modulus was compared to the product of estimated body mass and femur length. The difference essentially disappears between the Late Pleistocene samples. Dolní Věstonice 35 and Pavlov 1 have moderately low values, but one needs to bear in mind that the femoral length of the former is approximate and, if estimated too long, will decrease the perceived level of robusticity. The one moderately high value from Dolní Věstonice is, again, the left femur of Dolní Věstonice 15.

Related to these issues of robusticity is the degree of anterior femoral curvature, for which the absolute subtense is the relevant variable of comparison (Shackelford & Trinkaus, 2002). The distributions of curvature subtense for the comparative samples (Figure 18.20) provide similar values across the samples; the two high outliers are Veneri 1 and Skhul 6, and the curvature of the latter may be exaggereated by postmortem deformation. As documented elsewhere (Shackelford & Trinkaus, 2002), most of these values cluster toward the upper limits of more recent human samples.

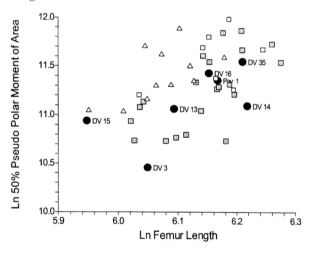

Figure 18.19 Bivariate plots of the midshaft pseudopolar moment of area versus femoral length (above) and midshaft (50%) polar section modulus versus body mass times femoral length (below). Solid circles: Dolní Věstonice and Pavlov specimens; gray squares: European earlier Upper Paleolithic specimens; open squares: Qafzeh-Skhul specimens; open triangles: Neandertals.

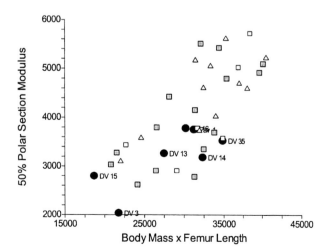

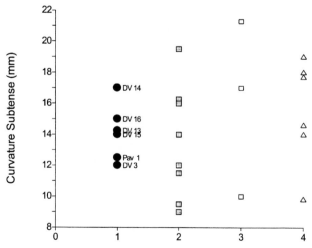

Figure 18.20 Plot of the anterior femoral curvature subtense for (1) Dolní Věstonice and Pavlov specimens ($N = 6$), (2) earlier Upper Paleolithic femora ($N = 9$), (3) Qafzeh-Skhul specimens ($N = 3$), and (4) Neandertals ($N = 6$).

The Distal Epiphyses

The distal epiphyses of the Dolní Věstonice and Pavlov femora provide little of note. They all exhibit, to the extent preserved, higher and broader lateral versus medial patellar margins and evenly elliptical condyles. None of the postero-proximal condylar surfaces exhibits the flattening interpreted (Trinkaus, 1975b) as a femoral squatting facet. These observations apply to at least one distal epiphysis for each of the associated skeletons, although the Dolní Věstonice 3 and 16 and Pavlov 1 distal femora provide the least data. In addition, Dolní Věstonice 13 exhibits a small left adductor tubercle, but none on the right, and bilaterally prominent lateral epicondyles. The Dolní Věstonice 15 distal femora have modest adductor tubercles, at least on the right as preserved. There is no asymmetry there beyond the normal trivial metric differences beween sides.

One feature that can be quantified on the femora of five of the individuals (all but Dolní Věstonice 3) is the bicondylar angle. They vary between 8° and 13°, and in this they are very similar to other Late Pleistocene femora (Figure 18.21) and to recent human samples (Tardieu & Trinkaus, 1994). Since this feature is developmentally plastic and only forms in response to normal weight bearing at the knee (Tardieu & Trinkaus, 1994), it indicates normal postural development for these individuals This is not surprising for four of these individuals, but it helps to confirm that Dolní Věstonice 15, both of whose femora have bicondylar angles of 11°, was able to support weight normally through the knee region.

The Patellae

Both patellae are well preserved for Dolní Věstonice 3 and 13–15, and the left patella remains for Dolní Věstonice 16 (Figures 18.22 to 18.26). The anterior surfaces are all gently marked for the quadriceps femoris tendon, and none of them exhibits enthesopathies for the tendon fibers. Only the Dolní Věstonice 16 patella has a clear and large vastus notch, although small ones are present on the Dolní Věstonice 3 patellae. There is nonetheless some variation on the posterior (articular) surfaces.

On the Dolní Věstonice 3 patellae, the medial facet is convex near the central ridge but exhibits a distinct concavity near its superomedial corner; the lateral facet is gently concave superoinferiorly and flat mediolaterally. The Dolní Věstonice 13 patellae have a marked medial displacement of the central ridge and a clear concavity only on the distolateral facet. The Dolní Věstonice 14 patellae show little concavity of the medial and lateral facets, although there is some on the left one.

Each of the Dolní Věstonice 15 facets is strongly asymmetrical with modest lateral concavity. There is an unusual proximodistal sulcus that runs along each medial facet near its medial margin. The right one is 6.5 mm wide and 15.0 mm long, whereas the left one is 7.0 mm wide and 14.5 mm long. These sulci are not reflected in any apparent changes in the patellar facets of the distal femora, and they are unlikely to have affected movement of the patellofemoral articulations. They are associated

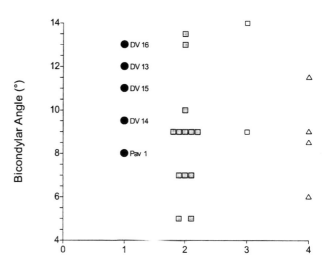

Figure 18.21 Plot of femoral bicondylar angles for (1) Dolní Věstonice and Pavlov specimens ($N = 5$), (2) Gravettian femora ($N = 13$), (3) Qafzeh-Skhul specimens ($N = 2$) and (4) Neandertals ($N = 4$).

Figure 18.22 Anterior (above) and posterior (below) views of the Dolní Věstonice 3 patellae. Scale in centimeters.

Figure 18.23 Anterior (above) and posterior (below) views of the Dolní Věstonice 13 patellae. Scale in centimeters.

Figure 18.24 Anterior (above) and posterior (below) views of the Dolní Věstonice 14 patellae. Scale in centimeters.

with normal bicondylar angles of the femora, and there are no alterations on the anterior (tendinous) surfaces that would suggest any changes in the quadriceps femoris insertions. They simply appear as anomalous alterations of the bones, those that probably would have been obscured in the living by the adjacent articular cartilage.

The Dolní Věstonice 16 patella has modest asymmetry of the facets and an even concavity of the lateral facet. It is also apparent in posterior view that the depth of its vastus notch is accentuated by a medial projection of articular bone.

It has been noted that Neandertal patellae have, on average, greater similarity of the medial and lateral facet breadths than do recent humans (Trinkaus, 1983b, 2000a), and this pattern appears to characterize earlier archaic humans (Carretero et al., 1999; Trinkaus, 2000a). A plot of Late Pleistocene medial versus lateral facet breadths (Figure 18.27) shows a broad distribution, with the Neandertals indeed having more similar medial and lateral facet breadths than the Upper Paleolithic comparative specimens, but the Qafzeh-Skhul specimens cluster with the Neandertals. The Dolní Věstonice specimens show considerable variation, with four of them falling in the middle of the Middle Paleolithic distribution (especially Dolní Věstonice 14) and one (Dolní Věstonice 13) having a very narrow medial facet. The functional significance of this variation remains unclear (Trinkaus, 2000a), especially in light of the consistently higher lateral patellar surfaces on the distal femora of all of these Late Pleistocene samples, including those of Dolní Věstonice.

Neandertals have also been noted to have relatively thick patellae (Trinkaus, 1983b), but it has been shown that, when appropriately scaled to body weight and the approximate load arm of that body weight at the knee (the perpendicular from the knee to the line between the hip

Figure 18.25 Anterior (above) and posterior (below) views of the Dolní Věstonice 15 patellae. Scale in centimeters.

The Lower Limb Remains

Figure 18.26 Anterior (left) and posterior (right) views of the Dolní Věstonice 16 left patella. Scale in centimeters.

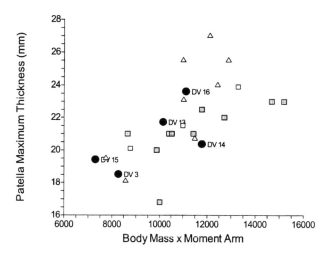

Figure 18.28 Bivariate plot of patellar maximum thickness versus body mass times its moment arm at the knee. Solid circles: Dolní Věstonice specimens; gray squares: European Gravettian specimens; open squares: Qafzeh-Skhul specimens; open triangles: Neandertals.

and ankle articulations), the differences largely disappear (Trinkaus & Rhoads, 1999). This is evident in Figure 18.28, in which a couple of Neandertals (Shanidar 1 and 4) do have relatively thick patellae, but the samples show considerable overlap. The Dolní Věstonice specimens scatter through the distribution, with those of Dolní Věstonice 14 being moderately thin and that of Dolní Věstonice 16 being relatively thicker. The Dolní Věstonice 15 patellae are by no means gracile for its size.

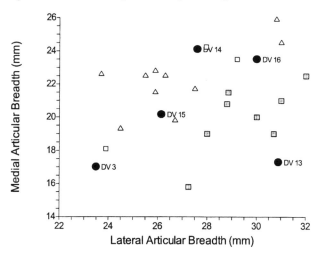

Figure 18.27 Bivariate plot of medial patellar facet breadth versus lateral facet breadth. Solid circles: Dolní Věstonice specimens; gray squares: European Gravettian specimens; open squares: Qafzeh-Skhul specimens; open triangles: Neandertals.

The Tibiae

Five of the associated skeletons preserve both tibiae (Pavlov 1 has only condylar fragments), but they vary markedly in preservation. Those of Dolní Věstonice 13 and 15 are essentially complete, with minor damage to the epiphyseal margins and associated restoration (Figures 18.30 and 18.32). The Dolní Věstonice 14 tibiae have largely complete diaphyses and distal epiphyses, but the proximal epiphyses were damaged and the condylar masses separated (Figure 18.31). The left one is probably too anteriorly placed, and the right one was set on backward; since they provide little information and their original orientations are unknown, they have been left in their incorrect positions. The Dolní Věstonice 16 tibiae largely lack their proximal ends (the tibial tuberosities are mostly present, and the right tibia has the trabecular core of the posterior medial condyle), all of the margins of the distal epiphyses are eroded, and the malleoli are absent. However, the diaphyses are largely intact (with left proximal damage), and the talar trochlear articulations are accurately placed (Figure 18.33). The least complete tibiae are those of Dolní Věstonice 3 since their proximal ends were extensively eroded and the distal ends are almost completely lacking (Figure 18.29). There is continuous bone to the middle of the left lateral condyle proximally, and the distal lateral capsular ligament attachment is preserved; it is on the basis of these landmarks that the lengths of the Dolní Věstonice 3 tibiae have been estimated (Trinkaus & Jelínek, 1997; Sládek et al., 2000).

These bones therefore provide reasonable proximal epiphyseal data mostly for Dolní Věstonice 13 and 15,

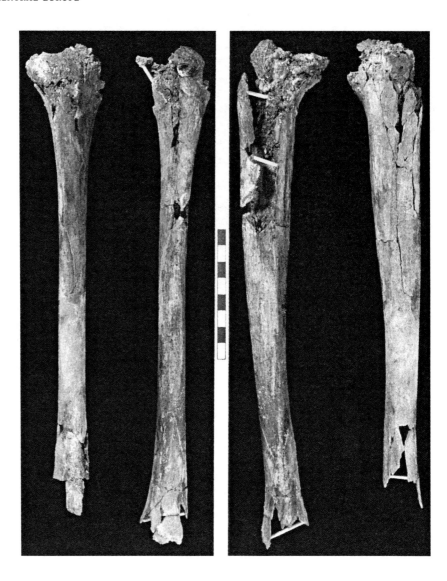

Figure 18.29 Anterior (left) and lateral (right) views of the Dolní Věstonice 3 tibiae. Scale in centimeters.

distal epiphyseal data for Dolní Věstonice 13 to 15, and diaphyseal data for all five individuals.

The Proximal Epiphyses

The tibiae of Dolní Věstonice 13 have modest intercondylar spines bilaterally, with a small pit just anterior. The lateral condyles round off posteriorly, whereas the left medial condyle (the right one is abraded) has a sharp posterior margin. The medial condyles are clearly concave in their middle portions, whereas the lateral ones are flat anteriorly. The tibial tuberosities are moderately rugose, with a strong distal raised transverse ridge on the left side but not on the right side. The fibular facets are smooth.

The Dolní Věstonice 14 proximal tibiae have modest medial condylar concavities and bilaterally distinct nubbins of bone on the tibial tuberosity, with an oblique edge running proximomedial to distolateral along each distal edge.

The intercondylar spines of the Dolní Věstonice 15 tibiae are small and rounded, with each bone having a small pit dorsal of the spines. The medial condylar concavity is present with a right angle posteriorly. The lateral condyles are gently convex anteroposteriorly, with little posterior marginal rounding. The fibular facets are slightly concave. The tibial tuberosities are distinct raised areas, but they are smooth proximally and taper off with vertical striations on their distal portions.

The Dolní Věstonice 16 tibiae exhibit broad and rugose tibial tuberosities with prominent anterior swellings that grade gradually onto the anterior crests distally. It should be noted that all of the Dolní Věstonice individuals have largely smooth anterior patellae, despite this variation in their tibial tuberosities.

The orientations and positions of their tibial condyles can be assessed for both sides of Dolní Věstonice 13 and 15. In the medial condylar retroversion angle (condylar plane relative to the perpendicular to the diaphyseal axis;

Figure 18.34), the Dolní Věstonice 15 angles of 10° are modest for a Late Pleistocene individual, and the values of 20° for Dolní Věstonice 13 are relatively high. However, both of them are within Late Pleistocene ranges of variation and are unexceptional for nonindustrial recent human samples (Trinkaus, 1975b). Both of these tibiae also exhibit the posterior displacement of the condyles relative to the diaphysis (and the tibial tuberosity) seen in nonindustrial human tibiae. The measure from the tibial tuberosity to the anteroposterior condylar middles, perpendicular to the diaphyseal axis, is an approximation of the quadriceps femoris moment arm, and it should therefore be scaled to body mass times the load arm for body mass at the knee (the perpendicular from the knee to the hip to ankle line) (Trinkaus & Rhoads, 1999). The Late Pleistocene samples are generally similar (Figure 18.35), a pattern previously documented for more temporally and geographically diverse Middle and Late Pleistocene samples (Trinkaus & Rhoads, 1999). The two Dolní Věstonice specimens are at the top of the range of variation.

The Diaphyses

The human tibial diaphysis exhibits considerable variation in cross-sectional shape and metric properties, some of which are related to patterns of remodeling in response to biomechanical strains (Lovejoy et al., 1976). In the Late Pleistocene, there is a general pattern in which the tibial cross sections change from generally amygdaloid in cross-sectional shape among late archaic humans to those in which there are distinct lateral concavities and more pronounced cresting among early modern humans. The Dolní Věstonice tibial diaphyses fall generally within the latter pattern, with individual variation.

The small Dolní Věstonice 3 diaphyses present anterior crests that are rounded and rugose for about 40 mm from each tibial tuberosity and then become sharply angled along the proximal and midshaft. The anterior crests are S-shaped, being convex medially along the proximal diaphyses and laterally convex distally. The medial diaphyseal surfaces are smooth and anteroposteriorly convex, rounding onto the posterior surface proximally and hav-

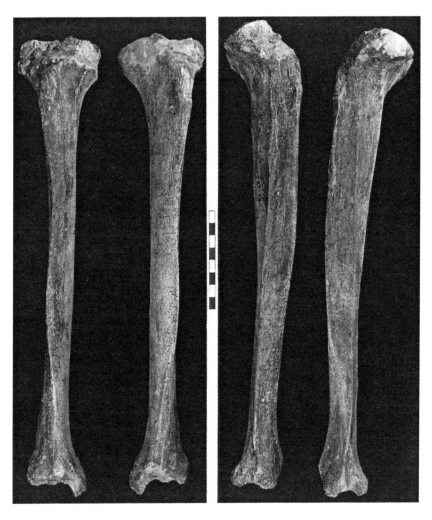

Figure 18.30 Anterior (left) and lateral (right) views of the Dolní Věstonice 13 tibiae. Scale in centimeters.

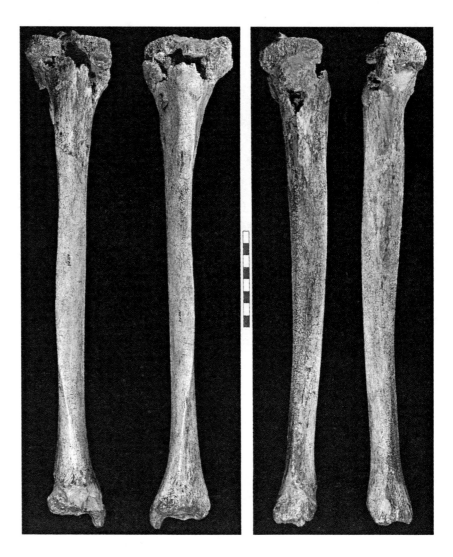

Figure 18.31 Anterior (left) and lateral (right) views of the Dolní Věstonice 14 tibiae. Scale in centimeters.

ing distinct angles to the posterior surfaces distally. The lateral surfaces are slightly concave anteroposteriorly where preserved proximally, flat through the midshafts, and then convex distally. The interosseous lines are very modest. The soleal lines are minimally rugose and not raised from the posterior surface. But there are prominent tibial pilasters, which emerge proximally from the soleal lines and are sufficiently distinct to form small sulci medially and laterally of them. The right one is prominent primarily in the nutrient foramen level, whereas the left one is prominent more distally.

The anterior crests of the Dolní Věstonice 13 tibiae have meandering paths, from laterally convex proximally, with a slight lateral lip adjacent to the tibial tuberosities (especially on the left), to medially convex down to midshaft, then laterally convex again, to traverse the distal shafts medially to the middles of the distal metaphyses. There are clear angles to the crests along their lengths. The interosseous lines form raised small crests with distinct angles down the lateral diaphyses. They never reach the posterolateral corner because of the distal extents of the tibial pilasters. Those pilasters are only suggestions of clear and separate pilasters at the nutrient foramen level, but all of the posterior surfaces of the diaphyses are extended ridges, or pilasters, from the soleal lines to the mid-distal diaphyses. The soleal lines themselves are minimally rugose and are evident primarily medial of the pilasters and proximal of midshaft. There is no evidence of a flexor line between the flexor digitorum longus and tibialis posterior origins.

Despite extensive root etching of their surfaces, the Dolní Věstonice 14 tibial diaphyses are well preserved. Their anterior crests have light S-curves, which are apparent mostly by the medial deviations of the crests on the distal diaphyses to reach the mediolateral middles of the metaphyses. There is a clear angle on the left crest, but more of a rounded margin on the right side. The anterolateral surfaces are distinctly concave anteroposteriorly, and the interosseous line presents a clear angle for the whole length of the right tibia; the less prominent interosseous line on the left bone may be due to the root etching of the surface. There is no clear evidence of a pilaster on either bone, but there are marked dorsal extensions of the diaphysis along the proximal two-thirds of the shafts. The soleal lines are modest, and there is a trace of a flexor line on each bone.

The Dolní Věstonice 15 anterior crests have a curvature similar to those seen on the other triple burial tibiae, with lateral convexities by the tibial tuberosities, as well as close to and distal of midshaft. They are well angled proximally and mid-distally, but in the midproximal level the anterior crests are rounded. The interosseous lines are sharp for the lengths of the bones, and they reach the posterior margins near midshaft. There is no real pilaster on either tibia, just a raised flexor line on the slightly convex posterior surface. The soleal lines are rugose along their medial edges but smooth proximally. The flexor lines are clearly marked on both sides, but there is a longer continuous extent of it on the right bone.

The Dolní Věstonice 16 tibial diaphyses have a less pronounced S-curve to the anterior crests, with rounded angles. The interosseous crests are prominent proximally, but they fade distally. The pilasters are gentle swellings among the proximal halves of the diaphyses that have no distinct medial and lateral boundaries. The soleal and flexor lines are clearly present, but the root etching makes their details uncertain.

These aspects of the diaphyseal subperiosteal morphology are evident in both the photographs of the bones (Figures 18.29 to 18.33) and the drawings (Figures 18.36 and 18.38) of their reconstructed cross sections at midshaft (50%) and mid-proximal shaft (65%). The anteroposterior versus mediolateral proportions of the diaphyses at midshaft and midproximal shaft can be assessed by both external diameters (the midproximal one being the basis of the cnemic index) and by perpendicular second moments of area (Figues 18.37 and 18.39). The one high outlier in both of the external diameter plots is Předmostí 3, and the lower Neandertal specimen in the proximal diameter comparison is La Ferrassie 2. The Dolní Věstonice specimens, including the small Dolní Věstonice 3 and Dolní Věstonice 15, are similar to the other specimens in these comparisons. Only the Dolní Věstonice 14 tibiae appear relatively expanded anteroposteriorly in the midshaft second moment of area comparison, but little more than Paviland 1 and Veneri 1.

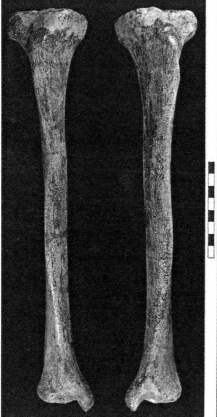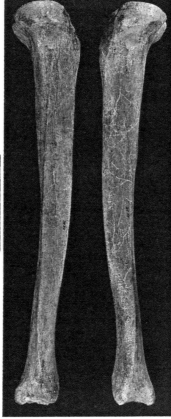

Figure 18.32 Anterior (left) and lateral (right) views of the Dolní Věstonice 15 tibiae. Scale in centimeters.

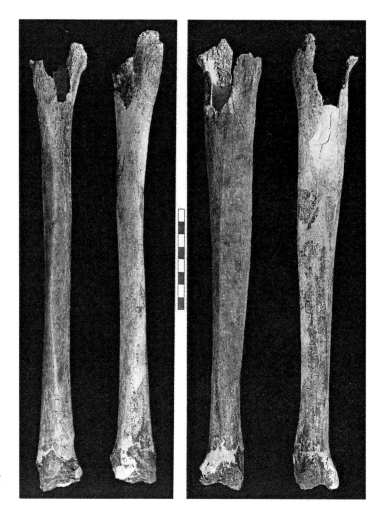

Figure 18.33 Anterior (left) and lateral (right) views of the Dolní Věstonice 16 tibiae. Scale in centimeters.

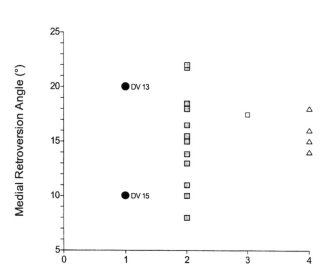

Figure 18.34 Plot of tibial retroversion angles for (1) Dolní Věstonice 13 and 15, (2) Gravettian specimens ($N = 12$), (3) Skhul 4, and (4) Neandertals ($N = 5$).

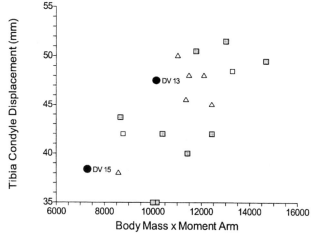

Figure 18.35 Bivariate plot of tibial condylar displacement versus body mass times its moment arm. Solid circles: Dolní Věstonice specimens; gray squares: European Gravettian specimens; open squares: Qafzeh-Skhul specimens; open triangles: Neandertals.

The Lower Limb Remains

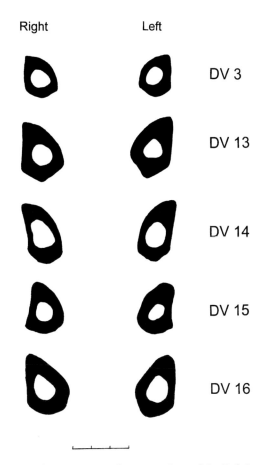

Figure 18.36 Reconstructed cross sections of the Dolní Věstonice 3 and 13–16 tibiae at 50% of length (midshaft level). Right tibiae are on the left, medial is toward the midline, anterior is above, and all are viewed distally. The endosteal contours are interpolated from parallax-corrected biplanar radiography. Scale in centimeters.

Comparisons of cross-sectional cortical area versus total subperiosteal area, as with the humeri and femora, provide little scatter across the Late Pleistocene samples, and all of the Dolní Věstonice specimens provide close to expected proportions (Figure 18.40). This holds even for the relatively small values for Dolní Věstonice 14, 15, and especially 3.

As with the femora, assessments of robusticity were undertaken by utilizing both external diameters and cross-sectional parameters (Figure 18.41). In the former, the cross section was modeled as a solid ellipse and the diameters used to compute polar moments of area; there is little overlap between the Neandertals, with their wide torsos and especially short tibiae (Chapter 12), and the more linear early modern humans. Only Dolní Věstonice 15 approaches the Neandertal distribution. When body proportions are taken into account by comparing the polar section modulus to estimated body mass times bone length, the differences between the reference samples disappear. In this, however, Dolní Věstonice 13 and 15 remain in the middle of the overall distribution, and Dolní Věstonice 3 and 14 appear relatively gracile.

The Distal Epiphyses

The distal epiphyses of Dolní Věstonice 13 present several features. There are no squatting facets, and the anterior medial malleolar surfaces turn only slightly medially, suggesting little hyperdorsiflexion of the talocrural joints. There is a clear midtrochlear ridge on each bone. The

Figure 18.37 Bivariate plot of anteroposterior versus mediolateral external diameters at the midshaft level (above) and maximum (anteroposterior) versus minimum (mediolateral) second moments of area at 50% of length (below). Solid circles: Dolní Věstonice specimens; gray squares: European Gravettian specimens; open squares: Qafzeh-Skhul specimens; open triangles: Neandertals.

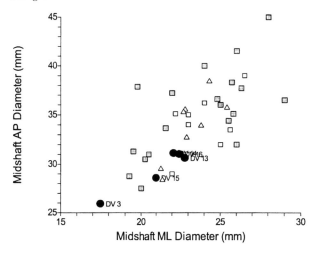

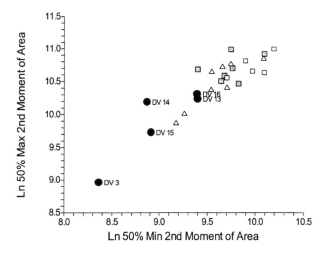

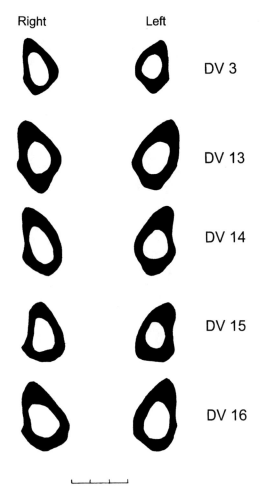

Figure 18.38 Reconstructed cross sections of the Dolní Věstonice 3 and 13–16 tibiae at 65% of length (midproximal shaft level). Right tibiae are on the left, medial is toward the midline, anterior is above, and all are viewed distally. The endosteal contours are interpolated from parallax-corrected biplanar radiography. Scale in centimeters.

plete separation of the trochlear and malleolar subchondral bone, with a small (6.0 mm wide) bridging of subchondral bone between them. The two surfaces on each side were therefore in one articular capsule, but the articular surfaces were largely separate.

Little remains of the Dolní Věstonice 16 distal tibiae, and the only feature of note is a prominent midtrochlear ridge on each tibia.

The Fibulae

The Dolní Věstonice human remains preserve major portions of twelve fibulae, five right and left pairs from the

Figure 18.39 Bivariate plot of anteroposterior versus mediolateral external diameters at the cnemic (nutrient foramen) index level (above) and maximum (anteroposterior) versus minimum (mediolateral) second moments of area at 65% of length (below). Solid circles: Dolní Věstonice specimens; gray squares: European Gravettian specimens; open squares: Qafzeh-Skhul specimens; open triangles: Neandertals.

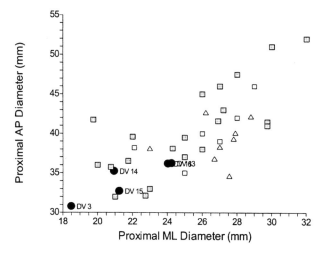

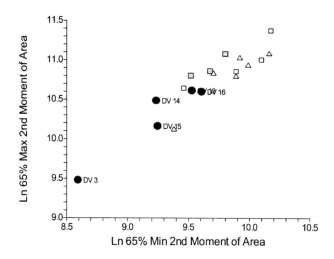

flexor hallucis longus bursal sulcus is smooth on the left but has a raised ridge on the right bone.

On the Dolní Věstonice 14 distal epiphysis, there is a prominent lateral squatting facet on the left tibia, 10.5 mm wide and 3.8 mm high (the right bone is eroded distoanteriorly). There is a small nubbin of bone on the left for the flexor hallucis longus bursa (the right side is damaged). The midtrochlear ridge is absent from the right bone and barely noticeable on the left side.

There is a damaged squatting facet on the left Dolní Věstonice 15 tibia, 6.5 mm high but of unknown breadth; the right bone has a trace of one, but its size is uncertain. The right flexor hallucis longus sulcus is eroded, but the left one has a medial ridge and a sulcus that is porous proximally and smooth distally. There is little midtrochlear ridge on either bone, but each exhibits an almost com-

The Lower Limb Remains

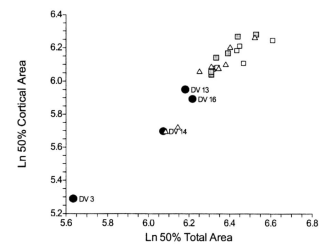

Figure 18.40 Bivariate plot of tibial midshaft (50%) cortical area versus total subperiosteal area. Solid circles: Dolní Věstonice specimens; gray squares: European Gravettian specimens; open squares: Qafzeh-Skhul specimens; open triangles: Neandertals.

The Diaphyses

All of the Dolní Věstonice and Pavlov fibular shafts are notable for their strong development of muscular and interosseous crests and the associated sulci, both ventrally and dorsally. This is particularly evident in midshaft cross sections of the Dolní Věstonice 13 to 16 plus Dolní Věstonice 42 and 48 fibulae (Figure 18.48). The Dolní Věstonice 3 fibulae have little rugosity of their muscle markings, but the interosseous crests are prominent, sharply angled, and curve laterally. The associated sulci reach depths of 3.0 mm on each side. Those of Dolní Věstonice 13 are similarly smooth, have relatively flat dorsal surfaces with little measurable concavity, but have prominent interosseous crests and associated anteromedial (or ventral) sulci. They

Figure 18.41 Bivariate plots of the midshaft pseudopolar moment of area versus tibial length (above) and midshaft (50%) polar section modulus versus body mass times tibial length (below). Solid circles: Dolní Věstonice specimens; gray squares: European earlier Upper Paleolithic specimens; open squares: Qafzeh-Skhul specimens; open triangles: Neandertals.

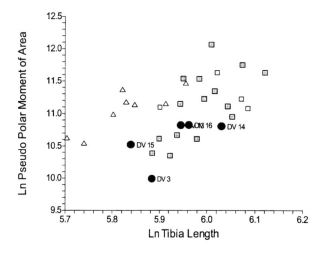

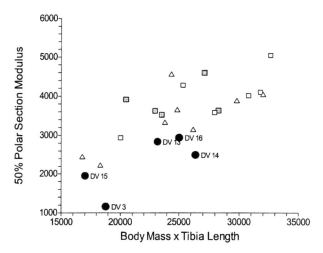

associated burials (Dolní Věstonice 3 and 13–16) plus two isolated right fibulae (Dolní Věstonice 42 and 48). In addition, there is a proximal diaphyseal segment from Pavlov 1 (Figures 18.42 to 18.47). However, all of the fibulae except those from the triple burial consist principally of diaphyseal sections, and only the three from the triple burial therefore provide either direct measurements of length (Dolní Věstonice 13 and 15) or lengths with minor estimation (Dolní Věstonice 14). The fibular lengths of Dolní Věstonice 3 and 16 could be estimated from their tibial lengths, but given the estimation already involved for their tibial lengths (Sládek et al., 2000), this has not been pursued.

A series of measurements, most of them diaphyseal, have been provided (Trinkaus & Jelínek, 1997; Sládek et al., 2000: 228–229). However, there is limited comparative data available, and it remains uncertain to what degree the diaphyseal diameters accurately quantify the complex morphology of these fibulae.

The Proximal Epiphyses

Dolní Věstonice 13 and 15 retain most of their fibular heads, but portions have been eroded and reconstructed. The Dolní Věstonice 14 fibulae possess the eroded cores of the fibular heads with small pieces of the tibial articular facets, and the Pavlov 1 fibula has the flare for the proximal epiphysis. There are no reliable osteometrics on them and little of morphological note. The Dolní Věstonice 15 fibular heads appear relatively bulbous (Figure 18.45), but those of Dolní Věstonice 13 are less so (Figure 18.43).

Figure 18.42 Anterior view of the Dolní Věstonice 3 right and left fibulae. Scale in centimeters.

are slightly asymmetrical, with the right having a deeper (5.0 mm) sulcus than the left (3.3 mm). The Dolní Věstonice 14 fibular midshafts are even more marked by crests and sulci since ventral sulci similar to those of Dolní Věstonice 13 (depths of 5.0 mm and 3.8 mm) are accompanied by dorsal sulci of about half of the ventral depths (3.2 mm and 1.8 mm). Its anterior crests also curve strongly laterally. The degree of muscle marking on the bones is obscured by root etching.

The Dolní Věstonice 15 midshaft sulci are smaller, being barely measurable dorsally (0.0 mm and 0.2 mm)

Figure 18.43 Anterior view of the Dolní Věstonice 13 right and left fibulae. Scale in centimeters.

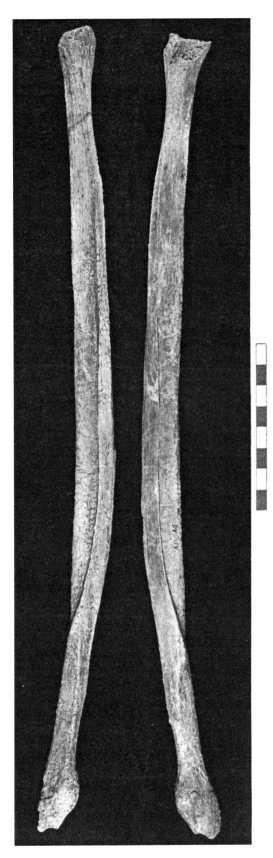

Figure 18.44 Anterior view of the Dolní Věstonice 14 right and left fibulae. Scale in centimeters.

Figure 18.45 Anterior view of the Dolní Věstonice 15 right and left fibulae. Scale in centimeters.

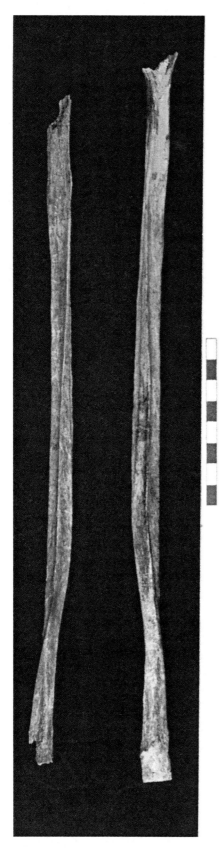

Figure 18.46 Anterior view of the Dolní Věstonice 16 right and left fibulae. Scale in centimeters.

and similar to those of Dolní Věstonice 3 ventrally (2.1 mm and 3.4 mm). The anterior crests do not curve laterally, and the bone surfaces are generally smooth. The Dolní Věstonice 16 fibulae have the smallest sulci (0.0 dorsally and 1.7 mm and 1.3 mm ventrally), and the anterior crests do not curve laterally. The isolated Dolní Věstonice 42 and 48 fibular midshafts are very similar in general shape to the other Dolní Věstonice fibulae, but their sulcal depths, as a measure of the degree of accentuation of the crests and sulci, are intermediate. They are 0.9 mm and 0.5 mm, respectively, dorsally and 3.8 mm and 2.9 mm ventrally. The ventral sulci are open, with no lateral curving of the anterior crest. The Pavlov 1 fibula is not complete to near midshaft, but the proximal shaft portion exhibits a very prominent anterior interosseous crest; comparison with the more complete Dolní Věstonice fibulae suggests a degree of crest development similar to the more pronounced of those specimens.

The fibulae of Neandertals all lack this degree of cresting and sulcal formation, with the various surfaces

Figure 18.47 Anterior views of the Dolní Věstonice 42 and 48 right fibulae and the Pavlov 1 left fibula. Scale in centimeters.

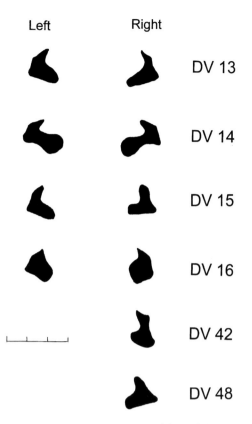

Figure 18.48 Midshaft cross sections of the Dolní Věstonice 13–16, 42, and 48 fibulae. The first four are oriented according to the parasagittal plane defined by their medial distal epiphyses (or close approximations thereof). The last two are oriented approximately by placing the interosseous crest anteriorly. Medial is toward the middle and anterior is above. Scale in centimeters.

being flat to convex around the circumferences (Heim, 1982b; Trinkaus, 1983b). Qafzeh-Skhul (McCown & Keith, 1939; Vandermeersch, 1981) and other Gravettian (Matiegka, 1938; Vallois & Billy, 1965; Trinkaus, 2000b) fibular diaphyses are similar to these Pavlovian ones in being strongly crested and grooved. Recent humans exhibit considerable variation in fibular diaphyseal cross-sectional morphology, with the degree of cresting present in these early modern human fibulae representing one end of the range of variation (Sprecher, 1932).

The Dolní Věstonice fibular shafts exhibit varying degrees of medial convexity. The subtenses from the medial diaphyseal chords vary from 7.0 mm and 6.0 mm for the Dolní Věstonice 15 fibulae, to 8.0 mm for the Dolní Věstonice 3 and 13 left fibulae, to 8.5 mm for each Dolní Věstonice 16 bone, to 15.5 mm and 16.5 mm for the two Dolní Věstonice 14 fibulae (Trinkaus & Jelínek, 1997; Sládek et al., 2000). The medial curves on most of the fibulae are produced by a combination of a medial bulge of the mid-distal diaphysis combined with an overall medial convexity of the mid-diaphyseal axis. However, in the Dolní Věstonice 15 fibulae, this is transformed into an S-curve resulting from the addition of a distal diaphyseal medial concavity (Figure 18.45). This S-curve reduced the total amount of curvature, resulting in Dolní Věstonice 15 having the smallest curvature subtenses, even though its curvature chords (267 mm and 260 mm) are larger than that of Dolní Věstonice 3 (235 mm). However, the longer Dolní Věstonice 13 left fibula and the Dolní Věstonice 16 fibulae (chords of 292 mm and 300 mm, respectively) also exhibit some degree of a distal medial diaphyseal concavity and a resultant S-curve, producing their more modest medial subtenses (Figures 18.43 and 18.46). The Dolní Věstonice 48 fibular shaft is not sufficiently complete to provide curvature measurements or indicate whether an S-curve was present (Figure 18.47); however, its marked medial bulge indicates a pattern of curvature similar to the other Dolní Věstonice fibulae.

In addition to these features, the Dolní Věstonice 14 fibulae exhibit deep fossae on the posterolateral neck leading onto short sulci, especially on the right side. The sulci are 44.0 mm and 33.6 mm long on the right and left sides, respectively, and they have breadths of 5.4 mm and 5.0 mm and depths of 2.3 mm and 1.0 mm respectively. They may well be associated with the soleus origin in that location.

The Distal Epiphyses

Distal epiphyses are present on the fibulae from the triple burial individuals only. The Dolní Věstonice 13 left fibula has a ridge of bone dorsally across the proximal margin of the lateral malleolar facet, and the malleolus projects minimally beyond the talar articulation. The digital fossae are deep (3.3 mm). The distal ends of the Dolní Věstonice 14 fibulae are notable only for their deep digital fossae (3.4 mm). The malleoli of the Dolní Věstonice 15 fibulae appear long relative to their talar facets (Chapter 19), and their digital fossae are even more pronounced than those of the previous two individuals (4.0 and 3.8 mm), especially considering the smaller size of Dolní Věstonice 15.

The Pedal Remains

Five of the associated skeletons (all except Pavlov 1) retain pedal remains, and three of these individuals (Dolní Věstonice 3, 15, and 16) have relatively large sets of bones for at least one foot (Table 18.1). In addition, there are several isolated foot bones from Dolní Věstonice II, which include the tarsal bones of Dolní Věstonice 39 and 46; the metatarsal bones of Dolní Věstonice 44, 47, and 49; and the hallucal proximal phalanx of Dolní Věstonice 52.

Table 18.1 Summary inventory of the Dolní Věstonice and Pavlov pedal remains.

	DV 3	DV 13	DV 14	DV 15	DV 16	DV Isolated
Talus	L	R		RL	RL	
Calcaneus	L	R	L	RL	RL	
Navicular	L		R	RL	RL	R
Cuboid	L		R		R	
Medial Cuneiform	L		RL		RL	L
Intermedial Cuneiform	L				RL	
Lateral Cuneiform	L			L	RL	
MT 1			RL	L	RL	
MT 2	L		RL	L	RL	L
MT 3	L		R	RL	RL	R
MT 4	L		R	RL	RL	
MT 5	L		RL	L	RL	R
PP 1			?		R	L
PP 2	?			R	R	
PP 3	L			R		
PP 4					R	
PP 5				R	R	
MP 2	L				R	
MP 3	L					
MP 4					R	
MP 5	L				R	
DP 1	L				R	
DP 2–4					R	
DP 5					R	
Sesamoids	LLL?			?		

R: right or probable right bone preserved; L: left or probable left bone preserved; ?: bone of uncertain side preserved. Multiple entries for the sesamoid bones indicate the number of bones preserved but with uncertainty as to the digits represented. For details on preservation, see Sládek et al. (2000).

However, these remains vary markedly in completeness and morphological utility (Sládek et al., 2000).

The Dolní Věstonice 3 foot remains, all probably from her left foot, retain all of the tarsals, four of the metatarsals, and a number of phalanges, but all of these tarsals except the talus have extensive plantar abrasion and crushing, and most of the metatarsal and phalangeal epiphyses are modified beyond utility. It is principally the talus (Figure 18.52) and some phalanges that provide comparative data. Dolní Věstonice 13 preserves only a talus and a calcaneus (Figure 18.53). Dolní Věstonice 14 has only five partial tarsal bones and a series of metatarsals with eroded bases and lacking the heads (Figure 18.57). The Dolní Věstonice 15 pedal remains are considerably more complete (Figures 18.49 & 18.54), even though many of the phalanges and some tarsals are absent and there is damage to some of the bones. It should be noted that the pedal bones of Dolní Věstonice 14 and 15 were mixed in situ and that they have been sorted according to preservation (the Dolní Věstonice 14 bones tend to be more extensively root-etched), size, and matching articulations. It is possible that a couple of the more distal bones have been misidentified as to individual, but it is unlikely that this will affect any of the considerations below.

The Dolní Věstonice 16 pedal skeletons are by far the most complete since they retain most of both tarsometatarsal skeletons and a number of the right phalanges (Figures 18.50 and 18.55). However, many of the edges are abraded, limiting some of the osteometrics and discrete trait observations from them.

The considerations below of these relatively abundant Dolní Věstonice pedal bones are concerned with the configurations of individual bones and some comparisons of articular and diaphyseal robusticity. It is thoroughly documented (e.g., Matiegka, 1938; McCown & Keith, 1939; Trinkaus, 1975a, 1983a; Vandermeersch, 1981; Heim, 1982b) that all of the Late Pleistocene human pedal remains are fully commensurate with a normal recent human bipedal gait. Issues of variation between samples relating to basic locomotion are therefore no longer pertinent. However, there may well be differences between these samples in more subtle aspects of the articulations and proportions (e.g, Rhoads & Trinkaus, 1977; Trinkaus & Hilton, 1996; Trinkaus, 2000c), and it is these considerations to which the Dolní Věstonice remains are related.

Overall Pedal Proportions

Two of the Dolní Věstonice pedal skeletons, those of Dolní Věstonice 15 (left) and Dolní Věstonice 16 (right), are sufficiently complete to permit assessment of their overall subtalar proportions (Figures 18.49 and 18.50). In order to do this, the preserved bones of each foot were articulated on a base of plasticene and manipulated until all of the articular facets were largely congruent. It is possible

Figure 18.49 Dorsal view of the articulated Dolní Věstonice 15 left foot. The connection between the talus and the calcaneus and the metatarsals is made through the navicular and lateral cuneiform bones, plus the intermetatarsal facets. Scale in centimeters.

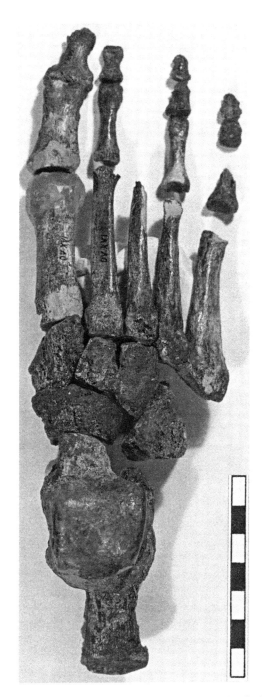

Figure 18.50 Dorsal view of the Dolní Věstonice 16 right foot. Scale in centimeters.

that this approach placed the tarsometatarsal skeletons in pronated positions, but it should not affect their longitudinal proportions. The Dolní Věstonice 16 pedal skeleton, despite erosion to a number of the elements, is sufficiently complete to require little estimation. The Dolní Věstonice 15 remains, however, lack the cuboid, medial cuneiform and intermedial cuneiform bones; it was therefore necessary to base the medial arch length on the articular connections through the lateral cuneiform bone and across the intermetatarsal facets of the three medial metatarsal bones. It is possible that the inferred length of the absent medial cuneiform bone is in error by 1 or 2 millimeters, but any such error will have little effect on the results.

The two positions in which the pedal skeleton is most stressed by ground reaction force during locomotion are heel-strike and heel-off (Stott et al., 1973; Cavanagh & Rodgers, 1984). It is the latter that is of interest since the maximum propulsive force through the foot occurs at that time. In particular, the anterior tarsometatarsal skeleton, from the talar trochlea to the area of ground contact near the metatarsal heads, serves initially as a continuation of the lower leg during the extension of the knee by quadriceps femoris and as a load arm for body mass during the contraction of triceps surae (soleus and gastrocnemius), producing talocrural plantarflexion. In both of these actions, the anterior tarsometatarsal skeleton serves as the load arm for body mass, and the total tarsometatarsal length (metatarsal heads to calcaneal tuberosity) serves as the power arm for triceps surae (whether stabilizing the talocrural joint against passive dorsiflexion or producing active plantarflexion). The subtalar skeleton therefore acts at heel-off as a second-class lever with the fulcrum at the metatarsal heads.

Consequently, measurements of the distance from the calcaneal tuberosity to the distal metatarsal 1 head and of each endpoint to the middle of the talar trochlea, all parallel to the plantar surface of the subtalar skeleton, were taken for Dolní Věstonice 15 and 16. These were compared to the limited set of sufficiently complete Late Pleistocene subtalar skeletons for whom data are available, Caviglione 1, Předmostí 3 (estimated from the scaled photograph in Matiegka, 1938), Qafzeh 9, La Ferrassie 1 and 2, Kiik-Koba 1 (from a cast), and Shanidar 1. The comparison of tarsometatarsal foot length (as the triceps surae moment arm) versus anterior tarsometatarsal length (as the body mass moment arm) shows little difference across the available Late Pleistocene data (Figure 18.51). The combination of body mass and its moment arm suggests a slight difference between the Middle and Upper Paleolithic samples, but the data are insufficient to provide confidence in such an inference. Primarily, the addition of body mass estimates makes the Dolní Věstonice 15 foot appear powerful, probably largely as a result of its diminutive body core.

The Tali

The tali of the four individuals preserving them, Dolní Věstonice 3, 13, 15, and 16, are generally similar to each other and to other early modern human tali (Figures 18.52 to 18.55). All except those of Dolní Věstonice 16, and to a

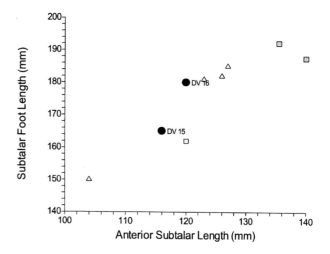

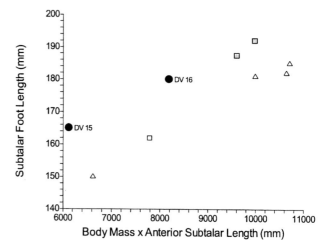

Figure 18.51 Bivariate plots of the triceps surae moment arm during heel-off (medial subtalar length) versus the pedal body weight moment arm at heel-off (midtalar trochlea to metatarsal 1 head) (above) and the former versus the body weight moment arm times body mass (below). Solid circles: Dolní Věstonice specimens; gray squares: European earlier Upper Paleolithic specimens; open squares: Qafzeh-Skhul specimens; open triangles: Neandertals.

lesser extent Dolní Věstonice 3, present largely complete sets of osteometrics and articular discrete traits.

On their trochlea, all of them have anterior extensions of their medial malleolar surfaces, all except Dolní Věstonice 15 have medial anterior extensions of the trochlea, and all except Dolní Věstonice 3 have lateral anterior extensions of the trochlea. Medial squatting facets are absent, but lateral ones are present on the Dolní Věstonice 3 and 13 tali and the Dolní Věstonice 15 right talus, being absent from the other Dolní Věstonice 15 talus and the better preserved right one of Dolní Věstonice 16. All of these features indicate frequent hyperdorsiflexion of the talocrural articulation, and they are common among Late

Figure 18.52 Dorsal view of the Dolní Věstonice 3 left talus and navicular bone. Scale in centimeters.

Pleistocene human tali (Trinkaus, 1975b, personal observation). Interestingly, lateral squatting facets are present on the distal tibiae of Dolní Věstonice 15 and 16 (as well as of Dolní Věstonice 14), but not on those of Dolní Věstonice 13.

The trochlear angles, which measure the degree of posterior convergence of the medial and lateral trochlear margins and reflect indirectly the characteristically human pattern of talocrural articular dynamics (Trinkaus, 1975a), are all well above zero (Table 18.2). They share

Figure 18.53 Dorsal and plantar views of the Dolní Věstonice 13 right talus and dorsal view of the right calcaneus. Scale in centimeters.

Figure 18.54 Dorsal and plantar views of the Dolní Věstonice 15 tali and dorsal view of the Dolní Věstonice 15 calcanei. Scale in centimeters.

this pattern with all recent and Late Pleistocene human tali.

The talar trochlea appears to change proportions during the Late Pleistocene. This was originally interpreted (Rhoads & Trinkaus, 1977) as a relative reduction in the overall size of the talar trochlea, probably reflecting decreases in habitual joint reaction forces. This is reinforced by a comparison of trochlear length to talar length (Figure 18.56), in which most of the high values are those of Neandertals, and the lower values, including all of the Dolní Věstonice specimens, are those of early modern humans. However, even in this, one early Upper Paleolithic specimen, the Aurignacian Mladeč 30 talus, and two Qafzeh-Skhul specimens, Qafzeh 3 and Skhul 5, have proportions among the Neandertals; the remainder of the Qafzeh-Skhul specimens are among the Gravettian

Figure 18.55 Dorsal and plantar views of the Dolní Věstonice 16 tali. Scale in centimeters.

specimens, and the long Aurignacian Fontana Nuova 4 talus has the relatively shortest trochlea.

However, if the geometric mean of the trochlear length and midbreadth is compared to talar length (Figure 18.56), the differences between the samples largely disappear. The Neandertals, on average, still have moderately large trochleae, but the differences are small. Those of Mladeč 30 and Skhul 5 remain high, but those of several early modern specimens, including Dolní Věstonice 3 and 13, overlap the Neandertal range of variation. The talar trochlear proportions in the Late Pleistocene are therefore affected by both the relative overall size of the trochlea and the length/breadth proportions of them. In this, the Dolní Věstonice tali on average have moderately large trochleae for Gravettian specimens but, similar to other Gravettian tali, relatively broad ones.

The neck angles of the Dolní Věstonice specimens are similar to those of other early modern and late archaic humans (Table 18.2). All of these values are well within recent human ranges of variation (Trinkaus, 1983a), and they indicate mediolaterally compact posterior tarsal regions. Similarly, their talar head torsion angles (measured relative to the transverse plane of the trochlea) are well within the ranges of variation of both other Late Pleistocene samples and recent humans (Table 18.2; Trinkaus, 1983a), indicating the pattern of stabilization of the midtarsal joint characteristic of recent humans.

Rounding of the sulcus tali margin is variably developed, being present on the Dolní Věstonice 3 talus but absent from those of Dolní Věstonice 13 and 15, whereas clear sulcus tali facets are present on the Dolní Věstonice 15 tali but absent from the other two (the Dolní Věstonice 16 tali do not preserve this region). All of the tali present complete fusion of the anterior and medial calcaneal facets, as do the tali of most other preindustrial human samples.

In addition to these comparative aspects of the Dolní Věstonice tali, there are a few aspects of the individual specimens of note. The Dolní Věstonice 13 talus has a ridge or step that extends across most of the posterior aspect of the medial calcaneal facet, and it borders a broad sulcus tali. The Dolní Věstonice 15 tali have a pronounced S-curve across the sulcus tali margins of the anterior and medial calcaneal facets, but the facets are fully fused. There is also a symmetrical crease along the posterior edge of each lateral malleolar surface that leads around onto the posterolateral corner of the trochlea; this has the effect of making the posterolateral trochlea plantarly prominent relative to the posterior calcaneal surface. The anterior ends of the medial malleolar anterior extensions are strongly deflected medially.

The Subtalar Skeletons

The Dolní Věstonice subtalar skeletons are variably preserved but nonetheless provide a variety of information about their pedal anatomies.

Table 18.2 Comparative talar articular angles.

	Trochlear Angle	Neck Angle	Torsion Angle
Dolní Věstonice 3	8°	24°	ca. 38°
Dolní Věstonice 13	ca. 4°	ca. 21°	35°
Dolní Věstonice 15	8° / 10°	ca. 19° / ca. 23°	33° / 32°
Dolní Věstonice 16	13° / 9°	ca. 19° / ca. 20°	—/—
Earlier Upper Paleolithic	11.7° ± 4.0° (5)	23.9° ± 2.7° (10)	32.9° ± 2.7° (5)
Qafzeh-Skhul	7.0° ± 3.1° (7)	25.8° ± 2.3° (6)	38.4° ± 9.0° (5)
Neandertals	6.5° ± 3.4° (11)	26.0° ± 4.0° (11)	40.4° ± 4.8° (11)

Right and left values are given for Dolní Věstonice 15 and 16 as available. Mean, standard deviation, and N are provided for comparative samples.

The Lower Limb Remains

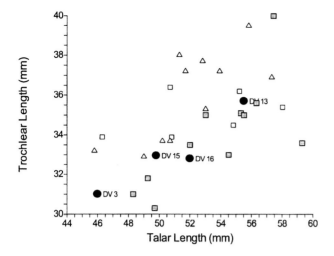

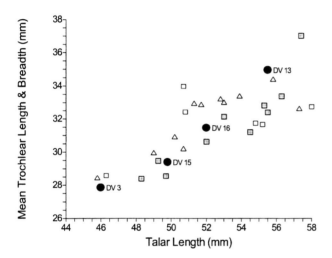

Figure 18.56 Bivariate plots of talar trochlear dimensions versus talar length: Trochlear length versus talar length (above); the geometric mean of the trochlear length and breadth versus talar length (below). Solid circles: Dolní Věstonice specimens; gray squares: European earlier Upper Paleolithic specimens; open squares: Qafzeh-Skhul specimens; open triangles: Neandertals.

The calcanei, to the extents preserved (best for Dolní Věstonice 15; worst for Dolní Věstonice 3 and 13), are compact and robust bones similar to those of most nonindustrial recent humans. Their anterior and medial talar facets are consistently completely fused, with only a suggestion of a posterolateral concavity between them (when adequately preserved: Dolní Věstonice 13, 15, and 16; Figures 18.53 and 18.54). Their sustentacula tali are moderately projecting, and the sufficiently preserved ones (Dolní Věstonice 13 and 15) exhibit only one flexor tendon sulcus (for flexor hallucis longus).

Despite the presence of cuboid and/or navicular bones for five individuals (Dolní Věstonice 3, 14, 15, 16 and 39), evidence for naviculocuboid facets are present only on the middle three individuals. Dolní Věstonice 16 exhibits matching facets on its right bones; however, Dolní Věstonice 14 lacks a facet on its right cuboid, and the Dolní Věstonice 15 left navicular lacks such a facet. Since they occur in about half of recent and Late Pleistocene specimens (Trinkaus, 1975a), this distribution is expected. The navicular tuberosities on Dolní Věstonice 15 are modest. In addition, the Dolní Věstonice 14 cuboid bone has a polished area in the peroneus longus sulcus, which might indicate the former presence of a sesamoid bone. Such a sesamoid bone has been identified for Dolní Věstonice 3, since it was found adherent to her plantar lateral cuboid bone.

The Dolní Věstonice 14 left medial cuneiform bone has a largely flat metatarsal 1 facet, but with some dorsoplantar torsion to the surface. The Dolní Věstonice 16 medial cuneiform bone is flat dorsally but distally convex plantarly, matching the mediolateral concavity of the metatarsal 1 base. The Dolní Věstonice 46 medial cuneiform bone has a strongly concave and twisted navicular facet, and its metatarsal 1 facet is largely flat mediolaterally, has dorsoplantar torsion, and has slight edge rounding on the mediodorsal half and the lateroplantar half. The intermedial cuneiform bones are unremarkable. The Dolní Věstonice 15 lateral cuneiform bone has a modest plantar tubercle, which does not extend plantarly below the metatarsal facet, and it has a prominent metatarsal 4 facet.

The metatarsal bones are variably preserved, with most of them having damaged heads; only the Dolní Věstonice 3 third and fifth metatarsals and the Dolní Věstonice 15 fourth and fifth metatarsals are sufficiently complete to provide any head diameters. The shafts present variably pronounced interosseous crests, with Dolní Věstonice 3 providing some of the more strongly marked ones. Even though none of them can be measured, all of the shafts have the progressively increasing torsion from the second to the fifth metatarsals that indicates well-formed, normal pedal arches (Figures 18.49, 18.50, 18.57, and 18.58). In addition, the base horizontal angles, in which a lower angle indicates a more medially directed tarsal facet, decrease steadily from the second to the fifth metatarsals.

The bases of the first metatarsal bones are variably concave mediolaterally, matching their distal medial cuneiform bones. The Dolní Věstonice 14 right metatarsal 1 is flat dorsally but slightly concave plantarly, and a piece of the left one shows a mediolateral concavity. The Dolní Věstonice 15 metatarsal 1 has a twisted and concave tarsal facet. The Dolní Věstonice 16 metatarsal 1 bases are similarly mediolaterally concave. The lateral bases of the Dolní Věstonice 14 and 15 first metatarsals lack any sign of a second metatarsal facet, and this is matched by the absence of such facets on the Dolní Věstonice 14 to 16

Figure 18.57 Dorsal view of the Dolní Věstonice 14 right anterior tarsals, metatarsals, and hallucal proximal phalanx. Scale in centimeters.

second metatarsals. However, such facets are absent on between 67.5% and 91.0% of individuals in five recent humans samples and on 61.5% of a late archaic human sample (Trinkaus, 1983a).

The plantar tuberosities of the Dolní Věstonice 15 metatarsals 2 to 4 are small (the bases of the Dolní Věstonice 3, 14, 16, 44, and 47 metatarsals are too eroded to provide information), and the peroneus longus tubercle on the proximolateral metatarsal 1 is small. Similarly, the proximal lateral tubercles of the Dolní Věstonice 3, 14–16, and 49 fifth metatarsals are modest in size and projection and are usually only partially differentiated from the cuboid facet and associated capsular attachments.

The Distal Halluces

The largely flat to slightly concave first tarsometetarsals articulations have already been described. Distally, the Dolní Vestonice 15 metatarsal 1 head has a lateral deviation of 6°, and the Dolní Věstonice 52 proximal hallucal phalanx has a lateral head deviation of 3° (the Dolní Věstonice 16 proximal phalanx head has no lateral deviation, but damage precludes certainty about this measurement). In addition, the bases of Dolní Věstonice 3 and 16 distal hallucal phalanges have lateral deviations of 6° and 11°. All of these lateral deviation angles are well within recent human ranges of variation (Trinkaus, 1975a; Meyer, 1979), and they are similar to those of other Late Pleis-

tocene archaic and early modern humans (Trinkaus, 1975a, personal observation; Courtaud, 1989). They indicate normal human hallux valgus (Barnicot & Hardy, 1955).

Measures of pedal robusticity are difficult, given the fragmentation and structural integration of the various components of the tarsometatarsal skeleton. However, the proximal hallucal phalanx is an important element of the propulsive phase of human locomotion, and therefore an assessment of its diaphyseal robusticity is appropriate for evaluating pedal robusticity. A pseudopolar moment of area was computed from the midshaft dorsoplantar and mediolateral diameters by using standard ellipse formulae and modeling the phalanges as solid beams. The resultant plot of this measure of diaphyseal rigidity (Figure 18.59) against phalanx length shows a scatter, with the late archaic humans having generally more robust phalanges than the early modern humans, despite some overlap between the samples. In this comparison, Dolní Věstonice 52 clusters with the Neandertals, and Dolní Věstonice 14 and 16 are in the middle of the early modern human distribution.

However, since the proximal hallucal phalanx is serving as a beam for the propulsion of body mass, its strength should preferably be scaled against bone length times body mass. Given the presence of several isolated phalanges, including Dolní Věstonice 52, this reduces the total sample size from twenty-four to sixteen. However, it also decreases the general scatter and produces more exten-

Figure 18.58 Dorsal views of Dolní Věstonice isolated metatarsals and hallucal phalanx. From left to right, the Dolní Věstonice 44 left metatarsal 2, the Dolní Věstonice 49 right metatarsal 5, and the Dolní Věstonice 52 left proximal hallucal phalanx. Scale in centimeters.

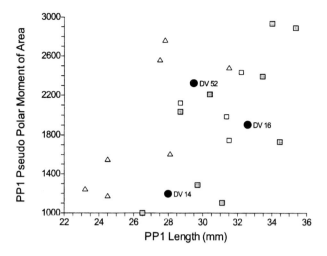

proximal phalanges, is since it has been shown (Trinkaus & Hilton, 1996) that there was a shift from mediolaterally broad to dorsoplantarly high diaphyses, at least between late archaic and recent humans. As a simple measure of this, an index of midshaft breadth to height was computed for the pooled digits 2 to 4 proximal phalanges. The resultant values are 81.2 and 88.6 for the Dolní Věstonice 15 and 16 second phalanges, 98.1 and 80.8 for the Dolní Věstonice 3 and 15 third phalanges, and 96.7 for the Dolní Věstonice 16 fourth phalanx. These compare with pooled means of 100.4 ($\pm$ 9.1; $N = 17$) for Gravettian humans, 107.0 ($\pm$ 6.8; $N = 12$) for Qafzeh-Skhul humans, and 116.7 ($\pm$ 8.8; $N = 17$) for Neandertals. The Dolní Věstonice values are all outside of the Neandertal range (minimum: 101.7), but they fall partially within the Qafzeh-Skhul (92.6–114.5) and Gravettian (85.1–116.7) ranges, with those of Dolní Věstonice 15 being the narrowest. This shift in diaphyseal proportions almost certainly reflects a general decrease in robusticity differentially affecting mediolateral dimensions (Trinkaus & Hilton, 1996).

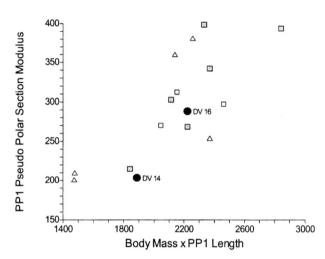

For a better assessment of proximal phalangeal hypertrophy, pseudopolar moments of area were computed as above and plotted against phalanx length and against phalanx length times body mass (Figure 18.60). In both of the comparisons, and especially in the latter, the Neandertal samples remain more robust than the early modern humans. The Dolní Věstonice specimens are all well within the early modern human distributions. This is in contrast with the patterns seen in their femoral, tibial, and hallucal proximal phalangeal diaphyses, as well as in assessments of gluteal tuberosity size and quadriceps femoris moment arms at the knee. In all of these other assessments of locomotor robusticity, there is little if any documentable difference between these Late Pleistocene human samples once the property in question is scaled appropriately, using moment arm and body mass. This implies that there was a reduction in the habitual levels of stress through the lateral toes with the emergence of early modern humans, a pattern that is not evident medially in the foot or proximally up the leg. Since the lateral toes are stressed primarily through ground reaction forces, generating traction at heel-off, it is likely that this indicates a protective layer between them and the substrate. This would not affect the rest of the leg and the hallux since they would be engaged in primary body support and propulsion, regardless of their mode of contact with the ground.

Figure 18.59 Bivariate plots of hallucal proximal phalanx pseudopolar moment of area versus length (above) and pseudopolar section modulus versus body mass times length (below). Solid circles: Dolní Věstonice specimens; gray squares: European Gravettian specimens; open squares: Qafzeh-Skhul specimens; open triangles: Neandertals.

sive overlap between the late archaic and early modern humans; Dolní Věstonice 14 and 16 remain in the middle of the overall distribution.

The distal hallucal phalanges of Dolní Věstonice 3 and 16 present few features of note. The distal tuberosities of both of them are very small. The latter has a marked plantar proximal pit associated with the flexor hallucis longus tendon, which leads distally onto a raised area for the insertion of the tendon.

The Lateral Toes

The scattered phalanges of the lateral toes provide few features of singular interest. Of primary concern are the proportions of the diaphyses of their middle (digits 2 to 4)

Summary

These considerations of the lower limb remains from Dolní Věstonice and Pavlov reveal an overall morphological pattern that is quite similar to that which is observed

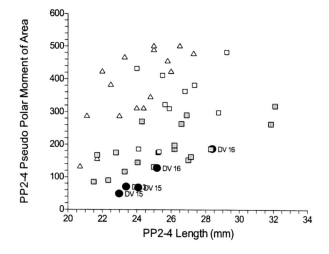

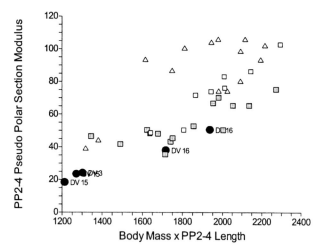

Figure 18.60 Bivariate plots of pedal proximal phalanx 2 to 4 pseudopolar moment of area versus length (above) and pseudopolar section modulus versus body mass times length (below). Solid circles: Dolní Věstonice specimens; gray squares: European Gravettian specimens; open squares: Qafzeh-Skhul specimens; open triangles: Neandertals.

in other Gravettian human remains. The femora all have pilasters, but as with other Gravettian specimens, the degree of separation of the pilaster from the remainder of the diaphysis is variable. The proximal femora exhibit variably prominent but universally present gluteal buttresses, being clearest on the Dolní Věstonice 3, 14, and 35 femora and least so on the Dolní Věstonice 16 and Pavlov 1 remains. All of them have third trochanters (although of variable size), but none exhibits hypotrochanteric fossae within their gluteal tuberosities. The femoral neck/shaft angles are mostly within the middle of the Gravettian (and normal hunter-gatherer) range, although Dolní Věstonice 3 has a high one and Dolní Věstonice 15 has a (possibly pathologically) low one.

The tibial diaphyses are marked by their consistent presence of at least some sharply angled diaphyseal "crests" (and variably demarcated tibial pilasters), and this is even more pronounced in the marked cresting and associated sulcal formation on all of their fibular diaphyses. The talar trochleae are similar in relative size to those of other Late Pleistocene humans, but as with other Gravettian tali, their trochleae tend to be shorter and broader than those of the late archaic and some early modern human tali.

The overall level of hypertrophy is similar to that seen in other Late Pleistocene human remains, as is evident in measures of femoral, tibial, and hallucal phalangeal diaphyseal robusticity and of femoral gluteal tuberosity development, although the Dolní Věstonice and other early modern human lateral pedal phalanges have relatively gracile shafts. However, the rugosities of most muscle markings and musculoligamentous tuberosities are modest, with the exceptions of some of those of Dolní Věstonice 16. The patellae and tibiae have degrees of relative development of the quadriceps femora moment arms similar to those of other Late Pleistocene humans, and similarly common proportions are seen in the tarsometatarsal skeletons of Dolní Věstonice 15 and 16 with respect to triceps surae.

Of further interest is the fact that Dolní Věstonice 15, despite the deformities of the right femur and the relative shortness of the femora, exhibits levels of robusticity and general morphology that conform entirely to what is observed in the other Gravettian specimens. Even if a slightly higher body mass estimate were employed to scale these reflections of lower limb hypertrophy (see Chapter 13), Dolní Věstonice 15 would remain well within the range of variation of other Pavlovian and Gravettian specimens. There is little, beyond the obvious femoral deformities and overall proportions, to separate it from the other specimens. Dolní Věstonice 15 kept up with the others of its social group and probably contributed his share of burden carrying and other economic activities.

Furthermore, despite a variety of changes in detailed morphology, including aspects of articular proportions and especially diaphyseal cross-sectional shape, there is little change in overall levels of lower limb robusticity between these Gravettian specimens and antecedent Middle Paleolithic late archaic and early modern humans. This suggests that, despite clear shifts in the manifestations of these skeletal features, the general levels of locomotor stress and burden carrying remained similar though this time period.

19

Skeletal and Dental Paleopathology

Erik Trinkaus, Simon W. Hillson, Robert G. Franciscus, and Trenton W. Holliday

The paleobiological analysis of the human remains from Dolní Věstonice and Pavlov involves a detailed consideration of the developmental and/or degenerative pathological lesions and anomalies on the remains, since these have the potential to provide information on both the natures and levels of the stress to which the individuals were subjected and their abilities to survive various forms of hardship. Many of the lesions have been referred to in previous publications (Jelínek, 1954, 1992; Vlček, 1991, 1997; Kuklík, 1992; Trinkaus & Jelínek, 1997; Sládek et al., 2000; Trinkaus et al., 2000c, 2001; Formicola et al., 2001), but none of those references has provided a complete list of the lesions or thorough diagnoses of them. In addition, most of the lesions have been mentioned in the preceding morphological chapters since they affect, to varying degrees, the morphological and functional anatomical interpretations of the remains. However, none of those previous citations here deals with the lesions specifically or provides diagnoses of their etiologies.

For these reasons, this chapter presents, by individual and by anatomical region, the lesions and other abnormalities on the remains. Also included are brief descriptions of several unusual aspects of the skeletal morphology or preservation; some of these features are clearly antemortem but may not be pathological, whereas others are or may be the result of peri- or postmortem alteration of the remains. For each individual, to the extent possible and warranted, the lesions and their interpretations are summarized to provide an overall assessment of the health status of the individual.

Dolní Věstonice 3

The initial description of the adult female Dolní Věstonice 3 (Jelínek, 1954) includes a brief description of the series of asymmetries of the cranium, which were attributed to congenital deformities and to the osteoarthritic degeneration of the left temporomandibular articulation. The following assessment builds upon those original remarks, as well as the subsequent assessment of the postcranial remains (Trinkaus & Jelínek, 1997).

The Cranium and Mandible

Neurocranial Lesion

A surface irregularity is located in the cranium 30 mm posterior to metopion in the midline that is 16.5 mm in width and 13.0 mm anteroposterior (Figures 8.1 and 8.27). It appears to be a healed antemortem lesion. It is very shallow, less than 1.0 mm in depth, and does not extend into the diploë. The surface appearance indicates remodeling activity. It does not appear to be a depressed fracture; instead, it resembles a reaction to a minor injury that damaged the pericranium and outer tabular bone in a nonuniform fashion. It was subsequently well healed.

Facial Modifications

As previously noted (Jelínek, 1954), there is a large degree of asymmetry in the orbital region of Dolní Věstonice

3 (Figure 8.1). The right orbit is square to round in shape, and it resembles the nonpathological orbits of the other Dolní Věstonice crania. The left orbit is normal in its medial aspect but vertically compressed in its lateral aspect, giving it a relatively more rounded appearance. The vertical compression is due to the relatively superior placement of the inferolateral border, which is also unusually posteriorly repositioned (compare Figures 8.9 and 8.21).

Internally, the right orbit contains a large amount of intact bone in its lateral, medial, superior, and inferior aspects, although it is covered with paint and filler to the point where details of anatomy are obscured. The left orbit also contains a large amount of internal bone structure; however, it is even more obscured by consolidant and paint. Nonetheless, some observations are relevant to the alteration of the left side of the face. First, the left internal orbit is mediolaterally compressed, and although its superior aspect is identical to the right, the inferior floor of the orbit (as reconstructed) is depressed vertically relative to the right. Moreover, in the area of the confluence of the frontal bone and lesser and greater wings of the sphenoid bone that forms the superior orbital fissure, bone is clearly preserved and appears to be remodeled.

On the left side, the medial aspect of the infraorbital region is inferiorly positioned (consistent with the vertically lowered orbital floor) relative to the right side, such that the anterior extent of the origin for masseter is ca. 5 mm lower than on the right side. Moreover, the lateral left cheek height is diminished relative to the right by 25% (17.5 mm vs. 23.2 mm). This diminished aspect is entirely relegated to the masseter origin area and suggests major resorption of the zygomatic bone because of the diminished, or even absence of, tension in masseter for some time.

In lateral view (Figure 8.9), it is clear that the left zygomatic bone is altered (albeit unfortunately painted over with virtually no surface detail visible). In addition to a more posterior placement of the lateral orbital margin than that of the right, there is substantial medial inflection at the junction of the base of the frontal process and zygomatic body. The body of the zygomatic is also substantially narrowed and somewhat superiorly positioned; this is most evident in the unnaturally superiorly elevated lateral inferior orbital margin on the left side. The zygomatic bone (especially the body) is taphonomically strong and unlikely to break postmortem. Also, if it does break postmortem, it is more likely to break across one of its processes than in the body of the bone.

Associated with these orbital and zygomatic alterations are changes in the left well-preserved temporomandibular joint (Figures 8.22 and 19.1). The right glenoid fossa appears completely normal, with a fossa that is deep (6.8 mm) relative to the anterior and posterior articular eminences and with no signs of remodeling (Chapter 8; Figure 8.22). The left glenoid fossa, however, was substantially remodeled antemortem (Figure 19.1). The anterior and posterior articular eminences have been completely remodeled away with only 1.1 mm maximum glenoid fossa depth. The remodeled surface has become a nearly circular and flat articular surface measuring 23.0 mm in width and length. The remodeled glenoid surface is nonuniform with irregular texture. There is a transverse line that nearly bisects the two halves of the expanded surface that probably corresponds to the original division between the anterior and posterior glenoid eminences.

The changes in the left glenoid fossa remodeling are related to a concomitant alteration to the left mandibular condylar process. In fact, the condyle is essentially absent, and only the remodeled basal aspect remains, mediolaterally narrowed, anteroposteriorly mushroomed out, and irregularly porous on the superior surface (Figures 19.2 and 19.3). As a result, the normal right condyle has an anteroposterior depth of 9.9 mm and a mediolateral breadth of 19.5 mm; the pseudocondyle on the left side measures 13.5 mm by 16.0 mm, respectively. Moreover, the normal right condyle is 34.7 mm above the postcanine alveolar plane, whereas the left pseudocondyle is only 26.1 mm above that plane (Figures 10.1 and 19.4). In posterior view, there is a clear line that runs around the posterior superior neck just inferior to the affected condyle.

Figure 19.1 Close-up view of the left temporal glenoid fossa of Dolní Věstonice 3, showing the substantial remodeling associated with the altered mandibular condyle. Note the absence of anterior and posterior articular eminences and the circular, flat, and enlarged articular surface. Note also the transverse line bisecting the two halves of the markedly expanded surface, which probably indicates the original division between the anterior and posterior glenoid eminences. Scale in centimeters.

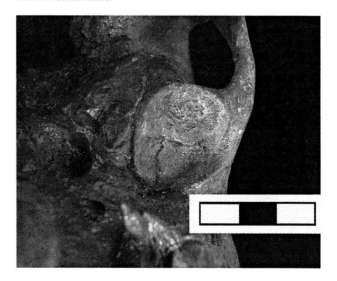

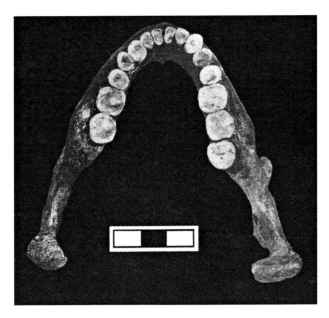

Figure 19.2 Superior view of the Dolní Věstonice 3 mandible, showing the altered left condyle compared to its normal right counterpart. Scale in centimeters.

In addition, the left coronoid process appears reduced compared to the right one (heights from the right alveolar plane of 32.4 mm vs. ca. 27.5 mm left; there is damage to the tip of the left coronoid process, making its original height uncertain), and the mandibular corpus height near the base of the ramus line on the external aspect is smaller on the left side than on the right side by 2.0 mm (19.0 mm vs. 21.0 mm). However, as noted in Chapter 10, corpus diameters more mesially are not markedly asymmetrical, and the minimum ramus breadths are essentially the same. In addition, masseter muscle markings on the affected side are still apparent and extend all the way anteriorly to the middle of the second molar.

Whereas the normal right infratemporal fossa area shows the usual configuration, the pathologically altered left side shows a shortening of the infratemporal fossa length in its superior aspect (left: 33.2 mm; right: 41.6 mm for the maximum anteroposterior passage for the temporal muscle; see Figure 8.22), and the left fossa is medially obliquely reoriented compared to the normal right side. This is associated with a reorientation of the anterior border of the infratemporal fossa (i.e., the confluence of the posterior maxillary zygomatic process, the external posterior wall of the maxillary sinus, and the anterior temporal surface of the zygomatic bone) on the left side, which is nearly horizontally rather than vertically oriented.

There also appears to be a compensatory misalignment of Dolní Věstonice 3's dental arcade relative to midline, such that the anterior dentition is oriented obliquely to the left, with each tooth antimere on the left more distally positioned than its right antimere (Figure 11.1). Concomitant asymmetries extend even to the usually nonmalleable nasal capsule, although these are less substantial (Chapter 8).

If the reconstruction of the left orbital and zygomatic region is remotely accurate, it is likely that distortion in the left orbit is secondary to the changes that involved the left side of the face, including the mandibular ramus, the zygomatic bone, and the internal orbit. Several secnarios are possible to explain these asymmetrical changes in the facial skeleton of Dolní Věstonice 3, but the most straightforward one is a traumatic impact to the left side of the face, which minimally fractured the left mandibular condylar process. Indeed, fracture of the mandibular condyle or condylar neck is a common result of blunt trauma to the anterior inferior face since it forces the mandibular condyle against the vertically strong lateral temporal bone (Wish-Baratz et al., 1992). There also may have been injuries to the zygomatic region that have become obscured through remodeling of the bone and/or reconstruction and painting of the region. Much of the facial aysmmetry not directly impacted by the inferred fracture of the condylar (and possibly zygomatic) areas, of both the maxilla and the mandible, are likely, in this scenario, to be the results of secondary remodeling in response to the fractures and the altered biomechanical environment of the left side of the face. The osteoarthritic

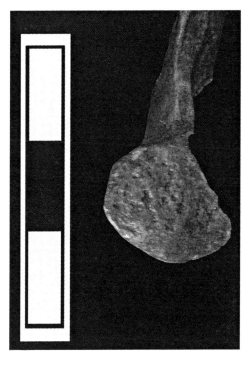

Figure 19.3 Close-up view of the altered left mandibular condyle of Dolní Věstonice 3 in superior view. Scale in centimeters.

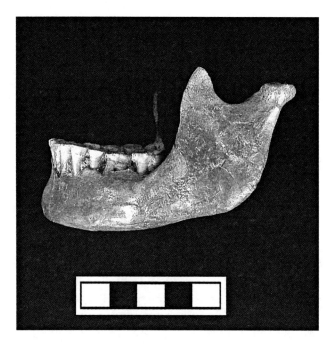

Figure 19.4 Left lateral view of the Dolní Věstonice 3 mandible. Note the reduced and remodeled condyle and the reduced coronoid process compared to the right (compare with Figure 10.1). Scale in centimeters

changes of the left glenoid fossa, rather than being simple temporomandibular joint degeneration from overloading (as originally suggested by Jelínek, 1954), are best seen as secondary to the loss of the mandibular condyle and the continued used of the mandible for mastication.

It is important to note that, despite the marked cranial asymmetry and left condylar loss, there appears to be a pattern of relatively even tooth wear (Table 11.8). First molar wear, for example, is largely symmetrical (Figure 11.1; Table 11.8), and second molar wear is moderately greater on the left side. However, the left lower third molar was congenitally absent, which may account for the greater left second molar wear. All of this implies that the condylar loss and associated changes healed sufficiently long before death so as to not affect dental wear, since the loss of the left condyle would normally imply that there was little joint reaction force at the left temporomandibular joint and hence little chewing on the right molar side, which would require joint reaction force at the contralateral balancing side (Shipman et al., 1985).

The available evidence therefore suggests that Dolní Věstonice 3 sustained a substantial blow to the left side of her face that fractured her left mandibular condylar process, possibly the zygomatic area, and produced changes in the left mandibular ramus, zygomatic, and orbital areas. The orbital modifications may have been sufficient to have caused difficulties in eye coordination from the changed positions of the ocular muscles. It probably precluded any normal level of bite force for a substantial period of time, which would have meant that she ingested a soft diet for some time. The damaged condyle probably resorbed, and the altered forces acting at the left temporomandibular joint resulted in the marked remodeling of the glenoid fossa. However, following healing, recruitment of chewing on both sides must have ensued, given the largely even wear gradient that resulted prior to death.

It is difficult to assess whether such an injury would have been the result of an accidental fall, close contact with a large animal, or interpersonal violence. Certainly, the very localized nature of the injury and the apparent absence of postcranial injury imply a very specific traumatic event and not the kind of minor injury responsible for the lesion on the frontal squamous.

Posterior Neurocranial Asymmetry

As mentioned in Chapter 8, in posterior view there are several asymmetries of the Dolní Věstonice 3 cranial vault. It is unclear whether these are the product of developmental abnormalities or postmortem deformations. Therefore, although they are mentioned in the cranial morphology chapter, they are considered further here.

The sagittal suture deviates to the left slightly as it approaches lambda. The left lambdoidal suture is more inferiorly positioned than the right one. The nuchal lines, asterion, and tip of the mastoid process all show a more inferior placement on the left side than on to the right side in occipital view (Figure 8.25). Moreover, there is also a small but distinct globular protrusion, or inferior bulging, of the left nuchal plane inferior to the superior nuchal line. Several questions arise with respect to the last feature in particular: (1) is it related to other adjacent external asymmetries? (2) Did it reflect a larger or bulging internal cerebellar fossa unrelated to the other external asymmetries? (3) If endocranial in origin, does it reflect an expansion of the left cerebellar lobe, or would the endocranial asymmetry be cerebral and therefore reflect posteroinferior positioning of the left occipital lobe? (4) Is it related to remodeling associated with the temporomandibular pathological alteration? (5) Is it primarily postmortem deformation, and consequently of little paleobiological interest?

The endocast of Dolní Věstonice 3 (Jelínek, 1954) indicates a degree of posterior bulging on the left posterior endocranial aspect in the same area. Further observations on the endocast by Jelínek include a right hemisphere that is somewhat larger than the left one, a right occipital lobe that is markedly shorter than the left one, a right frontal lobe that is broader and longer than the left one, a right cerebellar area that is smaller and flatter than the left side, and an associated asymmetry of the temporal lobes.

According to Jelínek, these asymmetries were most likely the result of a congenitally present plagiocephaly. "Plagiocephaly," however, is a largely descriptive term and clinically is employed to describe a variety of asymmetries and malformations of the neurocranial vault of varying etiologies. Moreover, since the endocast is a mold of the endocranium and not of the brain itself, a postmortem-induced asymmetry cannot be ruled out, especially given the mediolateral flattening of the cranium in situ (Klíma, 1963) and its extensive reconstruction from multiple pieces (Chapter 8). If the condition was antemortem, there is no clear evidence that the cerebral asymmetries were directly related to the left-side facial lesions discussed above. Finally, even if the hemispheric and lobe asymmetries present in Dolní Věstonice 3 are real rather than a postmortem artifact, it is not clear whether they lie beyond normal (i.e., nonpathological) ranges of variation.

The Alveolar Process and Dentition

Most of the common pathological conditions affecting the alveolar process are age-progressive, and in the Dolní Věstonice and Pavlov remains they particularly involve continuous eruption (or supereruption) of the teeth associated with advancing occlusal wear. This has been shown to be particularly rapid in populations with rapidly wearing teeth (Whittaker et al., 1982, 1985; Levers & Darling, 1983; Danenberg et al., 1991), and it causes more root to be exposed than would be considered normal in dental practice today. In radiographs of living patients, it is expected that 1 to 2 mm of root will be showing between the base of the crown and the crest of the alveolar process (the so-called CEJ-AC height; see Table 19.1; Whaites, 1992; Goaz & White, 1994; Wood & Goaz, 1997).

Dolní Věstonice 3, who would probably have been an adult of early middle age (Chapter 6), has considerable occlusal wear and consequently shows somewhat pronounced exposure of the tooth roots in places, but the degree is not great (Figures 8.9, 8.21; 10.1, and 11.1). It is, however, difficult to see if there are any localized losses of bone because of the postmortem damage to the bones; the dark coating of lacquer; extensive damage, especially to the buccal side of the alveolar processes; and the large amount of reconstruction. There is nothing visible on the surface of the bone or teeth that would suggest any deeper pattern of bone loss.

The teeth of Dolní Věstonice 3 do not show any convincing evidence of dental caries. The upper left third molar central fossa exhibits a dark stain and a slight widening of the fissures, but the radiograph does not show an associated radiolucency in the dentine. There is thus no supporting evidence to indicate that this represents a carious lesion, especially given the coating of the Dolní Věstonice 3 remains with preservative (contra Sládek et al., 2000).

The teeth of Dolní Věstonice 3 are heavily worn so that even if there were originally developmental defects of dental enamel (DDE), also referred to as dental enamel hypoplasias (DEH), they would have been largely worn away during the lifetime of the individual. The only remaining evidence is very slight furrow-form (Hillson &

Table 19.1 Maximum root exposure, measured as cement/enamel junction–alveolar crest (CEJ-AC) height in millimeters, giving the highest value for the buccal or lingual side.

		Left								Right							
		M3	M2	M1	P2	P1	C	I2	I1	I1	I2	C	P1	P2	M1	M2	M3
DV 3	Upper	—	4	3	1	1	—	—	4	2	2	2	2	2	3	3	—
	Lower		4	3	3	—	—	—	—	—	2	2	1	2	2	2	2
DV 13	Upper	—	—	2		2	1		2				1	2	1	—	
	Lower	—	2	2	2	2	1	2	3	3	2	1	2	1	2	2	—
DV 14	Upper	—	—	1	1	1	2	1	1	1	1	1	1	—	—	—	—
	Lower	—	0	1	1	2	1	2	—	—	2	2	—	—			
DV 15	Upper	—	2	2	1	1	1	2	2	2	2	2	2	1	2	—	—
	Lower	—	2	2	1	2	2			2	2	2	2	1	—	—	
DV 16	Upper	7	6	3	3	3	3	—			3	3	—	—	—	—	
	Lower		4	5	6?	3	5	—	3	2	2	4	3	3	3	4	—
Pav 1	Upper		6	3	3	3	—	—			—	—	—				
	Lower	4	3	4	—	—		—	—					2	3	4	
Pav 2	Upper		—	—	2								2	—	—		
Pav 3	Lower											2	—	—	3		

The measurement technique follows Hillson (2000). Where the crown is worn away completely, the measurement is taken from the worn attrition facet instead of the cement/enamel junction. Empty space: absence of the tooth;—: tooth and/or alveolar crest is too damaged to provide an accurate measurement or the tooth is not fully erupted into reasonably normal occlusion.

Bond, 1997) defects on the lower canines in the cervical region of the crown, together with what seems to be a corresponding defect on the second premolars (Table 19.2). All, however, are very difficult to see. Their position suggests a growth disruption around 5 years of age (Hillson, 1996; Reid & Dean, 2000).

The Axial Skeleton

The axial skeleton of Dolní Věstonice 3 is poorly preserved, but what elements still exist show no signs of degenerative changes or other lesions. None of the preserved vertebral elements from the cervical, thoracic, or lumbar regions has osteophytic lipping, nor is pathological wedging evident in the (albeit poorly preserved) thoracic column.

The Appendicular Skeleton

The appendicular skeleton of Dolní Věstonice 3 is remarkably free of lesions, given her moderately advanced age at death. Even though a number of the articular subchondral bone surfaces are missing or badly damaged, the ones that are preserved show almost no evidence of osteoarthritis. There is a minor degree of subchondral bone alteration in the weight-bearing area of each acetabulum, but it is likely to have been subclinical in extent. Neither the relatively well-preserved right femoral head nor the partial left femoral head shows any associated subchondral bone degeneration.

At the same time, the diaphyses of the humeri, femora, and tibiae, which should reflect any serious asymmetry in differential limb use (Trinkaus et al., 1994), exhibit very low levels of asymmetry (Tables 16.3, 19.4, and 19.5). Therefore, it is unlikely that the marked asymmetries in the skull affected limb use in Dolní Věstonice 3.

Very few of the metaphyseal regions of Dolní Věstonice 3 are sufficiently intact to reveal possible transverse (Harris) lines (Table 19.3), and her middle age makes it likely that any that were present in childhood would have been obscured by remodeling. It is therefore not surprising that none of the preserved metaphyseal regions exhibit transverse lines.

Summary

The Dolní Věstonice 3 skeleton seems to have survived to middle age with few skeletal and dental lesions except for one very marked traumatic injury. The skeletal remains, although fragmentary in portions, preserve one of the minor

Table 19.2 Distribution of dental enamel hypoplasias, or developmental defects of the dental enamel, in the Dolní Věstonice permanent tooth sample.

	DV 3		DV 13		DV 14		DV 15		DV 16		DV 31		DV 32	DV 33	DV 37
	R	L	R	L	R	L	R	L	R	L	R	L	L	L	
Upper															
I1	Abs	Abs		Abs	—	—	Pres	Pres							
I2	Pres	Pres					—	—	Pres	Pres		—			
C	Pres	Pres		Pres	—	—	Pres	Pres		—	—				
P1	Abs	Abs		Pres	—	—	—	—	Pres		—	—			
P2	Abs	Abs	Abs		—	—	—	—	Abs		—	—			
M1	Abs	Abs	Abs	Abs	—	—	Pres	Pres		—	—				
M2	Abs	Abs	Pres	Pres	—	—	Pres	Pres		—	—				
M3	Abs	Abs	Abs	Abs	—	—	Abs	Abs	Abs	Abs				Pres	
Lower															
I1	Abs	Abs	Abs	Abs	—	—	Pres					—			
I2	Abs	—	Abs	Abs	Pres	Pres	Pres					—			
C	Pres	Pres	Pres	Pres	Pres	Pres	Pres	Pres				—			
P1	Abs	—	Pres	Pres	Pres	—	—	Pres				—			
P2	Pres	Pres	Pres	Pres	—	—	Abs	Abs		—	—				
M1	Abs	Abs	Abs	Abs	—	Abs	Pres	Pres		—	—				
M2	—	—	Pres	Pres	Abs	Abs	Pres	Pres		—	—	Abs			Pres
M3	Abs		Abs	Abs	—	—	—	—		—	Abs				

Lesions are listed only as present (Pres) or absent (Abs) on sufficiently complete crowns;—indicates an insufficiently intact crown, either because of attrition or damage, or one that is sufficiently embedded in the alveolus to prevent proper observations; a blank indicates absence of the tooth. For details on the lesions, see discussions for the individual specimens. Because of insufficient development and/or preservation, the upper I1 and three M1s of Dolní Věstonice 36 and the lower M3 of Dolní Věstonice 38 are not included.

Table 19.3 Distribution of transverse (Harris) lines on the appendicular remains of the associated Dolní Věstonice and Pavlov human remains.

	DV 3		DV 13		DV 14		DV 15		DV 16		Pav 1	
	R	L	R	L	R	L	R	L	R	L	R	L
Humerus												
Proximal	—	—	Abs	Abs	—	—	Abs	Abs	—	—	—	—
Distal	—	—	Abs	Abs	Abs	Abs	Abs	Abs	Abs	Abs	Abs	Abs
Ulna												
Proximal	—	—	Abs	Abs	Abs	Abs	Abs	Abs	Abs	Abs	Abs	Abs
Distal	—	—	Abs	Abs	—	—	Abs	Abs	Abs	Abs	—	—
Radius												
Proximal	Abs	—	Abs	Abs	Abs	Abs	Abs	Abs	Abs	Abs	Pres	—
Distal	Abs	—	Abs	Abs	—	—	Abs	Abs	Abs	Abs	Abs	—
MC 1	—	—					Abs		—			
MC 2	Abs	—	Abs	Abs	Abs		—	—	Abs	—		
MC 3	—	—	—	Abs			Abs		Abs	Abs	—	
MC 4	—	—							—	—		
MC 5	—	—							Abs	—		
Femur												
Proximal	—	—	Abs	Abs	Abs	Abs	Abs	Abs	—	—		
Distal	Abs	—	Abs	Abs	Abs	Abs	Abs	Abs	—	—	—	—
Tibia												
Proximal	Abs	—	Abs	Abs	—	—	Pres	Pres	—	—		
Distal	—	—	Abs	Abs	Abs	Abs	Abs	Abs	—	—		
Fibula												
Distal	—	—	Abs	Abs	Abs	Abs	Abs	Abs	Abs	Abs		
MT 1	—	—			Pres	—	Abs	Abs	Pres	—		
MT 2	—	—			—	—	Pres	Abs	—	—		
MT 3	—	—			—	—	—	Abs	—	—		
MT 4	—	—			—	—	Abs	—	Abs	—		
MT 5	—	Abs			—	—	—	Abs	—	—		

Lines are listed only as present (Pres) or absent (Abs) on sufficiently complete bones;—indicates an insufficiently intact bone because of damage; a blank indicates absence of the bone. For details on the lesions, see discussions for the individual specimens.

neurocranial lesions that seem to be ubiquitous among the Late Pleistocene human populations, minimal evidence of developmental stress, little skeletal degeneration, and then a suite of changes of the facial skeleton that can be attributed to a single blow to the left side of the face. It is not clear when during her life this injury occurred, and she may well have had some assistance with food preparation during the healing process. Yet once healed, she appears to have led a normal life, with little or no loss of function. It is also possible, as suggested by Klíma (1963, 1983), that a small mammoth ivory human face with a marked asymmetry may be a portrait of Dolní Věstonice 3.

Dolní Věstonice 11/12

Even though they were given separate individual numbers in the field and they remain separately numbered as Dolní Věstonice 11 and 12 (Sládek et al., 2000; Chapter 8), it was recognized from the beginning (Klíma, 1987b; see also Vlček, 1991) that Dolní Věstonice 11 and 12 derive from one individual (Figure 8.29). The convincing evidence for this association is the presence of two sides of a traumatic lesion on the Dolní Věstonice 12 supraorbital piece (where it is obvious) and the Dolní Věstonice 11 frontal squamous (where it is less apparent). It is principally this lesion and other changes in the calotte of Dolní Věstonice 11 that are described here together.

Just lateral of the midline on the frontal bone of Dolní Věstonice 11, there is a beveling of the external half of the bony edge down to the postmortem break. This represents the superior edge of the major anterior frontal trauma on Dolní Věstonice 12. The edge of healed bone is 17 mm wide and up to 7 mm superoinferior. It has rounded edges of diploë with vascular canals, and as such it is distinct from the adjacent broken bone edge.

This lesion edge on the Dolní Věstonice 11 calotte matches up with the remains of a healed massive depressed trauma to the frontal squamous, just lateral of the midline to the right lateral break of the bone (Figures 8.29 to 8.31). The edges are all rounded, and they were long since healed over at death. The arching edge medially is concave laterally, and it has a raised step edge that reached a maximum step height of 2.3 mm. The step is a result of both a medially raised edge and the depression of the middle of the injury. The raised edge tapers off laterally and inferiorly for a mediolateral length of 17.5 mm. It ends laterally above the supraorbital sulcus. The scarring of the bone continues to the postmortem break, but comparison with the Dolní Věstonice 11 frontal bone indicates that it did not continue much beyond the preserved area. There is a vascular pit in the middle of the depressed area, 1.1 mm wide and 5.8 mm long. The depression of the middle of the injured area resulted in an inward (endocranial) displacement of the internal table 1 to 2 mm over an area of at least 17 mm by 17 mm. The postmortem broken edge is thin (ca. 1.0 mm), and there is no sign of diploë.

On the left side of Dolní Věstonice 12 along the postmortem break, the external table is gone superiorly near midline, exposing underlying diploë. However, to its left, there is a beveled edge to the left lateral break, preserved for at least 5 mm. This gives the impression of being related to postmortem erosion of the midline external table, but laterally it results in an edge that is less than 1 mm thick and shows no exposed diploë. The peak of the edge is 5 mm above the top of the left frontal sinus. This appears to be an abnormal thinning of the left frontal squamous, suggesting a second injury to the frontal squamous. Alternatively, but much less likely, it is the result of senile bone thinning.

In addition to the "shared" frontal depressed fracture, the Dolní Věstonice 11 calotte exhibits several less pronounced lesions. One of these, the one on the anteromedial right parietal might be traumatic, as suggested by Vlček (1991), but a traumatic etiology for any of these is hard to establish.

There is a modestly depressed area along the sagittal suture just posterior of the coronal suture on the right parietal (Figure 8.36). It measures 11.3 mm mediolaterally and 14.5 mm anteroposteriorly, with clear medial and anterior margins and the deepest portion along the medial edge. The posterior and lateral edges fade smoothly into the exocranial surface. The anterior edge is ca. 9 mm from the coronal suture, and the medial edge is ca. 3.5 mm to the sagittal suture. The depression was recorded as an injury by Vlček (1991), but its etiology is uncertain.

The external occipital protuberance is a prominent lip 16.2 mm wide and 8.7 mm in external height, with a sulcus ca. 1 mm beneath it. It is an enlarged enthesopathy rather than merely a pronounced external occipital protuberance (Figure 8.33). The superior nuchal region has depressions on the right side just lateral of the midline. They might represent antemortem irregularities of the superior nuchal area, but they are most likely the product of postmortem erosion of the surface bone.

Whatever the ultimate etiologies of the parietal and occipital abnormalities might be, it is apparent that Dolní Věstonice 11/12, like Dolní Věstonice 3, sustained a massive blow to the facial region. This injury must have caused considerable discomfort, internally and externally, and it is likely to have resulted in significant blood loss. Yet, by the time of the death of Dolní Věstonice 11/12, the bone and its adjacent soft tissue had completely healed over, leaving a massive surface scar. However, given the isolated nature of these neurocranial pieces, it is difficult to assess the possible behavioral contexts of the traumatic event.

Dolní Věstonice 13

The paleopathology of the young adult Dolní Věstonice 13 has received little attention beyond the observation by Vlček (1991) of a couple of minor traumatic lesions on the cranial vault.

The Cranium

There is an antemortem surface alteration in the cranium located 40 mm superior to the right superomedial orbital region between the right frontal eminence and midline in the vicinity of metopion (Figure 8.37). It is a surface irregularity that is ca. 20 mm in width and 8 mm in height. It is very shallow and did not extend into the diploë. To some extent, root etching obscures the delineation of the lesion, but the surface appearance indicates remodeling activity. It does not appear to be a depressed fracture, since it involves only the external table. It is likely to be the result of a minor trauma that damaged the pericranium and outer tabular bone in a nonuniform fashion. Regardless of its etiology, it was subsequently well healed.

In superior view, there is another surface irregularity on the right parietal bone, located ca. 14 mm lateral to bregma and ca. 12 mm posterior to the coronal suture, that also appears to be an antemortem healed lesion. The lesion is ca. 12 mm in width and 11 mm in anteroposterior length (Figure 8.46). Whether the lesion extended into the diploë is difficult to assess, but the surface appearance indicates healing. It may have been a depressed fracture or, alternatively, as noted for the previous lesion, damage to the pericranium and outer tabular bone that was later healed.

Along the superior nuchal line, starting 3.5 mm posterior of the left mastoid process, there is a raised area of new bone, 20 mm long and 6.5 mm wide, angled medially and tapering laterally (Figure 8.43). It appears to be an irregularity of the muscle attachment and is probably associated with a minor injury to the lateral posterior cranium.

Although whether they are pathological or taphonomical/excavation damage in origin is uncertain, there are two breaks, one on each lateral parietal bone posterior to the coronal suture (Figures 8.40 and 8.41). Each is an anteroposteriorly aligned break; the left one is 27.7 mm long and ca. 9.5 mm wide, and the right is 25.0 mm long and ca. 7.5 mm wide. They are more or less symmetrically placed along the sides of the neurocranium ca. 120 mm in arc length from each other. Each is located in the middle of a break line along the parietal sutures, and they may have been the postmortem impact points that broke the parietal along these subsequent fractures. However, certain oddities are associated with them that bear mentioning.

The break on the left side cleaves through the outer table and diploë, but it barely penetrates the inner table as minute fenestra. The anterior portion of the break appears to show a beveled fracture pattern usually associated with fresh perimortem "green-stick" fractures. However, fracture lines radiating out anteriorly and posteriorly from this break are more indicative of associated breakage due to loss of the organic component of bone and thus brittle breakage associated with postmortem damage. There is an additional oddity, however—in all other places on this cranium and those of the other individuals in the triple burial, break lines that did not join were filled in with consolidant to produce a smooth surface. This particular damaged area in Dolní Věstonice 13, and the similar one located in almost the same area on the right side of this cranium about 23 mm posterior to the coronal suture and about midway between the sagittal and squamosal suture, are conspicuous in that the preparator left these damaged areas unfilled. These breaks may well be the result of the postmortem vertical compression of the neurocranium from sediment pressure (Figures 4.5 and 4.6), as with much of the damage to the Dolní Věstonice 13 to 15 crania (Figures 4.6 to 4.8; Chapter 4). However, their unusual pattern relative to the other obviously postmortem damage remains enigmatic.

The occipital condyles of Dolní Věstonice 13 are well preserved and show some asymmetry. The maximum lengths of both condyles are similar (right: 18.3 mm; left: 19.5 mm). However, the width of the left condyle is expanded (right: 11.1 mm; left: 15.0 mm). This condylar asymmetry is associated with some asymmetries in the cervical vertebrae (Chapter 14). Unlike the uncertainty regarding the foramen magnum orientation (Chapter 8), these condylar asymmetries and some of the cervical vertebral ones are antemortem and not due to reconstruction efforts. Their etiology is uncertain, and they may be related to the kinds of developmental irregularities that are evident more caudally in the axial skeleton (see below) and are present in the vertebrae of other Dolní Věstonice individuals (Chapter 14). Cervical trauma would probably cause serious articular disruption, which should be evident on the bones; no such lesions are present.

The Alveolar Process and Dentition

A CEJ-AC measurement (see above) of 1 to 2 mm would be regarded as normal for living young adults, and this is the case for most teeth in Dolní Věstonice 13. In the mandible, however, the roots of the first incisors are more exposed than those of the cheek teeth (Table 19.1). This is not due to a reduction in the height of the alveolar process, but it can be explained in terms of the wear pattern (Chapter 11). Dolní Věstonice 13 was slightly older than the other two triple burial individuals, and he has generally more worn teeth, with an occlusal attrition gradient in which the first incisors are the most worn and the third molars the least worn (Table 11.8). This is reflected in the CEJ-AC heights.

Each of the lower first molars of Dolní Věstonice 13 exhibits a prominent buccal pit, which is darkly stained, but this does not appear to be a carious lesion (Figure 11.3). There is no radiographic evidence of dentine demineralization underneath. There are no other candidates for carious lesions on his teeth.

Modest furrow-form developmental defects of the dental enamel are present in the cervical third of the crown in the upper canine and both of the upper first premolars and second molars (Table 19.2). The matching positions are consistent with a growth disturbance at 4 to 5 years of age (Hillson, 1996; Reid & Dean, 2000).

The Axial Skeleton

The young adult age at death for the Dolní Věstonice 13 skeleton leads to the expectation of few degenerative changes in the axial skeleton, and this is the case. The vertebrae are free from osteophytes, indicating that the anterior longitudinal ligament and similar stabilizing ligaments associated with the vertebral column had not ossified, as is often the case in active individuals of a more advanced age. There are only two unusual, and hence potentially pathological, features of the Dolní Věstonice 13 axial skeleton.

The first is that its tenth and eleventh thoracic vertebrae were dorsally wedged (i.e., their dorsal body heights are shorter than their ventral body heights), an unusual state for these elements (Chapter 14). In contrast, the twelfth thoracic vertebra has a higher dorsal than ventral body height, and all of the lumbar vertebrae exhibit the morphology expected for their respective levels. Thus, it does not appear that this subtle shift in lower thoracic curvature was echoed in the lumbar region or that the individual had a pathologically accentuated lumbar lordosis.

The second unusual feature of the Dolní Věstonice 13 axial skeleton is a small (1.4 mm in height × 1.8 mm in breadth) pit found on the articular surface of the head of the right twelfth rib. This pit is seemingly antemortem, and thus it could be indicative of damage to the subchondral bone of the vertebrocostal joint. Yet, there is no bony reaction evident within or near the pit, nor is eburnation noted in the articular area surrounding it. This feature is therefore most likely nonpathological.

The Appendicular Skeleton

The Dolní Věstonice 13 appendicular remains are generally excellently preserved, with most of the long bone diaphyses and epiphyses largely or entirely intact. The hand and foot bones are less complete but still preserve a few elements. There is no evidence of healed traumatic injuries on the limb bones, and as might be expected in a young adult, there is little evidence of degenerative joint disease (or osteoarthritis). Indeed, almost all of the articulations exhibit smooth and dense subchondral bone with even margins, and there are no enthesopathies on the various tendinous attachments.

In the context of this, there is one very unusual articulation. The distal trochlea of the left second proximal hand phalanx (identified by its base morphology) exhibits advanced osteoarthritis. There is little involvement of the palmar portion of the trochlea or of the radial and ulnar sides of the distal portion of the phalanx, but the distal and dorsal margins of the trochlea are involved, including osteophytic growths along the edges of the trochlea, especially radially, and advanced porous degeneration of the subchondral bone (Figure 19.5). This advanced articular degeneration is striking, given both the young adult age of Dolní Věstonice 13 and the complete absence of degenerative lesions on the remainder of the appendicular skeleton, including the nineteen other preserved metacarpal and hand phalangeal articulations (Figure 16.55). No middle phalanx with a matching proximal articular degeneration is preserved for Dolní Věstonice 13.

The advanced state of degeneration of this proximal manual interphalangeal articulation, in the context of the otherwise healthy limbs of Dolní Věstonice 13, indicates that either it was being frequently overloaded mechanically or that it had sustained a traumatic disruption of the joint leading to abnormal loading patterns across the trochlea.

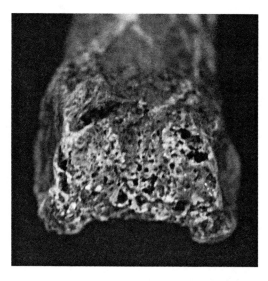

Figure 19.5 Dorsodistal view of the Dolní Věstonice 13 left second proximal hand phalanx, showing the osteoarthritic degeneration of the trochlea.

Given the absence of other pathological alterations of the phalanx or its associated metacarpal, it seems likely that a very localized form of habitual mechanical surcharge was responsible for the osteoarthritis.

A very similar pattern of interphalangeal degeneration is present on the right hand bones of Dolní Věstonice 15, but in that case both the second and fifth proximal interphalangeal articulations are involved (see below). However, the presence of a fifth middle phalanx, probably from the same hand, for Dolní Věstonice 13, suggests that the proximal interphalangeal osteoarthritis was localized to the second digit. Because of the more complete set of osteoarthritic degenerations present on Dolní Věstonice 15, the possible etiology of these interphalangeal degenerations is discussed below.

All of the long bones of Dolní Věstonice 13 except one fibula preserve both of their metaphyseal regions, and none of them shows transverse lines (Table 19.3). The same applies to the three retained metacarpals.

Summary

The Dolní Věstonice 13 remains, as those of a young adult, indicate an individual who experienced only minor levels of stress during his approximately two-and-one-half decades of life. There was modest systemic stress in the middle of the first decade. The other lesions are idiopathic and include three minor neurocranial vault injuries and the localized hand interphalangeal osteoarthritis. The parietal bone breaks are of uncertain origin and may well be postmortem, and the cervical and occipital condyle asymmetries suggest principally a developmental irregularity in this region.

Dolní Věstonice 14

The late adolescent remains of Dolní Věstonice 14 have received little paleopathological interest. The only mentioned possibility of a lesion concerns the pattern of breakage of the posterior neurocranium (Figure 8.52), which was described by Vlček (1991) as resulting from a perimortem traumatic blow to the back of the head, causing death in the process. Considerations of the general taphonomy of the triple burial (Chapter 4) and the marked postmortem deformation of the neurocranium (Chapter 8) make it likely that these fractures were postmortem, resulting from sediment pressure on the remains. No antemortem cranial or mandibular lesions were noted on this young individual.

The Auditory Ossicles

In the auditory ossicles of Dolní Věstonice 14, the lenticular process of the left incus is missing, which was destroyed antemortem by *otitis media*. In addition, the apex of the *crus longum incudis* without the lenticular process shows damage caused by inflammation during the life of the individual.

The Alveolar Process and Dentition

The state of the alveolar process of Dolní Věstonice 14 is fully compatible with its adolescent age (Chapter 6) and modest occlusal wear (Chapter 11). All of the preserved portions of the alveolar process have 1 to 2 mm of root showing between the base of the crown and the crest of the alveolar process (Table 19.1), the normal pattern in unworn or minimally worn teeth of living young dental patients (see above).

There is no evidence of dental caries on the teeth of Dolní Věstonice 14. Many teeth are marked by a postmortem roughening, which makes it difficult to see developmental enamel defects. It is possible, however, to make out moderate furrow-form defects on the midthird of the lower canines and first premolars and on the cervical third of the lower second incisors (Table 19.2). These defects suggest a growth disruption at about 3 years of age (Hillson, 1996; Reid & Dean, 2000).

The Axial Skeleton

The late adolescent age at death for the Dolní Věstonice 14 specimen suggests that its axial skeleton should be relatively free from degenerative changes. This is indeed the case; there are no osteophytes on the vertebrae, nor is any vertebral element pathologically wedged. Likewise, no lesions are noted on the ribs—this despite the fact that all of these elements are reasonably well preserved.

The Appendicular Skeleton

The appendicular remains of Dolní Věstonice 14 are largely free of any lesions, although the extensively root-etched nature of most of the diaphyses and damage to several of the epiphyseal regions mean that minor pathological alterations may not be preserved. The only possible lesion on the limb remains is a minor crest of bone on the left distal clavicle. On the deltoid attachment, in the anterior (concave) surface of the distal curvature, there is a raised spine leading onto the rugose muscle origin area (Figure 16.3). The spine is 5.8 mm long and projects 1.6 mm. It is more than a normal rugosity for the deltoid muscle, not necessarily pathological, and it is best seen as a minor enthesopathy for this muscle.

The metaphyseal regions of Dolní Věstonice 14 are variably preserved and damaged, but none of the sufficiently intact long bone metaphyses exhibits a transverse line (Table 19.3). However, the right first metatarsal has a faint line 4.8 mm from the proximal articular surface, which does not cross the entire proximal metatarsal bone. If this represents a transverse line from a growth arrest period, it would have occurred relatively late in development but would not have been severe.

Summary

The adolescent Dolní Věstonice 14, as with other individuals from these sites, sustained some minor systemic stress around weaning time, and he appears to have suffered from an ear canal inflammation. Otherwise, the remains show little evidence of either developmental or degenerative stress; the small clavicular spine for the deltoid attachment is probably related to minor muscular irritation, and the possible transverse line in the metatarsal to a minor period of stress during adolescence.

Dolní Věstonice 15

Ever since it was excavated (Klíma, 1987b), it has been noted that the Dolní Věstonice 15 skeleton presents a series of abnormalities. There has since been considerable discussion of the paleopathology of this individual (e.g., Klíma, 1987a; Vlček, 1991; Jelínek, 1992; Kuklík, 1992; Formicola et al., 2001; Trinkaus et al., 2001). These comments have been focused largely on the etiology of the set of lesions, but they have included a couple of attempts to document the abnormalities, with some discussion as to which features represent pathological alterations as opposed to either normal morphological variation or postmortem damage to the bones. The following presentation is oriented principally at documenting the natures of the apparent lesions and distinguishing them from normal morphological variation. There has been a series of attempts to provide an overall diagnosis of the suite of lesions, of which the attempt by Formicola et al. (2001) is the most thorough. And there has been a consideration of the general level of activity of Dolní Věstonice 15, as indicated by its appendicular robusticity, in the context of its pathological lesions (Trinkaus et al., 2001; see Chapters 16 and 18). Much of this has been combined with issues regarding its gender (Chapter 7).

From these past considerations, it has become apparent that the ultimate "true" diagnosis of the suite of lesions on Dolní Věstonice 15 may be difficult to ascertain. However, it is hoped that the description presented here and the associated assessment of the lesions will provide a foundation for such interpretations in the future.

The Cranium and Mandible

Given the abnormalities of the appendicular skeleton of Dolní Věstonice 15, it is tempting to view irregularities of the skull as related directly or indirectly to the growth irregularities of the limbs. It is therefore important to review possible craniomandibular lesions from the perspectives of both normal morphological variation and possible pathological configurations.

Neurocranial Modifications

There is a surface irregularity on the frontal bone to the left of the midline ca. 24 mm anterior to bregma; its maximum length is ca. 12 mm, and its maximum width is ca. 9 mm. The area is very shallow, with rough irregular edges, and shows antemortem remodeling. It appears to be the product of a well-healed minor injury that was probably of little significance.

There is another small surface irregularity on the right parietal bone ca. 27 mm to the right of midline and ca. 50 mm posterior to bregma; it is ca. 7 mm in maximum diameter, including the diffuse healed edge. It is very shallow, involving only the outer tabular bone, and it appears to have been a very minor, healed depressed fracture rather than a surface injury because of the gently circular appearance of its healed edge.

In addition, a 12.9 mm vertical groove is present in the area where the squamomastoid suture on the right temporal bone is usually found (Figure 8.60). It could represent the remains of the suture, but it has the appearance of a healed antemortem injury instead. Its status remains uncertain.

Approximately 20 mm to the right of the sagittal suture near the most posterior aspect in superior view, there is a small surface irregularity that superficially appears to be a benign button osteoma. Close inspection indicates that it is a sutural fragment that became dislodged and then was glued onto the cranial surface.

Facial Aspects

There is a large opening into the left frontal sinus region in the vicinity of the superomedial orbital corner (Figure 8.56). Internally, the enclosed, roughly circular space is bounded medially by a wall near midline, separating this sinus from the right-side internal structure. The opening is 11.4 mm wide anteriorly and 18.2 mm anteroposteriorly, and it invades internally into virtually all of the superciliary arch. The internal space is visible in CT scans and radiographs, and it appears to be a nonpathological frontal sinus that is limited to the left side (i.e., frontal sinus asymmetry with little to no development on the right side or central aspect). In this feature, Dolní Věstonice 15 is similar to both Dolní Věstonice 13 and 14, both of which had unilateral development of the left-side frontal sinus. There are no other features either internally or externally in Dolní Věstonice 15 to suggest that this feature is pathological. The anterior opening is the result of postmortem damage.

As noted in Chapter 8, the lateral aspect of the infraorbital region in Dolní Věstonice 15 is oriented in the coronal plane on both sides, and the medial facies adjacent to the piriform aperture is oriented in a parasagittal plane on either side, showing the typically modern human two-plane configuration. In Dolní Věstonice 15, the angulation is particularly acute, with a deeply depressed infraorbital region. Moreover, there is a sharp vertical sill on the lateral aspect of the right medial infraorbital facies that gives way to a recessed posterior area just medial to where the canine fossa is usually found (Figure 8.38). This morphology appears on the left medial infraorbital aspect as well, although the vertical sill is not so sharp on the right side. These bilaterally expressed depressed areas give the infraorbital region a deeply excavated appearance that may be perceived as pathological.

This infraorbital arrangement is similar to the concavity found in Dolní Věstonice 14, although that one is less extreme. It is also found in a more muted expression in Dolní Věstonice 13 (Figure 8.38). Such deeply excavated infraorbital regions as are found in Dolní Věstonice 15 are also found in recent humans and are usually associated with extremely everted maxillary frontal processes (Franciscus, personal observation). Interestingly, this infraorbital morphology has been noted in the Early Pleistocene Dmanisi 2700 (Vekua et al., 2002), where it is described as "a distinct pit behind the canine jugum." Its pronounced expression in Dolní Věstonice 15 and lesser expression in Dolní Věstonice 13 and 14 should therefore not be considered pathological.

The anteromedial infracondylar area at the top of the neck on both sides of the mandible shows a moderately excavated pterygoid fovea, and the left one shows a small area of raised remodeling in the lateral aspect of the fovea; it is very minor and probably was of little consequence.

As a result of these considerations, there appears to be little if anything on the cranium and mandible (as opposed to the dentition; see below) that can be related to the individual's appendicular abnormalities. Only the two minor traumatic lesions on the neurocranial vault appear pathological.

The Alveolar Process and Dentition

The modest level of occlusal attrition on the Dolní Věstonice 15 dentition is accompanied by a normal distance (for a young man) between the base of the crown and the crest of the alveolar process, for all of the teeth (Table 19.1).

Dolní Věstonice 15 has the most prominent developmental enamel defects among the Dolní Věstonice and Pavlov human remains (Figure 19.6; Table 19.2). The first molars are marked by a sharp groove running around the circumference of each crown. In the upper molars, this coincides with prominent buccal pits to make a T-shaped impression on the buccal side. It represents a disturbance at around 2 years of age (Hillson, 1996). A defect in this position would be expected to match with the middle part of the first incisor crowns, but they are too worn to see this. It would also be expected to match with the occlusal third of the canine, and it is possible to see some evidence of a defect at this point in the upper canines, in spite of the abrasion. There are also less prominent furrow-form defects in the cervical part of the upper and lower canines, first premolars, and second molars. This would represent an additional growth disturbance at around 5 years of age (Hillson, 1996; Reid & Dean, 2000).

Most of the Dolní Věstonice 15 teeth show no evidence of carious lesions. The only possible candidate for such a lesion is on the lower left first molar, where the prominent buccal pit intersects with the marked developmental defect (Figure 19.6). The defect is filled with material that may be dental calculus or sediment. The point of intersection of the buccal pit and the developmental defect is darkly stained, and this might represent a carious lesion. Developmental defects are known to predispose to caries, so it might be expected. In this case, however, the radiographic appearance of the tooth is normal, with no evidence of a radiolucency in the underlying dentine. The presence of caries is therefore not confirmed.

Overall Body Size

The stature and body mass estimates for Dolní Věstonice 15 have been presented and justified in Chapter 13. From those considerations, the estimated stature is ca. 159 cm, and its estimated body mass is ca. 53 kg. As discussed in Chapter 13, for an earlier Upper Paleolithic female these values would be modest but not exceptional. However, given the almost certain male gender of Dolní Věstonice 15 (Chapter 7), these values are unusually low. The stature estimate is 2.73 standard deviations below a European earlier Upper Paleolithic male mean (174.2 ± 5.6 cm; $N = 18$) and below the minimum male estimate of 165 cm (Barma Grande 4). The body mass estimate is 3.18 standard deviations below the mean body mass of the same male sample (70.5 ± 5.6 kg; $N = 16$) and substantially

Figure 19.6 Dolní Věstonice 15, marked hypoplastic dental enamel defects. The lower left canine (left) shows a marked furrow-form defect, representing a growth disturbance at about 4 to 5 years of age, and this is matched by a less marked defect in the cervical part of the first premolar. The lower left first molar (right) shows a very sharp groove defect, representing a disturbance at about 2 years of age. Deposits of calculus fill the defect, and they are darkly stained in the position of the buccal pit, but this does not appear to represent a carious lesion.

lower than the minimum value (62 kg for the Arene Candide IP adolescent) for that sample. It is therefore clear that Dolní Věstonice 15 sustained a developmental abnormality that prevented him from achieving a normal body length and mass for an earlier Upper Paleolithic male.

The Axial Skeleton

Even though Dolní Věstonice 15 was a young adult at the time of his death (Chapter 6), one cannot necessarily assume that the axial skeleton would be free from degenerative changes, given the length asymmetry present in the femora (see below). However, there are few, if any, obvious degenerative changes in the vertebral column. For example, no osteophytes are evident along the body margins, with the possible exception of a slight lip on the twelfth thoracic vertebra that runs from the posterolateral body margin onto the pedicle. The other potentially pathological features include a hole on the ventral face of the second lumbar vertebra that is too large and too ventral to be a nutrient foramen (Chapter 14). A second, larger hole is evident on the dorsal caudal surface of the L2 body, and it may be a Schmorl's node.

There is little evidence of pathological dorsoventral wedging in the Dolní Věstonice 15 vertebrae. Potential cases include the T3, the ventral height of which is lower than its dorsal body height. Although this is in the expected direction, the magnitude is greater than expected (Chapter 14). Likewise, neither dorsal nor ventral body height is preserved for the element, but the T6 also appears to have ventral wedging. In contrast, L1 is unusual in that its dorsal body height is much greater than its ventral height (the more typical pattern is a slight decrease from dorsal to ventral height). None of these elements is so unusual as to warrant a "pathological" designation; rather, they tend to fall near the ninetieth percentiles for recent humans (Chapter 14).

The length asymmetry of the femora, in combination with the minimal (1 mm) difference in tibial length (see below), should have produced pelvic tilting when Dolní Věstonice 15 was standing bipedally and upright. Combining the 16.0 mm difference in femoral bicondylar lengths with the 0.5 mm difference in tibial biomechanical lengths produces a difference in lower limb lengths of 15.5 mm. The combination of the interacetabular distance (97.0 mm) with half of each mean femoral head diameter (46.1 mm and 47.4 mm) provides an interweight-bearing distance of 143.8 mm. The resultant angle of pelvic tilt in a stationary bipedal stance with straight lower limbs, all else being equal, should therefore be 6.2°.

In order for the upper trunk, shoulder girdle, and head to be in normal balance, this pelvic tilting should have produced a right convex scoliosis of the vertebral column. However, all of the thoracic bodies are essentially symmetrical, although the L2 body is slightly taller on the left side rather than on the expected right side (the L3 to L5 bodies are very incomplete). The S1 body cranial surface is essentially perpendicular to the craniocaudal axis of the sacrum in a coronal plane (Chapter 17). If there was scoliosis, it must therefore have been produced by asymmetries in the damaged L3 to L5 bodies or solely within the intervertebral disks.

The spinous processes deviate to the right slightly on T8–T10 and T12 and more clearly so on T2, T4, L2, L4, and L5. The C7 spinous process is twisted clockwise (when viewed dorsally), whereas the T12 and L2 twist counterclockwise. The T7 right superior articular facet is in a slightly more cranial position than the left one, and the L3 and L4 articular facets are slightly asymmetrical in their dorsoventral positions on the neural arches. There are therefore unusual patterns in these vertebrae, although a skeletal diagnosis of scoliosis is disputable.

The sacrum of Dolní Věstonice 15 has been described as long and narrow. As noted in its description (Chapter 17), the apparently unusual length of the sacrum is the product of fusion of the first coccygeal vertebra (Cx1) on to the caudal end of the S5, an infrequent but not exceptional variant among recent humans. Moreover, the apparent narrowness of the S1 is the result of the small overall pelvic size since it is not proportionately small when compared to bi-iliac breadth estimates; nor are the alae especially narrow. It is therefore not considered as pathological here.

The Appendicular Skeleton: Lower Limb Proportions and Deformities

As mentioned in Chapter 18, the right femur of Dolní Věstonice 15 exhibits an abnormal anterior curvature of the proximal diaphysis and associated alterations of the proximal epiphyseal region (Figure 18.4). The left femur, however, appears to be closer to the expected morphology and proportions and was therefore considered morphologically among the other femora from Dolní Věstonice and Pavlov. The leg and foot remains below the femoral midshafts, however, show little in the way of abnormalities, even though some of their aspects could be related to the deformities of the more proximal lower limbs. For these reasons, the proximal femoral configurations of Dolní Věstonice 15, particularly of the right side, are described in detail here, and more distal aspects of the limbs are discussed as they might relate to the widespread deformities of this individual.

As discussed in Chapter 13, the lower limbs of Dolní Věstonice 15, in particular his femora, are relatively short. Indices of the more normal left femoral bicondylar length and the average tibial maximum length to skeletal trunk height (81.7 and 73.2, respectively) are 2.41 and 1.59 standard deviations, respectively, below Gravettian means (94.7 ± 5.4; N = 12; and 81.8 ± 5.4; N = 8). The index of the right bicondylar length to skeletal trunk height is 78.3, which is 3.04 standard deviations below the Gravettian mean. Viewed differently, the left bicondylar length of 383.0 mm is 86.5% of the estimated length of 443 mm based on the mean ratio of femoral bicondylar length to skeletal trunk height for Gravettian remains (Chapter 13). The percentage of this estimated length for the right femur then becomes 82.8%. The tibial maximum lengths are 89.9% and 89.7% of the estimated length of 382.5 mm based on the mean Gravettian bone length to skeletal trunk height indices. The relative tibial length could be considered a normal variant, but the femoral length is exceptional.

This foreshortening of the femora, especially of the right one, with its curvature subtense of 20 mm, was undoubtedly produced in part by the pronounced proximal anterior diaphyseal curvature of the bones, but that alone is insufficient to explain the femoral longitudinal abbreviation. This is reinforced by the fact that the anterior curvature subtense for the left femur of 14 mm is within 1 standard error of the mean (13.6 ± 2.8 mm; N = 15) of an earlier Upper Paleolithic sample (including the other specimens from Pavlov and Dolní Věstonice). It could also be argued that the low femoral neck/shaft angles of Dolní Věstonice 15 (right: 111°; left: 113°), compared to a Gravettian mean (including the other Dolní Věstonice femora) of 121.0° (± 5.7°; N = 22), contribute to this femoral foreshortening. If the Dolní Věstonice 15 femora had neck/shaft angles identical to the Gravettian mean, the increase in femoral length can be approximated by multiplying the tangent of the difference in angle (10° on the right and 8° on the left) by the measured anatomical biomechanical neck lengths of 47.0 mm and 51.0 mm, respectively. This would yield increases of 8.3 mm on the right side and 7.2 mm on the left side. The resultant bicondylar lengths of ca. 375 mm and ca. 390 mm would still be very short. The resulting indices relative to skeletal trunk height (80.0 and 83.2) still remain 2.72 and 2.13 standard deviations below the Gravettian mean. Finally, the distal bicondylar angles of the femora (both 11°) are normal for both Upper Paleolithic and recent humans (Chapter 18; Tardieu & Trinkaus, 1994). The Dolní Věstonice 15 femora are therefore simply short, and the abnormalities of the proximal epiphyses do not account for their abbreviation.

It should be noted, before the detailed description of the femoral abnormalities, that the osseous tissue of the femora appears, macroscopically, to be fully normal. This is apparent in the smooth and dense subperiosteal diaphyseal bone; the relatively smooth and regular ligamentous and muscular attachment areas, with an absence of enthesopathies (especially around the trochanters); and the normal and smooth subchondral bone of most of the femoral articulations (Figures 18.4, 19.7, and 19.8). There is some subchondral bone degeneration of the weight-bearing areas of the femoral heads and associated acetabulae and on the distal femora (see below). In addition, radiographically (Figure 19.9), the cortical diaphyseal bone is even and dense, with clear and modestly trabeculated endosteal margins. This is combined proximally with the completely normal arrangements of the proximal trabecular bundles through the necks, trochanters, and heads (Figure 19.10). There is no evidence of porosity externally or radiographically, and the distributions of cortical and trabecular bone (given the overall abnormal curvatures) are as expected for human femora.

The principal abnormality of the right femur is a diaphyseal angular deformity (ca. 20°) in the sagittal plane relative to the dorsal condyles, centered ca. 130 mm distal to the proximal head and 35.4% of bicondylar length

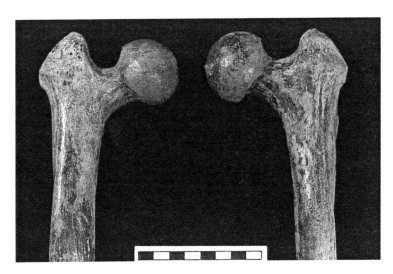

Figure 19.7 Anterior view of the Dolní Věstonice 15 right and left proximal femora, showing the low femoral neck/shaft angles and the relatively high position of the right greater trochanter. Scale in centimeters.

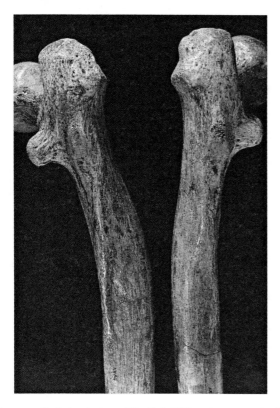

Figure 19.8 Lateral view of the Dolní Věstonice 15 proximal right (on the left) and left (on the right) femora, illustrating the marked anterior convexity and twisting of the right proximal diaphysis.

from the proximal end (Figures 18.4 and 19.8). The angle is clearest in the anterior profile since the posterior angulation is bridged partly by the pilaster leading onto the gluteal buttress. The curvature accentuates the rounded anterior buttress leading up to the greater trochanter, but there is no effect on the dorsal muscle markings. Endosteally, there is a thickening of the posterior cortical bone, but it is generally similar to the one seen on the left femur, with its less pronounced sagittal curvature (Figure 19.9). This parasagittal angular deformity is accompanied by an axial deformity, which leads to a retroversion of the proximal epiphysis ca. 5° relative to the left femur. It results in a lateral protrusion of the gluteal buttress and in a stronger, but still moderate, twisting of the gluteal tuberosity laterally.

These changes in the right proximal diaphysis result in an increase in the meric index (anteroposterior/mediolateral diameter) to 89.5, compared to the one of 80.8 on the left side. However, there is little change in the ratio of the subtrochanteric anatomically oriented second moments of area, with the right one providing an I_x / I_y ratio of 0.947 and the left one having one of 0.953. Yet, the right second moments of area are larger, with the polar moment of area being 3.8% greater than the left one.

The relatively, but not extraordinarily, low neck/shaft angles of the Dolní Věstonice 15 femora (Chapter 18) may well be associated with unusual levels of hip joint reaction force during the first decade of life, since the degree of normal decrease is proportional to the habitual level of loading of the joint (Houston & Zaleski, 1967; see clinical references in Anderson & Trinkaus, 1998). One of the consequences of these low neck/shaft angles is the apparent proximal protrusions of the greater trochanters (Figure 19.7). Both maximum trochanteric lengths are greater than the maximum articular lengths, being 7 mm longer on the right and 2 mm longer on the left. However, when the femora are placed in anatomical (bicondylar) position, the right bicondylar trochanteric length is 3 mm shorter and the left one is 8 mm shorter than their respective bicondylar interarticular lengths.

At the same time, there is a series of asymmetries of the proximal femora, some of which have been mentioned. The asymmetry in femoral length is marked, although some degree of asymmetry is present in other, apparently developmentally normal specimens. The asymmetry value of 4.27 for Dolní Věstonice 15 is approached by those of Dolní Věstonice 14 and Předmostí 10 (assuming that the latter is not a typographical error in Matiegka, 1938) and by a couple of recent humans (Table 19.4). The asymmetry in its neck/shaft angle (2°) is unexceptional,

Figure 19.9 Mediolateral radiograph of the Dolní Věstonice 15 right (R) and left femora.

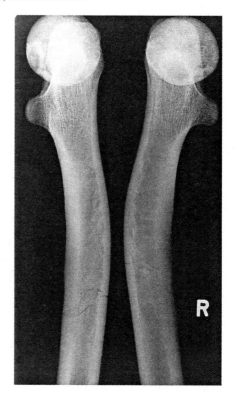

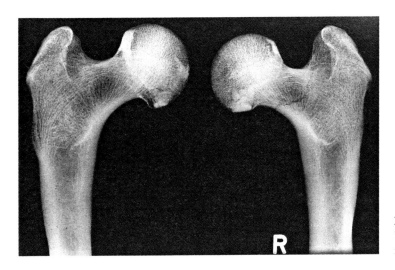

Figure 19.10 Posteroanterior radiographs of the Dolní Věstonice 15 right (R) and left proximal femora.

falling on the recent human median and in the middle of the Late Pleistocene range of variation. The right femoral head is smaller than the left one, but the degree of asymmetry is not exceptional. It is approached by Dolní Věstonice 13 and Předmostí 4, 10, and 14 and matched by Skhul 4; the very high values for Barma Grande 2 and Veneri 1 may be artifacts of postmortem damage, but at least the former individual exhibits high levels of asymmetry elsewhere in the appendicular skeleton (Churchill & Formicola, 1997; Tables 16.3 and 19.4). Few comparative data are available for midshaft polar moment of area asymmetry, and Dolní Věstonice 15 has a rather high value for it. Yet, both Skhul 5 and La Ferrassie 2 also exhibit relatively high values.

In the context of these unusual features of the Dolní Věstonice 15 proximal femora and midshafts, it is surprising that there are very few lower limb features distal of the femoral midshafts that can be considered pathological. The distal femoral epiphyses are completely normal in their shapes and orientations (Figure 18.4). The patellae are moderately asymmetrical in their overall dimensions but similar to those of other fossil and recent humans (Table 19.5). The medial and lateral facets are normally asymmetrical within each patella, even though each exhibits a curious proximodistal sulcus that runs along each medial facet near its medial margin (Figure 18.25). The tibiae are essentially the same lengths, as are the associated fibulae (articular lengths of 324.0 mm and 323.5 mm and of 327.0 mm, and 329.0 mm respectively). The asymmetry in the tibial midshaft polar moments of area is only slightly greater than those of Dolní Věstonice 13 and 14, and it is exceeded by those of Skhul 4 and La Ferrassie 1 and 2 (Table 19.5). The tibiae (Figure 18.32) have relatively sharp and protruding anterior crests along their proximal two-thirds, but they are only slightly more exaggerated in this feature than the tibiae of Dolní Věstonice 13. The fibulae (Figure 18.45) have a pronounced S-curve of the distal diaphysis, which is merely an exaggeration of the normal curvature and not markedly different from that seen in Dolní Věstonice 16. The pedal remains (Chapter 18) present little of note, and the talar trochleae are quite symmetrical, much more so than those of Dolní Věstonice 16, Arene Candide IP, Pataud 26.228, Předmostí 14, Veneri 2, several Middle Paleolithic specimens, and many recent humans (Table 19.5).

The only other feature of the lower limb remains of Dolní Věstonice 15 that appears unusual, as noted by Formicola et al. (2001), is the relative elongation of the distal fibulae. The interarticular lengths of the fibulae fit the tibiae appropriately for normal proximal and distal articulations, and the malleolar facets are appropriate for the talar lateral malleolar facets. However, the distal ends of the malleoli appear relatively long. This is reflected in ratios of 1.037 and 1.026 between the maximum and articular lengths. The one Upper Paleolithic specimen providing comparative data, Dolní Věstonice 13, has a ratio of 1.008.

It needs to be emphasized, nonetheless, that several features of the lower limbs indicate that Dolní Věstonice 15 had assumed normal bipedal postural and locomotor loading patterns. The arrangements of the femoral head and neck trabecular bundles are as expected, and the medial trabecular angles of 134° and 140° are close to those expected for its low neck/shaft angles (Trinkaus et al., 1991). The femoral bicondylar angles of 11° each are normal (Tardieu & Trinkaus, 1994). The tibial medial retroversion angles of 10° are normal for an active, nonurban population (Trinkaus, 1975b). And the patterns of torsion of the metatarsals indicate the formation of normal pedal arches (Trinkaus, 1975a). All of these features are developmentally plastic and only assume their characteristic mature human patterns with appropriate and sufficient loading from normal human bipedal posture and locomotion during immature life.

Table 19.4 Lower limb crural metric asymmetry for Dolní Věstonice and Pavlov, available Gravettian and Middle Paleolithic data, and pooled recent human samples (there are no asymmetry data for Aurignacian human remains).

	Femur Bicondylar Length	Femur Head Anteroposterior Diameter	Femur Neck/Shaft Angle	Femur Midshaft Polar Moment of Area
Dolní Věstonice and Pavlov				
Dolní Věstonice 3	—	—	—	0.73
Dolní Věstonice 13	0.56	2.11	1°	0.02
Dolní Věstonice 14	2.79	1.77	2°	0.97
Dolní Věstonice 15	4.27	2.77	2°	3.53
Dolní Věstonice 16	—	0.00	1°	0.43
Pavlov 1	—	—	—	1.16
Gravettian				
Arene Candide IP	0.22	1.02	7°	—
Barma Grande 1	0.94	—	—	—
Barma Grande 2	—	(6.37)	7°	—
Barma Grande 6	—	0.21	—	—
Fanciulli 4	0.00	—	—	—
Paglicci 25	0.00	1.16	3°	—
Předmostí 3	1.03	0.00	4°	—
Předmostí 4	0.12	2.11	3°	—
Předmostí 9	1.12	0.00	0°	—
Předmostí 10	3.62	2.11	5°	—
Předmostí 14	1.33	2.15	0°	—
Veneri 1	—	(6.16)	—	—
Veneri 2	0.41	0.41	5°	—
Qafzeh-Skhul				
Qafzeh 9	—	—	1°	—
Skhul 4	0.82	2.75	11°	1.80
Skhul 5	—	—	—	3.05
Neandertals				
La Ferrassie 1	—	0.92	2°	0.80
La Ferrassie 2	—	—	—	2.27
Neandertal 1	0.48	1.00	3°	—
Spy 2	—	0.00	6°	—
Recent Humans				
Median	0.53	0.00	2°	—
95th percentile	1.77	2.60	9°	—
Maximum	3.92	5.13	13°	—
N	96	153	150	—

Femur neck/shaft angle asymmetry is the difference between the sides. The others are the absolute value of the difference between the sides divided by the mean of the two values, expressed as a percentage. The fourth roots of the polar moments of area are employed to linearize the values. Recent human data are pooled from Hirai and Tabata (1928), Kiyono and Hirai (1928), and Ishisawa (1931); because of the tendency to assume bilateral symmetry in the lower limbs, asymmetry data are not available for recent human and other earlier Upper Paleolithic femoral midshafts. Values in parentheses are estimated.

Table 19.5 Lower limb subcrural metric asymmetry for Dolní Věstonice remains, available Gravettian and Middle Paleolithic data, and pooled recent human samples (there are no asymmetry data for Aurignacian human remains).

	Patella Maximum Breadth	Patella Maximum Thickness	Tibia Maximum Length	Tibia Midshaft Polar Moment of Area	Talus Trochlear Breadth
Dolní Věstonice					
Dolní Věstonice 3	—	—	—	0.89	—
Dolní Věstonice 13	—	2.76	0.13	2.09	—
Dolní Věstonice 14	4.45	0.49	—	2.45	—
Dolní Věstonice 15	3.28	3.09	0.29	2.55	0.76
Dolní Věstonice 16	—	—	0.51	1.24	4.31
Gravettian					
Arene Candide IP	8.51	0.00	0.53	—	3.17
Barma Grande 2	1.94	1.25	0.45	—	0.76
Fanciulli 4	2.06	0.00	1.34	—	—
Paglicci 25	0.00	2.35	1.70	—	1.94
Pataud 26.228	—	—	—	—	4.04
Předmostí 3	—	—	0.95	—	0.00
Předmostí 4	—	—	0.27	—	—
Předmostí 9	—	—	0.27	—	0.00
Předmostí 10	—	—	—	—	0.00
Předmostí 14	—	—	0.51	—	3.39
Veneri 1	6.49	16.9	0.50	—	0.82
Veneri 2	—	3.77	0.51	—	4.26
Qafzeh-Skhul					
Qafzeh 8	—	—	—	—	0.72
Qafzeh 9	—	—	—	—	0.32
Skhul 4	1.48	1.67	—	3.04	1.74
Neandertals					
La Ferrassie 1	—	—	—	3.52	3.82
La Ferrassie 2	—	5.52	—	2.91	1.65
Kiik-Koba 1	—	—	—	—	2.60
La Quina 1	—	—	—	—	1.77
Regourdou 1	—	—	—	—	0.37
Recent Humans					
Median	2.30	0.00	0.61	0.21	0.00
95th percentile	7.41	7.99	2.35	0.30	7.12
Maximum	8.70	12.77	8.80	8.00	8.33
N	88	93	90	71	81

The values are the absolute value of the difference between the sides divided by the mean of the two values, expressed as a percentage. The fourth roots of the polar moments of area are employed to linearize the values. Recent human data are pooled from Hirai & Tabata (1928), Kiyono & Hirai (1928), Ishisawa (1931) and Trinkaus et al. (1994).

Furthermore, as documented in Chapter 18, a variety of lower limb reflections of robusticity during development and/or maturity place Dolní Věstonice 15 among the more robust of the earlier Upper Paleolithic Europeans. Those comparisons assume that the bone tissue of his limbs was histologically normal, an interpretation that is supported by the external and radiographic appearance of the bones. And given the weight-bearing nature of the lower limbs, they are all scaled to estimates of body mass and appropriate measures of beam and/or moment arm length, which take into account the unusual body proportions of Dolní Věstonice 15. The inference from these data is that Dolní Věstonice 15, despite his developmental abnormalities, was similar to other earlier Upper Paleolithic humans in his levels and general patterns of locomotion.

The Appendicular Skeleton: Upper Limb Proportions and Deformities

As mentioned in Chapter 16, the upper limb remains of Dolní Věstonice 15 present a series of deformities, lesions, and asymmetries. These abnormalities are scattered through the shoulder and arm bones in a pattern that makes their interpretation difficult, and indeed, they have been considered by Formicola et al. (2001) as largely secondary to more primary abnormalities in the lower limbs.

The Dolní Věstonice 15 clavicular lengths and overall scapular dimensions are quite symmetrical. Given preservation, the only possible comparisons are the clavicular conoid lengths and the scapular infraspinatus breadths, but the former differ by only 1 mm (109 mm vs. 108 mm) and the latter (given estimation on the left side) are essentially the same. However, the long bones of the arms are distinctly different in lengths.

The comparison of humeral maximum lengths (Table 19.6) provides a very modest 0.99 asymmetry for Dolní Věstonice 15, but it is confused by the medial deviation of the right distal epiphysis, which decreased the distal projection of the distal medial trochlear flange. A comparison of articular lengths (to the distal capitulum) provides an asymmetry value of 3.12, and the intermediate biomechanical length to the lateral trochlear flange furnishes an asymmetry value of 1.99. Among the Late Pleistocene specimens furnishing reliable estimates of both humeral lengths (Table 19.6), only Kebara 2 has a humeral maximum length asymmetry value exceeding the Dolní Věstonice 15 biomechanical length asymmetry figure, and none of them matches the one for his articular lengths. The latter is matched, however, by some recent human maximum length asymmetry values, and it is only close to the ninety-fifth percentile of that pooled sample.

In the Dolní Věstonice 15 radii, the right bones are also longer, but in this case the deformity influencing the difference is on the left side. Percentage asymmetry values, using articular lengths, provide values for Late Pleistocene humans that are generally modest, and Dolní Věstonice 15 has a level of asymmetry that exceeds most of them. However, Fanciulli 4 has a similar level of radial length asymmetry, and among recent humans there are a few individuals, all beyond the ninety-fifth percentile, that match or surpass Dolní Věstonice 15 (Table 19.6).

The cubital articulations appear morphologically normal, despite the medial deviation of the right distal humeral epiphysis. There is nothing in any of them that would be out of place in the range of variation of Late Pleistocene early modern human arm bones. However, there is some degree of asymmetry in several aspects of them The humeral distal articular breadth asymmetry value of 7.14 is well above those of all other Late Pleistocene specimens except Barma Grande 2, and it is near the maximum value for the recent human pooled sample. This is joined by a very high asymmetry value (14.17) for radial head/neck length, with the right side being markedly longer (Table 19.6). Yet, as noted in Chapter 16 (Figure 16.48), for each of the radii the proportions between head/neck length and articular length are similar to those of other earlier Upper Paleolithic specimens, including the other Pavlovian ones.

Despite these size asymmetries, the articular dimensions of the proximal ulnae are very similar (Table 19.7). The radial heads are insufficiently preserved to measure asymmetry, but the average of the radial neck diameters provides an asymmetry value for Dolní Věstonice 15 that is in the middle of the values for Dolní Věstonice 13, 14, and 16 (Table 19.7). The marked asymmetry in the cubital articulations of Dolní Věstonice 15, therefore, appears to be principally in humeral distal articular breadth, especially in the radial head/neck length, and the latter is principally due to a foreshortening of the left radius overall.

The most marked abnormalities of the upper limbs, however, are the above-mentioned curvature deformities of the distal right humeral diaphysis and of the left radial and ulnar diaphyses. The former (Figures 19.11 and 19.12) is a clear medial angulation of the distal humeral diaphysis. It is located ca. 75 to 80 mm proximal to the distal capitulum (ca. 25% of bone length from the distal end), and there is an angle in the shaft of ca. 10°. The plane in which the angulation occurs is very close (±5°) to the mediolateral axis of the cubital articulation, with no perceptible deformity in the parasagittal plane of the bone. This mediolateral deformity results in a cubital angle of 82°, as opposed to the normal left one of 92°, indicating that the deformity is strictly diaphyseal and not epiphyseal.

The bone tissue appears normal, with smooth subperiosteal bone, sharp endosteal boundaries, and few tra-

Table 19.6 Asymmetry of humeral and radial dimensions, computed as the absolute value of the difference between the two sides divided by the mean of the two values, expressed as a percentage.

	Humerus Maximum Length	Radius Articular Length	Humerus Distal Articular Breadth	Radius Head/Neck Length
Dolní Věstonice				
Dolní Věstonice 13	1.20	0.82	1.57	3.65
Dolní Věstonice 14	—	—	0.68	9.77
Dolní Věstonice 15	0.99/3.12/1.99	2.67	7.14	14.17
Dolní Věstonice 16	—	—	—	7.50
Gravettian				
Arene Candide IP	—	0.44	—	—
Barma Grande 2	1.34	—	7.19	—
Cro-Magnon 4294	0.94	—	3.44	—
Fanciulli 4	0.54	2.61	3.43	3.16
Paglicci 25	1.54	—	0.71	5.20
Pataud 26.230	—	—	0.49	—
Předmostí 3	1.68	1.14	0.00	—
Předmostí 4	1.87	0.84	4.88	—
Předmostí 9	0.61	0.00	4.65	—
Předmostí 10	0.64	—	5.13	—
Předmostí 14	0.30	0.00	0.00	—
La Rochette 1	—	—	3.87	—
Sunghir 1	—	1.13	—	—
Qafzeh-Skhul				
Skhul 5	0.26	—	—	—
Neandertals				
La Chapelle-aux-Saints 1	—	—	0.66	—
La Ferrassie 1	0.89	—	—	4.30
Kebara 2	2.19	1.21	—	—
La Quina 5	—	—	1.72	—
Regourdou 1	—	—	—	3.69
Spy 2	—	—	0.89	—
Recent Humans				
Median	1.08	0.92	1.74	2.88
95th percentile	3.16	2.46	5.00	8.46
Maximum	4.19	3.68	8.71	17.96
N	128	148	97	89

The pooled recent human samples are those used in Trinkaus et al. (1994), Churchill (1994), and Churchill and Formicola (1997), but the percentage asymmetry is calculated differently; data are largely from Churchill (personal communication). The three values for the Dolní Věstonice 15 humeral lengths are for the maximum lengths to the distal medial trochleae, the articular lengths to the mid-distal capitulae, and the biomechanical lengths to the distal lateral trochleae, respectively.

Table 19.7 Asymmetry of the ulnar and radial proximal epiphyseal dimensions, computed as the absolute value of the difference between the two sides divided by the mean of the two values, expressed as a percentage.

	Ulna Olecranon Length	Ulna Olecranon Depth	Ulna Coronoid Height	Radius Mean Neck Diameters
Dolní Věstonice & Pavlov				
Dolní Věstonice 13	2.60	0.58	6.11	1.64
Dolní Věstonice 14	1.00	3.41	(0.00)	10.40
Dolní Věstonice 15	3.21	1.04	4.75	3.10
Dolní Věstonice 16	—	0.54	—	2.24
Pavlov 1	0.49	8.27	—	—
Gravettian				
Fanciulli 4	—	—	7.94	—
Paglicci 25	—	—	1.95	—
Neandertals				
Regourdou 1	—	—	—	4.63
Shanidar 6	—	—	—	4.06
Spy 2	—	—	1.73	—

The radial neck diameter asymmetry is based on the average of the neck anteroposterior and mediolateral diameters.

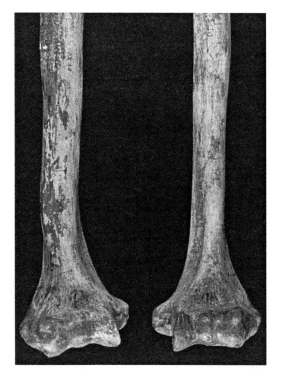

Figure 19.11 Anterior view of the Dolní Věstonice 15 distal humeri, showing the abnormal curvature of the distal right humerus and the clear diaphyseal asymmetry of the bones with the larger right diaphysis.

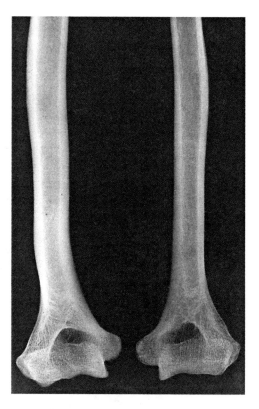

Figure 19.12 Anteroposterior radiograph of the Dolní Věstonice 15 distal humeri.

beculae at the level of the deformity. There is no evidence of a callus in the subperiosteal surface bone, in the even contours of the endosteal cortical bone, or in any bridging trabeculae across the medullary cavity. If the curvature was the result of a healed fracture, the callus must have been completely remodeled into normal diaphyseal cortical bone, something that would be unusual unless the deformity occurred very early in development and then persisted despite all of the other growth changes.

There is some mediolateral expansion of the diaphysis near the curvature, since the 35% anteroposterior to mediolateral second moments of area ratio (I_x/I_y) is 0.86. This compares with a ratio of 0.97 for the Dolní Věstonice 15 left humerus, as well as values of 1.47, 1.25, 1.54, 1.16, 1.31, and 1.13 for the Dolní Věstonice 3, 13, 14, 16, and 41 and Pavlov 1 right humeri, respectively. A non-Pavlovian earlier Upper Paleolithic right humeral sample provides a mean ratio of 1.33 (± 0.16; N = 8; pooled earlier Upper Paleolithic sample: 1.32 ± 0.15; N = 14), and similar values are available for Neandertal (1.23 ± 0.15; N = 9) and Qafzeh-Skhul (1.29 ± 0.23; N = 6) right humeri. The left I_x/I_y value of 0.97 compares to values of 1.29 ± 0.16 (N = 16) for a pooled earlier Upper Paleolithic sample, 1.12 ± 0.20 (N = 4) for Qafzeh-Skhul and 1.31 ± 0.14 (N = 7) for Neandertals. Both of the Dolní Věstonice 15 humeri are therefore unusually mediolaterally, as opposed to anteroposteriorly, expanded; the right one is 3.03 standard deviations from the pooled earlier Upper Paleolithic right humeral sample, and the left side is 1.99 standard deviations from a similar sample for the left humerus.

The right ulna appears to be normal, but the left one exhibits irregularities of the cortical bone in the mid-distal diaphysis (Figures 16.40, 16.41, and 19.13). It has a bony callus formation along the interosseous crest, ca. 30 mm long, 9 mm wide, and centered 96 mm from the distal ulnar head (or 38% of its maximum length from the distal end). It disrupts the interosseous crest, and it increases the mediolateral diameter to 15.2 mm, compared to 13.4 mm at midshaft. Radiographically, there is some thickening of the anterior and lateral cortical bone in the affected area, and there are a few bridging trabeculae across the medullary canal. The normal S-curve of the diaphysis remains, but its lateral extent near midshaft has been exaggerated. This appears to be a long-since healed fracture of the ulnar diaphysis, which fused with a slight deformity of the bone. It is likely that it occurred before longitudinal growth of the bone had ceased and that it is therefore correlated with the abbreviation of the left forearm.

This moderately common fracture of the ulnar diaphysis (Lovejoy & Heiple, 1981; Ortner & Putschar, 1981) is associated with an unusual dorsal convexity of the left radius (Figures 16.41 and 19.13). The left radius exhibits an angulation of the proximal diaphysis within the sag-

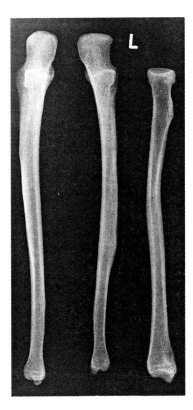

Figure 19.13 Anteroposterior radiographs of the Dolní Věstonice 15 right and left (L) ulnae and mediolateral radiograph of the Dolní Věstonice 15 left radius.

ittal plane of the bone. The apex of the angle is 85 mm distal of the head (or 36.5% of the maximum length from the head), and the resultant angle is ca. 8°. It is associated with a torsional change in the orientation of the proximal head and tuberosity, with the tuberosity more anteriorly placed than normal. However, as with the right humerus, the cortical bone through the affected area appears normal, with clear endosteal margins and no unusual trabeculation.

The Appendicular Skeleton: Lower Limb Osteoarthritis

As mentioned above, the articulations of the Dolní Věstonice 15 lower limbs are generally in excellent condition, without evidence of degeneration. However, both of the hip articulations and the distal femoral surfaces exhibit evidence of moderate localized osteoarthritis, in the form of both porosity and the laying down of new bone on the subchondral bone.

The left acetabulum presents a distinct area of new bone laid down on the lunate surface just below the anterior inferior iliac spine. The new bone area is ca. 12 mm anteroposterior and at least 15 mm mediolateral, and there is mild porosity around the new bone. In addition, there is an area of distinctive porous subchondral bone on the superolateral lunate surface, ca. 6 mm wide. These changes match those on the left femoral head, on which the entire proximal surface is covered with a fine layer 0.1 to 0.3 mm thick of poorly organized new bone, with minimal porosity. It does not change the curvature of the head. The new bone has a distinct irregular margin with a tiny step posteriorly, whereas there is just a change in texture anteriorly. The altered surface bone, measured as chords, covers an area 43.5 mm anteroposteriorly and 28.5 mm mediolaterally.

The right hip joint also presents some subchondral surface alterations, but they consist entirely of porosity, without the laying down of new bone. On the femoral head, the porous area is across the superior surface, rounding onto the medial and dorsal sides to the level of the fovea capitis. Measured as chords, the porous area is ca. 29 mm anteroposterior by ca. 28 mm mediolateral. The femoral head changes match an area of moderate subchondral porosity on the right acetabulum just below the anterior inferior iliac spine, up to 21.5 mm wide and ca. 17.5 mm anteroposterior.

Most of the femoral condylar subchondral bone is normal. However, there is an area of irregular subchondral bone greater than 16.5 mm anteroposterior (damage precludes knowing its full anteroposterior extent) and the full width of the condyle, located on the distoposterior curve of the right lateral condyle. There is similar but less pronounced remodeling of the left distoposterior lateral condyle. The tibial condyles do not show similar changes, suggesting that the alterations were restricted to the contact between the femoral condyles and the menisci.

Given the young adult age of Dolní Věstonice 15, these mild osteoarthritic changes suggest some overloading of the joints on a regular basis. It is possible that changes in the hip articulations are related to the abnormalities of the proximal femora, with greater loads being placed on the left side because of the more pronounced deformities on the right hip (but see discussion below). The trivial changes in the distal femora are more difficult to explain, given the otherwise normal appearance, configurations, and orientations of the distal femoral epiphyses.

The Appendicular Skeleton: Upper Limb Ossifications and Osteoarthritis

The left clavicle of Dolní Věstonice 15 exhibits a curious enlargement of the conoid tubercle (Figures 16.4). Although projecting conoid tubercles are known in other Late Pleistocene and recent human clavicles, and some level of metric asymmetry at this level is not uncommon, the degree of tubercle asymmetry in Dolní Věstonice 15 is moderately high and its direction appears unusual. For example, the asymmetry value (difference divided by the bilateral mean) in the dorsoventral conoid diameter is 20.8, whereas it is 4.3 in Dolní Věstonice 14, 22.2 in Dolní Věstonice 13, and ca. 27.3 in Pavlov 1. However, in all four of these individuals the right humeral diaphysis is more robust, indicating right-handedness; the three other Pavlovian specimens providing both conoid dorsoventral diameters have larger right clavicular diameters, whereas in Dolní Věstonice 15 it is the left clavicle that exhibits the distinctly larger conoid dorsoventral diameter.

In addition, the deltoid attachment within the left acromial curve of Dolní Věstonice 15 exhibits a small bony crest (Figure 16.4), generally similar to the one on the Dolní Věstonice 14 clavicle. It is 7.7 mm long, projects anteroinferiorly (or ventrocaudally), and is up to 2.3 mm thick.

Even though most of the well-preserved articular surfaces of the Dolní Věstonice 15 upper limb remains are normal, there is evidence of highly localized osteoarthritis.

The right glenohumeral articulation exhibits a focal area of degeneration (Figure 19.14). On the scapula, there is a patch of thin dense bone laid down on the subchondral bone in the middle of the glenoid fossa, ca. 11 mm high and ca. 9 mm wide. More inferiorly on the glenoid surface, there is some subchondral osteoarthritic bone surface erosion. On the directly medial aspect of the associated humeral head—medial relative to the transverse axis of the distal articulation and anteromedial relative to the head mediolateral axis, given humeral torsion (head retroversion) of 21°—there is a layer of thin dense bone 16–17 mm in diameter on top of the subchondral bone. The middle area of the patch (ca. 9 mm in diameter) has porous openings with rounded edges that communicate with the trabecular spaces. These two areas of alteration, which involved both subchondral bone erosion (particularly on the humerus) and new bone, are clearly associated aspects of the same osteoarthritic lesion. If these two lesions are placed together, the humeral shaft is oriented in the coronal plane at an angle of ca. 90° to the spine of the scapula. If this represents focal degeneration from biomechanical overloading of the joint in a given position, this orientation indicates that there was persistent loading of the joint when the humerus was in a parasagittal plane of the body.

In addition, there is advanced osteoarthritis of the second and fifth proximal interphalangeal articulations of the right hand. On the head of the proximal phalanx 2 (securely identified by its base morphology), the radial mid-distal to dorsal side of the trochlea is beveled off radially, with exposure and rounding of the trabeculae and a slight polish of the bone. The radioulnar middle of the trochlea has a deposition of extra bone in the mid-distal to dorsal sulcus, and there is slight remodeling of the dorsodistal portion of the ulnar side of the trochlea (Figure 19.15). This is

Figure 19.14 Medial view of the Dolní Věstonice 15 right humeral head and lateral view of its matching scapular glenoid fossa, showing the osteoarthritis with porosity and eburnation on the inferomedial humeral head and the osteoarthritis with porosity and new bone in the inferomiddle of the glenoid fossa. The two images are approximately the same scale.

accompanied by changes in the base of the second middle phalanx. All of its proximal articular surface is eroded, with islands of the original subchondral bone remaining; the edges were abraded postmortem, but the surface appears to have been broadly eroded antemortem.

Similarly, on the head trochlea of the proximal phalanx 5 (digit identified by base morphology and relative size), there is marked subchondral bone erosion all through the midtrochlear sulcus, extending from side to side across the articulation. There is slight polish of the radial and ulnar sides of the trochlea. This matches erosion of the middle phalanx 5 subchondral bone across all of the radial side of the proximal articular facet, across the middorsopalmar crest but not onto the dorsal or palmar tips of the crest, and onto the middle but not the edges of the ulnar facet.

Despite the advanced nature of these osteoarthritic degenerations of the right glenohumeral joint and of the

Figure 19.15 Dorsodistal views of the Dolní Věstonice 15 second (left) and fifth (right) right proximal hand phalanges, showing the marked osteoarthritis of the trochleae.

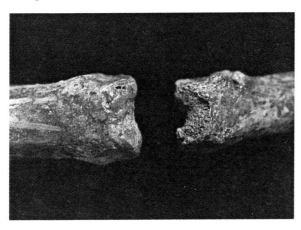

right second and fifth proximal interphalangeal articualtions, none of the other right upper limb articular surfaces and none of the preserved left upper limb articulations exhibits any degeneration. All of the arm articulations are preserved on at least one side, and many of the hand articulations (mostly on the right side) remain. This, combined with the young age of the individual, makes it unlikely that these articular changes are the product of age-accumulated wear on the joints. Moreover, one would expect the abnormally oriented right cubital articulations to have been susceptible to unusual loading patterns. However, they are otherwise healthy, as are those of the left elbow and wrist associated with the radial and ulnar deformities. This implies that these osteoarthritic changes are the product of localized, repetitive overloading of these three joints.

The glenohumeral degeneration indicates that the arm was in a parasagittal plane of the body during loading, such as would be assumed when carrying a heavy object in the hand alongside of the body or when dragging a heavy object behind oneself. The resultant glenohumeral joint reaction force should be largely perpendicular to the glenoid surface (Apreleva et al., 2000). The presence of proximal interphalangeal degenerations on the second and fifth digits, but not on the third and fourth, implies overloading of these joints when either a strap or a proximally convex handle was placed across the ulnar digits, thereby placing the maximum load on the second and fifth digits.

It is not possible on the basis of these biomechanical inferences to specify the actual activity that would produce these articular changes, but one activity that might generate such focal loading of these joints would be the dragging of a heavy load on a sled or skin across the ground behind oneself, with a strap that went across the flexed fingers, as in the hook grip of Napier (1956).

Such a behavior has been documented ethnographically for historic subarctic Native Americans (Gillespie, 1981; Savishinsky & Hara, 1981) and could well have been practiced by Dolní Věstonice 15. Although possibly localized to the second digit of the left hand, the osteoarthritis of Dolní Věstonice 13 (see above) suggests that a similar behavior may have been practiced by more Pavlovian individuals than just Dolní Věstonice 15.

Appendicular Skeleton: Transverse Lines

Despite the abundance of developmental abnormalities on the Dolní Věstonice 15 skeleton and the pronounced dental enamel hypoplasias, there is only a little evidence for transverse lines in the metaphyseal regions of his long bones, hands, or feet (Table 19.3). Both of the proximal tibiae, within the trabeculae of the metaphyseal regions, exhibit faint radiodensity lines that transverse only part of the metaphyses. On the right tibia, they are located 25.5, 28.3, and 30.8 mm from the condylar articular surface, and on the left they are 25.4 and 31.1 mm from the articular surface. In addition, the left first metatarsal has a radiodensity line near midshaft, 23.7 mm from the tarsal articulation, and one within the proximal trabeculae 4.0 mm from the tarsal surface. The midshaft first metatarsal line may be a remnant of the earlier childhood stress episodes evident in the tooth lesions, but the other ones are likely to have occurred closer to full maturity, given the proximities to the articular surfaces.

Summary and Interpretations

It has been evident since shortly after its discovery that Dolní Věstonice 15 suffered from some form of congenital dysplasia. This is particularly evident in his small body size and relatively short lower limbs, but the other deformities lend credence to such an interpretation. The etiological interpretation of the suite of lesions and abnormalities, as well as the possible determination of what form of dysplasia was responsible for the abnormalities, however, initially requires an assessment of which lesions are primary, which are secondary, and which are incidental to the dysplasia.

In reverse order, it is apparent that the right upper limb osteoarthritic lesions should be considered as incidental to the dysplasia. In a young adult individual, they must indicate localized articular biomechanical overloading, possibly related to the carrying and/or dragging of loads on the landscape. It might be argued that they, along with the right humeral diaphyseal deformity, could be due to the habitual use of some form of support, such as a crutch, to compensate for abnormalities in walking patterns induced by the right proximal femoral deformities. However, the basically normal developmental and functional anatomy, plus the symmetry, of the lower limbs distal of the midfemoral diaphyses indicates that Dolní Věstonice 15 did not have any deficiencies in his bipedal gait. It would therefore be surprising if he used a crutch or any other locomotor aid.

Other lesions that were probably incidental to the dysplasia are the minor traumatic lesions on the frontal and parietal bones. As with the minor neurocranial traumatic lesions on the other Pavlovian individuals (and Pleistocene human specimens in general; see below), these lesions are best seen as evidence of the generally high risk of minor injuries among these populations, evidence of which would preserve skeletally more readily on the subcutaneous cranial bone.

A possible secondary lesion is the deformity of the left forearm, the apparent healed fracture of the ulnar diaphysis and the associated deformity of the radial diaphysis from the resultant unusual biomechanical loading. One can conceive of Dolní Věstonice 15, with some impairment of locomotor function (if such existed), being more likely to stumble, fall, and fracture a forearm. Yet, falls are normally associated with Colles's fractures of the distal radial diaphysis and secondarily of the distal ulna once the radius has yielded. An ulnar midshaft fracture, or a "parry fracture," would more commonly be associated with a defensive block, such as in warding off a blow or shielding the head. It may therefore be incidental to the more systemic abnormalities of Dolní Věstonice 15.

In the same context, the angular deformity of the distal right humerus may be either secondary or incidental to more general problems. However, difficulties in ascertaining the proximate etiology of the deformity make assessments of its ultimate etiology ambiguous. If it is somehow related to the proximal femoral bowing, it could be seen as part of the dysplasia. However, it is unclear how unilateral, localized bowing of a distal humerus would be related to localized bowing of a proximal femur. Alternatively, it could be the product of an infantile fracture that healed with an angular deformity, which then had the callus completely remodeled through the subsequent two decades. If this were the case, it could be either secondary to impaired function in a very young individual or incidental to the rest of the lesions. Immature humeral fractures from physical activities tend to be metaphyseal (proximal or distal) (Kocher et al., 2000), and the most common mid-distal diaphyseal fractures from activities such as throwing tend to be external rotation spiral fractures (Allen, 1984; Ogawa & Yoshida, 1998), an unlikely cause of the simple medial deviation of the right distal humerus. If the fracture were across the distal metaphysis, it is conceivable that it healed with a deformity that, subsequently through growth, came to be located in the distal diaphysis. However, the distal epiphyseal region is entirely normal from the olecranon fossa to the distal articulations, and a fracture through the metaphysis might be expected to alter at least the olecranon fossa and the associated pillars. Moreover, the right forearm and hand have no evidence of trauma, and fractures from defensive activities or falls are likely to have affected those bones as well (see above). The etiology of this deformity therefore remains uncertain, as does its relationship to the other abnormalities of this individual.

Probably secondary lesions, some of which may be related to the general level of annual or seasonal availability of resources, are the indicators of nonspecific systemic stress—the developmental enamel defects and the transverse lines. These include indications of moderate stress episodes during adolescence (transverse lines) and around weaning (enamel defects) on Dolní Věstonice 15. The more severe developmental lesions, those on his first molars, represent some of the most pronounced developmental enamel defects known for a Pleistocene human. Since they should have formed at around 2 years postnatal, they are likely to be due to serious stress related to the more pronounced developmental abnormalities of Dolní Věstonice 15. In fact, since they occur neither in the immediately postnatal period nor around weaning, assuming that it was between 3 and 5 years of age (Dettwyler, 1995), they should have been related to a pronounced episode of stress that marshaled the individual's resources for survival at the expense of tooth formation and (probably) overall growth.

Other possible secondary lesions include osteoarthritis of the hip and knee articulations. The latter, the degeneration of the bilateral posterodistal lateral femoral condyles, is difficult to understand since it occurs in a region that would have been loaded principally in flexion of the knees. It is possible that the hip osteoarthritis was produced by the low femoral neck/shaft angles and twisting of the right proximal femur. Yet, the deformity did not directly affect either the femoral heads or the acetabulae, the low neck/shaft angles would not increase joint reaction forces across the hip articulations, and there is no sign of osteoarthritis in any of the weight-bearing portions of the more distal lower limb articulations. It may be related to the apparently consistently greater diaphyseal robusticity of the Dolní Věstonice 15 appendicular skeleton, suggesting that he had to work harder to accomplish the same tasks as the other members of his social group. Could this overloading have started to adversely affect his femoral joints?

Also possibly secondary to the femoral deformities are the minor variations of the lower thoracic and lumbar vertebrae. As noted above, the difference in femoral lengths would have produced, when standing, a pelvic tilt of about 6° to the right. However, one of the normal aspects of human locomotion is a lateral pelvic tilting during the stance phase, which can routinely achieve levels of 6° in either direction (Gard & Childress, 1997). Compounding these 6°s with the anatomical 6° tilt of Dolní Věstonice 15 would produce a total of 12° of pelvic tilt, which should affect the lateral curvature of the lumbar vertebral segment. However, not only does pelvic tilting have little overall effect on trunk vertical displacement (Gard & Childress, 1997), but it is likely that Dolní Věstonice 15 would have engaged in asymmetrical pelvic tilting, given the slight right tilting from the shorter right femur, and thereby would have reduced any unusual lateral flexion of the lumbar vertebral column. This would fit with the absence of skeletal evidence for scoliosis in the lumbar and more caudal thoracic vertebrae. Although one cannot rule out the possibility that the unusual variations of the lumbar vertebrae, especially the right deviations of some of the lower thoracic and lumbar spinous processes, are secondary to the femoral deformities, it is difficult to argue for either scoliosis of this individual or a direct connection between his vertebral variations and the lower limb abnormalities.

As a consequence, the primary abnormalities of this unfortunate individual appear to be (1) the small body size for a Gravettian male; (2) the relative narrowing of the pelvis, but not any abnormal overall proportions of the pelvis; (3) the marked relative shortening of the femora; (4) the less pronounced shortening of the tibiae and fibulae; and (5) the abnormal anterior bowing and torsion of the right femur.

The small overall body size can be seen in large part as a consequence of pronounced systemic stress during early childhood, as is evident in the marked dental enamel defects on the first molars, indicating a relatively long period of growth arrest at about 2 years of age. This was followed by less pronounced periods of growth inhibition in the early juvenile years, as indicated by more moderate but still readily evident dental enamel defects, as well as some continued stress episodes during adolescence, evident in the lower limb transverse lines. Cumulatively, these periods may have been sufficient to place Dolní Věstonice 15 on a growth trajectory well below that of his genetic potential, producing as an adult the reduced stature evident in the long bones and vertebrae. These observations, however, do not explain whether the periods of growth arrest were due to environmental and subsistence stress or medical difficulties specific to Dolní Věstonice 15.

The small overall dimensions of the pelvis, femora, tibiae, and fibulae are harder to explain, since all but the femora are essentially normal in their external and internal morphological configurations. Moreover, even the femora have all the appearance of having normal skeletal tissue, normal trabecular patterns (albeit altered in curvature by the right proximal femoral deformities), and fully normal proximal and distal articular surfaces. There is nothing in these bones to indicate histological malformation, diaphyseal or epiphyseal, thereby eliminating chondrodysplasia punctata (Formicola et al., 2001) as a valid diagnosis of the abnormalities (W. H. McAlister, personal communication).

Any differential diagnosis of these abnormalities must therefore incorporate both the detailed proportional and morphological changes of these bones (and possibly some of the others discussed above) with the apparent histological normality of the skeletal tissue. It must also differentiate between clearly pathological skeletal configurations and morphological patterns that are unusual but nonetheless within the ranges of "normal." This can be combined with the considerations above that many of the abnormalities of Dolní Věstonice 15 can be interpreted either as related to some form of a systemic dysplasia or as isolated lesions, incidental to the dysplasia although possibly secondarily related to it through some increased level of susceptibility or risk. And finally, the interpretation of the set of lesions and form of dysplasia of Dolní Věstonice 15 may be beyond paleopathological diagnosis because of both the plethora of dysplasias known clinically (e.g., Sillence et al., 1979; Wynne-Davies et al., 1985) and the heterogeneous and mosaic nature of many of the abnormalities encountered in living individuals.

Consequently, rather than attempt to identify the specific form of dysplasia of Dolní Věstonice 15 and to construct a scenario that accommodates most of the apparent lesions, we are leaving the ultimate diagnosis open, hopefully with sufficient documentation to allow individuals with the inclination for such pursuits to investigate it further. It may be possible to link the plethora of unusual features (orthodontic problems, arm deformities, small body and leg size, and osteoarthritis and fractures) to a few developmental anomalies. Or we may be left with an individual with a unique combination of the "normal," the "unusual," and the "abnormal."

At the same time, a few points need to be emphasized. First, Dolní Věstonice 15 survived his deformities and an apparently severe childhood stress episode until early adulthood. Second, although we do not know the cause of death, there is no indication that it was related to his deformities, especially given the contemporaneous deaths of the otherwise healthy Dolní Věstonice 13 and 14. Third, all of the developmentally sensitive epiphyseal regions indicate the attainment of a normal bipedal gait with normally functioning lower limbs. Fourth, the level of robusticity of the femoral and tibial diaphyses, lower limb muscular attachment areas, and lower limb muscular power arms—all plastic and all at or above the average levels of strength of Gravettian humans—indicate a level of locomotion and burden carrying similar to those of other members of the social group. And fifth, the localized osteoarthritis of the right shoulder and hand emphasize the active economic participation of this individual in the subsistence activities of the social group. Despite his problems, Dolní Věstonice 15 was a survivor, and his social group accorded him respect in the ritual burial of his remains (Chapter 4).

Dolní Věstonice 16

The elderly male skeleton, Dolní Věstonice 16, preserves major portions of his cranial, mandibular, and appendicular skeleton, plus much of its dentition and alveoli. A pair of lesions of the frontal squamous and a possible fracture of the maxillae (see below) were mentioned by Svoboda and Vlček (1991) and Vlček (1991), but detailed pathological considerations of the dentoalveolar region and of the remainder of the skeleton have not been undertaken. This description therefore details these previously noted lesions and complements them with additional observations.

The Cranium and Mandible

The Dolní Věstonice 16 cranium has a series of minor nondentoalveolar lesions, plus one more significant lesion. The former are described first, and then the more important one is considered.

Neurocranial Modifications

There is an antemortem surface modification located ca. 30 mm anterior to bregma, nearly in the midline (Figure 8.75), which appears to be a healed lesion. It is 17.5 mm anteroposteriorly, ca. 12.5 mm mediolaterally, and 2.0 mm in depth. The actual area of involvement is slightly larger than is apparent without sighting at an oblique angle to the lesion because it is well healed, especially around its peripheral borders. It might represent a healed depressed fracture; however, there are no clear lines indicating tabular fracture, and the lesion does not seem to have extended into the diploë, suggesting a substantial surface trauma.

Approximately 23 mm anterolateral to this lesion there is a smaller lesion ca. 8.5 mm in length and 6.5 mm in width. It is shallower and less well defined, and it is healed. Whether it was a small compression fracture or scrape is unknown, but it was of little consequence.

About 40 mm superior to the right orbital margin on the frontal bone and 19.3 mm medial to the right temporal line, there is a straight anteroposteriorly oriented groove that is 14.5 mm in length. It is located in the general area that frontal grooves, marking the passage of branches of the supraorbital nerve posteriorly into the scalp, are found in varying frequencies. There is no such groove on the left side, although the trait can be unilateral in expression (Hauser & DeStefano, 1989). However, because the trait is frequently found bilaterally, the groove may also represent a healed injury (Figures 8.66 and 8.69). If the latter, it was minor.

In addition, there is a wide, shallow groove coursing along the left superior nuchal line at an oblique angle ca. 20 mm lateral to the midline (Figure 8.71). It is ca. 25 mm in length, deepest anterolaterally, and broadest posteromedially. There is also a swelling along the inferior aspect of the groove. It is possible that the groove is a healed injury across the muscle insertion line.

Much of the frontal bone exhibits very fine surface porosity. This is particularly evident in the supraorbital region, but it extends to much of the frontal squama and the parietals along the midline. It is unclear whether this is the result of postmortem erosion or a degenerative effect such as senile porosity.

Facial Modifications

These minor lesions, or possible lesions, on the neurocranial vault are accompanied by a more marked alteration of the midfacial skeleton (Chapter 8; Figures 8.65 to 8.69). The left lower orbital rim has been markedly altered. Whether this is due to healed trauma or some combination of postmortem deformation and subsequent reconstructive efforts is not grossly obvious. Yet, several aspects of the infraorbital and zygomatic region, as well as mandibular occlusion and tooth wear, provide important clues. Principally, the left maxilla is superiorly positioned at the midline by ca. 4 mm relative to the right maxilla, as well as anteriorly displaced by ca. 4 mm. There are five reasons for suspecting that this is a growth abnormality or, less likely, the result of healed trauma to the maxilla.

First, close inspection of the alveoli between the central incisors indicates that there is fusion of the intermaxillary suture at the most anterior aspect and not a reconnection following fossilization damage. Second, there has been clear remodeling of the anterior nasal spine such that it has become obliquely reoriented. Third, the subnasoalveolar clivus is superoinferiorly shortened with a central margin of the inferior piriform aperture that appears abnormally lowered. Fourth, the two sides of the maxillary dentition in the transverse plane are aligned with completely symmetrical maxillary and mandibular wear along the entire dental arcade (Figures 11.8 and 11.9). This indicates that any misalignment of the maxillae occurred long enough before death for tooth attrition and/or supereruption in the maxillae and mandible to accommodate it. The mandible cannot be completely occluded, although it is quite close, probably because the contact points of the face at both frontozygomatic sutures and the right frontonasal suture are not correct, and both the height and the relative anteroposterior hinging could be off slightly; a minor correction in these defects would produce near perfect dental occlusion. Fifth, the palate of Dolní Věstonice 16 is rather deep.

As a first interpretation, it is possible that Dolní Věstonice 16 had a cleft palate. This would explain the long-term remodeling and symmetrical dental occlusion despite the misaligned maxillae. Possibly consistent with the cleft palate hypothesis is that Dolní Věstonice 16 has a very deep palate compared to the other Dolní Věstonice individuals. Moreover, although damage and restoration prevent confirmation that this is antemortem and not postmortem, the midline of the palate posterior to the incisive foramen is patent and may represent the remnant of an incompletely formed palatal midline (Figure 8.70).

The second possibility for Dolní Věstonice 16's maxillary misalignment is trauma. In this case, the maxillae were forcibly separated along the incompletely fused intermaxillary suture or, more likely, given the early fusion of this suture (about the ninth week in utero), fractured after fusion. If so, it is difficult to see how a traumatic blow sufficient to separate the two maxillae could have occurred without causing serious damage to other fragile bones of the midface. It is possible that the healed bone from such damage has been lost, given the postmortem damage to the facial skeleton and restoration of portions of the facial skeleton (Chapter 8), but further evidence of such trauma would be expected and is not present. Therefore, trauma seems less likely than some form of developmental abnormality, such as cleft palate, to explain the misalignment and associated changes of the midfacial skeleton.

The Alveolar Process and Dentition

Dolní Věstonice 16 shows extensive evidence of localized alveolar bone loss, in contrast with the other specimens preserving the alveolar process (Figures 19.16 and 19.17). It represents an older adult, rather later in middle age than Dolní Věstonice 3 or Pavlov 1 (Chapter 6), and the teeth are a great deal more worn. Little of the crown is left in any tooth anterior of the second molar, and the upper third molars remain minimally worn only because of the impaction and congenital absence of the lower third molars (Chapter 11). Most of the teeth show 3 mm or more of exposed root (measuring between the wear facet and the alveolar crest if the crown has been worn away; Table 19.1). This can be interpreted to a large extent as part of the process of continuous eruption. The overall height of the alveolar process (if both the lingual and buccal sides are considered together) does not seem to have been reduced anywhere, although there has been some localized antemortem loss of bone in places—largely to the buccal side. The buccal plate of cortical bone is thinned through-

Figure 19.16 Dolní Věstonice 16, pathology of the alveolar process in the upper jaw. The anterior part of the alveolar process (left) shows a large cavity to accommodate the root of the upper left first incisor. Next to this is the worn remnant of the upper left second incisor, with its root swollen by hypercementosis, suggesting that the cavity for the first incisor represents bone remodeling around a similarly swollen root. This part of the specimen is heavily reconstructed, but it is apparent that the buccal plate of the alveolar process was already greatly thinned. The upper left first molar (right) has a large periapical cavity in the alveolar process, centered on its mesiobuccal root. This probably represents a granuloma or cyst.

out both alveolar processes, but this is part of the usual remodeling process as the teeth continue to erupt and their sockets gradually migrate upward and mesially through the jaw (Hillson, 2000, 2001).

The roots of the upper incisors were swollen with hypercementosis (an irregular build up of cement common in very worn teeth) because, although the teeth themselves are not preserved, the form of the surviving sockets shows that the bone of the alveolar process has remodeled to accommodate this. Between the canines and incisors, there is some evidence of localized loss of bone to the interdental plates. The alveolar process around them is reconstructed, and it is difficult to record its morphology. However, where the interdental plates are exposed, they are porotic and irregularly broken down. This may represent the effects of periodontal disease. Today, many people suffer from an intermittent inflammation of the gums and tooth-supporting tissues in response to the presence of large bacterial communities in dental plaque. Inflammation of the deeper supporting tissues (periodontitis) causes a loss of attachment of the tooth root in its socket and this, in turn, leads to loss of bone from the crest of the alveolar process (Hillson, 1996, 2000). Initially, it affects particularly the bone lining the tooth sockets so that a trenchlike lesion results, running around the tooth. Later, there is a more general remodeling in which the whole height of the alveolar process is reduced.

In the case of Dolní Věstonice 16, this overall alveolar process reduction does not seem to have occurred, but the small signs of more localized bone loss described above are characteristic of periodontitis. In particular, the lower right canine shows a trenchlike zone of bone loss running around the buccal and distal sides of the tooth. The pattern of bone loss can be scored (Kerr, 1991). The

Figure 19.17 Dolní Věstonice 16, pathology of the alveolar process in the lower jaw. The lower right first molar (left) has a large periapical cavity in the alveolar process, centered on its distal root. All of the incisors (right) also have associated periapical cavities, some of which were filled with glue during reconstruction. All the periapical cavities most likely represent granulomata.

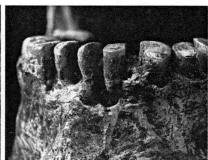

most affected tooth sockets fit Kerr's score of 3 (acute episode of periodontitis) and 4 (a quiescent phase following an acute episode of periodontitis). It is, however, very difficult to record such patterns of bone loss in fossil material, where there are often postmortem changes and the strong continuous eruption associated with rapid wear may mimic the effects of a general loss to the height of the alveolar process.

Several teeth in Dolní Věstonice 16 also show evidence of bone loss deep inside the alveolar process, around the apex of the root. The lower incisors, lower right first molar, and upper left first molar all have bony cavities in the periapical region, exposed by the thinning of the buccal plate described above. Dental radiographs do not show such bony cavities associated with any other teeth, and the defects around the lower left canine and premolars are probably due to gaps left in reconstruction (Chapter 11).

The periapical cavity associated with the upper left first molar is around the mesiobuccal root. It is about 10 mm in diameter, but it is asymmetrically distributed around the root apex, with its radius from the side of the root varying between 3 and 7 mm. In the radiograph, the cavity is difficult to make out because it is obscured by a variation in density caused by the reconstruction. The wall is relatively smooth, and this collection of features suggests it was a granuloma or cyst (Dias & Tayles, 1997; Hillson, 2000). Most small periapical cavities (visible in the living as radiolucencies in radiographs) represent low-level chronic inflammatory lesions known as granulomata (Wood & Goaz, 1997). They are usually painless, and patients are unaware of their existence. Abscesses are much more rarely associated with radiolucencies, and instead make themselves apparent by the drainage of pus. This seems an unlikely diagnosis in this case. Larger radiolucencies, over 15 mm or so in diameter, usually represent noninflammatory lesions called radicular cysts. A cyst is a large fluid-filled cavity, which develops in most cases from a chronic inflammatory lesion, such as a granuloma, that has resolved. The size of Dolní Věstonice 16's upper first molar radiolucency is nearly large enough to be a cyst, but it is more likely to have represented a granuloma.

In the lower right first molar in Dolní Věstonice 16, the periapical cavity is centered on the distal root. It is smaller than the cavity in the upper molar, at around 7 mm in diameter, but it has a similarly smooth bony wall inside. Once again, it probably represents a granuloma, exposed by thinning of the buccal plate during remodeling of the mandible. In addition, there is a large periapical cavity around the roots of the lower left first and second incisors and another around the lower right first and second incisors. Both are smooth-walled and about 7 to 8 mm in maximum diameter, and once more they probably represent granulomata.

In all of these teeth—the upper left first molar, lower right first molar, and all the lower incisors—the pulp must have been exposed to infection. The pulp would have died, and the bacteria, or the products of the pulp inflammation, would have emerged from the apical foramen of the root, to result in inflammation around the apex. The most common cause of such pulp exposure in living people is dental caries, and there is no sign of caries in Dolní Věstonice 16. The occlusal attrition facets of the upper first molar show smooth, unbroken expanses of dentine, with darker patches of secondary dentine where it has filled the root canals. There is little sign of a canal running down the mesiobuccal root in the radiograph, but somehow bacteria or the products of inflammation must have made their way to the apex. In the lower right first molar, the floor of the pulp chamber is open, with narrow apertures leading into the root canals, which can be seen clearly in the radiograph. Here, the pulp must have died before it was filled with secondary dentine, and the exposure must therefore have occurred at an earlier stage in the life history of the tooth. The pulp chamber and root canals are so little occluded by secondary dentine that it must have been at a point when the tooth was relatively little worn. One likely explanation is that there may have been a tooth fracture, as seen in Pavlov 1 (Chapter 11). The radiograph of the lower incisors associated with periapical cavities shows that some retained open root canals, although others did not. Once again, those with open root canals indicate a much earlier exposure of the pulp. The form of the alveolar process in individuals like Dolní Věstonice 16 is therefore a complex mix of several processes—continuous tooth eruption, remodeling to accommodate the changing form of worn tooth crowns, periodontal disease, and periapical inflammation. The form seen at death represents an accumulation of events, some of which occurred at considerably younger ages.

As noted above, there is no evidence of caries in the heavily worn teeth of Dolní Věstonice 16. In addition, any evidence of developmental defects of the enamel, which might have been present in early life, had been worn away long before the death of the individual. The one exception is the upper third molars, which do not show any such defects (Table 19.2).

The Axial Skeleton

The axial skeleton preserves only a series of vertebral elements from C1 to T1, a portion of an eleventh rib, and a dozen larger rib fragments. Yet despite this poor state of preservation, many degenerative changes are evident. In fact, all of the vertebral elements from C4 to T1 have degenerative changes reflective of osteoarthritis. The caudal surface of the C4 body is flattened on the left and has an osteophyte 8.5 mm in length. The caudal surface of the

C5 body has two osteophytic growths along its dorsal margin, and the underside of its anterior process evinces a flattened pseudoarthrosis for the C6 uncinate process. Osteoarthritic bony lips are evident on the left caudal margin and right dorsal cranial margin of the C6 body, and an osteophyte is present on the C6 right inferior articular facet. Last, the body of C7 has osteophytic lipping along its left caudal/dorsal margin, and lateral osteophytic lipping is evident on both the cranial and caudal surfaces of the T1 body. Even though it is not possible to assess these vertebral osteoarthritic degenerations in the context of the entire vertebral column for Dolní Věstonice 16 because of preservation, the occurrence of marked degenerations in C4 to T1 is a common distribution among recent humans (Bridges, 1994; Dawson & Trinkaus, 1997) related to the high level of vertebral mobility in that vertebral column segment (Levangie & Norkin, 2001).

The Appendicular Skeleton: Articular Alterations

Even though the appendicular skeleton of Dolní Věstonice 16 retains major portions of all of the long bones and a number of hand and foot bones, many of the long bone epiphyses are heavily damaged, and there was postmortem erosion to some of the surfaces, especially the fragile cortical bone overlying trabeculae in the articulations and around the epiphyses. However, a few observations can be made about his appendicular articular health.

Despite the relatively advanced age of Dolní Věstonice 16, there is remarkably little evidence for degeneration of the appendicular articular subchondral bone on those surfaces that are preserved. There are some changes of the acetabular lunate surface fragments, which involve increased porosity of the subchondral bone. Yet, the bone has also undergone minor postmortem alteration, and it is possible that these changes do not reflect antemortem degeneration of the surfaces. The porosity is not associated with new bone deposition, altered curvature of the articulations, or polishing of the surfaces. Therefore, although mention of it as a potential minor alteration in a primary weight-bearing location is appropriate, it appears unlikely that it reflects actual antemortem degeneration.

In contrast, both of the trapezioscaphoid articulations exhibit marked osteoarthritis (Figure 19.18). This is evident in matching altered facets on the right scaphoid and trapezium bones and in a similar one on the left scaphoid bone (the left trapezium is not preserved). The degenerations consist of extensive erosion of the surfaces, such that there is marked porosity, exposure of underlying trabeculae, and polishing (eburnation) across most of the altered facets (the last is more apparent on the left because of better bone preservation). The altered facets on the scaphoid bones are at distinct angles to the trapezoid facets, producing a clear angulation ridge running dorsopalmarly across the distal bone. The facets remain biconvex but are altered from their original contours. The matching facet on the right trapezium is concave dorsopalmarly with a sharp angle between it and the trapezoid facet. There is similarly even porosity with eburnation across the scaphoid facet of the more distal carpal bone.

This bilateral alteration of the trapezioscaphoid facets would normally be seen as one consequence of pollically related osteoarthritis, a common phenomenon in older recent humans. However, these are the only facets in all of the Dolní Věstonice 16 hand remains that exhibit any degeneration. In particular, all of the other intercarpal facets of the scaphoid bones and of the right trapezium are completely normal. There is no alteration of the shape or surface texture of the right first carpometacarpal articulation, the most common site of hand osteoarthritic degeneration (Brunelli, 1999). The metacarpophalangeal and interphalangeal articulations of the pollex are normal. Twenty-three of the right and left ulnar carpometacarpal, metacarpophalangeal, proximal interphalangeal, and distal interphalangeal articulations are sufficiently preserved on at least one surface (out of a possible thirty-two from both hands, or 71.9% of the originally present ulnar finger facets), and all of them are completely normal. In addition, both lunate bones, one trapezoid bone, both capitate bones, and both hamate bones are largely present, and none of them exhibits any antemortem degenerations.

Figure 19.18 From left to right, the Dolní Věstonice 16 left scaphoid bone, right scaphoid bone, and right trapezium bone, illustrating the marked osteoarthritic degeneration of the trapezioscaphoid facets with bone loss, porosity, and eburnation. Scale in millimeters.

The left radius retains its carpal facets, and these are also normal. Indeed, the only other pathological alteration on any of these bones is a small osteophyte on the right scaphoid bone along the dorsal margin of the capitate facet, but it did not affect the facet proper. Consequently, the trapezioscaphoid facets of Dolní Věstonice 16 are very unusual in exhibiting marked degeneration in the context of wrist and hand articulations that are completely normal.

Given the complex mechanical interrelationships between the various intercarpal and other hand articulations, it is unlikely that one would get marked, bilaterally symmetrical degeneration of the trapezioscaphoid facets without any osteological evidence of degeneration on the adjacent articulations. These changes are therefore not referable simply to mechanical overloading in the context of manipulative behaviors, combined with accumulated damage over the lifetime of the individual. Moreover, Dolní Věstonice 16 exhibits relatively elevated humeral diaphyseal asymmetry, with the right side more robust (Table 16.3); its asymmetry value of 9.55 is close to a Gravettian average of 10.32, below those of Dolní Věstonice 13 and 14, above that of Dolní Věstonice 3, and similar to that of Pavlov 1. It is also slightly above the maximum value for a large mixed sample of recent humans. Consequently, Dolní Věstonice 16 was distinctly right-handed, and any normal biomechanical overloading of the hand articulations would be expected to differentially affect the right hand.

An alternative etiology is a congenital, bilaterally symmetrical laxity of the trapezioscaphoid articulation (V. Glogovac, personal communication). In this interpretation, laxity of the articulations would have promoted abnormal loading of the articular surfaces, leading to the degenerations. Therefore, these changes in the Dolní Věstonice 16 hand remains should not be seen as part of normal wear-and-tear degeneration from activity levels but as a consequence of a congenital abnormality that predisposed him to this probably seriously painful degeneration.

The Appendicular Skeleton: Muscular Attachment Alterations

Each proximal ulna of Dolní Věstonice 16 presents a small but distinct enthesopathy for the dorsal insertion of the triceps brachii tendon on the dorsal olecranon process (Figure 16.43). There are two possible interpretations of these tendinous ossifications. They could result from localized, mechanically based tendon irritation, or they could be part of a systemic disorder that occurs mostly in older individuals, hyperostotic disease or diffuse idiopathic skeletal hyperostosis (DISH) (Arlet & Mazières, 1985; Utsinger, 1985; Crubézy & Trinkaus, 1992). Hyperostotic disease characteristically involves bridging ossifications between vertebral bodies combined with the bilateral formation of enthesopathies minimally at the insertions for triceps brachii on the ulnae, quadriceps femoris on the patellae, and triceps surae on the calcanei. The vertebral bodies of Dolní Věstonice 16 are not sufficiently preserved to determine whether any bridging ossifications were present, but the left patella lacks any enthesopathy, the right calcaneal tuberosity lacks enthesopathies (even though the tuberosity was damaged plantarly), and the only other osteophyte on the appendicular remains is the small one adjacent to the capitate facet on the right scaphoid bone. Despite the incompleteness of some critical portions of the skeleton, the absence of evidence in this case appears to support the interpretation of a localized irregularity.

Combined with these ulnar ossifications, each radius exhibits a patch of new bone on the dorsal side of each interosseous crest in the resultant sulcus. The right one is 20.5 mm long and the left one is 11.0 mm long. They involve new bone laid down within the sulcus and do not extend to the margins of the interosseous crests. The area of involvement is approximately the origin of abductor pollicis longus. Curiously, as with the trapezioscaphoid osteoarthritis and the ulnar enthesopathies, the alteration is largely bilaterally symmetrical, despite the marked asymmetry evident in the humeral diaphyses.

The Appendicular Skeleton: Transverse Lines

Only a minority of the metaphyseal regions of the Dolní Věstonice 16 limb bones are sufficiently intact to indicate whether transverse lines were present (Table 19.3), and only one of them, the right first metatarsal, has any indication of lines. That one is 19.1 mm from the proximal articular surface, and it is modest.

Summary

Dolní Věstonice 16, as the oldest individual in the Pavlovian sample, has the greatest accumulation of minor lesions, along with a couple of unusual ones. The neurocranial vault lesions are more of the minor alterations from subcutaneous bone disruptions. The dentoalveolar lesions appear to be entirely due to the advanced degree of occlusal wear of the teeth, and the vertebral alterations are similarly age-related. It is also likely that the ulnar enthesopathies, and possibly the irritations of the radial diaphyses, are a combination of age and activity levels. The unusual features are the facial asymmetry, most likely due to a developmental abnormality but possibly to an infantile fracture, and the bilateral trapezioscaphoid osteoarthritis, apparently due to idiopathic joint laxity.

Dolní Věstonice 17

The fragmentary pieces of immature parietal bone that make up Dolní Věstonice 17 exhibit no pathological lesions.

Dolní Věstonice 36

There are no pathological lesions apparent in the associated immature dentition of Dolní Věstonice 36.

Dolní Věstonice Isolated Dental Remains: Dolní Věstonice 31–33, 37, and 38

The isolated teeth of Dolní Věstonice 33 and 37 have furrow-form developmental enamel defects (Table 19.2). Dolní Věstonice 33 is an anomalous tooth of uncertain identification, so it is not possible to estimate an age for the growth disruption. The defect on the lower second molar of Dolní Věstonice 37 is in the cervical third of the crown, suggesting an age of 5 to 6 years for the growth disruption (Hillson, 1996). The lower third molar of Dolní Věstonice 38 exhibits hypercementosis of the roots, associated with occlusal attrition (Chapter 11), but it does not otherwise exhibit pathological lesions. The Dolní Věstonice 31 and 32 teeth do not exhibit any abnormalities.

Dolní Věstonice Isolated Postcranial Remains: Dolní Věstonice 34, 35, and 39–52

The isolated rib fragments; humeral, femoral, and fibular diaphyses; proximal radius, manual phalanges, and pedal bones that derive from the various localities of Dolní Věstonice I and II (Dolní Věstonice 34, 35, and 39–52) present no pathological lesions. This may not be surprising for the various nonarticular pieces present (all but the hand and foot remains), but the absence of any degenerative lesions on the hand and foot remains is worthy of note.

Pavlov 1

Even though the Pavlov 1 associated skeleton has been described in some detail (Vlček, 1997), the only mentions of pathological lesions concern the dentition and the associated alveoli. These are reconsidered here, along with paleopathological assessment of the other preserved elements.

The Cranium and Mandible

In the cranium, there is a small depressed area ca. 9 mm in diameter on the external posteroinferior aspect of the right parietal bone, with the center 12 mm from the lambdoid suture (Figure 8.81). There is growth of small bony spicules within the depression. The center of the lesion is ca. 60 mm from lambda. Also, on the right parietal, there is a broad depression ca. 10 to 15 mm in diameter, centered 20 mm from the coronal suture, 54 mm from the sagittal suture, and 26 mm above the temporal line. It is uncertain whether this lesion represents a long-since healed minor trauma to the external table, a secondary effect of the swellings along the sutures (see below), or normal variation of cranial contours.

There is an oblique line 60 mm long and less than 5 mm wide across the external left parietal bone. The anterior end is 16.5 mm from the coronal suture and 43.5 mm from the sagittal suture; the posterior end is 67 mm from the coronal suture and 78 mm from the sagittal suture. The posteroinferior end abuts against the temporal line, and it has slightly raised bone. The anterior portion is more of a depressed groove, and it is probably a long-since healed minor scalp wound.

There is a depressed area on the posterior shoulder of the articular eminence of the left temporal bone, covering its lateral two-thirds and bordered anteriorly by a vascular groove (Figure 8.82). There is minimal alteration of the subchondral bone, but there is a distinct, raised medial margin for the area. The width of the depressed area (13.3 mm) is little more than half the breadth of the articular eminence (24.0 mm). This change is associated with two depressions on the surface of the left mandibular condyle due to bone remodeling but not active osteoarthritis of the joint. The medial depression is 5.6 mm high and 10.6 mm wide, whereas the lateral one is 4.6 mm by 6.4 mm. These associated temporomandibular joint alterations represent more of a remodeling process, probably secondary to dental attrition, rather than active degeneration.

In addition to the above lesions, there is a series of swellings on the external (but not internal) surfaces of the calvarium, mostly along the sutures. It is unclear whether these are pathological in nature or the result of senile changes in an older individual. In addition to the comments on the external surfaces noted below, internal/external radiographs show slight increases in density along the swellings but no other obvious changes.

The first swelling is along the area where the metopic suture would have been located. The swelling starts 41 mm above nasion and gradually expands to a breadth of 48 mm ca. 20 mm anterior of bregma. This swelling is sufficient to create visible sulci bilaterally along its margins. Along the left coronal suture, there is a very small to questionably present swelling ca. 48 mm from bregma. On the right side, there is a prominent swelling up to 38 to 40 mm wide anteroposteriorly, extending ca. 38 mm from bregma. It then tapers off ca. 55 mm from bregma along the right coronal suture. Bilateral swelling is evident all along the sagittal suture (Figure 8.85), especially between 7 mm and 42 mm from bregma, where it reaches a maximum breadth of 24 mm. It then decreases to a more modest breadth of ca. 18 mm until close to lambda. Along the last 29 mm before lambda, there are separate right and left swellings, 11 mm wide on the right, 9 mm wide on the left, with a maximum breadth across the two of ca. 29 mm.

The pattern of swelling along the lambdoid suture is slightly distorted by postmortem separation along the suture and reconstruction with filler, but the effect on the pattern is small. There is a posterior displacement of the occipital relative to the parietal bones (also seen in Dolní Věstonice 16), with bony swellings along the medial halves of the two sides of the suture. The right side extends ca. 56 mm from lambda, reaching 25 mm wide, with clear sulci above and below the swelling. The left side extends ca. 60 mm from lambda and reaches 23 mm wide, with sulci on both sides. The swellings are mostly on the occipital bones, but they are also present on the parietal sides of the suture.

The mandible of Pavlov 1 has small lingual alveolar exostoses, involving two small bumps by the left second molar, one by the left first molar distal root, and one each by the left first premolar and canine. There are none on the right.

The Alveolar Process and Dentition

Pavlov 1 was an older middle-aged adult (Chapter 6), with relatively marked occlusal dental attrition and somewhat more root exposure (up to 4 mm) than in the younger individuals (Table 19.1). However, the limited preservation of the alveolar process makes full assessment of its status difficult.

The most likely candidate for a dental cavity is in the central fossa of the occlusal surface of the lower left third molar (Figure 19.19). This has a slight brown stain, which is often associated with the development of caries in the occlusal fissures of molars. A cavity in the occlusal surface would usually be regarded as a well-developed lesion, which would be likely to affect the underlying dentine (Hillson, 2001). Thus a radiolucency representing dentine demineralization might be expected, but none is visible in the radiograph of this tooth, thereby making the stain likely to be due to processes other than dental caries.

There is a furrow-form developmental dental enamel defect on the preserved portion of the lower right canine, indicating a formation at about 4 years of age. There are no such defects on the other remaining teeth, but most of them are too worn to provide information (Table 19.8).

Figure 19.19 Pavlov 1, lower left third molar. There is a large, dark-stained cavity at the center of the fissures in the occlusal surface. This may represent a carious lesion, but it is not associated with a radiolucency in the radiograph of this tooth.

The Axial Skeleton

No pathological conditions are noted on the fragmentary thoracic vertebrae and isolated rib fragment belonging to Pavlov 1.

The Appendicular Skeleton

The Pavlov 1 skeleton retains only a few of the appendicular articulations (both scapular glenoid fossae, one humeral head, one distal humerus, both ulnar trochlea, one distal radius, one proximal metacarpal, and two distal femora) and a number of diaphyses. However, despite the relatively advanced age at death of this individual, the appendicular remains are almost devoid of lesions. The only possible lesions are irregular ossifications along the glenohumeral articular margins (Chapter 16). The ventral margin of the right scapular glenoid fossa has a pair of protuberances, a smaller one caudally and a larger one cranially, that produce an irregular margin that would have extended into the glenoid labrum (Figure 16.17). On the inferior medial right humeral head there is a raised area 5.7 mm wide across 25.0 mm (Figure 16.25); it is subchondral but apparently not articular, since an unusual degree of adduction of the joint would be necessary to bring it into contact with the inferior glenoid surface. Both of these changes may be related to some minor, but not pathological, alterations of the glenoid labrum and articular capsule.

Table 19.8 Distribution of dental enamel hypoplasias, or developmental defects of the dental enamel, in the Pavlov permanent tooth sample.

	Pav 1		Pav 2		Pav 3	Pav 5		Pav 20	Pav 22	Pav 23	Pav 24	Pav 25	Pav 28
	R	L	R	L	R	R	L	L	L?	L	R	R	L
Upper													
I1	—												
I2									Pres				
C	—	—											
P1	—	—	Abs										
P2	—	—	Abs	Abs									
M1	—	—	Abs	Abs									
M2	—	—	Pres	Pres									
M3								Abs					
Lower													
I1						Abs	Abs			Pres	Abs	Abs	
I2	—	—											
C	Pres					Pres							
P1		—				Pres							
P2	—	—				Pres							
M1	—	—				Abs							
M2	Abs	Abs											
M3		Abs											Abs

Lesions are listed only as present (Pres) or absent (Abs) on sufficiently complete crowns; — indicates an insufficiently intact crown, because of either attrition or damage, or one that is sufficiently embedded in the alveolus to prevent proper observations; a blank indicates absence of the tooth. For details on the lesions, see discussions for the individual specimens.

Most of the Pavlov 1 long bones are insufficiently preserved to permit assessment of transverse lines. However, even though the distal humeri, proximal ulnae, and distal radii do not exhibit lines, the right proximal radius shows two partially bridging lines at the level of the radial tuberosity, which are interpreted as remnants of transverse lines from the radial head metaphysis (Table 19.3). They are located 27.5 mm and 32.5 mm from the middle of the head articular surface.

Summary

Pavlov 1, as with most of the other Pavlovian individuals, exhibits several minor neurocranial vault traumatic lesions. However, despite his middle age at death, he appears to have had little else in the way of pathological lesions. The dentoalveolar and temporomandibular changes appear to be secondary to dental occlusal wear. The curious swellings on the neurocranium may be related to developmental or aging processes along the sutures, and hence are not pathological. And the bony spurs around the right glenohumeral joint do not appear to have affected the articulation proper.

Pavlov 2

The prominent maxillary torus, or alveolar exostosis, of Pavlov 2 is discussed in Chapters 8 and 11 and by Vlček (1997). It is strongly developed from the mid–first molar, is largest by the second molar, and then continues past the absent third molar to the retromolar alveolus (Figure 19.20). Its maximum superoinferior height is 7.4 mm by the second molar, and it extends ca. 2.5 mm lingually out from the side of the alveolar process. It is unusual in being lingual since maxillary tori are normally buccal, but otherwise it is not exceptional.

This growth is accompanied by only 1 to 2 mm of exposed root, where it is possible to measure the teeth (Table 19.1). Since Pavlov 2 probably died in early middle age (Chapter 6), continuous eruption is unlikely to have been advanced.

There is no evidence for dental caries in the Pavlov 2 teeth. There is a slight furrow-form developmental defect on one part of the cervix of the upper right first molar but not on the upper left first molar (Table 19.8). Similarly, a small furrow-form defect on the upper left second molar is not matched on the upper right second molar. This

Skeletal and Dental Paleopathology

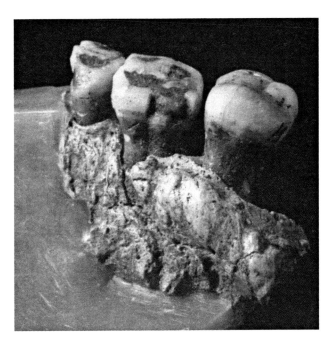

Figure 19.20 Pavlov 2, alveolar torus on the lingual side of the alveolar process, along the molar tooth row.

might be taken as evidence that the two halves of the dentition are not, in fact, from the same individual, but they are so slightly developed that this is a difficult case to argue.

Pavlov 3

The Pavlov 3 mandibular piece has 2 to 3 mm of exposed root, once the apparent postmortem displacement of teeth from their sockets is taken into account (Table 19.1). This is not unusual for the degree of dental attrition present on the teeth. There are no dental caries on the teeth, but fine furrow-form defects are present in the cervical part of the canine and premolars (Table 19.8). These are consistent with a growth disruption between 4 and 5 years of age.

Pavlov 4

The Pavlov 4 mandibular fragment is too incomplete to provide paleopathological information. The minimally preserved portions appear normal.

Pavlov Isolated Dental Remains: Pavlov 5–28

The twenty-six teeth represented by Pavlov 5 to 28 are largely free of pathological lesions (Tables 19.8 and 19.9).

Three of the teeth, Pavlov 21, 22, and 28, exhibit varying degrees of hypercementosis of the roots. Two of them have developmental defects of the enamel. Pavlov 22, a permanent upper second incisor, has a marked furrow-form defect of enamel hypoplasia in the cervical third of the crown, corresponding to a growth disruption between 4 and 5 years of age (Reid & Dean, 2000). Pavlov 23, a permanent lower first incisor, exhibits a defect in the cervical half of the crown that occurred between 2 and 4 years of age (Reid & Dean, 2000).

Summary and Interpretation

Alveolar Process and Dental Abnormalities

The Dolní Věstonice and Pavlov dental and associated alveolar remains reveal a generally low level of lesions, especially once the probable ages at death, degrees of dental attrition, and systemic abnormalities of Dolní Věstonice 15 are taken into account.

There are no confirmed cases of dental caries. The few suggestions of caries are not confirmed with radiographic examination, and postmortem staining is a more likely explanation for the features observed. This pattern fits with observed frequencies of caries in Late Pleistocene European samples. Among earlier Upper Paleolithic Europeans, they have been identified (but not radiographically confirmed) only in Cro-Magnon 4, Les Rois R50-4, and Les Rois R51-15 (Brennan, 1991; Trinkaus et al., 2000b); they are known from only six Middle Paleolithic Eurasian humans (Sognnaes, 1956; Boydstun et al., 1988; Lalueza et al., 1993; Tillier et al., 1995; Trinkaus et al., 2000b; Lebel & Trinkaus, 2002a), and in a more recent European foraging sample they remain at very lower percentages (Brennan, 1991), similar to those of recent arctic populations (Pedersen, 1966; Mayhall, 1978; Costa, 1980). Given the likely diets high in animal products, as indicated by the faunal remains (Musil, 1994, 1997, 2002; West, 2001), general environmental reconstructions (Svobodová, 1991a; Svoboda et al., 2002a), and the stable isotope analyses of European Gravettian humans (Richards et al., 2001), this low incidence of dental caries should not be surprising.

Similarly, with the exception of Dolní Věstonice 16 and the alveolar process abnormalities associated with his advanced dental attrition, the level of alveolar abnormalities of these individuals is relatively low. What root exposure is present, as indicated by the heights between the cement/enamel junctions and the alveolar crests (Table 19.1), is appropriate for the degrees of dental occlusal attrition evident on the teeth.

Table 19.9 Distribution of dental enamel hypoplasias, or developmental defects of the dental enamel, in the Dolní Věstonice and Pavlov deciduous tooth sample.

	DV 36		Pav 6	Pav 7	Pav 8	Pav 9	Pav 11	Pav 16	Pav 17	Pav 18	Pav 19
	R	L	L	L	R	R	R	R	R	L	L
Upper											
di1	Abs										Abs
di2								Abs			
dc											
dm1		Abs	—			Abs					
dm2		Abs	Abs								
Lower											
di1	Abs										
di2									Abs	Abs	
dc											
dm1											
dm2		Abs		Abs	Abs	Abs					

Lesions are listed only as present (Pres) or absent (Abs) on sufficiently complete crowns; — indicates an insufficiently intact crown, because of either attrition or damage, or one that is sufficiently embedded in the alveolus to prevent proper observations; a blank indicates absence of the tooth. For details on the lesions, see discussions for the individual specimens. Dolní Věstonice 27 and Pavlov 10 and 12–15 are not included because of preservation.

Developmental Stress Indicators

The principal indicator of periods of developmental systemic stress in these human remains is developmental defects of the dental enamel, or dental enamel hypoplasias. The Dolní Věstonice and Pavlov remains exhibit a moderate frequency of such lesions, since they are present on at least one tooth of each of Dolní Věstonice 3, 13–15, 33, and 37 and on Pavlov 1–3, 22, and 23 (Tables 19.2 and 19.8). However, there are none on the numerous deciduous teeth of Dolní Věstonice 36 and from Pavlov (Table 19.9), and almost all of those present on the permanent teeth are very minor. The one exception to this pattern is the marked defects on all four first molars of Dolní Věstonice 15, which indicate a marked stress episode and should be related to the pronounced developmental abnormalities of that individual.

Almost all of the developmental defects of the enamel appear to have occurred at the age of 4 to 5 years postnatal (Dolní Věstonice 3, 13, 15, and 37 and Pavlov 1–3 and 22). A couple of them occurred slightly earlier (Dolní Věstonice 14 and Pavlov 23), and the marked defects on the first molars of Dolní Věstonice 15 formed at about 2 years of age. It is tempting to attribute the defects that formed between the ages of about 3 to 5 years to the dietary stress associated with weaning, given the three to four years of nursing common in recent traditional societies and probably longer in mobile foraging populations (Dettwyler, 1995). The shift to a variable supply of foraged food, without the body reserves to effectively bridge over lean periods, could induce such minor developmental lesions. However, there is a large number of teeth calcifying during this time period (incisors, canines, premolars, and second molars), and hence there may simply be a higher probability that one of them will exhibit an enamel defect from whatever stress episodes may occur during this developmental stage. Yet, whether these defects are related to weaning stress or fluctuating stress during development, they indicate both a relatively consistent pattern of stress during this time period among these Pavlovian people and, given the modest development of most of the defects, few serious episodes (with the exception of the first molar one of Dolní Věstonice 15). This pattern is common among recent human groups.

Systematic data on dental hypoplasias for large samples of European earlier Upper Paleolithic humans are not available, but the few reports (e.g., Brennan, 1991; Trinkaus et al., 2002a) available, as well as personal observations, indicate a generally low level of such lesions and a predominance of relatively minor defects (although this observation would be accentuated by the heavy wear). Interestingly, only one of the Dolní Věstonice or Pavlov third molars has a defect (Tables 19.2 and 19.8), indicating that the periods of stress that are recorded are mostly during the early to middle of the first decade of life. This contrasts with the pattern seen in the Neandertals, in which the incidence of defects increases steadily with the later-forming molars (Ogilvie et al., 1989).

This generally low level of developmental defects of the enamel correlates with the rarity of transverse (Harris) lines on the long bones, metacarpals, and metatarsals (Table 19.3) The older ages at death of three individuals, Dolní Věstonice 3 and 16 and Pavlov 1, may partially account for the dearth of such lines on these remains, since those that occurred earlier in development are likely to have been remodeled away during later development and adulthood. However, minor defects evident as partial bridging of the medullary cavity or transverse densities in the trabecular bone, all assessed radiographically, are present on the first metatarsals of Dolní Věstonice 14–16, on the proximal tibiae of Dolní Věstonice 15, and on the proximal radius of Pavlov 1. Those near the middles of the Dolní Věstonice 15 and 16 first metatarsals probably occurred earlier in development, but the others appear as indications of modest stress during later development.

Developmental Abnormalities

The Dolní Věstonice and Pavlov sample may have a relatively elevated level of developmental abnormalities, depending in part on the interpretation of several features. Dolní Věstonice 15 was affected during development by a form of dysplasia, even if a secure diagnosis of the form of the disorder cannot be easily provided. Dolní Věstonice 3, if her posterior cranial fossa asymmetry is not due to postmortem deformation, could have had altered neurological developmental patterns and an associated growth response of the occipital bone; this might be considered a developmental abnormality. The facial asymmetry of Dolní Věstonice 16 could have been due to trauma or, alternatively and more likely, to some form of facial growth disruption similar to cleft palate. And Pavlov 1 exhibits curious swellings along several of the sutures, which could have been developmental in nature.

The last three cases, if indeed they are considered as developmental abnormalities, may have had little effect on the survivability of the individuals. However, the severity of the growth arrest of Dolní Věstonice 15, evident in the first molar enamel defects and probably in his overall stature, suggests considerable social input to get him through the crisis.

Evidence of Trauma

At the same time, there is almost ubiquitous evidence of traumatic injury among these people. Only the late adolescent remains of Dolní Věstonice 14 do not have any traumatic lesions. All of the others that are sufficiently complete, Dolní Věstonice 3, 11/12, 13, 15, and 16, plus Pavlov 1, have healed minor injuries to the cranial vault. Most of them are probably the result of surface disruptions of the scalp and pericranium with associated bony reactions of the external table and the diploë. A couple of them might reflect minor impacts on the vault. Such minor head injuries are common among Pleistocene archaic *Homo* remains (Berger & Trinkaus, 1995, and references therein), and they are present although never systematically recorded among earlier Upper Paleolithic modern humans (e.g., Szombathy, 1925; Matiegka, 1934; Trinkaus et al., 2002a). It is possible that they are simply a reflection of a generally elevated risk of minor superficial injuries to these Pleistocene foraging populations. In this, minor injuries to the trunk and limbs would largely remain within the soft tissue and leave no paleontological trace, whereas those to the scalp would result in skeletal lesions because of the subcutaneous proximity of the pericranium and the associated external cranial tables.

At the same time, two of the Dolní Věstonice individuals, 3 and 11/12, sustained serious facial traumatic injuries. Dolní Věstonice 3 had the left side of her face rearranged, and Dolní Věstonice 11/12 had his left forehead pushed in. Clearly they collided with resistant objects with considerable force. As mentioned above, it is always difficult to determine whether such injuries are the result of accidents, retaliation by injured prey animals, or interpersonal violence. And in the last case, it is only readily apparent when the lesion could only have been inflicted by a distinctively human object, as in the case of Saint-Césaire 1 (Zollikofer et al., 2002), or a portion of the weapon is retained within the lesion, as in the case of Fanciulli 2 (Henry-Gambier, 2001). Moreover, if interhuman violence was involved, it remains indeterminate whether it was the result of a short-term disagreement, a premeditated act, or an accident with a weapon. Yet, all of these individuals survived and assumed normal function. That survival, as with the infantile difficulties of Dolní Věstonice 15, is likely to have invoked some social support, although the degree in which it occurred is uncertain.

Infection

The only clear case of an infection, other than those possibly associated with the alveolar alterations of Dolní Věstonice 16, is the left auditory canal otitis media of Dolní Věstonice 14 (Chapter 9). Despite the abundance of trauma and other abnormalities, infection affecting the skeleton appears to have been rare among these Gravettian people.

Summary

The human remains from Dolní Věstonice I and II and Pavlov I exhibit a suite of pathological disorders and morphological variants that push the limits of "normal." In addition to the considerations of their comparative morphology presented in Chapters 8 to 18, these assessments of their lesions help to fill out their paleobiology, providing a window on their developmental and degenerative stress levels, their susceptibility to injury, their idiopathic variations, and their abilities to survive considerable strains. It is indeed their survivability that has provided us with this perspective on their biology and behavior.

20

The Paleobiology of the Pavlovian People

Erik Trinkaus and Jiří A. Svoboda

These paleoanthropological considerations of the human remains from the sites of Dolní Věstonice I and II and Pavlov I, in their archeological and taphonomic contexts, provide a modest window into the biocultural patterns of these Pavlovian or central European early Gravettian human populations. Even though the sample size remains small and a number of the elements sustained significant postmortem modification, their detailed analysis and the ongoing analyses of their associated archeological remains make it possible to make a few observations concerning these populations.

Given their proximity in time and, especially, space, it is likely that these human remains are a sampling of a regional lineage of early Gravettian human populations occupying and exploiting the central European landscape between approximately 27,000 and 25,000 radiocarbon years B.P., with the majority of the more complete remains from burials dating to between 26,000 and 25,000 years B.P. (Chapter 3). Only the direct date on the isolated and unprovenienced Dolní Věstonice 35 femur falls outside of these ranges, and it remains uncertain to what extent that later date is due to contamination. The populations were generally very mobile but, as the evidence of large settlements shows, were also suited to periods of sedentary life. They are represented by the human remains from the sites on the Pavlovské Hills and from Grub/Kranawetberg and Předmostí and by the slightly more recent remains from Brno-Francouzská, Willendorf, and perhaps Dolní Věstonice 35. The Dolní Věstonice and Pavlov sample will therefore be treated as representing one regional lineage over a relatively short period of human evolutionary time.

Variability

From a human skeletal and dental morphological perspective, one of the more striking aspects of these remains is their degree of variability. Some of this variability is due to sexual differences between the female Dolní Věstonice 3 and the male Pavlov 1 and Dolní Věstonice 11/12–16. Some of the variability can also be seen as that between the pathological Dolní Věstonice 15 individual and the rest of the sample, but the previous and following assessments of this variability take that into account. There are nonetheless a number of aspects of their morphological range that cannot be explained on the basis of sexual dimorphism and/or pathology.

There is considerable variation in adult stature among the male remains, with Dolní Věstonice 13 being rather small and Dolní Věstonice 14 (plus Dolní Věstonice 35) being among the tallest known Gravettian specimens (Chapter 13). Similarly, even though all of these remains exhibit relatively high crural indices and modest body breadths, among the two nonpathological individuals preserving bi-iliac breadth (Dolní Věstonice 13 and 14) the latter is markedly linear whereas the former is relatively broad for a Gravettian human (Chapter 12).

In male cranial morphology, three of the individuals, Dolní Věstonice 11/12 and 16 and, especially, Pavlov 1, possess strong development of cranial superstructures, particularly in the supraorbital region (Chapter 8). In addition, they exhibit a general hypertrophy of neurocranial features. Yet, whereas Dolní Věstonice 16 has strong development of the facial skeleton, Pavlov 1 does not particularly (Chapters 8 and 10). At the same time, the three male crania from the triple burial, Dolní Věstonice 13–15, all have comparatively gracile crania (Chapter 8), even though their features are overwhelmingly "male" (Chapter 7). If found isolated, those three crania might be diagnosed as female, particularly in the context of the other three male crania from these sites and of the robust male crania from Brno-Francouzská, Cro-Magnon, and Předmostí and from the earlier sites of Mladeč and Oase. There is an age at death difference between these more gracile crania from the triple burial and the three more robust crania (Chapter 6), but most of these features are established in their mature morphology by the end of adolescence, and it is unlikely that age alone explains the degree of variability.

In addition to contrasts in skull hypertrophy, there is considerable variation among the remains in midsagittal neurocranial curvature, development of an occipital bun, foramen magnum shape, infraorbital concavity, and the degree of separation of the supraorbital elements.

There is little in the absolute dimensions of the dental remains or their proportions along the dental arcades that would not be expected in a normal population, but there is a noticeable amount of variation in the anterior dentitions of those individuals that preserve substantial portions of the crowns. In particular, the incisors and canines of Dolní Věstonice 13, Pavlov 5, and particularly Dolní Věstonice 14 exhibit essentially no shoveling, whereas Dolní Věstonice 15 possesses a moderately marked development of these features (Chapter 11). There is also a fair amount of variability in discrete traits of the molars, although it tends to follow a scattered pattern across the individual teeth and associated dentitions.

Perhaps more noticeable in the dental variation is the presence of anomalous and/or supernumerary teeth, including the one below the lower premolars of Dolní Věstonice 15 and the unusual Dolní Věstonice 33 and Pavlov 21 molars (Chapter 11). In addition, the Dolní Věstonice 3 and 16 mandibles exhibit congenital absence of their lower left third molars, and two isolated anterior deciduous teeth, Pavlov 15 and 19, may indicate congenital absence (or anomalous eruption position) of their permanent successors by their degrees of wear and absence of root resorption.

There are also a few nonmetric features of the postcrania that deserve note in this context. The axillary borders of the scapular show marked variation, with the full range from the dorsal sulcus configuration to the ventral sulcus being present (Chapter 16). Moreover, two of the individuals, Dolní Věstonice 3 and 14, have asymmetrical axillary borders. Although all of them are modest in size, there is considerable variation in the pattern of their pectoralis major tuberosities on their proximal humeri (Chapter 16).

Most of the patellae follow normal recent human patterns of articular form, but those of Dolní Věstonice 15 have an unusual sulcus that runs vertically along the medial margin of each medial articular facet (Chapter 18). Furthermore, the talar trochlear and malleolar facets on the distal tibiae of Dolní Věstonice 15 are almost completely separate. It is possible that these lower limb anomalies are related to his dysplasia, but the absence of functionally related deformities distal of the midfemoral diaphyses implies that they are more likely to be the products of normal variation.

Even though these Pavlovian human remains exhibit these various aspects of skeletal and dental variation, there is at least one unusual feature that they seem to share. This feature is the marked angulation of the dorsal scapular spine relative to its border with the supraspinatus/infraspinatus surface (Chapter 16). This arrangement is manifest mostly as a caudal displacement of the vertebral end of the dorsal scapular spine. It is present on all of the sufficiently complete scapulae from this sample, and it appears to be absent or rare (and less pronounced when present) in other Gravettian samples. Moreover, the three individuals from the triple burial exhibit a distinctive arrangement of the infraorbital area (Chapter 8), a pattern that seems to morphologically link them together.

Phylogenetic Issues

These comments concerning morphological variability implicitly assume that these traits are epigenetic, or have a strong genetic component in their development. Traits that are known to be developmentally or degeneratively plastic have not been considered. By the same criteria, it should be possible to employ morphological aspects of the Dolní Věstonice and Pavlov human remains to assess their phylogenetic affinities (despite the caveats presented in Chapter 1).

By any criteria, it is apparent that these Pavlovian humans are "modern humans." In other words, their overall morphological gestalt is principally that evident among recent (Holocene) human populations, and there are few, if any, of their characteristics that are not readily observed in the skeletal and dental remains of these recent human populations. At the same time, some of their features—particularly as related to craniofacial robusticity, overall mandibular form and symphyseal morphology, some aspects of dental morphology, and body proportions—are more commonly found among roughly contemporaneous earlier Upper Paleolithic populations in Europe than among Holocene populations of the same geographical region. These morphological aspects are shared with the (now lost) Předmostí sample, the fragmentary Willendorf remains, and the incomplete Brno-Francouzská skeleton. More distantly, they are also found in the Gravettian samples from Cro-Magnon, Lagar Velho, Pataud, Paviland, and La Rochette to the west; Arene Candide, Baousso da Torre, Barma Grande, Caviglione, Fanciulli, Paglicci, and Veneri to the south; and Sunghir to the east. They are also generally similar, although a bit less robust on average, to the preceding Aurignacian or pre-28,000 years B.P. remains from Cioclovina, Mladeč, Muierii, Oase, La Quina and Les Rois.

Given the available sample sizes, relative chronologies (the southern European remains tend to be more recent on average), and geographical dispersal of the remains through the fluctuating paleoenvironments of oxygen isotope stage 3, it is not possible to reject a null hypothesis of a widely dispersed, highly mobile, and hence interconnected, fluctuating Gravettian human population across Europe between about 27,000 and 20,000 years B.P. The levels and patterns of variability in human anatomy, especially given the small sample sizes available for any one region for a given millennium, can be easily subsumed within expected minor individual and interpopulational variation across the time and space involved.

At the same time, there is abundant archeological evidence, in the form of artifact typology, artistic styles, patterns of personal ornamentation, and burial practices, that suggests that these populations were culturally interconnected across Europe, from central Russia in the east, to southern Italy in the south, to Portugal in the southwest, and to Wales in the northwest. Regional differences in the framework of this "Gravettian mosaic" are mainly observed in the foraging subsistence patterns and in the related technologies, but both of them may be explained, at least partly, as adaptive responses to the variability in landscapes and resources (both dietary and technological) across Europe. By the same criteria, the human skeletal remains are in complete agreement with such an interconnected network of European Gravettian human populations, affected by individual and some regional patterns of variation.

When these Gravettian human remains are compared to the preceding pre-28,000-year B.P. "modern" humans from Europe, there is little basis to reject a null hypothesis of general human population continuity. The pre-28,000-year B.P. human fossil sample has diminished markedly in recent years as a result of the redating of specimens to anywhere from the Gravettian (Cro-Magnon and La Rochette), to the Magdalenian (Zlatý kůn/Koněprusy), and to the Holocene (Hahnöfersand, Velika Pećina, Vogelherd, and Svatý Prokop). To this sample have been added the remains from Oase, currently the oldest known "modern" humans in Europe at about 35,000 years B.P. The primary difference between the Pavlovian humans remains or the larger Gravettian human sample and these pre-28,000-year B.P. remains is a general decrease in robusticity and the loss (or diminution) of several archaic human features. These minor changes, although supporting the ongoing evolution of human anatomy after the establishment of "modern" human populations across Europe between 35,000 to 30,000 years B.P., provide little ground for inferring any significant population replacement or even gene flow into the northwestern Old World with the advent of the Gravettian.

It is also inevitable that one should address the issue of possible genetic continuity with late Neandertal populations in Europe, given the presence of archaic human features in the pre-28,000-year B.P. European "modern" human sample. Since these are traits that appear to have been lost in the likely ancestral African and Southwest Asian earliest modern human populations that predate the earliest modern humans in Europe (e.g., Omo-Kibish, Qafzeh, and Skhul), the most probable source for these archaic human characteristics is the European Neandertal population. Indeed, as discussed in Chapter 1, the morphology of these scattered earliest "modern" human remains in Europe is one of the sources of evidence for some degree of admixture as early "modern" humans dispersed out of Africa and across Europe sometime after 40,000 years B.P.

The question remains whether there is further evidence in the Pavlovian human remains from central Europe, evidence that supports some degree of prior admixture with regional Neandertal populations. There is evidence in Gravettian human remains for such admixture, in the Lagar Velho skeleton, but it is from a region in which the human biological (and archeological) transition took place relatively late. In general, there is little indication of any Neandertal ancestry in the Dolní Věstonice and Pavlov human remains, with most of the apparently archaic features in some individuals reflecting merely a level of craniofacial robusticity that falls toward the more hypertrophied end of recent human ranges of variation (Chapters 8 and 10). There are nonetheless two features that are arguably evidence of such admixture, body proportions and maxillary dental shoveling.

The body proportions of the Dolní Věstonice and Pavlov human remains, as with those of other European Gravettian fossils, are most similar to the body proportions of modern equatorial Africans. Yet, they are in some respects intermediate between those of recent temperate (European) and equatorial (African) populations (Chapter 12). These results have been used to argue for human population dispersal and/or gene flow from Africa into Europe around the time of modern human appearance in Europe, but the less equatorial body proportions of these Gravettian remains suggests that there was some

European influence on their gene pools. That "European influence" could take one or both of two forms. It could be the result of short-term selection from the colder climates of interpleniglacial Europe, although various sources of information suggest that human body proportions, especially with the level of cultural thermal buffering available in Upper Paleolithic, are unlikely to change markedly within the time frame (maximally 10,000 years) available since early modern humans dispersed throughout Europe. Alternatively, it could result from some modest degree of admixture with local Neandertal populations, whose consistent polar, or even hyperpolar, body proportions could have the effect of making those European early modern humans appear less tropical in their body proportions. This is the pattern that is evident in the Lagar Velho Gravettian skeleton, although with respect to limb segment proportions in that individual as opposed to principally body breadths in the Dolní Věstonice sample.

Moderate maxillary incisor and canine shoveling is present in the Dolní Věstonice 15 dentition, even though it is absent in that of Dolní Věstonice 14 and probably also of Dolní Věstonice 13 (the other maxillary anterior teeth are too worn to provide accurate indications of the presence or degrees of shoveling). The degree of shoveling in Dolní Věstonice 15 does not reach that seen in most Neandertal specimens, but it is matched by that in some Neandertals, in some of the teeth of the earlier Aurignacian Les Rois sample, and in some other Gravettian specimens. It is therefore not unusual in a European early modern human. However, the best available sample of the presumed early modern human ancestor for European earlier Upper Paleolithic modern humans, that from the sites of Qafzeh and Skhul, lacks any noticeable shoveling of the maxillary dentition, arguing for an absence of shoveling in the earliest modern humans. If this is indeed the case, the appearance of anterior dental shoveling in European early modern humans, however muted compared to the pattern common among Neandertals, is best explained as the product of some degree of Neandertal ancestry among those earliest modern humans in Europe. Indeed, the variability in the degree of shoveling, whether in the Les Rois sample, the Dolní Věstonice sample, or across Aurignacian and Gravettian humans generally, would argue for the kind of genetic variability produced by interpopulational admixture. Alternatively, one would have to argue for another or an additional source of modern human ancestry in Europe, aside from that represented by the Qafzeh-Skhul sample, to explain the variable persistence of this dental trait in earlier Upper Paleolithic non-Neandertal Europeans.

These two aspects of the morphology of the Dolní Věstonice and Pavlov human remains could therefore be used to support some degree, however modest, of Neandertal genetic contribution to the populations of early modern humans in Europe that persisted to the time of these Pavlovian people. Yet, by themselves, they are not sufficient to establish the presence or degrees of such admixture, and the overwhelming morphological pattern of these remains is indeed modern. It is other data and analyses, to which these traits may provide support, in combination with the empirical, methodological, and theoretical demise of a strict replacement interpretation, that serve to establish some degree of Neandertal to early modern human admixture in Europe.

Manipulative Behavior

The archeological remains from the Dolní Věstonice and Pavlov sites testify abundantly to the manipulative skill and behavioral diversity of these Pavlovian human populations. The lithic and osteological technology, ceramics, and evidence of textiles all support the interpretation that their daily manipulative behaviors were greatly aided by an elaborate technology.

Their upper limb remains contribute to this impression (Chapter 16). In general, the upper limb bones of these Pavlovian humans, as well as those of other Gravettian human remains, provide evidence of a generally low level of habitual loading on the upper limb for a Late Pleistocene human sample. This is evident to a modest degree in diaphyseal strength measures, but it is most apparent in muscle attachments. The small humeral pectoralis major attachments provide the best indication of this gracility, but it is reflected in other aspects of their upper limbs, including radial diaphyseal curvature and ulnar pronator quadratus crests.

In addition, the associated dentitions of Dolní Věstonice 3, Dolní Věstonice 13 to 16, and Pavlov 1 and 3 provide some indication of the degree to which they were using their anterior teeth for manipulative behavior (Chapter 11). This is reflected in the degree of attrition of those anterior teeth, beyond what would be expected from the level of wear on the postcanine dentition. Dolní Věstonice 13 and 15 possess slightly more pronounced anterior dental wear than might be expected, given their young ages at death and moderate posterior dental attrition. A similar but less pronounced pattern is evident in the less worn dentition of Dolní Věstonice 14, and the incomplete dentition of Pavlov 3 implies a similar pattern. The three older adults with associated teeth, Pavlov 1 and Dolní Věstonice 3 and 16, exhibit little differential anterior versus posterior tooth wear, and it is primarily among older individuals that any differential wear should be most evident. It would be difficult to argue, on the basis of the wear patterns across these dentitions, that these Pavlovian humans were using their anterior teeth for more than routine mastication and occasional manipulative activities, similar to most recent hunter-gatherer human populations.

Mobility Levels

Analyses of the lower limb remains, particularly femora, tibiae, and patellae, of Late Pleistocene humans have shown that there is little change in general levels of locomotor hypertrophy prior to the advent of sedentism in the terminal Pleistocene or early Holocene. Once appropriately scaled for baseline loads of body mass times its relevant load arm (or beam length for diaphyses), it is not possible to differentiate the locomotor robusticity of late archaic humans, Middle Paleolithic early "modern" humans, or earlier Upper Paleolithic humans. It is only in some aspects of diaphyseal shape, particularly of the femur, that these samples contrast, reflecting patterns of locomotion and perhaps body proportions more than levels of mobility. The data presented here on Pavlovian human remains and Late Pleistocene comparative samples reinforce these interpretations (Chapters 17 and 18). The Pavlov 1 and Dolní Věstonice 3, 13, 14, 16, and 35 lower limb remains all conform to this pattern, despite size, sex, and some individual variation. These were highly mobile populations.

The high, or greater than many Holocene, levels of lower limb robusticity are complemented by the degree to which most of these Pavlovian femoral diaphyses are anteroposteriorly reinforced. The consistent presence of a pilaster (in all but some of the Předmostí femora) and the generally greater anteroposterior than mediolateral measures of midshaft bending rigidity (only Pavlov 1 is an exception) combine to indicate the frequent presence of longer distance mobility (as opposed to more localized movement on the landscape).

There are several indirect ways to address the question of mobility by archeological methods (Chapter 2). The first comes from simple stylistic comparisons across northern Eurasia. Technological and morphological similarities among such artifacts and symbols as the earlier Gravettian microliths or the later Gravettian female figurines are so striking that they suggest long-distance relationships between groups or, at least, sufficient awareness of other groups and their styles to require considerable movement on the landscape. Even more convincing are the lithic raw material imports. We do not know exactly what forms of social organization (self-supplying trips or simple forms of "trade") and which technical method of transport (simple carrying or, perhaps, profiting from the Moravian riverine network) stood behind the movement of lithic materials. However, the very presence of lithics originating from hundreds of kilometers away at the Dolní Věstonice and Pavlov sites is eloquent evidence of mobility of the population who transported them.

The importance of mobility to these populations is also implied by two paleopathological aspects of the sample. First, the Dolní Věstonice 15 skeleton, despite the small body size and readily apparent deformities of the individual, exhibits levels of appendicular robusticity that are similar to or greater than those of the other Gravettian individuals. He not only kept up with the other members of the social group, but he did so with what appears to be a greater expenditure of effort reflected in his level of femoral and tibial robusticity. Second, both Dolní Věstonice 13 and 15 exhibit curious, idiopathic patterns of advanced upper limb osteoarthritis, of hand proximal interphalangeal articulations on both individuals and the right glenohumeral articulations of Dolní Věstonice 15. These degenerations are located where joints would be loaded, particularly when carrying or dragging heavy objects for prolonged periods of time. Burden carrying, as part of hunter-gatherer mobility, was sufficiently important to induce these lesions in young adults with little other evidence of articular degeneration.

In the context of this high level of mobility inferred from both the archeological and human paleontological records for the Pavlovian, there is an apparent contradiction. This anomaly involves some of the female figurines associated with these early Gravettian humans. The most obvious of these is the ceramic female figurine from Dolní Věstonice I, but it is joined by others, including the one from the nearby site of Willendorf, as well as figurines from across Europe during the Gravettian. A significant percentage of these depictions are of women who are markedly obese. Some of the artworks may have been intended to illustrate pregnancy, and there is artistic license of form and exaggeration. However, the Dolní Věstonice and Willendorf figurines are sufficiently accurate depictions of the normal distribution of excess adipose tissue in women to illustrate that these Gravettian populations were familiar with human obesity. In the context of the data summarized above on locomotor robusticity, paleopathological indications of mobility, and the archeological reflections that indicate population movement, it becomes difficult to understand how a Pavlovian individual could obtain such excess weight or engage in serious long-distance mobility. It is also in contrast to the paleopathological evidence for at least some periods of systemic stress during development, most likely from seasonal food shortages.

It is possible that it was the rare individual who attained such levels of corpulence, but that would make it difficult to explain the presence of such figurines from the

Atlantic to northern Asia. It is possible, alternatively, that the mobility indicated by their skeletons and archeological remains was associated with prolonged periods of relative sedentism in particularly rich environments, locales that provided a sufficient and stable food supply to allow at least some individuals to achieve impressive levels of adipose tissue. This would be in agreement with the archeological evidence from the large hunters' settlements that indicate some level of sedentism (Chapter 2). It may also be supported by the unusual number of, and associations between, shed deciduous teeth at these sites (Chapter 11), suggesting some duration of the occupations by the children. These inferences therefore may add a further dimension to our perceptions of Gravettian mobility, one combining pronounced movement with periods of relative sedentism.

Stress and Survival

The level of preservation of the Dolní Věstonice and Pavlov human remains and the effort here to document the developmental and degenerative lesions on these bones and teeth (Chapter 19) make it possible to assess the levels and natures of both the stress to which these individual were subjected and their ability to survive these insults. In this sense, it is possible to go beyond previous paleopathological analyses of Upper Paleolithic human remains, which either focus on the abnormalities of individual specimens from a differential diagnosis perspective or plot the incidences of limited varieties of lesions across larger samples.

Indications of mild to moderate levels of stress during development, presumably from seasonal periods of food scarcity, are indicated by the relatively common occurrences of dental enamel hypoplasias, or developmental defects of the dental enamel, on the permanent teeth of these samples. They are not present on all of the teeth of the affected individuals, and they occur on only 55.6% ($N = 18$) of the individuals that preserve at least one sufficiently intact permanent tooth crown (more complete and less worn dentitions of most of the individuals would certainly raise this percentage). They are present on only one of the later-forming third molars, that of Dolní Věstonice 33 ($N = 9$ individuals), and they are absent from the deciduous teeth ($N = 10$ individuals). Moreover, with the exception of Dolní Věstonice 15, all of the enamel hypoplasias reflect minor disruptions of the crown formation process, suggesting short-term insults and/or low stress levels.

This moderate and not unusual level of dental enamel hypoplasias is supported by the rarity of transverse (Harris) lines on the associated postcranial remains. Some of these lines may have been lost from the less complete and older Dolní Věstonice 3 and 16 and Pavlov 1 remains, but low frequency of them in the late adolescent to young adult skeletons from the triple burial reinforces this pattern.

At the same time, evidence of trauma follows an interesting pattern. With the exception of the long bone deformities of Dolní Věstonice 15, one of which (the left ulna diaphysis) is almost certainly due to a fracture, the limb remains are free of either direct (through healed fractures) or indirect (through abnormal patterns of lower limb osteoarthritis) evidence of injuries. The skeletal evidence for trauma is therefore restricted almost exclusively to the cranium and mandible. This involves a series of minor superficial injuries to the neurocranial vaults of Dolní Věstonice 13 and 16 and Pavlov 1, such as are common among Paleolithic human remains. It may include a maxillary fracture of Dolní Věstonice 16, but it does include a serious facial trauma to Dolní Věstonice 3 and a major frontal injury to Dolní Věstonice 11/12.

A predominance of skull injuries is unusual among Holocene human populations, but it is common among other Paleolithic populations. Yet, when associated skeletons are available, and not just isolated craniofacial elements, the head injuries are normally associated with upper limb injuries, across the samples if not necessarily within individual skeletons. The higher levels of upper body traumatic lesions, all of which indicate some survival, may be due to a higher risk of injury to the head and arms. Alternatively, and perhaps more appropriately, given the emphasis on mobility among these Paleolithic populations, there is underrepresentation of serious lower limb injuries. Any traumatic event that inhibited mobility is likely not to be preserved in the human paleontological record, since those individuals would be left behind and their remains scavenged and destroyed. In the context of this pattern, therefore, it is not the frequent presence of craniofacial injuries in the Dolní Věstonice and Pavlov remains that is striking but the absence of upper limb injuries in all but the systemically pathological Dolní Věstonice 15.

Combined with these variably serious traumatic alterations are the pronounced developmental abnormalities of Dolní Věstonice 15 and probably 16. The case of Dolní Věstonice 15, despite the ongoing uncertainties regarding the differential diagnosis of the suite of abnormalities on the remains, represents a major developmental abnormality through some form of dysplasia combined with survival of those insults. In fact, the very pronounced dental enamel hypoplasias on the first molars of Dolní Věstonice 15, which occurred at about 2 years of age, indicate a serious growth arrest, almost certainly related to the dysplasia. It is not known when during development the deformities of the right humerus and femur occurred or the fracture and deformity of the left forearm, but it is likely that they took place at various times during at least the first decade of life. They would have altered both locomotor and manipulative function. More likely than a facial fracture, the abnormalities of the Dolní Věstonice 16 facial skeleton probably indicate that he was born with a form of cleft palate. This would have resulted in a serious deformity of the midfacial region. Yet, both of these individuals survived these developmental difficulties and were able to engage in reasonably normal behavior. What is not known is the extent to which the developmental abnormalities of Dolní Věstonice 15 and 16 may have been combined with other anatomical or physiological abnormalities, those that do not leave evidence on the skeletal remains but may have affected their general well-being.

The survival of major developmental insults by Dolní Věstonice 15 and 16 is joined by the survival of Dolní Věstonice 3 and 11/12 of their serious anterior cranial traumatic injuries. All three of these individuals testify to their survival abilities within their social contexts, a pattern not unexpected, given the complex social integration indicated by site organizations, the elaborate ritual burials (Chapter 4), and the abundance of art and body decoration at these Pavlovian sites.

Conclusion

The human remains from the Pavlovian sites of Dolní Věstonice I and II and of Pavlov I, in their rich archeological context, therefore provide a window onto the biology and behavior of these foraging populations of more than twenty-five millennia ago. As early, robustly built modern humans, they provide an evolutionary link between earlier and later populations in Europe. But just as important, if not more so, they provide a rather complex and still incompletely understood picture of the successes and stresses, the mobility and sedentism, and the cares and caring of these human populations.

References

Aamodt, A., Lund-Larsen, J., Eine, J., Andersen, E., Benum, P., & Schnell Husby, O. (1997). *In vivo* measurements show tensile axial strain in the proximal lateral aspect of the human femur. *Journal of Orthopaedic Research* 15:927–931.

Absolon, K. (1929). New finds of fossil human skeletons in Moravia. *Anthropologie (Prague)* 7:79–89.

Absolon, K. (1935). Ein Anhängsel aus einem fossilen Menschenzahn. *Zeitschrift für Rassenkunde* 1:317.

Absolon, K. (1938a). *Die Erforschung der diluvialen Mammutjägerstation von Unter-Wisternitz an den Pollauer Bergen in Mähren. Arbeitsbericht über das erste Grabungsjahr 1924.* Brno: Polygrafia.

Absolon, K. (1938b). *Die Erforschung der diluvialen Mammutjäger-Station von Unter-Wisternitz in Mähren. Arbeitsbericht über das zweite Jahr 1925.* Brno: Polygrafia.

Absolon, K. (1945). *Die Erforschung der diluvialen Mammutjägerstation von Unter-Wisternitz an den Pollauer Bergen in Mären. Arbeitsbericht über das dritte Jahr 1926.* Brno: Polygrafia.

Absolon, K., & Klíma, B. (1977). *Předmostí. Ein Mammutjägerplatz in Mähren.* Prague: Academia.

Aiello, L., & Dean, C. (1990). *An Introduction to Human Evolutionary Anatomy.* New York: Academic Press.

Alciati, G., Coppa, A., Dalmeri, G., Giacobini, G., Lanzinger, M., & Macchiarelli, R. (1997). Reperti di eta' epigravettiana dal Riparo Dalmeri: Casualita' o pratiche rituali? (abstract). *Thirty-third Riunione Scientifica of the Istituto Italiano di Preistoria e Protostoria*, p. 36.

Alexeeva, T.I., Bader, N.O., Munchaev, R.M., Buzhilova, A.P., Kozlovskaya, M.V., & Mednikova, M.B. (Eds.). (2000). *Homo Sungirensis. Upper Palaeolithic Man: Ecological and Evolutionary Aspects of the Investigation.* Moscow: Scientific World.

Allen, M.E. (1984). Stress fracture of the humerus. A case study. *American Journal of Sports Medicine* 12:244–245.

Alt, K.W., Pichler, S., Vach, W., Klíma, B., Vlček, E., & Sedlmeier, J. (1997). Twenty-five thousand-year-old burial from Dolni Věstonice: An ice-age family? *American Journal of Physical Anthropology* 102:123–131.

Alt, K.W., & Türp, J.C. (1998). Hereditary dental anomalies. In K.W. Alt, F.W. Rösing, & M. Teschler-Nicola (eds.), *Dental Anthropology. Fundamentals, Limits and Prospects.* Vienna: Springer-Verlag, pp. 95–128.

Anderson, J.Y., & Trinkaus, E. (1998). Patterns of sexual, bilateral and inter-populational variation in human femoral neck-shaft angles. *Journal of Anatomy* 192:279–285.

Angel, J.L. (1972). A Middle Palaeolithic temporal bone from Darra-i-Kur, Afghanistan. In L. Dupree, (ed.), *Prehistoric Research in Afghanistan (1959–1966). Transactions of the American Philosophical Society* 62:54–56.

Antl-Weiser, W., & Fladerer, F. (2004). Outlook to the East: The 25 ka BP Gravettian Grub/Kranawetberg campsite (Lower Austria). In J. Svoboda & L. Sedláčková (eds.), *The Gravettian along the Danube. Dolní Věstonice Studies 11.* Brno: Archeologický ústav AV ČR, pp. 116–130.

Apreleva, M., Parsons, I.M., IV, Warner, J.J.P., Fu, F.H., & Woo, S.L.Y. (2000). Experimental investigation of reaction forces at the glenohumeral joint during active abduction. *Journal of Shoulder and Elbow Surgery* 9:409–417.

Arensburg, B. (1977). New Upper Palaeolithic human remains from Israel. *Eretz-Israel* 13:208–215.

Arensburg, B. (1991). The vertebral column, thoracic cage, and hyoid bone. In O. Bar Yosef & B. Vandermeersch (eds.), *Le Squelette Moustérien de Kébara 2.* Paris: Éditions du C.N.R.S., pp. 113–146.

Arensburg, B., Harell, M., & Nathan, H. (1981). The human middle ear ossicles: Morphometry and taxonomic implications. *Journal of Human Evolution* 10:199–205

Arensburg, B., & Nathan, H. (1972). A propos de deux

References

osselets de l'oreille moyenne d'un neandertaloide trouves à Qafzeh (Israel). *L'Anthropologie* 76:301–307.

Arensburg, B., & Tillier, A.M. (1983). A new Mousterian child from Qafzeh (Israel): Qafzeh 4a. *Bulletin et Mémoires de la Société d'Anthropologie de Paris*, série XIII,10:61–69.

Arlet, J., & Mazières, B. (1985) La maladie hyperostosique. *Revue de Médicine Interne* 6:553–564.

Arsuaga, J.L., Martínez, I., Gracia, A., & Lorenzo, C. (1997). The Sima de los Huesos crania (Sierra de Atapuerca, Spain). A comparative study. *Journal of Human Evolution* 33:219–281.

Arsuaga, J.L., Lorenzo, C., Carretero, J.M., Gracia, A., Martínez, I., García, N., Bermúdez de Castro, J.M., & Carbonell, E. (1999). A complete human pelvis from the Middle Pleistocene of Spain. *Nature* 399:255–258.

Aujoulat, N., Geneste, J.M., Archambeau, C., Delluc, M., Duday, H., & Henry-Gambier, D. (2002). Le grotte ornée de Cussac—Le Buisson-de-Cadouin (Dordogne): Premières observations. *Bulletin de la Société Préhistorique Française* 99:129–153.

Baba, H., & Endo, B. (1982). Postcranial skeleton of the Minatogawa man. In H. Suzuki & K. Hanihara (eds.), *The Minatogawa Man: The Upper Pleistocene Man from the Island of Okinawa*. University Museum, University of Tokyo, Bulletin 19:61–195.

Bader, O. (1998). *Upper Palaeolithic Site of Sungir (Graves and Environment)* (in Russian). Moscow: Scientific World.

Badoux, D.M. (1965). Probabilité d'une différenciation due au climat chez les Néandertaliens d'Europe. *L'Anthropologie* 69:75–82.

Bailey, S.E. (2002). *Neandertal Dental Morphology: Implications for Modern Human Origins*. Ph.D. thesis, Arizona State University, Tempe.

Barnicot, N.A., & Hardy, R.H. (1955). The position of the hallux in West Africans. *Journal of Anatomy* 89:355–361.

Bass, W.M. (1987). *Human Osteology: A Laboratory and Field Manual*, 3rd ed. Columbia: Missouri Archaeological Society Special Publication No. 2.

Baume, L.J., Horowitz, H.S., Summers, C.J., Dirks, O.B., Brown, W.A.B., Carlos, J.P., Cohen, L.K., Freer, T.J., Harvold, E.P., Moorrees, C.F.A., Satzmann, J.A., Schmuth, G., Solow, B., & Taatz, H. (1970). A method for measuring occlusal traits. Developed by the F.D.I. Commission on Classification and Statistics for Oral Conditions (COCSTOC) Working Group 2 on Dentofacial Anomalies, 1969–1972. *International Dental Journal* 23:530–537.

Berger, T.D., & Trinkaus, E. (1995). Patterns of trauma among the Neandertals. *Journal of Archaeological Science* 22:841–852.

Bermúdez de Castro, J.M., Arusaga, J.L., & Pérez, P.J. (1997). Interproximal grooving in the Atapuerca-SH hominid dentitions. *American Journal of Physical Anthropology* 102:369–376.

Billy, G. (1972). L'évolution humaine au paléolithique supérieur. *Homo* 23:2–12.

Billy, G. (1975). Étude anthropologique des restes humains de l'Abri Pataud. In H.L. Movius Jr. (ed.), *Excavation of the Abri Pataud, Les Eyzies (Dordogne)*. Bulletin of the American School of Prehistoric Research 30:201–261.

Bogin, B., & Rios, L. (2003). Rapid morphological change in living humans: Implications for modern human origins. *Comparative Biochemistry and Physiology* 136:71–84.

Borgogini Tarli, S.M., Fornaciari, G., & Palma di Cesnola, A. (1980). Restes humains des niveaux Gravettiens de la Grotte Paglicci (Rignano Garganico): Contexte archéologique, étude anthropologique et notes de paléopathologie. *Bulletins et Mémoires de la Société d'Anthropologie de Paris*, Série XII, 7:125–152.

Boule, M. (1911–1913). L'homme fossile de La Chapelle-aux-Saints. *Annales de Paléontologie* 6:111–172; 7:21–56, 85–192; 8:1–70.

Boydstun, S.B., Trinkaus, E., & Vandermeersch, B. (1988). Dental caries in the Qafzeh 3 early modern human (abstract). *American Journal of Physical Anthropology* 75:188–189.

Brain, C.K. (1981). *The Hunters or the Hunted? An Introduction to African Cave Taphonomy*. Chicago: University of Chicago Press.

Bråten, M., Terjesen, T., & Rossvoll, I. (1992). Femoral anteversion in normal adults. Ultrasound measurements in 50 men and 50 women. *Acta Orthopaedica Scandinavica* 63:29–32.

Bräuer, G. (1988). Osteometrie. In R. Knussman (ed.), *Anthropologie I*. Stuttgart: Fischer Verlag, pp. 160–232.

Bräuer, G. (2001). The "Out-of-Africa" model and the question of regional continuity. In P.V. Tobias, M.A. Raath, J. Moggi-Cecchi, & G.A. Doyle (eds.), *Humanity from African Naissance to Coming Millennia*. Florence: Firenze University Press; Johannesburg: Witwatersrand University Press, pp. 183–189.

Bräuer, G., & Broeg, H. (1998). On the degree of Neandertal-modern continuity in the earliest Upper Palaeolithic crania from the Czech Republic: Evidence from nonmetrical features. In K. Omoto & P.V. Tobias (eds.), *The Origins and Past of Modern Humans—Towards Reconciliation*. Singapore: World Scientific, pp. 106–125.

Brekelmans, W.A.M., Poort, H.W., & Slooff, T.J.J.H. (1972). A new method to analyze the mechanical behaviour of skeletal parts. *Acta Orthopaedica Scandinavica* 43:301–317.

Brennan, M.U. (1991). *Health and Disease in the Middle and Upper Paleolithic of Southwestern France: A Bioarcheological Study*. Ph.D. thesis, New York University, New York.

Bridges, P.S. (1994). Vertebral arthritis and physical activities in the prehistoric southeastern United States. *American Journal of Physical Anthropology* 93:67–91.

Brothwell, D.R. (1963). *Digging up Bones*. London and Oxford: British Museum and Oxford University Press.

Brothwell, D.R. (1981). *Digging up Bones*, 2nd ed. London and Oxford: British Museum and Oxford University Press.

Brown, K.A.R., Davies, E.N., & McNally, D.S. (2002). Angular distribution of vertebral trabeculae in modern humans, chimpanzees and the Kebara 2 Neanderthal. *Journal of Human Evolution* 43:189–205.

Brown, T., & Molnar, S. (1990). Interproximal grooving and task activity in Australia. *American Journal of Physical Anthropology* 81:545–553.

Brunelli, G.R. (1999). Stability of the first carpometacarpal joint. In P. Brüser & A. Gilbert (eds.), *Finger Bone and Joint Injuries*. London: Martin Dunitz, pp.167–173.

Brůžek, J. (1991). *Fiabilité des procédés de détermination du sexe à partir de l'os coxal. Implications à l'étude du dimorphisme sexuel de l'Homme fossile*. Thèse de Doctorat, Muséum National d'Histoire Naturelle (Institut de Paléontologie Humaine), Paris.

Brůžek, J. (1992). Fiabilité des fonctions discriminantes dans la détermination sexuelle de l'os coxal. Critiques et propositions. *Bulletins et Mémoires de la Société d'Anthropologie de Paris* ns4:67–104.

Brůžek, J. (2002). A method for visual determination of sex, using the human hip bone. *American Journal of Physical Anthropology* 117:157–168.

Brůžek, J., Murail, P., Houët, F., & Clevenot, E. (1994). Inter and intra observer errors in pelvic measurements and its implication for the method of sex determination. *Anthropologie (Brno)* 32:215–223.

Brůžek, J., Castex, D., & Majó, T. (1996). Évaluation des caractères morphologiques de la face sacro-pelvienne de l'os coxal. Proposition d'une nouvelle méthode de diagnose sexuelle. *Bulletins et Mémoires de la Société d'Anthropologie de Paris* ns8:479–490.

Budinoff, L.C., & Tague, R.G. (1990). The anatomical and developmental bases for the ventral arc of the human pubis. *American Journal of Physical Anthropology* 82:73–79.

Buikstra, J.E., & Ubelaker, D.H. (1994). *Standards for Data Collection from Human Skeletal Remains*. Arkansas Archaeological Survey Report (Fayetteville) 44.

Byers, S.N. (2002). *Introduction to Forensic Anthropology*. Boston: Allyn & Bacon.

Campbell, T.D. (1925). *Dentition and Palate of the Australian Aboriginal*. Adelaide: University of Adelaide Press.

Carretero, J.M., Lorenzo, C., & Arsuaga, J.L. (1999). Axial and appendicular skeleton of *Homo antecessor*. *Journal of Human Evolution* 37:459–499.

Carter, D.R., & Beaupré, G.S. (2001). *Skeletal Function and Form*. Cambridge UK: Cambridge University Press.

Cavanagh, P.R., & Rodgers, M.M. (1984). Pressure distribution underneath the human foot. In S.M. Perren & E. Schneider (eds.), *Biomechanics: Current Interdisciplinary Research*. Boston: Martinus Nijhoff, pp. 85–95.

Chilardi, S., Frayer, D.W., Gioia, P., Macchiarelli, R., & Mussi, M. (1996). Fontana Nuova di Ragusa (Sicily, Italy): Southernmost Aurignacian site in Europe. *Antiquity* 70:553–563.

Churchill, S.E. (1994). *Human Upper Body Evolution in the Eurasian Later Pleistocene*. Ph.D. thesis, University of New Mexico, Albuquerque.

Churchill, S.E. (1996). Neandertal scapular axillay border morphology revisited (abstract). *American Journal of Physical Anthropology Supplement* 22:85.

Churchill, S.E. (1998). Cold adaptation, heterochrony, and Neandertals. *Evolutionary Anthropology* 7:46–61.

Churchill, S.E., & Formicola, V. (1997). A case of marked bilateral asymmetry in the upper limb of an Upper Palaeolithic male from Barma Grande (Liguria), Italy. *International Journal of Osteoarcheology* 7:18–38.

Churchill, S.E., & Smith, F.H. (2000). A modern human humerus from the early Aurignacian of Vogelherdhöhle (Stetten, Germany). *American Journal of Physical Anthropology* 112:251–273.

Churchill, S.E., & Trinkaus, E. (1990). Neandertal scapular glenoid morphology. *American Journal of Physical Anthropology* 83:147–160.

Churchill, S.E., Pearson, O.M., Grine, F.E., Trinkaus, E., & Holliday, T.W. (1996). Morphological affinities of the proximal ulna from Klasies River Mouth Main Site: Archaic or modern? *Journal of Human Evolution* 31:213–237.

Churchill, S.E., Formicola, V., Holliday, T.W., Holt, B.M., & Schumann, B.A. (2000). The Upper Palaeolithic population of Europe in an evolutionary perspective. In W. Roebroeks, M. Mussi, J. Svoboda, & K. Fennema (eds.), *Hunters of the Golden Age*. Leiden: University of Leiden Press, pp. 31–57.

Condemi, S. (1992). *Les Hommes Fossiles de Saccopastore et Leurs Relations Phylogénétiques*. Paris: CNRS Éditions.

Coon, C.S. (1962). *The Origin of Races*. New York: Alfred A. Knopf.

Corruccini, R.S. (1991). Anthropological aspects of orofacial and occlusal variations and anomalies. In M.A. Kelley & C.S. Larsen (eds.), *Advances in Dental Anthropology*. New York: Wiley-Liss, pp. 295–323.

Costa, R.L., Jr. (1980). Incidence of caries and abscesses in archeological Eskimo skeletal samples from Point Hope and Kodiak Island, Alaska. *American Journal of Physical Anthropology* 52:501–514.

Courtaud, P. (1989). Deux os du pied provenant des niveaux moustériens de la Grotte de Kébara (Israël). *Bulletins et Mémoires de la Société d'Anthropologie de Paris* ns1:45–58.

Crevecoeur, I., & Trinkaus, E. (2004). From the Nile to the Danube: A comparison of the Nazlet Khater 2 and Oase 1 early modern human mandibles. *Anthropologie (Brno)* 42:229–239.

Crubézy, E., & Trinkaus, E. (1992). Shanidar 1: A case of hyperostotic disease (DISH) in the Middle Paleolithic. *American Journal of Physical Anthropology* 89:411–420.

Čuk, T., Leben-Seljak, P., & Štefančič, M. (2001). Lateral asymmetry of human long bones. *Variability and Evolution* 9:19–32.

Daegling, D.J. (2002). Estimation of torsional rigidity in primate long bones. *Journal of Human Evolution* 43:229–239.

Danenberg, P.J., Hirsch, R.S., Clarke, N.G., Leppard, P.I., & Richards, L.C. (1991). Continuous tooth eruption in Australian aboriginal skulls. *American Journal of Physical Anthropology* 85:305–312.

Davies, T.G.H., & Pedersen, P.O. (1955). The degree of attrition of the deciduous teeth and the first permanent molars of primitive and urbanised Greenland natives. *British Dental Journal* 99:35–43.

Dawson, J.E., & Trinkaus, E. (1997). Vertebral osteoarthritis of the La Chapelle-aux-Saints 1 Neanderthal. *Journal of Archaeological Science* 24:1015–1021.

DeStefano, G.F., & Hauser, G. (1991). Epigenetic traits of the Circeo I skull. In M. Piperno & G. Scichilone (eds.), *The Circeo 1 Neandertal Skull: Studies and Documentation.* Rome: I.P.Z.S. Libreria Dello Stato, pp. 273–299.

Dettwyler, K.A. (1995). A time to wean: The hominid blueprint for the natural age of weaning in modern human populations. In P. Stuart-Macadam & K.A. Dettwyler (eds.), *Breastfeeding: Biocultural Perspectives.* New York: Aldine de Gruyter, pp. 39–73.

De Villiers, H. (1968). *The Skull of the South African Negro. A Biometrical and Morphological Study.* Johannesburg: University of Witwatersrand Press.

Dias, G., & Tayles, N. (1997). 'Abscess cavity'—a misnomer. *International Journal of Osteoarchaeology* 7:548–554.

Dittner, C.B. (1976). *The Morphology of the Axillary Border of the Scapula with Special Reference to the Neandertal Problem.* M.A. thesis, University of Tennessee, Knoxville.

Dobson, S.D., & Trinkaus, E. (2002). Cross-sectional geometry and morphology of the mandibular symphysis in Middle and Late Pleistocene *Homo*. *Journal of Human Evolution* 43:67–87.

Drozdová, E. (2001). Zhodnocení znovuobjeveného fragmentu lidské dolní čelisti č. 21 z Předmostí u Přerova. *Archeologické rozhledy* 53:452–460.

Duarte, C., Maurício, J., Pettitt, P.B., Souto, P., Trinkaus, E., van der Plicht, H., & Zilhão, J. (1999). The early Upper Paleolithic human skeleton from the Abrigo do Lagar Velho (Portugal) and modern human emergence in Iberia. *Proceedings of the National Academy of Sciences USA* 96:7604–7609.

Duda, G.N., Schneider, E., & Chao, E.Y.S. (1997). Internal forces and moments in the femur in walking. *Journal of Biomechanics* 30:933–941.

Duda, G.N., Heller, M., Albinger, J., Schulz, O., Schneider, E., & Claes, L. (1998). Influence of muscle forces on femoral strain distributions. *Journal of Biomechanics* 31:841–846.

Dujardin V. (2003). Sondages à La Quina aval (Gardes-le-Pontaroux, Charente). *Antiquités Nationales* 33:21–26.

Dunlap, S.S. (1979). Sex, parity and the preauricular sulcus (abstract). *American Journal of Physical Anthropology* 50:434–435.

Emmerling, E., Geer, H., & Klíma, B. (1993). Ein Mondkalenderstab aus Dolní Věstonice. *Quartär* 43/44:151–162.

Endo, B., & Kimura, T. (1970). Postcranial skeleton of the Amud man. In H. Suzuki & F. Takai (eds.), *The Amud Man and His Cave Site.* Tokyo: Academic Press, pp. 231–406.

Eschman, P.N. (1992). *SLCOMM Version 1.6.* Albuquerque: Eschman Archeological Services.

Fanning, E.A. (1961). A longitudinal study of tooth formation and root resorption. *New Zealand Dental Journal* 57:202–217.

Feldesman, M.R., & Lundy, J.K. (1988). Stature estimates for some African Plio-Pleistocene hominids. *Journal of Human Evolution* 17:583–596.

Ferembach, D. (1976). Les restes humains de la grotte de Dar-es-Soltane 2 (Maroc) Campagne 1975. *Bulletins et Mémoires de la Société d'Anthropologie de Paris*, Série XIII, 3:183–193.

Formicola, V. (1990). The triplex burial of Barma Grande (Grimaldi, Italy). *Homo* 39:130–143.

Formicola, V. (1991). Interproximal grooving: Different appearances, different etiologies. *American Journal of Physical Anthropology* 86:85–86.

Formicola, V. (2003). More is not always better: Trotter and Gleser's equations and stature estimates of Upper Paleolithic European samples. *Journal of Human Evolution* 45:239–243.

Formicola, V., & Franceschi, M. (1996). Regression equations for estimating stature from long bones of Early Holocene European samples. *American Journal of Physical Anthropology* 100:83–88.

Formicola, V., & Giannecchini, M. (1999). Evolutionary trends of stature in Upper Paleolithic and Mesolithic Europe. *Journal of Human Evolution* 36:319–333.

Formicola, V., Pontrandolfi, A., & Svoboda, J. (2001). The Upper Paleolithic triple burial of Dolní Věstonice: Pathology and funerary behavior. *American Journal of Physical Anthropology* 115:372–379.

Fraipont, J., & Lohest, M. (1887). La race humaine de Néanderthal ou de Canstadt en Belgique. Recherches ethnographiques sur des ossements humains, découvertes dans des dépôts quaternaires d'une grotte à Spy et détermination de leur âge géologique. *Archives de Biologie* 7:587–757.

Franciscus, R.G. (1989). Neandertal mesosterna and noses: Implications for activity and biogeographical patterning (abstract). *American Journal of Physical Anthropology* 78:223.

Franciscus, R.G. (1995). *Nasal Morphology in the Western Old World Later Pleistocene and the Origins of Modern Humans.* Ph.D. thesis, University of New Mexico, Albuquerque.

Franciscus, R.G. (2002). The midfacial morphology. In J. Zilhão & E. Trinkaus (eds.), *Portrait of the Artist as a Child. The Gravettian Human Skeleton from the Abrigo do Lagar Velho and Its Archeological Context.* Trabalhos de Arqueologia 22:297–311.

Franciscus, R.G. (2003a). Comparing internal nasal fossa dimensions and classical measures of the external nasal skeleton in recent humans: Inferences for respiratory airflow dynamics and climatic adaptation (abstract). *American Journal of Physical Anthropology Supplement* 36:96–97.

Franciscus, R.G. (2003b). Internal nasal floor configuration in *Homo* with special reference to the evolution of Neandertal facial form. *Journal of Human Evolution* 44:701–729.

Franciscus, R.G., & Churchill, S.E. (2002). The costal skeleton of Shanidar 3 and a reappraisal of Neandertal thoracic morphology. *Journal of Human Evolution* 42:303–356.

Franciscus, R.G., & Holliday, T.W. (1992). Hindlimb skeletal allometry in Plio-Pleistocene hominids with special reference to AL 288-1 ("Lucy"). *Bulletins et Mémoires de la Société d'Anthropologie de Paris* ns4:5–20.

Franciscus, R.G., & Trinkaus, E. (1995). Determinants of retromolar space presence in Pleistocene *Homo* mandibles. *Journal of Human Evolution* 28:577–595.

Frayer, D.W. (1980). Sexual dimorphism and cultural evolution in the Late Pleistocene and Holocene of Europe. *Journal of Human Evolution* 9:399–415.

Fully, G. (1956). Une nouvelle méthode de determination de la taille. *Annals de Médicine Légale* 36:266–273.

Gaillard, J. (1960). Détermination du sexe d'un os coxal fragmentaire. *Bulletins et Mémoires de la Société d'Anthropologie de Paris*, Série IX, 1:225–267.

García Sánchez, M. (1986). Estudio preliminar de los restos Neandertalenses del Boquete de Zafarraya (Alcaucin, Malaga). In *Homenaje a Luis Siret*. Malaga: Consejeria de Cultura de la Junta de Andalucia, pp. 49–56.

Gard, S.A., & Childress, D.S. (1997). The effect of pelvic tilting on the vertical displacement of the trunk during normal walking. *Gait & Posture* 5:233–238.

Genet-Varcin, E., & Miquel, M. (1967). Contribution à l'étude du squelette Magdalénien de l'Abri Lafaye à Bruniquel (Tarn-et-Garonne). *L'Anthropologie* 71:467–478.

Genovés, S. (1954). The problem of the sex of certain fossil hominids, with special reference to the Neandertal skeletons from Spy. *Journal of the Royal Anthropological Institute* 84:131–144.

Giffin, E.B. (1990). Gross spinal anatomy and limb use in living and fossil reptiles. *Paleobiology* 16:448–458.

Giffin, E.B. (1992). Functional implications of neural canal anatomy in recent and fossil marine carnivores. *Journal of Morphology* 214:357–374.

Gill, G.W., Hughes, S.S., Bennet, S.M., & Gilbert, B.M. (1988). Racial identification from the midfacial skeleton with special reference to American Indians and Whites. *Journal of Forensic Sciences* 33:92–99.

Gillespie, B.C. (1981). Mountain Indians. In J. Helm, (ed.), *Handbook of North American Indians 6: Subarctic*. Washington, DC: Smithsonian Institution, pp. 326–337.

Glanville, E.W. (1969). Nasal shape, prognathism and adaptation in man. *American Journal of Physical Anthropology* 30:29–38.

Goaz, P.W., & White, S.C. (1994). *Oral Radiology. Principles and Interpretation*. St Louis: C.V. Mosby.

Grahnén, H. (1956). Hypodontia in the permanent dentition. *Odontologisk Revy* 7 (Supplement 3).

Grine, F.E., Jungers, W.L., Tobias, P.V., & Pearson, O.M. (1995). Fossil *Homo* femur from Berg Aukas, northern Namibia. *American Journal of Physical Anthropology* 97:151–185.

Gustafson, G., & Koch, G. (1974). Age estimation up to 16 years of age based on dental development. *Odontologisk Revy* 25:297–306.

Guthrie, D., & Kolfschoten, T. van (2000). Neither warm and moist, nor cold and arid: The ecology of the Mid Upper Paleolithic. In W. Roebroeks, M. Mussi, J. Svoboda, & K. Fennema (eds.), *Hunters of the Golden Age*. Leiden: Leiden University Press, pp. 13–20.

Haavikko, K. (1973). The physiological resorption of the roots of deciduous teeth in Helsinki children. *Proceedings of the Finnish Dental Society* 69:93–98.

Haesaerts, P., Damblon, F., Bachner, M., & Trnka, G. (1996). Revised stratigraphy and chronology of the Willendorf II sequence, Lower Austria. *Archaeologia Austriaca* 80:25–42.

Hager, L.D. (1989). *The Evolution of Sex Differences in the Hominid Bony Pelvis*. Ph.D. thesis, University of California, Berkeley.

Haile-Selassie, Y., Asfaw, B., & White, T.D. (2004). Hominid cranial remains from Upper Pleistocene deposits at Aduma, Middle Awash, Ethiopia. *American Journal of Physical Anthropology* 123:1–10.

Hambücken, A. (1995). Étude du degrée de robustesse des os longs du membre supérieur des Néandertaliens. *Bulletins et Mémoires de la Société d'Anthropologie de Paris* ns7:37–47.

Hamilton, M.E. (1975). *Variation among Five Groups of Amerindians in the Magnitude of Sexual Dimorphism of Skeletal Size*. Ph.D. thesis, University of Michigan, Ann Arbor.

Hauser, G., & DeStefano, G.F. (1989). *Epigenetic Variants of the Human Skull*. Stuttgart: E. Schweizerbart'sche Verlagsbuchhandlung.

Heim, J.L. (1976). Les hommes fossiles de La Ferrassie I: Le gisement. Les squelettes adultes (crâne et squelette du tronc). *Archives de l'Institut de Paléontologie Humaine* 35:1–331.

Heim, J.L. (1982a). *Les Enfants Néandertaliens de La Ferrassie*. Paris: Masson.

Heim, J.L. (1982b). Les hommes fossiles de La Ferrassie II: Les squelettes adultes (squelette des membres). *Archives de l'Institut de Paléontologie Humaine* 38:1–272.

Heim, J.L., & Granat, J. (1995). La mandibule de l'enfant néandertalien de Malarnaud (Ariège). Une nouvelle approche anthropologique par la radiographie et la tomodensitométrie. *Anthropologie et Préhistoire* 106:75–96.

Henry-Gambier, D. (2001). *Le Sépulture des Enfants de Grimaldi (Baoussé-Roussé, Italie). Anthropologie et Palethnologie Funéraire des Populations de la Fin du Paléolithique Supérieur*. Paris: Éditions du Comité des travaux historiques et scientifiques. Réunion des Musées Nationaux.

Henry-Gambier, D., Bružek, J., Murail, P., & Houët, F. (2002). Estimation du sexe du squelette magdalénien de St. Germain-La-Rivière (Gironde, France). *Paléo* 14:205–212.

Henry-Gambier, D., Maureille, B., & White, R. (2004). Vestiges humains des niveaux de l'Aurignacien ancien du site de Brassempouy (Landes). *Bulletins et Mémoires de la Société d'Anthropologie de Paris*, 16:49–87.

Hershkovitz, I., Speirs, M.S., Frayer, D., Nadel, D., Wish-Baratz, S., & Arensburg, B. (1995) Ohalo II H2: A 19,000-year-old skeleton from a water-logged site at the sea of Galilee, Israel. *American Journal of Physical Anthropology* 96:215–234.

Hillson, S.W. (1996). *Dental Anthropology*. Cambridge UK: Cambridge University Press.

Hillson, S.W. (2000). Dental pathology. In M.A. Katzenberg & S.R. Saunders (eds.), *Biological Anthropology of the Human Skeleton*. New York: John Wiley, pp. 249–286.

Hillson, S.W. (2001). Recording dental caries in archaeological human remains. *International Journal of Osteoarchaeology* 11:249–289.

Hillson, S.W., & Bond, S. (1997). Relationship of enamel hypoplasia to the pattern of tooth crown growth: A discussion. *American Journal of Physical Anthropology* 104:89–104.

Hillson, S.W., & Fitzgerald, C. (2003). Tooth size variation and dental reduction in Europe, the Middle East and North Africa between 120,000 and 5000 BP (abstract). *American Journal of Physical Anthropology Supplement* 36:114.

Hillson, S.W., & Trinkaus, E. (2002). Comparative dental crown metrics. In J. Zilhão & E. Trinkaus (eds.), *Portrait of the Artist as a Child. The Gravettian Human Skeleton from the Abrigo do Lagar Velho and Its Archeological Context*. Trabalhos de Arqueologia 22:356–364.

Hinton, R.J. (1981). Form and patterning of anterior tooth wear among aboriginal human groups. *American Journal of Physical Anthropology* 54:555–564.

Hirai, T., & Tabata, T. (1928). Anthropologische Untersuchungen über das Skelett der rezenten Japaner. Die untere Extremität. *Journal of the Anthropological Society of Nippon Supplement* 43:1–176.

Hlusko, L.J. (2003). The oldest hominid habit? Experimental evidence for toothpicking with grass stalks. *Current Anthropology* 44:738–741.

Holliday, T.W. (1995). *Body Size and Proportions in the Late Pleistocene Western Old World and the Origins of Modern Humans*. Ph.D. thesis, University of New Mexico, Albuquerque.

Holliday, T.W. (1997a). Body proportions in Late Pleistocene Europe and modern human origins. *Journal of Human Evolution* 32:423–447.

Holliday, T.W. (1997b). Postcranial evidence of cold adaptation in European Neandertals. *American Journal of Physical Anthropology* 104:245–258.

Holliday, T.W. (1999). Brachial and crural indices of European Late Upper Paleolithic and Mesolithic humans. *Journal of Human Evolution* 36:549–566.

Holliday, T.W. (2000a). Body proportions of the Paviland 1 skeleton. In S. Aldhouse-Green (ed.), *Paviland Cave and the "Red Lady." A Definitive Report*. Bristol: Western Academic and Specialist Press, pp. 199–204.

Holliday, T.W. (2000b). Evolution at the crossroads: Modern human emergence in western Asia. *American Anthropologist* 102:54–68.

Holliday, T.W., & Ruff, C.B. (1997). Ecogeographical patterning and stature prediction in fossil hominids: Comment on M.R. Feldesman and R.L. Fountain. *American Journal of Physical Anthropology* 103:137–140.

Holliday, T.W., & Ruff, C.B. (2001). Relative variation in human proximal and distal limb segment lengths. *American Journal of Physical Anthropology* 116:26–33.

Hollimon, S.E. (2001). The gendered peopling of North America: Addressing the antiquity of systems of multiple genders. In N.S. Price (ed.), *The Archaeology of Shamanism*. London: Routledge, pp. 123–134.

Holt, B.M. (1999). *Biomechanical Evidence of Decreased Mobility in Upper Paleolithic and Mesolithic Europe*. Ph.D. thesis, University of Missouri, Columbia.

Holt, B.M., Mussi, M., Churchill, S.E., & Formicola, V. (2000). Biological and cultural trends in Upper Palaeolithic Europe. *Rivista di Antropologia* 78:179–192.

Horowitz, L.K., & Smith, P. (1988). The effects of striped hyaena activity on human remains. *Journal of Archaeological Science* 15:471–481.

Hoshi, H. (1961). On the preauricular groove in the Japanese pelvis. *Okajima Folia Anatomica Japan* 37:259–269.

Houët, F. (2001). Limites de variation, distance (position) probabiliste et écart réduit ajusté. *Paléo* 13:195–200.

Houët, F., Brůžek, J., & Murail, P. (1999). Computer program for sex diagnosis of the human pelvic bone based on a probabilistic approach (abstract). *American Journal of Physical Anthropology Supplement* 28:155.

Houghton, P. (1974). The relationship of the pre-auricular groove of the ilium to pregnancy. *American Journal of Physical Anthropology* 41:381–389.

Houghton, P. (1975). The bony imprint of pregnancy. *Bulletin of the New York Academy of Medicine* 51:655–661.

Houston, C.S., & Zaleski, W.A. (1967). The shape of vertebral bodies and femoral necks in relation to activity. *Radiology* 89:59–66.

Howells, W.W. (1973). Cranial variation in man. *Papers of the Peabody Museum* 67:1–259.

Howells, W.W. (1989). Skull shapes and the map. *Papers of the Peabody Museum* 79:1–189.

Hublin, J.J. (1978). *Le Torus Occipital Transverse et les Structures Associees: Évolution dans le Genre* Homo. Thèse de Troisième Cycle, Université de Paris VI, Paris.

Hublin, J.J. (2000). Modern-nonmodern hominid interactions: A Mediterranean perspective. In O. Bar-Yosef & D. Pilbeam (eds.), *The Geography of Neandertals and Modern Humans in Europe and the Greater Mediterranean*, Peabody Museum Bulletin 8:157–182.

Hublin, J.J., Barroso Ruiz, C., Medina Lara, P., Fontugne, M., & Reyss, J.L. (1995). The Mousterian site of Zafarraya (Andalucia, Spain): Dating and implications on the Palaeolithic peopling process of western Europe. *Comptes rendus de l'Académie des Sciences de Paris* 321:931–937.

Hublin, J.J., Spoor, F., Braun, M., Zonneveld, F., & Condemi, S. (1996). A late Neanderthal associated with Upper Palaeolithic artefacts. *Nature* 381:224–226.

Hublin, J.J., Trinkaus, E., & Stefan, V.H. (1998). The Mousterian human remains from Zafarraya (Andalucia, Spain). *American Journal of Physical Anthropolology Supplement* 26:122–123.

Ishisawa, M. (1931). Anthropologische Untersuchungen über das Skelett der Yoshiko-Steinzeitmenschen. III. Teil. Die unteren Extremitäten. *Journal of the Anthropological Society of Nippon* 46:1–192.

Jabbour, R.S., Richards, G.D., & Anderson, J.Y. (2002).

Mandibular condyle traits in Neanderthals and other *Homo*: A comparative, correlative, and ontogenetic study. *American Journal of Physical Anthropology* 119: 144–155.

Jelínek, J. (1953). Nález zubů fosilního člověka v Dolních Věstonicích. *Časopis Moravského Musea v Brně* 38:180–190.

Jelínek, J. (1954). Nález fosilního člověka Dolní Věstonice III. *Anthropozoikum* 3:37–92.

Jelínek, J. (1989). Upper Paleolithic Gravettian population in Moravia. In G. Giacobini (ed.), *Hominidae*. Milan: Editoriale Jaca Book, pp. 443–448.

Jelínek, J. (1992). New Upper Palaeolithic burials from Dolni Vestonice. In M. Toussaint (ed.), *Cinq Millions d'Années: L'Aventure Humaine*. Études et Recherches Archéologiques de l'Université de Liège 56:207–228.

Jelínek, J., Pelíšek, J., & Valoch, K. (1959). Der fossile Mensch Brno II. *Anthropos (Brno)* 9:5–30.

Jellema, L.M., Latimer, B., & Walker A. (1993). The rib cage. In A. Walker & R. Leakey (eds.), *The Nariokotome Homo erectus Skeleton*. Cambridge, MA: Harvard University Press, pp. 294–325.

Jidoi, K., Nara, T., & Dodo, Y. (2001). Bony bridging of the mylohyoid groove of the human mandible. *Anthropological Science* 108:345–370.

Jungers, W.L. (1991). Scaling of postcranial joint size in hominoid primates. *Human Evolution* 6:391–399.

Jungwirth, J., & Strouhal, E. (1972). Jungpaläolithische menschliche Skelettreste von Krems-Hundssteig in Niederösterreich. *Homo* 23:100–113.

Katzmarzyk, P.T., & Leonard, W.R. (1998). Climatic influences on human body size and proportions: Ecological adaptations and secular trends. *American Journal of Physical Anthropology* 106:483–503.

Kerr, N.W. (1991). Prevalence and natural history of periodontal disease in Scotland—the mediaeval period (900–1600 AD). *Journal of Periodontal Research* 26:346–354.

Kieser, J.A. (1990). *Human Adult Odontometrics*. Cambridge UK: Cambridge University Press.

Kieser, J.A., Preston, C.B., & Evans, W.G. (1983). Skeletal age at death: An evaluation of the Miles method of ageing. *Journal of Archaeological Science* 10:9–12.

Kiyono, K., & Hirai, T. (1928). Anthropologische Untersuchungen über das Skelett der Steinzeit Japaner. IV. Die untere Extremität. *Journal of the Anthropological Society of Nippon Supplement* 43:303–494.

Klaatsch, H., & Lustig, W. (1914). Morphologie der paläolithische Skelettreste des mittleren Aurignacien der Grotte von La Rochette, Dep. Dordogne. *Archiv für Anthropologie* 41:81–126.

Klíma, B. (1950). Objev diluviálního hrobu v Dolních Věstonicích. *Časopis Moravského muzea* 35:216–232.

Klíma, B. (1959). Objev paleolitického pohřbu v Pavlově. *Archeologické rozhledy* 11:305–316.

Klíma, B. (1963). *Dolní Věstonice*. Prague: Nakladatelství Československé Akademie Věd.

Klíma, B. (1977). Malaja poluzemljanka na paleoliticheskoj stojanke Pavlov v Chekhoslovakii. In N.D. Praslov (ed.), *Problemy paleolita Vostochnoj i Centralnoj Evropy*. Leningrad: Nauka, pp. 144–148.

Klíma, B. (1981). Der mittlere Teil der paläolithischen Station bei Dolní Věstonice. *Památky archeologické* 72:5–92.

Klíma, B. (1983). *Dolní Věstonice*. Prague: Academia.

Klíma, B. (1987a). Das jungpaläolithische Massengrab von Dolní Věstonice. *Quartär* 37/38:53–62.

Klíma, B. (1987b). Une triple sépulture du Pavlovien à Dolní Věstonice, Tchécoslovaquie. *L'Anthropologie* 91:329–334.

Klíma, B. (1987c). Zachraňovací výzkum nad cihelnou u Dolních Věstonic (okr. Břeclav). *Přehled výzkumů* 1985:16–18.

Klíma, B. (1988). A triple burial from the Upper Paleolithic of Dolní Věstonice, Czechoslovakia. *Journal of Human Evolution* 16:831–835.

Klíma, B. (1989). El arte del Gravetiense. In G. Albrecht, G. Bosinski, R. Feustel, J. Hahn, B. Klíma, & H.J. Müller-Beck (eds.), *Los comienzos del arte en Europa Central*. Madrid: Museo Arqueológico Nacional, pp. 36–43.

Klíma, B. (1990). Der pleistozäne Mensch aus Dolní Věstonice. *Památky archeologické* 81:5–16.

Klíma, B. (1995). *Dolní Věstonice II. Ein Mammutjägerrastplatz und seine Bestattungen*. Études et Recherches Archéologiques de l'Université de Liège 73/Dolní Věstonice Studies 3. Brno: Archeologický ústav AV ČR.

Klíma, B. (1997). Grabungsgeschichte, Stratigraphie und Fundumstände. In J. Svoboda (ed.), *Pavlov I, Northwest. The Upper Paleolithic Burial and Its Settlement Context*. Dolní Věstonice Studies 4, Brno: Archeologický ústav AV ČR, pp. 13–51.

Klíma, B. (2001). Die Kjökkenmöddinge Nr. 5–8 von Dolní Věstonice. In B. Ginter, B. Drobniewicz, B. Kazior, M. Nowak, & M. Poltowicz (eds.), *Problems of the Stone Age in the Old World*. Kraków: Uniwersytet Jagiellonski, pp. 173–193.

Klíma, B., Kukla, J., Ložek, V., & de Vries, H. (1962). Stratigraphie des Pleistozäns und Alter des paläolithischen Rastplatzes in der Ziegelei von Dolní Věstonice (Unter-Wisternitz). *Anthropozoikum* 11:93–145.

Kobayashi, K. (1967). Trend in the length of life based on human skeletons from prehistoric to modern times in Japan. *Journal of the Faculty of Science, University of Tokyo, Section V. Anthropology* 3:107–162.

Kocher, M.S., Waters, P.M., & Micheli, L.J. (2000). Upper extremity injuries in the paediatric athlete. *Sports Medicine* 30:117–135.

Konigsberg, L.W., Hens, S.M., Jantz, L.M., & Jungers, W.L. (1998). Stature estimation and calibration: Bayesian and maximum likelihood perspectives in physical anthropology. *Yearbook of Physical Anthropology* 41:65–92.

Kovanda, J. (1991). Molluscs from the section with the skeleton of Upper Palaeolithic man at Dolní Věstonice. In J. Svoboda (ed.), *Dolní Věstonice II Western Slope*. Études et Recherches Archéologiques de l'Université de Liège 54:89–96.

Kozlovskaya, M.V., & Mednikova, M.B. (2000). Catalogue of photos and tables on materials from Sunghirian graves 1 and 2. In T.I. Alexeeva, N.O. Bader, R.M. Munchaev,

A.P. Buzhilova, M.V. Kozlovskaya, & M.B. Mednikova (eds.), *Homo Sungirensis. Upper Palaeolithic Man: Ecological and Evolutionary Aspects of the Investigation*. Moscow: Scientific World, pp. 85–144.

Kozlowski, J.K. (1986). The Gravettian in central and eastern Europe. *Advances in World Archaeology* 5:131–200.

Králík, M., Novotný, V., & Oliva, M. (2002). Fingerprint on the venus of Dolní Věstonice I. *Anthropologie (Brno)* 40:107–113.

Krantz, G.S. (1982). The fossil record of sex. In R.L. Hall (ed.), *Sexual Dimorphism in* Homo sapiens. *A Question of Size*. New York: Praeger, pp. 85–105.

Krogman, W.M. (1962). *The Human Skeleton in Forensic Medicine*. Springfield, IL: C.C. Thomas.

Krogman, W.M., & Iscan, M.Y. (1986). *The Human Skeleton in Forensic Medicine*. Springfield, IL: C.C. Thomas.

Kuklík, M. (1992). Úvaha k nálezům z mladopaleolitického trojhrobu v Dolních Věstonicích z pohledu genetika. In E. Vlček (ed.), *Lovci Mamutů z Dolních Věstonic*. Sborník Národního Muzea v Praze (Acta Musei Nationales Pragae) 48B:148–151.

Kunter, M., & Wahl, J. (1992). Das Femurfragment eines Neandertalers aus der Stadelhöhle des Hohlensteins im Lonetal. *Fundberichte aus Baden-Württemberg* 17:111–124.

Lahr, M.M. (1996). *The Evolution of Modern Human Diversity*. Cambridge UK: Cambridge University Press.

Lalueza, C., Pérez-Pérez, A., Chimenos, E., Maroto, J., & Turbón, D. (1993). Estudi radiogràfic i microscòpic de la mandíbula de Banyoles: Patologies i estat de conservació. In J. Maroto (ed.), *La Mandíbula de Banyoles en el Context dels Fossils Humans del Pleistocè*. Centre d'Investigacions Arqueològiques de Girona Sèrie Monogràfica 13:135–144.

Latimer, B., & Ward, C.V. (1993). The thoracic and lumbar vertebrae. In A. Walker & R. Leakey (eds.), *The Nariokotome* Homo erectus *Skeleton*. Cambridge, MA: Harvard University Press, pp. 266–293.

Lavelle, C.L.B., & Moore, W.J. (1973). The incidence of agenesis and polygenesis in the primate dentition. *American Journal of Physical Anthropology* 38:671–680.

Lebel, S., & Trinkaus, E. (2002a) A carious Neandertal molar from the Bau de l'Aubesier, Vaucluse, France. *Journal of Archaeological Science* 28:555–557.

Lebel, S., & Trinkaus, E. (2002b). Middle Pleistocene human remains from the Bau de l'Aubesier. *Journal of Human Evolution* 43:659–685.

Legoux, P. (1975). Présentation des dents des restes humains de l'Abri Pataud. In H.L. Movius, Jr. (ed.), *Excavation of the Abri Pataud, Les Eyzies (Dordogne)*. American School of Prehistoric Research Bulletin 30: 262–305.

Leicher, H. (1928). Die Vererbung anatomischer Varianten der Nase, ihrer Nebenhöhlen und des Gehörorganes. In O. Korner (ed.), *Die Ohrenheilkunde der Gegenwart und ihre Grenzgebiete in Einzeldarstellungen*. München: J.F. Bergman.

Lengsfeld, M., Kaminsky, J., Merz, B., & Franke, R.P. (1996). Sensitivity of femoral strain pattern analyses to resultant and muscle forces at the hip joint. *Medical Engineering and Physics* 18:70–78.

Levangie, P.K., & Norkin, C.C. (2001). *Joint Structure and Function*, 3rd ed. Philadelphia: F.A. Davis.

Levers, B.G.H., & Darling, A.I. (1983). Continuous eruption of some adult human teeth of ancient populations. *Archives of Oral Biology* 28:401–408.

Lieberman, D.E., & Crompton, A.W. (1998). Responses of bone to stress: Constraints on symmorphosis. In E.R. Wiebel, C.R. Taylor, & L. Bolis (eds.), *Principles of Animal Design. The Optimization and Symmorphosis Debate*. Cambridge UK: Cambridge University Press, pp. 78–86.

Lisoněk, P. (1992). Ossicula auditus mladopaleolitických lovců mamutů z Dolních Věstonic. In E. Vlček (ed.), *Lovci mamutů z Dolních Věstonic*. Sborník Národního Muzea v Praze (Acta Musei Nationales Pragae) 48B:65–67.

Liversidge, H.M. (1994). Accuracy of age estimation from developing teeth of a population of known age (0 to 5.4 years). *International Journal of Osteoarchaeology* 4:37–46.

Llunggren, A.E. (1979). Clavicular function. *Acta Orthopaedica Scandinavica* 50:261–268.

Löhr, P. (1894). Ueber den *Sulcus praeauricularis* des Darmbeins und änhliche Furchen anderen Knochen. *Anatomischer Anzeiger* 9:521–536.

Lorenzo, C., Arsuaga, J.L., & Carretero, J.M. (1999). Hand and foot remains from the Gran Dolina Early Pleistocene site (Sierra de Atapuerca, Spain). *Journal of Human Evolution* 37:501–522.

Lovejoy, C.O. (1985). Dental wear in the Libben population: Its functional pattern and role in the determination of adult skeletal age at death. *American Journal of Physical Anthropology* 68:47–56.

Lovejoy, C.O., & Heiple, K.G. (1981). The analysis of fractures in skeletal populations with an example from the Libben site, Ottawa County, Ohio. *American Journal of Physical Anthropology* 55:529–541.

Lovejoy, C.O., Heiple, K.G., & Burstein, A.H. (1973). The gait of *Australopithecus*. *American Journal of Physical Anthropology* 38:757–779.

Lovejoy, C.O., Burstein, A.H., & Heiple, K.G. (1976). The biomechanical analysis of bone strength: A method and its application to platycnemia. *American Journal of Physical Anthropology* 44:489–505.

Lovejoy, C.O., Meindl, R.S., Mensforth, R.P., & Barton, T.J. (1985a). Multifactorial determination of skeletal age at death: A method and blind tests of its accuracy. *American Journal of Physical Anthropology* 68:1–14.

Lovejoy, C.O., Meindl, R.S., & Pryzbeck, T.R. (1985b). Chronological metamorphosis of the auricular surface of the ilium: A new method for the determination of adult skeletal age at death. *American Journal of Physical Anthropology* 68:15–28.

Lumley, M.A. de (1973). Anténéandertaliens et Néandertaliens du bassin méditerranéen occidental européen. *Études du Quaternaires* 2:1–626.

Macho, G. (1991). Anthropological evaluation of left-right differences in the femur of southern African populations. *Anthropologischer Anzeiger* 49:207–217.

MacLarnon, A. (1993). The vertebral canal. In A. Walker & R. Leakey (eds.), *The Nariokotome* Homo erectus *Skeleton*. Cambridge MA: Harvard University Press, pp. 359–390.

Makowsky, A. (1892). Der diluviale Mensch im Löss von Brünn. *Mitteilungen der Anthropologischen Gesellschaft Wien* 22:73–84.

Mallegni, F., & Palma di Cesnola, A. (1994). Les restes humains découverts dans les niveaux Gravettiens de la Grotte Paglicci (Rignano Garganico, Pouilles, Italie). *Anthropologie (Brno)* 32:45–57.

Mallegni, F., & Trinkaus, E. (1997). A reconsideration of the Archi 1 Neandertal mandible. *Journal of Human Evolution* 33:651–668.

Mallegni, F., Bertoldi, F., & Manolis, S.K. (1999). The Gravettian female human skeleton from Grotta Paglicci, south Italy. *Homo* 50:127–148.

Mallegni, F., Bertoldi, F., & Manolis, S.K. (2000). Paleobiology of two Gravettian skeletons from Veneri cave (Parabita, Puglia, Italy). *Homo* 51:235–257.

Malý, J. (1939). Lebky fosilního člověka v Dolních Věstonicích. *Anthropologie (Prague)* 17:171–190.

Manzi, G., Gracia, A., & Arsuaga, J.L. (2000). Cranial discrete traits in the Middle Pleistocene humans from Sima de los Huesos (Sierra de Atapuerca, Spain). Does hypostosis represent any increase in "ontogenetic stress" along the Neanderthal lineage? *Journal of Human Evolution* 38:411–446.

Martin, H. (1923). *L'Homme Fossile de La Quina*. Paris: Librairie Octave Doin.

Martin, R. (1928). *Lehrbuch der Anthropologie*, 2nd ed. Jena: Fischer Verlag.

Martin, R.B. (2003). Functional adaptation and fragility of the skeleton. In S.C. Agarwal & S.D. Stout (eds.), *Bone Loss and Osteoporosis: An Anthropological Perspective*. New York: Kluwer Academic/Plenum, pp. 121–138.

Maška, K.J. (1895). Diluviální člověk v Předmostí. *Časopis Vlastivědného spolku musejního Olomouc* 12:4–7.

Mason, S., Hather, J., & Hillman, G. (1994). Preliminary investigation of the plant macro-remains from Dolní Věstonice II and its implications for the role of plant foods in Palaeolithic and Mesolithic Europe. *Antiquity* 68:48–57.

Matiegka, J. (1934). *Homo předmostensis. Fosilní člověk z Předmostí na Moravě I. Lebky*. Prague: Česká Akademie Věd a Umění.

Matiegka, J. (1938). *Homo Předmostensis. Fosilní člověk z Předmostí na Moravě II. Ostatní části kostrové*. Prague: Česká Akademie Věd a Umění.

Mayhall, J.T. (1978). Canadian Inuit caries experience 1969–73. *Journal of Dental Research* 54:1245.

McCown, T.D., & Keith, A. (1939). *The Stone Age of Mount Carmel II: The Fossil Human Remains from the Levalloiso-Mousterian*. Oxford: Clarendon Press.

McHenry, H.M. (1992). Body size and proportions in early hominids. *American Journal of Physical Anthropology* 87:407–431.

McLeish, R.D., & Charnley, J. (1970). Abduction forces in the one-legged stance. *Journal of Biomechanics* 3:191–209.

Mednikova, M., & Trinkaus, E. (2001) Femoral midshaft diaphyseal cross-sectional geometry of the Sunghir 1 and 4 Gravettian human remains. *Anthropologie (Brno)* 39:135–141.

Mehta, C., & Patel, N. (1999). *StatXact 4 for Windows*. Cambridge, MA: Cytel Software.

Meindl, R.S., & Lovejoy, C.O. (1985). Ectocranial suture closure: A revised method for the determination of skeletal age at death based on the lateral-anterior sutures. *Journal of Physical Anthropology* 68:57–66.

Mercier, N., Valladas, H., Joron, J.L., Reyes, J.L., Lévêque, F., & Vandermeersch, B. (1991). Thermoluminescence dating of the late Neanderthal remains from Saint-Césaire. *Nature* 351:737–739.

Meyer, M. (1979). A comparison of hallux abducto valgus in two ancient populations. *Journal of the American Podiatry Association* 69:65–68.

Miles, A.E.W. (1958). The assessment of age from the dentition. *Proceedings of the Royal Society of Medicine* 51:1057–1060.

Miles, A.E.W. (1962). Assessment of the ages of a population of Anglo-Saxons from their dentitions. *Proceedings of the Royal Society of Medicine* 55:881–886.

Miles, A.E.W., & Grigson, C. (1990). *Colyer's Variations and Diseases of the Teeth of Animals*, rev. ed. Cambridge UK: Cambridge University Press.

Milner, G.R., & Larsen, C.S. (1991). Teeth as artifacts of human behavior: intentional mutilation and accidental modification. In M.A. Kelley & C.S. Larsen (eds.), *Advances in Dental Anthropology*. New York: Wiley-Liss, pp. 357–378.

Milunsky, J.M., Maher, T.A., & Metzenberg, A.B. (2003). Molecular, biochemical, and phenotypic analysis of a hemizygous male with a severe atypical phenotype for X-linked dominant Conradi-Hunermann-Happle Syndrome and a mutation in EBP. *American Journal of Medical Genetics* 116A:249–254.

Mincer, H.H., Harris, E.F., & Berryman, H.E. (1993). The A.B.F.O. study of third molar development and its use as an estimator of chronological age. *Journal of Forensic Sciences* 38:379–390.

Moerman, M.L. (1982). Growth of birth canal in adolescent girls. *American Journal of Obstetrics and Gynecology* 143:528–532.

Moorrees, C.F.A. (1957). *The Aleut Dentition*. Cambridg, MA: Harvard University Press.

Moorrees, C.F.A., Fanning, E.A., & Hunt, E.E. (1963a). Age variation of formation stages for ten permanent teeth. *Journal of Dental Research* 42:1490–1502.

Moorrees, C.F.A., Fanning, E.A., & Hunt, E.E. (1963b). Formation and resorption of three deciduous teeth in children. *American Journal of Physical Anthropology* 21:205–213.

Münter, A.H. (1936). A study of the lengths of the long bones of the arms and legs in man, with special reference to Anglo-Saxon skeletons. *Biometrika* 28:258–294.

Murail, P., Brůžek, J., & Braga, J. (1999). A new approach to sexual diagnosis in past populations. Practical adjustments from van Vark's procedure. *International Journal of Osteoarchaeology* 9:39–53.

Murray, W.M., Buchanan, T.S., & Delp, S.L. (2002). Scaling of peak moment arms of elbow muscles with upper extremity bone dimensions. *Journal of Biomechanics* 35: 19–26.

Musgrave, J.H. (1970). *An Anatomical Study of the Hands of Pleistocene and Recent Man.* Ph.D. thesis, University of Cambridge, Cambridge, UK.

Musil, R. (1994). Hunting game from the culture layer at Pavlov. In J. Svoboda (ed.), *Pavlov I, Excavations 1952–53*. Études et Recherches Archéologiques de l'Université de Liège 66/Dolní Věstonice Studies 2. Brno: Archeologický ústav AV ČR, pp. 170–196.

Musil, R. (1997). Hunting game analysis. In J. Svoboda (ed.), *Pavlov I, Northwest. The Upper Paleolithic Burial and Its Settlement Context.* Dolní Věstonice Studies 4. Brno: Archeologický ústav AV ČR, pp. 443–468.

Musil, R. (2002). Přírodní prostředí jako ekonomická báze paleolitických lovců. In J. Svoboda, P. Havlíček, V. Ložek, J. Macoun, R. Musil, A. Přichystal, H. Svobodová, & E. Vlček, *Paleolit Moravy a Slezska—The Paleolithic of Moravia and Silesia.* Dolní Věstonice Studies 8, Brno: Archeologický ústav AV ČR, pp. 52–66.

Nadel, D., & Hershkovitz, I. (1991). New subsistence data and human remains from the earliest Levantine Epipalaeolithic. *Current Anthropology* 32:631–635.

Nagurka, M.L., & Hayes, W.C. (1980). An interactive graphics package for calculating cross-sectional properties of complex shapes. *Journal of Biomechanics* 13:59–64.

Napier, J.R. (1956). The prehensile movements of the human hand. *Journal of Bone and Joint Surgery* 38B:902–913.

Nepstad-Thornberry, T.N., Whitelaw D.C., & Van Gerven, D.P. (2003). Sex determination from the human hip bone: A response to Bruzek. *American Journal of Physical Anthropology Supplement* 36:157.

Newman, M.T. (1953). The application of ecological rules to the racial anthropology of the aboriginal New World. *American Anthropologist* 55:311–327.

Niewoehner, W.A. (2000). *The Functional Anatomy of Late Pleistocene and Recent Human Carpometacarpal and Metacarpophalangeal Articulations.* Ph.D. thesis, University of New Mexico, Albuquerque.

Niewoehner, W.A. (2001). Behavioral inferences from the Skhul/Qafzeh early modern human hand remains. *Proceedings of the National Academy of Sciences USA* 98:2979–2984.

Niewoehner, W.A., Weaver, A.H., & Trinkaus, E. (1997). Neandertal capitate-metacarpal articular morphology. *American Journal of Physical Anthropology* 103:219–233.

Nordborg, M. (1998). On the probability of Neanderthal ancestry. *American Journal of Human Genetics* 63:1237–1240.

Novotný, V. (1975). Discriminantanalyse der Geschlechtsmerkmale auf dem Os coxae beim Menschen. *Thirteenth Congress, Czechoslovak Anthropological Association, Brno.*

Novotný, V. (1981). *Pohlavní rozdíly a identifikace pohlaví pánevní kosti (Sex Differences and Identification of Sex in Pelvic Bone).* Ph.D. thesis, Masarykova Univerzita, Brno.

Novotný, V. (1983). Sex differences of pelvis and sex determination in paleoanthropology. *Anthropologie (Brno)* 21:65–72.

Novotný, V. (1992). Pánev a sexuální dimorfismus lovců z Dolních Věstonic. In E. Vlček (ed.), *Lovci mamutů z Dolních Věstonic.* Sborník Národního Muzea v Praze (Acta Musei Nationales Pragae) 48B:152–163.

Novotný, V. (2003). Enigmatic skeleton DV XV (Upper Paleolithic, Dolní Věstonice), south Moravia. In J. Brůžek, B. Vandermeersch, & M.D. Garralda (eds.), *Changements Biologiques et Culturels en Europe de la Fin du Paléolithique Moyen au Néolithique.* Talence: Laboratoire d'Anthropologie des Populations du Passé, Université de Bordeaux 1, pp.129–143.

Novotný, V., & Brůžek, J. (1999). Methodological aspects of fossil sex assessment: Revision of Gravettian finds from Dolní Věstonice, Moravia (Czech Republic). *Fourth International Anthropological Congress of Aleš Hrdlička, Prague.*

Nowell, G.W. (1978). An evaluation of the Miles method of ageing using the Tepe Hissar dental sample. *American Journal of Physical Anthropology* 49:271–276.

Ogawa, K., & Yoshida, A. (1998). Throwing fracture of the humeral shaft. An analysis of 90 patients. *American Journal of Sports Medicine* 26:242–246.

Ogilvie, M.D., Curran, B.K., & Trinkaus, E. (1989) The incidence and patterning of dental enamel hypoplasias among the Neandertals. *American Journal of Physical Anthropology* 79:25–41.

Ogilvie, M.D., Hilton, C.E., & Ogilvie, C.D. (1998). Lumbar anomalies in the Shanidar 3 Neandertal. *Journal of Human Evolution* 35:597–610.

Oliva, M. (1998). Geografie moravského gravettienu. *Památky archeologické* 89:39–63.

Oliva, M. (2000). The Brno II Upper Paleolithic burial. In W. Roebroeks, M. Mussi, J. Svoboda, & K. Fennema (eds.), *Hunters of the Golden Age.* Leiden: Leiden University Press, pp. 143–153.

Olivier, G. (1951–1956). Anthropologie de la clavicule. *Bulletins et Mémoires de la Société d'Anthropologie de Paris,* Série 10, 2:67–99, 121–157; 3:269–279; 4:90–100; 5:35–56, 144–153; 6:283–302; 7:225–261, 404–447.

Opravil, E. (1994). The vegetation. In J. Svoboda (ed.), *Pavlov I, Excavations 1952–53.* Études et Recherches Archéologiques de l'Université de Liège 66/Dolní Věstonice Studies. 2, Brno: Archeologický ústav AV ČR, pp. 163–167.

Ortner, D.J., & Putschar, W.G.J. (1981). Identification of pathological conditions in human skeletal remains. *Smithsonian Contributions to Anthropology* 28:1–479.

Otte, M. (1981). *Le Gravettien en Europe Centrale.* Bruge: De Tempel.

Paoli, G., Parenti, R., & Sergi, S. (1980). Gli scheletri mesolitici della Caverna delle Arene Candide (Liguria). *Memorie dell'Istituto Italiano de Paleontologia Umana* ns3:33–154.

Pap, I., Tillier, A.M., Arensburg, B., & Chech, M. (1996). The Subalyuk Neanderthal remains (Hungary): A re-examination. *Annales Historico-Naturales Musei Nationalis Hungarici* 88:233–270.

Pearson, O.M. (1997). *Postcranial Morphology and the Origin of Modern Humans*. Ph.D. thesis, State University of New York, Stony Brook.

Pearson, O.M. (2000). Activity, climate and postcranial robusticity. Implications for modern human origins and scenarios of adaptive change. *Current Anthropology* 41:569–607.

Pedersen, D.R., Brand, R.A., & Davy, D.T. (1997). Pelvic muscle and acetabular contact forces during gait. *Journal of Biomechanics* 30:959–965.

Pedersen, P.O. (1966). Nutritional aspects of dental caries. *Odontologisk Revy* 17:91–100.

Pettitt, P.B., & Bader, N.O. (2000). Direct AMS radiocarbon dates for the Sungir mid Upper Palaeolithic burials. *Antiquity* 74:269–270.

Pettitt, P.B., Richards, M., Maggi, R., & Formicola, V. (2003). The Gravettian burial known as the Prince ("Il Principe"): New evidence for his age and diet. *Antiquity* 77:15–19.

Pettitt, P.B., & Trinkaus, E. (2000). Direct radiocarbon dating of the Brno 2 Gravettian human remains. *Anthropologie (Brno)* 38:149–150.

Phenice, T.W. (1969). A newly developed visual method of sexing the os pubis. *American Journal of Physical Anthropology* 30:297–301.

Ponce de León, M.S., & Zollikofer, C.P.E. (1999). New evidence from Le Moustier 1: Computer-assisted reconstruction and morphometry of the skull. *Anatomical Record* 254:474–489.

Přichystal, A. (2002). Zdroje kamenných surovin. In J. Svoboda, P. Havlíček, V. Ložek, J. Macoun, R. Musil, A. Přichystal, H. Svobodová, & E. Vlček, *Paleolit Moravy a Slezska—The Paleolithic of Moravia and Silesia*. Dolní Věstonice Studies 8. Brno: Archeologický ústav AV ČR, pp. 67–76.

Pusch, C.M., & Bachmann, L. (2004). Spiking of contemporary human template DNA with ancient DNA extracts induces mutations under PCR and generates nonauthentic mitochondrial sequences. *Molecular Biology and Evolution* 21:957–964.

Radlauer, C. (1908). Beiträge zur Anthropologie des Kreuzbeines. *Gegenbaurs Morphologisches Jahrbuch* 38:323–447.

Rak, Y. (1991). The pelvis. In O. Bar Yosef & B. Vandermeersch (eds.), *Le Squelette Moustérien de Kébara 2*. Paris: Éditions du C.N.R.S., pp. 147–156.

Rak, Y. (1998). Does any mousterian cave present evidence of two hominid species? In T. Akazawa, K. Aoki, & O. Bar-Yosef (eds.), *Neandertals and Modern Humans in Western Asia* New York: Plenum, pp. 353–366.

Reid, D.J., & Dean, M.C. (2000). The timing of linear hypoplasias on human anterior teeth. *American Journal of Physical Anthropology* 113:135–140.

Relethford, J.H. (2001). *Genetics and the Search for Modern Human Origins*. New York: Wiley-Liss.

Rhoads, J.G., & Trinkaus, E. (1977). Morphometrics of the Neandertal talus. *American Journal of Physical Anthropology* 46:29–44.

Rhoads, M.L., & Franciscus, R.G. (1996). Mandibular notch crest orientation in Neandertals and recent humans (abstract). *American Journal of Physical Anthropology Supplement* 22:196.

Rice, W.R. (1989). Analyzing tables of statistical tests. *Evolution* 43:223–225.

Richards, G.D., Jabbour, R.S., & Anderson, J.Y. (2003). Medial mandibular ramus: Ontogenetic, idiosyncratic, and geographic variation in recent *Homo*, great apes, and fossil hominids. *British Archaeological Reports International Series* 1138.

Richards, M.P., Pettitt, P.B., Stiner, M.C., & Trinkaus, E. (2001) Stable isotope evidence for increasing dietary breadth in the European mid–Upper Paleolithic. *Proceedings of the National Academy of Sciences USA* 98:6528–6532.

Roberts, D.F. (1978). *Climate and Human Variability*. 2nd ed. Menlo Park, CA: Cummings.

Roebroeks, W., Mussi, M., Svoboda, J., & Fennema, K. (Eds.). (2000). *Hunters of the Golden Age*. Leiden: University of Leiden Press.

Rohlmann, A., Mössner, U., Bergmann, G., & Kölbel, R. (1982). Finite-element-analysis and experimental investigation of stresses in a femur. *Journal of Biomedical Engineering* 4:241–246.

Rosas, A. (2001). Occurrence of Neanderthal features in mandibles from the Atapuerca-SH site. *American Journal of Physical Anthropology* 114:74–91.

Rosenberg, K.R., Lu, Z., & Ruff, C.B. (1999). Body size, body proportions and encephalization in the Jinniushan specimen (abstract). *American Journal of Physical Anthropology Supplement* 28:235.

Rosenberg, K., & Trevathan, W. (2002). Birth, obstetrics and human evolution. *British Journal of Obstetrics and Gynaecology* 109:1199–1206.

Rougier, H. (2003). *Étude Descriptive et Comparative de Biache-Saint-Vaast 1 (Biache-Saint-Vaast, Pas-de-Calais, France)*. Thèse de Doctorat, Université de Bordeaux 1, Talence.

Ruff, C.B. (1990). Body mass and hindlimb bone cross-sectional and articular dimensions in anthropoid primates. In J. Damuth & B.J. MacFadden (eds.), *Body Size in Mammalian Paleobiology: Estimation and Biological Implications*. New York: Cambridge University Press, pp. 119–149.

Ruff, C.B. (1991). Climate and body shape in human evolution. *Journal of Human Evolution* 21:81–105.

Ruff, C.B. (1993). Climatic adaptation and hominid evolution: The thermoregulatory imperative. *Evolutionary Anthropology* 2:53–60.

Ruff, C.B. (1994). Morphological adaptation to climate in modern and fossil hominids. *Yearbook of Physical Anthropology* 37:65–107.

Ruff, C.B. (1995). Biomechanics of the hip and birth in early *Homo*. *American Journal of Physical Anthropology* 98:527–574.

Ruff, C.B. (2000a). Biomechanical analyses of archaeological human skeletal samples. In M.A. Katzenburg & S.R. Saunders (eds.), *Biological Anthropology of the Human Skeleton*. New York: Alan R. Liss, pp. 71–102.

Ruff, C.B. (2000b). Body size, body shape, and long bone strength in modern humans. *Journal of Human Evolution* 38:269–290.

References

Ruff, C.B. (2000c). Prediction of body mass from skeletal frame size in elite athletes. *American Journal of Physical Anthropology* 113:507–517.

Ruff, C.B., & Hayes, W.C (1983a). Cross-sectional geometry of Pecos Pueblo femora and tibiae—a biomechanical investigation. I. Method and general patterns of variation. *American Journal of Physical Anthropology* 60:359–381.

Ruff, C.B., & Hayes, W.C (1983b). Cross-sectional geometry of Pecos Pueblo femora and tibiae – a biomechanical investigation. II. Sex, age, and side differences. *American Journal of Physical Anthropology* 60:383–400.

Ruff, C.B., Scott, W.W., & Liu, A.Y.C. (1991). Articular and diaphyseal remodeling of the proximal femur with changes in body mass in adults. *American Journal of Physical Anthropology* 86:397–413.

Ruff, C.B., Walker, A., & Trinkaus, E. (1994). Postcranial robusticity in *Homo*, III: Ontogeny. *American Journal of Physical Anthroplogy* 93:35–54.

Ruff, C.B., Trinkaus, E., Walker, A., & Larsen, C.S. (1993). Postcranial robusticity in *Homo*, I: Temporal trends and mechanical interpretations. *American Journal of Physical Anthropology* 91:21–53.

Ruff, C.B., Walker, A., & Trinkaus, E. (1994). Postcranial robusticity in *Homo*, III: Ontogeny. *American Journal of Physical Anthropology* 93:35–54.

Ruff, C.B., Trinkaus, E., & Holliday, T.W. (1997). Body mass and encephalization in Pleistocene *Homo*. *Nature* 387:173–176.

Ruff, C.B., Trinkaus, E., & Holt, B. (2000). Lifeway changes as shown by postcranial skeletal robustness (abstract). *American Journal of Physical Anthropology Supplement* 30:266.

Ruff, C.B., Trinkaus, E., & Holliday, T.W. (2002) Body proportions and size. In J. Zilhão & E. Trinkaus (eds.), *Portrait of the Artist as a Child. The Gravettian Human Skeleton from the Abrigo do Lagar Velho and its Archeological Context*. Trabalhos de Arqueologia 22:365–391.

Rybicki, E.F., Simonen, F.A., & Weis, E.B., Jr. (1972). On the mathematical analysis of stress in the human femur. *Journal of Biomechanics* 5:203–215.

Rybníčková, E., & Rybníček, K. (1991). The environment of the Pavlovian—Palaeoecological results from Bulhary, South Moravia. In J. Kovar-Eder (ed.), *Palaeovegetational Development in Europe, Pan-European Palaeobotanical Conference Vienna*. Vienna: Naturhistorisches Museum Wien, pp. 73–79.

Sakaue, K. (1997). Bilateral asymmetry of the humerus in Jomon people and modern Japanese. *Anthropological Science* 105:231–246.

Sakura, H. (1970). Dentition of the Amud man. In H. Suzuki & F. Takai (eds.), *The Amud Man and His Cave Site*. Tokyo: Academic Press of Japan, pp. 207–229.

Santa Luca, A.P. (1978). A re-examination of presumed Neandertal-like fossils. *Journal of Human Evolution* 7:619–636.

Savishinsky, J.S., & Hara, H.S. (1981). Hare. In J. Helm (ed.), *Handbook of North American Indians 6: Subarctic*. Washington, DC: Smithsonian Institution, pp. 314–325.

Scheuer, L., & Black, S. (2000). *Developmental Juvenile Osteology*. San Diego: Academic Press.

Schmitz, R.W., Serre, D., Bonani, G., Feine, S., Hillgruber, F., Krainitzki, H., Pääbo, S., & Smith, F.H. (2002). The Neandertal type site revisited: Interdisciplinary investigations of skeletal remains from the Neander Valley, Germany. *Proceedings of the National Academy of Sciences USA* 99:13342–13347.

Schour, I., & Massler, M. (1941). The development of the human dentition. *Journal of the American Dental Association* 28:1153–1160.

Schour, I., & Massler, M. (1944). *Development of the Human Dentition*. Chicago: American Dental Association.

Schulter-Ellis, F.P., Schmidt, D.J., Hayek, L.C., & Craig, J. (1983). Determination of sex with a discriminant analysis of new pelvic bone measurements: Part I. *Journal of Forensic Sciences* 28:169–180.

Schultz, A.H. (1930). The skeleton of the trunk and limbs of higher Primates. *Human Biology* 2:303–438.

Schumann, B.A. (1995). *Biological Evolution and Population Change in the European Upper Palaeolithic*. Ph.D. thesis, University of Cambridge, Cambridge, UK.

Scott, G.R., & Turner, C.G., II (1988). Dental anthropology. *Annual Review of Anthropology* 17:99–126.

Šefčáková, A., Mizera, I., & Thurzo, M. (1999). New human fossil remains from Slovakia. The skull from Moča (Late Upper Paleolithic, South Slovakia). *Bulletin Slovenskej Antropologickej Spoločnosti* 2:55–63.

Sekiguchi, M., Yabuki, S., Satoh, K., & Kikuchi, S. (2004). An anatomic study of the sacral hiatus: a basis for successful caudal epidural block. *Clinical Journal of Pain* 20:51–54.

Senut, B. (1981). *L'Humérus et ses Articulations chez les Hominidés Plio-Pleistocènes*. Paris: Editions du C.N.R.S.

Sergi, S. (1942–1946). L'uomo di Saccopastore (Il cranio del secondo paleantropo di Saccopastore). *Paleontographia Italica* 42:25–164.

Sergi, S. (1944). Craniometria e craniografia del primo paleantropo di Saccopastore. *Richerche di Morfologia* 20–21:733–791.

Sergi, S. (1974). *Il Cranio Neandertaliano del Monte Circeo (Circeo I)*. Rome: Accademia Nazionale dei Lincei.

Sergi, S., Parenti, R., & Paoli, G. (1974). Il giovane paleolitico della Caverna delle Arene Candide. *Studi di Paletnologia, Paleoantropologia, Paleontologia e Geologia del Quaternario* 2:13–38.

Serre, D., Langaney, A., Chech, M., Teschler-Nicola, M., Paunovič, M., Mennecier, P., Hofreiter, M., Possnert, G., & Pääbo, S. (2004) No evidence of Neandertal mtDNA contribution to early modern humans. *Public Library of Science Biology* 2:313–317.

Shackelford, L.L., & Trinkaus, E. (2002). Late Pleistocene human femoral diaphyseal curvature. *American Journal of Physical Anthropology* 118:359–370.

Shipman, P., Walker, A., & Bichell, D. (1985). *The Human Skeleton*. Cambridge, MA: Harvard University Press.

Sillence, D.O., Horton, W.A., & Rimoin, D.L. (1979). Morphologic studies in the skeletal dysplasias. *American Journal of Pathology* 96:813–870.

Sjøvold, T. (1984). A report on the heritability of some cranial measurements and non-metric traits. In G.N. van Vark & W.W. Howells (eds.), *Multivariate Statistics in Physical Anthropology*. Dordrecht: D. Reidel, pp. 223–246.

Sjøvold, T. (1988). Geschlechtsdiagnose am Skelett. In R. Knussmann (ed.), *Anthropologie Handbuch der vergleichenden Biologie des Menschen*. Stuttgart: Gustav Fischer Verlag, pp. 444–480.

Škrdla, P., & Lukáš, M. (2000). A contribution to the question of the geographical setting of Pavlovian localities in Moravia. *Přehled výzkumů* 41:21–33.

Sládek, V. (2000). *Évolution des Hominidés en Europe Centrale durant le Pléistocène Supérieur: Origines des Hommes Anatomiquement Modernes*. Thése de Doctorat, Université de Bordeaux 1, Talence.

Sládek, V., Trinkaus, E., Hillson, S.W., & Holliday, T.W. (2000). *The People of the Pavlovian: Skeletal Catalogue and Osteometrics of the Gravettian Fossil Hominids from Dolní Věstonice and Pavlov*, Dolní Věstonice Studies 5. Brno: Archeologický ústav AV ČR.

Sládek, V., Šefčáková, A., & Brůžek, J. (2001). Sex dimorphism among the Early Upper Paleolithic hominids from central Europe: Cranial and pelvic metric variation (abstract). *Journal of Human Evolution* 43:A23.

Sládek, V., Trinkaus, E., Šefčáková, A., & Halouzka, R. (2002). Morphological affinities of the Šal'a 1 frontal bone. *Journal of Human Evolution* 43:787–815.

Slome, D. (1929). The osteology of a Bushman tribe. *Annals of the South African Museum* 24:33–60.

Smith, B.H. (1984). Patterns of molar wear in hunter-gatherers and agriculturalists. *American Journal of Physical Anthropology* 63:39–56.

Smith, F.H. (1976). The Neandertal remains from Krapina. A descriptive and comparative study. *Department of Anthropology, University of Tennessee, Report of Investigations* 15:1–359.

Smith, F.H. (1978). Evolutionary significance of the mandibular foramen area in Neandertals. *American Journal of Physical Anthropology* 48:523–532.

Smith, F.H., & Ranyard, G.C. (1980). Evolution of the supraorbital region in Upper Pleistocene fossil hominids from south-central Europe. *American Journal of Physical Anthropology* 53:589–610.

Smith, F.H., Falsetti, A.B., & Donnelly, S.M. (1989). Modern human origins. *Yearbook of Physical Anthropology* 32:35–68.

Smith, F.H., Trinkaus, E., Pettitt, P.B., Karavanić, I., & Paunović, M. (1999). Direct radiocarbon dates for Vindija G_1 and Velika Pećina Late Pleistocene hominid remains. *Proceedings of the National Academy of Sciences USA* 96:12281–12286.

Smith, R.J. (1980). Rethinking allometry. *Journal of Theoretical Biology* 87:97–111.

Soffer, O. (1989). Storage, sedentism and the Eurasian Palaeolithic record. *Antiquity* 63:719–732.

Soffer, O. (2000). Gravettian technologies in social contexts. In W. Roebroeks, M. Mussi, J. Svoboda, & K. Fennema (eds.), *Hunters of the Golden Age*. Leiden: Leiden University Press, pp. 59–75.

Sognnaes, R.F. (1956). Histologic evidence of developmental lesions in teeth originating from Paleolithic, prehistoric, and ancient man. *American Journal of Pathology* 32:547–577.

Sokal, R.R., & Rohlf, F.J. (1981). *Biometry*, 2nd ed. San Francisco: Freeman.

Spoor, F. (2002). The auditory ossicles. In J. Zilhão & E. Trinkaus (eds.), *Portrait of the Artist as a Child. The Gravettian Human Skeleton from the Abrigo do Lagar Velho and Its Archeological Context*. Trabalhos de Arqueologia 22:293–296.

Sprecher, H. (1932). *Morphologische Untersuchungen an der Fibula des Menschen unter Berücksichtigung anderer Primaten*. Zürich: Art. Institut Orell Füssli.

Steele, D.G., & Bramblett, C.A. (1988). *The Anatomy and Biology of the Human Skeleton*. College Station: Texas A&M University Press.

Stefan, V.H., & Trinkaus, E. (1998a). Discrete trait and dental morphometric affinities of the Tabun 2 mandible. *Journal of Human Evolution* 34:443–468.

Stefan, V.H., & Trinkaus, E. (1998b). La Quina 9 and Neandertal mandibular variability. *Bulletins et Mémoires de la Société d'Anthropologie de Paris* ns10:293–324.

Stewart, T.D., Tiffany, M., Angel, J.L., & Kelley, J.O. (1986). Description of the human skeleton. In F. Wendorf, R. Schild & A.E. Close (eds.), *The Wadi Kubbaniya Skeleton: A Late Paleolithic Burial from Southern Egypt*. Dallas: Southern Methodist University, pp. 49–70.

Stini, W.A. (1969). Nutritional stress and growth: Sex differences in adaptive response. *American Journal of Physical Anthropology* 31:417–426.

Stott, J.R.R., Hutton, W.C., & Stokes, I.A.F. (1973). Forces under the foot. *Journal of Bone and Joint Surgery* 55B:335–344.

Streeter, M., Stout, S.D., Trinkaus, E., Stringer, C.B., Roberts, M.B., & Parfitt, S.A. (2001). Histomorphometric age assessment of the Boxgrove 1 tibial diaphysis. *Journal of Human Evolution* 40:331–338.

Stringer C.B. (1990). British Isles. In R. Orban (ed.), *Hominid Remains: An Update. British Isles and Eastern Germany*. Brussels: Université Libre de Bruxelles, pp. 1–40.

Stringer, C.B. (2001). What happened to the Neandertals? *General Anthropology* 7:4–5.

Sutcliffe, A.J. (1970). Spotted hyaena: Crusher, gnawer, digester and collector of bones. *Nature* 227:1110–1113.

Suwa, G., White, T.D., & Howell, F.C. (1996). Mandibular postcanine dentition from the Shungura Formation, Ethiopia: Crown morphology, taxonomic allocations, and Plio-Pleistocene hominid evolution. *American Journal of Physical Anthropology* 101:247–282.

Suzuki, H. (1970). The skull of the Amud man. In H. Suzuki & F. Takai (eds.), *The Amud Man and His Cave Site*. Tokyo: University of Tokyo, pp. 123–206.

Suzuki, T., Kusumoto, A., Fujita, H., & De Shi, C. (1995). The fourth molar in a mandible found in a Jomon skeleton in Japan. *International Journal of Osteoarchaeology* 5:174–180.

Svoboda, J. (1988). A new male burial from Dolní Věstonice. *Journal of Human Evolution* 16:827–830.

Svoboda, J. (1989). Další objev paleolitického hrobu v Dolních Věstonicích. *Archeologické rozhledy* 41:233–242.

Svoboda, J. (Ed.). (1991). *Dolní Věstonice II Western Slope*. Études et Recherches Archéologiques de l'Université de Liège 54.

Svoboda, J. (Ed.). (1994). *Pavlov I, Excavations 1952–53*. Études et Recherches Archéologiques de l'Université de Liège 66/Dolní Věstonice Studies 2. Brno: Archeologický ústav AV ČR.

Svoboda, J. (1997a). Lithic industries of the 1957 area. In J. Svoboda (ed.), *Pavlov I, Northwest. The Upper Paleolithic Burial and Its Settlement Context*. Dolní Věstonice Studies 4. Brno: Archeologický ústav AV ČR, pp. 179–209.

Svoboda, J. (Ed.) (1997b). *Pavlov I, Northwest. The Upper Paleolithic Burial and Its Settlement context*. Dolní Věstonice Studies 4. Brno: Archeologický ústav AV ČR.

Svoboda, J. (1997c). Symbolisme gravettien en Moravie: Espace, temps et formes. *Bulletin de la Société préhistorique de l'Ariége-Pyrenées* 52:87–104.

Svoboda, J. (2000a). The depositional context of the Early Upper Paleolithic human fossils from the Koněprusy (Zlatý kůn) and Mladeč caves, Czech Republic. *Journal of Human Evolution* 38:523–536.

Svoboda, J. (2000b). *Předmostí*. Olomouic: Archeologické památky střední Moravy 1.

Svoboda, J. (2001a). Analysis of the large hunter's settlements: Excavation at Předmostí in 1992. *Archeologické rozhledy* 53:431–443.

Svoboda, J. (2001b). Analysis of the large hunter's settlements: Spatial structure and chronology of the site Dolní Věstonice II–IIa. *Památky archeologické* 92:74–97.

Svoboda, J. (2001c). The Pavlov site and the Pavlovian: A large hunter's settlement in a context. *Praehistoria* 2:97–115.

Svoboda, J. (2003). The Gravettian of Moravia: Landscape, settlement, and dwellings. In S.A. Vasil´ev, O. Soffer, & J. Kozlowski (eds.), *Perceived Landscapes and Built Environments*, British Archaeological Reports S1122:121–129.

Svoboda, J.A., & Bar-Yosef, O. (Eds.). (2003). *Stránská skála. Origins of the Upper Paleolithic in the Brno Basin, Moravia, Czech Republic*. American School of Prehistoric Research Bulletin 47, Dolní Věstonice Studies 10.

Svoboda, J.A., & Sedláčková, L. (Eds.). (2004). *The Gravettian along the Danube*. Dolní Věstonice Studies 11. Brno: Archeologický ústav AV ČR.

Svoboda, J., & Vlček, E. (1991). La nouvelle sépulture de Dolni Vestonice (DV XVI), Tchécoslovaquie. *L'Anthropologie* 95:323–328.

Svoboda, J., Škrdla, P., & Jarošová, L. (1993). Analyse einer Siedlungsfläche von Dolní Věstonice. *Archäologisches Korrespondenzblatt* 23:393–404.

Svoboda, J., Ložek, V., & Vlček, E. (1996). *Hunters between East and West. The Paleolithic of Moravia*. New York: Plenum.

Svoboda, J., Havlíček, P., Ložek, V., Macoun, J., Musil, R., Přichystal, A., Svobodová, H., & Vlček, E. (2002a). *Paleolit Moravy a Slezska*, 2nd ed. Dolní Věstonice Studies 8. Brno: Archeologický ústav AV ČR.

Svoboda, J.A., van der Plicht, J., & Kuželka, V. (2002b). Upper Palaeolithic and Mesolithic human fossils from Moravia and Bohemia (Czech Republic): Some new ^{14}C dates. *Antiquity* 76:957–962.

Svoboda, J., Péan, S., & Wojtal, P. (2005). Mammoth deposits and subsistence practices during the mid Upper Palaeolithic in Central Europe. *Quaternary International* 126–128:209–221.

Svobodová, H. (1991a). The pollen analysis of Dolní Věstonice II—western slope. In J. Svoboda (ed.), *Dolní Věstonice II Western Slope*. Études et Recherches Archéologiques de l'Université de Liège 54:75–88.

Svobodová, H. (1991b). Pollen analysis of the Upper Paleolithic triple burial at Dolní Věstonice. *Archeologické rozhledy* 43:505–510.

Swartz, S.M. (1990). Curvature of the forelimb bones of anthropoid primates: Overall allometric patterns and specialization in suspensory species. *American Journal of Physical Anthropology* 83:477–498.

Szilvassy, J. (1982). Zur Variation, Entwicklung und Vererbung der Stirnhöhlen. *Annales der Naturhistorisches Museum Wien* 84:97–125.

Szilvassy, J. (1986). Eine neue Methode zur intraserialen Analyse von Gräberfeldern. *Mitteilungen der Berliner Gesellschaft für Anthropologie, Ethnologie und Urgeschichte* 7:49–62.

Szombathy, J. (1925). Die diluvialen Menschenreste aus der Fürst-Johanns-Höhle bei Lautsch in Mähren. *Die Eiszeit* 2:1–26, 73–95.

Tague, R.G. (1989). Variation in pelvic size between males and females. *American Journal of Physical Anthropology* 80:59–71.

Tague, R.G. (1992). Sexual dimorphism in the human bony pelvis, with a consideration of the Neandertal pelvis from Kebara Cave, Israel. *American Journal of Physical Anthropology* 88:1–21.

Tanner, J.M., Hayashi, T., Preece, M.A., & Cameron, N. (1982). Increase in length of leg relative to trunk in Japanese children and adults from 1957–1977: Comparison with British and with Japanese Americans. *Annals of Human Biology* 9:411–423.

Tardieu, C., & Trinkaus, E. (1994) The early ontogeny of the human femoral bicondylar angle. *American Journal of Physical Anthropology* 95:183–195.

Taylor, M.E., Tanner, K.E., Freeman, M.A.R., & Yettram, A.L. (1996). Stress and strain distribution within the intact femur: Compression or bending? *Medical Engineering and Physics* 18:122–131.

Templeton, A.R. (2002) Out of Africa again and again. *Nature* 416:45–51.

Teschler-Nicola, M., & Trinkaus, E. (2001). Human remains from the Austrian Gravettian: The Willendorf femoral diaphysis and mandibular symphysis. *Journal of Human Evolution* 40:451–465.

Thieme, F.P., & Schull, W.J. (1957). Sex determination from the skeleton. *Human Biology* 29:242–273.

Thoma, A. (1984). Morphology and affinities of the Nazlet Khater man. *Journal of Human Evolution* 13:287–296.

Tillier, A.M. (1999). *Les Enfants Moustériens de Qafzeh*.

Interprétation Phylogénétique et Paléoauxologique. Paris: CNRS Éditions.

Tillier, A.M., Arensburg, B., & Duday, H. (1989). La mandibule et les dents du Néanderthalien de Kebara (Homo 2), Mont Carmel, Israël. *Paléorient* 15:39–58.

Tillier, A.M., Arensburg, B., Rak, Y., & Vandermeersch, B. (1995). Middle Palaeolithic dental caries: New evidence from Kebara (Mount Carmel, Israel). *Journal of Human Evolution* 29:189–192.

Trinkaus, E. (1975a). *A Functional Analysis of the Neandertal Foot*. Ph.D. thesis, University of Pennsylvania, Philadelphia.

Trinkaus, E. (1975b). Squatting among the Neandertals: A problem in the behavioral interpretation of skeletal morphology. *Journal of Archaeological Science* 2:327–351.

Trinkaus, E. (1976a). The evolution of the hominid femoral diaphysis during the Upper Pleistocene in Europe and the Near East. *Zeitschrift für Morphologie und Anthropologie* 67:291–319.

Trinkaus, E. (1976b). The morphology of European and Southwest Asian Neandertal pubic bones. *American Journal of Physical Anthropology* 44:95–104.

Trinkaus, E. (1977). A functional interpretation of the axillary border of the Neandertal scapula. *Journal of Human Evolution* 6:231–234.

Trinkaus, E. (1980). Sexual differences in Neanderthal limb bones. *Journal of Human Evolution* 9:377–397.

Trinkaus, E. (1981). Neanderthal limb proportions and cold adaptation. In C.B. Stringer (ed.), *Aspects of Human Evolution*. London: Taylor & Francis, pp. 187–219.

Trinkaus, E. (1983a). Functional aspects of Neandertal pedal remains. *Foot and Ankle* 3:377–390.

Trinkaus, E. (1983b). *The Shanidar Neandertals*. New York: Academic Press.

Trinkaus, E. (1984a). Does KNM-ER 1481A establish *Homo erectus* at 2.0 myr BP? *American Journal of Physical Anthropology* 64:137–139.

Trinkaus, E. (1984b). Neandertal pubic morphology and gestation length. *Current Anthropology* 25:509–514.

Trinkaus, E. (1987). The Neandertal face: Evolutionary and functional perspectives on a recent hominid face. *Journal of Human Evolution* 16:429–443.

Trinkaus, E. (1989). Olduvai Hominid 7 trapezial metacarpal 1 articular morphology: Contrasts with recent humans. *American Journal of Physical Anthropology* 80:411–416.

Trinkaus, E. (1992). Morphological contrasts between the Near Eastern Qafzeh-Skhul and late archaic human samples: Grounds for a behavioral difference? In T. Akazawa, K. Aoki, & T. Kimura (eds.), *The Evolution and Dispersal of Modern Humans in Asia*. Tokyo: Hokusen-Sha, pp. 277–294.

Trinkaus, E. (1993a). Femoral neck-shaft angles of the Qafzeh-Skhul early modern humans, and activity levels among immature Near Eastern Middle Paleolithic hominids. *Journal of Human Evolution* 25:393–416.

Trinkaus, E. (1993b). A note on the KNM-ER 999 hominid femur. *Journal of Human Evolution* 24:493–504.

Trinkaus, E. (1995). Neanderthal mortality patterns. *Journal of Archaeological Science* 22:121–142.

Trinkaus, E. (1997a) Appendicular robusticity and the paleobiology of modern human emergence. *Proceedings of the National Academy of Sciences USA* 94:13367–13373.

Trinkaus, E. (1997b). Cross-sectional geometry of the long bone diaphyses of Pavlov 1. In J. Svoboda (ed.), *Pavlov I—Northwest. The Upper Paleolithic Burial and Its Settlement Context*. Dolní Věstonice Studies 4. Brno: Archeologický ústav AV ČR, pp. 155–166.

Trinkaus, E. (2000a). Human patellar articular proportions: Recent and Pleistocene patterns. *Journal of Anatomy (London)* 196:473–483.

Trinkaus, E. (2000b). The human remains from Paviland Cave: Late Pleistocene and Holocene human remains from Paviland Cave. In S.H.R. Aldhouse-Green (ed.), *Paviland Cave and the 'Red Lady': A Definitive Report*. Bristol: Western Academic and Specialist Press, pp. 141–199.

Trinkaus, E. (2000c). The "Robusticity Transition" revisited. In C.B. Stringer, R.N.E. Barton, & C. Finlayson (eds.), *Neanderthals on the Edge*. Oxford: Oxbow Books, pp. 227–236.

Trinkaus, E. (2001). Paleobiological perspectives on the Early Upper Paleolithic human transition in the northwestern Old World. *Bulletins et Mémoires de la Société d'Anthropologie de Paris* ns13:311–322.

Trinkaus, E. (2002a). The cranial morphology. In J. Zilhão & E. Trinkaus (eds.), *Portrait of the Artist as a Child. The Gravettian Human Skeleton from the Abrigo do Lagar Velho and Its Archeological Context*. Trabalhos de Arqueologia 22:256–286.

Trinkaus, E. (2002b). The mandibular morphology. In J. Zilhão & E. Trinkaus (eds.), *Portrait of the Artist as a Child. The Gravettian Human Skeleton from the Abrigo do Lagar Velho and Its Archeological Context*. Trabalhos de Arqueologia 22:312–325.

Trinkaus, E. (2003). Neandertal faces were not long; modern human faces are short. *Proceedings of the National Academy of Sciences USA* 100:8142–8145.

Trinkaus, E. (2004). Eyasi 1 and the suprainiac fossa. *American Journal of Physical Anthropology* 124:28–32.

Trinkaus, E., & Churchill, S.E. (1988). Neandertal radial tuberosity orientation. *American Journal of Physical Anthropology* 75:15–21.

Trinkaus, E., & Churchill, S.E. (1999). Diaphyseal cross-sectional geometry of Near Eastern Middle Paleolithic humans: The humerus. *Journal of Archaeological Science* 26:173–184.

Trinkaus, E., & Hilton, C.E. (1996). Neandertal pedal proximal phalanges: Diaphyseal loading patterns. *Journal of Human Evolution* 30:399–425.

Trinkaus, E., & Jelínek, J. (1997). Human remains from the Moravian Gravettian: The Dolní Věstonice 3 postcrania. *Journal of Human Evolution* 33:33–82.

Trinkaus, E., & Pettitt, P.B. (2000). The Krems-Hundssteig "Gravettian" human remains are Holocene. *Homo* 51:258–260.

Trinkaus, E., & Rhoads, M.L. (1999). Neandertal knees: Power lifters in the Pleistocene? *Journal of Human Evolution* 37:833–859.

References

Trinkaus, E., & Ruff, C.B. (1999a). Diaphyseal cross-sectional geometry of Near Eastern Middle Paleolithic humans: The femur. *Journal of Archaeological Science* 26:409–424.

Trinkaus, E., & Ruff, C.B. (1999b). Diaphyseal cross-sectional geometry of Near Eastern Middle Paleolithic humans: The tibia. *Journal of Archaeological Science* 26:1289–1300.

Trinkaus, E., & Ruff, C.B. (2000). Comment on: O.M. Pearson, "Activity, climate, and postcranial robusticity. Implications for modern human origins and scenarios of adaptive change." *Current Anthropology* 41:598.

Trinkaus, E., & Villemeur, I. (1991). Mechanical advantages of the Neandertal thumb in flexion: A test of an hypothesis. *American Journal of Physical Anthropology* 84:249–260.

Trinkaus, E., & Zilhão, J. (2002). Phylogenetic implications. In J. Zilhão & E. Trinkaus (eds.), *Portrait of the Artist as a Child. The Gravettian Human Skeleton from the Abrigo do Lagar Velho and Its Archeological Context.* Trabalhos de Arqueologia 22:497–518.

Trinkaus, E., Churchill, S.E., Villemeur, I., Riley, K.G., Heller, J.A., & Ruff, C.B. (1991) Robusticity *versus* shape: The functional interpretation of Neandertal appendicular morphology. *Journal of the Anthropological Society of Nippon* 99:257–278.

Trinkaus, E., Churchill, S.E., & Ruff, C.B. (1994). Postcranial robusticity in *Homo*, II: Humeral bilateral asymmetry and bone plasticity. *American Journal of Physical Anthropology* 93:1–34.

Trinkaus, E., Stringer, C.B., Ruff, C.B., Hennessy, R.J., Roberts, M.B., & Parfitt, S.A. (1999a). Diaphyseal cross-sectional geometry of the Boxgrove 1 Middle Pleistocene human tibia. *Journal of Human Evolution* 37:1–25.

Trinkaus, E., Jelínek, J., & Pettitt, P.B. (1999b). Human remains from the Moravian Gravettian: The Dolní Věstonice 35 femoral diaphysis. *Anthropologie (Brno)* 37:167–175.

Trinkaus, E., Churchill, S.E., Ruff, C.B., & Vandermeersch, B. (1999c). Long bone shaft robusticity and body proportions of the Saint-Césaire 1 Châtelperronian Neandertal. *Journal of Archaeological Science* 26:753–773.

Trinkaus, E., Lebel, S., & Bailey, S.E. (2000a). Middle Paleolithic and recent human dental remains from the Bau de l'Aubesier, Monieux (Vaucluse). *Bulletins et Mémoires de la Société d'Anthropologie de Paris* ns12:207–226.

Trinkaus, E., Smith, R.J., & Lebel, S. (2000b) Dental caries in the Aubesier 5 Neandertal primary molar. *Journal of Archaeological Science* 27:1017–1021.

Trinkaus, E., Svoboda, J., West, D.L., Sládek, V., Hillson, S.W., Drozdová, E., & Fišáková, M. (2000c). Human remains from the Moravian Gravettian: Morphology and taphonomy of isolated elements from the Dolní Věstonice II site. *Journal of Archaeological Science* 27:1115–1132.

Trinkaus, E., Formicola, V., Svoboda, J., Hillson, S.W., & Holliday, T.W. (2001). Dolní Věstonice 15: Pathology and persistence in the Pavlovian. *Journal of Archaeological Science* 28:1291–1308.

Trinkaus, E., Hillson, S.W., & Santos Coelho, J.M. (2002a). Paleopathology. In J. Zilhão & E. Trinkaus (eds.), *Portrait of the Artist as a Child. The Gravettian Human Skeleton from the Abrigo do Lagar Velho and Its Archeological Context.* Trabalhos de Arqueologia 22:489–495.

Trinkaus, E., Ruff, C.B., Esteves, F., Santos Coelho, J.M., Silva, M., & Mendonça, M. (2002b). The lower limb remains. In J. Zilhão & E. Trinkaus (eds.), *Portrait of the Artist as a Child. The Gravettian Human Skeleton from the Abrigo do Lagar Velho and its Archeological Context.* Trabalhos de Arqueologia 22:425–465.

Trinkaus, E., Milota, Ş., Rodrigo, R., Gherase, M., & Moldovan, O. (2003a). Early modern human cranial remains from the Peştera cu Oase, Romania. *Journal of Human Evolution* 45:245–253.

Trinkaus, E., Moldovan, O., Milota, Ş., Bîlgăr, A., Sarcina, L., Athreya, S., Bailey, S.E., Rodrigo, R., Gherase, M., Higham, T., Bronk Ramsey, C., & van der Plicht, J. (2003b). An early modern human from the Peştera cu Oase, Romania. *Proceedings of the National Academy of Sciences USA* 100:11231–11236.

Trinkaus, E., Smith, F.H., Stockton, T.C., & Shackelford, L.L. (2005). The human postcranial remains from Mladeč. In M. Teschler-Nicola (ed.), *Early Modern Humans at the Moravian Gate: The Mladeč Caves and Their Remains.* Vienna: Springer Verlag (in press).

Trotter, M., & Gleser, G.C. (1952). Estimation of stature from long bones of American whites and negroes. *American Journal of Physical Anthropology* 10:463–514.

Trotter, M., & Lanier, P.F (1945). Hiatus canalis sacralis in American whites and negroes. *Human Biology* 17:368–381.

Tschentscher, F., Capelli, C., Geisert, H., Krainitzki, H., Schmitz, R.W., & Krings, M. (2000). Mitochondrial DNA sequences from the Neanderthals. In J. Orschiedt & G.C. Weniger (eds.), *Neanderthals and Modern Humans—Discussing the Transition: Central and Eastern Europe from 50.000—30.000 B.P.*, Wissenschaftliche Schriften der Neanderthal Museum 2:303–314.

Turner, C.G., II, Nichol, C.R., & Scott, G.R. (1991). Scoring procedures for key morphological traits of the permanent dentition: The Arizona State University Dental Anthropology System. In M.A. Kelley & C.S. Larsen (eds.), *Advances in Dental Anthropology.* New York: Wiley-Liss, pp. 13–31.

Twiesselmann, F. (1958). Les ossements humains du Gîte Mésolithique d'Ishango. In J. de Heinzelin (ed.), *Exploration du Parc National Albert.* Brussels: Institute des Parcs Nationaux du Congo Belge 5:1–125.

Twiesselmannn, F. (1961). Le fémur néanderthalien de Fond-de-Forêt (Province de Liège). *Mémoire de l'Institut Royal des Sciences Naturelles de Belgique* 149:1–164.

Twiesselmann, F. (1973). Évolution des dimensions et de la forme de la mandibule, du palais et des dents de l'homme. *Annales de Paléontologie (Vertébrés)* 59:173–277.

Twiesselmann, F., & Brabant, H. (1967). Les dents et les maxillaires de la population d'âge Franc de Coxyde (Belgique). *Bulletin du Groupement International pour la Recherche Scientifique en Stomatologie* 10:5–180.

Ubelaker, D.H. (1978). *Human skeletal remains: Excavation, analysis, interpretation*. Chicago: Aldine.

Ubelaker, D.H., Phenice, T.W., & Bass, W.M. (1969). Artificial interproximal grooving of the teeth in American Indians. *American Journal of Physical Anthropology* 30: 145–149.

Ungar, P.S., Grine, F.E., Teaford, M.F., & Pérez-Pérez, A. (2001). A review of interproximal wear grooves on fossil hominin teeth with new evidence from Olduvai Gorge. *Archives of Oral Biology* 46:285–292.

Utsinger, P.D. (1985). Diffuse idiopathic skeletal hyperostosis. *Clinics in Rheumatic Diseases* 11:325–351.

Vallois, H.V. (1928–1946). L'omoplate humaine: Étude anatomique et anthropologique. *Bulletins et Mémoires de la Société d'Anthropologie de Paris*, Série 7, 9:129–168; 10:110–191; Série 8, 3:3–153; Série 9, 7:16–100.

Vallois, H.V., & Billy, G. (1965). Nouvelles recherches sur les hommes fossiles de l'Abri de Cro-Magnon. *L'Anthropologie* 69:47–74, 249–272.

Vallois, H.V., & Roche J. (1958). La mandibule acheuléenne de Témara, Maroc. *Comptes rendus de l'Académie des Sciences Paris* 246:3113–3116.

Valoch, K. (1996). *Le Paléolithique en Tchéquie et en Slovaquie*. Grenoble: J. Millen.

Vančata, V. (2003). Sexual dimorphism in body size and shape in Pavlovian Upper Paleolithic group: A population approach. *Anthropologie (Brno)* 41:213–240

Van Den Bogert, A., Read, L., & Nigg, B.M. (1999). An analysis of hip joint loading during walking, running, and skiing. *Medicine & Science in Sports & Exercise* 31:131–142.

Vandermeersch, B. (1981). *Les Hommes Fossiles de Qafzeh (Israël)*. Paris: Éditions du C.N.R.S.

Vandermeersch, B. (1991). La ceinture scapulaire et les membres supérieures. In O. Bar-Yosef & B. Vandermeersch (eds.), *Le Squelette Moustérien de Kébara 2*. Paris: Éditions du C.N.R.S. pp. 157–178.

Vandermeersch, B., & Trinkaus, E. (1995). The postcranial remains of the Régourdou 1 Neandertal: The shoulder and arm remains. *Journal of Human Evolution* 28:439–476.

Vanhaeren, M., & d'Errico, F. (2002). The body ornaments associated with the burial. In J. Zilhão & E. Trinkaus (eds.), *Portrait of the Artist as a Child. The Gravettian Human Skeleton from the Abrigo do Lagar Velho and its Archeological Context*. Trabalhos de Arqueologia 22:154–186.

Vekua, A., Lordkipanidze, D., Rightmire, G.P., Agusti, J., Ferring, R., Maisuradze, G., Mouskhelishvili, A., Nioradze, M., Ponce de León, M., Tappen, M., Tvalchrelidze, M., & Zollikofer, C. (2002). A new skull of early *Homo* from Dmanisi, Georgia. *Science* 297:85–89.

Verneau, R. (1906). *Les Grottes de Grimaldi (Baoussé-Roussé), Tome II—Fascicule I. Anthropologie*. Monaco: Imprimérie de Monaco.

Verpoorte, A. (2000). Pavlov-reflexes and Pompeii-premise. *Archeologické rozhledy* 52:577–595.

Verpoorte, A. (2001). *Places of Art, Traces of Fire*. Archaeological Studies, Leiden University 8/Dolní Věstonice Studies 6. Brno: Archeologický ústav AV ČR.

Villemeur, I. (1994). *La Main des Néandertaliens*. Paris: C.N.R.S. Éditions.

Vlček, E. (1951). Otisky papilárních linií mladodiluviálního člověka z Dolních Věstonic. *Zprávy Anthropologické Společnosti* 4:90–94.

Vlček, E. (1961). Pozůstatky mladopleistocenního člověka z Pavlova. *Památky Archeologické* 52:46–56.

Vlček, E. (1967). Der Jungpleistozäne Menschenfund aus Svitávka in Mähren. *Anthropos (Brno)* 19:262–270.

Vlček, E. (1971). Czechoslovakia. In K.P. Oakley, B.G. Campbell, & T.I. Molleson (eds.), *Catalogue of Fossil Hominids II: Europe*. London: British Museum (Natural History), pp. 47–64.

Vlček, E. (1991). Die Mammutjäger von Dolní Věstonice. *Archäologie und Museum*, 22.

Vlček, E. (Ed.). (1992). *Lovci mamutů z Dolních Věstonic*. Sborník Národního Muzea v Praze (Acta Musei Nationales Pragae) 48B.

Vlček, E. (1997). Human remains from Pavlov and the biological anthropology of the Gravettian human population of South Moravia. In J. Svoboda (ed.), *Pavlov I–Northwest*. Dolní Věstonice Studies 4. Brno: Archeologický ústav AV ČR, pp. 53–153.

Vlček, E., & Šmahel, Z. (2002). Roentgencraniometric analysis of skulls of mammoth hunters from Dolní Věstonice. *Acta Chirurgiae Plasticae* 44:136–141.

Walker, A., & Leakey, R. (1993). The postcranial bones. In A. Walker & R. Leakey (eds.), *The Nariokotome Homo erectus Skeleton*. Cambridge, MA: Harvard University Press, pp. 95–160.

Walker, M.J. (2001). Excavations at Cueva Negra del Estrecho del Río Quípar and Sima de las Palomas del Cabezo Gordo: Two sites in Murcia (south-east Spain) with Neanderthal skeletal remains, Mousterian assemblages and late Middle to early Upper Pleistocene fauna. In S. Milliken & J. Cook (eds.), *A Very Remote Period Indeed. Papers on the Paleolithic Presented to Derek Roe*. Oxford: Oxbow Books, pp. 153–159.

Wall, J.D. (2000). Detecting ancient admixture in humans using sequence polymorphism data. *Genetics* 154:1271–1279.

Warren, E. (1897). An investigation on the variability of the human skeleton: with especial reference to the Naqada race. *Philosophical Transactions of the Royal Society* 189B: 135–227.

Weidenreich, F. (1943). The skull of *Sinanthropus pekinensis*. A comparative study on a primitive hominid skull. *Palaeontologia Sinica* 10D:1–485.

Weisl, H. (1954). The ligaments of the sacro-iliac joint examined with particuliar reference to their function. *Acta Anatomica* 20:201–213.

West, D. (2001). Analysis of the fauna recovered from the 1986/1987 excavations at Dolní Věstonice II, western slope. *Památky archeologické* 92:98–123.

Whaites, E. (1992). *Essentials of Dental Radiography and Radiology*. Edinburgh: Churchill Livingstone.

White, T.D., & Folkens, P.A. (2000). *Human Osteology*, 2nd ed. San Diego: Academic Press.

White, T.D., Asfaw, B., DeGusta, D., Gilbert, H., Richards,

G.D., Suwa, G., & Howell, F.C. (2003) Pleistocene *Homo sapiens* from Middle Awash, Ethiopia. *Nature* 423:742–747.

Whittaker, D.K., Parker, J.H., & Jenkins, C. (1982). Tooth attrition and continuing eruption in a Romano-British population. *Archives of Oral Biology* 27:405–409.

Whittaker, D.K., Molleson, T., Daniel, A.T., Williams, J.T., Rose, P., & Resteghini, R. (1985). Quantitative assessment of tooth wear, alveolar-crest height and continuing eruption in a Romano-British population. *Archives of Oral Biology* 30:493–501.

Wish-Baratz, S., Arensburg, B., & Alter, Z. (1992). Anatomical relationships and superior reinforcement of the TMJ mandibular fossa. *Journal of Craniomandibular Disorders: Facial & Oral Pain* 6:171–175.

Wolpoff, M.H. (1999). *Paleoanthropology*, 2nd ed. New York: McGraw-Hill.

Wolpoff, M.H., Smith, F.H., Malez, M., Radovčić, J., & Rukavina, D. (1981). Upper Pleistocene human remains from Vindija Cave, Croatia, Yugoslavia. *American Journal of Physical Anthropology* 54:499–545.

Wolpoff, M.H., Hawks, J., Frayer, D.W., & Hunley, K. (2001). Modern human ancestry at the peripheries: A test of the replacement theory. *Science* 291:293–297.

Woo, T.L., & Morant, G.M. (1934). A biomechanical study of the "flatness" of the facial skeleton in man. *Biometrika* 26:196–250.

Wood, N.K., & Goaz, P.W. (1997). *Differential Diagnosis of Oral and Maxillofacial Lesions*. St Louis: Mosby.

Wynne-Davies, R., Hall, C.M., & Apley, A.G. (1985). *Atlas of Skeletal Dysplasias*. New York: Churchill Livingstone.

Youm, Y., Dryer, R.F., Thambyrajah, K., Flatt, A.E., & Sprague, B.L. (1979). Biomechanical analyses of forearm pronation-supination and elbow flexion-extension. *Journal of Biomechanics* 12:245–255.

Zaajer, T. (1866). Untersuchungen über die Form des Beckens javanischen Frauen. *Naturrk Verhandel Holland Maatch Vetensch Harlem* 24:1–42.

Zilhão, J., & Trinkaus, E. (2002). Social implications. In J. Zilhão & E. Trinkaus (eds.), *Portrait of the Artist as a Child. The Gravettian Human Skeleton from the Abrigo do Lagar Velho and its Archeological Context*. Trabalhos de Arqueologia 22:519–539.

Zollikofer, C.P.E., Ponce de León, M.S., Vandermeersch, B., & Lévêque, F. (2002). Evidence for interpersonal violence in the St. Césaire Neanderthal. *Proceedings of the National Academy of Sciences USA* 99:6444–6448.

Index

Agenesis, dental, 32, 39, 183, 184, 197, 217, 218
Aggsbach, 6, 8
Allen's Rule, 227
Alveolar exostosis, 204, 453, 454
Alveolar morphology
   Dolní Věstonice 3, 76
   Dolní Věstonice 13, 98, 100
   Dolní Věstonice 14, 110
   Dolní Věstonice 15, 122, 124
   Dolní Věstonice 16, 134, 135, 137
Amud, 63, 170, 176, 199, 327, 347, 388
Attrition, approximal, 179
Arcy-sur-Cure (Hyène), 170
Arcy-sur-Cure (Renne), 3
Arene Candide, 60, 461
   Axial skeleton, 283, 292, 293
   Body size and proportions, 224, 234, 432
   Lower limb, 381, 386, 388, 393, 435–437
   Skull, 63
   Upper limb, 327, 340, 439
Art, 22
ASUDAS (Arizona State University Dental Anthropology System), 179–182
Asymmetry, 171, 348–350, 421–424, 427, 430, 432, 434–440, 447, 451, 457
Aubesier, 171, 177
Axillary border, 340–342

Banyoles, 170, 173
Baousso da Torre, 60, 224, 234, 327, 461

Barma Grande, 36, 60, 461
   Axial skeleton, 230, 231
   Body size and proportions, 224, 233, 234, 237, 238, 431
   Lower limb, 381, 390, 435–437
   Skull, 63, 69, 170
   Upper limb, 327, 349, 350, 352, 361, 438, 439
Berg Aukas, 386
Bergmann's Rule, 227, 235
Biache, 153–155
Bi-iliac breadth, 226, 228, 229, 235, 236, 240, 379
Bisitun, 327
Body mass, 235, 236, 238, 240, 381, 431, 432
Boršice, 7, 8
Brachial index, 225–227
Brno, 7, 8, 9, 11, 15, 63, 74, 152, 177, 220, 459, 460
Broken Hill, 387
Bruniquel, 199
Buccal wear facets, 179, 184–187, 189–193, 198, 202–205, 208, 220–222
Burial(s), 12, 13, 15–26, 62

Cabezo Gordo, 3
Calculus, 187, 201, 205, 209
Canines, 431, 453
   Mandibular, 180, 184, 188, 191, 195, 197, 204, 207, 215, 219, 221, 222
   Maxillary, 180, 184, 187, 190, 196, 203, 207, 214, 216, 219, 221, 222
Caries (dental), 179, 184, 187, 423, 427, 431, 453–455
Carpal bones, 39, 366–368, 450, 451

Caviglione, 49, 60, 170, 224, 234, 327, 368, 370, 381, 412, 461
Chondrodysplasia punctata, 47, 446
Cioclocina, 461
Circeo. *See* Guattari
Clavicle, 35, 36, 38, 329–334, 429, 438, 442
Cleft palate, 447, 465
Combe Capelle, 63, 71
Congenitally missing teeth. *See* Agenesis, dental
Cro-Magnon, 50, 51, 60, 460, 461
   Axial skeleton, 319, 322, 323
   Body size and proportions, 224, 234, 240
   Dentition, 217, 455
   Lower limb, 378, 381, 388
   Skull, 63, 170, 173
   Upper limb, 327, 350, 439
Crown diameters, 214, 215
Crown morphology, 179
Crural index, 225–227

Dar-es-Soltane, 171, 173
Darra-i-Kur, 153–155
Deciduous canines, 40, 43, 185, 209, 210, 218, 222
Deciduous incisors, 40, 43, 199, 210, 211, 218, 222
Deciduous molars (premolars), 34, 42, 43, 197, 200, 207–209, 222
Deciduous teeth, retention, 210, 211
Deltoideus, 333, 334, 344–348, 371, 429
Dental development score, 179, 186, 187, 190, 197, 200, 201, 205, 206, 211–213

Developmental defects of dental enamel, 179, 186–188, 190–193, 198, 201, 204, 205, 207, 212, 423, 424, 427, 429, 431, 449, 452, 454, 456, 464
Diastema, 216
Dmanisi, 430
Dolní Věstonice I, 4, 7, 8, 9, 10–12, 27, 28
Dolní Věstonice II, 4, 7, 8, 9, 10–13, 29
Dolní Věstonice III, 8, 9, 10
Dolní Věstonice 1, 12, 19, 25, 27, 47,
Dolní Věstonice 2, 12, 19, 25, 27, 47, 65, 74
Dolní Věstonice 3, 12, 15–18, 27, 29, 31–33, 47
  Axial skeleton, 33, 243, 245–247, 249–253, 255, 257, 259–265, 267, 268, 270–273, 276, 278, 280, 281, 295, 296, 424
  Body size and proportions, 224, 225, 227–232, 236–239
  Cranium, 31, 32, 57–59, 63–75, 82, 84, 85, 151, 152, 419–423
  Dentition, 32, 33, 82, 156, 157, 176, 181–184, 213, 214, 215, 217, 219–221, 223, 423, 424
  Lower limb, 48, 53–56, 61, 374, 377, 383–384, 386, 388–395, 397–399, 403–406, 410, 412–415, 417, 418
  Mandible, 58, 59, 82, 156, 157, 171, 172, 176, 177, 420–423
  Paleopathology, 64, 67, 69–71, 79, 81, 82, 85, 156, 157, 171, 177, 419–425
  Postcranium, 18
  Upper limb, 33, 329, 334, 336, 338, 340–344, 348, 350–355, 358, 361, 362, 364–370
Dolní Věstonice 4, 12, 18, 19, 27
Dolní Věstonice 5, 19, 27
Dolní Věstonice 6, 19, 27
Dolní Věstonice 7, 19, 27
Dolní Věstonice 8, 19, 26, 27
Dolní Věstonice 9, 19, 27, 33, 34, 184, 185
Dolní Věstonice 10, 19, 34, 185, 220, 222, 223
Dolní Věstonice 11/12, 24, 25, 29, 34, 47, 57–59
  Cranium, 65, 66, 74, 75, 86, 89, 90
  Paleopathology, 86, 87, 425, 426
Dolní Věstonice 13, 13, 19–22, 25, 29, 34, 35, 47
  Axial skeleton, 34, 242, 243, 245–248, 250, 252–257, 260–270, 272–277, 279–282, 284–293, 295–303, 305–307, 309, 311–314, 316, 317, 319–326, 427, 428
  Body size and proportions, 224, 225, 227–232, 236–239
  Cranium, 34, 57–59, 63–66, 68–71, 73–75, 90–92, 98, 103, 104, 151, 426, 427
  Dentition, 34, 157–160, 176, 180–182, 185–188, 214–217, 219–223, 427
  Lower limb, 35, 48, 49, 53–56, 61, 374–379, 383, 384, 388–400, 402–406, 409, 410, 412–415
  Mandible, 58, 59, 82, 92, 102, 157–160, 172, 175–177
  Paleopathology, 98, 100, 253, 254, 365–366, 426–428
  Upper limb, 35, 329, 330, 332–338, 340–345, 347–355, 357–368, 370, 428
Dolní Věstonice 14, 13, 19–22, 25, 29, 30, 35–37, 47
  Axial skeleton, 36, 242, 244, 246–252, 254, 256–258, 260–263, 265, 266, 268, 270–273, 275, 277–282, 285–293, 295–310, 313–323, 429
  Body size and proportions, 224, 227, 228, 236–239
  Cranium, 35, 57–59, 63–66, 68–71, 73–75, 105, 110, 113–116, 151, 153–155, 429
  Dentition, 35, 36, 107, 160–162, 176, 180–182, 188–191, 214–216, 219–223, 429
  Lower limb, 36, 49, 50, 53–56, 61, 374–379, 384–386, 388–398, 400, 403–407, 409, 410, 416, 417
  Mandible, 58, 59, 82, 105, 113, 160–162, 176, 177
  Paleopathology, 155, 333, 429
  Upper limb, 36, 429, 330–338, 340–345, 347–353, 355, 356, 358–368, 370
Dolní Věstonice 15, 13, 19–22, 25, 29, 30, 37, 38, 47
  Axial skeleton, 37, 242, 244–246, 248, 249, 251–256, 258–271, 273, 275–282, 284, 286–293, 295, 297–302, 304, 306, 308–311, 313, 315–323, 325, 326, 432
  Body size and proportions, 224, 225, 227–232, 238–240, 431, 432
  Cranium, 37, 57–60, 63–66, 68–75, 116, 122, 126–128, 151–155, 430, 431
  Dentition, 37, 125, 162–165, 176, 180–182, 191–195, 214–223, 431
  Lower limb, 38, 50–56, 61, 374–379, 376, 385, 386, 387–399, 401–407, 409–417, 432–438, 441, 442
  Mandible, 58–60, 82, 119, 125, 162–165, 175–177
  Paleopathology, 128, 273, 278, 338, 342, 343, 354, 356, 357, 362, 365–366, 372, 387, 390, 395, 418, 429–446
  Upper limb, 38, 330–342, 344, 346–368, 370, 438–444
Dolní Věstonice 16, 13, 22–24, 25, 29, 30, 38, 39, 47
  Axial skeleton, 39, 242, 244–246, 249, 251, 253, 255–257, 259, 317, 319, 322, 323, 326, 449, 450
  Body size and proportions, 224, 225, 227, 228, 236–239
  Cranium, 38, 39, 57–60, 63–66, 68–71, 73, 75, 128, 135, 140–142, 151, 446, 447
  Dentition, 39, 140, 165–168, 176, 181, 195–197, 214–220, 222, 223, 447–449
  Lower limb, 39, 52–56, 61, 385, 387–397, 401–405, 408–417
  Mandible, 58–60, 82, 139, 140, 165–168, 176, 177
  Paleopathology, 128–132, 137, 142, 253, 256, 259, 339, 354, 366, 446–451
  Upper limb, 330–334, 336–342, 344, 346–354, 357–364, 366–371, 450, 451
Dolní Věstonice 17, 24, 29, 40, 142, 452
Dolní Věstonice 23, 19
Dolní Věstonice 24, 19, 27
Dolní Věstonice 25, 19, 26, 27
Dolní Věstonice 26, 19, 27
Dolní Věstonice 27, 19, 27, 40, 197, 198, 222, 223
Dolní Věstonice 28, 19, 27
Dolní Věstonice 29, 19, 27
Dolní Věstonice 30, 19, 27
Dolní Věstonice 31, 19, 27, 40, 182, 198, 215, 221, 452
Dolní Věstonice 32, 19, 27, 40, 198, 215, 452
Dolní Věstonice 33, 13, 24, 29, 40, 198, 199, 218, 452
Dolní Věstonice 34, 24, 29, 41, 364, 366, 371
Dolní Věstonice 35, 10, 11, 19, 27, 41, 61, 240, 241, 388, 390–394
Dolní Věstonice 36, 13, 24, 29, 40, 179, 181, 182, 199–201, 223, 452
Dolní Věstonice 37, 19, 27, 41, 182, 201, 215, 452
Dolní Věstonice 38, 19, 27, 41, 182, 201

Dolní Věstonice 39, 13, 24, 29, 41, 409, 415
Dolní Věstonice 40, 24, 29, 41, 389
Dolní Věstonice 41, 24, 29, 41, 349, 350, 352
Dolní Věstonice 42, 24, 29, 41, 408, 409
Dolní Věstonice 43, 24, 29, 41, 389, 390
Dolní Věstonice 44, 24, 29, 41, 409, 416
Dolní Věstonice 45, 24, 29, 41
Dolní Vostonice 46, 24, 29, 41, 409, 415
Dolní Věstonice 47, 13, 24, 29, 41, 409
Dolní Věstonice 48, 24, 29, 41, 408, 409
Dolní Věstonice 49, 13, 24, 29, 41, 409, 416
Dolní Věstonice 50, 24, 29, 41, 352, 356, 359
Dolní Věstonice 51, 24, 29, 41
Dolní Věstonice 52, 24, 29, 41, 409, 416, 417
Dolní Věstonice 53, 24, 29, 41, 364, 366, 371

Enthesopathy, 451
Epiphyseal fusion, 35–38
Exfoliation, 40, 42, 43, 179, 218, 222

Facial size and proportions
  Dolní Věstonice 3, 72, 76, 77
  Dolní Věstonice 11/12, 87, 88
  Dolní Věstonice 13, 95, 98
  Dolní Věstonice 14, 107, 110
  Dolní Věstonice 15, 119, 122
  Dolní Věstonice 16, 135
  Pavlov 1, 145, 147
Fanciulli, 36, 60, 436–440, 457, 461
  Axial skeleton, 231
  Body size and proportions, 224, 233, 234, 237, 238
  Lower limb, 378, 379, 381, 436, 437
  Skull, 63, 170
  Upper limb, 327, 350, 357, 438–440
Fédération Dentaire Internationale, 214
Feldhofer. See Neandertal
Femur, 10, 225–228, 230, 231, 463
  Dolní Věstonice 3, 33, 236, 383, 384, 388–395
  Dolní Věstonice 13, 35, 236–239, 383, 384, 386–395
  Dolní Věstonice 14, 36, 236–239, 383–395
  Dolní Věstonice 15, 38, 236–240, 385–395, 432–436, 441, 442
  Dolní Věstonice 16, 236–239, 384, 385, 387–395
  Dolní Věstonice 35, 19, 41, 240, 241, 383, 388, 390–394

  Dolní Věstonice 40, 41, 383, 389
  Dolní Věstonice 43, 41, 383, 388, 389
  Pavlov 1, 236–239, 388, 389–395
Fibula, 35, 36, 38, 41, 404–409, 435
Figurines, 12, 463, 464
Fond-de-Forêt, 381
Fontana Nuova, 381, 414
Fontéchevade, 327
Forbes' Quarry, 63
Frontal morphology
  Dolní Věstonice 3, 65, 72, 78
  Dolní Věstonice 11/12, 86, 87
  Dolní Věstonice 13, 92, 95, 96, 99, 103
  Dolní Věstonice 14, 105, 107
  Dolní Věstonice 15, 116, 117, 119, 120
  Dolní Věstonice 16, 129, 132, 135
  Pavlov 1, 143, 145

Gibraltar. See Forbes' Quarry
Glenoid fossa
  Scapula, 338–340, 442, 443, 453
  Temporal bone, 420
Gravettian, 13, 15
Grooves, approximal, 212, 220–223
Grotte-des-Enfants. See Fanciulli
Grub-Kranawetberg, 10, 459
Guattari, 63, 170, 173

Hortus, 381
Humerus, 35, 36, 38, 41, 230, 236, 237, 342–352, 438–441
Hypercementosis, 41, 43, 44, 185, 196, 211
Hypoplasias, dental enamel. See Developmental defects of dental enamel

Impaction, 37, 39, 216, 217
Incisors, 34, 42, 43, 448, 462
  Mandibular, 180, 184, 188, 191, 195, 197, 204, 207, 212, 213, 215, 219, 221, 222
  Maxillary, 180, 184, 185, 187, 190, 193, 196, 200, 203, 212, 214, 219, 221, 222
Incus, 154, 155, 429
Infraorbital area
  Dolní Věstonice 3, 69, 70, 74
  Dolní Věstonice 13, 93, 94, 98, 100
  Dolní Věstonice 14, 106, 108, 109, 112
  Dolní Věstonice 15, 118, 122, 124
  Dolní Věstonice 16, 130, 134, 136, 137
  Pavlov 1, 144, 146, 147
Interpleniglacial, 6

Ishango, 234, 236
Isturitz, 170

Jarošov, 7, 8
Jomon, 219

Kebara
  Axial skeleton, 319, 324–326
  Body size, 236
  Lower limb, 378, 379, 437
  Skull, 170, 173
  Upper limb, 327, 332, 350, 369, 438, 439
Kent's Cavern, 3
Kiik-Koba, 381, 412
Koněprusy, 15
Krapina, 216, 327, 332, 333, 340, 368, 387
Krems-Hundssteig, 10
Krems-Wachtberg, 8
Kubbaniya, 234

Lagar Velho, 153, 154, 173, 233, 461
La Chaise, 170
La Chapelle-aux-Saints, 63, 74, 224, 228, 229, 231, 322, 323, 327, 350, 381, 439
La Ferrassie
  Axial skeleton, 322, 323
  Body size and proportions, 224, 231
  Lower limb, 381, 401, 412, 435
  Skull, 63, 74, 153–155, 170, 172, 173, 176
  Upper limb, 327, 332, 347, 350, 362, 369, 439
La Naulette, 170, 176, 216
Langenlois, 8
La Quina, 3, 63, 74, 170, 327, 381, 439, 461
La Rochette, 224, 234, 240, 327, 381, 439, 461
Les Rois, 452, 461, 462
Limb-trunk proportions, 230, 231

Malarnaud, 216, 218
Malleus, 153, 154
Mammoth, 7, 17, 18, 24
Mandible, 25, 42, 58–60, 156–178, 183, 420–422, 431, 453, 455
Maxilla, 25, 42
  Dolní Věstonice 3, 69–71, 74, 76, 81, 82
  Dolní Věstonice 13, 93–95, 98, 100, 103
  Dolní Věstonice 14, 94, 106, 107, 110, 113
  Dolní Věstonice 15, 94, 118, 119, 122, 124, 126, 430
  Dolní Věstonice 16, 130–132, 134, 136, 137, 140, 447

Maxilla (*continued*)
  Pavlov 1, 144–147, 149
  Pavlov 2, 150, 454
Metacarpals, 34, 35, 37, 364, 366–370
Metatarsals, 33, 35, 36, 41, 410, 415, 416, 435
Minatogawa, 234, 236, 240, 387
Mladeč, 3, 13, 460, 461
  Body size, 234, 237
  Lower limb, 381, 393, 413, 414
  Skull, 63, 72, 94, 96, 119, 152
  Upper limb, 327, 347, 350
Molars, 32–34, 35, 37, 39–41, 43, 44, 431, 448, 449, 453
  Mandibular, 182, 184, 185, 187, 190, 194, 197, 198, 200, 201, 203, 206, 213, 215, 217–222
  Maxillary, 181, 183, 186, 189, 192, 196, 198, 202, 205, 211, 214, 221, 222
Montmaurin, 171, 177
Moorrees and colleagues dental development score. *See* Dental development score
Muierii, 461

Nahal Ein Gev, 63, 234, 236, 240
Nasal morphology
  Dolní Věstonice 3, 70, 71, 74, 76
  Dolní Věstonice 13, 94, 95, 98, 100
  Dolní Vostonice 14, 106, 107, 110, 112
  Dolní Věstonice 15, 118, 119, 122, 124
  Dolní Vostonice 16, 130, 131, 134, 137
  Pavlov 1, 144, 147
Nazlet Khater, 171, 173, 234, 236, 240
Neandertal, 3, 63, 322, 323, 327, 332, 381
Neandertals, 3, 4, 5, 461
  Body size, 236–239
  Dentition, 213–215
  Lower limb, 377, 379, 387, 396, 408, 417
  Skull, 153, 154
  Upper limb, 335, 340, 341, 347, 355, 357, 361, 366–370

Oase, 3, 63, 170, 172, 177, 460, 461
Obercassel, 51, 319, 322, 323
Occipital morphology
  Dolní Věstonice 3, 74, 79
  Dolní Věstonice 13, 97, 100–102
  Dolní Věstonice 14, 108, 111–113
  Dolní Věstonice 15, 120, 124, 125
  Dolní Věstonice 16, 133, 137–139
  Pavlov 1, 146, 148
Occlusal facet scratches, 184, 220
Occlusal wear stages, 179

Occlusion, 214–219
Ochre, 16, 18, 19, 21, 24, 265
Ohalo, 63, 170, 234, 236, 238, 368, 387
Omo, 218
Omo-Kibish, 461
Orbital morphology
  Dolní Věstonice 3, 66, 67, 69, 74
  Dolní Věstonice 11/12, 87
  Dolní Věstonice 13, 93, 97, 98, 100
  Dolní Věstonice 14, 106, 108, 111, 112
  Dolní Věstonice 15, 117, 118, 121, 122
  Dolní Věstonice 16, 130, 134, 136
  Pavlov 1, 143, 144, 146
Osteoarthritis, 253, 256, 259, 273, 421, 422, 424, 428, 432, 441–444, 449–451
Ostuni, 233
Over-eruption, 41, 217

Paglicci, 49, 50, 60, 461
  Axial skeleton, 283, 292, 293
  Body size and proportions, 224, 230, 233, 234
  Dentition, 199
  Lower limb, 381, 439
  Skull, 63, 170
  Upper limb, 327, 350, 369
Pair-non-Pair, 327, 436, 437
Palate, 81, 82, 103, 113, 126, 140, 149
Parabita. *See* Veneri
Parietal morphology
  Dolní Věstonice 3, 72, 73, 78
  Dolní Věstonice 11/12, 87–89
  Dolní Věstonice 13, 96, 97
  Dolní Věstonice 14, 107, 108, 110
  Dolní Věstonice 15, 120, 122
  Dolní Věstonice 16, 132, 133, 135, 136
  Pavlov 1, 145, 147
Pataud, 461
  Body size, 234
  Dentition, 216, 219
  Lower limb, 435, 437
  Upper limb, 327, 337, 350, 370, 439
Patella, 395–397, 435, 437
Paviland, 49, 51, 60, 234, 327, 381, 401, 461
Pavlov I, 4, 8–10, 13, 14, 28
Pavlov II, 9, 10
Pavlov 1, 14, 24, 25, 28, 30, 41, 42
  Axial skeleton, 41, 42, 242, 246, 259, 260, 286, 287, 289, 293, 295
  Body size and proportions, 224, 227, 236–239
  Cranium, 41, 57–60, 63–66, 68, 69, 74, 75, 142, 143, 149–152, 452, 453

Dentition, 41, 142, 168–170, 182, 201–204, 213–215, 217, 219, 220, 222, 223, 453
Lower limb, 61, 385, 388–395, 408
Mandible, 58–60, 142, 168–170, 172, 176, 177, 453
Paleopathology, 342, 452–454
Upper limb, 331–334, 337, 339–342, 344, 346, 348–354, 356, 358–360, 362, 363, 453
Pavlov 2, 14, 25, 28
  Dentition, 42, 150, 181, 204–206, 211, 213, 214, 219, 220, 222
  Maxilla, 150
  Paleopathology, 454, 455
Pavlov 3, 14, 25, 28, 42, 455
  Dentition, 42, 170, 182, 206, 207, 213, 215, 219, 222
  Mandible, 170, 174, 176
  Paleopathology, 454, 455
Pavlov 4, 14, 25, 28, 42, 170
Pavlov 5, 28, 42, 207, 215, 222
Pavlov 6, 28, 42, 43, 207, 208, 220–223
Pavlov 7, 28, 43, 208, 222, 223
Pavlov 8, 28, 43, 208, 222, 223
Pavlov 9, 28, 43, 208, 209, 222, 223
Pavlov 10, 28, 43, 185, 208, 209
Pavlov 11, 28, 43, 209, 222
Pavlov 12, 28, 43, 209, 222
Pavlov 13, 28, 43, 209, 210
Pavlov 14, 28, 43, 209, 210, 222
Pavlov 15, 28, 43, 210, 218
Pavlov 16, 28, 43, 210, 222
Pavlov 17, 28, 43, 210, 211, 222
Pavlov 18, 28, 43, 211, 222
Pavlov 19, 28, 43, 211, 218
Pavlov 20, 28, 43, 181, 211, 214, 222
Pavlov 21, 28, 43, 211, 212, 218, 221, 223, 455
Pavlov 22, 28, 43, 44, 212, 214, 455
Pavlov 23, 28, 43, 44, 212, 215, 455
Pavlov 24, 28, 44, 212
Pavlov 25, 28, 43, 44, 212, 213, 215, 222
Pavlov 26, 28, 44, 213
Pavlov 27, 28, 44, 213
Pavlov 28, 28, 42, 44, 181, 182, 213, 215, 455
Pavlovian, 3, 4, 6, 7, 9
Pectoralis major tuberosity, 344–347, 349
Pelvis, 235, 236, 238
  Dolní Věstonice 3, 48, 53, 55, 56, 374
  Dolní Věstonice 13, 35, 48–50, 53–56, 373–377
  Dolní Věstonice 14, 36, 37, 49–51, 53–56, 374–377
  Dolní Věstonice 15, 22, 53–56, 240, 374–377, 441, 442

Dolní Věstonice 16, 24, 39, 52, 55, 56, 376, 377
Pendants, 15
Perforated teeth, 15, 19, 21, 26, 213
Periapical inflammation, 196
Periodontitis, 448, 449
Petřkovice, 7
Phalanges
  Foot, 33, 41
  Hand, 33, 41, 364–366, 369–371, 428, 442, 443
Pofi, 381
Pollex, 368–370
Předmostí, 7, 8, 15, 50, 60, 460
  Axial skeleton, 283, 292, 293
  Body size and proportions, 224, 230, 233, 234, 240
  Cranium, 63, 94, 152
  Dentition, 179, 218, 220
  Lower limb, 381, 388, 390–392, 401, 412, 435–437
  Mandible, 170, 173, 177
  Upper limb, 327, 337, 368, 439
Premolars
  Mandibular, 182, 184, 187, 191, 194, 197, 204, 206, 207, 215, 216, 219–222
  Maxillary, 183, 187, 188, 193, 196, 202, 205, 214, 219–222
Pterion, 73, 97, 99, 108, 111, 120, 133, 145

Qafzeh, 461, 462
  Body size, 234, 237, 238
  Cranium, 63, 65, 67, 70, 71, 73, 94, 104, 117, 141, 150, 152, 153–155
  Dentition, 216
  Lower limb, 381, 386, 388, 390, 409, 412, 413, 417, 436, 437
  Mandible, 170, 171–174
  Upper limb, 327, 340, 358, 361, 368, 370

Radius, 33, 35, 36, 41, 230, 352–364, 438–440
Regourdou, 170, 283, 292, 325–326, 327, 332, 361, 381, 437, 439, 440
Respiratory area, 323–325
Rib agenesis, 272
Ribs, 34, 36, 41, 272, 295–325, 321–322
Rochers-de-Villeneuve, 381
Root resorption, 34, 42, 43, 179, 185, 208–211
Rotation, dental, 215

Saccopastore, 63
Sacrum, 377–379
Saint-Césaire, 3, 63, 170, 327, 340, 371, 387, 457

Santa Croce, 381
Scapula, 334–342
Schmorl's node, 278
Scoliosis, 432
Shanidar, 69, 319, 324–327, 350, 362, 378, 379, 381, 412, 440
Skhul, 461, 462
  Body size, 233, 234, 236, 238
  Cranium, 63, 65, 67, 70–74, 94, 96, 104, 117, 119, 141, 150, 152
  Dentition, 216
  Lower limb, 381, 386, 388, 390, 393, 409, 413, 414, 417, 435–437
  Mandible, 170, 171, 173, 174
  Upper limb, 327, 340, 350, 358, 361, 362
Smith occlusal wear stages. See Occlusal wear stages
Spadzista, 7
Sphenoid morphology
  Dolní Věstonice 3, 79
  Dolní Věstonice 13, 102
  Dolní Věstonice 14, 113
  Dolní Věstonice 15, 125
  Dolní Věstonice 16, 139
  Pavlov 1, 148
Spinous process, 288–291
Spy, 63, 170, 327, 350, 355, 381, 394, 436, 439, 440
Spytihněv, 7
Stadel, 381
Stature, 228, 234–241
Sternum, 34, 318, 319, 325, 326
Stillfried-Grub, 8
Subalyuk, 171, 381
Subnasalalveolar clivus
  Dolní Věstonice 3, 71, 72, 76
  Dolní Věstonice 13, 95, 98, 100
  Dolní Věstonice 14, 107, 110
  Dolní Věstonice 15, 119, 122, 124
  Dolní Věstonice 16, 131, 132, 134, 135, 137
  Pavlov 1, 145
Sunghir, 60, 170, 233, 234, 238, 327, 381, 439, 461
Supernumerary teeth, 199, 211, 212
Supraorbital morphology
  Dolní Věstonice 3, 65, 66
  Dolní Věstonice 11/12, 87
  Dolní Věstonice 13, 92, 93
  Dolní Věstonice 14, 105, 106
  Dolní Věstonice 15, 117
  Dolní Věstonice 16, 129, 130
  Pavlov 1, 143
Sutures, 31, 32, 34, 35, 37–39, 41
Svitávka, 10

Tabun, 63, 171, 177, 325–327, 340, 368, 387

Talus, 412–414, 437
Tarsal bones, 41, 410–416
Témara, 173
Temporal morphology
  Dolní Věstonice 3, 73, 78, 79
  Dolní Věstonice 13, 97, 99, 100, 102, 103
  Dolní Věstonice 14, 108, 111, 113
  Dolní Věstonice 15, 120, 121, 123, 125, 126
  Dolní Věstonice 16, 133, 136, 139, 140
Tibia, 35, 36, 230, 231, 234, 397–404, 432, 435, 437
Torus, alveolar. See Alveolar exostosis
Transverse lines, 424, 425, 428, 429, 444, 451, 457
Trauma, 457, 464
  Dolní Věstonice 3, 64, 67, 69, 74, 156, 157, 419–422
  Dolní Věstonice 11/12, 86, 87, 425, 426
  Dolní Věstonice 13, 426
  Dolní Věstonice 15, 346, 356, 357, 386, 430, 440, 441, 444
  Dolní Věstonice 16, 128, 129, 137, 446, 447
  Pavlov 1, 452
Trunk height
  Partial, 225
  Skeletal, 225, 230, 231, 240, 433

Ulna, 35, 36, 38, 352–364, 438–441, 444, 451

Veneri, 60, 224, 234, 381, 401, 435–437, 461
Vertebrae
  Cervical, 33, 39, 242–257, 292–293, 432, 449, 450, 451
  Lumbar, 33, 37, 273–282, 284–293, 432
  Thoracic, 33, 37, 39, 257–273, 284–293, 427, 432, 449, 450
Vindija, 3, 63, 171, 173

Willendorf, 6, 8, 10, 170, 234, 381, 459, 463
Willendorf-Kostenkian, 7, 9, 10

Zafarraya, 3, 171
Zygomatic morphology
  Dolní Věstonice 3, 74, 78, 79, 81
  Dolní Věstonice 13, 97, 100
  Dolní Věstonice 14, 108, 111
  Dolní Věstonice 15, 121, 123, 124
  Dolní Věstonice 16, 133, 134, 136
  Pavlov 1, 146